LONDON MATHEMATICAL SOCIETY LECTURE NOTE SERIES

Managing Editor: Professor N.J. Hitchin, Mathematical Institute,
University of Oxford, 24–29 St Giles, Oxford OX1 3LB, United Kingdom

The titles below are available from booksellers, or from Cambridge University Press at www.cambridge.org

100 Stopping time techniques for analysts and probabilists, L. EGGHE
105 A local spectral theory for closed operators, I. ERDELYI & WANG SHENGWANG
107 Compactification of Siegel moduli schemes, C.-L. CHAI
109 Diophantine analysis, J. LOXTON & A. VAN DER POORTEN (eds)
113 Lectures on the asymptotic theory of ideals, D. REES
116 Representations of algebras, P.J. WEBB (ed)
119 Triangulated categories in the representation theory of finite-dimensional algebras, D. HAPPEL
121 Proceedings of *Groups - St Andrews 1985*, E. ROBERTSON & C. CAMPBELL (eds)
128 Descriptive set theory and the structure of sets of uniqueness, A.S. KECHRIS & A. LOUVEAU
130 Model theory and modules, M. PREST
131 Algebraic, extremal & metric combinatorics, M.-M. DEZA, P. FRANKL & I.G. ROSENBERG (eds)
141 Surveys in combinatorics 1989, J. SIEMONS (ed)
144 Introduction to uniform spaces, I.M. JAMES
146 Cohen-Macaulay modules over Cohen-Macaulay rings, Y. YOSHINO
148 Helices and vector bundles, A.N. RUDAKOV *et al*
149 Solitons, nonlinear evolution equations and inverse scattering, M. ABLOWITZ & P. CLARKSON
150 Geometry of low-dimensional manifolds 1, S. DONALDSON & C.B. THOMAS (eds)
151 Geometry of low-dimensional manifolds 2, S. DONALDSON & C.B. THOMAS (eds)
152 Oligomorphic permutation groups, P. CAMERON
153 L-functions and arithmetic, J. COATES & M.J. TAYLOR (eds)
155 Classification theories of polarized varieties, TAKAO FUJITA
158 Geometry of Banach spaces, P.F.X. MÜLLER & W. SCHACHERMAYER (eds)
159 Groups St Andrews 1989 volume 1, C.M. CAMPBELL & E.F. ROBERTSON (eds)
160 Groups St Andrews 1989 volume 2, C.M. CAMPBELL & E.F. ROBERTSON (eds)
161 Lectures on block theory, BURKHARD KÜLSHAMMER
163 Topics in varieties of group representations, S.M. VOVSI
164 Quasi-symmetric designs, M.S. SHRIKANDE & S.S. SANE
166 Surveys in combinatorics, 1991, A.D. KEEDWELL (ed)
168 Representations of algebras, H. TACHIKAWA & S. BRENNER (eds)
169 Boolean function complexity, M.S. PATERSON (ed)
170 Manifolds with singularities and the Adams-Novikov spectral sequence, B. BOTVINNIK
171 Squares, A.R. RAJWADE
172 Algebraic varieties, GEORGE R. KEMPF
173 Discrete groups and geometry, W.J. HARVEY & C. MACLACHLAN (eds)
174 Lectures on mechanics, J.E. MARSDEN
175 Adams memorial symposium on algebraic topology 1, N. RAY & G. WALKER (eds)
176 Adams memorial symposium on algebraic topology 2, N. RAY & G. WALKER (eds)
177 Applications of categories in computer science, M. FOURMAN, P. JOHNSTONE & A. PITTS (eds)
178 Lower K- and L-theory, A. RANICKI
179 Complex projective geometry, G. ELLINGSRUD *et al*
180 Lectures on ergodic theory and Pesin theory on compact manifolds, M. POLLICOTT
181 Geometric group theory I, G.A. NIBLO & M.A. ROLLER (eds)
182 Geometric group theory II, G.A. NIBLO & M.A. ROLLER (eds)
183 Shintani zeta functions, A. YUKIE
184 Arithmetical functions, W. SCHWARZ & J. SPILKER
185 Representations of solvable groups, O. MANZ & T.R. WOLF
186 Complexity: knots, colourings and counting, D.J.A. WELSH
187 Surveys in combinatorics, 1993, K. WALKER (ed)
188 Local analysis for the odd order theorem, H. BENDER & G. GLAUBERMAN
189 Locally presentable and accessible categories, J. ADAMEK & J. ROSICKY
190 Polynomial invariants of finite groups, D.J. BENSON
191 Finite geometry and combinatorics, F. DE CLERCK *et al*
192 Symplectic geometry, D. SALAMON (ed)
194 Independent random variables and rearrangement invariant spaces, M. BRAVERMAN
195 Arithmetic of blowup algebras, WOLMER VASCONCELOS
196 Microlocal analysis for differential operators, A. GRIGIS & J. SJÖSTRAND
197 Two-dimensional homotopy and combinatorial group theory, C. HOG-ANGELONI *et al*
198 The algebraic characterization of geometric 4-manifolds, J.A. HILLMAN
199 Invariant potential theory in the unit ball of C^n, MANFRED STOLL
200 The Grothendieck theory of dessins d'enfant, L. SCHNEPS (ed)
201 Singularities, JEAN-PAUL BRASSELET (ed)
202 The technique of pseudodifferential operators, H.O. CORDES
203 Hochschild cohomology of von Neumann algebras, A. SINCLAIR & R. SMITH
204 Combinatorial and geometric group theory, A.J. DUNCAN, N.D. GILBERT & J. HOWIE (eds)
205 Ergodic theory and its connections with harmonic analysis, K. PETERSEN & I. SALAMA (eds)
207 Groups of Lie type and their geometries, W.M. KANTOR & L. DI MARTINO (eds)
208 Vector bundles in algebraic geometry, N.J. HITCHIN, P. NEWSTEAD & W.M. OXBURY (eds)
209 Arithmetic of diagonal hypersurfaces over finite fields, F.Q. GOUVÉA & N. YUI
210 Hilbert C*-modules, E.C. LANCE
211 Groups 93 Galway / St Andrews I, C.M. CAMPBELL *et al* (eds)
212 Groups 93 Galway / St Andrews II, C.M. CAMPBELL *et al* (eds)

214 Generalised Euler-Jacobi inversion formula and asymptotics beyond all orders, V. KOWALENKO *et al*
215 Number theory 1992–93, S. DAVID (ed)
216 Stochastic partial differential equations, A. ETHERIDGE (ed)
217 Quadratic forms with applications to algebraic geometry and topology, A. PFISTER
218 Surveys in combinatorics, 1995, PETER ROWLINSON (ed)
220 Algebraic set theory, A. JOYAL & I. MOERDIJK
221 Harmonic approximation, S.J. GARDINER
222 Advances in linear logic, J.-Y. GIRARD, Y. LAFONT & L. REGNIER (eds)
223 Analytic semigroups and semilinear initial boundary value problems, KAZUAKI TAIRA
224 Computability, enumerability, unsolvability, S.B. COOPER, T.A. SLAMAN & S.S. WAINER (eds)
225 A mathematical introduction to string theory, S. ALBEVERIO *et al*
226 Novikov conjectures, index theorems and rigidity I, S. FERRY, A. RANICKI & J. ROSENBERG (eds)
227 Novikov conjectures, index theorems and rigidity II, S. FERRY, A. RANICKI & J. ROSENBERG (eds)
228 Ergodic theory of $\mathbf{Z}^d$ actions, M. POLLICOTT & K. SCHMIDT (eds)
229 Ergodicity for infinite dimensional systems, G. DA PRATO & J. ZABCZYK
230 Prolegomena to a middlebrow arithmetic of curves of genus 2, J.W.S. CASSELS & E.V. FLYNN
231 Semigroup theory and its applications, K.H. HOFMANN & M.W. MISLOVE (eds)
232 The descriptive set theory of Polish group actions, H. BECKER & A.S. KECHRIS
233 Finite fields and applications, S. COHEN & H. NIEDERREITER (eds)
234 Introduction to subfactors, V. JONES & V.S. SUNDER
235 Number theory 1993–94, S. DAVID (ed)
236 The James forest, H. FETTER & B. GAMBOA DE BUEN
237 Sieve methods, exponential sums, and their applications in number theory, G.R.H. GREAVES *et al*
238 Representation theory and algebraic geometry, A. MARTSINKOVSKY & G. TODOROV (eds)
240 Stable groups, FRANK O. WAGNER
241 Surveys in combinatorics, 1997, R.A. BAILEY (ed)
242 Geometric Galois actions I, L. SCHNEPS & P. LOCHAK (eds)
243 Geometric Galois actions II, L. SCHNEPS & P. LOCHAK (eds)
244 Model theory of groups and automorphism groups, D. EVANS (ed)
245 Geometry, combinatorial designs and related structures, J.W.P. HIRSCHFELD *et al*
246 p-Automorphisms of finite p-groups, E.I. KHUKHRO
247 Analytic number theory, Y. MOTOHASHI (ed)
248 Tame topology and o-minimal structures, LOU VAN DEN DRIES
249 The atlas of finite groups: ten years on, ROBERT CURTIS & ROBERT WILSON (eds)
250 Characters and blocks of finite groups, G. NAVARRO
251 Gröbner bases and applications, B. BUCHBERGER & F. WINKLER (eds)
252 Geometry and cohomology in group theory, P. KROPHOLLER, G. NIBLO, R. STÖHR (eds)
253 The q-Schur algebra, S. DONKIN
254 Galois representations in arithmetic algebraic geometry, A.J. SCHOLL & R.L. TAYLOR (eds)
255 Symmetries and integrability of difference equations, P.A. CLARKSON & F.W. NIJHOFF (eds)
256 Aspects of Galois theory, HELMUT VÖLKLEIN *et al*
257 An introduction to noncommutative differential geometry and its physical applications 2ed, J. MADORE
258 Sets and proofs, S.B. COOPER & J. TRUSS (eds)
259 Models and computability, S.B. COOPER & J. TRUSS (eds)
260 Groups St Andrews 1997 in Bath, I, C.M. CAMPBELL *et al*
261 Groups St Andrews 1997 in Bath, II, C.M. CAMPBELL *et al*
262 Analysis and logic, C.W. HENSON, J. IOVINO, A.S. KECHRIS & E. ODELL
263 Singularity theory, BILL BRUCE & DAVID MOND (eds)
264 New trends in algebraic geometry, K. HULEK, F. CATANESE, C. PETERS & M. REID (eds)
265 Elliptic curves in cryptography, I. BLAKE, G. SEROUSSI & N. SMART
267 Surveys in combinatorics, 1999, J.D. LAMB & D.A. PREECE (eds)
268 Spectral asymptotics in the semi-classical limit, M. DIMASSI & J. SJÖSTRAND
269 Ergodic theory and topological dynamics, M.B. BEKKA & M. MAYER
270 Analysis on Lie groups, N.T. VAROPOULOS & S. MUSTAPHA
271 Singular perturbations of differential operators, S. ALBEVERIO & P. KURASOV
272 Character theory for the odd order theorem, T. PETERFALVI
273 Spectral theory and geometry, E.B. DAVIES & Y. SAFAROV (eds)
274 The Mandlebrot set, theme and variations, TAN LEI (ed)
275 Descriptive set theory and dynamical systems, M. FOREMAN *et al*
276 Singularities of plane curves, E. CASAS-ALVERO
277 Computational and geometric aspects of modern algebra, M. D. ATKINSON *et al*
278 Global attractors in abstract parabolic problems, J.W. CHOLEWA & T. DLOTKO
279 Topics in symbolic dynamics and applications, F. BLANCHARD, A. MAASS & A. NOGUEIRA (eds)
280 Characters and automorphism groups of compact Riemann surfaces, THOMAS BREUER
281 Explicit birational geometry of 3-folds, ALESSIO CORTI & MILES REID (eds)
282 Auslander-Buchweitz approximations of equivariant modules, M. HASHIMOTO
283 Nonlinear elasticity, Y. FU & R. W. OGDEN (eds)
284 Foundations of computational mathematics, R. DEVORE, A. ISERLES & E. SÜLI (eds)
285 Rational points on curves over finite fields, H. NIEDERREITER & C. XING
286 Clifford algebras and spinors 2ed, P. LOUNESTO
287 Topics on Riemann surfaces and Fuchsian groups, E. BUJALANCE, A.F. COSTA & E. MARTÌNEZ (eds)
288 Surveys in combinatorics, 2001, J. HIRSCHFELD (ed)
289 Aspects of Sobolev-type inequalities, L. SALOFF-COSTE
290 Quantum groups and Lie Theory, A. PRESSLEY (ed)
291 Tits buildings and the model theory of groups, K. TENT (ed)
292 A quantum groups primer, S. MAJID
293 Second order partial differential equations in Hilbert spaces, G. DA PRATO & J. ZABCZYK
294 Introduction to the theory of operator spaces, G. PISIER

London Mathematical Society Lecture Note Series. 309

Corings and Comodules

Tomasz Brzezinski
University of Wales Swansea

Robert Wisbauer
Heinrich Heine Universität Düsseldorf

CAMBRIDGE
UNIVERSITY PRESS

CAMBRIDGE
UNIVERSITY PRESS

University Printing House, Cambridge CB2 8BS, United Kingdom

One Liberty Plaza, 20th Floor, New York, NY 10006, USA

477 Williamstown Road, Port Melbourne, VIC 3207, Australia

4843/24, 2nd Floor, Ansari Road, Daryaganj, Delhi - 110002, India

79 Anson Road, #06-04/06, Singapore 079906

Cambridge University Press is part of the University of Cambridge.

It furthers the University's mission by disseminating knowledge in the pursuit of education, learning and research at the highest international levels of excellence.

www.cambridge.org
Information on this title: www.cambridge.org/9780521539319

First published 2003

A catalogue record for this publication is available from the British Library

Library of Congress Cataloging in Publication data
Brzezinski, Tomasz, 1966–
Corings and Comodules / Tomasz Brzezinski, Robert Wisbauer.
p. cm. – (London Mathematical Society lecture note series; 309)
Includes bibliographical references and index.
ISBN 0-521-53931-5 (pbk.)
1. Corings (Algebra) 2. Comodules. I. Wisbauer, Robert, 1941–
II. Title. III. Series.
QA251.5.B79 2003
512'.4–dc21 2003046040

ISBN 978-0-521-53931-9 Paperback

Contents

Preface

Corings and comodules are fundamental algebraic structures that can be thought of as both dualisations and generalisations of rings and modules. Corings were introduced by Sweedler in 1975 as a generalisation of coalgebras and as a means of presenting a semi-dual version of the Jacobson-Bourbaki Theorem, but their origin can be traced back to 1968 in the work of Jonah on cohomology of coalgebras in monoidal categories. In the late seventies they resurfaced under the name of bimodules over a category with a coalgebra structure, BOCSs for short, in the work of Rojter and Kleiner on algorithms for matrix problems. For a long time, essentially only two types of examples of corings truly generalising coalgebras were known – one associated to a ring extension, the other to a matrix problem. The latter example was also studied in the context of differential graded algebras and categories. This lack of examples hindered the full appreciation of the fundamental role of corings in algebra and obviously hampered their progress in general coring theory.

On the other hand, from the late seventies and throughout the eighties and nineties, various types of Hopf modules were studied. Initially these were typically modules and comodules of a common bialgebra or a Hopf algebra with some compatibility condition, but this evolved to modules of an algebra and comodules of a coalgebra with a compatibility condition controlled by a bialgebra. In fact, even the background bialgebra has been shown now to be redundant provided some relations between a coalgebra and an algebra are imposed in terms of an entwining. The progress and interest in such categories of modules were fuelled by the emergence of quantum groups and their application to physics, in particular gauge theory in terms of principal bundles and knot theory.

By the end of the last century, M. Takeuchi realised that the compatibility condition between an algebra and a coalgebra known as an entwining can be recast in terms of a coring. From this, it suddenly became apparent that various properties of Hopf modules, including entwined modules, can be understood and more neatly presented from the point of view of the associated coring. It also emerged, on the one hand, that coring theory is rich in many interesting examples and, on the other, that – based on the knowledge of Hopf-type modules – there is much more known about the general structure of corings than has been previously realised. It also turned out that corings might have a variety of unexpected and wide-ranging applications, to topics in noncommutative ring theory, category theory, Hopf algebras, differential graded algebras, and noncommutative geometry. In summary, corings appear to offer a new, exciting possibility for recasting known results in a unified general manner and for the development of ring and module theory from a completely different point of view.

As indicated above, corings can be viewed as generalisations of coalgebras, the latter an established and well-studied theory, in particular over fields. More precisely, a coalgebra over a commutative ring R can be defined as a coalgebra in the monoidal category of R-modules – a notion that is well known in general category theory. On the other hand, an A-coring is a coalgebra in the monoidal category of (A, A)-bimodules, where A is an arbitrary ring. With the emergence of quantum groups in the works of Drinfeld [110], Jimbo [135] and Woronowicz [214], new interest arose in the study of coalgebras, mainly those with additional structures such as bialgebras and Hopf algebras, because of their importance in various applications. In the majority of books on Hopf algebras and coalgebras, such as the now classic texts of Sweedler [45] and Abe [1] or in the more recent works of Montgomery [37] and Dăscălescu, Năstăsescu and Raianu [14] together with texts motivated by quantum group theory (e.g., Lusztig [30]; Majid [33, 34]; Chari and Pressley [11]; Shnider and Sternberg [43]; Kassel [25]; Klimyk and Schmüdgen [26]; Brown and Goodearl [7]), coalgebras are considered over fields. The vast variety of applications and new developments, in particular in ring and module theory, manifestly show that there is still a need for a better understanding of coalgebras over arbitrary commutative rings, as a preliminary step towards the theory of corings. Let us mention a few aspects of particular interest to the classical module and ring theory.

There are parts of module theory over algebras A that provide a perfect setting for the theory of comodules. Given any left A-module M, denote by $\sigma[M]$ the full subcategory of the category ${}_A\mathbf{M}$ of left A-modules that is subgenerated by M. This is the smallest Grothendieck subcategory of ${}_A\mathbf{M}$ containing M. Internal properties of $\sigma[M]$ strongly depend on the module properties of M, and there is a well-established theory that explores this relationship. Although, in contrast to ${}_A\mathbf{M}$, there need not be projectives in $\sigma[M]$, its Grothendieck property enables the use of techniques such as localisation and various homological methods in $\sigma[M]$. Consequently, one can gain a very good understanding of the *inner properties* of $\sigma[M]$. On the other hand, by definition, $\sigma[M]$ is closed under direct sums, submodules and factor modules in ${}_A\mathbf{M}$, and so it is a *hereditary pretorsion class* in ${}_A\mathbf{M}$. If $\sigma[M]$ is also closed under extensions in ${}_A\mathbf{M}$, it is a (*hereditary*) *torsion class*. Torsion theory then provides many characterisations of the *outer properties* of $\sigma[M]$, that is, the behaviour of $\sigma[M]$ as a subclass of ${}_A\mathbf{M}$.

Both the inner and outer properties of the categories of type $\sigma[M]$ are important in the study of coalgebras and comodules. If C is a coalgebra over a commutative ring R, then the dual C^* is an R-algebra and C is a left and right module over C^*. The link to the module theory mentioned above is provided by the basic observation that the category $\mathbf{M}^C$ of right C-comodules is subgenerated by C, and there is a faithful functor from $\mathbf{M}^C$

to the category ${}_{C^*}\mathbf{M}$ of left C^*-modules. Further properties of $\mathbf{M}^C$ and of this relationship depend on the properties of C as an R-module. First, if ${}_RC$ is (locally) projective, then $\mathbf{M}^C$ is the same as the category $\sigma[{}_{C^*}C]$, the full subcategory of ${}_{C^*}\mathbf{M}$ subgenerated by C. In this case, results for module categories of type $\sigma[M]$ can be transferred directly to comodules, with no new proofs required. This affords a deeper understanding of old results about comodules (for coalgebras over fields) and provides new proofs that readily apply to coalgebras over rings. For example, the inner properties of $\sigma[M]$ reveal the internal structure and decomposition properties of C, while the outer properties allow one to study when $\mathbf{M}^C$ is closed under extensions in ${}_{C^*}\mathbf{M}$. Second, if ${}_RC$ is flat, then, although $\mathbf{M}^C$ is no longer a full subcategory of ${}_{C^*}\mathbf{M}$, it is still an Abelian (Grothendieck) category, and the pattern of proofs for $\sigma[M]$ can often be followed literally to prove results for comodules. Third, in the absence of any condition on the R-module structure of ${}_RC$, the category $\mathbf{M}^C$ may lack kernels, and monomorphisms in $\mathbf{M}^C$ need not be injective maps. However, techniques from module theory may still be applied, albeit with more caution. For example, typical properties for Hopf algebras H do not need any restriction on ${}_RH$. Finally, it turns out that practically all of the traditional results, in the case of when R is a field, remain true over QF rings provided ${}_RC$ is flat.

Various properties of coalgebras over a commutative ring R extend to A-corings, provided one carefully addresses the noncommutativity of the base ring A in the latter case. For example, given an A-coring $\mathcal{C}$, one can consider three types of duals of $\mathcal{C}$, namely, the right A-module, the left A-module, or the (A,A)-bimodule dual. In each case one obtains a ring, but, since these rings are no longer isomorphic to each other, one has to carefully study relations between them. This then transfers to the study of categories of comodules of a coring and their relationship to various possible categories of modules over dual rings.

The aim of the present book is to give an introduction to the general theory of corings and to indicate their numerous applications. We would like to stress the role of corings as one of the most fundamental algebraic structures, which, in a certain sense, lie between module theory and category theory. From the former point of view, they give a unifying and general framework for rings and modules, while from the latter they are concrete realisations of adjoint pairs of functors or comonads. We start, however, with a description of the theory of coalgebras over commutative rings and their comodules. Therefore, the first part of the book should give the reader a feeling for the typical features of coalgebras over rings as opposed to fields, thus, in the first instance, filling a gap in the existing literature. It is also a preparation for the second part, which is intended to provide the reader with a reference to the wide tapestry of known results on the structure of corings, possible applications and developments. It is not our aim to give a complete picture of this rapidly developing theory. Instead we would like to indicate what is known and what can be done in this new, emerging field. Thus we provide an overview of known and, by now, standard results about corings scattered in the existing established literature. Furthermore, we outline various aspects of corings studied very recently by several authors in a number of published papers as well as preprints still awaiting formal publication. We believe, however, that a significant number of results included in this book are hereby published for the first time. It is our hope that the present book will become a reference and a starting point for further progress in this new exciting field. We also believe that the first part of the book, describing coalgebras over rings, may serve as a textbook for a graduate course on coalgebras and Hopf algebras for students who would like to specialise in algebra and ring theory.

A few words are in order to explain the structure of the book. The book is primarily intended for mathematicians working in ring and module theory and related subjects, such as Hopf algebras. We believe, however, that it will also be useful for (mathematically oriented) mathematical physicists, in particular those who work with quantum groups and noncommutative geometry. In the main text, we make passing references to how abstract constructions may be seen from their point of view. Moreover, the attention of noncommutative geometers should be drawn, in particular to the construction of connections in Section 29.

The book also assumes various levels of familiarity with coalgebras. The reader who is not familiar with coalgebras should start with Chapter 1. The reader who is familiar with coalgebras and Hopf algebras can proceed directly to Chapter 3 and return to preceding chapters when prompted. For the benefit of readers who are not very confident with the language of categories or with the structure of module categories, the main text is supplemented by an appendix in which we recall well-known facts about categories in general as well as module categories. This is done explicitly enough to provide a helpful guidance for the ideas employed in the main part of the text. Also included in the appendix are some new and less standard items that are used in the development of the theory of comodules in the main text.

It is a great pleasure to acknowledge numerous discussions with and comments of our friends and collaborators, in particular Jawad Abuhlail, Khaled Al-Takhman, Kostia Beidar, Stefaan Caenepeel, Alexander Chamrad-Seidel, John Clark, José Gómez-Torrecillas, Piotr Hajac, Lars Kadison, Christian Lomp, Shahn Majid, Claudia Menini, Gigel Militaru, Mike Prest and Blas Torrecillas. Special thanks are extended to Gabriella Böhm, who helped us to clarify some aspects of weak Hopf algebras. Tomasz Brzeziński thanks the UK Engineering and Physical Sciences Research Council for an Advanced Research Fellowship.

Notations

tw	the twist map $\mathsf{tw} : M \otimes_R N \to N \otimes_R M$, 40.1
$\mathrm{Ke}\, f$ $(\mathrm{Coke} f)$	the kernel (cokernel) of a linear map f
$\mathrm{Im}\, f$	the image of a map f
I, I_X	the identity morphism for an object X
A	algebra over a commutative ring R, 40.2
μ, μ_A	product of A as a map $A \otimes_R A \to A$, 40.2
ι, ι_A	the unit of A as a map $R \to A$, 40.2
$Z(A)$, $\mathrm{Jac}(A)$	the centre and the Jacobson radical of A
$\mathrm{Alg}_R(-,-)$	R-algebra maps
M_A $({}_AM)$	right (left) A-module M, 40.4
ϱ_M $({}_M\varrho)$	the A-action for a right (left) A-module M, 40.4
$\mathbf{M}_A$ $({}_A\mathbf{M})$	the category of right (left) A-modules, 40.4
$\mathrm{Hom}_A(-,-)$	homomorphisms of right A-modules, 40.4
${}_A\mathrm{Hom}(-,-)$	homomorphisms of left A-modules
${}_A\mathbf{M}_B$	the category of (A,B)-bimodules, 40.9
${}_A\mathrm{Hom}_B(-,-)$	homomorphisms of (A,B)-bimodules, 40.9
$\sigma[M]$	the full subcategory of $\mathbf{M}_A$ $({}_A\mathbf{M})$ of modules subgenerated by a module M, 41.1
C $(\mathcal{C})$	coalgebra over R (coring over A), 1.1, 17.1
Δ, Δ_C	the coproduct of C as map $C \to C \otimes_R C$, 1.1, 17.1
ε, ε_C	the counit of C as map $C \to R$, 1.1
$\underline{\Delta}$, $\underline{\Delta}_{\mathcal{C}}$	the coproduct of $\mathcal{C}$ as map $\mathcal{C} \to \mathcal{C} \otimes_A \mathcal{C}$, 17.1
$\underline{\varepsilon}$, $\underline{\varepsilon}_{\mathcal{C}}$	the counit of $\mathcal{C}$ as map $\mathcal{C} \to A$, 17.1
C^*	the dual (convolution) algebra of C, 1.3
$\mathcal{C}^*$, ${}^*\mathcal{C}$, ${}^*\mathcal{C}^*$	the right, left, and bi-dual algebras of $\mathcal{C}$, 17.8
ϱ^M $({}^M\varrho)$	the coaction of a right (left) comodule M, 3.1, 18.1
$\mathbf{M}^C$ $(\mathbf{M}^{\mathcal{C}})$	the category of right comodules over C $(\mathcal{C})$, 3.1, 18.1
$\mathrm{Hom}^C(-,-)$	the colinear maps of right C-comodules, 3.3
${}^{\mathcal{C}}\mathrm{Hom}(-,-)$	the colinear maps of left $\mathcal{C}$-comodules, 18.3
$\mathrm{End}^C(M)$	endomorphisms of a right C-comodule M, 3.12
${}^{\mathcal{C}}\mathrm{End}(M)$	endomorphisms of a left $\mathcal{C}$-comodule M, 18.12
${}^C\mathrm{Hom}^D(-,-)$	(C,D)-bicomodule maps, 11.1
$\mathrm{Rat}^{\mathcal{C}}(M)$	the rational comodule of a left ${}^*\mathcal{C}$-module M, 7.1, 20.1
${}^{\mathcal{C}}\mathbf{M}^{\mathcal{D}}$	the category of $(\mathcal{C},\mathcal{D})$-bicomodules, 22.1

Chapter 1

Coalgebras and comodules

Coalgebras and comodules are dualisations of algebras and modules. In this chapter we introduce the basic definitions and study several properties of these notions. The theory of coalgebras over fields and their comodules is well presented in various textbooks (e.g., Sweedler [45], Abe [1], Montgomery [37], Dăscălescu, Năstăsescu and Raianu [14]). Since the tensor product behaves differently over fields and rings, not all the results for coalgebras over fields can be extended to coalgebras over rings. Here we consider base rings from the very beginning, and part of our problems will be to find out which module properties of a coalgebra over a ring are necessary (and sufficient) to ensure the desired properties. In view of the main subject of this book, this chapter can be treated as a preliminary study towards corings. Also for this reason we almost solely concentrate on those properties of coalgebras and comodules that are important from the module theory point of view. The extra care paid to module properties of coalgebras will pay off in Chapter 3.

Throughout, R denotes a commutative and associative ring with a unit.

1 Coalgebras

Intuitively, a coalgebra over a ring can be understood as a dualisation of an algebra over a ring. Coalgebras by themselves are equally fundamental objects as are algebras. Although probably more difficult to understand at the beginning, they are often easier to handle than algebras. Readers with geometric intuition might like to think about algebras as functions on spaces and about coalgebras as objects that encode additional structure of such spaces (for example, group or monoid structure). The main aim of this section is to introduce and give examples of coalgebras and explain the (dual) relationship between algebras and coalgebras.

1.1. Coalgebras. An *R-coalgebra* is an R-module C with R-linear maps

$$\Delta : C \to C \otimes_R C \quad \text{and} \quad \varepsilon : C \to R,$$

called *(coassociative) coproduct* and *counit*, respectively, with the properties

$$(I_C \otimes \Delta) \circ \Delta = (\Delta \otimes I_C) \circ \Delta, \text{ and } (I_C \otimes \varepsilon) \circ \Delta = I_C = (\varepsilon \otimes I_C) \circ \Delta,$$

which can be expressed by commutativity of the diagrams

$$\begin{array}{ccc} C & \xrightarrow{\Delta} & C\otimes_R C \\ {\scriptstyle\Delta}\downarrow & & \downarrow{\scriptstyle I_C\otimes\Delta} \\ C\otimes_R C & \xrightarrow{\Delta\otimes I_C} & C\otimes_R C\otimes_R C \end{array} \qquad \begin{array}{ccc} C & \xrightarrow{\Delta} & C\otimes_R C \\ {\scriptstyle\Delta}\downarrow & \searrow^{I_C} & \downarrow{\scriptstyle \varepsilon\otimes I_C} \\ C\otimes_R C & \xrightarrow[I_C\otimes\varepsilon]{} & C\,. \end{array}$$

A coalgebra (C,Δ,ε) is said to be *cocommutative* if $\Delta = \mathsf{tw}\circ\Delta$, where

$$\mathsf{tw}: C\otimes_R C \to C\otimes_R C, \quad a\otimes b\mapsto b\otimes a,$$

is the twist map (cf. 40.1).

1.2. Sweedler's Σ-notation. For an elementwise description of the maps we use the Σ-*notation*, writing for $c\in C$

$$\Delta(c) = \sum_{i=1}^{k} c_i\otimes \tilde{c}_i = \sum c_{\underline{1}}\otimes c_{\underline{2}}.$$

The first version is more precise; the second version, introduced by Sweedler, turnes out to be very handy in explicit calculations. Notice that $c_{\underline{1}}$ and $c_{\underline{2}}$ do not represent single elements but families $c_1,\ldots,c_k$ and $\tilde{c}_1,\ldots,\tilde{c}_k$ of elements of C that are by no means uniquely determined. Properties of $c_{\underline{1}}$ can only be considered in context with $c_{\underline{2}}$. With this notation, the coassociativity of Δ is expressed by

$$\sum\Delta(c_{\underline{1}})\otimes c_{\underline{2}} = \sum c_{\underline{1}\underline{1}}\otimes c_{\underline{1}\underline{2}}\otimes c_{\underline{2}} = \sum c_{\underline{1}}\otimes c_{\underline{2}\underline{1}}\otimes c_{\underline{2}\underline{2}} = \sum c_{\underline{1}}\otimes\Delta(c_{\underline{2}}),$$

and, hence, it is possible and convenient to shorten the notation by writing

$$\begin{array}{rcl} (\Delta\otimes I_C)\Delta(c) = (I_C\otimes\Delta)\Delta(c) & = & \sum c_{\underline{1}}\otimes c_{\underline{2}}\otimes c_{\underline{3}}, \\ (I_C\otimes I_C\otimes\Delta)(I_C\otimes\Delta)\Delta(c) & = & \sum c_{\underline{1}}\otimes c_{\underline{2}}\otimes c_{\underline{3}}\otimes c_{\underline{4}}, \end{array}$$

and so on. The conditions for the counit are described by

$$\sum\varepsilon(c_{\underline{1}})c_{\underline{2}} = c = \sum c_{\underline{1}}\varepsilon(c_{\underline{2}}).$$

Cocommutativity is equivalent to $\sum c_{\underline{1}}\otimes c_{\underline{2}} = \sum c_{\underline{2}}\otimes c_{\underline{1}}$.

R-coalgebras are closely related or dual to algebras. Indeed, the module of R-linear maps from a coalgebra C to any R-algebra is an R-algebra.

1.3. The algebra $\mathrm{Hom}_R(C,A)$. *For any R-linear map $\Delta: C\to C\otimes_R C$ and an R-algebra A,* $\mathrm{Hom}_R(C,A)$ *is an R-algebra by the* convolution product

$$f*g = \mu\circ(f\otimes g)\circ\Delta, \quad i.e., \quad f*g(c) = \sum f(c_{\underline{1}})g(c_{\underline{2}}),$$

for $f,g\in$ $\mathrm{Hom}_R(C,A)$ *and $c\in C$. Furthermore,*

(1) *Δ is coassociative if and only if* $\mathrm{Hom}_R(C,A)$ *is an associative R-algebra, for any R-algebra A.*

(2) *C is cocommutative if and only if* $\mathrm{Hom}_R(C,A)$ *is a commutative R-algebra, for any commutative R-algebra A.*

(3) *C has a counit if and only if* $\mathrm{Hom}_R(C,A)$ *has a unit, for all R-algebras A with a unit.*

Proof. (1) Let $f,g,h \in \mathrm{Hom}_R(C,A)$ and consider the R-linear map

$$\tilde{\mu} : A \otimes_R A \otimes_R A \to A, \quad a_1 \otimes a_2 \otimes a_3 \mapsto a_1 a_2 a_3.$$

By definition, the products $(f * g) * h$ and $f * (g * h)$ in $\mathrm{Hom}_R(C,A)$ are the compositions of the maps

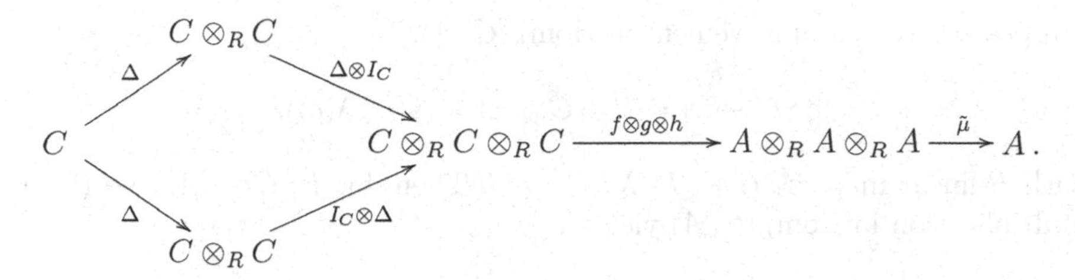

It is obvious that coassociativity of Δ yields associativity of $\mathrm{Hom}_R(C,A)$.

To show the converse, we see from the above diagram that it suffices to prove that, (at least) for one associative algebra A and suitable $f,g,h \in \mathrm{Hom}_R(C,A)$, the composition $\tilde{\mu} \circ (f \otimes g \otimes h)$ is a monomorphism. So let $A = T(C)$, the tensor algebra of the R-module C (cf. 15.12), and $f = g = h$, the canonical mapping $C \to T(C)$. Then $\tilde{\mu} \circ (f \otimes g \otimes h)$ is just the embedding $C \otimes C \otimes C = T_3(C) \to T(C)$.

(2) If C is cocommutative and A is commutative,

$$f * g\,(c) = \sum f(c_{\underline{1}})g(c_{\underline{2}}) = \sum g(c_{\underline{1}})f(c_{\underline{2}}) = g * f\,(c),$$

so that $\mathrm{Hom}_R(C,A)$ is commutative. Conversely, assume that $\mathrm{Hom}_R(C,A)$ is commutative for any commutative A. Then

$$\mu \circ (f \otimes g)(\Delta(c)) = \mu \circ (f \otimes g)(\mathsf{tw} \circ \Delta(c)).$$

This implies $\Delta = \mathsf{tw} \circ \Delta$ provided we can find a commutative algebra A and $f,g \in \mathrm{Hom}_R(C,A)$ such that $\mu \circ (f \otimes g) : C \otimes_R C \to A$ is injective. For this take A to be the symmetric algebra $\mathcal{S}(C \oplus C)$ (see 15.13). For f and g we choose the mappings

$$C \to C \oplus C, \;\; x \mapsto (x,0), \quad C \to C \oplus C, \;\; x \mapsto (0,x),$$

composed with the canonical embedding $C \oplus C \to \mathcal{S}(C \oplus C)$.

With the canonical isomorphism $h : \mathcal{S}(C) \otimes \mathcal{S}(C) \to \mathcal{S}(C \oplus C)$ (see 15.13) and the embedding $\lambda : C \to \mathcal{S}(C)$, we form $h^{-1} \circ \mu \circ (f \otimes g) = \lambda \otimes \lambda$. Since $\lambda(C)$ is a direct summand of $\mathcal{S}(C)$, we obtain that $\lambda \otimes \lambda$ is injective and so $\mu \circ (f \otimes g)$ is injective.

(3) It is easy to check that the unit in $\mathrm{Hom}_R(C, A)$ is

$$C \xrightarrow{\varepsilon} R \xrightarrow{\iota} A, \quad c \mapsto \varepsilon(c)1_A.$$

For the converse, consider the R-module $A = R \oplus C$ and define a unital R-algebra

$$\mu : A \otimes_R A \to A, \quad (r, a) \otimes (s, b) \mapsto (rs, rb + as).$$

Suppose there is a unit element in $\mathrm{Hom}_R(C, A)$,

$$e : C \to A = R \oplus C, \quad c \mapsto (\varepsilon(c), \lambda(c)),$$

with R-linear maps $\varepsilon : C \to R$, $\lambda : C \to C$. Then, for $f : C \to A$, $c \mapsto (0, c)$, multiplication in $\mathrm{Hom}_R(C, A)$ yields

$$f * e : C \to A, \quad c \mapsto (0, (I_C \otimes \varepsilon) \circ \Delta(c)).$$

By assumption, $f = f * e$ and hence $I_C = (I_C \otimes \varepsilon) \circ \Delta$, one of the conditions for ε to be a counit. Similarly, the other condition is derived from $f = e * f$.

Clearly ε is the unit in $\mathrm{Hom}_R(C, R)$, showing the uniqueness of a counit for C. □

Note in particular that $C^* = \mathrm{Hom}_R(C, R)$ is an algebra with the convolution product known as the *dual* or *convolution algebra* of C.

Notation. From now on, C (usually) will denote a coassociative R-coalgebra (C, Δ, ε), and A will stand for an associative R-algebra with unit (A, μ, ι).

Many properties of coalgebras depend on properties of the base ring R. The base ring can be changed in the following way.

1.4. Scalar extension. *Let C be an R-coalgebra and S an associative commutative R-algebra with unit. Then $C \otimes_R S$ is an S-coalgebra with the coproduct*

$$\tilde{\Delta} : C \otimes_R S \xrightarrow{\Delta \otimes I_S} (C \otimes_R C) \otimes_R S \xrightarrow{\simeq} (C \otimes_R S) \otimes_S (C \otimes_R S)$$

and the counit $\varepsilon \otimes I_S : C \otimes_R S \to S$. If C is cocommutative, then $C \otimes_R S$ is cocommutative.

Proof. By definition, for any $c \otimes s \in C \otimes_R S$,

$$\tilde{\Delta}(c \otimes s) = \sum (c_{\underline{1}} \otimes 1_S) \otimes_S (c_{\underline{2}} \otimes s).$$

It is easily checked that $\tilde{\Delta}$ is coassociative. Moreover,

$$(\varepsilon \otimes I_S \otimes I_{C \otimes_R S}) \circ \tilde{\Delta}(c \otimes s) = \sum \varepsilon(c_{\underline{1}}) c_{\underline{2}} \otimes s = c \otimes s,$$

and similarly $(I_{C \otimes_R S} \otimes \varepsilon \otimes I_S) \circ \tilde{\Delta} = I_{C \otimes_R S}$ is shown.

Obviously cocommutativity of Δ implies cocommutativity of $\tilde{\Delta}$. □

To illustrate the notions introduced above we consider some examples.

1.5. R as a coalgebra. The ring R is (trivially) a coassociative, cocommutative coalgebra with the canonical isomorphism $R \to R \otimes_R R$ as coproduct and the identity map $R \to R$ as counit.

1.6. Free modules as coalgebras. Let F be a free R-module with basis $(f_\lambda)_\Lambda$, Λ any set. Then there is a unique R-linear map

$$\Delta : F \to F \otimes_R F, \quad f_\lambda \mapsto f_\lambda \otimes f_\lambda,$$

defining a coassociative and cocommutative coproduct on F. The counit is provided by the linear map $\varepsilon : F \to R$, $f_\lambda \longmapsto 1$.

1.7. Semigroup coalgebra. Let G be a semigroup. A coproduct and counit on the semigroup ring $R[G]$ can be defined by

$$\Delta_1 : R[G] \to R[G] \otimes_R R[G], \ g \mapsto g \otimes g, \quad \varepsilon_1 : R[G] \to R, \ g \mapsto 1.$$

If G has a unit e, then another possibility is

$$\Delta_2 : R[G] \to R[G] \otimes_R R[G], \quad g \mapsto \begin{cases} e \otimes e & \text{if } g = e, \\ g \otimes e + e \otimes g & \text{if } g \neq e. \end{cases}$$

$$\varepsilon_2 : R[G] \to R, \quad g \mapsto \begin{cases} 1 & \text{if } g = e, \\ 0 & \text{if } g \neq e. \end{cases}$$

Both Δ_1 and Δ_2 are coassociative and cocommutative.

1.8. Polynomial coalgebra. A coproduct and counit on the polynomial ring $R[X]$ can be defined as algebra homomorphisms by

$$\begin{aligned} \Delta_1 : R[X] &\to R[X] \otimes_R R[X], & X^i &\mapsto X^i \otimes X^i, \\ \varepsilon_1 : R[X] &\to R, & X^i &\mapsto 1, \quad i = 0, 1, 2, \ldots. \end{aligned}$$

or else by

$$\begin{aligned} \Delta_2 : R[X] &\to R[X] \otimes_R R[X], & 1 \mapsto 1,\ X^i &\mapsto (X \otimes 1 + 1 \otimes X)^i, \\ \varepsilon_2 : R[X] &\to R, & 1 \mapsto 1,\ X^i &\mapsto 0, \quad i = 1, 2, \ldots. \end{aligned}$$

Again, both Δ_1 and Δ_2 are coassociative and cocommutative.

1.9. Coalgebra of a projective module. Let P be a finitely generated projective R-module with dual basis $p_1, \ldots, p_n \in P$ and $\pi_1, \ldots, \pi_n \in P^*$. There is an isomorphism

$$P \otimes_R P^* \to \mathrm{End}_R(P), \quad p \otimes f \mapsto [a \mapsto f(a)p],$$

and on $P^* \otimes_R P$ the coproduct and counit are defined by

$$\Delta : P^* \otimes_R P \to (P^* \otimes_R P) \otimes_R (P^* \otimes_R P), \quad f \otimes p \mapsto \sum_i f \otimes p_i \otimes \pi_i \otimes p,$$

$$\varepsilon : P^* \otimes_R P \to R, \quad f \otimes p \mapsto f(p).$$

By properties of the dual basis,

$$(I_{P \otimes_R P^*} \otimes \varepsilon)\Delta(f \otimes p) = \sum_i f \otimes p_i \pi_i(p) = f \otimes p,$$

showing that ε is a counit, and coassociativity of Δ is proved by the equality

$$(I_{P \otimes_R P^*} \otimes \Delta)\Delta(f \otimes p) = \sum\nolimits_{i,j} f \otimes p_i \otimes \pi_i \otimes p_j \otimes \pi_j \otimes p = (\Delta \otimes I_{P \otimes_R P^*})\Delta(f \otimes p).$$

The dual algebra of $P^* \otimes_R P$ is (anti)isomorphic to $\mathrm{End}_R(P)$ by the bijective maps

$$(P^* \otimes_R P)^* = \mathrm{Hom}_R(P^* \otimes_R P, R) \simeq \mathrm{Hom}_R(P, P^{**}) \simeq \mathrm{End}_R(P),$$

which yield a ring isomorphism or anti-isomorphism, depending from which side the morphisms are acting.

For $P = R$ we obtain $R = R^*$, and $R^* \otimes_R R \simeq R$ is the trivial coalgebra. As a more interesting special case we may consider $P = R^n$. Then $P^* \otimes_R P$ can be identified with the matrix ring $M_n(R)$, and this leads to the

1.10. Matrix coalgebra. Let $\{e_{ij}\}_{1 \leq i,j \leq n}$ be the canonical R-basis for $M_n(R)$, and define the coproduct and counit

$$\Delta : M_n(R) \to M_n(R) \otimes_R M_n(R), \quad e_{ij} \mapsto \sum\nolimits_k e_{ik} \otimes e_{kj},$$

$$\varepsilon : M_n(R) \to R, \quad e_{ij} \mapsto \delta_{ij}\,.$$

The resulting coalgebra is called the *(n, n)-matrix coalgebra over R*, and we denote it by $M_n^c(R)$.

Notice that the matrix coalgebra may also be considered as a special case of a semigroup coalgebra in 1.7.

From a given coalgebra one can construct the

1.11. Opposite coalgebra. Let $\Delta : C \to C \otimes_R C$ define a coalgebra. Then

$$\Delta^{\mathsf{tw}} : C \xrightarrow{\Delta} C \otimes_R C \xrightarrow{\mathsf{tw}} C \otimes_R C, \quad c \mapsto \sum c_{\underline{2}} \otimes c_{\underline{1}},$$

where tw is the twist map, defines a new coalgebra structure on C known as the *opposite* coalgebra with the same counit. The opposite coalgebra is denoted by C^{cop}. Note that a coalgebra C is cocommutative if and only if C coincides with its opposite coalgebra (i.e., $\Delta = \Delta^{\mathsf{tw}}$).

1.12. Duals of algebras. Let (A, μ, ι) be an R-algebra and assume ${}_RA$ to be finitely generated and projective. Then there is an isomorphism

$$A^* \otimes_R A^* \to (A \otimes_R A)^*, \quad f \otimes g \mapsto [a \otimes b \mapsto f(a)g(b)],$$

and the functor $\mathrm{Hom}_R(-, R) = (-)^*$ yields a coproduct

$$\mu^* : A^* \to (A \otimes_R A)^* \simeq A^* \otimes_R A^*$$

and a counit (as the dual of the unit of A)

$$\varepsilon := \iota^* : A^* \to R, \quad f \mapsto f(1_A).$$

This makes A^* an R-coalgebra that is cocommutative provided μ is commutative. If ${}_RA$ is not finitely generated and projective, the above construction does not work. However, under certain conditions the *finite dual* of A has a coalgebra structure (see 5.7).

Further examples of coalgebras are the tensor algebra 15.12, the symmetric algebra 15.13, and the exterior algebra 15.14 of any R-module, and the enveloping algebra of any Lie algebra.

1.13. Exercises

Let $M_n^c(R)$ be a matrix coalgebra with basis $\{e_{ij}\}_{1 \leq i,j \leq n}$ (see 1.10). Prove that the dual algebra $M_n^c(R)^*$ is an (n, n)-matrix algebra.

(Hint: Consider the basis of M^* dual to $\{e_{ij}\}_{1 \leq i,j \leq n}$.)

References. Abuhlail, Gómez-Torrecillas and Wisbauer [50]; Bourbaki [5]; Sweedler [45]; Wisbauer [210].

2 Coalgebra morphisms

To discuss coalgebras formally, one would like to understand not only isolated coalgebras, but also coalgebras in relation to other coalgebras. In a word, one would like to view coalgebras as objects in a category.[1] For this one needs the notion of a *coalgebra morphism.* Such a morphism can be defined as an R-linear map between coalgebras that respects the coalgebra structures (coproducts and counits). The idea behind this definition is of course borrowed from the idea of an algebra morphism as a map respecting the algebra structures. Once such morphisms are introduced, relationships between coalgebras can be studied. In particular, we can introduce the notions of a *subcoalgebra* and a *quotient coalgebra.* These are the topics of the present section.

2.1. Coalgebra morphisms. Given R-coalgebras C and C', an R-linear map $f : C \to C'$ is said to be a *coalgebra morphism* provided the diagrams

$$\begin{array}{ccc} C & \xrightarrow{f} & C' \\ {\scriptstyle\Delta}\downarrow & & \downarrow{\scriptstyle\Delta'} \\ C \otimes_R C & \xrightarrow{f\otimes f} & C' \otimes_R C' , \end{array} \qquad \begin{array}{ccc} C & \xrightarrow{f} & C' \\ & {\scriptstyle\varepsilon}\searrow & \downarrow{\scriptstyle\varepsilon'} \\ & & R \end{array}$$

are commutative. Explicitly, this means that

$$\Delta' \circ f = (f \otimes f) \circ \Delta, \quad \text{and} \quad \varepsilon' \circ f = \varepsilon,$$

that is, for all $c \in C$,

$$\sum f(c_{\underline{1}}) \otimes f(c_{\underline{2}}) = \sum f(c)_{\underline{1}} \otimes f(c)_{\underline{2}}, \quad \text{and} \quad \varepsilon'(f(c)) = \varepsilon(c).$$

Given an R-coalgebra C and an S-coalgebra D, where S is a commutative ring, a *coalgebra morphism* between C and D is defined as a pair (α, γ) consisting of a ring morphism $\alpha : R \to S$ and an R-linear map $\gamma : C \to D$ such that

$$\gamma' : C \otimes_R S \to D, \quad c \otimes s \mapsto \gamma(c)s,$$

is an S-coalgebra morphism. Here we consider D as an R-module (induced by α) and $C \otimes_R S$ is the scalar extension of C (see 1.4).

As shown in 1.3, for an R-algebra A, the contravariant functor $\mathrm{Hom}_R(-, A)$ turns coalgebras to algebras. It also turns coalgebra morphisms into algebra morphisms.

---

[1]The reader not familiar with category theory is referred to the Appendix, §38.

2.2. Duals of coalgebra morphisms. *For R-coalgebras C and C', an R-linear map $f : C \to C'$ is a coalgebra morphism if and only if*

$$\mathrm{Hom}(f, A) : \mathrm{Hom}_R(C', A) \to \mathrm{Hom}_R(C, A)$$

is an algebra morphism, for any R-algebra A.

Proof. Let f be a coalgebra morphism. Putting $f^* = \mathrm{Hom}_R(f, A)$, we compute for $g, h \in \mathrm{Hom}_R(C', A)$

$$\begin{aligned} f^*(g * h) &= \mu \circ (g \otimes h) \circ \Delta' \circ f = \mu \circ (g \otimes h) \circ (f \otimes f) \circ \Delta \\ &= (g \circ f) * (h \circ f) \;=\; f^*(g) * f^*(h). \end{aligned}$$

To show the converse, assume that f^* is an algebra morphism, that is,

$$\mu \circ (g \otimes h) \circ \Delta' \circ f = \mu \circ (g \otimes h) \circ (f \otimes f) \circ \Delta,$$

for any R-algebra A and $g, h \in \mathrm{Hom}_R(C', A)$. Choose A to be the tensor algebra $T(C)$ of the R-module C and choose g, h to be the canonical embedding $C \to T(C)$ (see 15.12). Then $\mu \circ (g \otimes h)$ is just the embedding $C \otimes_R C \to T_2(C) \to T(C)$, and the above equality implies

$$\Delta' \circ f = (f \otimes f) \circ \Delta,$$

showing that f is a coalgebra morphism. □

2.3. Coideals. The problem of determining which R-submodules of C are kernels of a coalgebra map $f : C \to C'$ is related to the problem of describing the kernel of $f \otimes f$ (in the category of R-modules $\mathbf{M}_R$). If f is surjective, we know that $\mathrm{Ke}\,(f \otimes f)$ is the sum of the canonical images of $\mathrm{Ke}\, f \otimes_R C$ and $C \otimes_R \mathrm{Ke}\, f$ in $C \otimes_R C$ (see 40.15). This suggests the following definition.

The kernel of a surjective coalgebra morphism $f : C \to C'$ is called a *coideal* of C.

2.4. Properties of coideals. *For an R-submodule $K \subset C$ and the canonical projection $p : C \to C/K$, the following are equivalent:*

(a) K is a coideal;

(b) C/K is a coalgebra and p is a coalgebra morphism;

(c) $\Delta(K) \subset \mathrm{Ke}\,(p \otimes p)$ and $\varepsilon(K) = 0$.

If $K \subset C$ is C-pure, then (c) is equivalent to:

(d) $\Delta(K) \subset C \otimes_R K + K \otimes_R C$ and $\varepsilon(K) = 0$.

If (a) holds, then C/K is cocommutative provided C is also.

Proof. (a) $\Leftrightarrow$ (b) is obvious.
(b) $\Rightarrow$ (c) There is a commutative exact diagram

$$\begin{array}{ccccccccc} 0 & \longrightarrow & K & \longrightarrow & C & \xrightarrow{p} & C/K & \longrightarrow & 0 \\ & & \downarrow & & \downarrow{\scriptstyle \Delta} & & \downarrow{\scriptstyle \bar{\Delta}} & & \\ 0 & \longrightarrow & \mathrm{Ke}\,(p\otimes p) & \longrightarrow & C\otimes_R C & \xrightarrow{p\otimes p} & C/K\otimes_R C/K & \longrightarrow & 0, \end{array}$$

where commutativity of the right square implies the existence of a morphism $K \to \mathrm{Ke}\,(p\otimes p)$, thus showing $\Delta(K) \subset \mathrm{Ke}\,(p\otimes p)$. For the counit $\bar{\varepsilon} : C/K \to R$ of C/K, $\bar{\varepsilon}\circ p = \varepsilon$ and hence $\varepsilon(K) = 0$

(c) $\Rightarrow$ (b) Under the given conditions, the left-hand square in the above diagram is commutative and the cokernel property of p implies the existence of $\bar{\Delta}$. This makes C/K a coalgebra with the properties required.

(c) $\Leftrightarrow$ (d) If $K \subset C$ is C-pure, $\mathrm{Ke}\,(p\otimes p) = C\otimes_R K + K\otimes_R C$. □

2.5. Factorisation theorem. *Let $f : C \to C'$ be a morphism of R-coalgebras. If $K \subset C$ is a coideal and $K \subset \mathrm{Ke}\, f$, then there is a commutative diagram of coalgebra morphisms*

$$\begin{array}{ccc} C & \xrightarrow{p} & C/K \\ & {\scriptstyle f}\searrow & \downarrow{\scriptstyle \bar{f}} \\ & & C'. \end{array}$$

Proof. Denote by $\bar{f} : C/K \to C'$ the R-module factorisation of $f : C \to C'$. It is easy to show that the diagram

$$\begin{array}{ccc} C/K & \xrightarrow{\bar{f}} & C' \\ {\scriptstyle \bar{\Delta}}\downarrow & & \downarrow{\scriptstyle \Delta'} \\ C/K\otimes_R C/K & \xrightarrow{\bar{f}\otimes\bar{f}} & C'\otimes_R C' \end{array}$$

is commutative. This means that $\bar{f}$ is a coalgebra morphism. □

2.6. The counit as a coalgebra morphism. *View R as a trivial R-coalgebra as in 1.5. Then, for any R-coalgebra C,*

(1) ε is a coalgebra morphism;

(2) if ε is surjective, then $\mathrm{Ke}\,\varepsilon$ is a coideal.

Proof. (1) Consider the diagram

$$\begin{array}{ccc} C & \xrightarrow{\varepsilon} & R \\ {\scriptstyle \Delta}\downarrow & & \downarrow{\scriptstyle \simeq} \\ C\otimes_R C & \xrightarrow{\varepsilon\otimes\varepsilon} & R\otimes_R R \end{array} \qquad \begin{array}{ccc} c & \longmapsto & \varepsilon(c) \\ \downarrow & & \downarrow \\ \sum c_{\underline{1}}\otimes c_{\underline{2}} & \longmapsto & \sum \varepsilon(c_{\underline{1}})\otimes\varepsilon(c_{\underline{2}}) . \end{array}$$

The properties of the counit yield

$$\sum \varepsilon(c_{\underline{1}}) \otimes \varepsilon(c_{\underline{2}}) = \sum \varepsilon(c_{\underline{1}})\varepsilon(c_{\underline{2}}) \otimes 1 = \varepsilon(\sum c_{\underline{1}}\varepsilon(c_{\underline{2}})) \otimes 1 = \varepsilon(c) \otimes 1,$$

so the above diagram is commutative and ε is a coalgebra morphism.

(2)This is clear by (1) and the definition of coideals. □

2.7. Subcoalgebras. An R-submodule D of a coalgebra C is called a *subcoalgebra* provided D has a coalgebra structure such that the inclusion map is a coalgebra morphism.

Notice that a pure R-submodule (see 40.13 for a discussion of purity) $D \subset C$ is a subcoalgebra provided $\Delta_D(D) \subset D \otimes_R D \subset C \otimes_R C$ and $\varepsilon|_D : D \to R$ is a counit for D. Indeed, since D is a pure submodule of C, we obtain

$$\Delta_D(D) = D \otimes_R C \cap C \otimes_R D = D \otimes_R D \subset C \otimes_R C,$$

so that D has a coalgebra structure for which the inclusion is a coalgebra morphism, as required.

From the above observations we obtain:

2.8. Image of coalgebra morphisms. *The image of any coalgebra map $f : C \to C'$ is a subcoalgebra of C'.*

2.9. Remarks. (1) In a general category $\mathbf{A}$, *subobjects* of an object A in $\mathbf{A}$ are defined as *equivalence classes of monomorphisms* $D \to A$. In the definition of subcoalgebras we restrict ourselves to *subsets* (or inclusions) of an object. This will be general enough for our purposes.

(2) The fact that – over arbitrary rings – the tensor product of injective linear maps need not be injective leads to some unexpected phenomena. For example, a submodule D of a coalgebra C can have two distinct coalgebra structures such that, for both of them, the inclusion is a coalgebra map (see Exercise 2.15(3)). It may also happen that, for a submodule V of a coalgebra C, $\Delta(V)$ is contained in the image of the canonical map $V \otimes_R V \to C \otimes_R C$, yet V has no coalgebra structure for which the inclusion $V \to C$ is a coalgebra map (see Exercise 2.15(4)). Another curiosity is that the kernel of a coalgebra morphism $f : C \to C'$ need not be a coideal in case f is not surjective (see Exercise 2.15(5)).

2.10. Coproduct of coalgebras. For a family $\{C_\lambda\}_\Lambda$ of R-coalgebras, put $C = \bigoplus_\Lambda C_\lambda$, the coproduct in $\mathbf{M}_R$, $i_\lambda : C_\lambda \to C$ the canonical inclusions, and consider the R-linear maps

$$C_\lambda \xrightarrow{\Delta_\lambda} C_\lambda \otimes C_\lambda \subset C \otimes C, \quad \varepsilon : C_\lambda \to R.$$

By the properties of coproducts of R-modules there exist unique maps

$$\Delta : C \to C \otimes_R C \text{ with } \Delta \circ i_\lambda = \Delta_\lambda, \quad \varepsilon : C \to R \text{ with } \varepsilon \circ i_\lambda = \varepsilon_\lambda.$$

(C, Δ, ε) is called the *coproduct* (or *direct sum*) of the coalgebras C_λ. It is obvious that the $i_\lambda : C_\lambda \to C$ are coalgebra morphisms.

C is coassociative (cocommutative) if and only if all the C_λ have the corresponding property. This follows – by 1.3 – from the ring isomorphism

$$\mathrm{Hom}_R(C, A) = \mathrm{Hom}_R(\textstyle\bigoplus_\Lambda C_\lambda, A) \simeq \prod_\Lambda \mathrm{Hom}_R(C_\lambda, A),$$

for any R-algebra A, and the observation that the left-hand side is an associative (commutative) ring if and only if every component in the right-hand side has this property.

Universal property of $C = \bigoplus_\Lambda C_\lambda$. *For a family $\{f_\lambda : C_\lambda \to C'\}_\Lambda$ of coalgebra morphisms there exists a unique coalgebra morphism $f : C \to C'$ such that, for all $\lambda \in \Lambda$, there are commutative diagrams of coalgebra morphisms*

$$\begin{array}{ccc} C_\lambda & \xrightarrow{i_\lambda} & C \\ & \searrow^{f_\lambda} & \downarrow f \\ & & C' . \end{array}$$

2.11. Direct limits of coalgebras. Let $\{C_\lambda, f_{\lambda\mu}\}_\Lambda$ be a direct family of R-coalgebras (with coalgebra morphisms $f_{\lambda\mu}$) over a directed set Λ. Let $\varinjlim C_\lambda$ denote the direct limit in $\mathbf{M}_R$ with canonical maps $f_\mu : C_\mu \to \varinjlim C_\lambda$. Then the $f_{\lambda\mu} \otimes f_{\lambda\mu} : C_\lambda \otimes C_\lambda \to C_\mu \otimes C_\mu$ form a directed system (in $\mathbf{M}_R$) and there is the following commutative diagram

$$\begin{array}{ccccc} C_\mu & \xrightarrow{\Delta_\mu} & C_\mu \otimes C_\mu & & \\ {\scriptstyle f_\mu}\downarrow & & \downarrow & \searrow^{f_\mu \otimes f_\mu} & \\ \varinjlim C_\lambda & \xrightarrow{\delta} & \varinjlim (C_\lambda \otimes C_\lambda) & \xrightarrow{\theta} & \varinjlim C_\lambda \otimes \varinjlim C_\lambda, \end{array}$$

where the maps δ and θ exist by the universal properties of direct limits. The composition

$$\Delta_{lim} = \theta \circ \delta : \varinjlim C_\lambda \to \varinjlim C_\lambda \otimes \varinjlim C_\lambda$$

turns $\varinjlim C_\lambda$ into a coalgebra such that the canonical map (e.g., [46, 24.2])

$$p : \textstyle\bigoplus_\Lambda C_\lambda \to \varinjlim C_\lambda$$

is a coalgebra morphism. The counit of $\varinjlim C_\lambda$ is the map ε_{lim} determined by the commutativity of the diagrams

$$\begin{array}{ccc} C_\mu & \xrightarrow{f_\mu} & \varinjlim C_\lambda \\ & \searrow^{\varepsilon_\mu} & \downarrow \varepsilon_{lim} \\ & & R . \end{array}$$

For any associative R-algebra A,

$$\mathrm{Hom}_R(\varinjlim C_\lambda, A) \simeq \varprojlim \mathrm{Hom}_R(C_\lambda, A) \subset \prod_\Lambda \mathrm{Hom}_R(C_\lambda, A),$$

and from this we conclude – by 1.3 – that the coalgebra $\varinjlim C_\lambda$ is coassociative (cocommutative) whenever all the C_λ are coassociative (cocommutative).

Recall that for the definition of the tensor product of R-algebras A, B, the twist map $\mathsf{tw} : A \otimes_R B \to B \otimes_R A$, $a \otimes b \mapsto b \otimes a$ is needed. It also helps to define the

2.12. Tensor product of coalgebras. Let C and D be two R-coalgebras. Then the composite map

$$C \otimes_R D \xrightarrow{\Delta_C \otimes \Delta_D} (C \otimes_R C) \otimes_R (D \otimes_R D) \xrightarrow{I_C \otimes \mathsf{tw} \otimes I_D} (C \otimes_R D) \otimes_R (C \otimes_R D)$$

defines a coassociative coproduct on $C \otimes_R D$, and with the counits ε_C of C and ε_D of D the map $\varepsilon_C \otimes \varepsilon_D : C \otimes_R D \to R$ is a counit of $C \otimes_R D$. With these maps, $C \otimes_R D$ is called the *tensor product coalgebra* of C and D. Obviously $C \otimes_R D$ is cocommutative provided both C and D are cocommutative.

2.13. Tensor product of coalgebra morphisms. *Let $f : C \to C'$ and $g : D \to D'$ be morphisms of R-coalgebras. The tensor product of f and g yields a coalgebra morphism*

$$f \otimes g : C \otimes_R D \to C' \otimes_R D'.$$

In particular, there are coalgebra morphisms

$$I_C \otimes \varepsilon_D : C \otimes_R D \to C, \quad \varepsilon_C \otimes I_D : C \otimes_R D \to D,$$

which, for any commutative R-algebra A, lead to an algebra morphism

$$\mathrm{Hom}_R(C, A) \otimes_R \mathrm{Hom}_R(D, A) \to \mathrm{Hom}_R(C \otimes_R D, A),$$

$$\xi \otimes \zeta \mapsto (\xi \circ (I_C \otimes \varepsilon_D)) * (\zeta \circ (\varepsilon_C \otimes I_D)),$$

where $*$ denotes the convolution product (cf. 1.3).

Proof. The fact that f and g are coalgebra morphisms implies commutativity of the top square in the diagram

$$\begin{array}{ccc}
C \otimes_R D & \xrightarrow{f \otimes g} & C' \otimes_R D' \\
\downarrow{\scriptstyle \Delta_C \otimes \Delta_D} & & \downarrow{\scriptstyle \Delta_{C'} \otimes \Delta_{D'}} \\
C \otimes_R C \otimes_R D \otimes_R D & \xrightarrow{f \otimes f \otimes g \otimes g} & C' \otimes_R C' \otimes_R D' \otimes_R D' \\
\downarrow{\scriptstyle I_C \otimes \mathsf{tw} \otimes I_D} & & \downarrow{\scriptstyle I_{C'} \otimes \mathsf{tw} \otimes I_{D'}} \\
C \otimes_R D \otimes_R C \otimes_R D & \xrightarrow{f \otimes g \otimes f \otimes g} & C' \otimes_R D' \otimes_R C' \otimes_R D',
\end{array}$$

while the bottom square obviously is commutative by the definitions. Commutativity of the outer rectangle means that $f \otimes g$ is a coalgebra morphism.

By 2.2, the coalgebra morphisms $C \otimes_R D \to C$ and $C \otimes_R D \to D$ yield algebra maps

$$\mathrm{Hom}_R(C, A) \to \mathrm{Hom}_R(C \otimes_R D, A), \quad \mathrm{Hom}_R(D, A) \to \mathrm{Hom}_R(C \otimes_R D, A),$$

and with the product in $\mathrm{Hom}_R(C \otimes_R D, A)$ we obtain a map

$$\mathrm{Hom}_R(C, A) \times \mathrm{Hom}_R(D, A) \to \mathrm{Hom}_R(C \otimes_R D, A),$$

which is R-linear and hence factorises over $\mathrm{Hom}_R(C, A) \otimes_R \mathrm{Hom}_R(D, A)$. This is in fact an algebra morphism since the image of $\mathrm{Hom}_R(C, A)$ commutes with the image of $\mathrm{Hom}_R(D, A)$ by the equalities

$$\begin{aligned}
&((\xi \circ (I_C \otimes \varepsilon_D)) * (\zeta \circ (\varepsilon_C \otimes I_D)))(c \otimes d) \\
&= \sum \xi \circ (I_C \otimes \varepsilon_D) \otimes \zeta \circ (\varepsilon_C \otimes I_D)(c_{\underline{1}} \otimes d_{\underline{1}} \otimes c_{\underline{2}} \otimes d_{\underline{2}}) \\
&= \sum \xi(c_{\underline{1}}\varepsilon(d_{\underline{1}}))\, \zeta(\varepsilon(c_{\underline{2}})d_{\underline{2}}) \\
&= \sum \xi(c_{\underline{1}}\varepsilon(c_{\underline{2}}), \zeta(\varepsilon(d_{\underline{1}})d_{\underline{2}}) \\
&= \xi(c)\, \zeta(d) = \zeta(d)\, \xi(c) \\
&= ((\zeta \circ (\varepsilon_C \otimes I_D)) * (\xi \circ (I_C \otimes \varepsilon_D)))(c \otimes d),
\end{aligned}$$

where $\xi \in \mathrm{Hom}_R(C, A)$, $\zeta \in \mathrm{Hom}_R(D, A)$ and $c \in C$, $d \in D$. □

To define the comultiplication for the tensor product of two R-coalgebras C, D in 2.12, the twist map $\mathsf{tw} : C \otimes_R D \to D \otimes_R C$ was used. Notice that any such map yields a formal comultiplication on $C \otimes_R D$, whose properties strongly depend on the properties of the map chosen.

2.14. Coalgebra structure on the tensor product. For R-coalgebras $(C, \Delta_C, \varepsilon_C)$ and $(D, \Delta_D, \varepsilon_D)$, let $\omega : C \otimes_R D \to D \otimes_R C$ be an R-linear map. Explicitly on elements we write $\omega(c \otimes d) = \sum d^\omega \otimes c^\omega$. Denote by $C \ltimes_\omega D$ the R-module $C \otimes_R D$ endowed with the maps

$$\begin{aligned}
\bar{\Delta} = (I_C \otimes \omega \otimes I_D) \circ (\Delta_C \otimes \Delta_D) &: \; C \otimes_R D \to (C \otimes_R D) \otimes_R (C \otimes_R D), \\
\bar{\varepsilon} = \varepsilon_C \otimes \varepsilon_D &: \; C \otimes_R D \to R.
\end{aligned}$$

Then $C \ltimes_\omega D$ is an R-coalgebra if and only if the following bow-tie diagram *is commutative (tensor over R):*

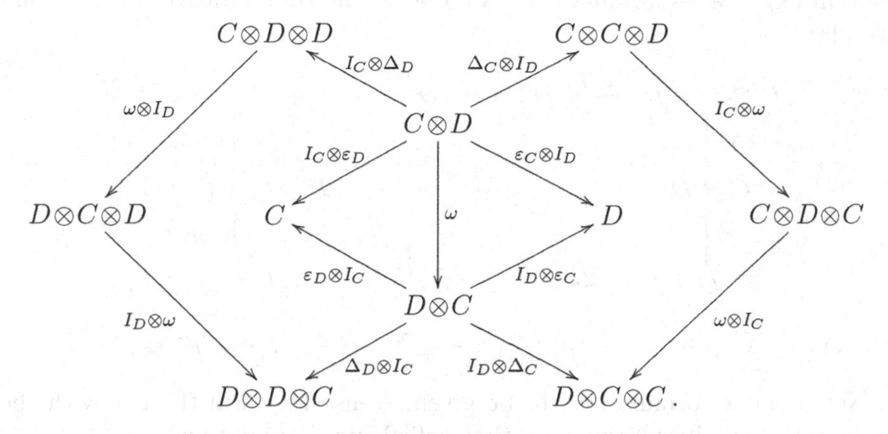

If this holds, the coalgebra $C \ltimes_\omega D$ is called a *smash coproduct* of C and D.

Proof. Notice that commutativity of the central trapezium (triangles) means

$$(I_D \otimes \varepsilon_C)\,\omega(c \otimes d) = \varepsilon_C(c)d, \quad (\varepsilon_D \otimes I_C)\,\omega(c \otimes d) = \varepsilon_D(d)c.$$

By definition, right counitality of $\bar{\varepsilon}$ requires $(I_{C\otimes_R D} \otimes_R \bar{\varepsilon}) \circ \bar{\Delta} = I_{C\otimes_R D}$, that is,

$$c \otimes d = \sum c_{\underline{1}} \otimes (I_D \otimes \varepsilon_C)\,\omega(c_{\underline{2}} \otimes d_{\underline{1}})\bar{\varepsilon}(d_{\underline{2}}) = \sum c_{\underline{1}} \otimes (I_D \otimes \varepsilon_C)\,\omega(c_{\underline{2}} \otimes d).$$

Applying $\varepsilon_C \otimes I_D$, we obtain the first equality (right triangle) for ω. Similarly, the second equality (left triangle) is derived. A simple computation shows that the two equalities imply counitality.

Coassociativity of $\bar{\Delta}$ means commutativity of the diagram

$$\begin{array}{ccccc}
C\otimes C\otimes D\otimes D & \xrightarrow{I\otimes\omega\otimes I} & C\otimes D\otimes C\otimes D & \xrightarrow{I\otimes I\otimes\Delta\otimes\Delta} & C\otimes D\otimes C\otimes C\otimes D\otimes D \\
\uparrow{\scriptstyle \Delta_C\otimes\Delta_D} & & & & \downarrow{\scriptstyle I\otimes I\otimes I\otimes\omega\otimes I} \\
C\otimes D & & (*) & & C\otimes D\otimes C\otimes D\otimes C\otimes D \\
\downarrow{\scriptstyle \Delta_C\otimes\Delta_D} & & & & \uparrow{\scriptstyle I\otimes\omega\otimes I\otimes I\otimes I} \\
C\otimes C\otimes D\otimes D & \xrightarrow{I\otimes\omega\otimes I} & C\otimes D\otimes C\otimes D & \xrightarrow{\Delta\otimes\Delta\otimes I\otimes I} & C\otimes C\otimes D\otimes D\otimes C\otimes D\,,
\end{array}$$

which is equivalent to the identity $(*)$

$$\sum c_{\underline{1}} \otimes d_{\underline{1}}{}^{\omega} \otimes c_{\underline{2}}{}^{\omega}{}_{\underline{1}} \otimes d_{\underline{2}\underline{1}}{}^{\bar{\omega}} \otimes c_{\underline{2}}{}^{\omega}{}_{\underline{2}}{}^{\bar{\omega}} \otimes d_{\underline{2}\underline{2}} = \sum c_{\underline{1}\underline{1}} \otimes d_{\underline{1}}{}^{\omega}{}_{\underline{1}}{}^{\bar{\omega}} \otimes c_{\underline{1}\underline{2}}{}^{\bar{\omega}} \otimes d_{\underline{1}}{}^{\omega}{}_{\underline{2}} \otimes c_{\underline{2}}{}^{\omega} \otimes d_{\underline{2}}.$$

Applying the map $\varepsilon_C \otimes I_C \otimes I_D \otimes I_D \otimes I_C \otimes \varepsilon_D$ to the last module in the diagram $(*)$ – or to formula $(*)$ – we obtain the commutative diagram and formula

$$\begin{array}{ccccc}
C\otimes D\otimes D & \xrightarrow{\omega\otimes I} & D\otimes C\otimes D & \xrightarrow{I\otimes\Delta_C\otimes I} & D\otimes C\otimes C\otimes D \\
\uparrow{\scriptstyle I\otimes\Delta_D} & & & & \downarrow{\scriptstyle I\otimes I\otimes\omega} \\
C\otimes D & & (**) & & D\otimes C\otimes D\otimes C \\
\downarrow{\scriptstyle \Delta_C\otimes I} & & & & \uparrow{\scriptstyle \omega\otimes I\otimes I} \\
C\otimes C\otimes D & \xrightarrow{I\otimes\omega} & C\otimes D\otimes C & \xrightarrow{I\otimes\Delta_D\otimes I} & C\otimes D\otimes D\otimes C\,,
\end{array}$$

$$(**)\qquad \sum d_{\underline{1}}{}^{\omega}\otimes c^{\omega}{}_{\underline{1}}\otimes d_{\underline{2}}{}^{\bar\omega}\otimes c^{\omega}{}_{\underline{2}}{}^{\bar\omega} = \sum d^{\omega}{}_{\underline{1}}{}^{\bar\omega}\otimes c_{\underline{1}}{}^{\bar\omega}\otimes d^{\omega}{}_{\underline{2}}\otimes c_{\underline{2}}{}^{\omega}.$$

Now assume formula $(**)$ to be given. Tensoring from the left with the coefficients $c_{\underline{1}}$ and replacing c by the coefficients $c_{\underline{2}}$ we obtain

$$\begin{array}{rcl}
\sum c_{\underline{1}}\otimes d_{\underline{1}}{}^{\omega}\otimes c_{\underline{2}}{}^{\omega}{}_{\underline{1}}\otimes d_{\underline{2}}{}^{\bar\omega}\otimes c_{\underline{2}}{}^{\omega}{}_{\underline{2}}{}^{\bar\omega} & = & \sum c_{\underline{1}}\otimes d^{\omega}{}_{\underline{1}}{}^{\bar\omega}\otimes c_{\underline{2}\underline{1}}{}^{\bar\omega}\otimes d^{\omega}{}_{\underline{2}}\otimes c_{\underline{2}\underline{2}}{}^{\omega} \\
 & = & \sum c_{\underline{1}\underline{1}}\otimes d^{\omega}{}_{\underline{1}}{}^{\bar\omega}\otimes c_{\underline{1}\underline{2}}{}^{\bar\omega}\otimes d^{\omega}{}_{\underline{2}}\otimes c_{\underline{2}}{}^{\omega}.
\end{array}$$

Now, tensoring with the coefficients $d_{\underline{2}}$ from the right and replacing d by the coefficients $d_{\underline{1}}$ we obtain formula $(*)$. So both conditions $(**)$ and $(*)$ are equivalent to coassociativity of $\bar\Delta$.

Commutativity of the trapezium yields a commutative diagram

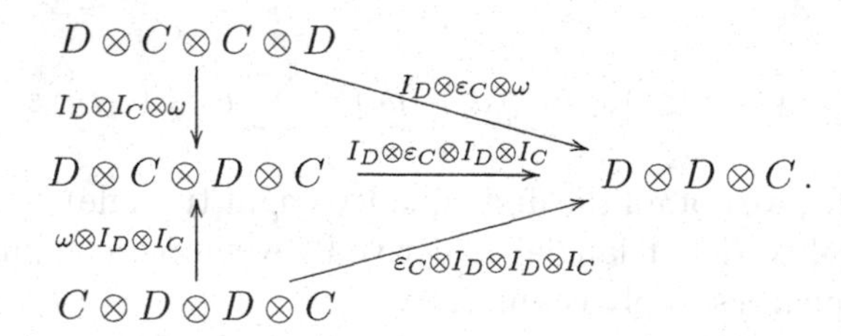

With this it is easy to see that the diagram $(**)$ reduces to the diagram

$$\begin{array}{ccccc}
C\otimes D\otimes D & & \xrightarrow{\omega\otimes I_D} & & D\otimes C\otimes D \\
\uparrow{\scriptstyle I_C\otimes\Delta_D} & & & & \downarrow{\scriptstyle I_C\otimes\omega} \\
C\otimes D & \xrightarrow{\omega} & D\otimes C & \xrightarrow{\Delta_D\otimes I_C} & D\otimes D\otimes C,
\end{array}$$

and a similar argument with $\varepsilon_D\otimes I_C$ yields the diagram

$$\begin{array}{ccccc}
C\otimes D & \xrightarrow{\omega} & D\otimes C & \xrightarrow{I_D\otimes\Delta_C} & D\otimes C\otimes C \\
\downarrow{\scriptstyle \Delta_C\otimes I_D} & & & & \uparrow{\scriptstyle \omega\otimes I_C} \\
C\otimes C\otimes D & & \xrightarrow{I_C\otimes\omega} & & C\otimes D\otimes C\,.
\end{array}$$

Notice that the two diagrams are the left and right wings of the bow-tie and hence one direction of our assertion is proven.

Commutativity of these diagrams corresponds to the equations

$$\sum d_{\underline{1}}{}^{\omega} \otimes d_{\underline{2}}{}^{\bar{\omega}} \otimes c^{\omega\bar{\omega}} = \sum d^{\omega}{}_{\underline{1}} \otimes d^{\omega}{}_{\underline{2}} \otimes c^{\omega}, \quad \sum d^{\omega} \otimes c^{\omega}{}_{\underline{1}} \otimes c^{\omega}{}_{\underline{2}} = \sum d^{\omega\bar{\omega}} \otimes c_{\underline{1}}{}^{\bar{\omega}} \otimes c_{\underline{2}}{}^{\omega},$$

and – alternatively – these can be obtained by applying $I_D \otimes \varepsilon_C \otimes I_D \otimes I_C$ and $I_D \otimes I_C \otimes \varepsilon_D \otimes I_C$ to equation $(**)$.

For the converse implication assume the bow-tie diagram to be commutative. Then the trapezium is commutative and hence $\bar{\varepsilon}$ is a counit. Moreover, the above equalities hold. Tensoring the first one with the coefficients $c_{\underline{1}}$ and replacing c by the coefficients $c_{\underline{2}}$ we obtain

$$\sum c_{\underline{1}} \otimes d_{\underline{1}}{}^{\omega} \otimes d_{\underline{2}}{}^{\tilde{\omega}} \otimes c_{\underline{2}}{}^{\omega\tilde{\omega}} = \sum c_{\underline{1}} \otimes d^{\omega}{}_{\underline{1}} \otimes d^{\omega}{}_{\underline{2}} \otimes c_{\underline{2}}{}^{\omega}.$$

Applying $\omega \otimes I_D \otimes I_C$ to this equation yields

$$\sum d_{\underline{1}}{}^{\omega\bar{\omega}} \otimes c_{\underline{1}}{}^{\bar{\omega}} \otimes d_{\underline{2}}{}^{\tilde{\omega}} \otimes c_{\underline{2}}{}^{\omega\tilde{\omega}} = \sum d^{\omega}{}_{\underline{1}}{}^{\bar{\omega}} \otimes c_{\underline{1}}{}^{\bar{\omega}} \otimes d^{\omega}{}_{\underline{2}} \otimes c_{\underline{2}}{}^{\omega}.$$

Now, tensoring the second equation with the coefficients $d_{\underline{2}}$ from the right, replacing d by the coefficients $d_{\underline{1}}$ and then applying $I_D \otimes I_C \otimes \omega$ yields

$$\sum d_{\underline{1}}{}^{\omega} \otimes c^{\omega}{}_{\underline{1}} \otimes d_{\underline{2}}{}^{\bar{\omega}} \otimes c^{\omega}{}_{\underline{2}}{}^{\bar{\omega}} = \sum d_{\underline{1}}{}^{\omega\bar{\omega}} \otimes c_{\underline{1}}{}^{\bar{\omega}} \otimes d_{\underline{2}}{}^{\tilde{\omega}} \otimes c_{\underline{2}}{}^{\omega\tilde{\omega}}.$$

Comparing the two equations we obtain $(**)$, proving the coassociativity of $\bar{\Delta}$. □

Notice that a dual construction and a dual bow-tie diagram apply for the definiton of a general product on the tensor product of two R-algebras A, B by an R-linear map $\omega' : B \otimes_R A \to A \otimes_R B$. A partially dual bow-tie diagram arises in the study of entwining structures between R-algebras and R-coalgebras (cf. 32.1).

2.15. Exercises

(1) Let $g : A \to A'$ be an R-algebra morphism. Prove that, for any R-coalgebra C,

$$\mathrm{Hom}(C, g) : \mathrm{Hom}_R(C, A) \to \mathrm{Hom}_R(C, A')$$

is an R-algebra morphism.

(2) Let $f : C \to C'$ be an R-coalgebra morphism. Prove that, if f is bijective then f^{-1} is also a coalgebra morphism.

(3) On the $\mathbb{Z}$-module $C = \mathbb{Z} \oplus \mathbb{Z}/4\mathbb{Z}$ define a coproduct

$$\Delta : C \to C \otimes_{\mathbb{Z}} C, \quad (1,0) \mapsto (1,0) \otimes (1,0),$$
$$(0,1) \mapsto (1,0) \otimes (0,1) + (0,1) \otimes (1,0).$$

On the submodule $D = \mathbb{Z} \oplus 2\mathbb{Z}/4\mathbb{Z} \subset C$ consider the coproducts

$$\begin{aligned}\Delta_1 : D \to D \otimes_{\mathbb{Z}} D, \quad & (1,0) \mapsto (1,0) \otimes (1,0),\\ & (0,2) \mapsto (1,0) \otimes (0,2) + (0,2) \otimes (1,0),\\ \Delta_2 : D \to D \otimes_{\mathbb{Z}} D, \quad & (1,0) \mapsto (1,0) \otimes (1,0),\\ & (0,2) \mapsto (1,0)\otimes(0,2) + (0,2)\otimes(0,2) + (0,2)\otimes(1,0).\end{aligned}$$

Prove that (D, Δ_1) and (D, Δ_2) are not isomorphic but the canonical inclusion $D \to C$ is an algebra morphism for both of them (Nichols and Sweedler [168]).

(4) On the $\mathbb{Z}$-module $C = \mathbb{Z}/8\mathbb{Z} \oplus \mathbb{Z}/2\mathbb{Z}$ define a coproduct

$$\Delta : C \to C \otimes_{\mathbb{Z}} C, \quad \begin{aligned}&(1,0) \mapsto 0,\\ &(0,1) \mapsto 4(1,0) \otimes (1,0)\end{aligned}$$

and consider the submodule $V = \mathbb{Z}(2,0) + \mathbb{Z}(0,1) \subset C$. Prove:

(i) Δ is well defined.

(ii) $\Delta(V)$ is contained in the image of $V \otimes_R V \to C \otimes_R C$.

(iii) $\Delta : V \to C \otimes_R C$ has no lifting to $V \otimes_R V$ (check the order of the preimage of $\Delta(0,1)$ in $V \otimes_R V$) (Nichols and Sweedler [168]).

(5) Let $C = \mathbb{Z} \oplus \mathbb{Z}/2\mathbb{Z} \oplus \mathbb{Z}$, denote $c_0 = (1,0,0)$, $c_1 = (0,1,0)$, $c_2 = (0,0,1)$ and define a coproduct

$$\Delta(c_n) = \sum_{i=0}^{n} c_i \otimes c_{n-i}, \quad n = 0,1,2.$$

Let $D = \mathbb{Z} \oplus \mathbb{Z}/4\mathbb{Z}$, denote $d_0 = (1,0)$, $d_1 = (0,1)$ and

$$\Delta(d_0) = d_0 \otimes d_0, \quad \Delta(d_1) = d_0 \otimes d_1 + d_1 \otimes d_0.$$

Prove that the map

$$f : C \to D, \quad c_0 \mapsto d_0,\ c_1 \mapsto 2d_1,\ c_2 \mapsto 0,$$

is a $\mathbb{Z}$-coalgebra morphism and $\Delta(c_2) \notin c_2 \otimes C + C \otimes c_2$ (which implies that $\operatorname{Ke} f = \mathbb{Z}c_2$ is not a coideal in C) (Nichols and Sweedler [168]).

(6) Prove that the tensor product of coalgebras yields the product in the category of cocommutative coassociative coalgebras.

(7) Let $(C, \Delta_C, \varepsilon_C)$ and $(D, \Delta_D, \varepsilon_D)$ be R-coalgebras with an R-linear mapping $\omega : C \otimes_R D \to D \otimes_R C$. Denote by $C \ltimes_\omega D$ the R-module $C \otimes_R D$ endowed with the maps $\bar{\Delta}$ and $\bar{\varepsilon}$ as in 2.14. The map ω is said to be *left* or *right conormal* if for any $c \in C$, $d \in D$,

$$(I_D \otimes \varepsilon_C)\,\omega(c \otimes d) = \varepsilon(c)d \quad \text{or} \quad (\varepsilon_D \otimes I_C)\,\omega(c \otimes d) = \varepsilon_D(d)c.$$

Prove:

(i) The following are equivalent:

(a) ω is left conormal;

(b) $\varepsilon_C \otimes I_D : C \ltimes_\omega D \to D$ respects the coproduct;

(c) $(I_{C \otimes_R D} \otimes_R \bar{\varepsilon}) \circ \bar{\Delta} = I_{C \otimes_R D}$.

(ii) The following are equivalent:

(a) ω is right conormal;

(b) $I_C \otimes \varepsilon_D : C \ltimes_\omega D \to C$ respects the coproduct;

(c) $(\bar{\varepsilon} \otimes_R I_{C \otimes_R D}) \circ \bar{\Delta} = I_{C \otimes_R D}$.

References. Caenepeel, Militaru and Zhu [9]; Nichols and Sweedler [168]; Sweedler [45]; Wisbauer [210].

3 Comodules

In algebra or ring theory, in addition to an algebra, one would also like to study its modules, that is, Abelian groups on which the algebra acts. Correspondingly, in the coalgebra theory one would like to study R-modules on which an R-coalgebra C *coacts*. Such modules are known as (right) C-comodules, and for any given C they form a category $\mathbf{M}^C$, provided morphisms or *C-comodule maps* are suitably defined. In this section we define the category $\mathbf{M}^C$ and study its properties. The category $\mathbf{M}^C$ in many respects is similar to the category of modules of an algebra, for example, there are Hom-tensor relations, there exist cokernels, and so on, and indeed there is a close relationship between $\mathbf{M}^C$ and the modules of the dual coalgebra C^* (cf. Section 4). On the other hand, however, there are several marked differences between categories of modules and comodules. For example, the category of modules is an Abelian category, while the category of comodules of a coalgebra over a ring might not have kernels (and hence it is not an Abelian category in general). This is an important (lack of) property that is characteristic for coalgebras over rings (if R is a field then $\mathbf{M}^C$ is Abelian), that makes studies of such coalgebras particularly interesting. The ring structure of R and the R-module structure of C play in these studies an important role, which requires careful analysis of R-relative properties of a coalgebra or both C- and R-relative properties of comodules.

As before, R denotes a commutative ring, $\mathbf{M}_R$ the category of R-modules, and C, more precisely (C, Δ, ε), stands for a (coassociative) R-coalgebra (with counit). We first introduce right comodules over C.

3.1. Right C-comodules. For $M \in \mathbf{M}_R$, an R-linear map $\varrho^M : M \to M \otimes_R C$ is called a *right coaction* of C on M or simply a *right C-coaction.* To denote the action of ϱ^M on elements of M we write $\varrho^M(m) = \sum m_{\underline{0}} \otimes m_{\underline{1}}$.

A C-coaction ϱ^M is said to be *coassociative* and *counital* provided the diagrams

$$\begin{array}{ccc} M & \xrightarrow{\varrho^M} & M \otimes_R C \\ {\scriptstyle \varrho^M}\downarrow & & \downarrow{\scriptstyle I_M \otimes \Delta} \\ M \otimes_R C & \xrightarrow{\varrho^M \otimes I_C} & M \otimes_R C \otimes_R C, \end{array} \qquad \begin{array}{ccc} M & \xrightarrow{\varrho^M} & M \otimes_R C \\ & {\scriptstyle I_M}\searrow & \downarrow{\scriptstyle I_M \otimes \varepsilon} \\ & & M \end{array}$$

are commutative. Explicitly, this means that, for all $m \in M$,

$$\sum \varrho^M(m_{\underline{0}}) \otimes m_{\underline{1}} = \sum m_{\underline{0}} \otimes \Delta(m_{\underline{1}}), \quad m = \sum m_{\underline{0}} \varepsilon(m_{\underline{1}}).$$

In view of the first of these equations we can shorten the notation and write

$$(I_M \otimes \Delta) \circ \varrho^M(m) = \sum m_{\underline{0}} \otimes m_{\underline{1}} \otimes m_{\underline{2}},$$

and so on, in a way similar to the notation for a coproduct. Note that the elements with subscript 0 are in M while all the elements with positive subscripts are in C.

An R-module with a coassociative and counital right coaction is called a *right C-comodule.*

Recall that any semigroup induces a coalgebra $(R[G], \Delta_1, \varepsilon_1)$ (see 1.7) and for this the comodules have the following form.

3.2. Graded modules. *Let G be a semigroup. Considering R with the trivial grading, an R-module M is G-graded (see 40.6) if and only if it is an $R[G]$-comodule.*

Proof. Let $M = \bigoplus_G M_g$ be a G-graded module. Then a coaction of $(R[G], \Delta_1, \varepsilon_1)$ on M is defined by

$$\varrho^M : M \longrightarrow M \otimes_R R[G], \quad m_g \mapsto m \otimes g.$$

It is easily seen that this coaction is coassociative and, for any $m \in M$,

$$(I_M \otimes \varepsilon_1)\varrho^M(m) = (I_M \otimes \varepsilon_1)(\sum_{g \in G} m_g \otimes g) = \sum_{g \in G} m_g = m\,.$$

Now assume that M is a right $R[G]$-comodule and for all $m \in M$ write $\varrho^M(m) = \sum_{g \in G} m_g \otimes g$. By coassociativity, $\sum_{g \in G}(m_g)_h \otimes h \otimes g = \sum_{g \in G} m_g \otimes g \otimes g$, which implies $(m_g)_h = \delta_{g,h} m_g$ and also $\varrho^M(m_g) = m_g \otimes g$. Then $M_g = \{m_g \mid m \in M\}$ is an independent family of R-submodules of M. Now counitality of M implies $m = (I_M \otimes \varepsilon_1)(\sum_{g \in G} m_g \otimes g) = \sum_{g \in G} m_g$, and hence $M = \bigoplus_G M_g$. $\square$

Maps between comodules should respect their structure, that is, they have to commute with the coactions. This leads to

3.3. Comodule morphisms. Let M, N be right C-comodules. An R-linear map $f : M \to N$ is called a *comodule morphism* or a *morphism of right C-comodules* if and only if the diagram

$$\begin{array}{ccc} M & \xrightarrow{\ f\ } & N \\ {\scriptstyle \varrho^M}\downarrow & & \downarrow{\scriptstyle \varrho^N} \\ M \otimes_R C & \xrightarrow{f \otimes I_C} & N \otimes_R C \end{array}$$

is commutative. Explicitly, this means that $\varrho^N \circ f = (f \otimes I_C) \circ \varrho^M$; that is, for all $m \in M$ we require

$$\sum f(m)_{\underline{0}} \otimes f(m)_{\underline{1}} = \sum f(m_{\underline{0}}) \otimes m_{\underline{1}}.$$

Instead of *comodule morphism* we also say *C-morphism* or *(C-)colinear map.* It is easy to see that the sum of two C-morphisms is again a C-morphism. In fact, the set $\mathrm{Hom}^C(M,N)$ of C-morphisms from M to N is an R-module, and it follows from the definition that it is determined by the exact sequence in $\mathbf{M}_R$,

$$0 \to \mathrm{Hom}^C(M,N) \to \mathrm{Hom}_R(M,N) \xrightarrow{\gamma} \mathrm{Hom}_R(M, N\otimes_R C),$$

where $\gamma(f) := \varrho^N \circ f - (f\otimes I_C)\circ \varrho^M$. Notice that it can also be determined by the pullback diagram

$$\begin{array}{ccc} \mathrm{Hom}^C(M,N) & \longrightarrow & \mathrm{Hom}_R(M,N) \\ \downarrow & & \downarrow{\scriptstyle \varrho^N\circ -} \\ \mathrm{Hom}_R(M,N) & \xrightarrow{(-\otimes I_C)\circ\varrho^M} & \mathrm{Hom}_R(M, N\otimes_R C)\,. \end{array}$$

Obviously the class of right comodules over C together with the colinear maps form an additive category. This category is denoted by $\mathbf{M}^C$.

3.4. Left C-comodules. Symmetrically, for an R-module M, *left C-coaction* is defined as an R-linear map ${}^M\varrho : M \to C\otimes_R M$. It is said to be coassociative and counital if it induces commutative diagrams

$$\begin{array}{ccc} M & \xrightarrow{{}^M\varrho} & C\otimes_R M \\ {\scriptstyle {}^M\varrho}\downarrow & & \downarrow{\scriptstyle \Delta\otimes I_M} \\ C\otimes_R M & \xrightarrow{I_C\otimes {}^M\varrho} & C\otimes_R C\otimes_R M, \end{array} \qquad \begin{array}{ccc} M & \xrightarrow{{}^M\varrho} & C\otimes_R M \\ & \underset{=}{\searrow} & \downarrow{\scriptstyle \varepsilon\otimes I_M} \\ & & M. \end{array}$$

For $m\in M$ we write ${}^M\varrho(m) = \sum m_{\underline{-1}}\otimes m_{\underline{0}}$, and coassociativity is expressed as $\sum m_{\underline{-1}}\otimes {}^M\varrho(m_{\underline{0}}) = \sum \Delta(m_{\underline{-1}})\otimes m_{\underline{0}} = \sum m_{\underline{-2}}\otimes m_{\underline{-1}}\otimes m_{\underline{0}}$, where the final expression is a notation. The axiom for the counit reads $m = \sum \varepsilon(m_{\underline{-1}})m_{\underline{0}}$.

An R-module with a coassociative and counital left C-coaction is called a *left C-comodule.* C-morphisms between left C-comodules M, N are defined symmetrically, and the R-module of all such C-morphisms is denoted by ${}^C\mathrm{Hom}\,(M,N)$. Left C-comodules and their morphisms again form an additive category that is denoted by ${}^C\mathbf{M}$.

An example of a left and right C-comodule is provided by C itself. In both cases coaction is given by Δ. Unless explicitly stated otherwise, C is always viewed as a C-comodule with this coaction. In this context Δ is often referred to as a left or right *regular coaction.*

In what follows we mainly study the category of right comodules. The corresponding results for left comodules can be obtained by left-right symmetry. Similarly to the case of coalgebras, several constructions for C-comodules build upon the corresponding constructions for R-modules.

3.5. Kernels and cokernels in $\mathbf{M}^C$. Let $f : M \to N$ be a morphism in $\mathbf{M}^C$. The cokernel g of f in $\mathbf{M}_R$ yields the exact commutative diagram

$$\begin{array}{ccccccc} M & \xrightarrow{f} & N & \xrightarrow{g} & L & \longrightarrow & 0 \\ \downarrow{\scriptstyle \varrho^M} & & \downarrow{\scriptstyle \varrho^N} & & & & \\ M \otimes_R C & \xrightarrow{f \otimes I_C} & N \otimes_R C & \xrightarrow{g \otimes I_C} & L \otimes_R C & \longrightarrow & 0, \end{array}$$

which can be completed commutatively in $\mathbf{M}_R$ by some $\varrho^L : L \to L \otimes_R C$ for which we obtain the diagram

$$\begin{array}{ccccc} N & \xrightarrow{\varrho^N} & N \otimes_R C & \overset{\varrho^N \otimes I_C}{\underset{I_N \otimes \Delta}{\rightrightarrows}} & N \otimes_R C \otimes_R C \\ \downarrow{\scriptstyle g} & & \downarrow{\scriptstyle g \otimes I_C} & & \downarrow{\scriptstyle g \otimes I_C \otimes I_C} \\ L & \xrightarrow{\varrho^L} & L \otimes_R C & \overset{\varrho^L \otimes I_C}{\underset{I_L \otimes \Delta}{\rightrightarrows}} & L \otimes_R C \otimes_R C\,. \end{array}$$

The outer rectangle is commutative for the upper as well as for the lower morphisms, and hence

$$(\varrho^L \otimes I_C) \circ \varrho^L \circ g = (I_L \otimes \Delta) \circ \varrho^L \circ g\,.$$

Now, surjectivity of g implies $(\varrho^L \otimes I_C) \circ \varrho^L = (I_L \otimes \Delta) \circ \varrho^L$, showing that ϱ^L is coassociative. Moreover,

$$(I_L \otimes \varepsilon) \circ \varrho^L \circ g = (I_L \otimes \varepsilon) \circ (g \otimes I_C) \circ \varrho^N = g,$$

which shows that $(I_L \otimes \varepsilon) \circ \varrho^L = I_L$. Thus ϱ^L is counital, and so it makes L a comodule such that g is a C-morphism. This shows that cokernels exist in the category $\mathbf{M}^C$.

Dually, for the kernel h of f in $\mathbf{M}_R$ there is a commutative diagram

$$\begin{array}{ccccccc} 0 & \longrightarrow & K & \xrightarrow{h} & M & \xrightarrow{f} & N \\ & & & & \downarrow{\scriptstyle \varrho^M} & & \downarrow{\scriptstyle \varrho^N} \\ 0 & \longrightarrow & K \otimes_R C & \xrightarrow{h \otimes I_C} & M \otimes_R C & \xrightarrow{f \otimes I_C} & N \otimes_R C\,, \end{array}$$

where the top sequence is always exact while the bottom sequence is exact provided f is C-pure as R-morphism (see 40.13). If this is the case, the diagram can be extended commutatively by a coaction $\varrho^K : K \to K \otimes_R C$, and (dual to the proof for cokernels) it can be shown that ϱ^K is coassociative and counital. Thus kernels of C-morphisms are induced from kernels in $\mathbf{M}_R$ provided certain additional conditions are imposed, for example, when C is flat as an R-module.

3.6. C-subcomodules. Let M be a right C-comodule. An R-submodule $K \subset M$ is called a *C-subcomodule of M* provided K has a right comodule structure such that the inclusion is a comodule morphism.

This definition displays a number of typical features of coalgebras over a ring as opposed to coalgebras over a field. In the case in which $R = F$ is a field, one defines a C-subcomodule of M as a subspace $K \subset M$ such that $\varrho^M(K) \subset K \otimes_F C \subset M \otimes_F C$. In the case of a general commutative ring R, however, the fact that K is an R-submodule of M does not yet imply that $K \otimes_R C$ is a submodule of $M \otimes_R C$, since the tensor functor is only right but not left exact. However, if K is a C-pure R-submodule of M, then $K \otimes_R C \subset M \otimes_R C$ as well, and K is a subcomodule of M provided $\varrho^M(K) \subset K \otimes_R C \subset M \otimes_R C$. By the same token, the kernel K in $\mathbf{M}_R$ of a comodule morphism $f : M \to N$ need not be a subcomodule of M unless f is a C-pure morphism (compare 3.5).

3.7. Coproducts in $\mathbf{M}^C$. Let $\{M_\lambda, \varrho^M_\lambda\}_\Lambda$ be a family of C-comodules. Put $M = \bigoplus_\Lambda M_\lambda$, the coproduct in $\mathbf{M}_R$, $i_\lambda : M_\lambda \to M$ the canonical inclusions, and consider the linear maps

$$M_\lambda \xrightarrow{\varrho^M_\lambda} M_\lambda \otimes_R C \subset M \otimes_R C.$$

Note that the inclusions i_λ are R-splittings, so that $M_\lambda \otimes_R C \subset M \otimes_R C$ is a pure submodule. By the properties of coproducts of R-modules there exists a unique coaction

$$\varrho^M : M \to M \otimes_R C, \text{ such that } \varrho^M \circ i_\lambda = \varrho^M_\lambda,$$

which is coassociative and counital since all the ϱ^M_λ are, and thus it makes M a C-comodule for which the $i_\lambda : M_\lambda \to M$ are C-morphisms with the following universal property:

Let $\{f_\lambda : M_\lambda \to N\}_\Lambda$ be a family of morphisms in $\mathbf{M}^C$. Then there exists a unique C-morphism $f : M \to N$ such that, for each $\lambda \in \Lambda$, the following diagram of C-morphisms commutes:

$$\begin{array}{ccc} M_\lambda & \xrightarrow{i_\lambda} & M \\ & {\scriptstyle f_\lambda}\searrow & \downarrow {\scriptstyle f} \\ & & N . \end{array}$$

Similarly to the coproduct, the direct limit of direct families of C-comodules is derived from the direct limit in $\mathbf{M}_R$. Both constructions are special cases of a more general observation on colimits of F-coalgebras in 38.25.

3.8. Comodules and tensor products. *Let M be in $\mathbf{M}^C$ and consider any morphism $f : X \to Y$ of R-modules. Then:*

(1) *$X \otimes_R M$ is a right C-comodule with the coaction*

$$I_X \otimes \varrho^M : X \otimes_R M \longrightarrow X \otimes_R M \otimes_R C,$$

and the map $f \otimes I_M : X \otimes_R M \to Y \otimes_R M$ is a C-morphism.

(2) *In particular, $X \otimes_R C$ has a right C-coaction*

$$I_X \otimes \Delta : X \otimes_R C \longrightarrow X \otimes_R C \otimes_R C,$$

and the map $f \otimes I_C : X \otimes_R C \to Y \otimes_R C$ is a C-morphism.

(3) *For any index set Λ, $R^{(\Lambda)} \otimes_R C \simeq C^{(\Lambda)}$ as comodules and there exists a surjective C-morphism*

$$C^{(\Lambda')} \to M \otimes_R C, \quad \textit{for some } \Lambda'.$$

(4) *The structure map $\varrho^M : M \to M \otimes_R C$ is a comodule morphism, and hence M is a subcomodule of a C-generated comodule.*

Proof. (1) and (2) are easily verified from the definitions.

(3) Take a surjective R-linear map $h : R^{(\Lambda')} \to M$. Then, by (2),

$$h \otimes I_C : R^{(\Lambda')} \otimes_R C \to M \otimes_R C$$

is a surjective comodule morphism.

(4) By coassociativity, ϱ^M is a comodule morphism (where $M \otimes_R C$ has the comodule structure from (1)). Note that ρ^M is split by $I_M \otimes \varepsilon$ as an R-module; thus M is a pure submodule of $M \otimes_R C$ and hence is a subcomodule. □

Similarly to the classical Hom-tensor relations (see 40.18), we obtain

3.9. Hom-tensor relations in $\mathbf{M}^C$. *Let X be any R-module.*

(1) *For any $M \in \mathbf{M}^C$, the R-linear map*

$$\varphi : \mathrm{Hom}^C(M, X \otimes_R C) \to \mathrm{Hom}_R(M, X), \quad f \mapsto (I_X \otimes \varepsilon) \circ f,$$

is bijective, with inverse map $h \mapsto (h \otimes I_C) \circ \varrho^M$.

(2) *For any $M, N \in \mathbf{M}^C$, the R-linear map*

$$\psi : \mathrm{Hom}^C(X \otimes_R M, N) \to \mathrm{Hom}_R(X, \mathrm{Hom}^C(M, N)), \; g \mapsto [x \mapsto g(x \otimes -)],$$

is bijective, with inverse map $h \mapsto [x \otimes m \mapsto h(x)(m)]$.

Proof. (1) For any $f \in \mathrm{Hom}^C(M, X \otimes_R C)$ the diagram

$$\begin{array}{ccccc} M & \xrightarrow{f} & X \otimes_R C & & \\ {\scriptstyle \varrho^M}\downarrow & & \downarrow{\scriptstyle I_X \otimes \Delta} & \searrow^{=} & \\ M \otimes_R C & \xrightarrow[f \otimes I_C]{} & X \otimes_R C \otimes_R C & \xrightarrow[I_X \otimes \varepsilon \otimes I_C]{} & X \otimes_R C \end{array}$$

is commutative, that is,

$$f = (I_X \otimes \varepsilon \otimes I_C) \circ (f \otimes I_C) \circ \varrho^M = (\varphi(f) \otimes I_C) \circ \varrho^M.$$

This implies that φ is injective.

Since ϱ^M is a C-morphism, so is $(h \otimes I_C) \circ \varrho^M$, for any $h \in \mathrm{Hom}_R(M, X)$. Therefore

$$\varphi((h \otimes I_C) \circ \rho^M) = (I_X \otimes \varepsilon) \circ (h \otimes I_C) \circ \varrho^M = h \circ (I_M \otimes \varepsilon) \circ \varrho^M = h,$$

implying that φ is surjective.

(2) The Hom-tensor relations for modules provide one with an isomorphism of R-modules,

$$\psi : \mathrm{Hom}_R(X \otimes_R M, N) \to \mathrm{Hom}_R(X, \mathrm{Hom}_R(M, N)). \qquad (*)$$

For any $x \in X$, by commutativity of the diagram

$$\begin{array}{ccc} M & \xrightarrow{x \otimes -} & X \otimes_R M \\ {\scriptstyle \varrho^M}\downarrow & & \downarrow{\scriptstyle I_X \otimes \varrho^M} \\ M \otimes_R C & \xrightarrow{(x \otimes -) \otimes I_C} & X \otimes_R M \otimes_R C, \end{array} \qquad \begin{array}{ccc} m & \longmapsto & x \otimes m \\ \downarrow & & \downarrow \\ \varrho^M(m) & \longmapsto & x \otimes \varrho^M(m), \end{array}$$

the map $x \otimes -$ is a C-morphism. Hence, for any $g \in \mathrm{Hom}^C(X \otimes_R M, N)$, the composition $g \circ (x \otimes -)$ is a C-morphism. On the other hand, there is a commutative diagram, for all $h \in \mathrm{Hom}_R(X, \mathrm{Hom}^C(M, N))$,

$$\begin{array}{ccc} X \otimes_R M & \longrightarrow & N \\ {\scriptstyle I_X \otimes \varrho^M}\downarrow & & \downarrow{\scriptstyle \varrho^N} \\ X \otimes_R M \otimes_R C & \longrightarrow & N \otimes_R C, \end{array} \qquad \begin{array}{ccc} x \otimes m & \longmapsto & h(x)(m) \\ \downarrow & & \downarrow \\ x \otimes \varrho^M(m) & \longmapsto & (h(x) \otimes I_C) \circ \varrho^M(m). \end{array}$$

This shows that $\psi^{-1}(h)$ lies in $\mathrm{Hom}^C(X \otimes_R M, N)$ and therefore implies that ψ in $(*)$ restricts to the bijective map $\psi : \mathrm{Hom}^C(X \otimes_R M, N) \to \mathrm{Hom}_R(X, \mathrm{Hom}^C(M, N))$, as required. □

For completeness we formulate the left-sided version of 3.9.

3.10. Hom-tensor relations in ${}^C\mathbf{M}$. *Let X be any R-module.*

(1) *For any left C-comodule M, there is an isomorphism*

$$\varphi' : {}^C\mathrm{Hom}\,(M, C \otimes_R X) \to \mathrm{Hom}_R(M, X), \quad f \mapsto (\varepsilon \otimes I_X) \circ f,$$

with inverse map $h \mapsto (I_C \otimes h) \circ {}^M\varrho$.

(2) *For any $M, N \in {}^C\mathbf{M}$, there is an isomorphism*

$$\psi' : {}^C\mathrm{Hom}\,(M \otimes_R X, N) \to \mathrm{Hom}_R(X, {}^C\mathrm{Hom}(M,N)), \quad g \mapsto [x \mapsto g(- \otimes x)],$$

with inverse map $h \mapsto [m \otimes x \mapsto h(x)(m)]$.

Unlike for A-modules (see 40.8), the R-dual of a right C-comodule need not be a left C-comodule unless additional conditions are imposed. To specify such sufficient conditions, first recall that, for a finitely presented R-module M and a flat R-module C, there is an isomorphism (compare 40.12)

$$\nu_M : C \otimes_R \mathrm{Hom}_R(M, R) \to \mathrm{Hom}_R(M, C), \quad c \otimes h \mapsto c \otimes h(-).$$

3.11. Comodules finitely presented as R-modules. *Let ${}_RC$ be flat and $M \in \mathbf{M}^C$ such that ${}_RM$ is finitely presented. Then $M^* = \mathrm{Hom}_R(M, R)$ is a left C-comodule by the structure map*

$${}^{M^*}\varrho : M^* \to \mathrm{Hom}_R(M, C) \simeq C \otimes_R M^*, \quad g \mapsto (g \otimes I_C) \circ \varrho^M.$$

Proof. The comodule property of M^* follows from the commutativity of the following diagram (with obvious maps), the central part of which arises from the coassociativity of C (tensor over R):

$$\begin{array}{ccccccc}
M^* & \xrightarrow{\simeq} & \mathrm{Hom}^C(M, C) & \longrightarrow & \mathrm{Hom}^C(M, C \otimes C) & \xrightarrow{\simeq} & C \otimes M^* \\
\downarrow & & \downarrow & & \downarrow & & \downarrow {\scriptstyle \Delta \otimes I_{M^*}} \\
C \otimes M^* & \xrightarrow{\simeq} & \mathrm{Hom}^C(M, C \otimes C) & \longrightarrow & \mathrm{Hom}^C(M, C \otimes C \otimes C) & \xrightarrow{\simeq} & C \otimes C \otimes M^* .
\end{array}$$

□

For $X = R$ and $M = C$, the isomorphism φ describes the comodule endomorphisms of C.

3.12. Comodule endomorphisms of C.

(1) *There is an algebra anti-isomorphism $\varphi : \mathrm{End}^C(C) \to C^*$, $f \mapsto \varepsilon \circ f$, with the inverse map $h \mapsto (h \otimes I_C) \circ \Delta$ and so $h \in C^*$ acts on $c \in C$ from the right by*

$$c \leftharpoonup h = (h \otimes I_C)\Delta(c) = \sum h(c_{\underline{1}})c_{\underline{2}}.$$

(2) *There is an algebra isomorphism* $\varphi' : {}^C\mathrm{End}(C) \to C^*$, $f \mapsto \varepsilon \circ f$, *with the inverse map* $h \mapsto (I_C \otimes h) \circ \Delta$ *and so* $h \in C^*$ *acts on* $c \in C$ *from the left by*

$$h \rightharpoonup c = (I_C \otimes h)\Delta(c) = \sum c_{\underline{1}} h(c_{\underline{2}}).$$

(3) *For any* $f \in C^*$ *and* $c \in C$,

$$\begin{aligned} \Delta(f \rightharpoonup c) &= \sum c_{\underline{1}} \otimes (f \rightharpoonup c_{\underline{2}}), \\ \Delta(c \leftharpoonup f) &= \sum (c_{\underline{1}} \leftharpoonup f) \otimes c_{\underline{2}}, \\ \Delta(f \rightharpoonup c \leftharpoonup g) &= \sum (c_{\underline{1}} \leftharpoonup g) \otimes (f \rightharpoonup c_{\underline{2}}), \\ \sum c_{\underline{1}} \otimes (c_{\underline{2}} \leftharpoonup f) &= \sum (f \rightharpoonup c_{\underline{1}}) \otimes c_{\underline{2}}. \end{aligned}$$

(4) φ *and* φ' *are homeomorphisms for the finite topologies (cf. 42.1).*

(5) *The coproduct* Δ *yields the embedding*

$$C^* \simeq \mathrm{Hom}^C(C, C) \to \mathrm{Hom}^C(C, C \otimes_R C) \simeq \mathrm{End}_R(C).$$

Proof. (1) By 3.9(1), φ is R-linear and bijective. Take any $f, g \in \mathrm{End}^C(C)$, recall that $(f \otimes I_C) \circ \Delta = \Delta \circ f$, and consider the convolution product applied to any $c \in C$,

$$\begin{aligned} (\varepsilon \circ f) * (\varepsilon \circ g)(c) &= \sum \varepsilon(f(c_{\underline{1}}))\, \varepsilon(g(c_{\underline{2}})) \\ &= \varepsilon \circ g\, [(\varepsilon \otimes I_C) \circ (f \otimes I_C) \circ \Delta(c)] \\ &= \varepsilon \circ g\, [(\varepsilon \otimes I_C) \circ \Delta \circ f(c)] \;=\; \varepsilon \circ (g \circ f)(c)\,. \end{aligned}$$

This shows that φ is an anti-isomorphism.

(2) For all $f, g \in {}^C\mathrm{End}\,(C)$, $(I_C \otimes g) \circ \Delta = \Delta \circ g$, and hence

$$\begin{aligned} (\varepsilon \circ f) * (\varepsilon \circ g)(c) &= \sum \varepsilon(f(c_{\underline{1}}))\, \varepsilon(g(c_{\underline{2}})) \\ &= \varepsilon \circ f\, [(I_C \otimes \varepsilon) \circ (I_C \otimes g) \circ \Delta(c)] \\ &= \varepsilon \circ g\, [(I_C \otimes \varepsilon) \circ \Delta \circ g(c)] = \varepsilon \circ (f \circ g)(c)\,. \end{aligned}$$

(3) By definition,

$$\begin{aligned} \Delta(f \rightharpoonup c) &= \Delta(\sum c_{\underline{1}} f(c_{\underline{2}}) \;=\; \sum c_{\underline{11}} \otimes c_{\underline{12}} f(c_{\underline{2}}) \\ &= \sum c_{\underline{1}} \otimes c_{\underline{21}} f(c_{\underline{22}}) \;=\; \sum c_{\underline{1}} \otimes (f \rightharpoonup c_{\underline{2}}). \end{aligned}$$

The remaining assertions are shown similarly.

(4) We show that both φ and φ^{-1} map open neighbourhoods of zero to open neighbourhoods of zero. For any $x_1, \ldots, x_k \in C$, writing $\Delta(x_i) = \sum x_{i\underline{1}} \otimes x_{i\underline{2}}$,

$$\{h \in C^* \mid h(x_{i\underline{1}}) = 0, i = 1, \ldots, k\} \subset \varphi(\{f \in \mathrm{End}^C(C) \mid f(x_i) = 0, i = 1, \ldots, k\}),$$

where the left-hand side denotes an open subset in C^* and

$$\{f \in \mathrm{End}^C(C) \mid f(x_i) = 0, i=1,\ldots,k\} \subset \varphi^{-1}(\{h \in C^* \mid h(x_i) = 0, i=1,\ldots,k\}),$$

with the left-hand side an open subset in $\mathrm{End}^C(C)$. This shows that φ is a homeomorphism.

(5) This follows from the Hom-tensor relations 3.9 for $M = C = X$. □

Notice that in 3.12(1) the comodule morphisms are written on the left of the argument. By writing morphisms of right comodules on the right side, we obtain an isomorphism between C^* and the comodule endomorphism ring.

The next theorem summarises observations on the category of comodules.

3.13. The category $\mathbf{M}^C$.

(1) The category $\mathbf{M}^C$ has direct sums and cokernels, and C is a subgenerator.

(2) $\mathbf{M}^C$ is a Grothendieck category provided that C is a flat R-module.

(3) The functor $-\otimes_R C : \mathbf{M}_R \to \mathbf{M}^C$ is right adjoint to the forgetful functor $(-)_R : \mathbf{M}^C \to \mathbf{M}_R$.

(4) For any monomorphism $f : K \to L$ of R-modules,

$$f \otimes I_C : K \otimes_R C \to L \otimes_R C$$

is a monomorphism in $\mathbf{M}^C$.

(5) For any family $\{M_\lambda\}_\Lambda$ of R-modules, $(\prod_\Lambda M_\lambda) \otimes_R C$ is the product of the $M_\lambda \otimes_R C$ in $\mathbf{M}^C$.

Proof. (1) The first assertions follow from 3.5 and 3.7. By 3.8(4), any comodule M is a subcomodule of the C-generated comodule $M \otimes_R C$.

(2) By 3.5, $\mathbf{M}^C$ has kernels provided C is a flat R-module. This implies that the intersection of two subcomodules and the preimage of a (sub)comodule is again a comodule. It remains to show that $\mathbf{M}^C$ has (a set of) generators. For any right C-comodule M, there exists a surjective comodule map $g : C^{(\Lambda)} \to M \otimes_R C$ (see 3.8). Then $L := g^{-1}(M) \subset C^{(\Lambda)}$ is a subcomodule. Furthermore, for any $m \in M$ there exist $k \in \mathbb{N}$ and an element x in the comodule $C^k \cap L \subset C^k$ such that $g(x) = m$. Therefore $m \in g(C^k \cap L)$. This shows that M is generated by comodules of the form $C^k \cap L$, $k \in \mathbb{N}$. Hence the subcomodules of C^k, $k \in \mathbb{N}$, form a set of generators of $\mathbf{M}^C$.

(3) For all $M \in \mathbf{M}^C$ and $X \in \mathbf{M}_R$, let $\varphi_{M,X}$ denote the isomorphism constructed in 3.9(1). We need to show that $\varphi_{M,X}$ is natural in M and X. First take any right C-comodule N and any $g \in \mathrm{Hom}^C(M, N)$. Then, for all

$f \in \mathrm{Hom}^C(N, X \otimes_R C)$,

$$\begin{aligned}(\varphi_{M,X} \circ \mathrm{Hom}^C(g, X \otimes_R C))(f) &= (I_X \otimes \varepsilon) \circ \mathrm{Hom}^C(g, X \otimes_R C)(f) \\ &= (I_X \otimes \varepsilon) \circ f \circ g \\ &= \mathrm{Hom}_R(g, X)((I_X \otimes \varepsilon) \circ f) \\ &= (\mathrm{Hom}_R(g, X) \circ \varphi_{N,X})(f).\end{aligned}$$

Similarly, take any R-module Y and $g \in \mathrm{Hom}_R(X, Y)$. Then, for any map $f \in \mathrm{Hom}^C(M, X \otimes_R C)$,

$$\begin{aligned}(\varphi_{M,Y} \circ \mathrm{Hom}^C(M, g \otimes I_C))(f) &= (I_Y \otimes \varepsilon) \circ (\mathrm{Hom}^C(M, g \otimes I_C)(f)) \\ &= (I_Y \otimes \varepsilon) \circ (g \otimes I_C) \circ f \\ &= (g \otimes \varepsilon) \circ f = g \circ (I_X \otimes \varepsilon) \circ f \\ &= (\mathrm{Hom}_R(M, g) \circ \varphi_{M,X})(f).\end{aligned}$$

This proves the naturality of φ and thus the adjointness property. Note that the unit of this adjunction is provided by the coaction $\varrho^M : M \to M \otimes_R C$, while the counit is $I_X \otimes \varepsilon : X \otimes_R C \to X$.

(4) Any functor that has a left adjoint preserves monomorphisms (cf. 38.21). Note that monomorphisms in $\mathbf{M}^C$ need not be injective maps, unless ${}_RC$ is flat.

(5) By (3), for all $X \in \mathbf{M}^C$ there are isomorphisms

$$\begin{aligned}\mathrm{Hom}^C(X, (\textstyle\prod_\Lambda M_\lambda) \otimes_R C) &\simeq \mathrm{Hom}_R(X, \textstyle\prod_\Lambda M_\lambda) \\ &\simeq \textstyle\prod_\Lambda \mathrm{Hom}_R(X, M_\lambda) \\ &\simeq \textstyle\prod_\Lambda \mathrm{Hom}^C(X, M_\lambda \otimes_R C).\end{aligned}$$

These isomorphisms characterise $(\prod_\Lambda M_\lambda) \otimes_R C$ as product of the $M_\lambda \otimes_R C$ in $\mathbf{M}^C$. □

3.14. C as a flat R-module. *The following are equivalent:*

(a) C is flat as an R-module;

(b) every monomorphism in $\mathbf{M}^C$ is injective;

(c) every monomorphism $U \to C$ in $\mathbf{M}^C$ is injective;

(d) the forgetful functor $\mathbf{M}^C \to \mathbf{M}_R$ respects monomorphisms.

Proof. (a) ⇒ (b) Consider a monomorphism $f : M \to N$. Since ${}_RC$ is flat, the inclusion $i : \mathrm{Ke}\, f \to M$ is a morphism in $\mathbf{M}^C$ (by 3.5) and $f \circ i = f \circ 0 = 0$ implies $i = 0$, that is, $\mathrm{Ke}\, f = 0$.

(b) ⇒ (c) and (b) ⇔ (d) are obvious.

(c) ⇒ (a) For every ideal $J \subset R$, the canonical map $J \otimes_R C \to R \otimes_R C$ is a monomorphism in $\mathbf{M}^C$ by 3.13(4), and hence it is injective by assumption. This implies that ${}_RC$ is flat (e.g., [46, 12.16]). □

3.15. $-\otimes_R C$ as a left adjoint functor. *If the functor $-\otimes_R C : \mathbf{M}_R \to \mathbf{M}^C$ is left adjoint to the forgetful functor $\mathbf{M}^C \to \mathbf{M}_R$, then C is finitely generated and projective as an R-module.*

Proof. As a right adjoint functor, the forgetful functor respects monomorphisms and products (see 38.21). Hence C is a flat R-module (by 3.14), and by 3.13, for any family $\{M_\lambda\}_\Lambda$ of R-modules there is an isomorphism

$$(\prod_\Lambda M_\lambda) \otimes_R C \simeq \prod_\Lambda (M_\lambda \otimes_R C).$$

By 40.17 this implies that C is a finitely presented R-module, and hence it is projective. □

3.16. Finiteness Theorem (1). *Assume C to be flat as an R-module and let $M \in \mathbf{M}^C$.*

(1) Every finite subset of M is contained in a subcomodule of M that is contained in a finitely generated R-submodule.

(2) If C is a Mittag-Leffler R-module (cf. 40.17), then every finite subset of M is contained in a subcomodule of M that is finitely generated as R-module.

Proof. (1) Obviously it is enough to prove this for a single element $m \in M$. Write $\varrho^M(m) = m_1 \otimes c_1 + \cdots + m_k \otimes c_k$ and $M' = \sum_i Rm_i$. Let N denote the kernel of the composition of the canonical maps

$$M \xrightarrow{\varrho^M} M \otimes_R C \longrightarrow (M \otimes_R C)/(M' \otimes_R C).$$

Then N is a C-comodule, $m \in N$, and $\varrho^N(N) \subset M' \otimes_R C$, implying

$$N \subset (I_{M'} \otimes \varepsilon)(M' \otimes_R C) \subset M'.$$

(2) For $m \in M$, let $\{M_\lambda\}_\Lambda$ denote the family of all R-submodules of M such that $\varrho^M(m) \in M_\lambda \otimes_R C$. Consider the commutative diagram

$$\begin{array}{ccccccc}
0 \longrightarrow & (\bigcap_\Lambda M_\lambda) \otimes_R C & \longrightarrow & M \otimes_R C & \longrightarrow & (\prod_\Lambda M/M_\lambda) \otimes_R C \\
& \downarrow & & \downarrow = & & \downarrow \varphi_C \\
0 \longrightarrow & \bigcap_\Lambda (M_\lambda \otimes_R C) & \longrightarrow & M \otimes_R C & \longrightarrow & \prod_\Lambda (M/M_\lambda \otimes_R C),
\end{array}$$

where φ_C is injective by the Mittag-Leffler property of C (see 40.17) and hence – by diagram lemmata – we obtain $(\bigcap_\Lambda M_\lambda) \otimes_R C = \bigcap_\Lambda (M_\lambda \otimes_R C)$.

Putting $M' = \bigcap_\Lambda M_\lambda$ and defining N as above, we can write $\varrho^M(m) = n_1 \otimes c_1 + \cdots + n_l \otimes c_l$, where all $n_i \in M'$. This implies $N \subset M' = \sum_{i=1}^k Rn_i \subset N$, and so $N = \sum_{i=1}^k Rn_i$ is finitely generated. □

Special cases of these finiteness properties are considered in 4.12 and 4.16.

Recall that a monomorphism $i : N \to L$ in $\mathbf{M}_R$ is a *coretraction* provided there exists $p : L \to N$ in $\mathbf{M}_R$ with $p \circ i = I_N$ (see 38.8).

3.17. Relative injective comodules. A right C-comodule M is said to be *relative injective* or *(C, R)-injective* if, for every C-comodule map $i : N \to L$ that is an R-module coretraction, and for every morphism $f : N \to M$ in $\mathbf{M}^C$, there exists a right C-comodule map $g : L \to M$ such that $g \circ i = f$. In other words, we require that every diagram in $\mathbf{M}^C$

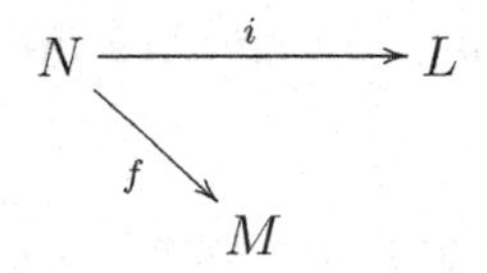

can be completed commutatively by some C-morphism $g : L \to M$, provided there exists an R-module map $p : L \to N$ such that $p \circ i = I_N$.

3.18. (C, R)-injectivity. *Let M be a right C-comodule.*

(1) The following are equivalent:

(a) M is (C, R)-injective;

(b) any C-comodule map $i : M \to L$ that is a coretraction in $\mathbf{M}_R$ is also a coretraction in $\mathbf{M}^C$;

(c) the coaction $\varrho^M : M \to M \otimes_R C$ is a coretraction in $\mathbf{M}^C$.

(2) For any $X \in \mathbf{M}_R$, $X \otimes_R C$ is (C, R)-injective.

(3) If M is (C, R)-injective, then, for any $L \in \mathbf{M}^C$, the canonical sequence

$$0 \longrightarrow \mathrm{Hom}^C(L, M) \xrightarrow{\ i\ } \mathrm{Hom}_R(L, M) \xrightarrow{\ \gamma\ } \mathrm{Hom}_R(L, M \otimes_R C)$$

splits in $\mathbf{M}_B$, where $B = \mathrm{End}^C(L)$ and $\gamma(f) = \varrho^M \circ f - (f \otimes I_C) \circ \varrho^L$ (see 3.3).

In particular, $\mathrm{End}^C(C) \simeq C^*$ is a C^*-direct summand in $\mathrm{End}_R(C)$.

Proof. (1) (a) $\Rightarrow$ (b) Suppose that M is (C, R)-injective and take $N = M$ and $f = I_M$ in 3.17 to obtain the assertion.

(b) $\Rightarrow$ (c) View $M \otimes_R C$ as a right C-comodule with the coaction $I_M \otimes \Delta$, and note that $\varrho^M : M \to M \otimes_R C$ is a right C-comodule map that has an R-linear retraction $I_M \otimes \varepsilon$. Therefore ϱ^M is a coretraction in $\mathbf{M}^C$.

(c) $\Rightarrow$ (a) Suppose there exists a right C-comodule map $h : M \otimes_R C \to M$ such that $h \circ \rho^M = I_M$, consider a diagram

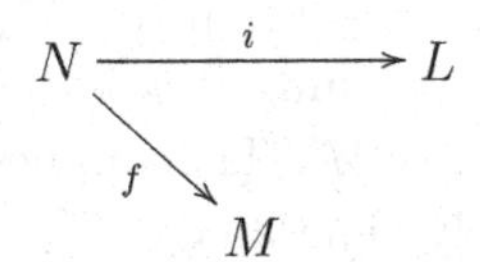

as in 3.17, and assume that there exists an R-module map $p : L \to N$ such that $p \circ i = I_N$. Define an R-linear map $g : L \to M$ as a composition

$$g : L \xrightarrow{\varrho^L} L \otimes_R C \xrightarrow{f \circ p \otimes I_C} M \otimes_R C \xrightarrow{h} M .$$

Clearly, g is a right C-comodule map as a composition of C-comodule maps. Furthermore,

$$\begin{aligned} g \circ i &= h \circ (f \circ p \otimes I_C) \circ \varrho^L \circ i = h \circ (f \circ p \circ i \otimes I_C) \circ \varrho^N \\ &= h \circ (f \otimes I_C) \circ \varrho^N \;=\; h \circ \varrho^M \circ f \;=\; f, \end{aligned}$$

where we used that both i and f are C-colinear. Thus the above diagram can be completed to a commutative diagram in $\mathbf{M}^C$, and hence M is (C, R)-injective.

(2) The coaction for $X \otimes_R C$ is given by $\varrho^{X \otimes_R C} = I_X \otimes \Delta$, and it is split by a right C-comodule map $I_X \otimes \varepsilon \otimes I_C$. Thus $X \otimes_R C$ is (C, R)-injective by part (1).

(3) Denote by $h : M \otimes_A C \to M$ the splitting map of ϱ^M in $\mathbf{M}^C$. Then the map

$$\mathrm{Hom}_R(L, M) \simeq \mathrm{Hom}^C(L, M \otimes_A C) \to \mathrm{Hom}^C(L, M), \quad f \mapsto h \circ (f \otimes I_C) \circ \varrho^L,$$

splits the first inclusion in $\mathbf{M}_B$, and the map

$$\mathrm{Hom}_R(L, M \otimes_A C) \to \mathrm{Hom}_R(L, M),\; g \mapsto h \circ g$$

yields a splitting map $\mathrm{Hom}_R(L, M \otimes_A C) \to \mathrm{Hom}_R(L, M)/\mathrm{Hom}^C(L, M)$, since for any $f \in \mathrm{Hom}_R(L, M)$,

$$h \circ \gamma(f) = f - h \circ (f \otimes I_C) \circ \varrho^L \in f + \mathrm{Hom}^C(L, M).$$

□

If ${}_R C$ is flat, $\mathbf{M}^C$ is a Grothendieck category by 3.13, so exact sequences are defined in $\mathbf{M}^C$ and we can describe

3.19. Exactness of the Hom^C-functors. *Assume ${}_R C$ to be flat and let $M \in \mathbf{M}^C$. Then:*

(1) $\mathrm{Hom}^C(-, M) : \mathbf{M}^C \to \mathbf{M}_R$ *is a left exact functor.*

(2) $\mathrm{Hom}^C(M, -) : \mathbf{M}^C \to \mathbf{M}_R$ *is a left exact functor.*

Proof. (1) From any exact sequence $X \to Y \to Z \to 0$ in $\mathbf{M}^C$ we derive the commutative diagram (tensor over R)

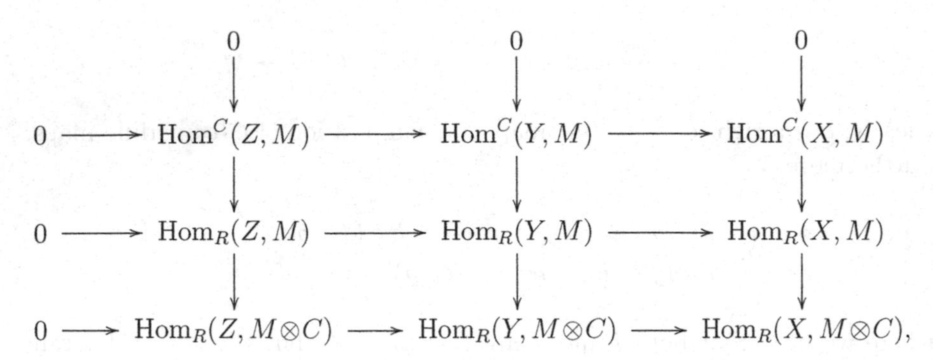

where the columns are exact by the characterisation of comodule morphisms (in 3.3). The second and third rows are exact by exactness properties of the functors Hom_R. Now the diagram lemmata imply that the first row is exact, too.

Part (2) is shown with a similar diagram. □

3.20. (C,R)-exactness. If $_RC$ is flat, exact sequences in $\mathbf{M}^C$ are called *(C,R)-exact* provided they split in $\mathbf{M}_R$. A functor on $\mathbf{M}^C$ is called *left (right) (C,R)-exact* if it is left (right) exact on short *(C,R)-exact* sequences. From properties of the functors given in 3.9 and 3.19, and from properties of (C,R)-injective comodules in 3.18, we immediately obtain the following characterisation of (C,R)-injective comodules in case $_RC$ is flat:

$M \in \mathbf{M}^C$ is (C,R)-injective if and only if $\mathrm{Hom}^C(-,M) : \mathbf{M}^C \to \mathbf{M}_R$ *is a (C,R)-exact functor.*

An object $Q \in \mathbf{M}^C$ is *injective in* $\mathbf{M}^C$ if, for any monomorphism $M \to N$ in $\mathbf{M}^C$, the canonical map $\mathrm{Hom}^C(N,Q) \to \mathrm{Hom}^C(M,Q)$ is surjective.

3.21. Injectives in $\mathbf{M}^C$. *Assume $_RC$ to be flat.*

(1) $Q \in \mathbf{M}^C$ is injective if and only if $\mathrm{Hom}^C(-,Q) : \mathbf{M}^C \to \mathbf{M}_R$ *is exact.*

(2) If $X \in \mathbf{M}_R$ is injective in $\mathbf{M}_R$, then $X \otimes_R C$ is injective in $\mathbf{M}^C$.

(3) If $M \in \mathbf{M}^C$ is (C,R)-injective and injective in $\mathbf{M}_R$, then M is injective in $\mathbf{M}^C$.

(4) C is (C,R)-injective, and it is injective in $\mathbf{M}^C$ provided that R is injective in $\mathbf{M}_R$.

(5) If $_RM$ is flat and N is injective in $\mathbf{M}^C$, then $\mathrm{Hom}^C(M,N)$ *is injective in $\mathbf{M}_R$.*

Proof. (1) The assertion follows by 3.19.

(2) This follows from the isomorphism in 3.9(1).

(3) Since M is R-injective, assertion (2) implies that $M \otimes_R C$ is injective in $\mathbf{M}^C$. Moreover, by 3.18, M is a direct summand of $M \otimes_R C$ as a comodule, and hence it is also injective in $\mathbf{M}^C$.

Part (4) is a special case of (2).

(5) This follows from the isomorphism in 3.9(2). □

An object $P \in \mathbf{M}^C$ is *projective in* $\mathbf{M}^C$ if, for any epimorphism $M \to N$ in $\mathbf{M}^C$, the canonical map $\mathrm{Hom}^C(P, M) \to \mathrm{Hom}^C(P, N)$ is surjective.

3.22. Projectives in $\mathbf{M}^C$. *Consider any $P \in \mathbf{M}^C$.*

(1) If P is projective in $\mathbf{M}^C$, then P is projective in $\mathbf{M}_R$.

(2) If ${}_RC$ is flat, the following are equivalent:

(a) P is projective in $\mathbf{M}^C$;

(b) $\mathrm{Hom}^C(P, -) : \mathbf{M}^C \to \mathbf{M}_R$ is exact.

Proof. (1) For any epimorphism $f : K \to L$ in $\mathbf{M}_R$, $K \otimes_R C \xrightarrow{f \otimes I_C} L \otimes_R C$ is an epimorphism in $\mathbf{M}^C$ and the projectivity of P implies the exactness of the top row in the commutative diagram

$$\begin{array}{ccccc} \mathrm{Hom}^C(P, K \otimes_R C) & \longrightarrow & \mathrm{Hom}^C(P, L \otimes_R C) & \longrightarrow & 0 \\ \downarrow \simeq & & \downarrow \simeq & & \\ \mathrm{Hom}_R(P, K) & \xrightarrow{\mathrm{Hom}(P,f)} & \mathrm{Hom}_R(P, L), & & \end{array}$$

where the vertical maps are the functorial isomorphisms from 3.9(1). From this we see that $\mathrm{Hom}(P, f)$ is surjective, proving that P is projective as an R-module.

(2) This follows from left exactness of $\mathrm{Hom}^C(P, -)$ described in 3.19. □

Note that, although there are enough injectives in $\mathbf{M}^C$, there are possibly no projective objects in $\mathbf{M}^C$. This remains true even if R is a field (see Exercise 8.12).

3.23. Tensor product and Hom^C. *Let ${}_RC$ be flat, and consider $M, N \in \mathbf{M}^C$ and $X \in \mathbf{M}_R$ such that*

(i) M_R is finitely generated and projective, and N is (C, R)-injective; or

(ii) M_R is finitely presented and X is flat in $\mathbf{M}_R$.

Then there exists a canonical isomorphism

$$\nu : X \otimes_R \mathrm{Hom}^C(M, N) \longrightarrow \mathrm{Hom}^C(M, X \otimes_R N), \quad x \otimes h \mapsto x \otimes h(-).$$

Proof. Consider the defining exact sequence for Hom^C (see 3.3),

$$(*)\quad 0 \longrightarrow \mathrm{Hom}^C(M,N) \longrightarrow \mathrm{Hom}_R(M,N) \longrightarrow \mathrm{Hom}_R(M, N\otimes_R C).$$

Tensoring with X_R yields the commutative diagram (tensor over R)

$$\begin{array}{ccccccc} 0 \longrightarrow & X\otimes\mathrm{Hom}^C(M,N) & \longrightarrow & X\otimes\mathrm{Hom}_R(M,N) & \longrightarrow & X\otimes\mathrm{Hom}_R(M,N\otimes C) \\ & \downarrow \nu & & \downarrow \simeq & & \downarrow \simeq \\ 0 \longrightarrow & \mathrm{Hom}^C(M,X\otimes N) & \longrightarrow & \mathrm{Hom}_R(M,X\otimes N) & \longrightarrow & \mathrm{Hom}_R(M,X\otimes N\otimes C), \end{array}$$

where the bottom row is exact (again by 3.3) and the vertical isomorphisms follow from the finiteness assumptions in (i) and (ii) (cf. 40.12).

If X is flat, the top row is exact. On the other hand, if N is (C,R)-injective, the sequence $(*)$ splits by 3.18, and hence the top row is exact, too. Therefore, in either case, the exactness of the diagram implies that ν is an isomorphism, as required. □

3.24. Bicomodules over C. An R-module M that is both a left and a right C-comodule is called a (C,C)-*bicomodule* if the diagram

$$\begin{array}{ccc} M & \xrightarrow{\varrho^M} & M\otimes_R C \\ {\scriptstyle {}^M\!\varrho}\downarrow & & \downarrow{\scriptstyle {}^M\!\varrho\otimes I_C} \\ C\otimes_R M & \xrightarrow{I_C\otimes\varrho^M} & C\otimes_R M\otimes_R C\,, \end{array}$$

where ${}^M\!\varrho$ is a left and ϱ^M is a right coaction, commutes.

An R-linear map $f: M\to N$ between two (C,C)-bicomodules is said to be a (C,C)-*bicomodule* or (C,C)-*bicolinear map* if it is both a left and a right comodule morphism. The set of these maps is denoted by ${}^C\mathrm{Hom}^C(M,N)$. The category whose objects are (C,C)-bicomodules and morphisms are the (C,C)-bicolinear maps is denoted by ${}^C\mathbf{M}^C$. By definition, any object, resp. morphism, in ${}^C\mathbf{M}^C$ is also an object, resp. morphism, in both ${}^C\mathbf{M}$ and $\mathbf{M}^C$, and the functor

$$C\otimes_R - \otimes_R C : \mathbf{M}_R \to {}^C\mathbf{M}^C, \quad M\mapsto C\otimes_R M\otimes_R C,$$

is simply the composition of the functors $C\otimes_R -$ and $-\otimes_R C$. Consequently, the proof of 3.9 can be extended to derive

3.25. Hom-tensor relations for bicomodules. *Let X be an R-module. For any $M\in {}^C\mathbf{M}^C$, the R-linear map*

$$\varphi : {}^C\mathrm{Hom}^C(M, C\otimes_R X\otimes_R C) \to \mathrm{Hom}_R(M,X), \quad f\mapsto (\varepsilon\otimes I_X\otimes\varepsilon)\circ f,$$

is bijective, with inverse map $h\mapsto (I_C\otimes h\otimes I_C)\circ(I_C\otimes\varrho^M)\circ {}^M\!\varrho$.

Several properties of comodules can be restricted to bicomodules.

3.26. The category ${}^C\mathbf{M}^C$. *Let C be an R-coalgebra.*

(1) *The category ${}^C\mathbf{M}^C$ has direct sums and cokernels, and $C \otimes_R C$ is a subgenerator.*

(2) *${}^C\mathbf{M}^C$ is a Grothendieck category provided C is a flat R-module.*

(3) *The functor $C \otimes_R - \otimes_R C : \mathbf{M}_R \to {}^C\mathbf{M}^C$ is right adjoint to the forgetful functor ${}^C\mathbf{M}^C \to \mathbf{M}_R$.*

(4) *For any monomorphism $f : K \to L$ of R-modules,*
$$I_C \otimes f \otimes I_C : C \otimes_R K \otimes_R C \to C \otimes_R L \otimes_R C$$
is a monomorphism in ${}^C\mathbf{M}^C$.

(5) *For any family $\{M_\lambda\}_\Lambda$ of R-modules, $C \otimes_R (\prod_\Lambda M_\lambda) \otimes_R C$ is the product of the family $\{C \otimes_R M_\lambda \otimes_R C\}_\Lambda$ in ${}^C\mathbf{M}^C$.*

Notice that ${}^C\mathbf{M}^C$ can also be considered as $\mathbf{M}^{C^{cop}\otimes_R C}$, the category of right comodules over the coalgebra $C^{cop} \otimes_R C$. With this identification it is obvious how to define *relative injective (C, C)-bicomodules* and *relative exact sequences in ${}^C\mathbf{M}^C$*.

3.27. Relative semismple coalgebras. An R-coalgebra C is said to be *left (C, R)-semisimple* or *left relative semisimple* if every left C-comodule is (C, R)-injective. *Right relative semisimple* coalgebras are defined similarly.

Furthermore, C is said to be *relative semisimple as a (C, C)-bicomodule* if every bicomodule is relative injective in ${}^C\mathbf{M}^C$. This means that $C^{cop} \otimes_R C$ is a right relative semisimple coalgebra.

If C is flat as an R-module, then C is left (C, R)-semisimple if and only if any (C, R)-splitting sequence of left C-comodules splits in ${}^C\mathbf{M}$. A similar characterisation holds for right (C, R)-semisimple coalgebras.

Interesting examples of relative semisimple coalgebras are provided by

3.28. Coseparable coalgebras. $C \otimes_R C$ can be viewed as a (C, C)-bicomodule with the left coaction ${}^{C\otimes_R C}\varrho = \Delta \otimes I_C$ and the right coaction $\varrho^{C\otimes_R C} = I_C \otimes \Delta$. Note that both of these coactions are split in $\mathbf{M}_R$ by a single (C, C)-bicomodule map $I_C \otimes \varepsilon \otimes I_C : C \otimes_R C \otimes_R C \to C \otimes_R C$. This means that $C \otimes_R C$ is a relative injective (C, C)-bicomodule.

On the other hand, C is a (C, C)-bicomodule by the left and right regular coaction Δ. Although Δ is split as a left C-comodule map by $I_C \otimes \varepsilon$ and as a right C-comodule map by $\varepsilon \otimes I_C$, C is not necessarily a relative injective (C, C)-bicomodule. Coalgebras that are relative injective (C, C)-bicomodules are of particular interest because they are dual to separable algebras. Thus, a coalgebra C is called a *coseparable* coalgebra if the structure map $\Delta : C \to C \otimes_R C$ splits as a (C, C)-bicomodule map. Explicitly this means that there exists a map $\pi : C \otimes_R C \to C$ with the properties

$$(I_C \otimes \pi) \circ (\Delta_C \otimes I_C) = \Delta_C \circ \pi = (\pi \otimes I_C) \circ (I_C \otimes \Delta_C) \quad \text{and} \quad \pi \circ \Delta_C = I_C.$$

The coseparability of C can be described equivalently as the separability of certain functors (cf. 38.18, 38.19 and 38.20 for the definition and discussion of separable functors).

3.29. Properties of coseparable coalgebras. *For an R-coalgebra C the following are equivalent:*

(a) *C is coseparable;*

(b) *there exists an R-linear map $\delta : C \otimes_R C \to R$ satisfying*

$$\delta \circ \Delta = \varepsilon \quad \text{and} \quad (I_C \otimes \delta) \circ (\Delta \otimes I_C) = (\delta \otimes I_C) \circ (I_C \otimes \Delta);$$

(c) *the forgetful functor $(-)_R : \mathbf{M}^C \to \mathbf{M}_R$ is separable;*

(d) *the forgetful functor ${}_R(-) : {}^C\mathbf{M} \to \mathbf{M}_R$ is separable;*

(e) *the forgetful functor ${}_R(-)_R : {}^C\mathbf{M}^C \to \mathbf{M}_R$ is separable;*

(f) *C is relative semisimple as a (C, C)-bicomodule;*

(g) *C is relative injective as a (C, C)-bicomodule.*

If these conditions are satisfied, then C is left and right (C, R)-semisimple.

Proof. (a) $\Rightarrow$ (b) Let $\pi : C \otimes_R C \to C$ be left and right C-colinear with $\pi \circ \Delta = I_C$ and define $\delta = \varepsilon \circ \pi : C \otimes_R C \to R$. Then $\delta \circ \Delta = \varepsilon \circ \pi \circ \Delta = \varepsilon$ and

$$\begin{aligned} (I_C \otimes \delta) \circ (\Delta \otimes I_C) &= (I_C \otimes \varepsilon) \circ (I_C \otimes \pi) \circ (\Delta \otimes I_C) \\ &= (I_C \otimes \varepsilon) \circ \Delta \circ \pi = \pi \\ &= (\varepsilon \otimes I_C) \circ (\pi \otimes I_C) \circ (I_C \otimes \Delta) \\ &= (\delta \otimes I_C) \circ (I_C \otimes \Delta). \end{aligned}$$

Thus $\delta = \varepsilon \circ \pi$ has all the required properties.

(b) $\Rightarrow$ (c) Given $\delta : C \otimes_R C \to R$ with the stated properties, for any $N \in \mathbf{M}^C$ define an R-linear map

$$\nu_N : N \otimes_R C \xrightarrow{\varrho^N \otimes I_C} N \otimes_R C \otimes_R C \xrightarrow{I_N \otimes \delta} N.$$

Explicitly, $\nu_N : n \otimes c \mapsto \sum n_{\underline{0}} \delta(n_{\underline{1}} \otimes c)$. The map ν_N is a right C-comodule map because – by the properties of δ – we obtain for all $n \in N$ and $c \in C$

$$\begin{aligned} \sum \nu_N(n \otimes c_{\underline{1}}) \otimes c_{\underline{2}} &= \sum n_{\underline{0}} \otimes \delta(n_{\underline{1}} \otimes c_{\underline{1}}) \cdot c_{\underline{2}} \\ &= \sum n_{\underline{0}} \otimes n_{\underline{1}} \cdot \delta(n_{\underline{2}} \otimes c) \\ &= \sum \nu_N(n \otimes c)_{\underline{0}} \otimes \nu_N(n \otimes c)_{\underline{1}}. \end{aligned}$$

Note that ν_N is a retraction for ϱ^N by the computation

$$\begin{aligned} \nu_N \circ \varrho^N &= (I_N \otimes \delta) \circ (\varrho^N \otimes I_C) \circ \varrho^N \\ &= (I_N \otimes \delta) \circ (I_N \otimes \Delta) \circ \varrho^N = (I_N \otimes \varepsilon) \circ \varrho^N = I_N. \end{aligned}$$

Now, define a functorial morphism $\Phi : \mathrm{Hom}_R((-)_R, (-)_R) \to \mathrm{Hom}^C(-,-)$ by assigning to any R-linear map $f : M \to N$, where $M, N \in \mathbf{M}^C$, the map

$$\Phi(f) : M \xrightarrow{\varrho^M} M \otimes_R C \xrightarrow{f \otimes I_C} N \otimes_R C \xrightarrow{\nu_N} N .$$

View $M \otimes_R C$ and $N \otimes_R C$ as right C-comodules via $I_M \otimes \Delta$ and $I_N \otimes \Delta$, respectively. Then the map $f \otimes I_C$ is right C-colinear, and so is $\Phi(f)$ as a composition of right C-colinear maps. If, in addition, f is a morphism in $\mathbf{M}^C$, then $(f \otimes I_C) \circ \varrho^M = \varrho^N \circ f$, and, since ν_N is a retraction for ϱ^N, it follows that $\Phi(f) = f$. This shows that the forgetful functor is separable.

(c) $\Rightarrow$ (a) Assume $(-)_R$ to be separable. Then there exists a functorial morphism $\nu_C : C \otimes_R C \to C$ in $\mathbf{M}^C$. Since the composition of $(-)_R$ with $- \otimes_R C$ preserves colimits and $C \otimes_R C$ is also a left C-comodule, we conclude that ν_C is also left C-colinear (cf. 39.7). Thus ν_C is a (C, C)-bicolinear splitting of Δ, and therefore C is a coseparable coalgebra.

(a) $\Leftrightarrow$ (d) Since the condition in (a) is symmetric the proof of (a) $\Leftrightarrow$ (c) applies.

(a) $\Rightarrow$ (e) The forgetful functor ${}_R(-)_R : {}^C\mathbf{M}^C \to \mathbf{M}_R$ is the composition of the forgetful functors ${}_R(-)$ and $(-)_R$, and hence it is separable (by 38.20).

(e) $\Rightarrow$ (c) Since the composition of the left and right forgetful functors is separable, then so is each one of these (by 38.20).

(e) $\Rightarrow$ (f) For a bicomodule N consider the maps $N \xrightarrow{{}^N\varrho} C \otimes_R N \xrightarrow{I \otimes \varrho^N} C \otimes_R N \otimes_R C$. By coseparability, both $I \otimes \varrho^N$ and ${}^N\varrho$ are split by morphisms from ${}^C\mathbf{M}^C$.

(g) $\Rightarrow$ (a) $\Delta : C \to C \otimes_R C$ is R-split, and hence it splits in ${}^C\mathbf{M}^C$.

If $(-)_R$ is a separable functor, it reflects retractions and hence C is (C, R)-semisimple in $\mathbf{M}^C$ and ${}^C\mathbf{M}$ (see 38.19). □

Remark. While the relative semisimplicity of ${}^C\mathbf{M}^C$ is sufficient (equivalent) to obtain the coseparability of C (by 3.29(f)), the relative semisimplicity of $\mathbf{M}^C$ need not imply the coseparability of C.

The forgetful functor is left adjoint to the tensor functor $- \otimes_R C$, and we may ask when the latter is separable. This is dual to the separability of the functor $A \otimes_R -$ for an R-algebra A considered in 40.22.

3.30. Separability of $- \otimes_R C$. *The following are equivalent for C:*

(a) $- \otimes_R C : \mathbf{M}_R \to \mathbf{M}^C$ *is separable;*

(b) $C \otimes_R - : \mathbf{M}_R \to {}^C\mathbf{M}$ *is separable;*

(c) $C \otimes_R - \otimes_R C : \mathbf{M}_R \to {}^C\mathbf{M}^C$ *is separable;*

(d) *there exists* $e \in C$ *with* $\varepsilon(e) = 1_R$.

Proof. Let ψ denote the counit of the adjoint pair $((-)_R, - \otimes_R C)$, that is, $\psi_R = I_R \otimes \varepsilon : R \otimes_R C \to R$.

(a) $\Rightarrow$ (d) By 38.24(2), ψ_R is split by some morphism $\nu_R : R \to C$. Putting $e = \nu_R(1_R)$ yields $\varepsilon(e) = \varepsilon \circ \nu_R(1_R) = 1_R$.

(d) $\Rightarrow$ (a) Suppose there exists $e \in C$ such that $\varepsilon(e) = 1_R$. For any $M \in \mathbf{M}_R$ define an R-linear map $\nu_M : M \to M \otimes_R C$, $m \mapsto m \otimes e$, for which we get $\psi \circ \nu_M(m) = m\varepsilon(e) = m$, that is, $\psi \circ \nu_M = I_M$. Moreover, for any $f \in \mathrm{Hom}_R(M, N)$ and $m \in M$ we compute

$$\nu_N \circ f(m) = f(m) \otimes e = (f \otimes I_C) \circ \nu_M(m),$$

so that ν_M is functorial in M.

The remaining implications follow by symmetry and the basic properties of separable functors (see 38.20). □

3.31. Exercises

Let C be an R-coalgebra with ${}_RC$ flat. Prove that the following are equivalent:

(a) every (C, R)-injective right C-comodule is injective in $\mathbf{M}^C$;

(b) every exact sequence in $\mathbf{M}^C$ is (C, R)-exact.

References. Caenepeel, Ion and Militaru [84]; Caenepeel, Militaru and Zhu [9]; Castaño Iglesias, Gómez-Torrecillas and Năstăsescu [91]; Doi [104]; Gómez-Torrecillas [122]; Larson [148]; Rafael [180]; Sweedler [45]; Wisbauer [210].

4 C-comodules and C^*-modules

Let C be again an R-coalgebra. As explained in 1.3, the dual R-module $C^* = \text{Hom}_R(C, R)$ is an associative algebra. As already mentioned at the beginning of the previous section, there is a close relationship between the comodules of C and the modules of C^*. More precisely, there is a faithful functor $\mathbf{M}^C \to {}_{C^*}\mathbf{M}$. It is therefore natural to ask when $\mathbf{M}^C$ is a full subcategory of the latter (i.e., when all the (left) C^*-linear maps between right C-comodules arise from (right) C-comodule morphisms) or when $\mathbf{M}^C$ is isomorphic to the category of left C^*-modules. This connection between the comodules of a coalgebra and the modules of a dual algebra allows one to relate comodules to much more familiar (from the classical ring theory) and often nicer (for example, Abelian) categories of modules. In this section we study this relationship, and in particular we introduce in 4.2 an important property of coalgebras termed the *α-condition*. Coalgebras that satisfy the α-condition have several nice module-theoretic properties that are revealed in a number of subsequent sections, in particular in Sections 7, 8 and 9.

4.1. C-comodules and C^*-modules.

(1) Any $M \in \mathbf{M}^C$ is a (unital) left C^*-module by

$$\rightharpoonup : C^* \otimes_R M \to M, \quad f \otimes m \mapsto (I_M \otimes f) \circ \varrho^M(m) = \sum m_{\underline{0}} f(m_{\underline{1}}).$$

(2) Any morphism $h : M \to N$ in $\mathbf{M}^C$ is a left C^*-module morphism, that is,

$$\text{Hom}^C(M, N) \subset {}_{C^*}\text{Hom}\,(M, N).$$

(3) There is a faithful functor from $\mathbf{M}^C$ to $\sigma[{}_{C^*}C]$, the full subcategory of ${}_{C^*}\mathbf{M}$ consisting of all C^*-modules subgenerated by C (cf. 41.1).

Proof. (1) By definition, for all $f, g \in C^*$ and $m \in M$, the actions $f \rightharpoonup (g \rightharpoonup m)$ and $(f * g) \rightharpoonup m$ are the compositions of the maps in the top and bottom rows of the following commutative diagram:

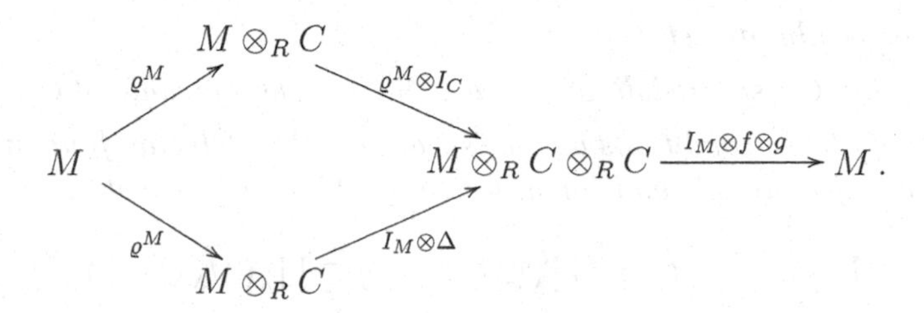

Clearly, for each $m \in M$, $\varepsilon \rightharpoonup m = m$, and thus M is a C^*-module.

(2) For any $h : M \to N$ in $\mathbf{M}^C$ and $f \in C^*$, $m \in M$, consider

$$\begin{aligned} h(f \rightharpoonup m) &= \sum h(m_{\underline{0}} f(m_{\underline{1}})) = (I_N \otimes f) \circ (h \otimes I_C) \circ \varrho^M(m) \\ &= (I_N \otimes f) \circ \varrho^N \circ h(m) = f \rightharpoonup h(m). \end{aligned}$$

This shows that h is a C^*-linear map.

(3) By 3.13, C is a subgenerator in $\mathbf{M}^C$ and hence all C-comodules are subgenerated by C as C^*-modules (by (1),(2)); thus they are objects in $\sigma[{}_{C^*}C]$, and hence (1)–(2) define a faithful functor $\mathbf{M}^C \to \sigma[{}_{C^*}C]$. □

Now, the question arises when $\mathbf{M}^C$ is a full subcategory of $\sigma[{}_{C^*}C]$ (or ${}_{C^*}\mathbf{M}$), that is, when $\mathrm{Hom}^C(M,N) = \mathrm{Hom}_{C^*}(M,N)$, for any $M, N \in \mathbf{M}^C$. In answering this question the following property plays a crucial role.

4.2. The α-condition. C is said to *satisfy the α-condition* if the map

$$\alpha_N : N \otimes_R C \to \mathrm{Hom}_R(C^*, N), \quad n \otimes c \mapsto [f \mapsto f(c)n],$$

is injective, for every $N \in \mathbf{M}_R$. By 42.10, the following are equivalent:

(a) C satisfies the α-condition;

(b) for any $N \in \mathbf{M}_R$ and $u \in N \otimes_R C$, $(I_N \otimes f)(u) = 0$ for all $f \in C^*$, implies $u = 0$;

(c) C is locally projective as an R-module.

In particular, this implies that C is a flat R-module, and that it is cogenerated by R.

The importance of the α-condition in the context of the category of comodules becomes clear from the following observations.

4.3. $\mathbf{M}^C$ as a full subcategory of ${}_{C^*}\mathbf{M}$. *The following are equivalent:*

(a) $\mathbf{M}^C = \sigma[{}_{C^*}C]$;

(b) $\mathbf{M}^C$ is a full subcategory of ${}_{C^*}\mathbf{M}$;

(c) for all $M, N \in \mathbf{M}^C$, $\mathrm{Hom}^C(M,N) = {}_{C^*}\mathrm{Hom}(M,N)$;

(d) ${}_RC$ is locally projective;

(e) every left C^*-submodule of C^n, $n \in \mathbb{N}$, is a subcomodule of C^n.

If (any of) these conditions are satisfied, then the inclusion functor $\mathbf{M}^C \to {}_{C^*}\mathbf{M}$ has a right adjoint, and for any family $\{M_\lambda\}_\Lambda$ of R-modules,

$$(\textstyle\prod_\Lambda M_\lambda) \otimes_R C \simeq \prod^C_\Lambda (M_\lambda \otimes_R C) \subset \prod_\Lambda (M_\lambda \otimes_R C),$$

where $\prod^C$ denotes the product in $\mathbf{M}^C$.

Proof. (a) ⇔ (b) ⇔ (c) follow by the fact that C is always a subgenerator of ${}^C\mathbf{M}$ (see 3.13) and the definition of the category $\sigma[{}_{C^*}C]$ (cf. 41.1).

(a) ⇒ (d) The equality obviously implies that monomorphisms in $\mathbf{M}^C$ are injective maps. Hence ${}_RC$ is flat by 3.13(4). For any $N \in \mathbf{M}_R$ we prove the injectivity of the map $\alpha_N : N \otimes_R C \to \mathrm{Hom}_R(C^*, N)$.

$\mathrm{Hom}_R(C^*, N)$ is a left C^*-module by (see 40.8)

$$g \cdot \gamma(f) = \gamma(f * g), \qquad \text{for } \gamma \in \mathrm{Hom}_R(C^*, N),\ f, g \in C^*,$$

and considering $N \otimes_R C$ as left C^*-module in the canonical way we have

$$\alpha_N(g \rightharpoonup (n \otimes c))(f) = \sum n\, f(c_{\underline{1}})g(c_{\underline{2}}) = n\, f * g(c) = [g \cdot \alpha_N(n \otimes c)](f),$$

for all $f, g \in C^*$, $n \in N$, and $c \in C$. So α_N is C^*-linear, and for any right C-comodule L there is a commutative diagram

$$\begin{array}{ccc} {}_{C^*}\mathrm{Hom}(L, N \otimes_R C) & \xrightarrow{\mathrm{Hom}(L,\alpha_N)} & {}_{C^*}\mathrm{Hom}(L, \mathrm{Hom}_R(C^*, N)) \\ \downarrow \simeq & & \downarrow \simeq \\ \mathrm{Hom}_R(L, N) & \xrightarrow{=} & \mathrm{Hom}_R(L, N). \end{array}$$

The first vertical isomorphism is obtained by assumption and the Hom-tensor relations 3.9, explicitly,

$${}_{C^*}\mathrm{Hom}(L, N \otimes_R C) = \mathrm{Hom}^C(L, N \otimes_R C) \simeq \mathrm{Hom}_R(L, N).$$

The second vertical isomorphism results from the canonical isomorphisms

$${}_{C^*}\mathrm{Hom}(L, \mathrm{Hom}_R(C^*, N)) \simeq \mathrm{Hom}_R(C^* \otimes_{C^*} L, N) \simeq \mathrm{Hom}_R(L, N).$$

This shows that $\mathrm{Hom}(L, \alpha_N)$ is injective for any $L \in \mathbf{M}^C$, and so, by 38.8, (the corestriction of) α_N is a monomorphism in $\mathbf{M}^C$ (see 38.8). Since ${}_RC$ is flat, this implies that α_N is injective (by 3.14).

(d) ⇒ (e) We show that, for right C-comodules M, any C^*-submodule N is a subcomodule. For this consider the map

$$\rho_N : N \to \mathrm{Hom}_R(C^*, N),\ n \mapsto [f \mapsto f \rightharpoonup n].$$

The inclusion $i : N \to M$ yields the commutative diagram with exact rows

$$\begin{array}{ccccccccc} 0 & \longrightarrow & N & \xrightarrow{i} & M & \xrightarrow{p} & M/N & \longrightarrow & 0 \\ & & & & \downarrow \varrho_M & & & & \\ 0 & \longrightarrow & N \otimes_R C & \xrightarrow{i \otimes I} & M \otimes_R C & \xrightarrow{p \otimes I} & M/N \otimes_R C & \longrightarrow & 0 \\ & & \downarrow \alpha_{N,C} & & \downarrow \alpha_{M,C} & & \downarrow \alpha_{M/N,C} & & \\ 0 & \longrightarrow & \mathrm{Hom}_R(C^*, N) & \xrightarrow{\mathrm{Hom}(C^*, i)} & \mathrm{Hom}_R(C^*, M) & \longrightarrow & \mathrm{Hom}_R(C^*, M/N), & & \end{array}$$

where $\mathrm{Hom}(C^*, i) \circ \rho_N = \alpha_{M,C} \circ \varrho_M \circ i$. Injectivity of $\alpha_{M/N,C}$ implies $(p \otimes I) \circ \varrho_M \circ i = 0$, and by the kernel property $\varrho_M \circ i$ factors through $N \to N \otimes_R C$, thus yielding a C-coaction on N.

(e) $\Rightarrow$ (a) First we show that every finitely generated C^*-module $N \in \sigma[_{C^*}C]$ is a C-comodule. There exist a C^*-submodule $X \subset C^n$, $n \in \mathbb{N}$, and an epimorphism $h : X \to N$. By assumption, X and the kernel of h are comodules and hence N is a comodule (see 3.5). So, for any $L \in \sigma[_{C^*}C]$, finitely generated submodules are comodules and this obviously implies that L is a comodule.

It remains to prove that, for $M, N \in \mathbf{M}^C$, any C^*-morphism $f : M \to N$ is a comodule morphism. $\mathrm{Im}\, f \subset N$ and $\mathrm{Ke}\, f \subset M$ are C^*-submodules and hence – as just shown – are subcomodules of N and M, respectively. Therefore the corestriction $M \to \mathrm{Im}\, f$ and the inclusion $\mathrm{Im}\, f \to N$ both are comodule morphisms and so is f (as the composition of two comodule maps).

For the final assertions, recall that the inclusion $\sigma[_{C^*}C] \to {}_{C^*}\mathbf{M}$ has a right adjoint functor (trace functor, see 41.1) and this respects products (cf. 38.21). So the isomorphism follows from the characterisation of the products of the $M_\lambda \otimes_R C$ in $\mathbf{M}^C$ (see 3.13). □

4.4. Coaction and C^*-modules. *Let $_RC$ be locally projective. For any R-module M, consider an R-linear map $\varrho : M \to M \otimes_R C$. Define a left C^*-action on M by*

$$\rightharpoonup : C^* \otimes_R M \to M, \quad f \otimes m \mapsto (I_M \otimes f) \circ \varrho(m).$$

Then the following are equivalent:

(a) ϱ is coassociative and counital;

(b) M is a unital C^*-module by $\rightharpoonup$.

Proof. The implication (a) $\Rightarrow$ (b) is shown in 4.1. Conversely, suppose that M is a unital C^*-module by $\rightharpoonup$, that is,

$$(f * g) \rightharpoonup m = f \rightharpoonup (g \rightharpoonup m), \text{ for all } f, g \in C^*,\ m \in M.$$

By the definition of the action $\rightharpoonup$, this means that

$$(I_M \otimes f \otimes g) \circ (I_M \otimes \Delta) \circ \varrho(m) = (I_M \otimes f \otimes g) \circ (\varrho \otimes I_C) \circ \varrho(m),$$

and from this $_RC$ locally projective implies $(I_M \otimes \Delta) \circ \varrho(m) = (\varrho \otimes I_C) \circ \varrho(m)$ (see 4.2), showing that ϱ is coassociative. Moreover, for any $m \in M$, $m = \varepsilon \rightharpoonup m = (I_M \otimes \varepsilon) \circ \varrho(m)$. □

By symmetry, there is a corresponding relationship between left C-comodules and right C^*-modules, which we formulate for convenience.

4.5. Left C-comodules and right C^*-modules.

(1) Any $M \in {}^C\mathbf{M}$ is a (unital) right C^*-module by

$$\leftharpoonup : M \otimes_R C^* \to M, \quad m \otimes f \mapsto (f \otimes I_M) \circ {}^M\varrho(m) = \sum f(m_{\underline{-1}}) m_{\underline{0}}.$$

(2) Any morphism $h : M \to N$ in ${}^C\mathbf{M}$ is a right C^*-module morphism, so

$$^C\mathrm{Hom}\,(M, N) \subset \mathrm{Hom}_{C^*}(M, N)$$

and there is a faithful functor ${}^C\mathbf{M} \to \sigma[C_{C^*}] \subset \mathbf{M}_{C^*}$.

(3) ${}_RC$ is locally projective if and only if ${}^C\mathbf{M} = \sigma[C_{C^*}]$.

Since C is a left and right C-comodule by the regular coaction (cf. 3.4), we can study the structure of C as a (C^*, C^*)-bimodule (compare 3.12).

4.6. C as a (C^*, C^*)-bimodule. *C is a (C^*, C^*)-bimodule by*

$$\rightharpoonup : C^* \otimes C \to C, \quad f \otimes c \mapsto f \rightharpoonup c = (I_C \otimes f) \circ \Delta(c),$$
$$\leftharpoonup : C \otimes C^* \to C, \quad c \otimes g \mapsto c \leftharpoonup g = (g \otimes I_C) \circ \Delta(c).$$

(1) For any $f, g \in C^*$, $c \in C$,

$$f * g\,(c) = f(g \rightharpoonup c) = g(c \leftharpoonup f).$$

(2) C is faithful as a left and right C^*-module.

(3) Assume C to be cogenerated by R. Then for any central element $f \in C^*$ and any $c \in C$, $f \rightharpoonup c = c \leftharpoonup f$.

(4) If C satisfies the α-condition, it is a balanced (C^*, C^*)-bimodule, that is,

$$_{C^*}End(C) = \mathrm{End}^C(C) \simeq C^* \simeq {}^C\mathrm{End}(C) = \mathrm{End}_{C^*}(C) \text{ and}$$

$$_{C^*}End_{C^*}(C) = {}^C\mathrm{End}^C(C) \simeq Z(C^*),$$

where morphisms are written opposite to scalars and $Z(C^*)$ denotes the centre of C^*. In this case a pure R-submodule $D \subset C$ is a subcoalgebra if and only if D is a left and right C^*-submodule.

Proof. The bimodule property is shown by the equalities

$$\begin{aligned}(f \rightharpoonup c) \leftharpoonup g &= (g \otimes I_C \otimes f) \circ ((\Delta \otimes I_C) \circ \Delta))(c) \\ &= (g \otimes I_C \otimes f) \circ ((I_C \otimes \Delta) \circ \Delta))(c) \quad = \quad f \rightharpoonup (c \leftharpoonup g).\end{aligned}$$

(1) From the definition it follows

$$\begin{aligned} f * g\,(c) = (f \otimes g) \circ \Delta(c) &= (f \otimes I_C) \circ (I_C \otimes g) \circ \Delta(c) = f(g \rightharpoonup c) \\ &= (I_C \otimes g) \circ (f \otimes I_C) \circ \Delta(c) = g(c \leftharpoonup f).\end{aligned}$$

(2) For $f \in C^*$, assume $f \rightharpoonup c = 0$ for each $c \in C$. Then applying (1) yields $f(c) = \varepsilon(f \rightharpoonup c) = 0$, and hence $f = 0$.

(3) For any central element $f \in C^*$, by (1),

$$g(c \leftharpoonup f) = f * g(c) = g * f(c) = g(f \rightharpoonup c),$$

for all $c \in C$, $g \in C^*$. Since C is cogenerated by R, this can only hold if, for all $c \in C$, $c \leftharpoonup f = f \rightharpoonup c$.

(4) The isomorphisms follow from 3.12, 4.3 and 4.5. Let $D \subset C$ be a pure R-submodule. If D is a subcoalgebra of C, then it is a right and left subcomodule and hence a left and right C^*-submodule. Conversely, suppose that D is a left and right C^*-submodule. Then the restriction of Δ yields a left and right C-coaction on D and, by 40.16,

$$\Delta(D) \subset D \otimes_R C \ \cap \ C \otimes D = D \otimes_R D\,,$$

proving that D is a subcoalgebra. □

4.7. When is $\mathbf{M}^C = {}_{C^*}\mathbf{M}$? *The following are equivalent:*

(a) $\mathbf{M}^C = {}_{C^*}\mathbf{M}$*;*

(b) the functor $- \otimes_R C : \mathbf{M}_R \to {}_{C^*}\mathbf{M}$ *has a left adjoint;*

(c) ${}_RC$ *is finitely generated and projective;*

(d) ${}_RC$ *is locally projective and* C *is finitely generated as right* C^**-module;*

(e) ${}^C\mathbf{M} = \mathbf{M}_{C^*}$*.*

Proof. (a) ⇒ (b) is obvious (by 3.13(3)).

(b) ⇒ (c) Since $- \otimes_R C$ is a right adjoint, it preserves monomorphisms (injective morphisms) by 38.21. Therefore, ${}_RC$ is flat. Moreover $- \otimes_R C$ preserves products, so for any family $\{M_\lambda\}_\Lambda$ in $\mathbf{M}_R$ there is an isomorphism

$$(\prod_\Lambda M_\lambda) \otimes_R C \simeq \prod_\Lambda (M_\lambda \otimes_R C),$$

which implies that ${}_RC$ is finitely presented (see 40.17) and hence projective.

(c) ⇒ (d) Clearly, projective modules are locally projective, and C finitely generated as an R-module implies that C is finitely generated as a right (and left) C^*-module.

(d) ⇒ (a) By 4.6, C is a faithful left C^*-module that is finitely generated as a module over its endomorphism ring C^*. This implies that C is a subgenerator in ${}_{C^*}\mathbf{M}$, that is, $\mathbf{M}^C = \sigma[{}_{C^*}C] = {}_{C^*}\mathbf{M}$ (see 41.7). □

Recall from 38.23 that a functor is said to be *Frobenius* when it has the same left and right adjoint. Recall also that an extension of rings is called a *Frobenius extension* when the restriction of scalars functor is a Frobenius functor (see 40.21 for more details).

4.8. Frobenius coalgebras. *The following are equivalent:*

(a) *the forgetful functor* $(-)_R : \mathbf{M}^C \to \mathbf{M}_R$ *is Frobenius;*

(b) ${}_RC$ *is finitely generated and projective, and* $C \simeq C^*$ *as left* C^**-modules;*

(c) ${}_RC$ *is finitely generated and projective, and there is an element* $e \in C$ *such that the map* $C^* \to C$, $f \mapsto f \rightharpoonup e$, *is bijective;*

(d) *the ring morphism* $R \to C^*$, $r \mapsto r\varepsilon$, *is a Frobenius extension.*

Proof. (a) $\Rightarrow$ (b) By 3.15, ${}_RC$ is finitely generated and projective, and 4.7 implies that $\mathbf{M}^C = {}_{C^*}\mathbf{M}$; so the forgetful functor $(-)_R : {}_{C^*}\mathbf{M} \to \mathbf{M}_R$ is Frobenius. Now 40.21 applies.

In view of 4.7, the remaining assertions also follow from 40.21. □

The comodules of the coalgebra associated to any finitely generated projective R-module are of fundamental importance.

4.9. Projective modules as comodules. *Let* P *be a finitely generated projective* R*-module with dual basis* $p_1, \ldots, p_n \in P$ *and* $\pi_1, \ldots, \pi_n \in P^*$. *Then* P *is a right* $P^* \otimes_R P$*-comodule with the coaction*

$$\varrho^P : P \to P \otimes_R (P^* \otimes_R P), \quad p \mapsto \sum_i p_i \otimes \pi_i \otimes p.$$

P *is a subgenerator in* $\mathbf{M}^{P^* \otimes_R P}$, *and there is a category isomorphism*

$$\mathbf{M}^{P^* \otimes_R P} \simeq \mathbf{M}_{\mathrm{End}_R(P)}.$$

The dual P^* *is a left* $P^* \otimes_R P$*-comodule with the coaction*

$${}^P\varrho : P \to (P^* \otimes_R P) \otimes_R P, \quad f \mapsto \sum_i f \otimes p_i \otimes \pi_i.$$

Proof. Coassociativity of ϱ^P follows from the equality

$$(I \otimes \Delta)\varrho^P(f \otimes p) = \sum_{i,j} f \otimes p_i \otimes \pi_i \otimes p_j \otimes \pi_j \otimes p = (\varrho^P \otimes I)\varrho^P(f \otimes p).$$

By properties of the dual basis, $(I_P \otimes \varepsilon)\varrho^P(p) = \sum_i p_i \pi_i(p) = p$, so that P is indeed a right comodule over $P^* \otimes_R P$. There exists a surjective R-linear map $R^n \to P^*$ that yields an epimorphism $P^n \simeq R^n \otimes P \to P^* \otimes_R P$ in $\mathbf{M}^{P^* \otimes_R P}$. So P generates $P^* \otimes_R P$ as a right comodule and hence is a subgenerator in $\mathbf{M}^{P^* \otimes_R P}$. Since $P^* \otimes_R P$ is finitely generated and projective as an R-module, the category isomorphism follows by 4.7.

A simple computation shows that P^* is a left comodule over $P^* \otimes_R P$. □

As a special case, for any $n \in \mathbb{N}$, R^n may be considered as a right comodule over the matrix coalgebra $M_n^c(R)$ (cf. 1.10).

For an algebra A, any two elements $a, b \in A$ define a subalgebra $aAb \subset A$, and for an idempotent $e \in A$, eAe is a subalgebra with a unit. Dually, one considers

4.10. Factor coalgebras. *Let $f, g, e \in C^*$ with $e * e = e$. Then:*

(1) $f \rightharpoonup C \leftharpoonup g$ is a coalgebra (without a counit) and there is a coalgebra morphism

$$C \to f \rightharpoonup C \leftharpoonup g, \quad c \mapsto f \rightharpoonup c \leftharpoonup g.$$

(2) $e \rightharpoonup C \leftharpoonup e$ is a coalgebra with counit e and coproduct

$$e \rightharpoonup c \leftharpoonup e \;\mapsto\; \sum e \rightharpoonup c_{\underline{1}} \leftharpoonup e \otimes e \rightharpoonup c_{\underline{2}} \leftharpoonup e.$$

The kernel of $C \to e \rightharpoonup C \leftharpoonup e$ is equal to $(\varepsilon - e) \rightharpoonup C + C \leftharpoonup (\varepsilon - e)$.

(3) If C is R-cogenerated, and e is a central idempotent, then $e \rightharpoonup C$ is a subcoalgebra of C.

Proof. (1) For any $f, g \in C^*$ consider the left, respectively right, comodule maps $L_f : C \to C$, $c \mapsto f \rightharpoonup c$, and $R_g : C \to C$, $c \mapsto c \leftharpoonup g$. Construct the commutative diagram

$$\begin{array}{ccccc} C & \xrightarrow{L_f} & C & \xrightarrow{R_g} & C \\ \downarrow{\scriptstyle \Delta} & & \downarrow{\scriptstyle \Delta} & & \downarrow{\scriptstyle \Delta} \\ C \otimes_R C & \xrightarrow{I_C \otimes L_f} & C \otimes_R C & \xrightarrow{R_g \otimes I_C} & C \otimes_R C, \end{array}$$

which leads to the identity $\Delta \circ R_g \circ L_f = (R_g \otimes L_f) \circ \Delta$. Putting $\delta := L_f \circ R_g = R_g \circ L_f$, we obtain the commutative diagram

$$\begin{array}{ccccc} C & \xrightarrow{\Delta} & C \otimes_R C & & \\ \downarrow{\scriptstyle \delta} & & \downarrow{\scriptstyle R_g \otimes L_f} & & \\ \delta(C) & \xrightarrow{\Delta} & R_g(C) \otimes_R L_f(C) & \xrightarrow{L_f \otimes R_g} & \delta(C) \otimes_R \delta(C). \end{array}$$

Thus $\Delta_\delta = (L_f \otimes R_g) \circ \Delta$ makes $\delta(C)$ a coalgebra. It is easily verified that

$$\begin{array}{ccc} C & \xrightarrow{\Delta} & C \otimes_R C \\ \downarrow{\scriptstyle \delta} & & \downarrow{\scriptstyle \delta \otimes \delta} \\ \delta(C) & \xrightarrow{\Delta_\delta} & \delta(C) \otimes_R \delta(C) \end{array}$$

is a commutative diagram, and hence δ is a coalgebra morphism.

(2) The form of the coproduct follows from (1). For $c \in C$, $\varepsilon(e \rightharpoonup c \leftharpoonup e) = \varepsilon * e(c \leftharpoonup e) = e(c)$ showing that e is the counit of $e \rightharpoonup C \leftharpoonup e$.

For $x \in C$, $e \rightharpoonup x \leftharpoonup e = 0$ implies $x \leftharpoonup e = (\varepsilon - e) \rightharpoonup (x \leftharpoonup e) \in (\varepsilon - e) \rightharpoonup C$, and so

$$x = x \leftharpoonup e + x \leftharpoonup (\varepsilon - e) \in (\varepsilon - e) \rightharpoonup C + C \leftharpoonup (\varepsilon - e).$$

This proves the stated form of the kernel.

(3) By 4.6, for a central idempotent e and $c \in C$, $e \rightharpoonup c \leftharpoonup e = e \rightharpoonup c$. Putting $f = g = e$ in (the proof of) (1) we obtain $\Delta_e(e \rightharpoonup C) \subset e \rightharpoonup C \otimes e \rightharpoonup C$. □

4.11. Idempotents and comodules. *Let $e \in C^*$ be an idempotent and consider the coalgebra $e \rightharpoonup C \leftharpoonup e$ (as in 4.10).*

(1) *For any $M \in \mathbf{M}^C$, $e \rightharpoonup M$ is a right $e \rightharpoonup C \leftharpoonup e$-comodule with the coaction*

$$e \rightharpoonup M \to e \rightharpoonup M \otimes_R e \rightharpoonup C \leftharpoonup e, \quad e \rightharpoonup m \mapsto \sum e \rightharpoonup m_{\underline{0}} \otimes e \rightharpoonup m_{\underline{1}} \leftharpoonup e.$$

(2) *For any $f : M \to N \in \mathbf{M}^C$, $f(e \rightharpoonup M) = e \rightharpoonup f(M)$, and so there is a covariant functor*

$$e \rightharpoonup - : \mathbf{M}^C \to \mathbf{M}^{e \rightharpoonup C \leftharpoonup e}, \quad M \mapsto e \rightharpoonup M.$$

(3) *For any $M \in \mathbf{M}^C$, M^* is a right C^*-module canonically and*

$$\mathrm{Hom}_R(e \rightharpoonup M, R) = (e \rightharpoonup M)^* \simeq M^* \cdot e.$$

(4) *The map $- \leftharpoonup e : e \rightharpoonup C \to e \rightharpoonup C \leftharpoonup e$ is a surjective right $e \rightharpoonup C \leftharpoonup e$-comodule morphism, and so $e \rightharpoonup C$ is a subgenerator in $\mathbf{M}^{e \rightharpoonup C \leftharpoonup e}$.*

(5) *$(e \rightharpoonup C \leftharpoonup e)^* \simeq e * C^* * e$, and hence there is a faithful functor $\mathbf{M}^{e \rightharpoonup C \leftharpoonup e} \to {}_{e*C^**e}\mathbf{M}$.*

(6) *If ${}_RC$ is locally projective, then $e \rightharpoonup C \leftharpoonup e$ is a locally projective R-module and*

$$\mathbf{M}^{e \rightharpoonup C \leftharpoonup e} = \sigma[{}_{e*C^**e}\, e \rightharpoonup C] = \sigma[{}_{e*C^**e}\, e \rightharpoonup C \leftharpoonup e].$$

Proof. (1), (3) and (4) are easily verified.

(2) By 4.1, right comodule morphisms are left C^*-morphisms.

(5) The isomorphism in (3) holds similarly for the right action of e on C and from this the isomorphism in (5) follows.

(6) Clearly direct summands of locally projectives are locally projective, and hence the assertion follows from (3) and 4.3. □

Even if C is not finitely generated as an R-module, it is (C^*, R)-finite as defined in 41.22 provided it satisfies the α-condition.

4.12. Finiteness Theorem (2). *Assume ${}_RC$ to be locally projective.*

(1) *Let $M \in \mathbf{M}^C$. Every finite subset of M is contained in a subcomodule of M that is finitely generated as an R-module.*

(2) *Any finite subset of C is contained in a (C^*, C^*)-sub-bimodule which is finitely generated as an R-module.*

(3) *Minimal C^*-submodules and minimal (C^*, C^*)-sub-bimodules of C are finitely generated as R-modules.*

Proof. (1) Since any sum of subcomodules is again a subcomodule, it is enough to show that each $m \in M$ lies in a subcomodule that is finitely generated as an R-module. Moreover, by the correspondence of subcomodules and C^*-submodules, this amounts to proving that the submodule $C^* \rightharpoonup m$ is finitely generated as an R-module. Writing $\varrho^M(m) = \sum_{i=1}^k m_i \otimes c_i$, where $m_i \in C^* \rightharpoonup m$, $c_i \in C$, we compute for every $f \in C^*$

$$f \rightharpoonup m = (I_M \otimes f) \circ \varrho^M(m) = \sum_{i=1}^k m_i \, f(c_i) \, .$$

Hence $C^* \rightharpoonup m$ is finitely generated by $m_1, \ldots, m_k$ as an R-module.

(2) It is enough to prove the assertion for single elements $c \in C$. By (1), $C^* \rightharpoonup c$ is generated as an R-module by some $c_1, \ldots, c_k \in C$. By symmetry, each $c_i \leftharpoonup C^*$ is a finitely generated R-module. Hence $C^* \rightharpoonup c \leftharpoonup C^*$ is a finitely generated R-module.

(3) This is an obvious consequence of (1) and (2). □

Now we turn our attention to those objects that are fundamental in any structure theory. A right C-comodule N is called *semisimple* (in $\mathbf{M}^C$) if every C-monomorphism $U \to N$ is a coretraction, and N is called *simple* if all these monomorphisms are isomorphisms (see 38.9). Semisimplicity of N is equivalent to the fact that every right C-comodule is N-injective (by 38.13). (Semi)simple left comodules and bicomodules are defined similarly.

The coalgebra C is said to be *left (right) semisimple* if it is semisimple as a left (right) comodule. C is called a *simple coalgebra* if it is simple as a (C, C)-bicomodule.

4.13. Semisimple comodules. *Assume that ${}_RC$ is flat.*

(1) Any $N \in \mathbf{M}^C$ is simple if and only if N has no nontrivial subcomodules.

(2) For $N \in \mathbf{M}^C$ the following are equivalent:

(a) N is semisimple (as defined above);

(b) every subcomodule of N is a direct summand;

(c) N is a sum of simple subcomodules;

(d) N is a direct sum of simple subcomodules.

Proof. (1) By 3.14, any monomorphism $U \to N$ is injective, and hence it can be identified with a subcomodule. From this the assertion is clear.

(2) ${}_RC$ flat implies that the intersection of any two subcomodules is again a subcomodule. Hence in this case the proof for modules (e.g., [46, 20.2]) can be transferred to comodules. □

4.14. Right semisimple coalgebras. *For C the following are equivalent:*

(a) *C is a semisimple right C-comodule;*

(b) *${}_RC$ is flat and every right subcomodule of C is a direct summand;*

(c) *${}_RC$ is flat and C is a direct sum of simple right comodules;*

(d) *${}_RC$ is flat and every comodule in $\mathbf{M}^C$ is semisimple;*

(e) *${}_RC$ is flat and every short exact sequence in $\mathbf{M}^C$ splits;*

(f) *${}_RC$ is projective and C is a semisimple left C^*-module;*

(g) *every comodule in $\mathbf{M}^C$ is (C-)injective;*

(h) *every comodule in $\mathbf{M}^C$ is projective;*

(i) *C is a direct sum of simple coalgebras that are right (left) semisimple;*

(j) *C is a semisimple left C-comodule.*

Proof. (a) $\Rightarrow$ (b) $\Rightarrow$ (c) $\Rightarrow$ (d) $\Rightarrow$ (e) Assume every monomorphism $i : U \to C$ to be a coretraction. Then i is in particular an injective map, and hence, by 3.14, ${}_RC$ is flat. Now the assertions follow by 4.13.

The implications (e) $\Rightarrow$ (g) and (e) $\Rightarrow$ (h) are obvious.

(h) $\Rightarrow$ (f) By 3.22, any projective comodule is projective as an R-module. In particular, C is a projective R-module, and hence $\mathbf{M}^C = \sigma[{}_{C^*}C]$ and all modules in $\sigma[{}_{C^*}C]$ are projective. This characterises C as a semisimple C^*-module (see 41.8).

The implication (f) $\Rightarrow$ (a) is obvious since $\mathbf{M}^C = \sigma[{}_{C^*}C]$.

(g) $\Rightarrow$ (a) This is shown in 38.13. Notice that, in view of (f), ${}_RC$ is projective, and hence the C-injectivity of any comodule N implies that N is injective in $\mathbf{M}^C = \sigma[{}_{C^*}C]$.

(f) $\Rightarrow$ (i) Let C be a left semisimple C^*-module. Let $\{E_i\}_I$ be a minimal representative set of simple C^*-submodules of C. Form the traces $D_i := \mathrm{Tr}_{C^*}(E_i, C)$. By the structure theorem for semisimple modules (see 41.8),

$$C \simeq \bigoplus_I D_i,$$

where the D_i are minimal fully invariant C^*-submodules. Considering C^* as an endomorphism ring acting from the right, this means that the D_i are minimal (C^*, C^*)-submodules. By 4.6, each D_i is a minimal subcoalgebra of C and every subcoalgebra of D_i is a subcoalgebra of C. So every D_i is a right semisimple simple coalgebra.

(i) $\Rightarrow$ (f) It follows from the proof (a) $\Rightarrow$ (f) that all simple comodules of C are projective as R-modules and hence ${}_RC$ is also projective. Now the assertion follows.

(f) $\Leftrightarrow$ (j) By 41.8, the semisimple module ${}_{C^*}C$ is semisimple over its endomorphism ring, that is, C_{C^*} is also semisimple. Since ${}^C\mathbf{M} = \sigma[C_{C^*}]$, the assertion follows from the preceding proof by symmetry. □

4.15. Simple coalgebras. *For C the following are equivalent:*

(a) *C is a simple coalgebra that is right (left) semisimple;*

(b) *${}_RC$ is projective and C is a simple (C^*, C^*)-bimodule containing a minimal left (right) C^*-submodule;*

(c) *C is a simple coalgebra and a finite-dimensional vector space over R/m, for some maximal ideal $m \subset R$.*

Proof. (a) $\Rightarrow$ (b) We know from 4.14 that ${}_RC$ is projective. Clearly a simple right subcomodule is a simple left C^*-submodule. Let $D \subset C$ be a (C^*, C^*)-sub-bimodule. Then it is a direct summand as a left C^*-module, and hence it is a subcoalgebra of C (by 4.6) and so $D = C$.

(b) $\Rightarrow$ (c) Let $D \subset C$ be a minimal left C^*-submodule. For any maximal ideal $m \subset R$, $mD \subset D$ is a C^*-submodule and hence $mD = 0$ or $mD = D$. Since D is finitely generated as an R-module (by 4.12), $mD = 0$ for some maximal $m \subset R$. Moreover, $mC = mD \leftharpoonup C^* = 0$, and so C is a finite-dimensional R/m-algebra.

(c) $\Rightarrow$ (a) is obvious. Notice that in this case $\mathbf{M}^C = {}_{C^*}\mathbf{M}$ (see 4.7). $\square$

The Finiteness Theorem 4.12 and the Hom-tensor relations 3.9 indicate that properties of R have a strong influence on properties of C-comodules.

4.16. Coalgebras over special rings. *Let ${}_RC$ be locally projective.*

(1) *If R is Noetherian, then C is locally Noetherian as a right and left comodule, and in $\mathbf{M}^C$ and ${}^C\mathbf{M}$ direct sums of injectives are injective.*

(2) *If R is perfect, then in $\mathbf{M}^C$ and ${}^C\mathbf{M}$ any comodule satisfies the descending chain condition on finitely generated subcomodules.*

(3) *If R is Artinian, then in $\mathbf{M}^C$ and ${}^C\mathbf{M}$ every finitely generated comodule has finite length.*

Proof. All these assertions are special cases of 41.22. $\square$

Notice that, over Artinian (perfect) rings R, ${}_RC$ is locally projective if and only if ${}_RC$ is projective (any flat R-module is projective).

In 42.3 the M-adic topology on the base ring and its relevance for the category $\sigma[M]$ are considered. Naturally for C-comodules, the C-adic topology on C^* is of importance.

4.17. The C-adic topology in C^*. Let ${}_RC$ be locally projective. Then the finite topology in $\mathrm{End}_R(C)$ induces the C-adic topology on C^* and the open left ideals determine right C-comodules.

Open left ideals. A filter basis for the open left ideals of C^* is given by

$$\mathcal{B}_C = \{\mathrm{An}_{C^*}(E) \,|\, E \text{ a finite subset of } C\},$$

where $\mathrm{An}_{C^*}(E) = \{f \in C^* \mid f \rightharpoonup E = 0\}$. The filter of all open left ideals is

$$\mathcal{F}_C = \{I \subset C^* \mid I \text{ is a left ideal and } C^*/I \in \mathbf{M}^C\}.$$

This is a bounded filter, that is, there is a basis of two-sided ideals

$$\mathcal{B}'_C = \{J \subset C^* \mid J \text{ is an ideal and } C^*/J \in \mathbf{M}^C\}.$$

Thus generators in $\mathbf{M}^C$ are given by, for example,

$$G = \bigoplus\{C^*/I \mid I \in \mathcal{B}_C\} \text{ and } G' = \bigoplus\{C^*/J \mid J \in \mathcal{B}'_C\}.$$

Closed left ideals. For a left ideal $I \subset C^*$ the following are equivalent:

(a) I is closed in the C-adic topology;

(b) $I = \mathrm{An}_{C^*}(W)$ for some $W \in \mathbf{M}^C$;

(c) C^*/I is cogenerated by some (minimal) cogenerator of $\mathbf{M}^C$;

(d) $I = \bigcap_\Lambda I_\lambda$, where all $C^*/I_\lambda \in \mathbf{M}^C$ and are finitely cogenerated (cocyclic).

Over QF rings. Let R be a QF ring.

(i) Any finitely generated left (right) ideal $I \subset C^*$ is closed in the C-adic topology.

(ii) A left ideal $I \subset C^*$ is open if and only if it is closed and C^*/I is finitely R-generated (= finitely R-cogenerated).

The first two parts are special cases of 42.3 and 41.22(5). Notice that, over a QF ring R, C is injective in $\mathbf{M}^C$ and ${}^C\mathbf{M}$ (by 3.21), and hence every finitely generated left (or right) ideal in (the endomorphism ring) C^* is closed in the C-adic topology (see 42.3).

We conclude this section by discussing a more general framework for the α-condition.

4.18. Pairings of algebras and coalgebras. A *pairing* (C, A) consists of an R-algebra A, an R-coalgebra C, and a bilinear form $\beta : C \times A \to R$ such that the map $\gamma : A \to C^*$, $a \mapsto \beta(-, a)$, is a ring morphism. The pairing (C, A) is called a *rational pairing* if, for any $N \in \mathbf{M}_R$, the map

$$\tilde{\alpha}_N : N \otimes_R C \to \mathrm{Hom}_R(A, N),\ n \otimes c \mapsto [a \mapsto \beta(c, a)n]$$

is injective.

The interest in rational pairings (C, A) arises from the fact that they allow one to identify C-comodules with A-modules. Observe that, for any R-coalgebra C, (C, C^*) is a pairing with the bilinear form induced by evaluation. This pairing is rational if and only if ${}_RC$ is locally projective. The relationship to more general pairings is given by the next theorem.

4.19. Rational pairings. *For a pairing (C, A) the following are equivalent:*

(a) *(C, A) is a rational pairing;*

(b) *${}_R C$ is locally projective and $\gamma(A)$ is a dense subalgebra of C^*;*

(c) *$\mathbf{M}^C = \sigma[{}_{C^*}C]$ and $\sigma[{}_{C^*}C] = \sigma[{}_A C]$;*

(d) *$\mathbf{M}^C = \sigma[{}_A C]$.*

Proof. Since $\gamma : A \to C^*$ is a ring morphism, any C-comodule has a left A-module structure (via C^*), and so there are faithful functors $\mathbf{M}^C \to {}_{C^*}\mathbf{M} \to {}_A\mathbf{M}$.

(a) $\Rightarrow$ (d) For any $M, N \in \mathbf{M}^C$ and $h \in \mathrm{Hom}_A(M, N)$, consider the diagram

$$\begin{array}{ccccc} M & \xrightarrow{\varrho^M} & M \otimes_R C & \xrightarrow{\tilde{\alpha}_M} & \mathrm{Hom}_R(A, M) \\ \downarrow{\scriptstyle h} & & \downarrow{\scriptstyle h \otimes I_C} & & \downarrow{\scriptstyle \mathrm{Hom}(A,h)} \\ N & \xrightarrow{\varrho^N} & N \otimes_R C & \xrightarrow{\tilde{\alpha}_N} & \mathrm{Hom}_R(A, N)\,, \end{array}$$

in which the right-hand square is always commutative. For $m \in M$ and $a \in A$, the outer paths yield $h(a \rightharpoonup m)$ and $a \rightharpoonup h(m)$, respectively, and so the outer rectangle is commutative since h is A-linear. By assumption, $\tilde{\alpha}_N$ is injective, and this implies that the left square is also commutative, thus proving that h is a C-comodule morphism. So ${}_A\mathrm{Hom}(M, N) = \mathrm{Hom}^C(M, N)$ and $\mathbf{M}^C = \sigma[{}_A C]$.

(b) $\Leftrightarrow$ (c) $\Leftrightarrow$ (d) There are embeddings $\mathbf{M}^C \subset \sigma[{}_{C^*}C] \subset \sigma[{}_A C]$. Moreover, we know that ${}_R C$ locally projective is equivalent to $\mathbf{M}^C = \sigma[{}_{C^*}C]$ (cf. 4.3), while $\gamma(A)$ dense in C^* is equivalent to $\sigma[{}_{C^*}C] = \sigma[{}_A C]$ (Density Theorem).

(b) $\Rightarrow$ (a) For $N \in \mathbf{M}_R$, consider the commutative diagram

$$\begin{array}{ccc} N \otimes_R C & \xrightarrow{\alpha_N} & \mathrm{Hom}_R(C^*, N) \\ \downarrow{\scriptstyle =} & & \downarrow{\scriptstyle \mathrm{Hom}(\gamma,N)} \\ N \otimes_R C & \xrightarrow{\tilde{\alpha}_N} & \mathrm{Hom}_R(A, N). \end{array}$$

Since ${}_R C$ is locally projective, α_N is injective. Assume $\sum_{i=1}^k n_i \otimes c_i \in \mathrm{Ke}\,\tilde{\alpha}_N$. By our density condition, for any $f \in C^*$ there exist $a \in A$ such that $f(c_i) = \beta(c_i, a)$, for all $i = 1, \ldots, k$. This implies $\sum f(c_i) n_i = \sum \beta(c_i, a) n_i = 0$ and hence $\sum_{i=1}^k n_i \otimes c_i \in \mathrm{Ke}\,\alpha_N = 0$. So $\tilde{\alpha}_N$ is injective for any $N \in \mathbf{M}_R$, proving that (C, A) is a rational pair. $\square$

For further literature on pairings of coalgebras and algebras over rings the reader is referred to [48], [49] and [121].

References. Abuhlail [48]; Abuhlail, Gómez-Torrecillas and Lobillo [49]; El Kaoutit, Gómez-Torrecillas and Lobillo [112]; Gómez-Torrecillas [121]; Radford [178]; Sweedler [45]; Wisbauer [210].

5 The finite dual of an algebra

As observed in 1.3, for any R-coalgebra C the dual module $C^* = \mathrm{Hom}_R(C, R)$ has an algebra structure. This rises the question of whether the dual module A^* of any R-algebra A has a coalgebra structure. As explained in 1.12, this is the case provided A_R is finitely generated and projective. In case A_R is not finitely generated and projective, we may try another way to associate a coalgebra to A^* by looking for a submodule $B \subset A^*$ such that $\mu_A^*(B) \subset B \otimes_R B$. We prepare this approach by some more general constructions related to monoids. To overcome technical problems with the tensor product, we will assume R to be Noetherian at crucial steps.

For any set S, maps $S \to R$ can be identified with a product of copies of R, $\mathrm{Map}(S, R) = R^S$, which may be considered as an R-algebra. Notice that, for a Noetherian (coherent) ring R, R^S is a flat R-module.

5.1. Lemma. *For a Noetherian ring R, the following map is injective:*

$$\pi : R^S \otimes_R R^T \to R^{S\times T}, \quad f \otimes g \mapsto [(s,t) \mapsto f(s)g(t)].$$

Proof. Any finitely generated submodule $M \subset R^S$ is finitely presented, and hence the restriction of π, $M \otimes_R R^T \simeq M^T \subset (R^S)^T \simeq R^{S\times T}$, is injective (see 40.17). Therefore π is injective. □

5.2. Maps on monoids. Let G be a monoid with product $\mu : G \times G \to G$ and neutral element e. Denote by $R[G]$ the monoid algebra over G. The algebra R^G is an $R[G]$-bimodule: The action of $x, y \in G$ on $f \in R^G$ is defined by $xfy(z) = f(yzx)$ for all $z \in G$. These actions are extended uniquely to make R^G an $R[G]$-bimodule. There is an R-linear map

$$\mu^\times : R^G \to R^{G\times G}, \quad f \mapsto [(x,y) \mapsto f(xy)].$$

For the maps $I_G \times \mu, \mu \times I_G : G \times G \times G \to G \times G$, the associativity of μ implies

$$(I_G \times \mu)^\times \circ \mu^\times = (\mu \times I_G)^\times \circ \mu^\times.$$

Furthermore, define $\alpha, \beta : R^{G\times G} \to R^G$ by $\alpha(h)(x) = h(x, e)$, and $\beta(h)(x) = h(e, x)$, for all $h \in R^{G\times G}$ and $x \in G$. Then

$$\alpha \circ \mu^\times = I_{R^G} = \beta \circ \mu^\times.$$

5.3. Subsets with finiteness conditions. Let G be a monoid and R a Noetherian ring. For any $R[G]$-sub-bimodule $B \subset R^G$ define subsets

$$\begin{aligned} {}^{\mathsf{f}}B &= \{b \in B \mid R[G]b \text{ is finitely generated as an } R\text{-module}\}, \\ B^{\mathsf{f}} &= \{b \in B \mid bR[G] \text{ is finitely generated as an } R\text{-module}\}, \end{aligned}$$

where obviously ${}^{\mathsf{f}}B$ is a left and B^{f} a right $R[G]$-submodule of B.

For an element $f \in R^G$, the following assertions are equivalent:

(a) $f \in {}^{\mathsf{f}}B$;

(b) $\mu^\times(f) \in \pi(B \otimes_R R^G)$;

(c) $\mu^\times(f) \in \pi({}^{\mathsf{f}}B \otimes_R R^G)$;

(d) $\mu^\times(f) \in \pi({}^{\mathsf{f}}B \otimes_R R^G) \cap \pi(R^G \otimes_R B^{\mathsf{f}})$;

(e) $f \in B$ and $R[G]fR[G]$ is finitely generated as an R-module;

(f) $f \in B^{\mathsf{f}}$.

As a consequence, $B^{\mathsf{f}} = {}^{\mathsf{f}}B$ and is an $R[G]$-sub-bimodule of B.

Proof. (a) $\Rightarrow$ (c) For $f \in {}^{\mathsf{f}}B$, $R[G]f$ is a finitely generated R-submodule of ${}^{\mathsf{f}}B$. Hence, there are $b_1, \ldots, b_n \in {}^{\mathsf{f}}B$ such that $R[G]f = \sum_{i=1}^n Rb_i$. For each $y \in G$, choose $f_1(y), \ldots, f_n(y) \in R$ such that $yf = \sum_{i=1}^n f_i(y)b_i$. Now, for $x, y \in G$,

$$\mu^\times(f)(x,y) = f(xy) = (yf)(x) = \sum_{i=1}^n f_i(y)b_i(x) = \pi(\sum_{i=1}^n b_i \otimes f_i)(x,y).$$

Thus, $\mu^\times(f) \in \pi({}^{\mathsf{f}}B \otimes_R R^G)$.

(c) $\Rightarrow$ (b) This is evident since ${}^{\mathsf{f}}B \subseteq B$.

(b) $\Rightarrow$ (f) Assume (b). First we prove $f \in B$. In fact, for $x \in G$,

$$f(x) = f(xe) = \mu^\times(f)(x,e) = \sum_{i=1}^n b_i(x)f_i(e),$$

where $\mu^\times(f) = \pi(\sum_{i=1}^n b_i \otimes f_i)$, $b_i \in B$ and $f_i \in R^G$. So $f = \sum_{i=1}^n f_i(e)b_i \in B$. Now, for any $y, x \in G$,

$$(fy)(x) = f(yx) = \mu^\times(f)(y,x) = \sum_{i=1}^n b_i(y)f_i(x) = \left(\sum_{i=1}^n b_i(y)f_i\right)(x).$$

Therefore, $fy \in \sum_{i=1}^n Rf_i$, and, since R is Noetherian, $fR[G]$ is finitely R-generated.

(f) $\Rightarrow$ (a) follows by symmetry.

(c) $\Leftrightarrow$ (d) Symmetric to (a) $\Rightarrow$ (c), (c) implies $\mu^\times(f) \in \pi(R^G \otimes_R B^{\mathsf{f}})$ and from this (d) follows.

(e) $\Rightarrow$ (a) is clear.

(a) $\Rightarrow$ (e) For $f \in {}^{\mathsf{f}}B$, let $b_1, \ldots, b_n \in B$ be such that $R[G]f = \sum_{i=1}^n Rb_i$. Now, $R[G]b_i \subset R[G]f$, whence $R[G]b_i$ is finitely generated, and thus $b_i \in {}^{\mathsf{f}}B$. We have already proved that (a) $\Rightarrow$ (f), so $b_i \in B^{\mathsf{f}}$. This means that $b_iR[G]$ is finitely generated as an R-module, and, therefore, $R[G]fR[G]$ is a finitely generated R-module. □

5.4. Coalgebra structure on B^{f}. *Let G be a monoid, R a Noetherian ring, and $B \subset R^G$ an $R[G]$-sub-bimodule. If B^{f} is pure as an R-submodule of R^G, then B^{f} is an R-coalgebra with the coproduct*

$$\Delta : B^{\mathsf{f}} \xrightarrow{\mu^\times} \pi(B^{\mathsf{f}} \otimes_R B^{\mathsf{f}}) \xrightarrow{\pi^{-1}} B^{\mathsf{f}} \otimes_R B^{\mathsf{f}}$$

and counit $\varepsilon : B^{\mathsf{f}} \to R$, $h \mapsto h(e)$.

Proof. By 5.3, for any $b \in B^{\mathsf{f}}$ and $\pi : R^G \otimes_R R^G \to R^{G\times G}$,

$$\mu^\times(b) \in \pi(R^G \otimes_R B^{\mathsf{f}}) \cap \pi(B^{\mathsf{f}} \otimes_R R^G) = \pi((R^G \otimes_R B^{\mathsf{f}}) \cap (B^{\mathsf{f}} \otimes_R R^G)) = \pi(B^{\mathsf{f}} \otimes_R B^{\mathsf{f}}),$$

where the last equality is justified by the intersection property of pure submodules (see 40.16). To show the coassociativity of the coproduct, consider the diagram

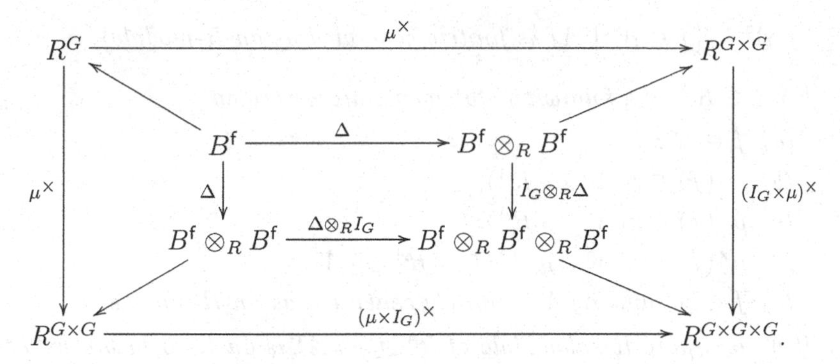

The outer rectangle is commutative by the associativity of μ (cf. 5.2). All the trapezia, built from obvious maps, are commutative. Since B^{f} is pure in R^G, the canonical map $B^{\mathsf{f}} \otimes_R B^{\mathsf{f}} \otimes_R B^{\mathsf{f}} \to R^{G\times G\times G}$ is injective. Therefore, the inner square is commutative, proving the coassociativity of Δ. □

As a special case in the above construction we may take $B = R^G$.

5.5. Representative functions. Let G be a monoid and R a Noetherian ring. The set

$$\mathcal{R}_R(G) = (R^G)^{\mathsf{f}} = \{f \in R^G \mid fR[G] \text{ is finitely generated as an } R\text{-module}\},$$

is called the set of *R-valued representative functions* on the monoid G. It follows from 5.4 that $\mathcal{R}_R(G)$ is a coalgebra provided that $\mathcal{R}_R(G)$ is a pure R-submodule in R^G. Notice that $\mathcal{R}_R(G)$ is also an R-subalgebra of R^G, which is compatible with this coalgebra structure (bialgebra; see 13.1).

Any algebra is a multiplicative monoid yielding the

5.6. Monoid ring of an algebra. Let A be an R-algebra. Then

$$A^* = \text{Hom}_R(A, R) \subset \text{Map}(A, R) = R^A.$$

R^A is a bimodule over the monoid ring $R[A]$, where $(r_a)_A \in R[A]$ acts on $f \in R^A$ by

$$((r_a)_A \cdot f)(b) = \textstyle\sum_A r_a f(ba), \quad (f \cdot (r_a)_A)(b) = \textstyle\sum_A r_a f(ab), \quad \text{for } b \in A.$$

A^* is an $R[A]$-sub-bimodule of R^A, and its $R[A]$-module structure coincides with the (A, A)-bimodule structure that reads for $a, b \in A$ and $f \in A^*$ as $(af)(b) = f(ba)$ and $(fa)(b) = f(ab)$, that is, $R[A]$-submodules of A^* are precisely A-submodules.

5.7. Finite dual of an algebra. *Let A be any algebra over a Noetherian ring R and put*

$$A^\circ = \{f \in A^* \mid Af \text{ is finitely generated as an } R\text{-module}\}.$$

(1) For $f \in R^A$, the following statements are equivalent:

(a) $f \in A^\circ$;
(b) $\mu^\times(f) \in \pi(A^* \otimes_R R^A)$;
(c) $\mu^\times(f) \in \pi(A^\circ \otimes_R R^A)$;
(d) $\mu^\times(f) \in \pi(A^\circ \otimes_R R^A) \cap \pi(R^A \otimes_R A^\circ)$;
(e) $f \in A^*$ and AfA is finitely generated as an R-module.

(2) If A° is a pure R-submodule of R^A, then A° is an R-coalgebra with the coproduct

$$\Delta : A^\circ \xrightarrow{\mu^\times} \pi(A^\circ \otimes_R A^\circ) \xrightarrow{\pi^{-1}} A^\circ \otimes_R A^\circ,$$

and counit $\varepsilon : A^\circ \to R$, $h \mapsto h(1_A)$. Furthermore, the canonical map

$$\phi : A \to A^{\circ *}, \quad a \mapsto [f \mapsto f(a)],$$

is an algebra morphism.

Proof. (1) It follows from 5.3 and 5.6 that $A^\circ = (A^*)^{\mathsf{f}}$; hence A° is an A-sub-bimodule of A^* and the given properties are equivalent.

(2) If A° is a pure R-submodule in R^A, the coalgebra structure follows from 5.4. To recall the construction observe that, for every $f \in A^\circ$, there is a unique $\sum_{i=1}^n f_i \otimes_R \tilde{f}_i \in A^\circ \otimes_R A^\circ$ such that $f(ab) = \sum_{i=1}^n f_i(a)\tilde{f}_i(b)$, for every $a, b \in A$, and the coproduct is

$$m_A^\times : A^\circ \to A^\circ \otimes_R A^\circ, \quad f \mapsto \sum_{i=1}^n f_i \otimes \tilde{f}_i = \sum f_{\underline{1}} \otimes f_{\underline{2}}.$$

Clearly ϕ is R-linear, and for $a, b \in A$ and $f \in A^\circ$

$$\phi(ab)(f) = f(ab) = \sum f_{\underline{1}}(a) f_{\underline{2}}(b) = [\phi(a) * \phi(b)](f),$$

showing that ϕ is an algebra morphism. □

There are further characterisations for the finite dual of an algebra. A submodule X of an R-module M is called *R-cofinite* if M/X is a finitely generated R-module.

5.8. Cofinite ideals. *Let R be Noetherian and A an R-algebra. The following statements are equivalent for $f \in A^*$:*

(a) $f \in A^\circ$;

(b) $\operatorname{Ke} f$ *contains an R-cofinite ideal of A;*

(c) $\operatorname{Ke} f$ *contains an R-cofinite left ideal of A;*

(d) $\operatorname{Ke} f$ *contains an R-cofinite right ideal of A.*

Proof. (a) ⇒ (b) If $f \in A^\circ$, then, by 5.7, AfA is finitely generated as an R-module. Let $f_1, \ldots, f_n$ be a set of generators and consider $I = \bigcap_{i=1}^n \operatorname{Ke} f_i$. Clearly I is an R-cofinite submodule of A and $I \subset \operatorname{Ke} f$. For $a, b \in A$ and $c \in I$, we observe $f_i(acb) = (bf_i a)(c) = 0$ since $bf_i a \in AfA = \sum_{i=1}^n Rf_i$. So $acb \in I$, showing that I is an ideal.

(b) ⇒ (c) This is obvious.

(c) ⇒ (a) Let I be an R-cofinite left ideal of A contained in $\operatorname{Ke} f$. Then $f \in \operatorname{Hom}_R(A/I, R)$, which implies $Af \subset \operatorname{Hom}_R(A/I, R)$. Since R is Noetherian, $\operatorname{Hom}_R(A/I, R)$ is a Noetherian R-module, and so Af is a finitely generated R-module, that is, $f \in A^\circ$.

(a) ⇔ (d) follows by symmetry. □

5.9. Pure submodules of products. *Let A be any algebra over a Noetherian ring R. Then, for an R-submodule $K \subset \operatorname{Hom}_R(A, R) = A^*$, the following are equivalent:*

(a) $K \subset R^A$ *is a pure submodule;*

(b) for any $N \in \mathbf{M}_R$, *the following map is injective:*

$$\tilde{\alpha}_{N,K} : N \otimes_R K \to \operatorname{Hom}_R(A, N), \; n \otimes k \mapsto [a \mapsto nk(a)].$$

Proof. First notice that $K \subset A^*$ implies that the image of $\tilde{\alpha}$ lies in $\operatorname{Hom}_R(A, N)$. Moreover, for any finitely generated R-module N, there is a commutative diagram

$$\begin{array}{ccc} N \otimes_R K & \xrightarrow{\tilde{\alpha}_{N,K}} & \operatorname{Hom}_R(A, N) \\ {\scriptstyle I_N \otimes i}\downarrow & & \downarrow \\ N \otimes_R R^A & \xrightarrow{\simeq} & N^A, \end{array}$$

where the right downward arrow is the inclusion map. From this we deduce that, for any $N \in \mathbf{M}_R$, the map $I_N \otimes i$ is injective if and only $\tilde{\alpha}_{N,K}$ is injective. □

For an algebra A over a Noetherian ring R, the condition that A° is pure in R^A implies that A° has a coalgebra structure (5.7), and also that the A°-comodules may be described as $A^{\circ *}$-modules (4.3). From 5.9 we can even derive that they are characterised by their A-module structure.

5.10. A°-comodules and A-modules. *Let A be an R-algebra with R Noetherian and A° pure in R^A. Then the category $\mathbf{M}^{A^\circ}$ of right A°-comodules is isomorphic to the category of left $A^{\circ *}$-modules subgenerated by A°, and also to the category of left A-modules subgenerated by A°, that is, there are category equivalences*

$$\mathbf{M}^{A^\circ} \simeq \sigma[{}_{A^{\circ *}}A^\circ] \simeq \sigma[{}_A A^\circ].$$

Proof. As mentioned above, the first isomorphism follows by 4.3. In view of 5.7(2) and 5.9, the canonical bilinear form $(A^\circ, A) \to R$, $(f, a) \mapsto f(a)$, makes (A°, A) a rational pairing, and hence the second isomorphism follows by 4.19. □

An A-module M is called (A, R)-finite if every finitely generated A-submodule is finitely generated as an R-module (see 41.22). Over QF rings R, the module A° determines all (A, R)-finite modules.

5.11. A° and (A, R)-finite modules. *Let A be any algebra over a QF ring R. Then A° is a subgenerator for all (A, R)-finite modules, that is, $\sigma[{}_A A^\circ]$ coincides with the class of all (A, R)-finite modules in ${}_A\mathbf{M}$.*

Proof. By 5.7, A° is (A, R)-finite as a left (and right) A-module. Now consider any cyclic A-module N that is finitely generated as an R-module. Then $N^* = \mathrm{Hom}_R(N, R)$ is a right A-module that is finitely generated as an R-module and hence is finitely generated as a right A-module, that is, there is an epimorphism $\phi : A^n \to N^*$ in $\mathbf{M}_A$. Applying $\mathrm{Hom}_R(-, R)$, one obtains a monomorphism $\phi^* : N^{**} \to (A^n)^*$ in ${}_A\mathbf{M}$. Since N^{**} is a finitely generated R-module, the image of ϕ^* is contained in $(A^\circ)^n$. As a QF ring, R cogenerates N, and hence the canonical map $N \to N^{**}$ is a monomorphism in ${}_A\mathbf{M}$. As a consequence, all (A, R)-finite modules are subgenerated by A°. □

5.12. Finite duals over QF rings. *Let A be an algebra over a QF ring R such that A° is projective as an R-module. Then all (A, R)-finite A-modules are A°-comodules.*

Proof. Since R is QF, the projective R-module A° is R-injective and hence is pure in R^A. By 5.7, A° has a coalgebra structure and $\mathbf{M}^{A^\circ} = \sigma[{}_A A^\circ]$ (by 5.10). Now the assertion follows from 5.11. □

5.13. Exercises

(1) Let A be an R-algebra and assume ${}_RA$ to be finitely generated and projective. Prove that the dual of the coalgebra A^* is isomorphic (as an algebra) to A.

(2) A ring R is *hereditary* if every ideal is projective. A Noetherian ring R is hereditary if and only if every submodule of an R-cogenerated module is flat (e.g., [46, 39.13]).

Let A be an algebra over a Noetherian ring R. Prove:

(i) If R is hereditary, then A° is pure in R^A.

(ii) If A is projective as an R-module, then A° is a pure R-submodule of A^* if and only if A° is pure in R^A.

(3) Let G be any monoid, R a Noetherian ring, and $R[G]$ the monoid algebra. Prove that, if $R[G]^\circ$ is R-pure in $R[G]^*$, then $R[G]^\circ$ and $\mathcal{R}_R(G)$ are isomorphic coalgebras (in fact bialgebras).

(4) An ideal I in the polynomial ring $R[X]$ is called *monic* if it contains a polynomial with leading coefficient 1. Prove:

(i) An ideal $I \subset R[X]$ is monic if and only if $R[X]/I$ is a finitely generated (free) R-module.

(ii) If R is Noetherian, then $R[X]^\circ$ is a pure R-submodule of $R[X]^*$ and $R[X]^\circ$ is a coalgebra with the coproduct

$$\Delta : R[X]^\circ \to R[X]^\circ \otimes_R R[X]^\circ, \quad \xi \mapsto [x^i \otimes x^j \mapsto \xi(x^{i+j}),\ i,j \geq 0],$$

and counit $\varepsilon : R[X]^\circ \to R,\ \xi \mapsto \xi(1)$.

Notice that $R[X]^\circ$ can be identified with the set of *linearly recursive sequences over* R (cf. [50]).

References. Abuhlail, Gómez-Torrecillas and Lobillo [49]; Abuhlail, Gómez-Torrecillas and Wisbauer [50]; Cao-Yu and Nichols [89].

6 Annihilators and bilinear forms

In this section we collect some useful technical notions that will help us study how certain substructures of an R-coalgebra C correspond to certain substructures of the associated dual algebra $C^* = \mathrm{Hom}_R(C, R)$. To facilitate such studies, consider the following definitions.

6.1. Annihilators. Let D be an R-submodule of C. The R-module

$$D^\perp := \{f \in C^* \mid f(D) = 0\} = \mathrm{Hom}_R(C/D, R) \subset C^*$$

is called the *annihilator* of D in C. For any subset $J \subset C^*$, denote

$$J^\perp := \bigcap \{\mathrm{Ke}\, f \mid f \in J\} \subset C\,.$$

Notice that $D \subset D^{\perp\perp}$ always and that $D = D^{\perp\perp}$ whenever C/D is cogenerated by R.

6.2. Properties of annihilators. *Let $D \subset C$ be an R-submodule.*

(1) If D is a left C^*-submodule of C, then $D^\perp$ is a right C^*-submodule.

(2) If D is a (C^*, C^*)-sub-bimodule of C, then $D^\perp$ is an ideal in C^*.

(3) If D is a coideal in C, then $D^\perp$ is a subalgebra of C^*.

Proof. (1),(2) If D is a left C^*-module, then for all $f \in D^\perp$ and $g \in C^*$,

$$f * g(D) = f(g \rightharpoonup D) \subset f(D) = 0\,,$$

where the first equality follows from 4.6. Therefore $f * g \in D^\perp$, as required. The left-side version and (2) are shown similarly.

(3) By definition $C \to C/D$ is a (surjective) coalgebra morphism, and as a consequence $\mathrm{Hom}_R(C/D, R) \to \mathrm{Hom}_R(C, R)$ is an injective algebra morphism. □

6.3. Kernels. *Let $J \subset C^*$ be an R-submodule.*

(1) If J is a right (left) ideal in C^*, then $J^\perp$ is a left (right) C^*-submodule of C.

(2) If J is an ideal in C^*, then $J^\perp$ is a (C^*, C^*)-sub-bimodule of C.

Assume R to be a semisimple ring. Then:

(3) If $J \subset C^*$ is a subalgebra, then $J^\perp$ is a coideal.

(4) $D \subset C$ is a coideal if and only if $D^\perp$ is a subalgebra.

Proof. (1) and (2) easily follow from 4.6.

(3) Let $J \subset C^*$ be a subalgebra and put $U = J^\perp$. Then $\varepsilon \in J$, and so $\varepsilon(U) = 0$. Furthermore, for any $f, g \in J$, $(f \otimes g)\Delta(U) = f * g(U) = 0$ and hence $(J \otimes_R J)\Delta(U) = 0$. Considering $J \otimes_R J$ as a submodule of $(C \otimes_R C)^*$ canonically, we know from linear algebra that $(J \otimes_R J)^\perp = J^\perp \otimes_R C + C \otimes_R J^\perp$ (R is semisimple) and this implies

$$\Delta(U) \subset U \otimes_R C + C \otimes_R U,$$

showing that U is a coideal.

(4) This follows from (3), 6.2(3), and $D = D^{\perp\perp}$. □

A useful technique for the investigation of coalgebras is provided by certain bilinear forms.

6.4. Balanced bilinear forms. Let $\beta : C \times C \to R$ be a bilinear form. Associated to β there are R-linear maps

$$\begin{array}{rcccccl} \bar{\beta}: & C \otimes_R C & \to & R, & c \otimes d & \mapsto & \beta(c, d), \\ \beta^l: & C & \to & C^*, & d & \mapsto & \beta(-, d), \\ \beta^r: & C & \to & C^*, & c & \mapsto & \beta(c, -)\,. \end{array}$$

β is said to be *C-balanced* if

$$\beta(c \leftharpoonup f, d) = \beta(c, f \rightharpoonup d), \text{ for all } c, d \in C,\ f \in C^*.$$

If ${}_RC$ is locally projective (cf. 4.2), the following are equivalent:

(a) *β is C-balanced;*

(b) $(I_C \otimes \bar{\beta}) \circ (\Delta \otimes I_C) = (\bar{\beta} \otimes I_C) \circ (I_C \otimes \Delta)$;

(c) *$\beta^l : C \to C^*$ is left C^*-linear, that is, for all $f \in C^*$ and $d \in C$,* $f * \beta^l(d) = \beta^l(f \rightharpoonup d)$;

(d) *$\beta^r : C \to C^*$ is right C^*-linear, that is, for all $f \in C^*$ and $c \in C$,* $\beta^r * f(c) = \beta^r(c \leftharpoonup f)$;

(e) *β factors through $\beta^* : C \otimes_{C^*} C \to R$.*

Notice that description (b) occurs in the characterisation of coseparable coalgebras (see 3.29).

Proof. (a) ⇔ (b) The condition in (b) explicitly reads, for all $c, d \in C$,

$$\sum c_{\underline{1}} \beta(c_{\underline{2}}, d) = \sum \beta(c, d_{\underline{1}}) d_{\underline{2}}\,. \qquad (*)$$

Since ${}_RC$ is locally projective, property 4.2(b) implies that $(*)$ is equivalent to the statement that, for all $c, d \in C$ and for all $f \in C^*$,

$$f(\sum c_{\underline{1}} \beta(c_{\underline{2}}, d)) = f(\sum \beta(c, d_{\underline{1}}) d_{\underline{2}})\,.$$

Now, since both f and β are R-linear, this is equivalent to

$$\beta(c\leftharpoonup f, d) = \beta(\sum f(c_{\underline{1}})c_{\underline{2}}, d) = \beta(c, \sum d_{\underline{1}} f(d_{\underline{2}})) = \beta(c, f\rightharpoonup d),$$

that is, β is C-balanced.

(a) $\Leftrightarrow$ (c) $\Leftrightarrow$ (d) are verified by similar arguments.

(a) $\Leftrightarrow$ (e) follows from the definition of the tensor product $C \otimes_{C^*} C$. □

6.5. Corollary. *Let ${}_RC$ be locally projective, let $\beta : C \times C \to R$ be a C-balanced bilinear form, and let $X \subset C$ be an R-submodule.*

(1) If X is a left C-subcomodule, then

$$X^{\perp_\beta} = \{d \in C \mid \beta(x, d) = 0 \text{ for all } x \in X\}$$

is a right C-comodule.

(2) If X is a right C-subcomodule, then

$${}^{\perp_\beta}X = \{c \in C \mid \beta(c, x) = 0 \text{ for all } x \in X\}$$

is a left C-subcomodule.

(3) In particular, $C^{\perp_\beta}$ is the kernel of β^l and ${}^{\perp_\beta}C$ is the kernel of β^r.

(4) Let R be Noetherian and X a finitely generated R-module. Then $C/X^{\perp_\beta}$ and $C/{}^{\perp_\beta}X$ are finitely generated as R-modules.

Proof. (1) For $d \in X^{\perp_\beta}$ and any $f \in C^*$, $\beta(x, f\rightharpoonup d) = \beta(x\leftharpoonup f, d) = 0$. This shows that $X^{\perp_\beta}$ is a left C^*-submodule and hence a right C-comodule by the α-condition.

(2) This is shown by a similar argument.

(3) The assertions follow immediately from the definitions.

(4) Obviously $X^{\perp_\beta}$ is the kernel of the map $C \to X^*, d \mapsto \beta(-, d)|_X$. Since R is Noetherian, X^* is a Noetherian R-module, and hence the submodules (isomorphic to) $C/X^{\perp_\beta}$ and $C/{}^{\perp_\beta}X$ are finitely generated. □

6.6. Nodegenerate bilinear forms. A bilinear form $\beta : C \times C \to R$ is called *left (right) nondegenerate* if β^l (resp. β^r) is injective.

A family of bilinear forms $\{\beta_\lambda : C \times C \to R\}_\Lambda$ is said to be *left (right) nondegenerate* if $\bigcap_\Lambda \operatorname{Ke} \beta^l_\lambda = 0$ (resp. $\bigcap_\Lambda \operatorname{Ke} \beta^r_\lambda = 0$).

Properties. *Let ${}_RC$ be locally projective.*

(1) The following are equivalent:

(a) there exists a left C^*-monomorphism $\gamma : C \to C^*$;

(b) there is a left nondegenerate C-balanced bilinear form β on C.

(2) *The following are equivalent:*

(a) *there exists a left C^*-monomorphism $\gamma : C \to (C^*)^\Lambda$;*

(b) *there exists a family of left nondegenerate C-balanced bilinear forms $\{\beta_\lambda\}_\Lambda$ on C.*

(3) *Assume that R is Noetherian and that the conditions in (2) hold. Then essential extensions of simple C^*-submodules of C are finitely generated as R-modules.*

Proof. (1) Let $\gamma : C \to C^*$ be any left C^*-linear map. Then

$$\beta : C \times C \to R, \quad (c,d) \mapsto \gamma(d)(c),$$

is a C-balanced bilinear form with $\beta^l(d) = \gamma(d)$. Clearly γ is injective if and only if β is left nondegenerate, and this proves the assertion.

(2) Let $\gamma : C \to (C^*)^\Lambda$ be a left C^*-monomorphism. Then

$$\gamma_\lambda : C \xrightarrow{\gamma} (C^*)^\Lambda \xrightarrow{\pi_\lambda} C^*$$

is a C^*-linear map and (as in (1)) there are C-balanced bilinear forms $\beta_\lambda : C \times C \to R$ with $\beta^l_\lambda = \gamma_\lambda$, for which

$$\bigcap\nolimits_\Lambda \mathrm{Ke}\,\beta^l_\lambda = \bigcap\nolimits_\Lambda \mathrm{Ke}\,\gamma_\lambda = \mathrm{Ke}\,\gamma.$$

From this the assertion follows immediately.

(3) Let S be a simple C^*-submodule of C with an essential extension $\tilde{S} \subset C$. There exists a bilinear form $\beta : C \times C \to R$ such that $\beta(x,S) \neq 0$, for some $x \in C$. Putting $X = x \leftharpoonup C^*$, we observe for $X^{\perp_\beta} = \{d \in C \mid \beta(X,d) = 0\}$ that $C/X^{\perp_\beta}$ is a finitely generated R-module (by 6.5).

Since $S \cap X^{\perp_\beta} = 0$, we also have $\tilde{S} \cap X^{\perp_\beta} = 0$ and there is a monomorphism $\tilde{S} \to C/X^{\perp_\beta}$. This implies that $\tilde{S}$ is a finitely generated R-module. □

6.7. Exercises

Let C be a matrix coalgebra with basis $\{e_{ij}\}_{1 \leq i,j \leq n}$ (see 1.10). Prove that

$$\beta : C \times C \to R, \quad (e_{ij}, e_{rs}) \mapsto \delta_{is}\delta_{jr},$$

determines a C-balanced, nondegenerate and symmetric bilinear form ([167]).

References. Doi [104]; Gómez-Torrecillas and Năstăsescu [123]; Heyneman and Radford [129]; Lin [152]; Miyamoto [160]; Nichols [167]; Sweedler [45]; Wischnewsky [213].

7 The rational functor

We know from Section 4 that there is a faithful functor from the category of right C-comodules to the category of left C^*-modules; we also know that $\mathbf{M}^C$ is a full subcategory of ${}_{C^*}\mathbf{M}$ if and only if the α-condition holds. Now we want to study an opposite problem. Suppose M is a left C^*-module; is it also a right C-comodule? If the answer is negative, does there exist a (maximal) part of M on which a right C-coaction can be defined? In other words, is it possible to define a functor ${}_{C^*}\mathbf{M} \to \mathbf{M}^C$, that selects the maximal part of a module that can be made into a comodule (hence acts as identity on all left C^*-modules that already are right C-comodules)? Such a functor exists and is known as the *rational functor*, provided that C satisfies the α-condition. Then $\mathbf{M}^C$ coincides with $\sigma[{}_{C^*}C]$, the full subcategory of ${}_{C^*}\mathbf{M}$ subgenerated by C (cf. 4.3), and the inclusion functor has a right adjoint $\mathcal{T}^C$ (see 41.1), which is precisely the rational functor. This functor is the topic of the present section.

Throughout (except for 7.9 and 7.10) we assume that ${}_RC$ is locally projective. We also use freely torsion-theoretic aspects of the category of modules and its subcategory $\sigma[M]$. The reader not familiar with those aspects is referred to Section 42.

7.1. Rational functor. For any left C^*-module M, define the *rational submodule*

$$\mathrm{Rat}^C(M) = \mathcal{T}^C(M) = \sum \{\mathrm{Im}\, f \mid f \in {}_{C^*}\mathrm{Hom}(U, M),\, U \in \mathbf{M}^C\},$$

where $\mathcal{T}^C$ is the trace functor ${}_{C^*}\mathbf{M} \to \sigma[C]$ (cf. 41.1). Clearly $\mathrm{Rat}^C(M)$ is the largest submodule of M that is subgenerated by C, and hence it is a right C-comodule. The induced functor (subfunctor of the identity)

$$\mathrm{Rat}^C : {}_{C^*}\mathbf{M} \to \mathbf{M}^C, \quad M \mapsto \mathrm{Rat}^C(M),$$

is called the *rational functor*. Since Rat^C is a trace functor, we know from 41.1 that it is right adjoint to the inclusion $\mathbf{M}^C \to {}_{C^*}\mathbf{M}$ and its properties depend on (torsion-theoretic) properties of the class $\mathbf{M}^C$ in ${}_{C^*}\mathbf{M}$. Of course $\mathrm{Rat}^C(M) = M$ for $M \in {}_{C^*}\mathbf{M}$ if and only if $M \in \mathbf{M}^C$, and $\mathbf{M}^C = {}_{C^*}\mathbf{M}$ if and only if ${}_RC$ is finitely generated (see 4.7).

7.2. Rational elements. Let M be a left C^*-module. An element $k \in M$ is said to be *rational* if there exists an element $\sum_i m_i \otimes c_i \in M \otimes_R C$, such that

$$fk = \sum_i m_i f(c_i), \text{ for all } f \in C^*.$$

This means that, from the diagram

$$\begin{array}{ccc} & M & \\ & \downarrow{\scriptstyle \psi_M} & \\ M \otimes_R C & \xrightarrow{\alpha_M} \mathrm{Hom}_R(C^*, M) \end{array} \qquad m \otimes c \longmapsto [f \mapsto mf(c)], \qquad \begin{array}{c} m \\ \downarrow \\ [f \mapsto fm], \end{array}$$

we obtain $\psi_M(k) = \alpha_M(\sum_i m_i \otimes c_i)$ (see 4.2). Since it is assumed that α_M is injective, the element $\sum_i m_i \otimes c_i$ is uniquely determined.

7.3. Rational submodule. *Let M be a left C^*-module.*

(1) *An element $k \in M$ is rational if and only if C^*k is a right C-comodule with $fk = f\rightharpoonup k$, for all $f \in C^*$.*

(2) $\mathrm{Rat}^C(M) = \{k \in M \mid k$ *is rational*$\}$.

Proof. (1) Let $k \in M$ be rational and $\sum_i m_i \otimes c_i \in M \otimes_R C$ such that $fk = \sum_i m_i f(c_i)$ for all $f \in C^*$. Put $K := C^*k$ and define a map

$$\varrho : K \to M \otimes_R C, \quad fk \mapsto \sum_i m_i \otimes f\rightharpoonup c_i.$$

For $f, h \in C^*$,

$$\alpha_M(\sum_i m_i \otimes f\rightharpoonup c_i)(h) = \sum_i m_i h(f\rightharpoonup c_i) = h * f\, k = h \cdot fk\,.$$

So the map ϱ is well defined since $fk = 0$ implies $\alpha_M(\sum_i m_i \otimes f\rightharpoonup c_i) = 0$, and hence $\sum_i m_i \otimes f\rightharpoonup c_i = 0$ by injectivity of α_M. Moreover, it implies that $\alpha_M \circ \varrho(K) \subset \mathrm{Hom}_R(C^*, K)$, and we obtain the commutative diagram with exact rows

$$\begin{array}{ccccccccc} & & & & K & & & & \\ & & & & \downarrow{\scriptstyle \varrho} & & & & \\ 0 & \longrightarrow & K \otimes_R C & \longrightarrow & M \otimes_R C & \longrightarrow & M/K \otimes_R C & \longrightarrow & 0 \\ & & \downarrow{\scriptstyle \alpha_K} & & \downarrow{\scriptstyle \alpha_M} & & \downarrow{\scriptstyle \alpha_{M/K}} & & \\ 0 & \longrightarrow & \mathrm{Hom}_R(C^*, K) & \longrightarrow & \mathrm{Hom}_R(C^*, M) & \longrightarrow & \mathrm{Hom}_R(C^*, M/K) & \longrightarrow & 0, \end{array}$$

where all the α are injective. By the kernel property we conclude that ϱ factors through some $\varrho^K : K \to K \otimes_R C$, and it follows by 4.4 that ϱ^K is coassociative and counital, thus making K a comodule. $\square$

As a first application we consider the rational submodule of C^{**}. The canonical map $\Phi_C : C \to C^{**}$ is a C^*-morphism, since, for all $c \in C$, $f, h \in C^*$,

$$\Phi_C(f\rightharpoonup c)(h) = h(f\rightharpoonup c) = \sum h(c_{\underline{1}})f(c_{\underline{2}}) = \Phi_C(c)(h * f) = f\Phi_C(c)(h).$$

Hence the image of Φ_C is a rational module. The next lemma shows that this is equal to the rational submodule of C^{**}.

7.4. Rational submodule of C^{**}. *$\Phi_C : C \to \mathrm{Rat}^C(C^{**})$ is an isomorphism.*

Proof. Local projectivity of ${}_RC$ implies that Φ_C is injective. Let $\varrho : \mathrm{Rat}^C(C^{**}) \to \mathrm{Rat}^C(C^{**}) \otimes_R C$ denote the comodule structure map. For $\gamma \in \mathrm{Rat}^C(C^{**})$ write $\varrho(\gamma) = \sum_i \gamma_i \otimes c_i$. Then, for any $f \in C^*$,

$$\gamma(f) = f \cdot \gamma(\varepsilon) = \sum_i f(c_i)\gamma_i(\varepsilon) = f(\sum_i \gamma_i(\varepsilon)c_i),$$

where $\sum_i \gamma_i(\varepsilon)c_i \in C$. So $\gamma \in \mathrm{Im}\, \Phi_C$, proving that Φ_C is surjective. □

The rational submodule of ${}_{C^*}C^*$ is a two-sided ideal in C^* and is called the *left trace ideal.* From the above observations and the Finiteness Theorem it is clear that $\mathrm{Rat}^C(C^*) = C^*$ if and only if ${}_RC$ is finitely generated.

Right rational C^*-modules are defined in a symmetric way, yielding the *right trace ideal* ${}^C\mathrm{Rat}(C^*)$, which in general is different from $\mathrm{Rat}^C(C^*)$.

7.5. Characterisation of the trace ideal. *Let $T = \mathrm{Rat}^C(C^*)$ be the left trace ideal.*

(1) Let $f \in C^*$ and assume that $f \rightharpoonup C$ is a finitely presented R-module. Then $f \in T$.

(2) If R is Noetherian, then T can be described as

$$\begin{aligned} T_1 &= \{f \in C^* \mid C^* * f \text{ is a finitely generated } R\text{-module}\}; \\ T_2 &= \{f \in C^* \mid \mathrm{Ke} f \text{ contains a right } C^*\text{-submodule } K, \text{ such that } \\ &\qquad C/K \text{ is a finitely generated } R\text{-module}\}; \\ T_3 &= \{f \in C^* \mid f \rightharpoonup C \text{ is a finitely generated } R\text{-module}\}. \end{aligned}$$

Proof. Assertion (1) and the inclusion $T \subset T_1$ in (2) follow from the Finiteness Theorem 4.12.

[$T_1 \subset T_2$]: For $f \in T_1$, let $C^* * f$ be finitely R-generated by $g_1, \ldots, g_k \in C^*$. Consider the kernel of $C^* * f$,

$$K := \bigcap \{\mathrm{Ke}\ h \mid h \in C^* * f\} = \bigcap_{i=1}^{k} \mathrm{Ke}\, g_i\,.$$

Clearly K is a right C^*-submodule of C. Moreover, all the $C/\mathrm{Ke}\, g_i$ are finitely generated R-modules, and hence

$$C/K \subset \bigoplus_{i=1}^{k} C/\mathrm{Ke}\, g_i$$

is a finitely generated R-module. This proves the inclusion $T_1 \subset T_2$.

[$T_2 \subset T_3$]: Let $f \in T_2$. Since $\Delta(K) \subset C \otimes_R K$, $f \rightharpoonup K = 0$ and $f \rightharpoonup C = f \rightharpoonup C/K$ is a finitely generated R-module, that is, $f \in T_3$.

$[T_3 \subset T]$: For $f \in T_3$, the rational right C^*-module $f \leftharpoonup C$ is a finitely presented R-module. Then, by 3.11, $(f \leftharpoonup C)^*$ is a rational left C^*-module. Since $\varepsilon(f \leftharpoonup c) = f(c)$ for all $c \in C$, we conclude $f \in (f \leftharpoonup C)^*$ and hence $f \in T$. □

7.6. $\mathbf{M}^C$ closed under extensions. *The following are equivalent:*

(a) *$\mathbf{M}^C$ is closed under extensions in ${}_{C^*}\mathbf{M}$;*

(b) *for every $X \in {}_{C^*}\mathbf{M}$, $\mathrm{Rat}^C(X/\mathrm{Rat}^C(X)) = 0$;*

(c) *there exists a C^*-injective $Q \in {}_{C^*}\mathbf{M}$ such that*

$$\mathbf{M}^C = \{N \in {}_{C^*}\mathbf{M} \mid {}_{C^*}\mathrm{Hom}(N, Q) = 0\}.$$

If R is QF, then (a)-(c) are equivalent to:

(d) *the filter of open left ideals $\mathcal{F}_C$ (cf. 4.17) is closed under products.*

Proof. The assertions follow from 42.14 and 42.15. □

Over a QF ring there is another finiteness condition that implies left exactness of Rat^C.

7.7. Corollary. *Let R be QF and $\mathcal{F}_C$ of finite type. Then $\mathbf{M}^C$ is closed under extensions in ${}_{C^*}\mathbf{M}$.*

Proof. The filter $\mathcal{F}_C$ is always bounded and by assumption of finite type. Hence it suffices to show that the product of an ideal $J \in \mathcal{F}_C$ and a finitely generated left ideal $I \in \mathcal{F}_C$ belongs to $\mathcal{F}_C$. Clearly, JI is finitely generated and hence closed. An epimorphism $(C^*)^n \to I$ yields the commutative exact diagram

$$\begin{array}{ccccc}
(C^*)^n & \longrightarrow & I & \longrightarrow & 0 \\
\downarrow & & \downarrow & & \\
(C^*/J)^n & \longrightarrow & I/JI & \longrightarrow & 0,
\end{array}$$

showing that I/JI is finitely R-generated. Now, in the exact sequence

$$0 \longrightarrow I/JI \longrightarrow C^*/JI \longrightarrow C^*/I \longrightarrow 0,$$

I/JI and C^*/I are finitely R-generated and so is C^*/JI. By 4.17, this implies that $JI \in \mathcal{F}_C$ and now the assertion follows from 7.6. □

7.8. $\mathbf{M}^C$ closed under essential extensions. *The following are equivalent:*

(a) *$\mathbf{M}^C$ is closed under essential extensions in ${}_{C^*}\mathbf{M}$;*

(b) *$\mathbf{M}^C$ is closed under injective hulls in ${}_{C^*}\mathbf{M}$;*

(c) *every C-injective module in $\mathbf{M}^C$ is C^*-injective;*

(d) *for every injective C^*-module Q, $\mathrm{Rat}^C(Q)$ is a direct summand in Q;*

(e) *for every injective C^*-module Q, $\mathrm{Rat}^C(Q)$ is C^*-injective.*

If (any of) these conditions hold, then $\mathbf{M}^C$ is closed under extensions.

Proof. This is a special case of 42.20. □

Before concentrating on properties of the trace ideal we consider density for any subalgebras of C^*. From the Density Theorem 42.2 we know that for any C-dense subalgebra $T \subset C^*$ the categories $\mathbf{M}^C$ and $\sigma[{}_TC]$ can be identified. For the next two propositions we need not assume a priori that C satisfies the α-condition.

7.9. Density in C^*. *For an R-submodule $U \subset C^*$ the following assertions are equivalent:*

(a) *U is dense in C^* in the finite topology (of R^C);*

(b) *U is a C-dense subset of C^* (in the finite topology of $\mathrm{End}_R(C)$).*

If C is cogenerated by R, then (a), (b) imply:

(c) $\mathrm{Ke}\, U = \{x \in C \mid u(x) = 0$ *for all* $u \in U\} = 0$.

If R is a cogenerator in $\mathbf{M}_R$, then (c) $\Rightarrow$ (b).

Proof. (a) $\Leftrightarrow$ (b) As shown in 3.12 the finite topologies in C^* and $\mathrm{End}^C(C)$ can be identified.

(a) $\Rightarrow$ (c) Let C be cogenerated by R. Then, for any $0 \neq x \in C$, there exists $f \in C^*$ such that $f(x) \neq 0$. Then, for some $u \in U$, $u(x) = f(x) \neq 0$, that is, $x \notin \mathrm{Ke}\, U$, and hence $\mathrm{Ke}\, U = 0$.

(c) $\Rightarrow$ (b) Let R be a cogenerator in $\mathbf{M}_R$. Let $f \in C^*$ and $x_1, \ldots, x_n \in C$. Suppose that

$$f \rightharpoonup (x_1, \ldots, x_n) \notin U \rightharpoonup (x_1, \ldots, x_n) \subset C^n.$$

Then there exists an R-linear map $g : C^n \to R$ such that

$$g(f \rightharpoonup (x_1, \ldots, x_n)) \neq 0 \text{ and } g(U \rightharpoonup (x_1, \ldots, x_n)) = 0.$$

For each $u \in U$ (by 4.6),

$$0 = \sum_i g_i(u \rightharpoonup x_i) = \sum_i u(x_i \leftharpoonup g_i) = u(\sum_i x_i \leftharpoonup g_i),$$

where $g_i : C \to C^n \xrightarrow{g} R$, and this implies $\sum_i x_i \leftharpoonup g_i = 0$ and

$$0 \neq g(f \rightharpoonup (x_1, \ldots, x_n)) = \sum_i g_i(f \rightharpoonup x_i) = \sum_i f(x_i \leftharpoonup g_i) = f(\sum_i x_i \leftharpoonup g_i) = 0,$$

contradicting the choice of g. □

Remark. Notice that, for a vector space V over a field R, a subspace $U \subset V^*$ is dense if and only if $\mathrm{Ke}\, U = 0$. This is the *density* criterion that is well known in the comodule theory for coalgebras over fields (e.g., [129], [152]).

7.10. Dense subalgebras of C^*. *For a subalgebra $T \subset C^*$ the following are equivalent:*

(a) *${}_RC$ is locally projective and T is dense in C^*;*

(b) *$\mathbf{M}^C = \sigma[{}_TC]$.*

If T is an ideal in C^*, then (a),(b) are equivalent to:

(c) *C is an s-unital T-module and C satisfies the α-condition.*

Proof. (a) $\Leftrightarrow$ (b) This was also observed in 4.19: there are embeddings $\mathbf{M}^C \subset \sigma[{}_{C^*}C] \subset \sigma[{}_TC]$. Now $\mathbf{M}^C = \sigma[{}_{C^*}C]$ is equivalent to the α-condition while $\sigma[{}_TC] = \sigma[{}_{C^*}C]$ corresponds to the density property.

(a) $\Leftrightarrow$ (c) By 42.6, for an ideal T the density property is equivalent to s-unitality of the T-module C. □

Combining the properties of the trace functor observed in 42.16 with the characterisation of dense ideals in 42.6, we obtain:

7.11. The rational functor exact. *Let $T = \mathrm{Rat}^C(C^*)$. The following statements are equivalent:*

(a) *the functor $\mathrm{Rat}^C : {}_{C^*}\mathbf{M} \to \mathbf{M}^C$ is exact;*

(b) *$\mathbf{M}^C$ is closed under extensions in ${}_{C^*}\mathbf{M}$ and the class*

$$\{X \in {}_{C^*}\mathbf{M} \mid \mathrm{Rat}^C(X) = 0\}$$

is closed under factor modules;

(c) *for every $N \in \mathbf{M}^C$ (with $N \subset C$), $TN = N$;*

(d) *for every $N \in \mathbf{M}^C$, the canonical map $T \otimes_{C^*} N \to N$ is an isomorphism;*

(e) *C is an s-unital T-module;*

(f) *$T^2 = T$ and T is a generator in $\mathbf{M}^C$;*

(g) *$TC = C$ and C^*/T is flat as a right C^*-module;*

(h) *T is a left C-dense subring of C^*.*

7.12. Corollary. *Assume that Rat^C is exact and let $T = \mathrm{Rat}^C(C^*) \subset C^*$.*

(1) *$\mathbf{M}^C$ is closed under small epimorphisms in ${}_{C^*}\mathbf{M}$.*

(2) *If P is finitely presented in $\mathbf{M}^C$, then P is finitely presented in ${}_{C^*}\mathbf{M}$.*

(3) *If P is projective in $\mathbf{M}^C$, then P is projective in ${}_{C^*}\mathbf{M}$.*

(4) *For any $M \in \mathbf{M}^C$, the canonical map ${}_{C^*}\mathrm{Hom}(C^*, M) \to {}_{C^*}\mathrm{Hom}(T, M)$ is injective.*

Proof. (1)–(3) follow from Corollary 42.17.

(4) By density, for every $f \in {}_{C^*}\mathrm{Hom}(C^*, M)$, $f(\varepsilon) = tf(\varepsilon) = f(t)$ for some $t \in T$, and hence $f(T) = 0$ implies $f(C^*) = 0$. □

The density of an ideal $S \subset C^*$ in the finite topology implies that C is s-unital both as left and right S-module and hence $\mathbf{M}^C = \sigma[{}_SC]$ and ${}^C\mathbf{M} = \sigma[C_S]$ (see 7.10). Therefore the exactness of Rat^C, that is, the density of $\mathrm{Rat}^C(C^*)$ in C^*, also has some influence on left C-comodules.

7.13. Corollary. *Assume that* Rat^C *is exact and let* $T = \mathrm{Rat}^C(C^*) \subset C^*$.

(1) For any $N \in {}^C\mathbf{M}$, *the canonical map* $\mathrm{Hom}_{C^*}(C^*, N) \to \mathrm{Hom}_{C^*}(T, N)$ *is injective.*

(2) ${}^C\mathrm{Rat}(C^*) \subset T$ *and equality holds if and only if* $T \in {}^C\mathbf{M}$.

Proof. (1) By the preceding remark, C is also s-unital as a right T-module and hence the proof of Corollary 7.12(4) applies.

(2) By the density of $T \subset C^*$, $X \leftharpoonup T = X$, for each $X \in {}^C\mathbf{M}$ (see 42.6). This implies

$$\mathrm{Hom}_{C^*}(X, C^*) = \mathrm{Hom}_{C^*}(X \leftharpoonup T, C^*) = \mathrm{Hom}_{C^*}(X, C^* * T) = \mathrm{Hom}_{C^*}(X, T);$$

hence ${}^C\mathrm{Rat}(C^*) \subset T$ and ${}^C\mathrm{Rat}(C^*) = T$ provided $T \in {}^C\mathbf{M}$. □

The assertion in 42.18 yields here:

7.14. Corollary. *Suppose that* $\mathbf{M}^C$ *has a generator that is locally projective in* ${}_{C^*}\mathbf{M}$. *Then* $\mathrm{Rat}^C : {}_{C^*}\mathbf{M} \to \mathbf{M}^C$ *is an exact functor.*

Except when ${}_RC$ is finitely generated (i.e., $\mathrm{Rat}^C(C^*) = C^*$) the trace ideal does not contain a unit element. However, if C is a direct sum of finitely generated left (and right) C^*-submodules, the trace ideal has particularly nice properties.

7.15. Trace ideal and decompositions. *Let* $T := \mathrm{Rat}^C(C^*)$ *and* $T' := {}^C\mathrm{Rat}(C^*)$.

(1) If C *is a direct sum of finitely generated right* C^**-modules, then* T *is* C*-dense in* C^* *and there is an embedding*

$$\gamma : T' \to \bigoplus_\Lambda T' * e_\lambda \subset T,$$

for a family of orthogonal idempotents $\{e_\lambda\}_\Lambda$ *in* T.

(2) If C *is a direct sum of finitely generated right* C^**-modules and of finitely generated left* C^**-modules, then* $T = T'$ *and* T *is a projective generator both in* $\mathbf{M}^C$ *and* ${}^C\mathbf{M}$.

Proof. (1) Under the given conditions there exist orthogonal idempotents $\{e_\lambda\}_\Lambda$ in C^* with $C = \bigoplus_\Lambda e_\lambda \rightharpoonup C$, where all $e_\lambda \rightharpoonup C$ are finitely generated right C^*-modules. By the Finiteness Theorem 4.12, the $e_\lambda \rightharpoonup C$ are finitely generated as R-modules, and they are R-projective as direct summands of C. Now it follows from 7.5(1) that $e_\lambda \in T$. Clearly C is an s-unital left T-module and hence the density property follows (see 42.6).

Consider the assignment $\gamma : T' \to \bigoplus_\Lambda T' * e_\lambda$, $t \mapsto \sum_\Lambda t * e_\lambda$. For any $t \in T'$, $t * C^*$ is finitely R-generated and so $t * e_\lambda = 0$ for almost all $\lambda \in \Lambda$. Hence γ is a well-defined map. Assume $\gamma(t) = 0$. Then, for any $c \in C$, $0 = t * e_\lambda(c) = t(e_\lambda \rightharpoonup c)$, for all $\lambda \in \Lambda$, implying $t = 0$.

(2) By symmetry, (1) implies $T = T'$ and so $T = \bigoplus_\Lambda T * e_\lambda$ and $T = \bigoplus_\Omega f_\omega * T$, where the $\{f_\omega\}_\Omega$ are orthogonal idempotents in C^*, and the $C \leftharpoonup f_\omega$ are finitely R-generated (hence $f_\omega \in T'$). Clearly each $T * e_\lambda$ is a projective left T-module and $f_\omega * T$ a projective right T-module. Now the density property implies that T is a projective generator both in $\mathbf{M}^C$ and in ${}^C\mathbf{M}$ (see 7.11). □

Notice that, in 7.15, $e_\lambda \in T'$ need not imply that $C \leftharpoonup e_\lambda$ is finitely R-generated, unless we know that R is Noetherian (see 7.5).

7.16. Decompositions over Noetherian rings. *Let R be Noetherian, $T = \mathrm{Rat}^C(C^*)$ and $T' = {}^C\mathrm{Rat}(C^*)$. Then the following are equivalent:*

(a) C_{C^*} and ${}_{C^*}C$ are direct sums of finitely generated C^*-modules;

(b) C_{C^*} is a direct sum of finitely generated C^*-modules and $T = T'$;

(c) ${}_{C^*}C$ is a direct sum of finitely generated C^*-modules and $T = T'$;

(d) $C = T \rightharpoonup C$ and $T = T'$ and is a ring with enough idempotents.

If these conditions hold, T is a projective generator both in $\mathbf{M}^C$ and in ${}^C\mathbf{M}$.

Proof. (a) ⇒ (b) follows by 7.15.

(b) ⇒ (d) Let $C = \bigoplus_\Lambda e_\lambda \rightharpoonup C$, with orthogonal idempotents $\{e_\lambda\}_\Lambda$ in C^*, where all $e_\lambda \rightharpoonup C$ are finitely R-generated. Then $e_\lambda \in T = T'$ and $T = \bigoplus_\Lambda T * e_\lambda$. For any $t \in T$, the module $t \rightharpoonup C$ is finitely R-generated (by 7.5) and so

$$t \rightharpoonup C \subset e_1 \rightharpoonup C \oplus \cdots \oplus e_k \rightharpoonup C, \text{ for some idempotents } e_i \in \{e_\lambda\}_\Lambda.$$

This implies $t = (e_1 + \cdots + e_k) * t \in \bigoplus_\Lambda e_\lambda * T$. So $\bigoplus_\Lambda T * e_\lambda = T = \bigoplus_\Lambda e_\lambda * T$, showing that T is a ring with enough idempotents.

(d) ⇒ (a) If $T = \bigoplus_\Lambda e_\lambda * T$, then

$$C = T \rightharpoonup C = \bigoplus_\Lambda e_\lambda \rightharpoonup C,$$

and $e_\lambda \in T$ implies that $e_\lambda \rightharpoonup C$ is finitely R-generated. So, by 7.15, T is dense in C^*, implying $C \leftharpoonup T = C$. Now symmetric arguments yield the decomposition of C as a direct sum of finitely R-generated left C^*-modules.

(c) ⇔ (a) The statement is symmetric to (d) ⇔ (a).

If the conditions hold, the assertion follows by 7.15. □

Fully invariant submodules of C that are direct summands are precisely subcoalgebras that are direct summands, and they are of the form $e \rightharpoonup C$, where e is a central idempotent in C^*. Hence 7.16 yields:

7.17. Corollary. *If R is Noetherian, the following are equivalent:*

(a) C is a direct sum of finitely generated subcoalgebras;

(b) C is a direct sum of finitely generated (C^*, C^*)-sub-bimodules;

(c) $T \rightharpoonup C = C$ and T is a ring with enough central idempotents.

7.18. Corollary. *Let R be Noetherian and C cocommutative. The following are equivalent:*

(a) C is a direct sum of finitely generated subcoalgebras;

(b) C is a direct sum of finitely generated C^*-submodules;

(c) $T \rightharpoonup C = C$ and T is a ring with enough idempotents.

Notice that, in the preceding results, the projectivity properties (of T) are derived from decompositions of C into finitely generated summands. In the next sections we will return to the problem of projectivity for comodules. One of the highlights will be the observation that, for coalgebras over QF rings, the exactness of Rat^C (see 7.11) is equivalent to the existence of enough projectives in $\mathbf{M}^C$ (see 9.6).

References. Heyneman and Radford [129]; Lin [152]; Wisbauer [210].

8 Structure of comodules

In this section we study further properties of a coalgebra C provided ${}_RC$ is locally projective. Since this means that the category of right C-comodules $\mathbf{M}^C$ is isomorphic to the category $\sigma[{}_{C^*}C]$ of C-subgenerated left C^*-modules (cf. 4.3), we can use the knowledge of module categories to derive properties of comodules. For example, we analyse projective objects in $\mathbf{M}^C$ and study properties of C motivated by module and ring theory.

Throughout this section we assume that C is an R-coalgebra with ${}_RC$ locally projective (see 4.2).

Let N be a right C-comodule. Then a C-comodule Q is said to be N-*injective* provided $\mathrm{Hom}^C(-, Q)$ turns any monomorphism $K \to N$ in $\mathbf{M}^C$ into a surjective map. We recall characterisations from 41.4.

8.1. Injectives in $\mathbf{M}^C$. *(1) For $Q \in \mathbf{M}^C$ the following are equivalent:*

(a) Q is injective in $\mathbf{M}^C$;

(b) the functor $\mathrm{Hom}^C(-, Q) : \mathbf{M}^C \to \mathbf{M}_R$ is exact;

(c) Q is C-injective (as left C^*-module);

(d) Q is N-injective for any (finitely generated) subcomodule $N \subset C$;

(e) every exact sequence $0 \to Q \to N \to L \to 0$ in $\mathbf{M}^C$ splits.

(2) Every injective object in $\mathbf{M}^C$ is C-generated.

(3) Every object in $\mathbf{M}^C$ has an injective hull.

A C-comodule P is N-*projective* if $\mathrm{Hom}^C(P, -)$ turns any epimorphism $N \to L$ into a surjective map. From 41.6 we obtain:

8.2. Projectives in $\mathbf{M}^C$. *(1) For $P \in \mathbf{M}^C$ the following are equivalent:*

(a) P is projective in $\mathbf{M}^C$;

(b) the functor $\mathrm{Hom}^C(P, -) : \mathbf{M}^C \to \mathbf{M}_R$ is exact;

(c) P is $C^{(\Lambda)}$-projective, for any set Λ;

(d) every exact sequence $0 \to K \to N \to P \to 0$ in $\mathbf{M}^C$ splits.

(2) If P is finitely generated and C-projective, then P is projective in $\mathbf{M}^C$.

Notice that projectives need not exist in $\mathbf{M}^C$. As observed in 3.22, projective objects in $\mathbf{M}^C$ (if they exist) are also projective in $\mathbf{M}_R$.

8.3. Cogenerator properties of C. *If C cogenerates all finitely C-generated left comodules, then the following are equivalent:*

(a) ${}_{C^*}C$ is linearly compact (see 41.13);

(b) C_{C^*} is C^*-injective.

If R is perfect, then (a),(b) are equivalent to:

(c) ${}_{C^*}C$ *is Artinian.*

Proof. The equivalence of (a) and (b) follows from 43.2. If R is perfect, C is semi-Artinian by 4.16, and hence, by 41.13, (a) implies that ${}_{C^*}C$ is Artinian. The implication (c) ⇒ (a) is trivial. □

Over a Noetherian ring R, C is left and right locally Noetherian as a C^*-module (by 4.16), and therefore we can apply 43.4 to obtain:

8.4. C as injective cogenerator in $\mathbf{M}^C$. *If R is Noetherian, then the following are equivalent:*

(a) *C is an injective cogenerator in $\mathbf{M}^C$;*

(b) *C is an injective cogenerator in ${}^C\mathbf{M}$;*

(c) *C is a cogenerator both in $\mathbf{M}^C$ and ${}^C\mathbf{M}$.*

8.5. C as injective cogenerator in $\mathbf{M}_{C^*}$. *If R is Artinian, then the following are equivalent:*

(a) *C is an injective cogenerator in $\mathbf{M}_{C^*}$;*

(b) *${}_{C^*}C$ is Artinian and an injective cogenerator in $\mathbf{M}^C$;*

(c) *C is an injective cogenerator in $\mathbf{M}^C$ and C^* is right Noetherian.*

If these conditions hold, then C^* is a semiperfect ring and every right C^*-module that is finitely generated as an R-module belongs to ${}^C\mathbf{M}$.

Proof. Since R is Artinian, C has locally finite length as a C^*-module.

(a) ⇒ (b) Assume C to be an injective cogenerator in $\mathbf{M}_{C^*}$. Then, by 8.4, C is an injective cogenerator in $\mathbf{M}^C$. Now 43.8 implies that ${}_{C^*}C$ is Artinian.

(b) ⇒ (a) and (b) ⇔ (c) follow again from 43.8.

Assume the conditions hold. C^* is f-semiperfect, being the endomorphism ring of a self-injective module (see 41.19). So $C^*/\mathrm{Jac}(C^*)$ is von Neumann regular and right Noetherian, and hence right (and left) semisimple. This implies that C^* is semiperfect.

Let $L \in \mathbf{M}_{C^*}$ be finitely generated as an R-module. Then L is finitely cogenerated as a C^*-module, and hence it is finitely cogenerated by C. This implies $L \in {}^C\mathbf{M}$. □

The decomposition of left semisimple coalgebras as a direct sum of (indecomposable) subcoalgebras (see 4.14) can be extended to more general situations. Recall that a relation on any family of (co)modules $\{M_\lambda\}_\Lambda$ is defined by setting (cf. 44.11)

$$M_\lambda \sim M_\mu \quad \text{if there exist nonzero morphisms } M_\lambda \to M_\mu \text{ or } M_\mu \to M_\lambda,$$

and the smallest equivalence relation determined by $\sim$ is given by

$$M_\lambda \approx M_\mu \quad \text{if there exist } \lambda_1, \ldots, \lambda_k \in \Lambda, \text{ such that } M_\lambda = M_{\lambda_1} \sim \cdots \sim M_{\lambda_k} = M_\mu .$$

8.6. σ-decomposition of coalgebras. *Let R be a Noetherian ring.*

(1) *There exist a σ-decomposition $C = \bigoplus_\Lambda C_\lambda$ and a family of orthogonal central idempotents $\{e_\lambda\}_\Lambda$ in C^* with $C_\lambda = C\leftharpoonup e_\lambda$, for each $\lambda \in \Lambda$.*

(2) *Each C_λ is a subcoalgebra of C, $C_\lambda^* \simeq C^* * e_\lambda$, $\sigma[_{C^*}C_\lambda] = \sigma[_{C_\lambda^*}C_\lambda]$, and*

$$\mathbf{M}^C = \bigoplus_\Lambda \sigma[_{C^*}C_\lambda] = \bigoplus_\Lambda \mathbf{M}^{C_\lambda}.$$

(3) *$\mathbf{M}^C$ is indecomposable if and only if, for any two injective uniform $L, N \in \mathbf{M}^C$, $L \approx N$ holds.*

(4) *Assume that R is Artinian. Then $\mathbf{M}^C$ is indecomposable if and only if, for any two simple $E_1, E_2 \in \sigma_{C^*}[C]$, $\widehat{E}_1 \approx \widehat{E}_2$ holds.*

Proof. (1),(2) By the Finiteness Theorem 4.12, C is a locally Noetherian C^*-module. Now the decomposition of $\mathbf{M}^C$ ($=\sigma[_{C^*}C]$) follows from 44.14. Clearly the resulting σ-decomposition of C is a fully invariant decomposition, and hence it can be described by central idempotents in the endomorphism ring ($= C^*$; see 44.1). Fully invariant submodules $C_\lambda \subset C$ are in particular R-direct summands in C and hence are subcoalgebras (by 4.6). It is straightforward to verify that $\mathrm{Hom}_R(C_\lambda, R) = C_\lambda^* \simeq C^* * e_\lambda$ is an algebra isomorphism. This implies $\sigma[_{C^*}C_\lambda] = \sigma[_{C_\lambda^*}C_\lambda] = \mathbf{M}^{C_\lambda}$.

(3) is a special case of 44.14(2).

(4) follows from 44.14(3). Notice that $\widehat{E}_1 \approx \widehat{E}_2$ can be described by extensions of simple modules (see 44.11). (The assertion means that the Ext quiver of simple modules in $\mathbf{M}^C$ is connected.) □

Transferring Corollary 44.7 we obtain:

8.7. Corollary. *Let C be a coalgebra with σ-decomposition $C = \bigoplus_\Lambda C_\lambda$. Then the left rational functor* Rat^C *is exact if and only if the left rational functors* Rat^{C_λ} *are exact, for each C_λ.*

Even for coalgebras C over fields there need not be any projective comodules in $\mathbf{M}^C$ (see example in 8.12). We discuss the existence of (enough) projectives in $\mathbf{M}^C$ and the projectivity of C in $\mathbf{M}^C$ or in $_{C^*}\mathbf{M}$.

Definition. A coalgebra C is called *right semiperfect* if every simple right comodule has a projective cover in $\mathbf{M}^C$. If $_RC$ is locally projective, this is obviously equivalent to the condition that every simple module in $\sigma[_{C^*}C]$ has a projective cover in $\sigma[_{C^*}C]$ (by 4.3), that is, $\mathbf{M}^C = \sigma[_{C^*}C]$ is a semiperfect category (see 41.16).

Notice that a right semiperfect coalgebra C need not be a semiperfect left C^*-module as defined in 41.14. The following characterisations can be shown.

8.8. Right semiperfect coalgebras. *The following are equivalent:*

(a) C is a right semiperfect coalgebra;

(b) $\mathbf{M}^C$ has a generating set of local projective modules;

(c) every finitely generated module in $\mathbf{M}^C$ has a projective cover.

If R is a perfect ring, then (a)-(c) are equivalent to:

(d) $\mathbf{M}^C$ has a generating set of finitely generated C-projective comodules.

Proof. The equivalence of (a), (b) and (c) follows from 41.16. If R is perfect, any finitely generated comodule is supplemented (see 41.22), and (a) $\Leftrightarrow$ (d) holds by 41.14. □

Semiperfect coalgebras over QF rings are described in 9.6. As an obvious application of 8.6 we obtain:

8.9. σ-decomposition of semiperfect coalgebras. *Let R be Noetherian and C with σ-decomposition $C = \bigoplus_\Lambda C_\lambda$. Then C is a right semiperfect coalgebra if and only if the C_λ are right semiperfect coalgebras, for all $\lambda \in \Lambda$.*

We finally turn to the question of when C itself is projective in $\mathbf{M}^C$ or ${}_{C^*}\mathbf{M}$. Since C is a balanced (C^*, C^*)-bimodule, we can use standard module theory to obtain some properties of C as a locally projective C^*-module.

8.10. C locally projective as C^*-module.

(1) If C is locally projective as a left C^*-module, then C is a generator in ${}^C\mathbf{M}$.

(2) If C is locally projective as a left and right C^*-module, then both Rat^C *and* ${}^C\mathrm{Rat}$ *are exact.*

Proof. (1) If ${}_{C^*}C$ is locally projective, then, by 42.10(g), C_{C^*} is a generator in $\sigma[C_{C^*}] = {}^C\mathbf{M}$.

(2) Assume that both ${}_{C^*}C$ and C_{C^*} are locally projective. Then, by (1), ${}_{C^*}C$ is a locally projective generator in $\sigma[{}_{C^*}C]$, and, by 7.14, Rat^C is an exact functor. Similar arguments show that ${}^C\mathrm{Rat}$ is exact. □

8.11. C projective in $\mathbf{M}^C$. *Assume that C is projective in $\mathbf{M}^C$.*

(1) If C^* is an f-semiperfect ring, or C is C-injective, then C is a direct sum of finitely generated left C^*-modules.

(2) If C^* is a right self-injective ring, then C is a generator in ${}^C\mathbf{M}$.

(3) If C^* is a semiperfect ring, then ${}_RC$ is finitely generated.

Proof. (1) This is a decomposition property of projective modules with f-semiperfect endomorphism rings (see 41.19). If C is self-injective, then C^* is f-semiperfect.

(2) As a self-injective ring, C^* is f-semiperfect. By 7.15, (1) implies that C is s-unital over the right trace ideal T', and so T' is a generator (by 7.11). Moreover, right injectivity of C^* implies that $T' = \text{Tr}({}^C\mathbf{M}, C^*) = \text{Tr}(C_{C^*}, C^*)$, and so C generates T' (see 42.7 for the definition of a trace).

(3) This follows from (1) and 41.19. □

8.12. Exercises

(1) Let C be a free R-module with basis $\{c_n \mid n = 0, 1, \ldots\}$ and define

$$\Delta : C \to C \otimes C,\ c_k \mapsto \sum_{i=0}^{k} c_i \otimes c_{k-i}, \quad \varepsilon : C \to R,\ c_k \mapsto \delta_{0,k}.$$

Prove ([45], [152]):

(i) (C, Δ, ε) is a coassociative coalgebra.

(ii) The nontrivial subalgebras of C are $\sum_{i=0}^{n} Rc_i$, $n \in \mathbb{N}$.

(iii) C^* is a power series ring $R[[X]]$ in one variable X (where $X(c_k) = \delta_{1,k}$).

(iv) There are no finitely generated projectives in $\mathbf{M}^C$.

(v) If R is a field, then $\mathbf{M}^C$ is the class of all torsion modules in ${}_{C^*}\mathbf{M}$.

(2) Let C be a free R-module with basis $\{g_k, d_k \mid k = 1, 2, \ldots\}$ and define

$$\begin{array}{lll} \Delta : C \to C \otimes C, & g_k \mapsto g_k \otimes g_k, & d_k \mapsto g_k \otimes d_k + d_k \otimes g_{k+1}, \\ \varepsilon : C \to R, & g_k \mapsto 1, & d_k \mapsto 0\,. \end{array}$$

(i) Prove that C is left and right semiperfect.

(ii) Show that C is projective as a left C^*-module but not as a right C^*-module.

Hint: Consider the elements in C^*:

$$g_k^*(g_i) := \delta_{ki},\ g_k^*(d_i) := 0 \ \text{ and } \ d_k^*(d_i) := \delta_{ki},\ d_k^*(g_i) := 0.$$

(iii) Replace Δ by $\Delta' : C \to C \otimes C$, $g_k \mapsto g_k \otimes g_k$, $d_k \mapsto g_1 \otimes d_k + d_k \otimes g_{k+1}$. Prove that (C, Δ') is a right semiperfect coalgebra that is not left semiperfect.

(From [152, Example 1], [123, Example 1.6], [152, Example 3].)

References. Allen and Trushin [53]; Al-Takhman [51]; Beattie, Dăscălescu, Grünenfelder and Năstăsescu [60]; García, Jara and Merino [116, 117]; Green [124]; Gómez-Torrecillas and Năstăsescu [123]; Kaplansky [24]; Lin [152]; Montgomery [161]; Shudo [189]; Shudo and Miyamoto [190]; Sweedler [45]; Vanaja [202]; Wisbauer [210].

9 Coalgebras over QF rings

Recall that a QF ring R is an Artinian injective cogenerator in $\mathbf{M}_R$. Over such rings coalgebras have particularly nice properties. In fact, we obtain essentially all structural properties known for coalgebras over fields. We consider R-coalgebras C with ${}_RC$ locally projective (see 4.2). If R is a QF ring, then this is equivalent to C being projective as an R-module (cf. 42.11).

9.1. Coalgebras over QF rings. *If R is a QF ring, then:*

(1) C is a (big) injective cogenerator in $\mathbf{M}^C$.

(2) Every comodule in $\mathbf{M}^C$ is a subcomodule of some direct sum $C^{(\Lambda)}$.

(3) C^* is an f-semiperfect ring.

(4) $K := \mathrm{Soc}_{C^*}C \trianglelefteq C$ and $\mathrm{Jac}(C^*) = \mathrm{Hom}_R(C/K, R)$.

(5) C^* is right self-injective if and only if C is flat as left C^*-module.

Proof. (1),(2) By 3.21, C is injective in $\mathbf{M}^C$. Over a QF ring R, every R-module M is contained in a free R-module $R^{(\Lambda)}$. This yields, for any right C-comodule, an injection $\varrho^M : M \to M \otimes_R C \subset R^{(\Lambda)} \otimes_R C \simeq C^{(\Lambda)}$.

(3) The endomorphism ring of any self-injective module is f-semiperfect (see 41.19).

(4) By 4.16, ${}_{C^*}C$ is locally of finite length and hence has an essential socle. By the Hom-tensor relations (see 3.9),

$$\mathrm{Jac}(C^*) = \mathrm{Hom}^C(C/K, C) \simeq \mathrm{Hom}_R(C/K, R).$$

(5) For any $N \in \mathbf{M}_{C^*}$, there is an isomorphism $\mathrm{Hom}_R(N \otimes_{C^*} C, R) \simeq \mathrm{Hom}_{C^*}(N, \mathrm{Hom}_R(C, R)) = \mathrm{Hom}_{C^*}(N, C^*)$ (cf. 40.18). So, if C^* is right self-injective, the functor $\mathrm{Hom}_R(- \otimes_{C^*} C, R) : \mathbf{M}_{C^*} \to \mathbf{M}_R$ is exact. Since R is a cogenerator in $\mathbf{M}_R$, this implies that $- \otimes_{C^*} C$ is exact, that is, ${}_{C^*}C$ is flat. Similar arguments yield the converse conclusion. □

By 44.8, for any injective cogenerator, fully invariant decompositions (coalgebra decompositions) are σ-decompositions (see discussion of decompositions in Section 44). Consequently, 8.6 yields:

9.2. σ-decomposition of C. *If R is a QF ring, then:*

(1) C has fully invariant decompositions with σ-indecomposable summands.

(2) Each fully invariant decomposition (= decomposition into coalgebras) is a σ-decomposition.

(3) C is σ-indecomposable if and only if C has no nontrivial fully invariant decomposition, that is, C^* has no nontrivial central idempotents.

(4) If C is cocommutative, then $C = \bigoplus_\Lambda \widehat{E}_\lambda$ is a fully invariant decomposition, where $\{E_\lambda\}_\Lambda$ is a minimal representing set of simple comodules in $\mathbf{M}^C$, and $\widehat{E}_\lambda$ denotes the injective hull of E_λ.

Proof. By 9.1, C is an injective cogenerator in $\sigma[_{C^*}C]$, and so (1), (2) and (3) follow from 44.8 and 8.6. In (4), C^* is a commutative algebra by assumption, and so the assertion follows from 43.7. □

Over QF rings there is a bijective correspondence between closed subcategories of $\mathbf{M}^C$ and (C^*, C^*)-sub-bimodules in C. However, the latter need not be pure R-submodules of C, and hence they may not be subcoalgebras. Recall that injectivity of C in $\mathbf{M}^C$ implies $\text{Tr}(\sigma[N], C) = \text{Tr}(N, C)$, for any $N \in \mathbf{M}^C$.

9.3. Correspondence relations. *Let R be a QF ring and $N \in \mathbf{M}^C$. Then:*

(1) $\sigma[N] = \sigma[\text{Tr}(N, C)]$.

(2) The map $\sigma[N] \mapsto \text{Tr}(N, C)$ *yields a bijective correspondence between closed subcategories of* $\mathbf{M}^C$ *and* (C^*, C^*)*-sub-bimodules of* C.

(3) $\sigma[N]$ *is closed under essential extensions (injective hulls) in* $\mathbf{M}^C$ *if and only if* $\text{Tr}(N, C)$ *is a* C^**-direct summand of* $_{C^*}C$. *In this case* $\text{Tr}(N, C)$ *is a subcoalgebra of* C.

(4) N *is a semisimple comodule if and only if* $\text{Tr}(N, C) \subset \text{Soc}(_{C^*}C)$.

(5) If R *is a semisimple ring, then* $\text{Tr}(N, C)$ *is a subcoalgebra of* C.

Proof. Since R is a QF ring, $_{C^*}C$ has locally finite length and is an injective cogenerator in $\mathbf{M}^C$. Hence (1)–(4) follow from 44.3. Furthermore, if R is semisimple, the (C^*, C^*)-sub-bimodule $\text{Tr}(N, C)$ is an R-direct summand in C and so is a subcoalgebra by 4.6. This proves assertion (5). □

Since over a QF ring R any R-coalgebra C is an injective cogenerator in $\mathbf{M}^C$ and $^C\mathbf{M}$ (by 9.1), the results from 8.5 simplify to the following.

9.4. C injective in $\mathbf{M}_{C^*}$. *If R is QF, the following are equivalent:*

(a) C *is injective in* $\mathbf{M}_{C^*}$;

(b) C *is an injective cogenerator in* $\mathbf{M}_{C^*}$;

(c) $_{C^*}C$ *is Artinian;*

(d) C^* *is a right Noetherian ring.*

Proof. In view of the preceding remark the equivalence of (b), (c) and (d) follows from 8.5. The implication (b) $\Rightarrow$ (a) is trivial, and (a) $\Rightarrow$ (c) is a consequence of 8.3. □

For finitely generated comodules, injectivity and projectivity in $\mathbf{M}^C$ may extend to injectivity, resp. projectivity, in $_{C^*}\mathbf{M}$.

9.5. Finitely presented modules over QF rings. *Let R be a QF ring and $M \in \mathbf{M}^C$.*

(1) If M is projective in $\mathbf{M}^C$, then M^* is C-injective as a right C^*-module and $\mathrm{Rat}^C(M^*)$ is injective in ${}^C\mathbf{M}$.

(2) If M is finitely generated as an R-module, then:

(i) if M is injective in $\mathbf{M}^C$, then M^* is projective in $\mathbf{M}_{C^*}$.

(ii) M is injective in $\mathbf{M}^C$ if and only if M is injective in ${}_{C^*}\mathbf{M}$.

(iii) M is projective in $\mathbf{M}^C$ if and only if M is projective in ${}_{C^*}\mathbf{M}$.

Proof. (1) Consider any diagram with exact row in ${}^C\mathbf{M}$,

$$\begin{array}{ccccc} 0 & \longrightarrow & K & \longrightarrow & N \\ & & \downarrow{\scriptstyle f} & & \\ & & M^*, & & \end{array}$$

where N is finitely generated as an R-module. Applying $(-)^* = \mathrm{Hom}_R(-, R)$ we obtain – with the canonical map $\Phi_M : M \to M^{**}$ – the diagram

$$\begin{array}{ccccc} M & \longrightarrow & M^{**} & & \\ & & \downarrow{\scriptstyle f^*} & & \\ N^* & \longrightarrow & K^* & \longrightarrow & 0\,, \end{array}$$

where the lower row is in $\mathbf{M}^C$ and hence can be extended commutatively by some right comodule morphism $g : M \to N^*$. Again applying $(-)^*$ - and recalling that the composition $M^* \xrightarrow{\Phi_{M^*}} M^{***} \xrightarrow{(\Phi_M)^*} M^*$ yields the identity (by 40.23) - we see that g^* extends f to N. This proves that M^* is N-injective for all modules $N \in {}^C\mathbf{M}$ that are finitely presented as R-modules.

In particular, by the Finiteness Theorem 4.12, every finitely generated C^*-submodule of C is finitely generated - hence finitely presented – as an R-module. So M^* is N-injective for all these modules, and hence it is C-injective as a right C^*-module (see 8.1). Notice that M^* need not be in ${}^C\mathbf{M}$ (not rational). It is straightforward to show that $\mathrm{Rat}^C(M^*)$ is an injective object in ${}^C\mathbf{M}$.

(2)(i) We know that $M \subset R^k$, for some $k \in \mathbb{N}$, and so there is a monomorphism in $\mathbf{M}^C$, $M \xrightarrow{\varrho^M} M \otimes_R C \longrightarrow R^k \otimes_R C \simeq C^k$, that splits in $\mathbf{M}^C$ and hence in ${}_{C^*}\mathbf{M}$ (by 4.1). So the dual sequence $(C^*)^k \to M^* \to 0$ splits in $\mathbf{M}_{C^*}$, and hence M^* is projective in $\mathbf{M}_{C^*}$.

(ii) Let M be injective in $\mathbf{M}^C$. Then M^* is projective in $\mathbf{M}_{C^*}$ (by (i)). Consider any monomorphism in $M \to X$ in ${}_{C^*}\mathbf{M}$. Then $X^* \to M^* \to 0$ is exact and splits in $\mathbf{M}_{C^*}$, and hence, in the diagram

$$\begin{array}{ccccc} 0 & \longrightarrow & M & \longrightarrow & X \\ & & \downarrow \simeq & & \downarrow \\ 0 & \longrightarrow & M^{**} & \longrightarrow & X^{**}, \end{array}$$

the bottom row splits in ${}_{C^*}\mathbf{M}$ and as a consequence so does the upper row, proving that M is injective in ${}_{C^*}\mathbf{M}$.

(iii) Let M be projective in $\mathbf{M}^C$. Since M^* is in ${}^C\mathbf{M}$ (by 3.11), we know from (1) that it is injective in ${}^C\mathbf{M}$. Now we conclude, by the right-hand version of (i), that $M \simeq M^{**}$ is projective in ${}_{C^*}\mathbf{M}$. □

As shown in 9.5, for coalgebras over QF rings, finitely generated projective modules in $\mathbf{M}^C$ are in fact projective in ${}_{C^*}\mathbf{M}$. This is the key to the fact that in this case right semiperfect coalgebras are characterised by the exactness of the left trace functor (so also by all the equivalent properties of the trace functor given in 7.11).

9.6. Right semiperfect coalgebras over QF rings. *Let R be QF and $T = \mathrm{Rat}^C(C^*)$. Then the following are equivalent:*

(a) C is a right semiperfect coalgebra;

(b) $\mathbf{M}^C$ has a generating set of finitely generated modules that are projective in ${}_{C^*}\mathbf{M}$;

(c) injective hulls of simple left *C-comodules are finitely generated as R-modules;*

(d) the functor $\mathrm{Rat}^C : {}_{C^*}\mathbf{M} \to \mathbf{M}^C$ is exact;

(e) T is left C-dense in C^*;

(f) $\mathrm{Ke}\,T = \{x \in C \,|\, T(x) = 0\} = 0$.

Proof. (a) ⇔ (b) If C is right semiperfect, there exists a generating set of finitely generated projective modules in $\mathbf{M}^C$ (see 8.8). By 9.5, all these are projective in ${}_{C^*}\mathbf{M}$. The converse conclusion is immediate.

(a) ⇒ (c) Let U be a simple left C-comodule with injective hull $U \to \widehat{U}$ in ${}^C\mathbf{M}$. Applying $\mathrm{Hom}_R(-, R)$ we obtain a small epimorphism in ${}_{C^*}\mathbf{M}$,

$$\widehat{U}^* \to U^* \to 0,$$

where U^* is a simple left C^*-module. Moreover, since R is QF, we know that $\widehat{U}$ is a direct summand of C_{C^*}, and so $\widehat{U}^*$ is a direct summand of C^*, and hence is projective in ${}_{C^*}\mathbf{M}$. By assumption there exists a projective cover $P \to U^*$ in $\mathbf{M}^C$. Since P is finitely generated as an$R$-module and projective in $\mathbf{M}^C$, it is also projective in ${}_{C^*}\mathbf{M}$ (by 9.5), and hence $\widehat{U}^* \simeq P$. So $\widehat{U}^*$ is finitely generated as an R-module and so is $\widehat{U}$.

(c) $\Rightarrow$ (a) Let $V \subset C$ be a simple left C^*-submodule. Then V^* is a simple right C^*-module in ${}^C\mathbf{M}$. Let $V^* \to K$ be its injective hull in ${}^C\mathbf{M}$. By assumption, K is a finitely generated R-module, and so K^* is a projective C^*-module (by 9.5) and $K^* \to V^{**} \simeq V$ is a projective cover in $\mathbf{M}^C$.

(b) $\Rightarrow$ (d) The assumption implies that $\mathbf{M}^C$ has a generator that is projective in ${}_{C^*}\mathbf{M}$, and the assertion follows from 7.14.

(d) $\Leftrightarrow$ (e) $\Leftrightarrow$ (f) These equivalences follow from 7.11 and 7.9.

(d) $\Rightarrow$ (c) Let $V \subset C$ be a simple left C^*-submodule. Then $U = V^*$ is a simple left C-comodule and there is a projective cover $\widehat{U}^* \to V$ in ${}_{C^*}\mathbf{M}$ (see proof (a) $\Rightarrow$ (c)). By 7.12, (d) implies that $\mathbf{M}^C$ is closed under small epimorphisms and hence $\widehat{U}^* \in \mathbf{M}^C$. □

The conditions on left C^*-modules (right C-comodules) posed in the preceding theorem imply remarkable properties of the left C-comodules.

9.7. Left side of right semiperfect coalgebras. *Let C be right semiperfect, R a QF ring and $T = \mathrm{Rat}^C(C^*)$. Then:*

(1) the injective hull of any $X \in {}^C\mathbf{M}$ is finitely R-generated, provided X is finitely R-generated.

(2) For every $X \in {}^C\mathbf{M}$ that is finitely R-generated, $\mathrm{Hom}_{C^*}(T, X) \simeq X$.

(3) For every $M \in {}^C\mathbf{M}$, the trace of $\mathbf{M}^C$ in M^* is nonzero.

(4) Any module in ${}^C\mathbf{M}$ has a maximal submodule and has a small radical.

Proof. (1) Let $X \in {}^C\mathbf{M}$ be finitely generated as an R-module. Then X has finite uniform dimension, and so its injective hull in ${}^C\mathbf{M}$ is a finite direct sum of injective hulls of simple modules, which are finitely generated by 9.6(c).

(2) By (1), the C-injective hull $\widehat{X}$ of X is finitely R-generated and hence is C^*-injective (see 9.5). So any $f \in \mathrm{Hom}_{C^*}(T, X)$ can be uniquely extended to some $h : C^* \to \widehat{X}$ and $h(\varepsilon) \in \widehat{X}$, which is s-unital over T (see 7.13). Hence

$$h(\varepsilon) \in h(\varepsilon) \cdot C^* = h(\varepsilon) \cdot T = h(T) = f(T) \subset X,$$

showing that $h \in \mathrm{Hom}_{C^*}(C^*, X) \simeq X$.

(3) For every simple submodule $S \subset M$ with injective hull $\widehat{S}$ in ${}^C\mathbf{M}$, there are commutative diagrams

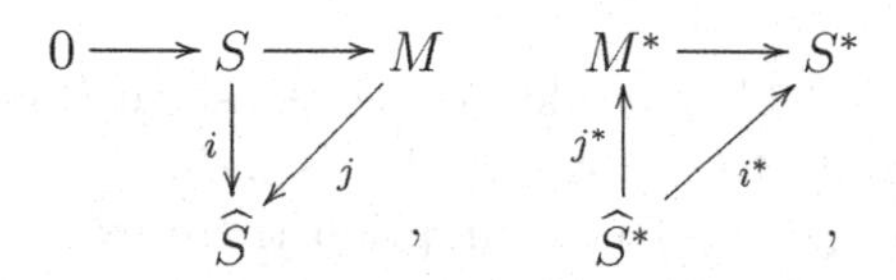

where i is injective and j is nonzero. By 3.11, $\widehat{S}^*$ belongs to $\mathbf{M}^C$ and so does its nonzero image under j^*.

(4) Let $M \in {}^C\mathbf{M}$. By (3), there exists a simple submodule $Q \subset M^*$ with $Q \in \mathbf{M}^C$. Then $\mathrm{Ke}\, Q = \{m \in M \mid Q(m) = 0\}$ is a maximal C^*-submodule of M. This shows that all modules in ${}^C\mathbf{M}$ have maximal submodules, and hence every proper submodule of M is contained in a maximal C^*-submodule. This implies that $\mathrm{Rad}(M)$ is small in M. □

9.8. Finiteness properties. *Let R be a QF ring.*

(1) If C is right semiperfect and there are only finitely many nonisomorphic simple right C-comodules, then ${}_RC$ is finitely generated.

(2) If C is right semiperfect and any two nonzero subalgebras have non-zero intersection (i.e., C is irreducible*), then ${}_RC$ is finitely generated.*

(3) ${}_RC$ is finitely generated if and only if $\mathbf{M}^C$ has a finitely generated projective generator.

(4) C^* is an algebra of finite representation type if and only if there are only finitely many nonisomorphic finitely generated indecomposable modules in $\mathbf{M}^C$.

Proof. (1) Since C_{C^*} is self-injective, the socle of C_{C^*} is a finitely generated R-module by 41.23. Hence $\mathrm{Soc}(C_{C^*})$ has finite uniform dimension, and since $\mathrm{Soc}(C_{C^*}) \trianglelefteq C$, C is a finite direct sum of injective hulls of simple modules in ${}^C\mathbf{M}$ that are finitely generated R-modules by 9.7.

(2) Under the given condition there exists only one simple right C-comodule (up to isomorphisms), and the assertion follows from (1).

(3) If ${}_RC$ is finitely generated, then $\mathbf{M}^C = {}_{C^*}\mathbf{M}$. Conversely, assume there exists a finitely generated projective generator P in $\mathbf{M}^C$. Then P is semiperfect and there are only finitely many simples in $\mathbf{M}^C$. Now (1) applies.

(4) One implication is obvious. Assume there are only finitely many non-isomorphic finitely generated indecomposables in $\mathbf{M}^C$. Since C is subgenerated by its finitely generated submodules, this implies that $\mathbf{M}^C$ has a finitely generated subgenerator. Now [46, 54.2] implies that there is a progenerator in $\mathbf{M}^C$, and hence ${}_RC$ is finitely generated by (3). □

Unlike in the case of associative algebras, right semiperfectness is a strictly one-sided property for coalgebras – it need not imply left semiperfectness (see example in 8.12). The next proposition describes coalgebras that are both right and left semiperfect.

9.9. Left and right semiperfect coalgebras. *Let R be a QF ring, $T = \mathrm{Rat}^C(C^*)$ and $T' = {}^C\mathrm{Rat}(C^*)$. The following are equivalent:*

(a) C is a left and right semiperfect coalgebra;

(b) all left C-comodules and all right C-comodules have projective covers;

(c) $T = T'$ and is dense in C^*;

(d) ${}_{C^*}C$ *and* C_{C^*} *are direct sums of finitely generated* C^**-modules.*

Under these conditions, T *is a ring with enough idempotents, and it is a generator in* $\mathbf{M}^C$.

Proof. (b) $\Rightarrow$ (a) is obvious.

(a) $\Rightarrow$ (b) By 41.16, all finitely generated projective modules in $\mathbf{M}^C$ are semiperfect in $\mathbf{M}^C$. According to 41.14 and 41.15, a direct sum of projective semiperfect modules in $\mathbf{M}^C$ is semiperfect provided it has a small radical. Since this is the case by 9.7, we conclude that every module in $\mathbf{M}^C$ has a projective cover. Similar arguments apply to the category ${}^C\mathbf{M}$.

(a) $\Leftrightarrow$ (c) This is obvious by the characterisation of exactness of the rational functor in 7.11 and 9.6.

(c) $\Leftrightarrow$ (d) follows from 7.15.

The final assertions follow from 7.15 and 7.11. □

For cocommutative coalgebras we can combine 9.9 with 9.2(4).

9.10. Cocommutative semiperfect coalgebras. *Let* R *be QF and* C *cocommutative. The following are equivalent:*

(a) C *is semiperfect;*

(b) C *is a direct sum of finitely generated* C^**-modules;*

(c) C *is a direct sum of finitely* R*-generated subcoalgebras;*

(d) *every uniform subcomodule (*C^**-submodule) of* C *is finitely* R*-generated.*

The trace functors combined with the dual functor $(-)^*$ define covariant functors ${}^C\mathrm{Rat} \circ (-)^* : \mathbf{M}^C \to {}^C\mathbf{M}$ and $\mathrm{Rat}^C \circ (-)^* : {}^C\mathbf{M} \to \mathbf{M}^C$. Over QF rings, these functors clearly are exact if and only if ${}^C\mathrm{Rat}$, respectively Rat^C, is exact, that is, C is left or right semiperfect. In this case they yield dualities between subcategories of $\mathbf{M}^C$ and ${}^C\mathbf{M}$.

9.11. Composition of Rat^C **and** $(-)^*$**.** *Let* R *be a QF ring. For any* $M \in {}^C\mathbf{M}$ *let* $\Phi_M : M \to M^{**}$ *denote the canonical map and* $i : \mathrm{Rat}^C(M^*) \to M^*$ *the inclusion. Consider the composition map*

$$\beta_M := i^* \circ \Phi_M : M \to M^{**} \to \left(\mathrm{Rat}^C(M^*)\right)^*.$$

If C *is right semiperfect, then* β_M *is a monomorphism and*

$$\beta_C : C \to {}^C\mathrm{Rat}\left(\mathrm{Rat}^C(C^*)\right)^*$$

is an isomorphism.

Proof. Let $U \subset M$ be any finitely generated subcomodule. Then $\beta_U : U \to U^{**}$ is an isomorphism, and, by the exactness of Rat^C, $\mathrm{Rat}^C(M^*) \to$

$\mathrm{Rat}^C(U^*)$ is surjective and hence $U^{**} \simeq \left(\mathrm{Rat}^C(U^*)\right)^* \subset \left(\mathrm{Rat}^C(M^*)\right)^*$. This implies that β_M is injective. On the other hand, $C \simeq {}^C\mathrm{Rat}(C^{**})$ by 7.4, and the map ${}^C\mathrm{Rat}(C^{**}) \to {}^C\mathrm{Rat}\left(\mathrm{Rat}^C(C^*)\right)^*$ is surjective, so that β_C is an isomorphism. □

9.12. Duality between left and right C^*-modules. *Let R be QF and $T = \mathrm{Rat}^C(C^*)$. Denote by $\sigma_f[{}_{C^*}C]$ (resp. $\sigma_f[T_T]$) the full subcategory of $\mathbf{M}^C$ (resp. ${}^C\mathbf{M}$) of C^*-modules that are submodules of finitely C-generated (resp. finitely T-generated) modules. Then the following assertions are equivalent:*

(a) *C is left and right semiperfect;*

(b) *the functors ${}^C\mathrm{Rat} \circ (-)^* : \mathbf{M}^C \to {}^C\mathbf{M}$ and $\mathrm{Rat}^C \circ (-)^* : {}^C\mathbf{M} \to \mathbf{M}^C$ are exact;*

(c) *the left and right trace ideals coincide and form a ring with enough idempotents, and ${}^C\mathrm{Rat} \circ (-)^* : \sigma_f[{}_{C^*}C] \to \sigma_f[T_T]$ defines a duality.*

Proof. (a) ⇔ (b) This follows from 9.9.

(b) ⇒ (c) The first assertion was noticed in 9.9, and by this the trace functor Rat^C is described by multiplication with T. It is obvious that the functor $(-)^*T$ transfers $\sigma_f[{}_{C^*}C]$ to $\sigma_f[T_T]$.

We know from 9.11 that $\beta_C : C \to T\left((C^*)T\right)^*$ is an isomorphism. It remains to show that $\beta_M : M \to T\left((M^*)T\right)^*$ is an isomorphism, for every $M \in \sigma_f[{}_{C^*}C]$, that is, M is reflexive with respect to ${}^C\mathrm{Rat} \circ (-)^*$. For this it is obviously enough to prove that submodules and factor modules of reflexive modules are again reflexive. To this end, consider an exact sequence in $\mathbf{M}^C$, $0 \to K \to L \to N \to 0$, in which L is reflexive. Then there is the following commutative diagram with exact rows:

$$\begin{array}{ccccccccc}
0 & \longrightarrow & K & \longrightarrow & L & \longrightarrow & N & \longrightarrow & 0 \\
 & & \downarrow{\scriptstyle \beta_K} & & \downarrow{\scriptstyle \beta_L} & & \downarrow{\scriptstyle \beta_N} & & \\
0 & \longrightarrow & T\left((K^*)T\right)^* & \longrightarrow & T\left((L^*)T\right)^* & \longrightarrow & T\left((N^*)T\right)^* & \longrightarrow & 0 .
\end{array}$$

By assumption, β_L is an isomorphism, and β_K, β_N are monomorphisms by 9.11. From the diagram properties we conclude that they are in fact isomorphisms, so both K and N are reflexive, as required.

(c) ⇒ (a) Clearly the functors transfer finitely generated R-modules to finitely generated R-modules. By our assumptions, the simples in $\sigma_f[T_T]$ have projective covers that are finitely generated as R-modules. This implies that C is a direct sum of finitely generated left C^*-modules and hence is left semiperfect. Finally, since C is reflexive by 9.11, $C = T \rightharpoonup C = \bigoplus_\Lambda e_\lambda \rightharpoonup C$ for idempotents $e_\lambda \in T$, where the $e_\lambda \rightharpoonup C$ are finitely generated by 7.5. □

Over a QF ring, projective comodules in $\mathbf{M}^C$ that are finitely generated as left C^*-modules are also projective in ${}_{C^*}\mathbf{M}$ (see 9.5). Moreover, any direct sum of copies of C is C-injective as a left and right C^*-module.

9.13. Projective coalgebras over QF rings. *If R is QF, the following are equivalent:*

(a) *C is a submodule of a free left C^*-module;*

(b) *C (or every right C-comodule) is cogenerated by C^* as a left C^*-module;*

(c) *there exists a family of left nondegenerate C-balanced bilinear forms $C \times C \to R$;*

(d) *in $\mathbf{M}^C$ every (indecomposable) injective object is projective;*

(e) *C is projective in $\mathbf{M}^C$;*

(f) *C is projective in ${}_{C^*}\mathbf{M}$.*

If these conditions are satisfied, then C is a left semiperfect coalgebra and C is a generator in ${}^C\mathbf{M}$.

Proof. (a) ⇔ (b) By 4.16, C is a direct sum of injective hulls of simple modules in $\mathbf{M}^C$. If C is cogenerated by C^*, then each of these modules is contained in a copy of C^*, and hence C is contained in a free C^*-module. Recall from 9.1 that C is a cogenerator in $\mathbf{M}^C$ and hence C^* cogenerates any $N \in \mathbf{M}^C$ provided it cogenerates C.

(b) ⇔ (c) This is shown in 6.6(2).

(c) ⇒ (f) Let U be a simple left C^*-submodule of C with injective hull $\widehat{U} \subset C$ in $\mathbf{M}^C$. Then $\widehat{U}$ is a finitely generated R-module by 6.6(3). Now we conclude from 9.5 that $\widehat{U}$ is injective in ${}_{C^*}\mathbf{M}$. Being cogenerated by C^*, we observe in fact that $\widehat{U}$ is a direct summand of C^*, and hence it is projective in ${}_{C^*}\mathbf{M}$. This implies that C is projective in ${}_{C^*}\mathbf{M}$.

(f) ⇒ (a) and (f) ⇒ (e) are obvious, and so is (d) ⇔ (e) (by 9.1).

(e) ⇒ (f) C is a direct sum of injective hulls $\widehat{U} \subset C$ of simple submodules $U \subset C$. By (e), $\widehat{U}$ is projective in $\mathbf{M}^C$. Since it has a local endomorphism ring, we know from 41.11 that it is finitely generated as a C^*-module and hence finitely generated as an R-module (by 4.12). Now we conclude from 9.5 that $\widehat{U}$ is projective in ${}_{C^*}\mathbf{M}$ and so is C.

Finally, assume these conditions hold. By the proof of 9.13, the injective hulls of simple modules in $\mathbf{M}^C$ are finitely generated R-modules. By 8.8, this characterises left semiperfect coalgebras, implying that the right trace ideal $T' := {}^C\mathrm{Rat}(C^*)$ is a generator in ${}^C\mathbf{M}$. Now, by 9.5(1), T' is injective in ${}^C\mathbf{M}$, and hence it is generated by C and therefore C is a generator in ${}^C\mathbf{M}$. □

9.14. Corollary. *Let R be QF and C projective in $\mathbf{M}^C$. Then the following are equivalent:*

(a) *${}_{C^*}C$ contains only finitely many nonisomorphic simple submodules;*

(b) *$\mathrm{Soc}({}_{C^*}C)$ is finitely generated as an R-module;*

(c) *C^* is a semiperfect ring;*

(d) ${}_RC$ *is finitely generated.*

Proof. (a) $\Rightarrow$ (b) Since ${}_{C^*}C$ is self-injective, 41.23 applies.

(b) $\Rightarrow$ (c) We know that C^* is f-semiperfect. Clearly $\text{Soc}({}_{C^*}C) \trianglelefteq C$, and hence C has a finite uniform dimension as a left C^*-module. This implies that C^* is semiperfect.

(c) $\Rightarrow$ (a) For any semiperfect ring there are only finitely many simple left (or right) modules (up to isomorphisms).

(c) $\Rightarrow$ (d) is shown in 8.11(3).

(d) $\Rightarrow$ (b) follows from the fact that R is Noetherian. □

From 9.1 we know that, over a QF ring R, C is always an injective cogenerator in $\mathbf{M}^C$. Which additional properties make C a projective generator?

9.15. C as a projective generator in $\mathbf{M}^C$. *Let R be QF and $T = \text{Rat}^C(C^*)$. The following are equivalent:*

(a) C is projective as left and right C-comodule;

(b) C is a projective generator in $\mathbf{M}^C$;

(c) C is a projective generator in ${}^C\mathbf{M}$;

(d) $C = TC$ and T has enough idempotents and is an injective cogenerator in $\mathbf{M}^C$.

Proof. (a) $\Rightarrow$ (b) This is obtained from 9.13 and 8.10.

(b) $\Rightarrow$ (a) By 9.13, C is projective as a left C^*-module and hence C^* is C-injective as a right C^*-module (by 9.5). To show that C is projective as a right C^*-module we show that C^* cogenerates C as a right C^*-module. For this it is enough to prove that each simple submodule $U \subset C_{C^*}$ is embedded in C^*. By 4.12, U is a finitely generated R-module. Clearly U^* is a simple module in $\mathbf{M}^C$, and hence there is a C^*-epimorphism $C \to U^*$. From this we obtain an embedding $U \simeq U^{**} \subset C^*$, which proves our assertion.

(a) $\Leftrightarrow$ (c) is clear by symmetry.

(a) $\Rightarrow$ (d) From the above discussion we know that C is a left and right semiperfect coalgebra. Hence T is a ring with enough idempotents and $\mathbf{M}^C = \sigma[{}_{C^*}T]$ by 9.9. Since C is projective, $C \subset T^{(\Lambda)}$, and hence T is a cogenerator in $\mathbf{M}^C$. T is injective in $\mathbf{M}^C$ by 9.5.

(d) $\Rightarrow$ (b) Since T is projective in $\mathbf{M}^C$, injective hulls of simple modules in $\mathbf{M}^C$ are projective, and so C is projective in $\mathbf{M}^C$. T is injective, and hence it is generated by C. By our assumptions T is a generator in $\mathbf{M}^C$ and so is C. □

In the situation of 9.15, T is an injective cogenerator in $\mathbf{M}^C$ and ${}^C\mathbf{M}$, and hence the functors $\text{Hom}_T(-,T)$ and ${}_T\text{Hom}(-,T)$ have properties similar to $(-)^*$. Following the arguments of the proof of 9.12, we obtain:

9.16. Dualities in case C is a projective generator. *Let R be QF and assume that C is a projective generator in $\mathbf{M}^C$. Let $T = \mathrm{Rat}^C(C^*) = {}^C\mathrm{Rat}(C^*)$, and let $\sigma_f[T_T]$ and $\sigma_f[{}_TT]$ be the categories of submodules of finitely T-generated modules. Then the following pair of functors defines a duality:*

(1) $\mathrm{Rat}^C \circ \mathrm{Hom}_T(-, C) : \sigma_f[T_T] \to \sigma_f[{}_TT]$,

(2) ${}^C\mathrm{Rat} \circ {}_T\mathrm{Hom}(-, T) : \sigma_f[{}_TT] \to \sigma_f[T_T]$.

In case C is finitely R-generated, $\mathbf{M}^C = {}_{C^*}\mathbf{M}$ and we obtain:

9.17. C as a projective generator in ${}_{C^*}\mathbf{M}$. *If R is QF, the following are equivalent:*

(a) C is a projective generator in ${}_{C^*}\mathbf{M}$;

(b) C is a generator in ${}_{C^*}\mathbf{M}$;

(c) C is a generator in $\mathbf{M}^C$ and ${}_RC$ is finitely generated;

(d) C^* is a QF algebra and ${}_RC$ is finitely generated.

Proof. (a) $\Rightarrow$ (b) is obvious.

(b) $\Rightarrow$ (c) As a generator in ${}_{C^*}\mathbf{M}$, C is finitely generated as a module over its endomorphism ring C^*, and hence ${}_RC$ is finitely generated.

(c) $\Rightarrow$ (d) Clearly C^* is left (and right) Artinian. By assumption, C is an injective generator in ${}_{C^*}\mathbf{M}$. This implies that C^* is self-injective and hence QF.

(d) $\Rightarrow$ (a) As a QF ring, C^* is an injective cogenerator in $\mathbf{M}^C = {}_{C^*}\mathbf{M}$. From this it is easy to see that C is a projective generator in ${}_{C^*}\mathbf{M}$. □

An R-coalgebra C is called *coreflexive* if every locally finite left C^*-module is rational.

9.18. Coreflexive coalgebras over QF rings. *Let C be an R-coalgebra with ${}_RC$ projective and R a QF ring. Then the following are equivalent:*

(a) C is coreflexive;

(b) every R-cofinite (left) ideal of C^* is closed in the C-adic topology;

(c) $C^{*\circ}$ is a right (left) C-comodule;

(d) the evaluation map $C \to C^{*\circ}$ is an isomorphism.

Proof. (a) $\Rightarrow$ (b) For a coreflexive coalgebra C, R-cofinite ideals of C^* are open in the C-adic topology, and open ideals are always closed in topological rings.

(b) $\Rightarrow$ (a) This follows by the fact that, for an R-cofinite closed ideal $I \subset C^*$, C^*/I is finitely cogenerated as a C^*-module and hence I is open.

(a) $\Leftrightarrow$ (c) By 5.11, all (C^*, R)-finite C^*-modules are subgenerated by $C^{*\circ}$. Hence they all are C-comodules provided $C^{*\circ}$ is a C-comodule.

(c) $\Leftrightarrow$ (d) This is a consequence of the isomorphism $C \simeq \mathrm{Rat}^C(C^{**})$ shown in 7.4. □

9.19. Remarks. As mentioned in the introduction to this section, coalgebras over QF rings share most of the properties of coalgebras over fields. Let us relate some of our observations to this special case.

Theorem 8.6 (and 9.2) extends decomposition results for coalgebras over fields to coalgebras over Noetherian (QF) rings. It was shown in [24] that any coalgebra C over a field is a direct sum of indecomposable coalgebras, and that, for C cocommutative, these components are even irreducible. In [161, Theorem 2.1] it is proved that C is a direct sum of link-indecomposable components. It is easy to see that the link-indecomposable components are simply the σ-indecomposable components of C. As outlined in [161, Theorem 1.7], this relationship can also be described by using the "wedge". In this context, another proof of the decomposition theorem is given in [190, Theorem]. We refer also to [116] and [117] for a detailed description of these constructions. In [190] an example is given of a σ-indecomposable coalgebra (over a field) with infinitely many simple comodules.

In [124], for every C-comodule M, the coefficient space $C(M)$ is defined as the smallest subcoalgebra $C(M) \subset C$ such that M is a $C(M)$-comodule. The definition heavily relies on the existence of an R-basis for comodules. In the more general correspondence theorem 9.3, $C(M)$ is replaced by $\text{Tr}(\sigma[M], C)$. For coalgebras over fields, $C(M)$ and $\text{Tr}(\sigma[M], C)$ coincide and 9.3 yields [124, 1.3d], [117, Proposition 7], and [161, Lemma 1.8]. Notice that in [124] closed subcategories in $\mathbf{M}^C$ are called *pseudovarieties*. The assertion in 9.3(3) was obtained in [165, Proposition 4.6].

For coalgebras over fields, 9.5(1) is shown in [152, Lemma 11] and (2) is proved in [104, Proposition 4]. The characterisations of semiperfect coalgebras in 9.6 are partly shown in [152, Theorem 10] and [123, Theorem 3.3]. The equivalence of (a) and (b) from 9.9 can be found in [152, Corollary 18].

In [53] semiperfect coalgebras are called *coproper coalgebras.* Their decomposition theorem [53, Theorem 1.11] follows from 8.9.

The exactness of the rational functor is also investigated in [189]. It is proven there, for example, that any irreducible coalgebra C with the exact rational functor is finite-dimensional (see 9.8) and that cocommutative coalgebras are semiperfect if and only if every irreducible subcoalgebra is finite-dimensional (see 9.10).

Projective coalgebras as considered in 9.13 are called *quasi-co-Frobenius* in [123], and some of the characterisations are given in [123, Theorem 1.3]. Moreover, characterisations of C as a projective generator in 9.15 are contained in [123, Theorem 2.6]. In this paper dualities similar to 9.12 and 9.16 also are considered (see [123, Theorems 3.5, 3.12]).

9.20. Exercises

(1) Let F be a free module over a QF ring R. Prove that an R-submodule $U \subset F^*$ is dense in the finite topology if and only if $\text{Ke}\, U = 0$. (Hint: 7.9)

(2) Let R be QF and ${}_RC$ projective. Denote by ${}_f\mathbf{M}^C$ (resp. ${}^C\mathbf{M}_f$) the full subcategory of $\mathbf{M}^C$ (resp. ${}^C\mathbf{M}$) consisting of finitely generated C^*-modules. Prove that $\mathrm{Hom}_R(-,R) : {}_f\mathbf{M}^C \to {}^C\mathbf{M}_f$ induces a duality of categories.

(3) A comodule $L \in {}^C\mathbf{M}$ is called *s-rational* if the injective envelope of L in ${}^C\mathbf{M}$ embeds in M^* as a right C^*-module, for some $M \in {}^C\mathbf{M}$. Let R be a QF ring and ${}_RC$ projective. Prove ([157]):

   (i) For a simple object $S \in \mathbf{M}^C$, the following are equivalent:

      (a) S is a quotient of a projective object of $\mathbf{M}^C$;

      (b) S is a quotient of an object of $\mathbf{M}^C$ that is a cyclic projective left C^*-module (and is finitely R-generated);

      (c) S^* is s-rational;

      (d) the injective envelope of S^* in ${}^C\mathbf{M}$ is finitely generated as an R-module.

   (ii) C is a right semiperfect coalgebra if and only if every simple object of ${}^C\mathbf{M}$ is s-rational.

(4) Let R be QF and assume C to be a projective generator in $\mathbf{M}^C$. Put $T = \mathrm{Rat}^C(C^*)$ and denote by T-mod and mod-T the categories of finitely generated left and right T-modules M, N with $TM = M$ or $N = NT$, respectively. Prove that ${}_T\mathrm{Hom}(-,T) : T\text{-mod}\to\text{mod-}T$ induces a duality.

(5) An R-algebra A is called *left almost Noetherian* if any R-cofinite left ideal is finitely generated as an A-module.
Let C be an R-coalgebra with ${}_RC$ projective and R a QF ring. Assume C^* to be left almost Noetherian. Prove that C is coreflexive and $\mathbf{M}^C$ is closed under extensions in ${}_{C^*}\mathbf{M}$.

(6) A comodule $M \in \mathbf{M}^C$ is called *cohereditary* if every factor comodule of M is injective. Let R be QF and assume ${}_RC$ to be projective. Prove that the following are equivalent:

   (a) C is cohereditary;

   (b) every injective comodule in $\mathbf{M}^C$ is cohereditary;

   (c) every indecomposable injective comodule in $\mathbf{M}^C$ is cohereditary;

   (d) every C-generated comodule in $\mathbf{M}^C$ is injective;

   (e) for every (simple) comodule in $\mathbf{M}^C$ the injective dimension is ≤ 1.

References. Allen and Trushin [53]; Al-Takhman [51]; García, Jara and Merino [116, 117]; Green [124]; Gómez-Torrecillas and Năstăsescu [123]; Heyneman and Radford [129]; Kaplansky [24]; Lin [152]; Menini, Torrecillas and Wisbauer [157]; Montgomery [161]; Năstăsescu and Torrecillas [165]; Năstăsescu, Torrecillas and Zhang [166]; Radford [176]; Shudo [189]; Shudo and Miyamoto [190]; Taft [194]; Wisbauer [210].

10 Cotensor product of comodules

In this section we dualise the definition and characterisations of tensor products of modules over algebras (see 40.10) to obtain the corresponding notion for comodules over coalgebras. In classical ring theory, a tensor product of modules is defined as a coequaliser and can be seen as a (bi)functor from the category(-ies) of modules to Abelian groups. Properties of modules such as purity, flatness or faithful flatness can be defined via the properties of the corresponding tensor functor. Dually, a cotensor product of comodules is defined as an equaliser and can be viewed as a (bi)functor from comodules to R-modules. The properties of this functor then determine whether a comodule is coflat, faithfully coflat or whether certain purity conditions are satisfied. In this section we study all such properties, as well as relations of the cotensor functor to the tensor functor. In these studies the fact that we consider coalgebras over a ring rather than coalgebras over a field plays a significant role.

As before, C denotes an R-coalgebra.

10.1. Cotensor product of comodules. For $M \in \mathbf{M}^C$ and $N \in {}^C\mathbf{M}$, the *cotensor product* $M \Box_C N$ is defined as the following equaliser in $\mathbf{M}_R$:

$$M \Box_C N \longrightarrow M \otimes_R N \underset{I_M \otimes {}^N\varrho}{\overset{\varrho^M \otimes I_N}{\rightrightarrows}} M \otimes_R C \otimes_R N,$$

or – equivalently – by the exact sequence in $\mathbf{M}_R$,

$$0 \longrightarrow M \Box_C N \longrightarrow M \otimes_R N \xrightarrow{\omega_{M,N}} M \otimes_R C \otimes_R N,$$

where $\omega_{M,N} := \varrho^M \otimes I_N - I_M \otimes {}^N\varrho$. It can also be characterised by the pullback diagram

$$\begin{array}{ccc} M \Box_C N & \longrightarrow & M \otimes_R N \\ \downarrow & & \downarrow {\scriptstyle \varrho^M \otimes I_N} \\ M \otimes_R N & \xrightarrow{I_M \otimes {}^N\varrho} & M \otimes_R C \otimes_R N . \end{array}$$

In particular, for the comodule C, by properties of the structure maps there are R-module isomorphisms

$$M \Box_C C = \varrho^M(M) \simeq M, \quad C \Box_C N = {}^N\varrho(N) \simeq N.$$

Proof. First observe that $\omega_{M,C} \circ \varrho^M = 0$ and hence $\varrho^M(M) \subset M \Box_C C$. Conversely, take any $\sum_i m^i \otimes c^i \in M \Box_C C$ and put $m = \sum_i m^i \varepsilon(c^i)$. Then

$$\varrho^M(m) = \sum_i m^i{}_{\underline{0}} \otimes m^i{}_{\underline{1}} \varepsilon(c^i) = \sum_i m^i \otimes c^i{}_{\underline{1}} \varepsilon(c^i{}_{\underline{2}}) = \sum_i m^i \otimes c^i,$$

showing that $M \Box_C C = \varrho^M(M)$. $\square$

10.2. Cotensor product of comodule morphisms. *Let $f : M \to M'$ and $g : N \to N'$ be morphisms of right, resp. left, C-comodules. Then there exists a unique R-linear map,*

$$f \Box g : \; M \Box_C N \longrightarrow M' \Box_C N',$$

yielding a commutative diagram

$$\begin{array}{ccccccc} 0 \longrightarrow & M \Box_C N & \longrightarrow & M \otimes_R N & \xrightarrow{\omega_{M,N}} & M \otimes_R C \otimes_R N \\ & \downarrow{\scriptstyle f \Box g} & & \downarrow{\scriptstyle f \otimes g} & & \downarrow{\scriptstyle f \otimes I_C \otimes g} \\ 0 \longrightarrow & M' \Box_C N' & \longrightarrow & M' \otimes_R N' & \xrightarrow{\omega_{M',N'}} & M' \otimes_R C \otimes_R N' . \end{array}$$

Proof. Since f and g are comodule morphisms,

$$\begin{aligned} (f \otimes I_C \otimes g) \circ \omega_{M,N} &= (f \otimes I_C) \circ \varrho^M \otimes g - f \otimes (I_C \otimes g) \circ \varrho^N \\ &= (\varrho^{M'} \circ f) \otimes g - f \otimes (\varrho^{N'} \circ g) \\ &= \omega_{M',N'} \circ (f \otimes g). \end{aligned}$$

This means that the right square is commutative and the kernel property yields the map stated. □

10.3. The cotensor functor. *For any $M \in \mathbf{M}^C$ there is a covariant functor*

$$\begin{array}{llll} M \Box_C - : {}^C\mathbf{M} \to \mathbf{M}_R, & N & \mapsto & M \Box_C N, \\ & f : N \to N' & \mapsto & I_M \Box f : M \Box_C N \to M \Box_C N'. \end{array}$$

Similarly, every left C-comodule N yields a functor $-\Box_C N : \mathbf{M}^C \to \mathbf{M}_R$.

To understand the exactness properties of the cotensor product as well as its relationship with the tensor product, a good understanding of the *purity of morphisms* is required. The reader is referred to 40.13 for these notions.

10.4. Exactness of the cotensor functor. *Let $M \in \mathbf{M}^C$ and ${}_RC$ be flat.*

(1) Consider an exact sequence $0 \longrightarrow N' \xrightarrow{f} N \xrightarrow{g} N''$ in ${}^C\mathbf{M}$. Assume

(i) M is flat as an R-module, or

(ii) the sequence is M-pure, or

(iii) the sequence is (C, R)-exact.

Then, cotensoring with M yields an exact sequence of R-modules,

$$0 \to M \Box_C N' \xrightarrow{I_M \Box f} M \Box_C N \xrightarrow{I_M \Box g} M \Box_C N''.$$

(2) The cotensor functor $M \Box_C - : {}^C\mathbf{M} \to \mathbf{M}_R$ is left (C, R)-exact, and it is left exact provided that M is flat as an R-module.

Proof. There is a commutative diagram with exact vertical columns:

$$\begin{array}{ccccccc}
 & & 0 & & 0 & & 0 \\
 & & \downarrow & & \downarrow & & \downarrow \\
0 & \longrightarrow & M\square_C N' & \xrightarrow{I_M \square f} & M\square_C N & \xrightarrow{I_M \square g} & M\square_C N'' \\
 & & \downarrow & & \downarrow & & \downarrow \\
0 & \longrightarrow & M \otimes_R N' & \xrightarrow{I_M \otimes f} & M \otimes_R N & \xrightarrow{I_M \otimes g} & M \otimes_R N'' \\
 & & \downarrow \omega_{M,N'} & & \downarrow \omega_{M,N} & & \downarrow \omega_{M,N''} \\
0 & \longrightarrow & M \otimes_R C \otimes_R N' & \xrightarrow{I \otimes I \otimes f} & M \otimes_R C \otimes_R N & \xrightarrow{I \otimes I \otimes g} & M \otimes_R C \otimes_R N''.
\end{array}$$

Under any of the given conditions, the second and third rows are exact, and hence the first row is exact by the Kernel-Cokernel Lemma. □

10.5. Direct limits and cotensor products. *Let $\{N_\lambda\}_\Lambda$ be a directed family in ${}^C\mathbf{M}$. Then, for any $M \in \mathbf{M}^C$,*

$$\varinjlim (M\square_C N_\lambda) \simeq M\square_C \varinjlim N_\lambda.$$

Proof. Since $M \otimes_R -$ commutes with direct limits, there is a commutative diagram with exact rows:

$$\begin{array}{ccccccc}
0 & \longrightarrow & \varinjlim (M\square_C N_\lambda) & \longrightarrow & \varinjlim (M \otimes_R N_\lambda) & \longrightarrow & \varinjlim (M \otimes_R C \otimes_R N_\lambda) \\
 & & \downarrow & & \downarrow \simeq & & \downarrow \simeq \\
0 & \longrightarrow & M\square_C \varinjlim N_\lambda & \longrightarrow & M \otimes_R \varinjlim N_\lambda & \longrightarrow & M \otimes_R C \otimes_R \varinjlim N_\lambda \, .
\end{array}$$

This yields the isomorphism stated. □

Associativity properties between tensor and cotensor products are of fundamental importance, and for this we show:

10.6. Tensor-cotensor relations. *Let $M \in \mathbf{M}^C$ and $N \in {}^C\mathbf{M}$. For any $W \in \mathbf{M}_R$ and the canonical C-comodule structures on $W \otimes_R M$ and $N \otimes_R W$ (cf. 3.8), there exist canonical R-linear maps*

$$\begin{aligned}
\tau_W : &\quad W \otimes_R (M\square_C N) \to (W \otimes_R M)\square_C N, \\
\tau'_W : &\quad (M\square_C N) \otimes_R W \to M\square_C (N \otimes_R W),
\end{aligned}$$

and the following are equivalent:

(a) *the map $\omega_{M,N} : M \otimes_R N \to M \otimes_R C \otimes_R N$ is W-pure;*

(b) *τ_W is an isomorphism;*

(c) *τ'_W is an isomorphism.*

Proof. With obvious maps there is a commutative diagram:

$$\begin{array}{ccccccc}
0 \longrightarrow & W\otimes_R(M\Box_C N) & \longrightarrow & W\otimes_R(M\otimes_R N) & \xrightarrow{I\otimes\omega_{M,N}} & W\otimes_R(M\otimes_R C\otimes_R N) \\
& \downarrow \tau_W & & \downarrow \simeq & & \downarrow \simeq \\
0 \longrightarrow & (W\otimes_R M)\Box_C N & \longrightarrow & (W\otimes_R M)\otimes_R N & \xrightarrow{\omega_{W\otimes M,N}} & (W\otimes_R M)\otimes_R C\otimes_R N\,,
\end{array}$$

where the bottom row is exact (by definition). If $\omega_{M,N}$ is a W-pure morphism, then the top row is exact and τ_W is an isomorphism by the diagram properties. On the other hand, if τ_W is an isomorphism, we conclude that the top row is exact and $\omega_{M,N}$ is a W-pure morphism. Similar arguments apply to τ'_W. $\square$

We consider two cases where the conditions in 10.6(1) are satisfied.

10.7. Purity conditions. *Let $M \in \mathbf{M}^C$ and $N \in {}^C\mathbf{M}$.*

(1) If the functor $M\Box_C -$ or $-{}_C\Box N$ is right exact, then $\omega_{M,N}$ is a pure morphism.

(2) If $M \in \mathbf{M}^C$ is (C,R)-injective and $N \in {}^C\mathbf{M}$, then the exact sequence

$$0 \longrightarrow M\Box_C N \longrightarrow M \otimes_R N \xrightarrow{\omega_{M,N}} M \otimes_R C \otimes_R N$$

splits in $\mathbf{M}_R$ (and hence is pure).

Proof. (1) Let $M\Box_C - : {}^C\mathbf{M} \to \mathbf{M}_R$ be right exact and $W \in \mathbf{M}_R$. From an exact sequence $F_2 \to F_1 \to W \to 0$ with free R-modules F_1, F_2, we obtain a commutative diagram:

$$\begin{array}{ccccccc}
(M\Box_C N)\otimes_R F_2 & \longrightarrow & (M\Box_C N)\otimes_R F_1 & \longrightarrow & (M\Box_C N)\otimes_R W & \longrightarrow & 0 \\
\downarrow \simeq & & \downarrow \simeq & & \downarrow \tau'_W & & \\
M\Box_C(N\otimes_R F_2) & \longrightarrow & M\Box_C(N\otimes_R F_1) & \longrightarrow & M\Box_C(N\otimes_R W) & \longrightarrow & 0\,,
\end{array}$$

where both sequences are exact. The first two vertical maps are isomorphisms since tensor and cotensor functors commute with direct sums. This implies that τ'_W is an isomorphism and the assertion follows by 10.6.

A similar proof applies when $-{}_C\Box N$ is exact.

(2) If N is (C,R)-injective, the structure map ${}^N\varrho : N \to C\otimes_R N$ is split by some comodule morphism $\lambda : C\otimes_R N \to N$. The image of the map

$$\beta = (I_M \otimes \lambda)\circ(\varrho^M \otimes I_N) : M\otimes_R N \longrightarrow M\otimes_R N$$

lies in $M\Box_C N$, since, for any $m\otimes n \in M\otimes_R N$,

$$\begin{aligned}
(\varrho^M \otimes I_N)\circ\beta(m\otimes n) &= \sum m_{\underline{0}} \otimes m_{\underline{1}} \otimes \lambda(m_{\underline{2}} \otimes n) \\
&= \sum m_{\underline{0}} \otimes (I_M \otimes \lambda)(\Delta \otimes I_N)(m_{\underline{1}} \otimes n) \\
&= \sum m_{\underline{0}} \otimes \varrho^N(\lambda(m_{\underline{1}} \otimes n)) \\
&= (I_M \otimes \varrho^N)\circ\beta(m\otimes n).
\end{aligned}$$

Furthermore, for any $x \in M\Box_C N$,

$$\beta(x) = (I_M \otimes \lambda) \circ (\varrho^M \otimes I_N)(x) = (I_M \otimes \lambda) \circ (I_M \otimes {}^N\!\varrho)(x) = x,$$

thus proving that $M\Box_C N$ is an R-direct summand of $M \otimes_R N$.

To show splitting in $M \otimes_R N$ consider the map

$$-(I_M \otimes \lambda) : M \otimes_R C \otimes_R N \longrightarrow M \otimes_R N .$$

For any element $m \otimes n \in M \otimes_R N$ we compute

$$\begin{aligned} -(I_M \otimes \lambda) \circ \omega_{M,N}(m \otimes n) &= -(I_M \otimes \lambda) \circ (\varrho^M \otimes I_N - I_M \otimes {}^N\!\varrho)(m \otimes n) \\ &= -(I_M \otimes \lambda) \circ (\varrho^M \otimes I_N)(m \otimes n) + m \otimes n \\ &= m \otimes n - \beta(m \otimes n) \\ &\in m \otimes n + M\Box_C N , \end{aligned}$$

which shows that $M \otimes_R N/M\Box_C N$ is isomorphic to a direct summand of $M \otimes_R C \otimes_R N$. A similar proof works if M is (C, R)-injective. □

10.8. Coflat comodules. Let ${}_RC$ be flat. A comodule $M \in \mathbf{M}^C$ is said to be *coflat* if the functor $M\Box_C - : {}^C\mathbf{M} \to \mathbf{M}_R$ is exact.

Notice that, for any coflat $M \in \mathbf{M}^C$, M is flat as an R-module. Indeed, since C is (C, R)-injective, 10.7 implies $M\Box_C(C \otimes_R -) \simeq (M\Box_C C) \otimes_R - \simeq M \otimes_R -$, from which the assertion is clear (${}_RC$ is assumed to be flat).

As a consequence of 10.5 we observe that direct sums and direct limits of coflat C-comodules are coflat.

A right C-comodule M is said to be *faithfully coflat* provided the functor $M\Box_C - : {}^C\mathbf{M} \to \mathbf{M}_R$ is exact and faithful. Recall that faithfulness means that the canonical map ${}^C\mathrm{Hom}(L, N) \to \mathrm{Hom}_R(M\Box_C L, M\Box_C N)$ is injective, for any $L, N \in {}^C\mathbf{M}$. Similar to the characterisation of faithfully flat modules (e.g., [46, 12.17]) we obtain:

10.9. Faithfully coflat comodules. *Let ${}_RC$ be flat. For $M \in \mathbf{M}^C$ the following are equivalent:*

(a) M is faithfully coflat;

(b) $M\Box_C - : {}^C\mathbf{M} \to \mathbf{M}_R$ is exact and reflects exact sequences (zero morphisms);

(c) M is coflat and $M\Box_C N \neq 0$, for any nonzero $N \in {}^C\mathbf{M}$.

Proof. (a) ⇒ (c) Assume $M\Box_C K = 0$ for any $K \in {}^C\mathbf{M}$. Then $I_M\Box I_K : M\Box_C K \to M\Box_C K$ is the zero map, and hence I_K is the zero map, that is, $K = 0$.

(c) $\Rightarrow$ (a) Assume for some $L \to N \in {}^C\mathbf{M}$ that the map $M\Box_C L \to M\Box_C N$ is zero. Then $M\Box_C L \to M\Box_C \mathrm{Im}\, f$ is surjective and zero, and so $\mathrm{Im}\, f = 0$ and f is zero.

(a) $\Leftrightarrow$ (b) This follows by standard arguments in category theory (e.g., [46, 12.17]). □

For coalgebras C with ${}_RC$ locally projective, the comodule categories can be considered as categories of C^*-modules (cf. Section 4). In this case the cotensor product can be described by C^*-modules.

10.10. Cotensor and Hom. *Let ${}_RC$ be locally projective and, for $M \in \mathbf{M}^C$ and $N \in {}^C\mathbf{M}$, consider $M \otimes_R N$ as a (C^*, C^*)-bimodule. Then*

$$M\Box_C N \simeq {}_{C^*}\mathrm{Hom}_{C^*}(C^*, M \otimes_R N).$$

Proof. Consider the following commutative diagram:

$$\begin{array}{ccccc}
M\Box_C N & \longrightarrow & M \otimes_R N & \xrightarrow{\simeq} & {}_{C^*}\mathrm{Hom}(C^*, M \otimes_R N) \\
\downarrow & & \downarrow {\scriptstyle \varrho^M \otimes I_N} & & \\
M \otimes_R N & \xrightarrow{I_M \otimes {}^N\varrho} & M \otimes_R C \otimes_R N & & \downarrow {\scriptstyle i_L} \\
\simeq \downarrow & & & \searrow {\scriptstyle \alpha_{M \otimes_R N}} & \\
\mathrm{Hom}_{C^*}(C^*, M \otimes_R N) & & \xrightarrow{\quad i_R \quad} & & \mathrm{Hom}_R(C^*, M \otimes_R N),
\end{array}$$

in which the internal rectangle (pullback) defines $M\Box_C N$, i_L and i_R denote the inclusion maps, and $\alpha_{M \otimes_R N}$ is defined as in 4.2. The pullback of i_L and i_R is simply the intersection

$${}_{C^*}\mathrm{Hom}(C^*, M \otimes_R N) \cap \mathrm{Hom}_{C^*}(C^*, M \otimes_R N) = {}_{C^*}\mathrm{Hom}_{C^*}(C^*, M \otimes_R N),$$

and we obtain a morphism $M\Box_C N \to {}_{C^*}\mathrm{Hom}_{C^*}(C^*, M \otimes_R N)$. From the fact that $\alpha_{M \otimes_R N}$ is injective and the pullback property of the inner rectangle, we obtain a morphism in the other direction and their composition yields the identity. □

Recall from 3.11 that, for ${}_RC$ flat and a right C-comodule N that is finitely presented as an R-module, $N^* := \mathrm{Hom}_R(N, R)$ is a left C-comodule. This yields the following

10.11. Hom-cotensor relation. *Let ${}_RC$ be flat and $M, L \in \mathbf{M}^C$ such that ${}_RM$ is flat and ${}_RL$ is finitely presented. Then there exists an isomorphism (natural in L)*

$$M\Box_C L^* \xrightarrow{\simeq} \mathrm{Hom}^C(L, M).$$

Proof. Consider the following commutative diagram with exact rows:

$$\begin{array}{ccccccc} 0 \longrightarrow & M\Box_C L^* & \longrightarrow & M\otimes_R L^* & \xrightarrow{\omega_{M,L^*}} & M\otimes_R C\otimes_R L^* \\ & \downarrow & & \downarrow\simeq & & \downarrow\simeq \\ 0 \longrightarrow & \mathrm{Hom}^C(L,M) & \longrightarrow & \mathrm{Hom}_R(L,M) & \xrightarrow{\gamma} & \mathrm{Hom}_R(L,M\otimes_R C), \end{array}$$

where $\omega_{M,L^*} = \varrho^M \otimes I_{L^*} - I_M \otimes \varrho_{L^*}$ and $\gamma(f) := \varrho^M \circ f - (f\otimes I_C)\circ \varrho^L$ (see 3.3). For $m\otimes f \in M\otimes_R L^*$ and $l\in L$,

$$\begin{array}{ccc} m\otimes f & \longmapsto & \varrho^M(m)\otimes f - m\otimes (f\otimes I_C)\circ\varrho^L \\ \downarrow & & \downarrow \\ [l\mapsto f(l)m] & \longmapsto & [l \;\mapsto\; f(l)\varrho^M(m) - \sum f(l_{\underline{0}})m\otimes l_{\underline{1}}], \end{array}$$

showing that the square is commutative. Now the diagram properties yield the isomorphism desired. □

10.12. Coflatness and injectivity. *Let $M\in \mathbf{M}^C$, and assume ${}_RM$ and ${}_RC$ to be flat.*

(1) Let $0\to L_1\to L_2\to L_3\to 0$ be an exact sequence in $\mathbf{M}^C$, where each L_i is finitely R-presented. If R is injective or the sequence is (C,R)-exact, then there is a commutative diagram:

$$\begin{array}{ccccccccc} 0 \longrightarrow & M\Box_C L_3^* & \longrightarrow & M\Box_C L_2^* & \longrightarrow & M\Box_C L_1^* & \longrightarrow 0 \\ & \downarrow\simeq & & \downarrow\simeq & & \downarrow\simeq & \\ 0 \longrightarrow & \mathrm{Hom}^C(L_3,M) & \longrightarrow & \mathrm{Hom}^C(L_2,M) & \longrightarrow & \mathrm{Hom}^C(L_1,M) & \longrightarrow 0. \end{array}$$

So the upper sequence is exact if and only if the lower sequence is exact.

(2) Let R be QF. Then

(i) M is coflat if and only if M is C-injective;

(ii) M is faithfully coflat if and only if M is an injective cogenerator in $\mathbf{M}^C$.

Proof. (1) Since the L_i are finitely R-presented, the isomorphisms are provided by 10.11.

(2) Let R be QF. Then ${}_RC$ is projective.

(i) Assume $M\Box_C -$ to be exact. Then, by (1), $\mathrm{Hom}^C(-,M)$ is exact with respect to all exact sequences $0\to I\to J\to J/I\to 0$ in $\mathbf{M}^C$, where $J\subset C$ is a finitely generated left C^*-submodule of C. This implies that M is C-injective as a C^*-module, and hence it is injective in $\mathbf{M}^C$ (see 8.1).

Now assume M to be C-injective. With R being a QF-ring, for any finitely generated left C-comodule K, $K \simeq K^{**}$ and $M\Box_C K \simeq \mathrm{Hom}^C(K^*, M)$ hold. Any exact sequence $0 \to K' \to K \to K'' \to 0$ in ${}^C\mathbf{M}$ is a direct limit of exact sequences $0 \to K'_\lambda \to K_\lambda \to K''_\lambda \to 0$ in ${}^C\mathbf{M}$, where all K'_λ, K_λ and K''_λ are finitely presented as R-modules. By C-injectivity of M and (1),

$$0 \to M\Box_C K'_\lambda \to M\Box_C K_\lambda \to M\Box_C K''_\lambda \to 0$$

is exact, for each λ. Now the direct limit yields an exact sequence in $\mathbf{M}_R$,

$$0 \to M\Box_C K' \to M\Box_C K \to M\Box_C K'' \to 0,$$

showing that M is coflat.

(ii) Let E be any simple right C-comodule. Clearly E is a finitely presented R-module and hence $M\Box_C E^* \simeq \mathrm{Hom}^C(E, M)$. The left side is nonzero provided M is faithfully coflat, and the right side is nonzero provided M is a cogenerator in $\mathbf{M}^C$. From this the assertion follows. □

Notice that, over arbitrary rings R, coflatness is not equivalent to injectivity: If ${}_RC$ is flat, then C is a coflat but need not be an injective C-comodule (unless R is injective).

References. Al-Takhman [51]; Doi [104]; Lin [152]; Milnor and Moore [159]; Sweedler [45].

11 Bicomodules

Let A and B be associative R-algebras. An R-module M that is a left A- and right B-module is called an (A,B)-*bimodule* provided the compatibility condition $(am)b = a(mb)$ holds, for all $a \in A$, $m \in M$ and $b \in B$. Similar (dual) notions can be considered for coalgebras. In this case we are interested in R-modules that are at the same time left comodules of one coalgebra and right comodules of a second coalgebra. In classical ring theory, given an (A,B)-bimodule M, one uses the tensor product to define an induction functor from the category of right A-modules to the category of right B-modules, $-\otimes_A M : \mathbf{M}_A \to \mathbf{M}_B$. To perform a similar construction for coalgebras, one uses the cotensor product. However, the corresponding functor from comodules of one coalgebra to comodules of the other is not always well defined. This is again a feature that is typical for coalgebras over rings, and in this section we study when such a *coinduction functor* is defined and when the cotensor product is associative.

11.1. Bicomodules (cf. 3.24). Let C, D be R-coalgebras, and M a right C-comodule and left D-comodule by $\varrho^M : M \to M\otimes_R C$ and ${}^M\varrho : M \to D\otimes_R M$, respectively. M is called a (D,C)-*bicomodule* if the diagram

$$\begin{array}{ccc} M & \xrightarrow{\varrho^M} & M\otimes_R C \\ {\scriptstyle {}^M\varrho}\downarrow & & \downarrow{\scriptstyle {}^M\varrho\otimes I_C} \\ D\otimes_R M & \xrightarrow{I_D\otimes\varrho^M} & D\otimes_R M\otimes_R C \end{array}$$

is commutative, that is, if ϱ^M is a left D-comodule morphism or – equivalently – ${}^M\varrho$ is a right C-comodule morphism.

Right C-comodules are left C^*-modules, and left D-comodules are right D^*-modules canonically. Hence, any (D,C)-bicomodule is a left C^*-module and a right D^*-module. The compatibility condition for bicomodules implies that every (D,C)-bicomodule is in fact a (C^*,D^*)-bimodule. This is proven by the following explicit computation for $f \in C^*$, $g \in D^*$, $m \in M$:

$$\begin{aligned} (f\rightharpoonup m)\leftharpoonup g &= (g\otimes I_M\otimes f)\circ(({}^M\varrho\otimes I_C)\circ\varrho^M))(m) \\ &= (g\otimes I_M\otimes f)\circ((I_D\otimes\varrho^M)\circ {}^M\varrho))(m) \\ &= f\rightharpoonup(m\leftharpoonup g). \end{aligned}$$

In particular, any coalgebra C is a (C,C)-bicomodule by the regular coactions (cf. 3.4) and hence a (C^*,C^*)-bimodule (see 4.6).

As outlined in 4.11, idempotents e in C^* induce for any right C-comodule M a right $e\rightharpoonup C\leftharpoonup e$ comodule $e\rightharpoonup M$. Now we look at the effect of this construction on bicomodules.

11.2. Idempotents and bicomodules. *For R-coalgebras C, D, let M be a (D, C)-bicomodule and $g^2 = g \in D^*$, $e^2 = e \in C^*$.*

(1) $e \rightharpoonup M$ is a $(D, e \rightharpoonup C \leftharpoonup e)$-bicomodule by the coactions

$$\begin{array}{lll} e \rightharpoonup M & \longrightarrow e \rightharpoonup M \otimes_R e \rightharpoonup C \leftharpoonup e, & e \rightharpoonup m \mapsto \sum e \rightharpoonup m_{\underline{0}} \otimes e \rightharpoonup m_{\underline{1}} \leftharpoonup e \\ e \rightharpoonup M & \longrightarrow D \otimes_R e \rightharpoonup M, & e \rightharpoonup m \mapsto \sum m_{\underline{-1}} \otimes e \rightharpoonup m_{\underline{0}}. \end{array}$$

(2) $M \leftharpoonup g$ is a $(g \rightharpoonup D \leftharpoonup g, C)$-bicomodule by the coactions

$$\begin{array}{lll} M \leftharpoonup g & \longrightarrow M \leftharpoonup g \otimes_R C, & m \leftharpoonup g \mapsto \sum m_{\underline{0}} \leftharpoonup g \otimes m_{\underline{1}} \\ M \leftharpoonup g & \longrightarrow g \rightharpoonup D \leftharpoonup g \otimes_R M \leftharpoonup g, & m \leftharpoonup g \mapsto \sum g \rightharpoonup m_{\underline{-1}} \leftharpoonup g \otimes m_{\underline{0}} \leftharpoonup g. \end{array}$$

(3) $(e \rightharpoonup M)^* \simeq M^* \cdot e$ as $(D^*, e * C^* * e)$-bimodules and $(M \leftharpoonup g)^* \simeq g \cdot M^*$ as $(g * D^* * g, C^*)$-bimodules.

Proof. The assertions in (1) and (2) are easy to verify.

(3) The first isomorphism is already given in 4.11(3). The check that it is a morphism in the given sense is left as an exercise. The second isomorphism is obtained by symmetric arguments. □

In general, for $M \in \mathbf{M}^C$ and $N \in {}^C\mathbf{M}$, the cotensor product $M \Box_C N$ is just an R-module. For bicomodules M, N, $M \Box_C N$ has a (bi)comodule structure under additional assumptions.

For a (D, C)-bicomodule M and any left C-comodule N with coaction ${}^N\varrho : N \to C \otimes_R N$, we consider $M \otimes_R N$ as a left D-comodule canonically, as in 3.8. Then the map

$$\omega_{M,N} : \varrho^M \otimes I_N - I_M \otimes {}^N\varrho : M \otimes_R N \to M \otimes_R C \otimes_R N$$

is a left D-comodule morphism. Hence its kernel $M \Box_C N$ is a D-subcomodule of $M \otimes_R N$, provided $\omega_{M,N}$ is a D-pure morphism (see 40.13 and 40.14).

11.3. Cotensor product of bicomodules. *Let B, C, D be R-coalgebras, M a (D, C)-bicomodule, $L \in \mathbf{M}^D$, and $N \in {}^C\mathbf{M}$.*

(1) $M \Box_C N$ is a left D-comodule, provided $\omega_{M,N}$ is D-pure.

(2) $L \Box_D M$ is a right C-comodule, provided $\omega_{L,M}$ is C-pure.

(3) If N is a (C, B)-bicomodule, then $M \Box_C N$ is a (D, B)-bicomodule, provided $\omega_{M,N}$ is D-pure and B-pure.

Notice that the above conditions are in particular satisfied when B, C and D are flat as R-modules. Furthermore, since C is always (C, R)-injective, the purity conditions are satisfied for the bicomodule C (see 10.7), and hence there is the following corollary of 11.3.

11.4. Cotensor product with C. *If $M \in \mathbf{M}^C$ and $N \in {}^C\mathbf{M}$, then $M \simeq M\Box_C C$ and $N \simeq C\Box_C N$ always as comodules.*

Coseparability of a coalgebra (cf. 3.28) is a sufficient condition for the cotensor product to be a functor between comodule categories. Indeed, as observed in 3.29, over a coseparable coalgebra any comodule is relative injective and hence the purity conditions 10.7 are satisfied and 11.3 implies:

11.5. Cotensor product over coseparable coalgebras. *Let B, C and D be R-coalgebras. If C is a coseparable coalgebra, then:*

(1) For any (B,C)-bicomodule M and $N \in {}^C\mathbf{M}$, $M\Box_C N$ is a left B-comodule.

(2) For any $M \in \mathbf{M}^C$ and a (C,D)-bicomodule N, $M\Box_C N$ is a right D-comodule.

(3) For any (B,C)-bicomodule M and a (C,D)-bicomodule N, $M\Box_C N$ is a (B,D)-bicomodule.

The comodule structure on cotensor products rises the question of

11.6. Associativity of the cotensor product. *Let C, D be R-coalgebras with ${}_RC$, ${}_RD$ flat, M a (D,C)-bicomodule, $L \in \mathbf{M}^D$, and $N \in {}^C\mathbf{M}$. Then*

$$(L\Box_D M)\Box_C N \simeq L\Box_D(M\Box_C N),$$

provided that the canonical maps yield isomorphisms

$$(L\Box_D M)\otimes_R N \simeq L\Box_D(M\otimes_R N) \quad \text{and} \quad L\otimes_R(M\Box_C N) \simeq (L\otimes_R M)\Box_C N\,.$$

These conditions are satisfied if, for example,

(i) L and N are flat as R-modules, or

(ii) L is coflat in $\mathbf{M}^D$, or

(iii) N is coflat in ${}^C\mathbf{M}$, or

(iv) M is coflat in ${}^D\mathbf{M}$ and $\mathbf{M}^C$, or

(v) M is (D,R)-injective and (C,R)-injective, or

(vi) L is (D,R)-injective and N is (C,R)-injective.

Proof. Since ${}_RC$ and ${}_RD$ are flat, we know that $L\Box_D M \in \mathbf{M}^C$ and $M\Box_C N \in {}^D\mathbf{M}$ (by 11.3). In the commutative diagram

$$\begin{array}{ccccccc}
0 \longrightarrow & (L\Box_D M)\Box_C N & \longrightarrow & (L\otimes_R M)\Box_C N & \longrightarrow & (L\otimes_R D\otimes_R M)\Box_C N \\
& \downarrow{\scriptstyle \psi_1} & & \downarrow{\scriptstyle \psi_2} & & \downarrow{\scriptstyle \psi_3} \\
0 \longrightarrow & L\Box_D(M\Box_C N) & \longrightarrow & L\otimes_R(M\Box_C N) & \longrightarrow & L\otimes_R D\otimes_R(M\Box_C N)\,,
\end{array}$$

the top row is exact since $\omega_{L,M}$ is N-pure (see 10.4), and the bottom row is exact by definition of the cotensor product. L-purity of $\omega_{M,N}$ implies that ψ_2 and ψ_3 are isomorphisms (since ${}_RD$ is flat), and so ψ_1 is an isomorphism.

It follows from 10.7 that each of the given conditions imply the necessary isomorphisms. □

In 11.14, an example is considered showing that (unlike the tensor product) the cotensor product need not be associative in general.

11.7. Cotensor product of coflat comodules. *Let C, D be R-coalgebras with ${}_RC$, ${}_RD$ flat, $L \in \mathbf{M}^D$ and M a (D,C)-bicomodule. If L is D-coflat and M is C-coflat, then $L\Box_D M$ is a coflat right C-comodule.*

Proof. By the flatness conditions, $L\Box_D M$ is a right C-comodule and $M\Box_C K$ is a left D-comodule, for any $K \in {}^C\mathbf{M}$ (see 11.3). Any exact sequence $0 \to K' \to K \to K'' \to 0$ in ${}^C\mathbf{M}$ yields a commutative diagram:

$$\begin{array}{ccccccccc}
0 & \longrightarrow & (L\Box_D M)\Box_C K' & \longrightarrow & (L\Box_D M)\Box_C K & \longrightarrow & (L\Box_D M)\Box_C K'' & \longrightarrow & 0 \\
 & & \downarrow \simeq & & \downarrow \simeq & & \downarrow \simeq & & \\
0 & \longrightarrow & L\Box_D(M\Box_C K') & \longrightarrow & L\Box_D(M\Box_C K) & \longrightarrow & L\Box_D(M\Box_C K'') & \longrightarrow & 0,
\end{array}$$

with vertical isomorphisms (by 11.6). By assumption, the bottom row is exact and hence the top row is exact, too. □

11.8. Coalgebra morphisms. Let $\gamma : C \to D$ be an R-coalgebra morphism (cf. 2.1). A right C-comodule N can be considered as a right D-comodule by

$$\varrho^N_\gamma = (I \otimes \gamma) \circ \varrho^N : N \to N \otimes_R C \to N \otimes_R D,$$

and morphisms of right C-comodules $f : N \to M$ are clearly morphisms of the corresponding D-comodules. Similarly, every left C-comodule has a left D-comodule structure. In particular, C itself is a left and right D-comodule and γ is a left and right D-comodule morphism.

Consider the following commutative diagram with exact rows:

$$\begin{array}{ccccc}
N\Box_D C & \longrightarrow & N \otimes_R C & \xrightarrow{\omega_{N,C}} & N \otimes_R D \otimes_R C \\
\downarrow {\scriptstyle I_N\Box\gamma} & & \downarrow {\scriptstyle I_N\otimes\gamma} & & \downarrow {\scriptstyle I_N\otimes I_D\otimes\gamma} \\
N\Box_D D & \longrightarrow & N \otimes_R D & \xrightarrow{\omega_{N,D}} & N \otimes_R D \otimes_R D,
\end{array}$$

where $\omega_{N,C} := (I_N \otimes \gamma) \circ \varrho^N \otimes I_C - I_N \otimes (\gamma \otimes I_C) \circ \Delta_C$ and $\omega_{N,D} := (I_N \otimes \gamma) \circ \varrho^N \otimes I_D - I_N \otimes \Delta_D$. Clearly $\omega_{N,C} \circ \varrho^N = 0$, and hence the image of ϱ^N is contained in $N\Box_D C$, and thus there are maps

$$N \xrightarrow{\varrho^N} N\Box_D C \xrightarrow{I_N\Box\gamma} N\Box_D D \simeq N$$

whose composition yields the identity. Since both ϱ^N and $I_N\Box\gamma$ are D-comodule morphisms, N is a direct summand of $N\Box_D C$ as a D-comodule. In particular, Δ_C may be considered as a (C,C)-bicolinear map $\Delta_C : C \to C\Box_D C$.

11.9. Coinduction and corestriction. Related to any R-coalgebra morphism $\gamma : C \to D$ there is the *corestriction functor*

$$(\ )_\gamma : \mathbf{M}^C \to \mathbf{M}^D, \quad (M, \varrho^M) \mapsto (M, (I_M \otimes \gamma)\circ \varrho^M)$$

(usually we simply write $(M)_\gamma = M$). Furthermore, if ${}_RC$ is flat, then one defines the *coinduction functor*

$$-\Box_D C : \mathbf{M}^D \to \mathbf{M}^C, \quad N \mapsto N\Box_D C.$$

Here $N\Box_D C$ is said to be *induced* by N.

Considering C as a (C,D)-bicomodule by $C \xrightarrow{\Delta} C\otimes_R C \xrightarrow{I_C\otimes\gamma} C\otimes_R D$, the corestriction functor is isomorphic to $-\Box_C C : \mathbf{M}^C \to \mathbf{M}^D$, and hence it is left exact provided ${}_RC$ is flat, and is right exact provided C is coflat as a left comodule.

We define the corestriction functor ${}_\gamma(-)$ similarly, and finally the corestriction functor ${}_\gamma(-)_\gamma$ as a composition of ${}_\gamma(-)$ with $(-)_\gamma$.

11.10. Hom-cotensor relation. *Let ${}_RC$ be flat, $N \in \mathbf{M}^C$ and $L \in \mathbf{M}^D$. Then the map*

$$\mathrm{Hom}^C(N, L\Box_D C) \to \mathrm{Hom}^D(N,L), \quad f \mapsto (I_N\Box\gamma)\circ f,$$

is a functorial R-module isomorphism with the inverse map $g \mapsto (g\otimes I_C)\circ\varrho^N$. Therefore corestriction is left adjoint to coinduction.

Proof. With the defining maps for comodule morphisms (see 3.3) we construct the following commutative diagram:

$$\begin{array}{ccccccc}
0 \longrightarrow & \mathrm{Hom}^C(N, L\Box_D C) & \longrightarrow & \mathrm{Hom}_R(N, L\Box_D C) & \xrightarrow{\beta_1} & \mathrm{Hom}_R(N, L\Box_D C\otimes_R C) \\
 & & & \downarrow {\scriptstyle (I_N\Box\gamma)\circ -} & & \downarrow {\scriptstyle (I_N\Box\gamma\otimes\gamma)\circ -} \\
0 \longrightarrow & \mathrm{Hom}^D(N,L) & \longrightarrow & \mathrm{Hom}_R(N, L) & \xrightarrow{\beta_2} & \mathrm{Hom}_R(N, L\otimes_R D),
\end{array}$$

where $\beta_1(f) := (\varrho^L\otimes I_C)\circ f - (f\otimes I_C)\circ\varrho^N$ and $\beta_2(g) := \varrho^L\circ g - (g\otimes I_D)\circ (I_N\otimes\gamma)\circ\varrho^N$. From this we obtain the desired map.

The inverse assignment is a C-comodule morphism $N \to L \otimes_R C$, and we have to show that its image lies in $L\Box_D C$. This follows from the commutativity of the diagram

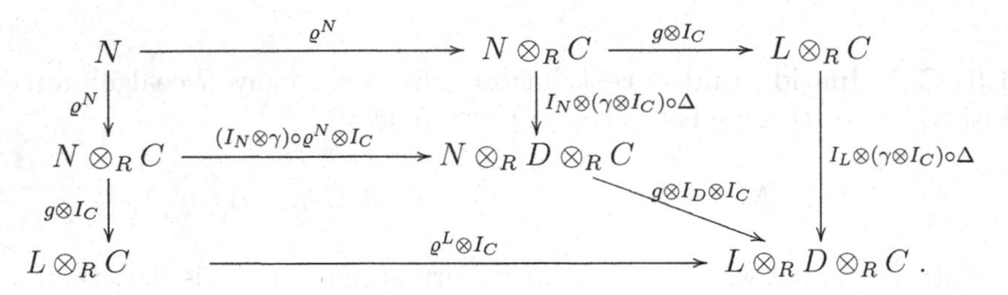

Direct calculation confirms that the compositions of these assignments yield the identity maps. □

Definitions. Let ${}_R C$ be flat. Given a coalgebra morphism $\gamma : C \to D$, a short exact sequence in $\mathbf{M}^C$ is called *(C, D)-exact* if it is splitting in $\mathbf{M}^D$. A right C-comodule N is called *(C, D)-injective* (resp. *(C, D)-projective*) if $\mathrm{Hom}^C(-, N)$ (resp. $\mathrm{Hom}^C(N, -)$) is exact with respect to (C, D)-exact sequences. The coalgebra C is called *right D-relative semisimple* or *right (C, D)-semisimple* if every right C-comodule is (C, D)-injective. Left-handed (in ${}^C\mathbf{M}$) and two-sided (in ${}^C\mathbf{M}^C$) versions of these notions are defined similarly.

11.11. (C, D)-injectivity. *Let ${}_R C$ be flat, $N \in \mathbf{M}^C$ and $\gamma : C \to D$ an R-coalgebra morphism.*

(1) If N is injective in $\mathbf{M}^D$, then $N\Box_D C$ is injective in $\mathbf{M}^C$.

(2) The following are equivalent:

(a) N is (C, D)-injective;

(b) every (C, D)-exact sequence splits in $\mathbf{M}^C$;

(c) the map $N \xrightarrow{\varrho^N} N\Box_D C$ splits in $\mathbf{M}^C$.

(3) If N is injective in $\mathbf{M}^D$ and (C, D)-injective, then N is injective in $\mathbf{M}^C$.

Proof. With the formalism from 11.8 and in view of 11.10, the proof of 3.18 can be adapted. □

Since the counit $\varepsilon : C \to R$ is a coalgebra morphism, one obtains the properties of (C, R)-exactness in 3.18 from 11.11 by setting $D = R$.

Definition. An R-coalgebra C is said to be *D-coseparable* if the map $\Delta_C : C \to C\Box_D C$ splits as a (C, C)-bicomodule morphism, that is, there exists a map $\pi : C\Box_D C \to C$ with the properties $\pi \circ \Delta_C = I_C$ and

$$(I_C \Box_D \pi) \circ (\Delta_C \Box_D I_C) = \Delta_C \circ \pi = (\pi \Box_D I_C) \circ (I_C \Box_D \Delta_C).$$

The characterisation of coseparable coalgebras in 3.29 can be extended to D-coseparable coalgebras.

11.12. D-coseparable coalgebras. *For a coalgebra morphism $\gamma : C \to D$, the following are equivalent:*

(a) *C is D-coseparable;*

(b) *there exists a (D, D)-bicolinear map $\delta : C\Box_D C \to D$ satisfying*

$$\delta \circ \Delta_C = \gamma \quad \text{and} \quad (I_C\Box_D\delta) \circ (\Delta_C\Box_D I_C) = (\delta\Box_D I_C) \circ (I_C\Box_D\Delta_C).$$

If ${}_RC$ is locally projective, (a) and (b) are also equivalent to:

(c) *the corestriction functor $(-)_\gamma : \mathbf{M}^C \to \mathbf{M}^D$ is separable;*

(d) *the corestriction functor ${}_\gamma(-) : {}^C\mathbf{M} \to {}^D\mathbf{M}$ is separable;*

(e) *the corestriction functor ${}_\gamma(-)_\gamma : {}^C\mathbf{M}^C \to {}^D\mathbf{M}^D$ is separable;*

(f) *C is D-relative semisimple as a (C, C)-bicomodule;*

(g) *C is D-relative injective as a (C, C)-bicomodule.*

If these conditions are satisfied, then all left or right C-comodules are (C, D)-injective.

Proof. (a) $\Rightarrow$ (b) Let $\pi : C\Box_D C \to C$ be left and right C-colinear with $\pi \circ \Delta_C = I_C$ and put $\delta = \gamma \circ \pi : C\Box_D C \to D$. Then $\delta \circ \Delta_C = \gamma \circ \pi \circ \Delta_C = \gamma$ and

$$\begin{aligned} (I_C\Box_D\delta) \circ (\Delta_C\Box_D I_C) &= (I_C\Box_D\gamma) \circ (I_C\Box_D\pi) \circ (\Delta_C\Box_D I_C) \\ &= (I_C\Box_D\gamma) \circ \Delta_C \circ \pi = \pi, \qquad \text{and} \\ (\delta\Box_D I_C) \circ (I_C\Box_D\Delta_C) &= (\gamma\Box_D I_C) \circ (\pi\Box_D I_C) \circ (I_C\Box_D\Delta_C) \\ &= (\gamma\Box_D I_C) \circ \Delta_C \circ \pi = \pi, \end{aligned}$$

as required.

(b) $\Rightarrow$ (a) Given $\delta : C\Box_D C \to D$ with the properties listed, define

$$\pi = (I_C\Box_D\delta) \circ (\Delta_C\Box_D I_C) : C\Box_D C \to C.$$

It is easily checked that π is left and right C-colinear. Furthermore,

$$\pi \circ \Delta_C = (I_C\Box_D\delta) \circ (\Delta_C\Box_D I_C) \circ \Delta_C = (I_C\Box_D\pi) \circ (I_C \otimes \Delta_C) \circ \Delta_C = I_C.$$

(b) $\Rightarrow$ (c) Given δ and $N \in \mathbf{M}^C$, define a map

$$\nu_N : N\Box_D C \xrightarrow{\simeq} N\Box_C C\Box_D C \xrightarrow{I_N\Box_D\delta} N.$$

Note that ν_N is a C-comodule map since so are the defining maps. Now define a functorial morphism $\Phi : \mathrm{Hom}^D((-)_\gamma, (-)_\gamma) \to \mathrm{Hom}^C(-,-)$ by assigning to any D-colinear map $f : M \to N$, where $M, N \in \mathbf{M}^C$, the map

$$\Phi(f) : M \xrightarrow{\varrho^M} M\Box_D C \xrightarrow{f\Box_D I_C} N\Box_D C \xrightarrow{\nu_N} N.$$

Again $\Phi(f)$ is C-colinear as a composition of C-colinear maps. Assume that f is already a morphism in $\mathbf{M}^C$. Then $(f\Box_D I_C)\circ \varrho^M = \varrho^N \circ f$, and it follows that $\Phi(f) = f$. This shows that the forgetful functor is separable.

Notice that the implication (b) $\Rightarrow$ (c) holds without any conditions on ${}_RC$. From now on we assume that ${}_RC$ is locally projective.

(c) $\Rightarrow$ (b) If $(-)_\gamma$ is a separable functor, then there exists a functorial morphism $\nu_C : C\Box_D C \to C$ in $\mathbf{M}^C$. Since $C\Box_D C$ is also in ${}_{C^*}\mathbf{M}$, we conclude from 39.5 that ν_C is left C^*-linear and hence is (C,C)-bicolinear.

(a) $\Leftrightarrow$ (d) Since the condition in (a) is symmetric, the proof of (a) $\Leftrightarrow$ (c) applies.

(a) $\Rightarrow$ (e) The forgetful functor ${}_\gamma(-)_\gamma : {}^C\mathbf{M}^C \to {}^D\mathbf{M}^D$ is the composition of the forgetful functors ${}_\gamma(-)$ and $(-)_\gamma$, and hence it is separable (by 38.20).

(e) $\Rightarrow$ (c) Since the composition of the left and right forgetful functors is separable, then so is (each) one of these (by 38.20).

(e) $\Rightarrow$ (f) For any (C,C)-bicomodule N, consider the map

$$N \xrightarrow{{}^N\varrho} C\Box_D N \xrightarrow{I_C\Box_D\varrho^N} C\Box_D(N\Box_D C).$$

By D-coseparability both $I_C\Box_D\varrho^N$ and ${}^N\varrho$ are split by morphisms from ${}^C\mathbf{M}^C$.

(f) $\Rightarrow$ (g) follows immediately from the definitions.

(g) $\Rightarrow$ (a) $\Delta : C \to C\Box_D C$ is (D,D)-split, and hence it splits in ${}^C\mathbf{M}^C$.

If $(-)_\gamma$ is a separable functor, it reflects retractions and hence all comodules in $\mathbf{M}^C$, ${}^C\mathbf{M}$ and ${}^C\mathbf{M}^C$ are D-relative injective (see 38.19). $\square$

Let A and S be R-algebras. For any (A,S)-bimodule M and a right S-module N, the R-module $\mathrm{Hom}_S(M,N)$ has a structure of a left A-module. Similarly, we may ask if, for any comodules $M \in {}^D\mathbf{M}^C$ and $N \in \mathbf{M}^C$, the R-module $\mathrm{Hom}^C(N,M)$ has a D-comodule structure. In case M is finitely R-presented, there is a Hom-cotensor relation $M\Box_C N^* \simeq \mathrm{Hom}^C(N,M)$ (see 10.11) and the left side is a left D-comodule provided ${}_RD$ is flat. More precisely, we can show:

11.13. Comodule structure on $\mathrm{Hom}^C(M,N)$. *Let ${}_RC$ be flat, $M \in {}^D\mathbf{M}^C$ and $N \in \mathbf{M}^C$ such that ${}_RM$ is flat and ${}_RN$ is finitely presented. If M is (C,R)-injective, or M is coflat in $\mathbf{M}^C$, or ${}_RD$ is flat, then*

$$M\Box_C N^* \simeq \mathrm{Hom}^C(N,M) \quad \textit{in } {}^D\mathbf{M}.$$

Proof. Since $\mathrm{Hom}_R(N,M) \simeq M\otimes_R N^*$ and $\mathrm{Hom}_R(N, M\otimes_R C) \simeq M\otimes_R C\otimes_R N^*$ (see 40.12), both modules are left D-comodules (induced by M). From 10.11 we obtain the commutative exact diagram

$$\begin{array}{ccccccc}
0 \longrightarrow & M\Box_C N^* & \longrightarrow & M\otimes_R N^* & \xrightarrow{\omega_{M,N^*}} & M\otimes_R C\otimes_R N^* \\
& \downarrow{\scriptstyle\beta_1} & & \downarrow{\scriptstyle\beta_2} & & \downarrow{\scriptstyle\beta_3} \\
0 \longrightarrow & \mathrm{Hom}^C(N,M) & \longrightarrow & \mathrm{Hom}_R(N,M) & \xrightarrow{\gamma} & \mathrm{Hom}_R(N, M\otimes_R C),
\end{array}$$

where β_1 is an R-isomorphism, while β_2, β_3 and γ are D-comodule morphisms (by definition of the comodule structures). Under any of the given conditions the sequences are D-pure in $\mathbf{M}_R$ (see 10.7). Hence the kernels of ω_{M,N^*} and γ are D-subcomdules and β_1 is a D-comodule morphism. □

11.14. Exercises

For some $0 \neq d \in \mathbb{Z}$, let $C = \mathbb{Z} \oplus \mathbb{Z}d$ be a $\mathbb{Z}$-coalgebra with structure maps Δ, ε given by $\Delta(1) = 1$, $\Delta(d) = d \otimes 1 + 1 \otimes d$, $\varepsilon(1) = 1$, $\varepsilon(d) = 0$. Prove (compare [125]):

(i) A $\mathbb{Z}$-module M is a (right) C-comodule if and only if there exists $\alpha \in \mathrm{End}_{\mathbb{Z}}(M)$ with $\alpha^2 = 0$.

(ii) Given $M, N \in \mathbf{M}_{\mathbb{Z}}$ and $\alpha \in \mathrm{End}_{\mathbb{Z}}(M)$, $\beta \in \mathrm{End}_{\mathbb{Z}}(N)$, with $\alpha^2 = 0$ and $\beta^2 = 0$, a C-colinear map $f : (M, \alpha) \to (N, \beta)$ is a $\mathbb{Z}$-morphism such that $\beta \circ f = f \circ \alpha$.

(iii) For C-comodules (M, α), (N, β),

$$(M,\alpha)\Box_C(N,\beta) = \{\sum m_i \otimes n_i \in M \otimes_{\mathbb{Z}} N \mid \sum \alpha(m_i) \otimes n_i = \sum m_i \otimes \beta(n_i)\}.$$

(iv) For some $n \in \mathbb{Z}$, consider the map $\gamma : \mathbb{Z}/n^2\mathbb{Z} \to \mathbb{Z}/n^2\mathbb{Z}$, $z \mapsto nz$, and show that

$$((\mathbb{Z},0)\Box_C(\mathbb{Z}/n^2\mathbb{Z},\gamma))\Box_C(\mathbb{Z}/n\mathbb{Z},0) \not\simeq (\mathbb{Z},0)\Box_C((\mathbb{Z}/n^2\mathbb{Z},\gamma)\Box_C(\mathbb{Z}/n\mathbb{Z},0)).$$

Hint: For the map $g : (\mathbb{Z}/n\mathbb{Z}, 0) \to (\mathbb{Z}/n^2\mathbb{Z}, 0)$, $z + n\mathbb{Z} \mapsto nz + n^2\mathbb{Z}$, consider $(I\Box_C I)\Box_C g$ and $I\Box_C(I\Box_C g)$ defined on the left and right side, respectively.

References. Al-Takhman [51]; Doi [104]; Grunenfelder and Paré [125].

12 Functors between comodule categories

The classical Eilenberg-Watts Theorem 39.4 states that colimit-preserving functors between module categories are essentially described by tensor functors. In particular, this is the background for the classical Morita theory of modules. The aim of this section is to study Eilenberg-Watts–type theorems for coalgebras, that is, to study when covariant functors between comodule categories can be considered as cotensor functors. These results are, in particular, applied to equivalences between comodule categories and thus lead to the Morita theory for coalgebras, known as the *Morita-Takeuchi theory*.

12.1. Functors between comodule categories. *Let C, D be R-coalgebras with ${}_RC$ and ${}_RD$ flat. Let $F : \mathbf{M}^C \to \mathbf{M}^D$ be an additive functor between the comodule categories that preserves kernels and colimits.*

(1) $F(C)$ is a (C, D)-bicomodule and there exists a functorial isomorphism $\nu : -\Box_C F(C) \to F$.

(2) For any $W \in \mathbf{M}_R$ and $N \in \mathbf{M}^C$,

$$W \otimes_R (N\Box_C F(C)) \simeq (W \otimes_R N)\Box_C F(C).$$

(3) $F(C)$ is a coflat left C-comodule, and, for all $W \in \mathbf{M}^D$, $N \in {}^D\mathbf{M}^C$,

$$W\Box_D(N\Box_C F(C)) \simeq (W\Box_D N)\Box_C F(C).$$

Proof. (1) Since all comodules are in particular R-modules, we learn from 39.3 that there is a functorial isomorphism

$$\Psi : - \otimes_R F(-) \to F(- \otimes_R -) \quad \text{of bifunctors} \quad \mathbf{M}_R \times \mathbf{M}^C \to \mathbf{M}^D.$$

Moreover, by 39.7, for the (C, C)-bicomodule C, $F(C)$ is a left C-comodule with the coaction

$${}^{F(C)}\varrho \; : \; F(C) \xrightarrow{F(\Delta_C)} F(C \otimes_R C) \xrightarrow{\Psi^{-1}_{C,C}} C \otimes_R F(C)\,.$$

With the defining equalisers for the cotensor product we obtain for any $M \in \mathbf{M}^C$ (recall $M \simeq M\Box_C C$) the commutative diagram

$$\begin{array}{ccccc}
M\Box_C F(C) & \longrightarrow & M \otimes_R F(C) & \overset{\varrho^M \otimes I_{F(C)}}{\underset{I_M \otimes {}^{F(C)}\varrho}{\rightrightarrows}} & M \otimes_R C \otimes_R F(C) \\
 & & \downarrow \Psi_{M,C} & & \downarrow \Psi_{M\otimes C,C} \\
F(M) & \longrightarrow & F(M \otimes_R C) & \overset{F(\varrho^M \otimes I_C)}{\underset{F(I_M \otimes \Delta)}{\rightrightarrows}} & F(M \otimes_R C \otimes_R C)\,,
\end{array}$$

where the top sequence is an equaliser by definition and the bottom sequence is an equaliser since F preserves kernels. From this we derive the isomorphism $\nu_M : M\Box_C F(C) \to F(M)$, which is functorial in M.

(2) This follows from the isomorphisms (tensor over R)

$$(W\otimes M)\Box_C F(C) \xrightarrow{\nu_{W\otimes M}} F(W\otimes M) \xrightarrow{\Psi^{-1}_{W,M}} W\otimes F(M) \xrightarrow{I\otimes\nu^{-1}_M} W\otimes(M\Box_C F(C)).$$

(3) Both $-\otimes_R C$ and F are left exact functors and so their composition $F(-\otimes_R C) \simeq -\otimes_R F(C)$ is also left exact, that is, $F(C)$ is a flat R-module. Since F preserves epimorphisms, so does $-\Box_C F(C)$, and hence $F(C)$ is coflat as a left C-comodule, and the associativity of the cotensor products follows from 11.6. □

12.2. Adjoint functors between comodule categories. *For R-flat coalgebras C, D, let (F,G) be an adjoint pair of additive functors, $F : \mathbf{M}^C \to \mathbf{M}^D$ and $G : \mathbf{M}^D \to \mathbf{M}^C$, with unit $\eta : I_{\mathbf{M}^C} \to GF$ and counit $\psi : FG \to I_{\mathbf{M}^D}$. If F preserves kernels and G preserves colimits, then:*

(1) $F(C)$ is a (C,D)-bicomodule and there exists a functorial isomorphism $\nu : -\Box_C F(C) \to F$.

(2) $G(D)$ is a (D,C)-bicomodule and there exists a functorial isomorphism $\mu : -\Box_D G(D) \to G$.

(3) For any $M \in \mathbf{M}^C$ and $M' \in {}^C\mathbf{M}$,

$$(M\Box_C F(C))\Box_D G(D) \simeq M\Box_C(F(C)\Box_D G(D)), \text{ and}$$
$$F(C)\Box_D(G(D)\Box_C M') \simeq (F(C)\Box_D G(D))\Box_C M'.$$

(4) There exist (C,C)-, resp. (D,D)-, bicomodule morphisms

$$\eta_C : C \to F(C)\Box_D G(D), \quad \psi_D : G(D)\Box_C F(C) \to D,$$

such that the following compositions yield identities:

$$F(C) \simeq C\Box_C F(C) \xrightarrow{\eta_C \Box I_{F(C)}} F(C)\Box_D G(D)\Box_C F(C) \xrightarrow{\psi_{F(C)}} F(C), \text{ and}$$
$$G(D) \xrightarrow{\eta_{G(D)}} G(D)\Box_C F(C)\Box_D G(D) \xrightarrow{\psi_D \Box I} D\Box_D G(D) \simeq G(D).$$

(5) There is an adjoint pair of functors (G', F'), where

$$G' = G(D)\Box_C - : {}^C\mathbf{M} \to {}^D\mathbf{M} \text{ and } F' = F(C)\Box_D - : {}^D\mathbf{M} \to {}^C\mathbf{M}.$$

Proof. (1),(2) Under the given conditions both F and G preserve kernels and colimits, and hence the assertions follow from 12.1.

(3) From 12.1 we also obtain the first isomorphism and the isomorphism

$$M \otimes_R (F(C)\Box_D G(D)) \simeq ((M \otimes_R F(C))\Box_D G(D)\,,$$

which, by 10.6, implies the isomorphism (here commutativity of R is crucial)

$$(F(C)\Box_D G(D)) \otimes_R M' \simeq F(C)\Box_D (G(D) \otimes_R M')\,.$$

Moreover, flatness of $F(C)$ induces an isomorphism

$$(F(C) \otimes_R G(D))\Box_C M' \simeq F(C) \otimes_R (G(D)\Box_C M')\,.$$

These isomorphisms imply associativity of the cotensor product (see 11.6).

(4) Notice that C is a (C,C)-bicomodule and GF preserves colimits, hence η_C is a (C,C)-bicomodule morphism by 39.7. Similar arguments apply to ψ_D, while the properties for the compositions are given in 38.21.

(5) To show that (G', F') is an adjoint pair, consider the maps

$$\begin{aligned} \eta'_M : &\quad M \simeq M\Box_C C \stackrel{I_M \Box \eta_C}{\longrightarrow} M\Box_C F(C)\Box_D G(D), \\ \psi'_N : &\quad G(D)\Box_C F(C)\Box_D N \stackrel{\psi_D \Box I_N}{\longrightarrow} D\Box_D N \simeq N\,, \end{aligned}$$

where $M \in \mathbf{M}^C$ and $N \in {}^D\mathbf{M}$. Applying the properties observed in (4) it is straightforward to show that these maps satisfy the conditions for a unit and a counit of an adjoint pair (see 38.21). □

12.3. Frobenius functors between comodule categories. *Let C, D be coalgebras that are flat as R-modules and (F,G) a Frobenius pair of additive functors, $F : \mathbf{M}^C \to \mathbf{M}^D$ and $G : \mathbf{M}^D \to \mathbf{M}^C$, that is, F is a left and right adjoint of G.*

(1) $F(C)$ is a (C,D)-bicomodule, $G(D)$ is a (D,C)-bicomodule, and there are functorial isomorphisms

$$-\Box_C F(C) \simeq F(-), \quad -\Box_D G(D) \simeq G(-).$$

(2) (G', F') is a Frobenius pair of functors where

$$G' = G(D)\Box_C - : {}^C\mathbf{M} \to {}^D\mathbf{M} \quad \text{and} \quad F' = F(C)\Box_D - : {}^D\mathbf{M} \to {}^C\mathbf{M}\,.$$

Proof. (1) Under the given conditions, F and G preserve limits and colimits (see 38.23), and hence the assertions follow from 12.2.

(2) In this case both (F,G) and (G,F) are adjoint pairs. By applying 12.2 to each one of them, we obtain adjoint pairs (G', F') and (F', G'), respectively. This means that (F', G') is a Frobenius pair, as required. □

Recall that equivalences between module categories can be described by the Hom and tensor functor (Morita Theorems). The next result shows that equivalences between comodule categories can be given by cotensor functors. This was observed by Takeuchi in [197] (for coalgebras over fields) and hence is called the *Morita-Takeuchi Theorem.* For coalgebras over rings one must beware of the fact that the cotensor product need not be associative.

12.4. Equivalences between comodule categories. *Let C, D be coalgebras that are flat as R-modules. The following are equivalent:*

(a) *there are functors $F : \mathbf{M}^C \to \mathbf{M}^D$ and $G : \mathbf{M}^D \to \mathbf{M}^C$ establishing an equivalence;*

(b) *there are functors $F' : {}^D\mathbf{M} \to {}^C\mathbf{M}$ and $G' : {}^C\mathbf{M} \to {}^D\mathbf{M}$ establishing an equivalence;*

(c) *there exist a (C, D)-bicomodule X and a (D, C)-bicomodule Y with bicomodule isomorphisms $\delta : C \to X\Box_D Y$, $\gamma : D \to Y\Box_C X$, such that*

$$(I_Y\Box_C\delta) \circ \varrho^Y = (\gamma\Box_D I_Y) \circ {}^Y\varrho, \quad (\delta\Box_D I_X) \circ {}^X\varrho = (I_X\Box_D\gamma) \circ \varrho^X,$$

X and Y are flat as R-modules, and the following pairs of morphisms

$$Y\otimes_R X \underset{I_Y\otimes {}^X\varrho}{\overset{\varrho^Y\otimes I_X}{\rightrightarrows}} Y\otimes_R C\otimes_R X, \quad X\otimes_R Y \underset{I_X\otimes {}^Y\varrho}{\overset{\varrho^X\otimes I_Y}{\rightrightarrows}} X\otimes_R D\otimes_R Y,$$

are pure in $\mathbf{M}_R$.

Proof. (a) $\Leftrightarrow$ (b) If the adjoint pair (F, G) gives an equivalence, then the unit and counit of the adjunction are isomorphisms. In particular,

$$\eta_C : C \to F(C)\Box_D G(D) \quad \text{and} \quad \psi_D : G(D)\Box_C F(C) \to D$$

are bimodule isomorphisms, and hence the functors

$$F' = F(C)\Box_D - : {}^D\mathbf{M} \to {}^C\mathbf{M} \quad \text{and} \quad G' = G(D)\Box_C - : {}^C\mathbf{M} \to {}^D\mathbf{M}$$

induce an equivalence (cf. 12.2). The converse follows by symmetry.

(a) $\Rightarrow$ (c) Putting $X = F(C)$ and $Y = G(D)$, we obtain the properties and constructions required from 12.2.

(c) $\Rightarrow$ (b) First observe that the given conditions imply associativity of the cotensor products, for any $M \in \mathbf{M}^C$, $N \in \mathbf{M}^D$,

$$(M\Box_C X)\Box_D Y \simeq M\Box_C(X\Box_D Y), \quad (N\Box_D Y)\Box_C X \simeq N\Box_D(Y\Box_C X).$$

For the functors $-\Box_C X : \mathbf{M}^C \to \mathbf{M}^D$ and $-\Box_D Y : \mathbf{M}^D \to \mathbf{M}^C$, the associativity properties above yield the functorial isomorphisms

$$M\Box_C X\Box_D Y \simeq M\Box_C C \simeq M \quad \text{and} \quad N\Box_D Y\Box_C X \simeq N\Box_D D \simeq N,$$

thus proving that they induce an equivalence. □

From 12.4 we know that functors describing equivalences between comodule categories are essentially cotensor functors. However, so far we have not investigated which properties of the comodules involved guarantee that the cotensor product yields an equivalence. One of the striking facts in the above observations is that $-\Box_D Y$ has a left adjoint, and we focus on this.

12.5. Quasi-finite comodules. Any $Y \in \mathbf{M}^C$ is called *quasi-finite* if the tensor functor $-\otimes_R Y : \mathbf{M}_R \to \mathbf{M}^C$ has a left adjoint. This left adjoint is called a *Cohom functor* and is denoted by $h_C(Y,-) : \mathbf{M}^C \to \mathbf{M}_R$. Thus, for each $M \in \mathbf{M}^C$, $W \in \mathbf{M}_R$ and a quasi-finite Y, there is a functorial isomorphism

$$\Phi_{M,W} : \mathrm{Hom}_R(h_C(Y,M),W) \to \mathrm{Hom}^C(M, W\otimes_R Y).$$

If C is flat as an R-module, then any quasi-finite comodule Y is flat as an R-module (since right adjoints respect monomorphisms).

As a left adjoint functor, $h_C(Y,-)$ respects colimits (see 38.21). By 39.3, this implies a functorial isomorphism

$$\Psi_{W,M} : W\otimes_R h_C(Y,M) \simeq h_C(Y, W\otimes_R M).$$

Moreover, if M is a (D,C)-bicomodule, then $h_C(Y,M)$ has a left D-comodule structure,

$$h_C(Y, {}^M\varrho) : h_C(Y,M) \to h_C(Y, D\otimes_R M) \simeq D\otimes_R h_C(Y,M),$$

such that the unit of the adjunction, $M \to h_C(Y,M)\otimes_R Y$, is a (D,C)-bicomodule morphism (see 39.7).

We know from the Hom-tensor relations 3.9 that $-\otimes_R C : \mathbf{M}_R \to \mathbf{M}^C$ is right adjoint to the forgetful functor $\mathbf{M}^C \to \mathbf{M}_R$, and hence C is a quasi-finite right (and left) C-comodule and the Cohom functor $h_C(C,-)$ is simply the forgetful functor.

For any quasi-finite $Y \in \mathbf{M}^C$ and $n \in \mathbb{N}$, a functorial isomorphism $\mathrm{Hom}_R(h_C(Y,-),-) \to \mathrm{Hom}^C(-,-\otimes_R Y)$ implies an isomorphism

$$\mathrm{Hom}_R(h_C(Y^n,-),-) \to \mathrm{Hom}^C(-,-\otimes_R Y^n),$$

showing that Y^n is again a quasi-finite C-comodule. Similarly it can be shown that any direct summand of a quasi-finite comodule is quasi-finite. So direct summands of C^n, $n \in \mathbb{N}$, provide a rich supply of quasi-finite right (and left) C-comodules.

12.6. Quasi-finite bicomodules. *Let Y be a (D,C)-bicomodule that is quasi-finite as a C-comodule and denote by $\eta : I_{\mathbf{M}^C} \to h_C(Y,-)\otimes_R Y$ the unit of the adjunction. Then there exists a unique D-comodule structure map $\varrho^{h_C(Y,M)} : h_C(Y,M) \to h_C(Y,M)\otimes_R D$ satisfying*

$$(I_{h_C(Y,M)} \otimes {}^Y\!\varrho) \circ \eta_M = (\varrho^{h_C(Y,M)} \otimes I_Y) \circ \eta_M \quad \text{and } \operatorname{Im}(\eta_M) \subset h_C(Y,M)\Box_D Y.$$

This yields a functor $h_C(Y,-) : \mathbf{M}^C \longrightarrow \mathbf{M}^D$.

Proof. For the C-colinear map

$$(I \otimes {}^Y\!\varrho) \circ \eta_M : M \longrightarrow h_C(Y,M)\otimes_R D \otimes_R Y,$$

there exists a unique R-linear map

$$\varrho^{h_C(Y,M)} : h_C(Y,M) \longrightarrow h_C(Y,M)\otimes_R D,$$

with $(I_{h_C(Y,M)}\otimes {}^Y\!\varrho)\circ\eta_M = (\varrho^{h_C(Y,M)}\otimes I_Y)\circ\eta_M$ (the preimage of the given map under Φ). It is straightforward to prove that this coaction makes $h_C(Y,M)$ a right D-comodule. The (defining) equality means that $\operatorname{Im}\eta_M$ lies in the equaliser of

$$h_C(Y,M)\otimes_R Y \overset{\varrho^{h_C(Y,M)}\otimes I_Y}{\underset{I_{h_C(Y,M)}\otimes {}^Y\!\varrho}{\rightrightarrows}} h_C(Y,M)\otimes_R D\otimes_R Y,$$

that is, $\operatorname{Im}(\eta_M) \subset h_C(Y,M)\Box_D Y$.

It remains to show that, for any $f : M \to M'$ in $\mathbf{M}^C$, the induced map $h_C(Y,f) : h_C(Y,M) \to h_C(Y,M')$ is D-colinear. First notice that the C-colinear map $\eta_{M'}\circ f : M \to h_C(Y,M')\otimes_R Y$ induces a unique R-linear map,

$$h_C(Y,f) : h_C(Y,M) \to h_C(Y,M'),$$

with $\eta_{M'}\circ f = (h_C(Y,f)\otimes I_Y)\circ\eta_M$. For the C-colinear map

$$(I_{h_C(Y,M)} \otimes {}^Y\!\varrho) \circ \eta_{M'} \circ f : M \longrightarrow h_C(Y,M')\otimes_R D\otimes_R Y,$$

there is a unique R-linear map $g : h_C(Y,M) \longrightarrow h_C(Y,M')\otimes_R D$, with $(g\otimes I_Y)\circ\eta_M = (I_{h_C(Y,M)}\otimes {}^Y\!\varrho)\circ\eta_{M'}\circ f$ (the preimage under Φ). We compute

$$\begin{aligned}
((\varrho^{h_C(Y,M')}\circ h_C(Y,f))\otimes I)\circ\eta_M &= (\varrho^{h_C(Y,M')}\otimes I)\circ(h_C(Y,f)\otimes I)\circ\eta_M\\
&= (\varrho^{h_C(Y,M')}\otimes I)\circ\eta_{M'}\circ f\\
&= (I\otimes {}^Y\!\varrho)\circ\eta_{M'}\circ f, \qquad \text{and}\\
((h_C(Y,f)\otimes I)\circ\varrho^{h_C(Y,M)}\otimes I)\circ\eta_M &= (h_C(Y,f)\otimes I\otimes I)\circ(\varrho^{h_C(Y,M)}\otimes I)\circ\eta_M\\
&= (h_C(Y,f)\otimes I\otimes I)\circ(I\otimes {}^Y\!\varrho)\circ\eta_M\\
&= (h_C(Y,f)\otimes {}^Y\!\varrho)\circ\eta_M\\
&= (I\otimes {}^Y\!\varrho)\circ(h_C(Y,f)\otimes I)\circ\eta_M\\
&= (I\otimes {}^Y\!\varrho)\circ\eta_{M'}\circ f,
\end{aligned}$$

where we have suppressed the subscripts of the identity maps $I_{h_C(Y,M)}$, and so on, to relieve the notation. By uniqueness of g we conclude

$$\varrho^{h_C(Y,M')} \circ h_C(Y,f) = (h_C(Y,f) \otimes I_Y) \circ \varrho^{h_C(Y,M)},$$

which shows that $h_C(Y,f)$ is D-colinear. $\square$

12.7. Cotensor functors with left adjoints. *For a (D,C)-bicomodule Y the following are equivalent:*

(a) $-\otimes_R Y : \mathbf{M}_R \to \mathbf{M}^C$ has a left adjoint (that is, Y is quasi-finite);

(b) $-\Box_D Y : \mathbf{M}^D \to \mathbf{M}^C$ has a left adjoint.

Proof. (b) $\Rightarrow$ (a) By the isomorphism $-\otimes_R Y \simeq (-\otimes_R D)\Box_D Y$, the functor $-\otimes_R Y$ is a composite of the functors $-\otimes_R D : \mathbf{M}_R \to \mathbf{M}^D$ and $-\Box_D Y : \mathbf{M}^D \to \mathbf{M}^C$. We know that $-\otimes_R D$ always has a left adjoint (forgetful functor). Hence the assertion follows from the fact that the composition of functors with left adjoints also has a left adjoint.

(a) $\Rightarrow$ (b) Let Y be quasi-finite as C-comodule with left adjoint $h_C(Y,-)$ and denote by $\eta : I_{\mathbf{M}^C} \to h_C(Y,-)\otimes_R Y$ the unit of the adjunction. By 12.6, there is a functor $h_C(Y,-) : \mathbf{M}^C \to \mathbf{M}^D$, and for any $M \in \mathbf{M}^C$, $\operatorname{Im} \eta_M \subset h_C(Y,M)\Box_D Y$. We claim that this functor is left adjoint to $-\Box_D Y$. For $M \in \mathbf{M}^C$ and $N \in \mathbf{M}^D$, define a map

$$\Phi'_{M,N} : \operatorname{Hom}^D(h_C(Y,M),N) \longrightarrow \operatorname{Hom}^C(M, N\Box_D Y), \quad f \mapsto (f\Box_D I_Y)\circ \eta_M,$$

which is just the restriction of the adjointness isomorphism Φ. There is a commutative diagram

$$\begin{array}{ccc}
\operatorname{Hom}^D(h_C(Y,M),N) & \xrightarrow{\Phi'_{M,N}} & \operatorname{Hom}^C(M,N\Box_D Y) \\
\downarrow & & \downarrow \\
\operatorname{Hom}_R(h_C(Y,M),N) & \xrightarrow{\Phi_{M,N}} & \operatorname{Hom}^C(M,N\otimes_R Y),
\end{array}$$

and it is left as an exercise to show that $\Phi'_{M,N}$ is an isomorphism. $\square$

12.8. Exactness of the Cohom functor. *For coalgebras C, D that are flat as R-modules, let Y be a (D,C)-comodule that is quasi-finite as a C-comodule.*

(1) The following are equivalent:

(a) $h_C(Y,-) : \mathbf{M}^C \to \mathbf{M}_R$ is exact;

(b) $W\otimes_R Y$ is injective in $\mathbf{M}^C$, for every injective R-module W;

(c) $h_C(Y,-) : \mathbf{M}^C \to \mathbf{M}^D$ is exact;

(d) $N\Box_D Y$ is injective in $\mathbf{M}^C$, for every injective comodule $N \in \mathbf{M}^D$.

If these conditions hold, then $h_C(Y,-) \simeq -\Box_C h_C(Y,C)$ as functors $\mathbf{M}^C \to \mathbf{M}_R$ (or $\mathbf{M}^C \to \mathbf{M}^D$), and therefore $h_C(Y,C)$ is a coflat left C-comodule.

(2) *If $h_C(Y,-)$ is exact, then the following are equivalent:*

  (a) *$h_C(Y,-)$ is faithful;*

  (b) *$h_C(Y,C)$ is a faithfully coflat left C-comodule.*

Proof. (1) (a) $\Rightarrow$ (b) and (c) $\Rightarrow$ (d) follow from the adjointness isomorphisms

$$\begin{aligned} \mathrm{Hom}_R(h_C(Y,-),W) &\simeq \mathrm{Hom}^C(-,W\otimes_R Y), \\ \mathrm{Hom}^D(h_C(Y,-),N) &\simeq \mathrm{Hom}^C(-,N\Box_D Y), \end{aligned}$$

since exactness of $h_C(Y,-)$ and $\mathrm{Hom}_R(-,W)$ (resp. $\mathrm{Hom}^D(-,N)$) implies exactness of their composition. This means that $W\otimes_R Y$ (resp. $N\Box_D Y$) is injective in $\mathbf{M}^C$.

(a) $\Rightarrow$ (c) This follows from the fact that any sequence of morphisms in $\mathbf{M}^D$ is exact if and only if it is exact in $\mathbf{M}_R$.

(b) $\Rightarrow$ (a) Since we know that $h_C(Y,-)$ is right exact, it remains to show that, for any monomorphism $f: K \to L$ in $\mathbf{M}^C$, $h = h_C(Y,f)$ is also a monomorphism. For any $W \in \mathbf{M}_R$ there is a commutative diagram

$$\begin{array}{ccc} \mathrm{Hom}_R(h_C(Y,L),W) & \xrightarrow{\simeq} & \mathrm{Hom}^C(L,W\otimes_R Y) \\ {\scriptstyle \mathrm{Hom}_R(h,W)}\downarrow & & \downarrow{\scriptstyle \mathrm{Hom}^C(f,W\otimes_R Y)} \\ \mathrm{Hom}_R(h_C(Y,K),W) & \xrightarrow{\simeq} & \mathrm{Hom}^C(K,W\otimes_R Y)\,. \end{array}$$

Let W be an injective cogenerator in $\mathbf{M}_R$. Then by assumption $W\otimes_R Y$ is injective in $\mathbf{M}^C$ and hence $\mathrm{Hom}^C(f,W\otimes_R Y)$ is surjective and so is $\mathrm{Hom}_R(h,W)$. By the cogenerator property of W this means that $h = h_C(Y,f)$ is injective.

(d) $\Rightarrow$ (c) This follows by an argument similar to the proof (b) $\Rightarrow$ (a).

(2) This is obvious by the isomorphism $h_C(Y,M) \simeq M\Box_C h_C(Y,C)$, for any $M \in \mathbf{M}^C$. $\square$

Property (1)(b) above motivates the following

Definition. A right C-comodule Y is called an *injector* if the tensor functor $-\otimes_R Y : \mathbf{M}_R \to \mathbf{M}^C$ respects injective objects. Left comodule injectors are defined similarly.

The Hom-tensor relations 3.9 imply that a coalgebra C is an injector as both a left and right comodule.

In the preceding propositions we considered quasi-finite C-comodules with an additional D-comodule structure. This is not a restriction because, for any quasi-finite C-comodule Y, there exists a coalgebra D (the dual of the algebra of C-colinear endomorphisms of Y) that makes Y a (D,C)-comodule. Now we outline the construction of such a coalgebra D.

Let $e_C(Y) = h_C(Y,Y)$. The unit of the adjunction yields a right C-comodule morphism,

$$(I_{e_C(Y)} \otimes \eta_Y) \circ \eta_Y : Y \longrightarrow e_C(Y) \otimes_R e_C(Y) \otimes_R Y \,.$$

By the adjunction isomorphism (tensor over R)

$$\Phi_{Y,e_C(Y)\otimes e_C(Y)} : \mathrm{Hom}_R(e_C(Y), e_C(Y)\otimes e_C(Y)) \simeq \mathrm{Hom}^C(Y, e_C(Y)\otimes e_C(Y)\otimes Y),$$

there exists a unique R-linear map

$$\Delta_e : e_C(Y) \to e_C(Y) \otimes_R e_C(Y), \text{ with } (I_{e_C(Y)} \otimes \eta_Y) \circ \eta_Y = (\Delta_e \otimes I_Y) \circ \eta_Y.$$

Moreover, by the isomorphism $\Phi_{Y,R} : \mathrm{Hom}_R(e_C(Y), R) \simeq \mathrm{Hom}^C(Y,Y)$, there exists an R-linear map

$$\varepsilon_e : e_C(Y) \to R, \text{ with } (\varepsilon_e \otimes I_Y) \circ \eta_Y = I_Y.$$

12.9. Coendomorphism coalgebra. *Let $Y \in \mathbf{M}^C$ be quasi-finite and put $e_C(Y) = h_C(Y,Y)$. Then $e_C(Y)$ is an R-coalgebra by the structure maps*

$$\Delta_e : e_C(Y) \to e_C(Y) \otimes_R e_C(Y), \quad \varepsilon_e : e_C(Y) \to R,$$

defined above. Furthermore, Y is an $(e_C(Y), C)$-bicomodule by the mapping $\eta_Y : Y \to e_C(Y) \otimes_R Y$, and there is a ring anti-isomorphism

$$e_C(Y)^* = \mathrm{Hom}_R(e_C(Y), R) \xrightarrow{\Phi_{Y,R}} \mathrm{End}^C(Y).$$

The coalgebra $e_C(Y)$ is known as the *coendomorphism coalgebra* of Y.

Proof. To show that ε_e is a counit for Δ_e, consider

$$\begin{aligned}
\Phi_{e_C(Y),e_C(Y)}((I_{e_C(Y)} \otimes \varepsilon_e) \circ \Delta_e) &= ((I_{e_C(Y)} \otimes \varepsilon_e) \circ \Delta_e \otimes I_Y) \circ \eta_Y \\
&= (I_{e_C(Y)} \otimes \varepsilon_e \otimes I_Y) \circ (\Delta_e \otimes I_Y) \circ \eta_Y \\
&= (I_{e_C(Y)} \otimes \varepsilon_e \otimes I_Y) \circ (I_{e_C(Y)} \otimes \eta_Y) \circ \eta_Y \\
&= \eta_Y, \\
\Phi_{e_C(Y),e_C(Y)}((\varepsilon_e \otimes I_{e_C(Y)}) \circ \Delta_e) &= ((\varepsilon_e \otimes I_{e_C(Y)}) \circ \Delta_e \otimes I_Y) \circ \eta_Y \\
&= (\varepsilon_e \otimes I_{e_C(Y)} \otimes I_Y) \circ (\Delta_e \otimes I_Y) \circ \eta_Y \\
&= (\varepsilon_e \otimes I_{e_C(Y)} \otimes I_Y) \circ (I_{e_C(Y)} \otimes \eta_Y) \circ \eta_Y \\
&= \eta_Y \,, \text{ and} \\
\Phi_{e_C(Y),e_C(Y)}(I_{e_C(Y)}) &= \eta_Y \,.
\end{aligned}$$

Now injectivity of Φ implies

$$(I_{e_C(Y)} \otimes \varepsilon_e) \circ \Delta_e = I_{e_C(Y)} = (\varepsilon_e \otimes I_{e_C(Y)}) \circ \Delta_e .$$

To prove coassociativity of Δ_e, we show that $\mathrm{Hom}_R(e_C(Y), A)$ with the convolution product is associative for any associative R-algebra A (cf. 1.3). For this consider the following R-module isomorphism:

$$\Phi' : \mathrm{Hom}_R(e_C(Y), A) \xrightarrow{\Phi_{Y,A}} \mathrm{Hom}^C(Y, A \otimes_R Y) \xrightarrow{\simeq} {}_A\mathrm{Hom}^C(A \otimes_R Y, A \otimes_R Y),$$

where the second isomorphism is given by an extension of scalars (see 40.20). This is in fact a ring anti-isomorphism since

$$\begin{aligned} \Phi'(f * g) = \Phi_{Y,A}((f \otimes g) \circ \Delta_e) &= ((f \otimes g) \circ \Delta_e \otimes I_Y) \circ \eta_Y \\ &= (f \otimes g \otimes I_Y) \circ (\Delta_e \otimes I_Y) \circ \eta_Y \\ &= (f \otimes g \otimes I_Y) \circ (I_{e_C(Y)} \otimes \eta_Y) \circ \eta_Y \\ &= (g \otimes I_Y) \circ \eta_Y \circ (f \otimes I_Y) \circ \eta_Y \\ &= \Phi'(g) \circ \Phi'(f) . \end{aligned}$$

Since $\mathrm{End}_A^C(A \otimes_R Y)$ is associative, we conclude that $\mathrm{Hom}_R(e_C(Y), A)$ is also associative. Therefore Δ_e is coassociative by 1.3.

By the definition of Δ_e there is a commutative diagram

$$\begin{array}{ccc} Y & \xrightarrow{\eta_Y} & e_C(Y) \otimes_R Y \\ \downarrow{\scriptstyle \eta_Y} & & \downarrow{\scriptstyle \Delta_e \otimes I_Y} \\ e_C(Y) \otimes_R Y & \xrightarrow{I \otimes \eta_Y} & e_C(Y) \otimes_R e_C(Y) \otimes_R Y , \end{array}$$

and the definition of ε_e shows that η_Y is a counital coaction. Since η_Y is C-colinear, Y is an $(e_C(Y), C)$-bicomodule. □

12.10. Equivalence with comodules over $e_C(Y)$. *Let C be a coalgebra that is flat as an R-module. Let Y be a right C-comodule that is quasi-finite, faithfully coflat and an injector, and let $e_C(Y)$ denote the coendomorphism coalgebra of Y. Then the functors*

$$-\Box_{e_C(Y)} Y : \mathbf{M}^{e_C(Y)} \to \mathbf{M}^C, \quad h_C(Y, -) : \mathbf{M}^C \to \mathbf{M}^{e_C(Y)},$$

where $h_C(Y, -)$ is the left adjoint to $- \otimes_R Y$, are inverse equivalences.

Proof. We prove that the conditions of 12.4(c) are satisfied. As shown in 12.9, Y is an $(e_C(Y), C)$-bicomodule and the image of $h_C(Y, -) : \mathbf{M}^C \to \mathbf{M}_R$ lies in $\mathbf{M}^{e_C(Y)}$ (see 12.6). Since Y is an injector, the functor $h_C(Y, -)$ is exact (by 12.8), and hence $h_C(Y,-) \simeq -\Box_C h_C(Y,C)$ (by 12.2) and so $h_C(Y,C)$ is coflat as a left C-comodule.

Given $M \in \mathbf{M}^C$, consider the right C-comodule morphism (notice that Y_R is flat),

$$I_M \Box_C \eta_C : M \simeq M \Box_C C \to M \Box_C (h_C(Y,C) \otimes_R Y) \simeq (M \Box_C h_C(Y,C)) \otimes_R Y.$$

There exists a unique right $e_C(Y)$-colinear morphism

$$\delta_M : h_C(Y,M) \to M \Box_C h_C(Y,C), \text{ with } (\delta_M \otimes_R I_Y) \circ \eta_M = I_M \Box_C \eta_C,$$

which is an isomorphism by exactness of $h_C(Y,-)$ (see 12.2). Since Y is a left $e_{C(Y)}$-comodule, the isomorphism $\delta_Y : e_C(Y) \to Y \Box_C h_C(Y,C)$ is in fact $(e_C(Y), e_C(Y))$-bicolinear by 39.7. Furthermore, since Y is a faithfully flat C-comodule, bijectivity of $I_M \Box_C \eta_C$ implies that $\eta_C : C \to h_C(Y,C) \Box_{e_C(Y)} Y$ is an isomorphism in $\mathbf{M}^C$. Since C is a bicomodule, this is a (C,C)-bicomodule morphism by 39.7.

To verify the purity conditions stated in 12.4(c)(2) we have to show that the canonical maps

$$\begin{aligned} W \otimes_R (h_C(Y,C) \Box_{e_C(Y)} Y) &\to (W \otimes_R h_C(Y,C)) \Box_{e_C(Y)} Y \text{ and} \\ W \otimes_R (Y \Box_C h_C(Y,C)) &\to (W \otimes_R Y) \Box_C h_C(Y,C) \end{aligned}$$

are isomorphisms for any $W \in \mathbf{M}_R$ (see 10.6). By 10.7, this follows from the coflatness of Y as a left $e_C(Y)$-comodule, and the coflatness of $h_C(Y,C)$ as a left C-comodule, respectively. □

12.11. Equivalences and coendomorphism coalgebra. *For R-coalgebras C, D that are flat as R-modules, let $F : \mathbf{M}^C \to \mathbf{M}^D$ be an equivalence with inverse $G : \mathbf{M}^D \to \mathbf{M}^C$. Then:*

(1) $G(D)$ is quasi-finite, an injector, and faithfully coflat both as a left D-comodule and as a right C-comodule;

(2) $F(C)$ is quasi-finite, an injector, and faithfully coflat both as a left C-comodule and as a right D-comodule;

(3) there are coalgebra isomorphisms

$$e_C(G(D)) \simeq D \simeq e_C(F(C)) \quad \textit{and} \quad e_D(G(D)) \simeq C \simeq e_D(F(C)).$$

Proof. The proofs of (1) and (2) are symmetric; thus we only need to prove (2). Since $F \simeq -\Box_C F(C)$ is an equivalence, it has a left adjoint and is exact and faithful. Hence $F(C)$ is quasi-finite as a right D-comodule and faithfully flat as a left C-comodule. Moreover, $F' = F(C) \Box_D - : {}^D\mathbf{M} \to {}^C\mathbf{M}$ is an equivalence (see 12.4) that implies that $F(C)$ is quasi-finite as a left C-comodule and faithfully flat as a right D-comodule.

For any injective $W \in \mathbf{M}_R$, by 3.9, $W \otimes_R C$ is injective in $\mathbf{M}^C$ and, by the properties of equivalences, $W \otimes_R F(C) \simeq F(W \otimes_R C)$ is injective in $\mathbf{M}^D$, that is, $F(C)$ is an injector as right D-comodule. Similarly we observe that $F'(D \otimes_R W) \simeq F(C)\Box_D(D \otimes_R W) \simeq F(C) \otimes_R W$ is injective in ${}^C\mathbf{M}$ and hence $F(C)$ is an injector as left C-comodule.

(3) Since $G \simeq -\Box_D G(D)$ is left adjoint to $F \simeq -\Box_C F(C)$,

$$e_D(F(C)) \simeq F(C)\Box_D G(D) \simeq GF(C) \simeq C.$$

The other isomorphisms are obtained similarly. □

The highlight of the characterisations of equivalences is the following version of the *Morita-Takeuchi Theorem.*

12.12. Equivalences between comodule categories (2). *For coalgebras C, D that are flat as R-modules, the following are equivalent:*

(a) the categories $\mathbf{M}^C$ and $\mathbf{M}^D$ are equivalent;

(b) the categories ${}^C\mathbf{M}$ and ${}^D\mathbf{M}$ are equivalent;

(c) there exists a (D, C)-bicomodule Y that is quasi-finite, faithfully coflat and an injector as a right C-comodule and $e_C(Y) \simeq D$ as coalgebras;

(d) there exists a (D, C)-bicomodule Y that is quasi-finite, faithfully coflat and an injector as a left D-comodule and $e_D(Y) \simeq C$ as coalgebras;

(e) there exists a (C, D)-bicomodule X that is quasi-finite, faithfully coflat and an injector as a right D-comodule and $e_D(X) \simeq C$ as coalgebras;

(f) there exists a (C, D)-bicomodule X that is quasi-finite, faithfully coflat and an injector as a left C-comodule and $e_C(X) \simeq D$ as coalgebras.

Proof. (a) $\Leftrightarrow$ (b) is shown in 12.4, and (c) $\Rightarrow$ (a) follows by 12.10.

(a) $\Rightarrow$ (c) Given an equivalence $F : \mathbf{M}^C \to \mathbf{M}^D$ with inverse $G : \mathbf{M}^D \to \mathbf{M}^C$, it was shown in 12.11 that $Y = G(D)$ satisfies the conditions stated.

The remaining implications follow by symmetry. □

For coalgebras over QF rings (in particular fields) quasi-finite comodules can be characterised by finiteness conditions; hence the terminology.

12.13. Quasi-finite comodules for coherent base ring. *Let R be a coherent ring, C an R-coalgebra that is flat as an R-module, and Y a quasi-finite right C-comodule. Then, for any $P \in \mathbf{M}^C$ that is finitely presented as an R-module, $\mathrm{Hom}^C(P, Y)$ is a finitely presented R-module.*

Proof. The coherence of R implies the flatness of R^Λ, for any index set Λ. Now 3.23 provides an isomorphism $R^\Lambda \otimes_R \mathrm{Hom}^C(P, Y) \simeq \mathrm{Hom}^C(P, R^\Lambda \otimes_R Y)$, and by quasi-finiteness of Y we obtain

$$\mathrm{Hom}^C(P, R^\Lambda \otimes_R Y) \simeq \mathrm{Hom}_R(h_C(Y, P), R^\Lambda) \simeq \mathrm{Hom}^C(P, Y)^\Lambda .$$

By 40.17, the resulting isomorphism $R^\Lambda \otimes_R \mathrm{Hom}^C(P,Y) \simeq \mathrm{Hom}^C(P,Y)^\Lambda$ implies that $\mathrm{Hom}^C(P,Y)$ is finitely presented as an R-module. □

12.14. Equivalences for QF base rings. *Let R be a QF ring, C an R-coalgebra that is projective as an R-module, and Y a quasi-finite right C-comodule. Then the following are equivalent:*

(a) Y is faithfully coflat and an injector in $\mathbf{M}^C$;

(b) Y is an injective cogenerator in $\mathbf{M}^C$.

If this is the case, there is an equivalence

$$-\Box_{e_C(Y)}Y : \mathbf{M}^{e_C(Y)} \to \mathbf{M}^C, \quad h_C(Y,-) : \mathbf{M}^C \to \mathbf{M}^{e_C(Y)},$$

and for any $M \in \mathbf{M}^C$,

$$h_C(Y,M) \simeq \varinjlim_\Lambda \mathrm{Hom}^C(M_\lambda, Y)^*,$$

where $\{M_\lambda\}_\Lambda$ denotes the family of all finitely generated subcomodules of M and $(-)^* = \mathrm{Hom}_R(-,R)$.

Proof. Recall from 10.12 that Y is injective (and a cogenerator) in $\mathbf{M}^C$ if and only if Y is a (faithfully) coflat C-comodule. Hence (a) implies (b). Conversely, suppose that (b) holds and consider any injective R-module W. Since R is QF, W is projective and hence

$$(W \otimes_R Y)\Box_C - \simeq W \otimes_R (Y\Box_C -) : \mathbf{M}^C \to \mathbf{M}_R.$$

Since $W \otimes_R (Y\Box_C -)$ is an exact functor, so is $(W \otimes_R Y)\Box_C -$. This means that $W \otimes_R Y$ is coflat and hence injective by the above remark. So Y is an injector and (a) holds.

It follows from 12.10 that $-\Box_{e_C(Y)}Y$ induces an equivalence with the inverse $h_C(Y,-)$. By the quasi-finiteness of Y,

$$h_C(Y,M_\lambda)^* \simeq \mathrm{Hom}_R(h_C(Y,M_\lambda),R) \simeq \mathrm{Hom}^C(M_\lambda,Y).$$

All M_λ are finitely presented R-modules, and hence the $\mathrm{Hom}^C(M_\lambda,Y)$ are finitely presented R-modules (by 12.13). This implies

$$h_C(Y,M_\lambda) \simeq h_C(Y,M_\lambda)^{**} = \mathrm{Hom}_R(h_C(Y,M_\lambda),R)^* \simeq \mathrm{Hom}^C(M_\lambda,Y)^*.$$

Since $h_C(Y,-)$ has a right adjoint, it commutes with direct limits, and therefore

$$h_C(Y,M) \simeq h_C(Y,\varinjlim_\Lambda M_\lambda) \simeq \varinjlim_\Lambda h_C(Y,M_\lambda) \simeq \varinjlim_\Lambda \mathrm{Hom}^C(M_\lambda,Y)^*,$$

as required. □

12.15. Idempotents and cotensor product. *Let $e^2 = e \in C^*$, assume ${}_RC$ to be flat, and let $e\rightharpoonup C \leftharpoonup e$ be the coalgebra in 4.10.*

(1) For any $M \in \mathbf{M}^C$ there is an $e\rightharpoonup C\leftharpoonup e$-comodule isomorphism

$$\gamma_M : e\rightharpoonup M \to M\Box_C e\rightharpoonup C, \quad e\rightharpoonup m \mapsto \sum m_{\underline{0}} \otimes e\rightharpoonup m_{\underline{1}}.$$

(2) $e\rightharpoonup C$ is a $(C, e\rightharpoonup C\leftharpoonup e)$-bicomodule and $C\leftharpoonup e$ is a $(e\rightharpoonup C\leftharpoonup e, C)$-bicomodule and there are bicomodule morphisms

$$\begin{array}{llll} \gamma : e\rightharpoonup C\leftharpoonup e & \longrightarrow C\leftharpoonup e\Box_C e\rightharpoonup C, & e\rightharpoonup c\leftharpoonup e & \mapsto \sum c_{\underline{1}}\leftharpoonup e \otimes e\rightharpoonup c_{\underline{2}}, \\ \delta : C & \longrightarrow e\rightharpoonup C\Box_{e\rightharpoonup C\leftharpoonup e} C\leftharpoonup e, & c & \mapsto \sum e\rightharpoonup c_{\underline{1}} \otimes c_{\underline{2}}\leftharpoonup e, \end{array}$$

where γ is an isomorphism.

(3) For any $K \in \mathbf{M}^{e\rightharpoonup C\leftharpoonup e}$, there are $e\rightharpoonup C\leftharpoonup e$-comodule isomorphisms

$$e\rightharpoonup(K\Box_{e\rightharpoonup C\leftharpoonup e} C\leftharpoonup e) \simeq K\Box_{e\rightharpoonup C\leftharpoonup e}(C\leftharpoonup e\Box_C e\rightharpoonup C) \simeq K.$$

(4) The functor $e\rightharpoonup -$ ($\simeq -\Box_C e\rightharpoonup C$) is left adjoint to the functor

$$-\Box_{e\rightharpoonup C\leftharpoonup e} C\leftharpoonup e : \mathbf{M}^{e\rightharpoonup C\leftharpoonup e} \to \mathbf{M}^C$$

by the isomorphism (for $M \in \mathbf{M}^C$, $K \in \mathbf{M}^{e\rightharpoonup C\leftharpoonup e}$)

$$\mathrm{Hom}^C(M, K\Box_{e\rightharpoonup C\leftharpoonup e} C\leftharpoonup e) \to \mathrm{Hom}^{e\rightharpoonup C\leftharpoonup e}(M\Box_C e\rightharpoonup C, K), \ f \mapsto f(e\rightharpoonup -),$$

with inverse map $h \mapsto (h\Box I_{C\leftharpoonup e}) \circ (I_M \Box \delta)$.

Proof. (1) To simplify notation, write eC for $e\rightharpoonup C$, Ce for $C\leftharpoonup e$, and eM for $e\rightharpoonup M$. From the defining diagram of $M\Box_C C$ we obtain by multiplication with e from the left

$$eM \xrightarrow{\varrho^M} M \otimes_R eC \underset{I_M \otimes \Delta}{\overset{\varrho^M \otimes I_{eC}}{\rightrightarrows}} M \otimes_R C \otimes_R eC,$$

where we have used that the maps involved are right C-comodule and hence left C^*-module morphisms. To show that this is an equaliser diagram we first observe that $\varrho^M|_{eM}$ is injective. Next, consider any $\sum_i m_i \otimes ec_i \in M \otimes_R eC$ that has the same image under $\varrho^M \otimes I_{eC}$ and $I_M \otimes \Delta$. Since $(I_C \otimes \varepsilon)\Delta|_{eC} = (\varepsilon \otimes I_C)\Delta|_{eC} = I_{eC}$, we obtain

$$\begin{array}{rcl} \sum_i m_i \otimes ec_i & = & (I_M \otimes I_C \otimes \varepsilon) \circ (I_M \otimes \Delta)(\sum_i m_i \otimes ec_i) \\ & = & (I_M \otimes I_C \otimes \varepsilon) \circ (\varrho^M \otimes I_{eC})(\sum_i m_i \otimes ec_i) \\ & = & (I_M \otimes I_C \otimes \varepsilon)(\sum_i m_{i\underline{0}} \otimes m_{i\underline{1}} \otimes ec_i) = \sum_i m_{i\underline{0}} \otimes m_{i\underline{1}} e(c_i) \\ & = & \sum_i m_{i\underline{0}} \otimes em_{i\underline{1}} e(c_i) = \sum_i \varrho^M(em_i e(c_i)), \end{array}$$

showing that this is an equaliser diagram, and hence – by definition of the cotensor product – $eM \simeq M\Box_C eC$.

(2) The bicomodule structures are special cases of 11.2, and it is easy to see that γ is an (eCe, eCe)-bimodule and δ is a (C, C)-bicomodule morphism. It follows from (1) that $\gamma = \gamma|_{Ce}$ is an isomorphism.

(3) The isomorphisms follow by (1), (2), and the fact that the cotensor product is associative by 11.6 since eC is a coflat left C-comodule.

(4) For every $f \in \mathrm{Hom}^C(M, K\Box_{eCe}Ce)$ and $m \in M$ we know that $f(m) = ef(m) = f(em) \in e(K\Box_{eCe}Ce) \simeq K$. The eCe-comodule morphisms

$$eM \simeq M\Box_C eC \xrightarrow{f\Box I_{eC}} K\Box_{eCe}Ce\Box_C eC \simeq K$$

show that the map $f \mapsto f(e\rightharpoonup -)$ is well defined. The inverse map is obtained by assigning to any $h \in \mathrm{Hom}^{eCe}(M\Box_C eC, K)$ the composition of C-comodule morphisms

$$M \simeq M\Box_C C \xrightarrow{I_M\Box\delta} M\Box_C eC\Box_{eCe}Ce \xrightarrow{h\Box I_{Ce}} K\Box_{eCe}Ce\,.$$

□

The choice of the idempotent e decides about further properties of the functors related to $e\rightharpoonup C$ and $C\leftharpoonup e$.

12.16. Idempotents and equivalences. *Let ${}_RC$ be flat. For an idempotent $e \in C^*$, the following are equivalent:*

(a) the functor $-\Box_C e\rightharpoonup C : \mathbf{M}^C \to \mathbf{M}^{e\rightharpoonup C\leftharpoonup e}$ is an equivalence;

(b) the functor $e\rightharpoonup C\Box_{e\rightharpoonup C\leftharpoonup e} - : {}^{e\rightharpoonup C\leftharpoonup e}\mathbf{M} \to {}^C\mathbf{M}$ is an equivalence;

(c) $\delta : C \to e\rightharpoonup C\Box_{e\rightharpoonup C\leftharpoonup e}C\leftharpoonup e$ (see 12.15) is an isomorphism;

(d) $e\rightharpoonup C$ is faithfully coflat as a left C-comodule.

If R is a QF ring, then (a)–(c) are equivalent to:

(e) δ (as in (c)) is injective;

(f) $e\rightharpoonup C$ is a cogenerator in ${}^C\mathbf{M}$.

In particular, if R is QF and $C^* * e * C^* = C^*$, then δ is injective.

Proof. Again we write eC for $e\rightharpoonup C$, and so on.

(a) ⇔ (b) This follows by 12.4 and 12.2.

(b) ⇔ (d) eC is a direct summand of C as a left C-comodule, and hence it is an injector. Clearly eC is quasi-finite and the assertion follows by 12.12.

(a) ⇔ (c) By the results in 12.15, $-\Box_C eC$ is an equivalence if and only if, for any $M \in \mathbf{M}^C$, $M \simeq (M\Box_C eC)\Box_{eCe}Ce$. In view of the associativity of the cotensor product in the given situation (see 11.6) the assertion is obvious.

(c) ⇒ (e) ⇒ (f) are obvious. Now suppose that R is a QF ring.

(f) ⇒ (d) Since eC is injective in ${}^C\mathbf{M}$, condition (e) implies that it is an injective cogenerator and hence is faithfully coflat and an injector in ${}^C\mathbf{M}$. Hence, by 12.12, eC induces an equivalence.

Finally, write $(-)^*$ for $\mathrm{Hom}_R(-, R)$, and consider the diagram with obvious maps

$$\begin{array}{ccc} C^* \otimes_R C^* & \longrightarrow & (eC \otimes_R Ce)^* \\ \downarrow & & \downarrow \delta^* \\ C^* * e \otimes_R e * C^* & \xrightarrow{\ *\ } & C^*, \end{array}$$

which is commutative since a straightforward computation shows that on both ways $f \otimes g \in C^* \otimes_R C^*$ is mapped to $f * e * g$. Therefore the equality $C^* * e * C^* = C^*$ implies that δ^* is surjective and hence δ is injective. □

Semiperfect rings A are called *basic* if $A/\mathrm{Jac}(A)$ is square-free as a left A-module (no distinct summands are isomorphic). Dually, one can consider

12.17. Basic coalgebras. *Let C be a coalgebra over a QF ring R with ${}_RC$ flat. Then the following are equivalent:*

(a) the left socle of C is square-free;

(b) C is the direct sum of pairwise nonisomorphic injective hulls of simple left comodules;

(c) C is the direct sum of pairwise nonisomorphic left comodules with local endomorphism rings;

(d) every simple left subcomodule of C is fully invariant;

(e) for every minimal (C^*, C^*)-sub-bimodule $U \subset C$, $\mathrm{Hom}_R(U, R)$ is a division ring;

(f) the right socle of C is square-free.

If R is a semisimple ring, then (a)-(f) are equivalent to:

(g) for every minimal subcoalgebra $U \subset C$, $\mathrm{Hom}_R(U, R)$ is a division ring.

Proof. (a) ⇔ (b) C is just the injective hull of its socle in ${}^C\mathbf{M}$, and simple comodules are isomorphic if and only if their injective hulls are.

(b) ⇔ (c) An injective comodule has an essential simple socle if and only if it has a local endomorphism ring.

(a) ⇒ (d) The image of any simple subcomodule $S \subset C$ is either zero or isomorphic to S and hence equal to S. Hence S is fully invariant in C.

(d) ⇒ (e) Any minimal (C^*, C^*)-sub-bimodule $U \subset C$ is the trace of a simple comodule $S \subset C$, and hence equal to S. By Schur's Lemma, $\mathrm{End}^C(U)$ is a division ring and is isomorphic to $\mathrm{Hom}^C(U, U) \simeq \mathrm{Hom}^C(U, C) \simeq \mathrm{Hom}_R(U, R)$.

(e) $\Rightarrow$ (a) Any minimal (C^*, C^*)-sub-bimodule of C is just a homogeneous component of the socle of C. Since its endomorphism ring is a division ring, it must be a simple subcomodule.

(e) $\Leftrightarrow$ (f) Condition (e) is left-right symmetric, and hence the assertion is obvious.

(f) $\Leftrightarrow$ (g) Over a semisimple ring all submodules are direct summands, and hence, by 4.6(4), (C^*, C^*)-sub-bimodules correspond to subcoalgebras. $\square$

The interest in basic coalgebras lies in the fact that over QF rings any coalgebra is Morita-Takeuchi equivalent to a basic coalgebra.

12.18. Basic coalgebra of a coalgebra. *Let C be a coalgebra over a QF ring R with ${}_RC$ flat. Then there exists an idempotent $e \in C^*$ such that $e\rightharpoonup C\leftharpoonup e$ is a basic coalgebra and*

$$-\Box_C e\rightharpoonup C : \mathbf{M}^C \to \mathbf{M}^{e\rightharpoonup C\leftharpoonup e}$$

is an equivalence. This $e\rightharpoonup C\leftharpoonup e$ is called the basic coalgebra of C.

Proof. As in the proof of 12.15 we write eC for $e\rightharpoonup C$, Ce for $C\leftharpoonup e$. Choose a family of pairwise orthogonal idempotents $\{e_\lambda\}_\Lambda$ such that the $e_\lambda C$ form an irredundant representing set of all injective hulls of simple left comodules. This is possible since C is an injective cogenerator in ${}^C\mathbf{M}$. The internal direct sum $\bigoplus_\Lambda e_\lambda C$ is injective and hence a direct summand in C, that is, it has the form eC for a suitable idempotent $e \in C^*$. Clearly eC is a cogenerator and hence induces an equivalence (see 12.16(f)). In particular, $Ce_\lambda \Box_C eC \simeq eCe_\lambda$ and hence $eCe_\lambda \simeq eCe_{\lambda'}$ if and only if $\lambda = \lambda'$.

Consider a right eCe-comodule decomposition $eCe \simeq \bigoplus_\Lambda eCe_\lambda$. All the $\mathrm{End}^{eCe}(eCe_\lambda) \simeq e_\lambda C^* e_\lambda \simeq \mathrm{End}^C(e_\lambda C)$ are local rings, and hence eCe is a basic coalgebra (see 12.17). $\square$

12.19. Projective modules and adjoint functors. *Let P be a finitely generated projective R-module with dual basis $p_1, \ldots, p_n \in P$, $\pi_1, \ldots, \pi_n \in P^*$. Then $C \otimes_R P$ is a direct summand of C^n and hence is quasi-finite. Furthermore, it is a (C, D)-bicomodule where $D = C \otimes_R (P^* \otimes_R P)$ is the tensor product of coalgebras. Considering $C \otimes_R P^*$ as a (D, C)-bicomodule, we obtain an adjoint pair of functors*

$$-\Box_C(C \otimes_R P) : \mathbf{M}^C \to \mathbf{M}^D, \quad -\Box_D(C \otimes_R P^*) : \mathbf{M}^D \to \mathbf{M}^C.$$

Proof. The coalgebra structure on $P^* \otimes_R P$ is introduced in 1.9, and the corresponding comodule structures of P and P^* are given in 4.9. The tensor product of coalgebras is defined in 2.12. There are bicomodule morphisms

$$\begin{array}{rcl} \gamma : C \otimes_R (P^* \otimes_R P) & \to & (C \otimes_R P^*)\Box_C(C \otimes_R P), \\ c \otimes f \otimes p & \mapsto & \sum_c c_{\underline{1}} \otimes f \otimes c_{\underline{2}} \otimes p, \\ \delta : C & \to & (C \otimes_R P)\Box_D(C \otimes_R P^*), \\ c & \mapsto & \sum_{i,c} c_{\underline{1}} \otimes p_i \otimes c_{\underline{2}} \otimes \pi_i, \end{array}$$

where γ is an isomorphism. For $M \in \mathbf{M}^C$ and $K \in \mathbf{M}^D$, the adjointness isomorphism is

$$\mathrm{Hom}^C(M, K\Box_D(C\otimes_R P^*)) \to \mathrm{Hom}^D(M\Box_C(C\otimes_R P), K),\ f \mapsto (I\Box\gamma)\circ(f\Box I),$$

with the inverse map $h \mapsto (h\Box I) \circ (I\Box\delta)$. All these assertions are verified by standard computations. □

Notice that the functors considered in 12.19 induce an equivalence if and only if δ is an isomorphism. This is the case when $P = R^n$, for any $n \in \mathbb{N}$. Then $P^*\otimes_R P$ is simply the matrix coalgebra $M_n^c(R)$ (cf. 1.10) and one obtains the coendomorphism coalgebra of C^n.

12.20. Matrix coalgebra and equivalence. *For any $n \in \mathbb{N}$, the coendomorphism coalgebra of C^n is the coalgebra $M_n^c(C) = C \otimes_R M_n^c(R)$ and there is an equivalence of categories*

$$-\Box_C C^n : \mathbf{M}^C \to \mathbf{M}^{M_n^c(C)}.$$

References. Al-Takhman [51]; Cuadra and Gómez-Torrecillas [102]; Doi [104]; Lin [150]; Takeuchi [197].

Chapter 2

Bialgebras and Hopf algebras

In classical algebraic geometry one thinks about commutative algebras as algebras of functions on spaces. If the underlying space is also a group, the corresponding algebra of functions becomes a coalgebra. Both coalgebra and algebra structures are compatible with each other in the sense that the coproduct and counit are algebra maps. This example motivates studies of coalgebras with a compatible commutative algebra structure. Allowing further generalisations, one considers coalgebras with noncommutative algebraic structures, initially over fields, but eventually over commutative rings. Such algebras with compatible coalgebra structures are known as *bialgebras* and *Hopf algebras* and often are referred to as *quantum groups*.

There are numerous textbooks and monographs on bialgebras, Hopf algebras and quantum groups (the latter mainly addressed to a physics audience), in particular, classic texts by Sweedler [45] or Abe [1], or more recent works (Montgomery [37], Dăscălescu, Năstăsescu and Raianu [14]), including the ones motivated by the quantum group theory (e.g., Lusztig [30], Majid [33, 34], Chari and Pressley [11], Shnider and Sternberg [43], Kassel [25], Klimyk and Schmüdgen [26], Brown and Goodearl [7]). In the majority of these texts it is assumed at the beginning that all algebras and coalgebras are defined over a field. By making this assumption, the authors are excused from not considering some of the module-theoretic aspects of the discussed objects. The aim of the present chapter is to glimpse at bialgebras and Hopf algebras, and study those properties that are significant from the point of view of ring and module theory, or which directly depend on the properties of algebras, coalgebras, and so on, as R-modules.

13 Bialgebras

In this section we are concerned with the compatibility of algebra and coalgebra structures on a given R-module. In particular, we define *bialgebras* and study their most elementary properties.

13.1. Bialgebras. An R-module B that is an algebra (B, μ, ι) and a coalgebra (B, Δ, ε) is called a *bialgebra* if Δ and ε are algebra morphisms or, equivalently, μ and ι are coalgebra morphisms. For Δ to be an algebra morphism one needs commutativity of the diagrams

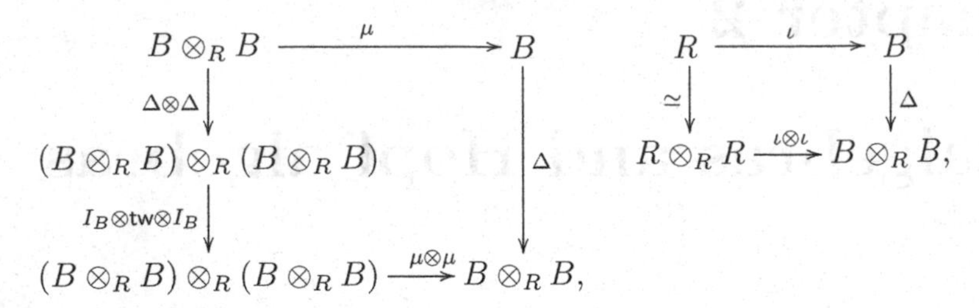

where tw denotes the twist map. Similarly, ε is an algebra morphism if and only if the following two diagrams

$$\begin{array}{ccc} B \otimes_R B & \xrightarrow{\mu} & B \\ {\scriptstyle \varepsilon\otimes\varepsilon}\downarrow & & \downarrow{\scriptstyle \varepsilon} \\ R \otimes_R R & \xrightarrow{\simeq} & R, \end{array} \qquad \begin{array}{ccc} & B & \\ {\scriptstyle \iota}\nearrow & & \searrow{\scriptstyle \varepsilon} \\ R & \xrightarrow{=} & R \end{array}$$

are commutative. The same set of diagrams makes μ and ι coalgebra morphisms. For the units $1_B \in B$, $1_R \in R$ and for all $a, b \in B$, the above diagrams explicitly mean that

$$\Delta(1_B) = 1_B \otimes 1_B, \qquad \varepsilon(1_B) = 1_R,$$
$$\Delta(ab) = \Delta(a)\Delta(b), \qquad \varepsilon(ab) = \varepsilon(a)\varepsilon(b).$$

Note that this implies that, in any R-bialgebra B, R is a direct summand of B as an R-module and hence B is a generator in $\mathbf{M}_R$. As first examples notice that, for any semigroup G, the semigroup coalgebra $R[G]$ is a bialgebra (see 1.7), in particular, the polynomial coalgebras (see 1.8) are bialgebras.

13.2. Decomposition. *Let B be a bialgebra over R.*

(1) $B = R1_B \oplus \mathrm{Ke}\,\varepsilon$ is a direct R-module decomposition.

(2) If the family $\{b_\lambda\}_\Lambda$, $b_\lambda \in B$, generates B as an R-module, then the family $\{b_\lambda - \varepsilon(b_\lambda)1_B\}_\Lambda$ generates $\mathrm{Ke}\,\varepsilon$ as an R-module.

Proof. (1) follows from the axioms for ε.

(2) Clearly, for each $\lambda \in \Lambda$, $b_\lambda - \varepsilon(b_\lambda)1_B \in \mathrm{Ke}\,\varepsilon$. Let $l = \sum r_\lambda b_\lambda \in \mathrm{Ke}\,\varepsilon$, that is, $\varepsilon(l) = \sum r_\lambda \varepsilon(b_\lambda) = 0$. This implies

$$l = \sum r_\lambda b_\lambda = \sum r_\lambda (b_\lambda - \varepsilon(b_\lambda)1_B,$$

thus proving the assertion. □

13.3. Bialgebra morphisms. An R-linear map $f : B \to B'$ of bialgebras is called a *bialgebra morphism* if f is both an algebra and a coalgebra morphism.

An R-submodule $I \subset B$ is a *sub-bialgebra* if it is a subalgebra as well as a subcoalgebra. I is a *bi-ideal* if it is both an ideal and a coideal.

Let $f : B \to B'$ be a bialgebra morphism. Then:

(1) *If f is surjective, then* $\operatorname{Ke} f$ *is a bi-ideal in B.*

(2) $\operatorname{Im} f$ *is a subcoalgebra of B'.*

(3) *For any bi-ideal $I \subset B$ contained in* $\operatorname{Ke} f$*, there is a commutative diagram of bialgebra morphisms*

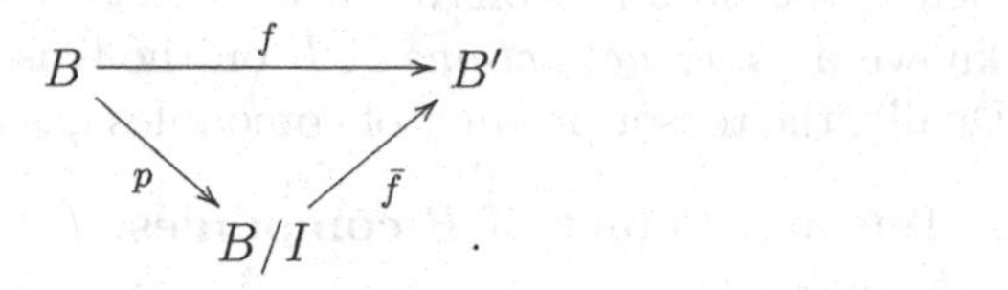

A remarkable feature of a bialgebra B is that the tensor product of B-modules is again a B-module. In other words, $\mathbf{M}_B$ (or ${}_B\mathbf{M}$) is a monoidal category with the tensor product $\otimes_R$ (cf. 38.31). This requires an appropriate definition of the action of B on the tensor product of modules, which we describe now. First, recall that an R-module N is a B-module if there is an algebra morphism $B \to \operatorname{End}_R(N)$.

13.4. Tensor product of B-modules. *Let K, L be right modules over an R-bialgebra B.*

(1) *$K \otimes_R L$ has a right B-module structure by the map*

$$B \xrightarrow{\Delta} B \otimes_R B \to \operatorname{End}_R(K) \otimes_R \operatorname{End}_R(L) \to \operatorname{End}_R(K \otimes_R L);$$

we denote this module by $K \otimes^b_R L$. The right action of B is given by

$$! : K \otimes_R L \otimes_R B \longrightarrow K \otimes_R L, \quad k \otimes l \otimes b \mapsto (k \otimes l)\Delta b,$$

where the product on the right side is taken componentwise, that is,

$$(k \otimes l) \,!\, b := (k \otimes l)\, \Delta b = \sum k b_{\underline{1}} \otimes l b_{\underline{2}}\,.$$

(2) *For any morphisms $f : K \to K'$, $g : L \to L'$ in $\mathbf{M}_B$, the tensor product map $f \otimes g : K \otimes^b_R L \to K' \otimes^b_R L'$ is a morphism in $\mathbf{M}_B$.*

Proof. (1) follows easily from the definitions. Assertion (2) is equivalent to the commutativity of the following diagram:

$$\begin{array}{ccc} K \otimes^b_R L \otimes_R B & \xrightarrow{f\otimes g\otimes I_B} & K' \otimes^b_R L' \otimes_R B \\ {\scriptstyle !}\downarrow & & \downarrow{\scriptstyle !} \\ K \otimes^b_R L & \xrightarrow{f\otimes g} & K' \otimes^b_R L', \end{array}$$

which follows immediately from B-linearity of f and g. □

Of course similar constructions apply to left B-modules K, L, in which case the left B-multiplication is given by

$$! : B \otimes_R K \otimes_R L \longrightarrow K \otimes_R L, \quad b \otimes k \otimes l \mapsto \Delta b(k \otimes l).$$

Explicitly, the product comes out as $b!(k \otimes l) = \sum b_{\underline{1}} k \otimes b_{\underline{2}} l$. The actions ! are known as *diagonal actions* of B on the tensor product of its modules.

Dually, the tensor product of comodules has a special comodule structure.

13.5. Tensor product of B-comodules. *Let K, L be right comodules over an R-bialgebra B.*

(1) $K \otimes_R L$ has a right B-comodule structure by the map (tensor over R)

$$\varrho^{K \otimes L} : K \otimes L \xrightarrow{\mathsf{tw}_{23} \circ (\varrho^K \otimes \varrho^L)} K \otimes L \otimes B \otimes B \xrightarrow{I_K \otimes I_L \otimes \mu} K \otimes L \otimes B,$$

where $\mathsf{tw}_{23} = I_K \otimes \mathsf{tw} \otimes I_B$. This comodule is denoted by $K \otimes^c_R L$. Thus, explicitly, for all $k \otimes l \in K \otimes^c_R L$,

$$\varrho^{K \otimes_R L}(k \otimes l) = \sum k_{\underline{0}} \otimes l_{\underline{0}} \otimes k_{\underline{1}} l_{\underline{1}}.$$

(2) For any morphisms $f : K \to K'$, $g : L \to L'$ in $\mathbf{M}^B$, the tensor product map $f \otimes g : K \otimes^c_R L \to K' \otimes^c_R L'$ is a morphism in $\mathbf{M}^B$.

Proof. (1) This is proved by computing for all $k \in K$, $l \in L$,

$$\begin{aligned}
(I_K \otimes I_L \otimes \Delta) \circ \varrho^{K \otimes_R L}(k \otimes l) &= \sum k_{\underline{0}} \otimes l_{\underline{0}} \otimes \Delta(k_{\underline{1}} l_{\underline{1}}) \\
&= \sum k_{\underline{0}} \otimes l_{\underline{0}} \otimes k_{\underline{11}} l_{\underline{11}} \otimes k_{\underline{12}} l_{\underline{12}} \\
&= \sum k_{\underline{00}} \otimes l_{\underline{00}} \otimes k_{\underline{01}} l_{\underline{01}} \otimes k_{\underline{11}} l_{\underline{11}} \\
&= (\varrho^{K \otimes_R L} \otimes I_B) \circ \varrho^{K \otimes_R L}(k \otimes l).
\end{aligned}$$

To prove (2), take any $k \in K$, $l \in L$ and compute

$$\begin{aligned}
\varrho^{K' \otimes_R L'} \circ (f \otimes g)(k \otimes l) &= \sum f(k)_{\underline{0}} \otimes g(l)_{\underline{0}} \otimes f(k)_{\underline{1}} g(l)_{\underline{1}} \\
&= \sum f(k_{\underline{0}}) \otimes g(l_{\underline{0}}) \otimes k_{\underline{1}} l_{\underline{1}} \\
&= (f \otimes g \otimes I_B) \circ \varrho^{K \otimes_R L}(k \otimes l).
\end{aligned}$$

This shows that $f \otimes g$ is a comodule morphism, as required. □

The coaction constructed in 13.5 is known as a *diagonal coaction* of a bialgebra B on the tensor product of its comodules.

In contrast to coalgebras, for a bialgebra B, any R-module K can be considered as B-comodule by $K \to K \otimes_R B$, $k \mapsto k \otimes 1_B$ (trivial coaction). In particular, the ring R is a right B-comodule, and this draws attention to those maps $B \to R$ that are comodule morphisms.

Definition. An element $t \in B^*$ is called a *left integral on B* if it is a left comodule morphism.

Recall that the rational part of B^* is denoted by $\mathrm{Rat}^B(B^*) = T$ and $\varrho^T : T \to T \otimes_R B$ denotes the corresponding coaction.

13.6. Left integrals on B. *Let B be an R-bialgebra and $t \in B^*$.*

(1) The following are equivalent:

(a) t is a left integral on B;

(b) $(I_B \otimes t) \circ \Delta = \iota \circ t$.

If B is cogenerated by R as an R-module, then (a) is equivalent to:

(c) For every $f \in B^*$, $f * t = f(1_B)t$.

(2) Assume that ${}_RB$ is locally projective.

(i) If $t \in T$, then t is a left integral on B if and only if $\varrho^T(t) = t \otimes 1_B$.

(ii) If R is Noetherian or if $t(B) = R$, then any left integral t on B is rational, that is, $t \in T$.

Proof. (1) (a) $\Leftrightarrow$ (b) The map t is left colinear if and only if there is a commutative diagram

$$\begin{array}{ccc} B & \xrightarrow{\ t\ } & R \\ {\scriptstyle\Delta}\big\downarrow & & \big\downarrow{\scriptstyle\iota} \\ B \otimes_R B & \xrightarrow{I_B \otimes t} & B \end{array} \qquad \begin{array}{ccc} b & \longmapsto & t(b) \\ \big\downarrow & & \big\downarrow \\ \Delta(b) & \longmapsto & (I_B \otimes t) \circ \Delta(b) = t(b)1_B. \end{array}$$

The commutativity of this diagram is expressed by condition (b).

(b) $\Leftrightarrow$ (c) For any $f \in B^*$ and $b \in B$,

$$\begin{array}{rcccl} f * t(b) & = & (f \otimes t) \circ \Delta(b) & = & f((I_B \otimes t) \circ \Delta(b)), \\ f(1_B)t(b) & = & f(t(b)1_B) & = & f(\iota \circ t(b)). \end{array}$$

From this (b) $\Rightarrow$ (c) is obvious. If B is cogenerated by R, then (c) $\Rightarrow$ (b).

(2)(i) If $t \in T$, that is, t is rational, then $f * t = (I_T \otimes f) \circ \varrho^T(t)$, for any $f \in B^*$, and (1)(c) implies

$$(I_T \otimes f)(\varrho^T(t)) = (I_T \otimes f)(t \otimes 1_B).$$

By local projectivity (α-condition; see 4.2) this means $\varrho^T(t) = t \otimes 1_B$. The converse conclusion is obvious.

(ii) By (1)(c), $B^* * t \subset R1_B$. If R is Noetherian, this implies that $B^* * t$ is finitely presented as an R-module, and by 7.5(2) this implies that the element $t \in \mathrm{Rat}^B(B^*) = T$. If $t(B) = R$, then $t \rightharpoonup B = (I_B \otimes t)\Delta(B) = \iota \circ t(B) = R1_B$ is finitely presented as an R-module and $t \in T$ by 7.5(1). □

References. Abe [1]; Dăscălescu, Năstăsescu and Raianu [14]; Montgomery [37]; Sweedler [45].

14 Hopf modules

Since an R-bialgebra B has both a coalgebra and an algebra structure, one can study R-modules that are both B-modules and B-comodules. Furthermore, since the coalgebra structure of B must be compatible with the algebra structure of B, one can require compatibility conditions for corresponding modules and comodules. This leads to the notion of *Hopf modules*, which, together with various generalisations (cf. Section 32), play an important role in representation theory of bialgebras, or indeed in classical module theory, in particular in the case of modules graded by groups. In this section we introduce Hopf modules, provide several constructions of such modules, study their category-theoretic aspects, and properties of their invariants and coinvariants. The latter can be viewed as a preparation for the Fundamental Theorem of Hopf algebras 15.5 in Section 15. We concentrate on the case when the module and comodule structures are given on the right side.

Throughout this section B denotes an R-bialgebra with product μ, coproduct Δ, unit map ι and counit ε.

14.1. B-Hopf modules. An R-module M is called a *right B-Hopf module* if M is

(i) a right B-module with an action $\varrho_M : M \otimes_R B \to M$,

(ii) a right B-comodule with a coaction $\varrho^M : M \to M \otimes_R B$,

(iii) for all $m \in M, b \in B$, $\varrho^M(mb) = \varrho^M(m)\Delta(b)$.

Condition (iii) means that $\varrho^M : M \to M \otimes^b_R B$ is B-linear and is equivalent to the requirement that the multiplication $\varrho_M : M \otimes^c_R B \to M$ is B-colinear, or to the commutativity of either of the diagrams

$$\begin{array}{ccc} M \otimes_R B & \xrightarrow{\varrho^M \otimes I_B} & M \otimes^b_R B \otimes_R B \\ {\scriptstyle \varrho_M} \downarrow & & \downarrow ! \\ M & \xrightarrow{\varrho^M} & M \otimes_R B\,, \end{array} \qquad \begin{array}{ccc} M \otimes^c_R B & \xrightarrow{\varrho^{M \otimes^c B}} & M \otimes^c_R B \otimes_R B \\ {\scriptstyle \varrho_M} \downarrow & & \downarrow {\scriptstyle \varrho_M \otimes I_B} \\ M & \xrightarrow{\varrho^M} & M \otimes_R B\,. \end{array}$$

An R-linear map $f : M \to N$ between right B-Hopf modules is a *Hopf module morphism* if it is both a right B-module and a right B-comodule morphism. Denoting these maps by

$$\mathrm{Hom}^B_B(M, N) = \mathrm{Hom}_B(M, N) \cap \mathrm{Hom}^B(M, N),$$

there are characterising exact sequences in $\mathbf{M}_R$,

$$0 \to \mathrm{Hom}^B_B(M, N) \to \mathrm{Hom}_B(M, N) \xrightarrow{\gamma} \mathrm{Hom}_B(M, N \otimes^b_R B),$$

where $\gamma(f) = \varrho^N \circ f - (f \otimes I_B) \circ \varrho^M$ or, equivalently,

$$0 \to \mathrm{Hom}^B_B(M, N) \to \mathrm{Hom}^B(M, N) \xrightarrow{\delta} \mathrm{Hom}^B(M \otimes^c_R B, N),$$

where $\delta(g) = \varrho_N \circ (g \otimes I_B) - g \circ \varrho_M$.

Left B-Hopf modules and the corresponding morphisms are defined similarly, and it is obvious that B is both a right and a left B-Hopf module with the action given by the product and the coaction given by the coproduct (regular coaction).

We give the following three motivating examples of right B-Hopf modules.

14.2. Trivial B-Hopf modules. *Let K be any R-module.*

(1) *$K \otimes_R B$ is a right B-Hopf module with the canonical structures*

$$I_K \otimes \Delta : K \otimes_R B \to (K \otimes_R B) \otimes_R B, \quad I_K \otimes \mu : (K \otimes_R B) \otimes_R B \to K \otimes_R B.$$

(2) *For any R-linear map $f : K \to K'$, the map $f \otimes I_B : K \otimes_R B \to K' \otimes_R B$ is a B-Hopf module morphism.*

Proof. We know that $K \otimes_R B$ is both a right B-module, and a comodule and the compatibility conditions are obvious from the properties of a bialgebra. It is clear that $f \otimes I_B$ is B-linear as well as B-colinear. □

14.3. B-modules and B-Hopf modules. *Let N be any right B-module.*

(1) *The right B-module $N \otimes_R^b B$ is a right B-Hopf module with the canonical comodule structure*

$$I_N \otimes \Delta : N \otimes_R^b B \to (N \otimes_R^b B) \otimes_R B, \quad n \otimes b \mapsto n \otimes \Delta b.$$

(2) *For any B-linear map $f : N \to N'$, the map $f \otimes I_B : N \otimes_R^b B \to N' \otimes_R^b B$ is a B-Hopf module morphism.*

(3) *The map*

$$\gamma_N : N \otimes_R B \to N \otimes_R^b B, \quad n \otimes b \mapsto (n \otimes 1_B)\Delta(b) = (n \otimes 1_B)!b$$

is a B-Hopf module morphism.

Proof. (1) To show that $I_N \otimes \Delta$ is B-linear one needs to check the commutativity of the following diagram:

$$\begin{array}{ccc} N \otimes_R^b B \otimes_R B & \xrightarrow{I_N \otimes \Delta \otimes I_B} & (N \otimes_R^b B) \otimes_R^b B \otimes_R B \\ \downarrow ! & & \downarrow ! \\ N \otimes_R^b B & \xrightarrow{I_N \otimes \Delta} & (N \otimes_R^b B) \otimes_R^b B\,. \end{array}$$

Evaluating this diagram at any $a, b \in B$ and $n \in N$ yields

$$\begin{aligned} (I_N \otimes \Delta)\left((n \otimes b)\Delta(a)\right) &= \sum n a_{\underline{1}} \otimes (b a_{\underline{2}})_{\underline{1}} \otimes (b a_{\underline{2}})_{\underline{2}} \\ &= \sum n a_{\underline{1}} \otimes b_{\underline{1}} a_{\underline{2}} \otimes b_{\underline{2}} a_{\underline{3}} \\ &= \left((I_N \otimes \Delta)(n \otimes b)\right) \Delta(a), \end{aligned}$$

by the multplicativity of Δ and the definition of the diagonal B-action on $(N \otimes^b_R B) \otimes^b_R B$ (cf. 13.4).

(2) It was shown in 3.8 that $f \otimes I_B$ is a comodule morphism, and we know from 13.4 that it is a B-module morphism.

(3) Clearly γ_N is B-colinear, and for any $c \in B$,

$$\gamma_N(n \otimes bc) = (n \otimes 1_B)\Delta(bc) = (n \otimes 1_B)(\Delta b)(\Delta c) = \gamma_N(n \otimes b)\Delta(c),$$

showing that γ_N is right B-linear. □

14.4. B-comodules and B-Hopf modules. *Let L be a right B-comodule.*

(1) The right B-comodule $L \otimes^c_R B$ is a right B-Hopf module with the canonical module structure

$$I_L \otimes \mu : L \otimes^c_R B \otimes_R B \to L \otimes^c_R B, \quad n \otimes b \otimes a \mapsto n \otimes ba.$$

(2) For any B-colinear map $f : L \to L'$, the map $f \otimes I_B : L \otimes^c_R B \to L' \otimes^c_R B$ is a B-Hopf module morphism.

(3) There is a B-Hopf module morphism

$$\gamma^L : L \otimes^c_R B \to L \otimes_R B, \quad l \otimes b \mapsto \varrho^L(l)(1_B \otimes b).$$

Proof. (1) To prove the colinearity of $I_L \otimes \mu$ one needs to show the commutativity of the diagram

$$\begin{array}{ccc} L \otimes^c_R B \otimes^c_R B & \xrightarrow{I_L \otimes \mu} & L \otimes^c_R B \\ \downarrow{\scriptstyle \varrho^{L \otimes^c B \otimes^c B}} & & \downarrow{\scriptstyle \varrho^{L \otimes^c B}} \\ L \otimes^c_R B \otimes^c_R B \otimes_R B & \xrightarrow{I_L \otimes \mu \otimes I_B} & L \otimes^c_R B \otimes_R B, \end{array}$$

which follows from the multiplicativity of Δ.

(2) Clearly $f \otimes I_B$ is B-linear, and, as shown in 13.5, it is also B-colinear.

(3) Obviously γ^L is right B-linear, and colinearity follows from the commutativity of the diagram (which is easily checked)

$$\begin{array}{ccc} L \otimes^c_R B & \xrightarrow{\gamma^L} & L \otimes_R B \\ \downarrow{\scriptstyle \varrho^{L \otimes B}} & & \downarrow{\scriptstyle I_L \otimes \Delta} \\ L \otimes^c_R B \otimes_R B & \xrightarrow{\gamma^L \otimes I} & L \otimes_R B \otimes_R B \end{array} \qquad \begin{array}{ccc} l \otimes b & \longmapsto & \sum l_{\underline{0}} \otimes l_{\underline{1}} b \\ \downarrow & & \downarrow \\ \sum l_{\underline{0}} \otimes b_{\underline{1}} \otimes l_{\underline{1}} b_{\underline{2}} & \longmapsto & \sum l_{\underline{0}} \otimes l_{\underline{1}} b_{\underline{1}} \otimes l_{\underline{2}} b_{\underline{2}}. \end{array}$$

This completes the proof. □

Right B-Hopf modules together with B-Hopf module morphisms form a category that is denoted by $\mathbf{M}^B_B$. This category is closed under direct sums and factor modules and has the following properties.

14.5. The category $\mathbf{M}_B^B$. *Let B be an R-bialgebra.*

(1) The right B-Hopf module $B \otimes_R^b B$ is a subgenerator in $\mathbf{M}_B^B$.

(2) The right B-Hopf module $B \otimes_R^c B$ is a subgenerator in $\mathbf{M}_B^B$.

(3) For any $M \in \mathbf{M}_B^B$, $N \in \mathbf{M}_B$,

$$\mathrm{Hom}_B^B(M, N \otimes_R^b B) \to \mathrm{Hom}_B(M,N),\ f \mapsto (I_N \otimes \varepsilon) \circ f,$$

is an R-module isomorphism with inverse map $h \mapsto (h \otimes I_B) \circ \varrho^M$.

(4) For any $M \in \mathbf{M}_B^B$, $N \in \mathbf{M}^B$,

$$\mathrm{Hom}_B^B(N \otimes_R^c B, M) \to \mathrm{Hom}^B(N,M),\ f \mapsto f(- \otimes 1_B),$$

is an R-module isomorphism with inverse map $h \mapsto \varrho_M \circ (h \otimes I_B)$.

(5) For any $K, L \in \mathbf{M}_R$,

$$\mathrm{Hom}_B^B(K \otimes_R B, L \otimes_R B) \to \mathrm{Hom}_R(K,L),\ f \mapsto (I_L \otimes \varepsilon) \circ f(- \otimes 1_B),$$

is an R-module isomorphism with inverse map $h \mapsto h \otimes I_B$.

Proof. (1) Let $M \in \mathbf{M}_B^B$. For a B-module epimorphism $f : B^{(\Lambda)} \to M$,

$$f \otimes I_B : B^{(\Lambda)} \otimes_R^b B \to M \otimes_R^b B$$

is an epimorphism in $\mathbf{M}_B^B$ (by 14.3), and so $M \otimes_R^b B$ is generated by

$$B^{(\Lambda)} \otimes_R^b B \simeq \left(B \otimes_R^b B\right)^{(\Lambda)}.$$

Moreover, $\varrho^M : M \to M \otimes_R^b B$ is a (B-splitting) Hopf module monomorphism, and so M is subgenerated by $B \otimes_R^b B$.

(2) For any $M \in \mathbf{M}_B^B$, there is a comodule epimorphism $B^{(\Lambda)} \to M \otimes_R B$, and from this we obtain a Hopf module epimorphism

$$(B \otimes_R^c B)^{(\Lambda)} \simeq B^{(\Lambda)} \otimes_R^c B \to (M \otimes_R B) \otimes_R^c B.$$

Moreover, there is a Hopf module monomorphism $\varrho^M \otimes I_B : M \otimes_R^c B \to (M \otimes_R B) \otimes_R^c B$ and a Hopf module epimorphism $M \otimes_R^c B \to M$, and hence M is subgenerated by $B \otimes_R^c B$.

(3) There is a commutative diagram with exact rows ($\otimes$ means $\otimes_R^b$),

$$\begin{array}{ccccccc}
0 \longrightarrow & \mathrm{Hom}_B^B(M, N\otimes B) & \longrightarrow & \mathrm{Hom}^B(M, N\otimes B) & \xrightarrow{\beta_1} & \mathrm{Hom}^B(M\otimes B, N\otimes B) \\
& \downarrow & & \downarrow{\scriptstyle (I_N\otimes\varepsilon)\circ -} & & \downarrow{\scriptstyle (I_N\otimes\varepsilon)\circ -} \\
0 \longrightarrow & \mathrm{Hom}_B(M,N) & \longrightarrow & \mathrm{Hom}_R(M,N) & \xrightarrow{\beta_2} & \mathrm{Hom}_R(M\otimes B, N),
\end{array}$$

where $\beta_1(f) = f \circ \varrho_M - \varrho_{N\otimes B} \circ (f \otimes I_B)$ and $\beta_2(g) = g \circ \varrho_M - \varrho_N \circ (g \otimes I_B)$. As shown in 3.9, the second and third vertical maps are isomorphisms and hence the first one is also an isomorphism.

(4) Consider the commutative diagram with exact rows ($\otimes$ for $\otimes^c_R$),

$$\begin{array}{ccccccc}
0 \longrightarrow & \mathrm{Hom}^B_B(N\otimes B, M) & \longrightarrow & \mathrm{Hom}_B(N\otimes B, M) & \xrightarrow{\gamma_1} & \mathrm{Hom}_B(N\otimes B, M\otimes B) \\
& \downarrow & & \downarrow {\scriptstyle -\circ(-\otimes 1_B)} & & \downarrow {\scriptstyle -\circ(-\otimes 1_B)} \\
0 \longrightarrow & \mathrm{Hom}^B(N, M) & \longrightarrow & \mathrm{Hom}_R(N, M) & \xrightarrow{\gamma_2} & \mathrm{Hom}_R(N, M\otimes B),
\end{array}$$

where $\gamma_1(f) = \varrho^M \circ f - (f \otimes I_B) \circ \varrho^{N\otimes B}$ and $\gamma_2(g) = \varrho^M \circ g - (g \otimes I_B) \circ \varrho^N$. The second and third vertical maps are isomorphisms and hence the first one is an isomorphism, too.

(5) View K as a trivial B-comodule. Then $K \otimes^c_R B \simeq K \otimes_R B$, and, by (4) and 3.9, $\mathrm{Hom}^B_B(K \otimes_R B, L \otimes_R B) \simeq \mathrm{Hom}^B(K, L \otimes_R B) \simeq \mathrm{Hom}_R(K, L)$, as required. □

14.6. $\mathbf{M}^B_B$ for B_R flat. *Let B be flat as an R-module and $M, N \in \mathbf{M}^B_B$. Then:*

(1) $\mathbf{M}^B_B$ is a Grothendieck category.

(2) The functor $\mathrm{Hom}^B_B(M, -) : \mathbf{M}^B_B \to \mathbf{M}_R$ is left exact.

(3) The functor $\mathrm{Hom}^B_B(-, N) : \mathbf{M}^B_B \to \mathbf{M}_R$ is left exact.

Proof. (1) For any morphism $f : M \to N$ in $\mathbf{M}^B_B$, $\mathrm{Ke}\, f$ is a B-submodule as well as a B-subcomodule (since B_R flat) and hence $\mathrm{Ke}\, f \in \mathbf{M}^B_B$.

(2) Any exact sequence $0 \to X \to Y \to Z$ in $\mathbf{M}^B_B$ induces the commutative diagram with exact columns

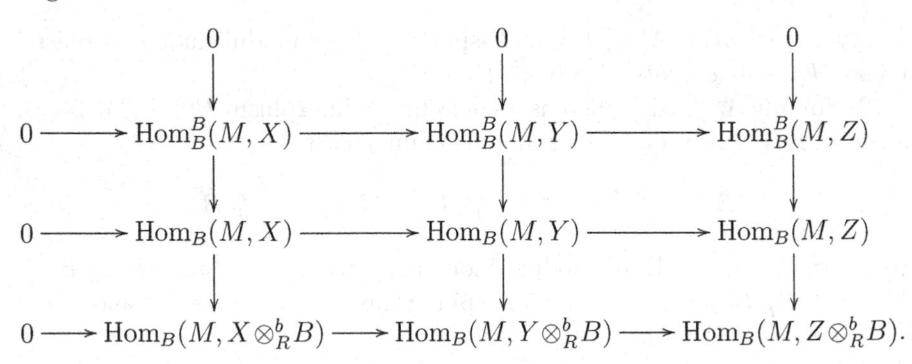

The columns are simply the defining sequences of $\mathrm{Hom}^B_B(M, -)$ in 14.1. The second and third rows are exact because of the left exactness of $\mathrm{Hom}_B(M, -)$ and $- \otimes_R B$. Now the diagram lemmata imply that the first row is exact.

(3) This is shown with a similar diagram that uses $- \otimes^c_R B$ instead of $- \otimes^b_R B$. □

14.7. Coinvariants of comodules. For $M \in \mathbf{M}^B$, the *coinvariants* of B in M are defined as

$$M^{coB} := \{m \in M \mid \varrho^M(m) = m \otimes 1_B\} = \mathrm{Ke}\,(\varrho^M - (- \otimes 1_B)).$$

This is clearly an R-submodule of M and there is an isomorphism

$$\mathrm{Hom}^B(R, M) \to M^{coB}, \quad f \mapsto f(1),$$

where R is considered as a B-comodule. In particular, this implies that $B^{coB} = R1_B$. Furthermore, for any R-module K,

$$\mathrm{Hom}^B(K, M) \simeq \mathrm{Hom}_R(K, M^{coB}),$$

where K is considered as a trivial B-comodule.

The last isomorphism follows by the fact that $f \in \mathrm{Hom}^B(K, M)$ is equivalent to the commutativity of the diagram

$$\begin{array}{ccc} K & \xrightarrow{f} & M \\ {\scriptstyle -\otimes 1_B}\downarrow & & \downarrow{\scriptstyle \varrho^M} \\ K \otimes_R B & \xrightarrow{f \otimes I_B} & M \otimes_R B \end{array} \qquad \begin{array}{ccc} k & \longmapsto & f(k) \\ \downarrow & & \downarrow \\ k \otimes 1_B & \longmapsto & f(k) \otimes 1_B = \varrho^M(f(k))\,. \end{array}$$

14.8. Coinvariants of Hopf modules. *For any $M \in \mathbf{M}_B^B$, the map*

$$\nu_M : \mathrm{Hom}_B^B(B, M) \to M^{coB}, \quad f \mapsto f(1_B),$$

is an R-module isomorphism with the inverse $\omega_M : m \mapsto [b \mapsto mb]$. Furthermore, the diagram

$$\begin{array}{ccc} \mathrm{Hom}_B^B(B, M) \otimes_R B & \longrightarrow & M \\ {\scriptstyle \nu_M \otimes I_B}\downarrow & & \| \\ M^{coB} \otimes_R B & \longrightarrow & M \end{array} \qquad \begin{array}{ccc} f \otimes b & \longmapsto & f(b) \\ \downarrow & & \| \\ f(1_B) \otimes b & \longrightarrow & f(1_B)b \end{array}$$

is commutative. In particular, $\mathrm{Hom}_B^B(B, B) \xrightarrow{\simeq} B^{coB} = R1_B$ is a ring isomorphism.

Proof. The isomorphism ν_M is obtained from the following commutative diagram of R-module maps with exact rows:

$$\begin{array}{ccccccc} 0 & \longrightarrow & \mathrm{Hom}_B^B(B, M) & \longrightarrow & \mathrm{Hom}_B(B, M) & \xrightarrow{\gamma_1} & \mathrm{Hom}_B(B, M \otimes_R B) \\ & & \downarrow & & \downarrow{\scriptstyle \simeq} & & \downarrow{\scriptstyle \simeq} \\ 0 & \longrightarrow & M^{coB} & \longrightarrow & M & \xrightarrow{\gamma_2} & M \otimes_R B, \end{array}$$

where $\gamma_1(f) = \varrho^M \circ f - (f \otimes I_B) \circ \Delta$, that is, the top row is the defining sequence of $\mathrm{Hom}_B^B(B, M)$, and $\gamma_2(m) = \varrho^M(m) - m \otimes 1_B$. □

14.9. Coinvariants of trivial Hopf modules.

(1) For any $K \in \mathbf{M}_R$, $\mathrm{Hom}_B^B(B, K \otimes_R B) \simeq K$ *as* R*-modules.*

(2) For all $L \in \mathbf{M}_R$ *and* $M \in \mathbf{M}_B^B$, *there are* R*-module isomorphisms*

$$\mathrm{Hom}_B^B(L \otimes_R B, M) \simeq \mathrm{Hom}_R(L, M^{coB}) \quad \text{and} \quad \mathrm{End}_B^B(B \otimes_R B) \simeq \mathrm{End}_R(B).$$

(3) There is an adjoint pair of functors

$$- \otimes_R B : \mathbf{M}_R \to \mathbf{M}_B^B, \quad \mathrm{Hom}_B^B(B, -) : \mathbf{M}_B^B \to \mathbf{M}_R,$$

and $\mathrm{Hom}_B^B(B, - \otimes_R B) \simeq I_{\mathbf{M}_R}$.

Proof. (1) Consider R as a B-comodule as in 14.5(4). Then the Hom-tensor relation 3.9(1) implies

$$\mathrm{Hom}_B^B(B, K \otimes_R B) \simeq \mathrm{Hom}_B^B(R \otimes_R^c B, K \otimes_R B) \simeq \mathrm{Hom}^B(R, K \otimes_R B) \simeq K.$$

(2) Combining 14.5(4) and 14.7, one obtains the chain of isomorphisms

$$\mathrm{Hom}_B^B(L \otimes_R B, M) \simeq \mathrm{Hom}^B(L, M) \simeq \mathrm{Hom}_R(L, M^{coB}).$$

(3) By 14.8, the adjointness is just an interpretation of the isomorphism in (2), and, by (1), the composition of the two functors is isomorphic to the identity functor on $\mathbf{M}_R$. □

14.10. Coinvariants and B-modules. *For any* $N \in \mathbf{M}_B$, *the map*

$$\nu'_{N \otimes B} : \mathrm{Hom}_B^B(B, N \otimes_R^b B) \to N, \; f \mapsto (I_N \otimes \varepsilon) \circ f(1_B),$$

is an R*-isomorphism with the inverse* $n \mapsto [b \mapsto \sum n b_{\underline{1}} \otimes b_{\underline{2}}]$. *Furthermore, the diagram*

$$\begin{array}{ccc} \mathrm{Hom}_B^B(B, N \otimes_R^b B) \otimes_R B & \longrightarrow & N \otimes_R^b B \\ {\scriptstyle \nu'_{N\otimes B} \otimes I_B} \downarrow & & \| \\ N \otimes_R B & \xrightarrow{\gamma_N} & N \otimes_R^b B \end{array} \qquad \begin{array}{ccc} g \otimes b & \longmapsto & g(b) \\ \downarrow & & \| \\ (I_N \otimes \varepsilon) g(1_B) \otimes b & \longmapsto & g(1_B)\Delta b, \end{array}$$

where γ_N *is described in 14.3(3), is commutative. This yields in particular*

$$(B \otimes_R^b B)^{coB} \simeq \mathrm{Hom}_B^B(B, B \otimes_R^b B) \simeq B,$$

and the commutative diagram

$$\begin{array}{ccc} \mathrm{Hom}_B^B(B, B \otimes_R^b B) \otimes_R B & \longrightarrow & B \otimes_R^b B \\ \downarrow & & \| \\ B \otimes_R B & \xrightarrow{\gamma_B} & B \otimes_R^b B \end{array} \qquad \begin{array}{ccc} h \otimes a & \longmapsto & h(a) \\ \downarrow & & \| \\ (I_B \otimes \varepsilon) h(1_B) \otimes a & \longmapsto & h(1_B)\Delta a. \end{array}$$

Proof. By 14.5, $\mathrm{Hom}_B^B(B, N \otimes_R^b B) \simeq \mathrm{Hom}_B(B, N) \simeq N$ and commutativity of the diagrams is shown by a straightforward computation. □

14.11. Invariants. Let A be an R-algebra A and $\varphi : A \to R$ a ring morphism. Considering R as a left A-module, one may ask for the A-morphisms from $R \to M$, where $M \in {}_A\mathbf{M}$. Define the *invariants* of M by

$$^AM = \{m \in M \mid am = \varphi(a)m \text{ for all } a \in A\}.$$

Then the map ${}_A\mathrm{Hom}(R, M) \to {}^AM$, $f \mapsto f(1)$, is an R-module isomorphism.

14.12. Invariants for bialgebras. For any bialgebra B, the counit ε is a ring morphism and hence induces a left and right B-module structure on R. Therefore, for any left B-module M, the *invariants of* M corresponding to ε come out as

$$^BM = \{m \in M \mid bm = \varepsilon(b)m \text{ for all } b \in B\}.$$

Furthermore, the map ${}_B\mathrm{Hom}(R, M) \to {}^BM$, $f \mapsto f(1)$, is an R-module isomorphism. The left invariants ${}^BB$ of B are called *left integrals in* B,

$${}_B\mathrm{Hom}\,(R, B) \simeq {}^BB = \{c \in B \mid bc = \varepsilon(b)c \text{ for all } b \in B\}.$$

Right invariants and *right integrals in* B are defined symmetrically .

On the other hand, for the dual algebra B^*, the map $\varphi : B^* \to R$, $f \mapsto f(1_B)$, is a ring morphism. Coinvariants of right B-comodules are closely related to invariants of left B^*-modules corresponding to φ.

14.13. Invariants and coinvariants. *Let B be a bialgebra that is locally projective as an R-module (cf. 4.2).*

(1) *For any $M \in \mathbf{M}^B$, ${}^{B^*}M = M^{coB}$.*

(2) *For the trace ideal $T = \mathrm{Rat}^B(B^*)$, ${}^{B^*}T = T^{coB}$.*

(3) *If B_R is finitely generated, then ${}^{B^*}B^* = (B^*)^{coB}$.*

Proof. (1) Let $m \in {}^{B^*}M$ and $f \in B^*$. From $f \rightharpoonup m = \sum m_{\underline{0}} f(m_{\underline{1}})$ we conclude

$$(I_M \otimes f)\varrho^M(m) = (I_M \otimes f)(m \otimes 1_B).$$

Now local projectivity of B implies that $\varrho_M(m) = m \otimes 1_B$, that is, $m \in M^{coB}$, as required. Conversely, take any $m \in M^{coB}$ and compute

$$f \rightharpoonup m = (I_M \otimes f)\varrho^M(m) = mf(1_B) = m\varphi(f).$$

This shows that $m \in {}^{B^*}M$, and therefore ${}^{B^*}M = M^{coB}$.

(2) From the definition of the trace ideal we know that $T \in \mathbf{M}^B$; hence the assertion follows from (1).

(3) If B_R is finitely generated and projective, then $T = B^*$ and the assertion follows from (2). □

For a bialgebra B the dual $B^* = \mathrm{Hom}_R(B, R)$ also has a natural (B, B)-bimodule structure with the following properties.

14.14. B-module structure of B^*. *B^* is a (B, B)-bimodule by*

$$\begin{array}{ll} \rightharpoondown : B \otimes_R B^* \to B^*, & b \otimes f \mapsto [c \mapsto f(cb)], \\ \leftharpoondown : B^* \otimes_R B \to B^*, & f \otimes b \mapsto [c \mapsto f(bc)], \end{array}$$

and for $a, b \in B$, $f, g \in B^*$,

$$a \rightharpoondown (f * g) = \sum (a_{\underline{1}} \rightharpoondown f) * (a_{\underline{2}} \rightharpoondown g).$$

Proof. For any $c \in B$, $[(a \rightharpoondown f) \leftharpoondown b](c) = f(bca) = [a \rightharpoondown (f \leftharpoondown b)](c)$, and

$$\begin{aligned} [a \rightharpoondown (f * g)](c) = (f * g)(ca) &= \sum f(c_{\underline{1}} a_{\underline{1}}) g(c_{\underline{2}} a_{\underline{2}}) \\ &= \sum (a_{\underline{1}} \rightharpoondown f)(c_{\underline{1}})\,(a_{\underline{2}} \rightharpoondown g)(c_{\underline{2}}) \\ &= \sum [(a_{\underline{1}} \rightharpoondown f) * (a_{\underline{2}} \rightharpoondown g)](c), \end{aligned}$$

as required. □

Every Hopf module $M \in \mathbf{M}_B^B$ is a right B-comodule, and hence it is a left B^*-module (in the canonical way). This yields an action of $B^{op} \otimes_R B^*$ on M,

$$B^{op} \otimes_R B^* \otimes_R M \to M, \;\; (a \otimes f) \otimes m \mapsto (a \otimes f) \varrho^M(m) = \sum m_{\underline{0}} a f(m_{\underline{1}}).$$

This action is obviously an R-linear map, but it does not make M a module with respect to the canonical algebra product in $B^{op} \otimes_R B^*$. On the other hand, there exists a different multiplication on $B^{op} \otimes_R B^*$ that makes M a module over the new algebra. Denote this product by "?". For all $a \in B$, $f, g \in B^*$, and $m \in M$, a product ? has to satisfy the associative law

$$\begin{aligned} [(a \otimes f) ? (b \otimes g)](m) = (a \otimes f)((b \otimes g)m) &= \sum (a \otimes f)(m_{\underline{0}} b g(m_{\underline{1}})) \\ &= \sum m_{\underline{0}} b_{\underline{1}} a\, f(m_{\underline{1}} b_{\underline{2}}) g(m_{\underline{2}}) \\ &= \sum m_{\underline{0}} b_{\underline{1}} a\, (b_{\underline{2}} \rightharpoondown f) * g(m_{\underline{1}}) \\ &= [\sum b_{\underline{1}} a \otimes (b_{\underline{2}} \rightharpoondown f) * g](m). \end{aligned}$$

From this we can see how the multiplication ? on $B^{op} \otimes_R B^*$ should be constructed in order to possess the desired properties.

14.15. Smash product $B^{op}\#B^*$. *Consider an algebra $B^{op}\#B^*$, which is isomorphic to the tensor product $B^{op}\otimes_R B^*$ as an R-module and has the product*

$$(a\#f)(b\#g) := ((\Delta b)(a\#f))\,(1_B\#g) = \sum b_{\underline{1}}a\#(b_{\underline{2}}\rightharpoondown f) * g,$$

where $a\#f = a\otimes f$ is the notation. Then $B^{op}\#B^*$ is an associative R-algebra with unit $1_B\#\varepsilon$, and the maps

$$\begin{array}{rcl} B^{op} & \to & B^{op}\#B^*, \quad a\mapsto a\#\varepsilon, \\ B^* & \to & B^{op}\#B^*, \quad f\mapsto 1_B\#f, \end{array}$$

are injective ring morphisms, making every left $B\#B^*$-module a right B-module and a left B^*-module. The algebra $B^{op}\#B^*$ is called a smash product.

Every $M\in \mathbf{M}^B_B$ is a left $B^{op}\#B^*$-module, and therefore $\mathbf{M}^B_B$ is embedded in $\sigma_{B^{op}\#B^*}[B\otimes^b_R B]\subset {}_{B^{op}\#B^*}\mathbf{M}$. If B_R is locally projective, then

$$\mathbf{M}^B_B = \sigma_{B\#B^*}[B\otimes^b_R B] = \sigma_{B\#B^*}[B\otimes^c_R B].$$

In particular, $\mathbf{M}^B_B = {}_{B^{op}\#B^*}\mathbf{M}$ provided that B_R is finitely generated and projective.

Proof. The first assertions are immediate consequences of the action considered above and the definition of the product $\#$. The local projectivity implies that the right B-comodule structures correspond to left B^*-module structures.

If B_R is finitely generated and projective, then there is a right coaction (see 3.11) $B^*\to \mathrm{End}_R(B)\simeq B^*\otimes_R B$, $g\mapsto (I_{B^*}\otimes g)\circ\Delta$. The map

$$B^*\otimes^c_R B\to B^{op}\#B^*, \quad f\otimes b\mapsto b\#f,$$

is an isomorphism of left $B^{op}\#B^*$-modules. Indeed, note that, for any $b,x\in B$ and $f,g\in B^*$,

$$\sum (b\rightharpoondown f)(g_{\underline{0}}(x)g_{\underline{1}}) = (b\rightharpoondown f)(I_B\otimes g)\Delta(x) = (b\rightharpoondown f)*g(x).$$

Using these identities we compute

$$\begin{array}{rcl} (a\#f)(g\otimes b) & = & (a\otimes f)\varrho^{B^*\otimes B}(g\otimes b) = \sum g_{\underline{0}}\otimes b_{\underline{1}}af(g_{\underline{1}}b_{\underline{2}}) \\ & \mapsto & \sum b_{\underline{1}}a\#(b_{\underline{2}}\rightharpoondown f)(g_{\underline{1}})g_{\underline{0}} = \sum b_{\underline{1}}a\#(b_{\underline{2}}\rightharpoondown f)*g \\ & = & (a\#f)(b\#g), \end{array}$$

that is, the map defined above is a morphism of left $B^{op}\#B^*$-modules. Clearly it is an isomorphism. Therefore, $B^{op}\#B^*\in\mathbf{M}^B_B$ and hence $\mathbf{M}^B_B = {}_{B^{op}\#B^*}\mathbf{M}$. □

References. Abe [1]; Dăscălescu, Năstăsescu and Raianu [14]; Lomp [153]; Montgomery [37]; Sweedler [45].

15 Hopf algebras

Bialgebras can be viewed as a generalisation of algebras of functions on monoids. Once one starts studying algebras of functions on groups, one immediately realises that the inverse in the group induces an R-module endomorphism of the corresponding bialgebra. The abstract version of such an endomorphism in a general noncommutative algebra is known as an *antipode*, and a bialgebra with an antipode is called a *Hopf algebra*. In this section we study Hopf algebras. In particular, we derive the Fundamental Theorem of Hopf algebras, which states that the category of Hopf modules of a Hopf algebra is equivalent to the category of R-modules. We also give a number of examples of Hopf algebras at the end of this section.

15.1. The ring $(\mathrm{End}_R(B), *)$. For any R-bialgebra B, $(\mathrm{End}_R(B), *)$ is an associative R-algebra with product, for $f, g \in \mathrm{End}_R(B)$,

$$f * g = \mu \circ (f \otimes g) \circ \Delta,$$

and unit $\iota \circ \varepsilon$, that is, $\iota \circ \varepsilon(b) = \varepsilon(b)1_B$, for any $b \in B$ (cf. 1.3). If B is commutative and cocommutative, then $(\mathrm{End}_R(B), *)$ is a commutative algebra.

Definitions. An element $S \in \mathrm{End}_R(B)$ is called a *left (right) antipode* if it is left (right) inverse to I_B with respect to the convolution product on $\mathrm{End}_R(B)$, that is, $S * I_B = \iota \circ \varepsilon$ (resp. $I_B * S = \iota \circ \varepsilon$). In case S is a left and right antipode, it is called an *antipode*. The corresponding conditions are

$$\mu \circ (S \otimes I_B) \circ \Delta = \iota \circ \varepsilon, \quad \mu \circ (I_B \otimes S) \circ \Delta = \iota \circ \varepsilon.$$

Explicitly, for all $b \in B$, an antipode S satisfies the following equalities:

$$\sum S(b_{\underline{1}})\, b_{\underline{2}} = \varepsilon(b)1_B = \sum b_{\underline{1}} S(b_{\underline{2}})\ .$$

Left and right antipodes need not be unique, whereas an antipode is unique whenever it exists. A bialgebra with an antipode is called a *Hopf algebra*.

Antipodes are related to the right Hopf module morphism (see 14.3, 14.10)

$$\gamma_B : B \otimes_R B \to B \otimes_R^b B,\ a \otimes b \mapsto (a \otimes 1_B)\Delta b = \sum a b_{\underline{1}} \otimes b_{\underline{2}}.$$

Notice that γ_B is also a left B-module morphism in an obvious way.

15.2. Existence of antipodes. *Let B be an R-bialgebra.*

(1) B has a right antipode if and only if γ_B has a left inverse in ${}_B\mathbf{M}$.

(2) If B has a left antipode, then γ_B has a right inverse in ${}_B\mathbf{M}$.

(3) γ_B is an isomorphism if and only if B has an antipode.

Proof. (1) If β is a left inverse of γ_B, for all $b \in B$, $1_B \otimes b = \beta \circ \gamma_B(1_B \otimes b) = \beta(\Delta b)$ holds. This implies that $\iota \circ \varepsilon(b) = (I_B \otimes \varepsilon) \circ \beta(\Delta b)$. Then $S = (I_B \otimes \varepsilon) \circ \beta(1_B \otimes -) : B \to B$ is a right antipode since

$$\mu \circ (I_B \otimes S) \circ \Delta(b) = \sum b_{\underline{1}}((I_B \otimes \varepsilon)\beta(1_B \otimes b_{\underline{2}})) = (I_B \otimes \varepsilon) \circ \beta(\Delta b) = \iota \circ \varepsilon(b),$$

where we used that β is left B-linear.

Now suppose that $S : B \to B$ is a right antipode. Then

$$\beta : B \otimes^b_R B \to B \otimes_R B, \quad a \otimes b \mapsto (a \otimes 1_B)(S \otimes I_B)(\Delta b) = \sum aS(b_{\underline{1}}) \otimes b_{\underline{2}},$$

is a left inverse of γ_B, since for any $b \in B$,

$$\begin{aligned}
\beta \circ \gamma_B(1_B \otimes b) = \beta(\Delta b) &= (\mu \otimes I_B) \circ (I_B \otimes S \otimes I_B) \circ (I_B \otimes \Delta)(\Delta b) \\
&= (\mu \otimes I_B) \circ (I_B \otimes S \otimes I_B) \circ (\Delta \otimes I_B)(\Delta b) \\
&= \sum \mu \circ (S \otimes I_B)(\Delta b_{\underline{1}}) \otimes b_{\underline{2}} \\
&= \sum \varepsilon(b_{\underline{1}}) 1_B \otimes b_{\underline{2}} \;=\; 1_B \otimes b.
\end{aligned}$$

(2) Let S be a left antipode, that is, $\mu \circ (S \otimes I_B)(\Delta b) = \iota \circ \varepsilon(b)$, for $b \in B$. Then

$$\beta : B \otimes^b_R B \to B \otimes_R B, \quad 1_B \otimes b \mapsto (S \otimes I_B)(\Delta b) = \sum S(b_{\underline{1}}) \otimes b_{\underline{2}},$$

is a right inverse of γ_B, since

$$\begin{aligned}
\gamma_B \circ \beta(1_B \otimes b) = \gamma_B((S \otimes I_B)(\Delta b)) &= \sum S(b_{\underline{1}}) b_{\underline{2}} \otimes b_{\underline{3}} \\
&= \sum \varepsilon(b_{\underline{1}}) 1_B \otimes b_{\underline{2}} \;=\; 1_B \otimes b.
\end{aligned}$$

(3) Suppose that γ_B is bijective. Take any $f \in \text{End}_R(B)$ and observe that if $\mu \circ (I_B \otimes f)(\Delta b) = 0$, for all $b \in B$, then $f = 0$. Indeed, any element in $B \otimes_R B$ can be written as a sum of elements of the form $(a \otimes 1_B)(\Delta b)$ and

$$\mu \circ (I_B \otimes f)((a \otimes 1_B)(\Delta b)) = a(\mu((I_B \otimes f)(\Delta b))) = 0,$$

implying $\mu(I_B \otimes f)(B \otimes_R B) = Bf(B) = 0$, and so $f = 0$, as claimed.

By (1), there exists a right antipode S, and for this we compute

$$\begin{aligned}
&\mu \circ (I_B \otimes \mu \circ (S \otimes I_B) \circ \Delta)(\Delta b) \\
&\quad = \mu \circ (I_B \otimes \mu) \circ (I_B \otimes S \otimes I_B) \circ (I_B \otimes \Delta)(\Delta b) \\
&\quad = \mu \circ (\mu \otimes I_B) \circ (I_B \otimes S \otimes I_B) \circ (\Delta \otimes I_B)(\Delta b) \\
&\quad = \sum \varepsilon(b_{\underline{1}}) b_{\underline{2}} \;=\; b \;=\; \mu \circ (I_B \otimes \iota \circ \varepsilon)(\Delta b).
\end{aligned}$$

By the preceding observation this implies $\mu \circ (S \otimes I_B) \circ \Delta = \iota \circ \varepsilon$, thus showing that S is also a left antipode. □

Definition. Let H be a Hopf algebra with antipode S. An R-submodule $J \subset H$ is called a *Hopf ideal* if J is a coideal, an ideal, and $S(J) \subset J$.

15.3. Factors by Hopf ideals. *For any Hopf ideal $J \subset H$, the factor module H/J is a Hopf algebra and the canonical map $H \to H/J$ is a bialgebra morphism.*

Proof. Clearly H/J is a factor algebra, and a factor coalgebra with counit $\bar{\varepsilon} : H/J \to R$ (see 2.4) satisfying the compatibility conditions and the projection $\overline{(\ )} : H \to H/J$ is a bialgebra map. Since $S(J) \subset J$, the map $S : H \to H$ induces a morphism

$$\bar{S} : H/J \to H/J, \quad \bar{h} \mapsto \overline{S(h)},$$

and for this we compute

$$\sum \bar{S}(\overline{h_{\underline{1}}})\overline{h_{\underline{2}}} = \overline{\sum S(h_{\underline{1}})h_{\underline{2}}} = \overline{\varepsilon(h)1_H} = \bar{\varepsilon}(\bar{h})1_{H/J}.$$

Similarly we get $\sum \overline{h_{\underline{1}}}\bar{S}(\overline{h_{\underline{2}}}) = \bar{\varepsilon}(\bar{h})1_{H/J}$, so that $\bar{S}$ is an antipode for H/J. □

15.4. Properties of antipodes. *Let H be a Hopf algebra with antipode S. Then:*

(1) S is an algebra anti-morphism, that is, for all $a, b \in H$, $S(ab) = S(b)S(a)$, and $S \circ \iota = \iota$.

(2) S is a coalgebra anti-morphism, that is, $\mathsf{tw} \circ (S \otimes S) \circ \Delta = \Delta \circ S$ and $\varepsilon \circ S = \varepsilon$.

(3) If S is invertible as a map, then, for any $b \in H$,

$$\sum S^{-1}(b_{\underline{2}})b_{\underline{1}} = \varepsilon(b)1_H = \sum b_{\underline{2}}S^{-1}(b_{\underline{1}}).$$

Proof. (1) Consider the convolution algebra $\tilde{H} := (\mathrm{Hom}_R(H \otimes_R H, H), \tilde{*})$ corresponding to the canonical coalgebra structure $\Delta_{H \otimes_R H}$ on $H \otimes_R H$ with the counit $\tilde{\varepsilon} = \varepsilon \otimes \varepsilon$. In particular, the unit in $\tilde{H}$ comes out as

$$\tilde{\iota} : H \otimes_R H \xrightarrow{\varepsilon \otimes \varepsilon} R \xrightarrow{\iota} H.$$

In addition to the product $\mu : H \otimes_R H \to H$, consider the R-linear maps

$$\nu : H \otimes_R H \to H, \ a \otimes b \mapsto S(b)S(a), \quad \rho : H \otimes_R H \to H, \ a \otimes b \mapsto S(ab).$$

To prove that S is an anti-multiplicative map, it is sufficient to show that $\rho \tilde{*} \mu = \mu \tilde{*} \nu = \tilde{\iota} \circ \tilde{\varepsilon}$ (the identity in $\tilde{H}$). By the uniqueness of inverse elements we are then able to conclude that $\nu = \rho$. Consider the R-linear maps

$$H \otimes_R H \xrightarrow{\Delta_{H \otimes H}} H \otimes_R H \otimes_R H \otimes_R H \underset{\mu \otimes \nu}{\overset{\rho \otimes \mu}{\rightrightarrows}} H \otimes_R H \xrightarrow{\mu} H.$$

Take any $a, b \in H$ and compute

$$\begin{aligned} a \otimes b \mapsto & \sum a_{\underline{1}} \otimes b_{\underline{1}} \otimes a_{\underline{2}} \otimes b_{\underline{2}} \\ & \stackrel{\rho\otimes\mu}{\longmapsto} \sum S(a_{\underline{1}}b_{\underline{1}})a_{\underline{2}}b_{\underline{2}} = S * I_H(ab) = \varepsilon(ab)1_H\,, \\ & \stackrel{\mu\otimes\nu}{\longmapsto} \sum a_{\underline{1}}b_{\underline{1}}S(b_{\underline{2}})\,S(a_{\underline{2}}) \\ & \quad = \sum a_{\underline{1}}S(a_{\underline{2}})\,\varepsilon(b) = \varepsilon(a)\varepsilon(b)1_H\,. \end{aligned}$$

Thus $\nu = \rho$, and S is an anti-multiplicative map, that is, $S(ab) = S(a)S(b)$. Furthermore, $1_H = \iota \circ \varepsilon(1_H) = (I_H * S)(1_H) = S(1_H)$, so that S is a unital map and hence an algebra anti-morphism.

(2) This is a dual statement to (1), and we use a similar technique as for the proof of (1). In this case consider the convolution algebra corresponding to H as a coalgebra and $H \otimes_R H$ as an algebra, $(\mathrm{Hom}_R(H, H \otimes_R H), \underline{*})$. Let $\nu := \mathsf{tw} \circ (S \otimes S) \circ \Delta$ and $\rho := \Delta \circ S$. Direct computation verifies that $\rho \underline{*} \Delta = \iota_H \circ \varepsilon_{H\otimes H} = \Delta \underline{*} \nu$. From this we conclude that $\rho = \nu$, so that S is an anti-comultiplicative map. Furthermore, for all $a \in H$, we know that $\varepsilon(\iota \circ \varepsilon(a)) = \varepsilon(a)$, and $\iota \circ \varepsilon(a) = \sum S(a_{\underline{1}})a_{\underline{2}}$. This implies

$$\varepsilon(a) = \varepsilon(\iota \circ \varepsilon(a)) = \sum \varepsilon\,(S(a_{\underline{1}}))\,\varepsilon(a_{\underline{2}}) = \varepsilon \circ S(a)\,,$$

hence S is a coalgebra anti-morphism, as stated.

(3) Apply S^{-1} to the defining properties of S. □

We now prove that Hopf algebras are precisely those R-bialgebras for which the category $\mathbf{M}_B^B$ is equivalent to $\mathbf{M}_R$. It is interesting to notice that this can be seen from a single isomorphism.

15.5. Fundamental Theorem of Hopf algebras. *For any R-bialgebra B the following are equivalent:*

(a) B is a Hopf algebra (that is, B has an antipode);

(b) $\gamma_B : B \otimes_R B \to B \otimes_R^b B$, $a \otimes b \mapsto (a \otimes 1_B)\Delta b$, is an isomorphism in $\mathbf{M}_B^B$;

(c) $\gamma^B : B \otimes_R^c B \to B \otimes_R B$, $a \otimes b \mapsto \Delta a(1_B \otimes b)$, is an isomorphism in $\mathbf{M}_B^B$;

(d) for every $M \in \mathbf{M}_B^B$, $M^{coB} \otimes_R B \to M$, $m \otimes b \mapsto mb$, is an isomorphism in $\mathbf{M}_B^B$;

(e) for every $M \in \mathbf{M}_B^B$, there is an isomorphism (in $\mathbf{M}_B^B$)

$$\varphi_M : \mathrm{Hom}_B^B(B, M) \otimes_R B \to M, \quad f \otimes b \mapsto f(b);$$

(f) $\varphi_{B\otimes B} : \mathrm{Hom}_B^B(B, B \otimes_R^b B) \otimes_R B \to B \otimes_R^b B$ is an isomorphism in $\mathbf{M}_B^B$;

(g) $\mathrm{Hom}_B^B(B, -) : \mathbf{M}_B^B \to \mathbf{M}_R$ is an equivalence (with inverse $- \otimes_R B$).

If B is flat as an R-module, then (a)-(g) are equivalent to:

(h) B is a (projective) generator in $\mathbf{M}_B^B$;

(i) B is a subgenerator in $\mathbf{M}_B^B$, and φ_M is injective for every $M \in \mathbf{M}_B^B$.

If B_R is locally projective, then (a) – (i) are equivalent to:

(j) B is a subgenerator in $\mathbf{M}_B^B$ and the image of $B^{op}\#B^* \to \mathrm{End}_R(B)$ is dense (for the finite topology).

For any Hopf module M over a Hopf algebra B, the coinvariants M^{coB} are an R-direct summand of M.

Proof. (a) ⇔ (b) was shown in 15.2, and by symmetry (see 14.4) the same proof implies (a)⇔(c). (b) ⇔ (f) is clear by 14.10.

(d) ⇔ (e) This follows from the commutative diagram in 14.8.

(a) ⇒ (d) For any $M \in \mathbf{M}_B^B$, consider $\phi : M \to M^{coB}, m \mapsto \sum m_{\underline{0}}S(m_{\underline{1}})$. The following equalities show that the image of ϕ is in M^{coB}:

$$\begin{aligned}\varrho^M(\phi(m)) = \varrho^M(\textstyle\sum m_{\underline{0}}S(m_{\underline{1}})) &= \textstyle\sum m_{\underline{0}}S(m_{\underline{3}}) \otimes m_{\underline{1}}S(m_{\underline{2}}) \\ &= \textstyle\sum m_{\underline{0}}S(m_{\underline{1}}) \otimes 1_B \;=\; \phi(m) \otimes 1_B.\end{aligned}$$

Now we show that the map

$$(\phi \otimes I_B) \circ \varrho^M : M \to M^{coB} \otimes_R B$$

is the inverse of the multiplication map $\varrho_M : M^{coB} \otimes_R B \to M$. For $m \in M$,

$$\varrho_M \circ (\phi \otimes I_B)(\varrho^M(m)) = \textstyle\sum \phi(m_{\underline{0}})m_{\underline{1}} = \sum m_{\underline{0}}S(m_{\underline{1}})m_{\underline{2}} = \sum m_{\underline{0}}\varepsilon(m_{\underline{1}}) = m.$$

On the other hand, for $n \otimes b \in M^{coB} \otimes_R B$,

$$\begin{aligned}(\phi \otimes I_B) \circ \varrho^M(nb) &= (\phi \otimes I_B)(\textstyle\sum nb_{\underline{1}} \otimes b_{\underline{2}}) \;=\; \sum \phi(nb_{\underline{1}}) \otimes b_{\underline{2}} \\ &= \textstyle\sum nb_{\underline{1}}\, S(b_{\underline{2}}) \otimes b_{\underline{3}} \;=\; \sum n\varepsilon(b_{\underline{1}}) \otimes b_{\underline{2}} \;=\; n \otimes b.\end{aligned}$$

(e) ⇒ (f) is trivial (take $M = B \otimes_R^b B$).

(e) ⇔ (g) From 14.9 we know $\mathrm{Hom}_B^B(B, - \otimes_R B) \simeq I_{R\mathbf{M}}$. Condition (f) induces $\mathrm{Hom}_B^B(B, -) \otimes_R B \simeq I_{\mathbf{M}_B^B}$, and the two isomorphisms characterise an equivalence between $\mathbf{M}_R$ and $\mathbf{M}_B^B$.

(g) ⇒ (h) Obviously (g) always implies that B is a generator in $\mathbf{M}_B^B$ and that B is projective in $\mathbf{M}_B^B$ (that is, $\mathrm{Hom}_B^B(B, -) : \mathbf{M}_B^B \to \mathbf{M}_R$ preserves epimorphisms).

Now suppose that ${}_RB$ is flat. Then $\mathbf{M}_B^B$ has kernels and $\mathrm{Hom}_B^B(B, -)$ is a left exact functor.

(h) ⇒ (i) Suppose that B is a generator in $\mathbf{M}_B^B$. Of course any generator is in particular a subgenerator. For any $M \in \mathbf{M}_B^B$, the set $\Lambda = \mathrm{Hom}_B(B, M)$ yields a canonical epimorphism

$$p : B^{(\Lambda)} \to M, \quad b_f \mapsto f(b).$$

Choosing $\Lambda' = \mathrm{Hom}_B(B, \mathrm{Ke}\, p)$ we form – with a similar map p' – the exact sequence in $\mathbf{M}_B^B$,

$$B^{(\Lambda')} \xrightarrow{p'} B^{(\Lambda)} \xrightarrow{p} M \longrightarrow 0\,.$$

Now apply $\mathrm{Hom}_B^B(B,-)$ to obtain the exact sequence

$$\mathrm{Hom}_B^B(B, B^{(\Lambda')}) \longrightarrow \mathrm{Hom}_B^B(B, B^{(\Lambda)}) \longrightarrow \mathrm{Hom}_B^B(B, M) \longrightarrow 0\,.$$

By the choice of Λ and Λ', this sequence is exact. Now tensor with $-\otimes_R B$ to obtain the commutative diagram with exact rows ($\otimes$ for $\otimes_R$),

$$\begin{array}{ccccccc}
\mathrm{Hom}_B^B(B, B^{(\Lambda')})\otimes B & \longrightarrow & \mathrm{Hom}_B^B(B, B^{(\Lambda)})\otimes B & \longrightarrow & \mathrm{Hom}_B^B(B,M)\otimes B & \longrightarrow & 0 \\
\downarrow \simeq & & \downarrow \simeq & & \downarrow \varphi_M & & \\
B^{(\Lambda')} & \longrightarrow & B^{(\Lambda)} & \longrightarrow & M & \longrightarrow & 0\,.
\end{array}$$

The first two vertical maps are bijective since $\mathrm{Hom}_B^B(B,-)$ commutes with direct sums. By the diagram properties this implies the bijectivity of φ_M.

(i) $\Rightarrow$ (f) Assume that B is a subgenerator in $\mathbf{M}_B^B$ and that φ_M is injective for all $M \in \mathbf{M}_B^B$. Then clearly φ_N is bijective for all B-generated objects N in $\mathbf{M}_B^B$ and M is a subobject of such an N. Choose an exact sequence $0 \to M \to N \to L$ in $\mathbf{M}_B^B$ where N and L are B-generated. Then clearly φ_N and φ_L are bijective and there is a commutative diagram with exact rows,

$$\begin{array}{ccccccc}
0 \longrightarrow & \mathrm{Hom}_B^B(B, M)\otimes_R B & \longrightarrow & \mathrm{Hom}_B^B(B, N)\otimes_R B & \longrightarrow & \mathrm{Hom}_B^B(B, L)\otimes_R B \\
 & \downarrow \varphi_M & & \downarrow \varphi_N & & \downarrow \varphi_L \\
0 \longrightarrow & M & \longrightarrow & N & \longrightarrow & L\,.
\end{array}$$

From this we conclude that φ_M is also bijective.

(h) $\Leftrightarrow$ (j) If B_R is locally projective, then, by 42.10(g), B is a generator in $\sigma[_{\mathrm{End}_R(B)}B]$. Moreover, $\mathbf{M}_B^B = \sigma[_{B^{op}\# B^*}B \otimes_R^b B]$. Now assume (h). Then $\mathbf{M}_B^B = \sigma[_{B^{op}\# B^*}B]$ and the density property follows by 43.12. On the other hand, given the density property and the subgenerating property of B, one has $\sigma[_{\mathrm{End}_R(B)}B] = \sigma[_{B^{op}\# B^*}B]$ and B is a generator in $\mathbf{M}_B^B$.

The R-linear map $\phi : M \to M^{coB}$ considered in the proof (a)$\Rightarrow$(d) splits the inclusion $M^{coB} \to M$, thus proving the final statement. □

Notice that parts of the characterisations in 15.5 apply to Hopf algebras that are not necessarily flat as R-modules (see 15.11 for such examples).

15.6. Finitely generated Hopf algebras. *For an R-bialgebra B with B_R finitely generated and projective, the following are equivalent:*

(a) B is a Hopf algebra;

(b) $\gamma_B : B \otimes_R B \to B \otimes^b_R B$ *is surjective;*

(c) B *has a left antipode;*

(d) $B^{op}\#B^* \simeq \mathrm{End}_R(B)$*;*

(e) B *is a generator in* ${}_{B^{op}\#B^*}\mathbf{M}$.

Proof. (a)$\Leftrightarrow$(b)$\Leftrightarrow$(c) follow from 15.2 and the fact that, for finitely generated projective R-modules, any surjective endomorphism is bijective.

(a) $\Rightarrow$ (d) As a generator in $\mathbf{M}^B_B = {}_{B^{op}\#B^*}\mathbf{M}$, B is a faithful $B^{op}\#B^*$-module and the density property of $B^{op}\#B^*$ (see 15.5) implies $B^{op}\#B^* \simeq \mathrm{End}_R(B)$.

(e) $\Leftrightarrow$ (d) Since B is a subgenerator in $\sigma[_{\mathrm{End}_R(B)}B]$, the assertion follows from 15.5(j).

(e) $\Rightarrow$ (a) Under the given conditions $\mathbf{M}^B_B = {}_{B^{op}\#B^*}\mathbf{M}$ (see 14.15) and the assertion again follows from the Fundamental Theorem 15.5. □

Clearly, if B is a finitely generated projective R-module, then $\mathbf{M}^B = {}_{B^*}\mathbf{M}$ has (enough) projectives and 13.6 implies the following corollary.

15.7. Existence of integrals. *Any Hopf algebra H with H_R finitely generated and projective has left (and right) integrals on H.*

15.8. Canonical isomorphisms. *Let H be a Hopf algebra with antipode S.*

(1) For any $N \in \mathbf{M}_H$, the Hopf module morphism (see 14.3)

$$\gamma_N : N \otimes_R H \to N \otimes^b_R H, \quad n \otimes h \mapsto \sum nh_{\underline{1}} \otimes h_{\underline{2}},$$

is invertible with the inverse,

$$\beta_N : N \otimes^b_R H \to N \otimes_R H, \quad m \otimes k \mapsto \sum mS(k_{\underline{1}}) \otimes k_{\underline{2}}.$$

(2) For any $L \in \mathbf{M}^B$, the Hopf module morphism (see 14.4)

$$\gamma^L : L \otimes^c_R H \to L \otimes_R H, \quad l \otimes h \mapsto \sum l_{\underline{0}} \otimes l_{\underline{1}}h,$$

is invertible with the inverse,

$$\beta^L : L \otimes_R H \to L \otimes^c_R H, \quad m \otimes k \mapsto \sum m_{\underline{0}} \otimes S(m_{\underline{1}})k.$$

Proof. (1) Take any $n \otimes h \in N \otimes_R H$ and $m \otimes k \in N \otimes^b_R H$ and compute

$$\begin{aligned}
\beta_N(\gamma_N(n \otimes h)) &= \beta_N(\textstyle\sum nh_{\underline{1}} \otimes h_{\underline{2}}) \\
&= \textstyle\sum nh_{\underline{1}}S(h_{\underline{2}}) \otimes h_{\underline{3}} = \sum n \otimes \varepsilon(h_{\underline{1}})h_{\underline{2}} = n \otimes h, \\
\gamma_N(\beta_N(m \otimes k)) &= \gamma_N(\textstyle\sum mS(k_{\underline{1}}) \otimes k_{\underline{2}}) \\
&= \textstyle\sum mS(k_{\underline{1}})k_{\underline{2}} \otimes k_{\underline{3}} = m \otimes k.
\end{aligned}$$

Thus β_N is the inverse of γ_N, as required.

(2) Take any $l \otimes h \in L \otimes^c_R H$ and $m \otimes k \in L \otimes_R H$ and compute

$$\begin{aligned}
\beta^L(\gamma^L(l \otimes h) &= \beta^L(\sum l_{\underline{0}} \otimes l_{\underline{1}} h) \\
&= \sum l_{\underline{0}} \otimes S(l_{\underline{1}}) l_{\underline{2}} h = \sum l_{\underline{0}} \varepsilon(l_{\underline{1}}) \otimes h = l \otimes h, \\
\gamma^L(\beta^L(m \otimes k) &= \gamma^L(\sum m_{\underline{0}} \otimes S(m_{\underline{1}}) k \\
&= \sum m_{\underline{0}} \otimes m_{\underline{1}} S(m_{\underline{2}}) k = \sum m_{\underline{0}} \varepsilon(m_{\underline{1}}) \otimes k = m \otimes k.
\end{aligned}$$

Thus β^L is the inverse of γ^L as required. □

15.9. Hom-tensor relations. *Let H be a Hopf algebra with antipode S, and $L, M, N \in \mathbf{M}_H$.*

(1) A right H-action on $\mathrm{Hom}_R(M, L)$ is defined by

$$(fh)(m) = \sum f(mS(h_{\underline{1}}))h_{\underline{2}}, \quad \text{for } h \in H,\ m \in M,\ f \in \mathrm{Hom}_R(M, L).$$

Denote $\mathrm{Hom}_R(M, L)$ with this H-module structure by $\mathrm{Hom}_R(M, L)^s$. Then there is a functorial isomorphism

$$\mathrm{Hom}_H(M \otimes^b_R N, L) \to \mathrm{Hom}_H(N, \mathrm{Hom}_R(M, L)^s), \ f \mapsto [n \mapsto f(- \otimes n)],$$

with the inverse $g \mapsto [m \otimes n \mapsto g(n)(m)]$.

(2) If S is invertible, then a right H-action on $\mathrm{Hom}_R(M, L)$ is defined by

$$(fh)(m) = \sum f(mS^{-1}(h_{\underline{2}}))h_{\underline{1}}, \quad \text{for } h \in H,\ m \in M,\ f \in \mathrm{Hom}_R(M, L).$$

Denote $\mathrm{Hom}_R(M, L)$ with this H-module structure by $\mathrm{Hom}_R(M, L)^t$. Then there is a functorial isomorphism

$$\mathrm{Hom}_H(M \otimes^b_R N, L) \to \mathrm{Hom}_H(M, \mathrm{Hom}_R(N, L)^t), \ f \mapsto [m \mapsto f(m \otimes -)],$$

with the inverse $g \mapsto [m \otimes n \mapsto g(m)(n)]$.

Proof. (1) Clearly, the defined right H-action is unital, and it is also associative since, for all $h, k \in H$ and $m \in M$,

$$\begin{aligned}
(f(hk))(m) &= \sum f\,(mS(h_{\underline{1}} k_{\underline{1}}))\, h_{\underline{2}} k_{\underline{2}} \\
&= \sum f\,(mS(k_{\underline{1}}) S(h_{\underline{1}}))\, h_{\underline{2}} k_{\underline{2}} = ((fh)k)(m).
\end{aligned}$$

The maps indicated yield the canonical isomorphism (see 40.18)

$$\mathrm{Hom}_R(M \otimes^b_R N, L) \to \mathrm{Hom}_R(N, \mathrm{Hom}_R(M, L)^s),$$

and it is left to show that H-morphisms on the left-hand side correspond to H-morphisms on the right-hand side. Let $m \otimes n \in M \otimes^b_R N$, $h \in H$, and $f \in \mathrm{Hom}_H(M \otimes^b_R N, L)$. Then

$$f(m \otimes nh) = \sum f(mS(h_{\underline{1}})h_{\underline{2}} \otimes nh_{\underline{3}}) = \sum f(mS(h_{\underline{1}}) \otimes n)h_{\underline{2}} = (fh)(m \otimes n).$$

This shows that $n \mapsto f(- \otimes n)$ is an H-linear map.

For $g \in \mathrm{Hom}_H(N, \mathrm{Hom}_R(M, L)^s)$ and $m \otimes n \in M \otimes^b_R N$, the map $m \otimes n \mapsto g(n)(m)$ is H-linear since, for any $h \in H$,

$$\begin{aligned}(m \otimes n)!h = \sum mh_{\underline{1}} \otimes nh_{\underline{2}} \quad \mapsto \quad & \sum g(nh_{\underline{2}})(mh_{\underline{1}}) = \sum (g(n)h_{\underline{2}})(mh_{\underline{1}}) \\ = \; & \sum g(n)(mh_{\underline{1}}S(h_{\underline{2}}))h_{\underline{3}} = g(n)(m)h.\end{aligned}$$

(2) The proof is similar to the proof of (1). □

We conclude this section by giving some general examples of bialgebras and Hopf algebras. More examples, for example, those coming from quantum groups, can be found in various monographs and textbooks on the subject (see the Preface or the introduction to this chapter).

15.10. Semigroup bialgebra. Let G be a semigroup with identity e. The *semigroup algebra* $R[G]$ is the R-module $R^{(G)}$ together with the maps (defined on the basis G and linearly extended)

$$\mu : R[G] \times R[G] \longrightarrow R[G], \ (g, h) \mapsto gh \ \text{ and } \iota : R \to R[G], \ r \mapsto re.$$

Since $R[G]$ is a free R-module, there are also linear maps (see 1.6)

$$\Delta : R[G] \longrightarrow R[G] \otimes_R R[G], \ g \mapsto g \otimes g, \ \text{ and } \varepsilon : R[G] \longrightarrow R, \ 1 \mapsto 1, g \mapsto 0.$$

It is easily seen from the definitions that Δ and ε are algebra morphisms.

If G is a group, then $S : R[G] \to R[G]$, $g \mapsto g^{-1}$, is an antipode, that is, in this case $R[G]$ is a Hopf algebra.

15.11. Polynomial Hopf algebra. As noticed in 1.8, for any commutative ring R, the polynomial algebra $R[X]$ is a coalgebra by

$$\begin{aligned}&\Delta_2 : R[X] \otimes_R R[X] \to R[X], \quad 1 \mapsto 1, \ X^i \mapsto (X \otimes 1 + 1 \otimes X)^i, \\ &\varepsilon_2 : R[X] \to R, \qquad\qquad\qquad\quad 1 \mapsto 1, \ X^i \mapsto 0, \quad i = 1, 2, \ldots.\end{aligned}$$

Together with the polynomial multiplication this yields a (commutative and cocommutative) bialgebra that is a Hopf algebra with antipode

$$S : R[X] \to R[X], \quad 1 \mapsto 1, \ X \to -X.$$

For any $a \in R$, denote by J the ideal in $R[X]$ generated by aX. Since

$$\Delta_2(aX) = 1 \otimes aX + aX \otimes 1, \ \varepsilon_2(aX) = 0 \ \text{ and } S(aX) = -aX,$$

it is easily seen that J is a Hopf ideal. Therefore $H = R[X]/J$ is a Hopf algebra over R. Notice that H need no longer be projective or flat as an R-module. In particular, if R is an integral domain and $0 \neq a \in R$, then $\mathrm{Hom}_R(R/aR, R) = 0$ and $H^* = \mathrm{Hom}(H, R) \simeq R$, and H-subcomodules of H do not correspond to H^*-submodules.

15.12. Tensor algebra of a module. For an R-module M, define

$$T_0(M) := R \quad \text{and} \quad T_n(M) := M \otimes \cdots \otimes M,\ n\text{-times}, n \geq 0.$$

Then $T(M) := \bigoplus_{n \geq 0} T_n(M)$ is an $\mathbb{N}$-graded algebra with the product

$$\begin{array}{rcl} T_n(M) \otimes_R T_k(M) & \longrightarrow & T_{n+k}(M), \\ (x_1 \otimes \cdots \otimes x_n) \otimes (y_1 \otimes \cdots \otimes y_k) & \longmapsto & x_1 \otimes \cdots \otimes x_n \otimes y_1 \otimes \cdots \otimes y_k, \end{array}$$

for all $n, k \in \mathbb{N}$. These give rise to an R-linear map $\mu : T(M) \otimes_R T(M) \to T(M)$. $(T(M), \mu)$ is called the *tensor R-algebra* over M. $T_0(M) = R$ is a subalgebra of $T(M)$ and the unit of R is the unit of $T(M)$. Moreover, $T_1(M) = M$ is a submodule of $T(M)$.

The tensor algebra is a cocommutative Hopf algebra.

This can be derived from the

Universal property of $T(M)$. *Let A be a unital associative R-algebra and let $f : M \to A$ be an R-linear map. Then there exists a unique R-algebra morphism $g : T(M) \to A$ such that $f = g|_M$. Since, for every R-algebra map $h : T(M) \to A$, the restriction $h|_M : M \to A$ is R-linear, there is in fact a bijective correspondence,*

$$\mathrm{Hom}_R(M, A) \to \mathrm{Alg}_R(T(M), A),$$

where $\mathrm{Alg}_R(T(M), A)$ *denotes all R-algebra maps $T(M) \to A$.*

Proof. For $n \geq 1$, the map $M^n \to A$, $(m_1, ..., m_n) \longmapsto f(m_1) \cdots f(m_n)$, is R-multilinear and hence it induces an R-linear map

$$g_n : T_n(M) \to A, \quad m_1 \otimes \cdots \otimes m_n \longmapsto f(m_1) \cdots f(m_n).$$

For $n = 0$ put $g_0 := \iota : R \to A, r \mapsto r1_A$. By the universal property of the direct sum this yields an R-linear map $g : T(M) \to A$, with $g|_{T_n(M)} = g_n$. It is easy to verify that g is an R-algebra morphism. □

To define comultiplication on $T(M)$, consider the maps, for $n \geq 1$,

$$\begin{array}{l} h_n : M^n \to T(M) \otimes_R T(M), \\ \quad (x_1, \cdots, x_n) \mapsto \displaystyle\sum_{0 \leq i \leq n} (x_1 \otimes \cdots \otimes x_i) \otimes (x_{i+1} \otimes \cdots \otimes x_n), \end{array}$$

which is R-multilinear and factorises over $\tilde{h}_n : T_n(M) \to T(M) \otimes_R T(M)$, yielding an R-linear map $\Delta : T(M) \to T(M) \otimes_R T(M)$. It is straightforward to show that this is a coassociative and cocommutative coproduct with counit

$$\varepsilon : T(M) \to R, \quad 1 \mapsto 1, \; z \mapsto 0, \; \text{for } z \in T_n(M), \; n \geq 1 .$$

Moreover, the algebra and coalgebra structures on $T(M)$ are compatible with each other, thus making $T(M)$ a bialgebra.

To define an antipode observe that, for the opposite algebra $T(M)^{op}$, the R-module map $M \to T(M)^{op}$, $m \mapsto -m$, can be extended to an algebra morphism $S : T(M) \to T(M)^{op}$ (by the universal property), which in turn can be considered as an algebra anti-morphism $S : T(M) \to T(M)$. By definition, for $m_1 \otimes \cdots \otimes m_n \in T_n(M)$,

$$S(m_1 \otimes \cdots \otimes m_n) = (-1)^n \, m_n \otimes \cdots \otimes m_1 ,$$

and from this we can deduce that S is an antipode for the bialgebra $T(M)$.

Other Hopf algebras can be derived from the Hopf algebra $T(M)$.

15.13. Symmetric algebra of a module. Let M be an R-module and $T(M)$ its tensor algebra. The *symmetric algebra of* M, denoted by $\mathcal{S}(M)$, is the factor algebra of $T(M)$ by the (two-sided) ideal J generated by the subset

$$\{x \otimes y - y \otimes x \mid x, y \in M\} \subset T(M).$$

Properties. *$\mathcal{S}(M)$ is a commutative graded algebra. $\mathcal{S}(M)$ contains R as a subring and M as a submodule.*

Proof. Denoting by $\bar{x}$ the image of $x \in M$ under $T(M) \to T(M)/J$, we see that $\mathcal{S}(M)$ is generated by $\{\bar{x} \mid x \in M\}$. It follows from the defining elements of J that $\bar{x}\bar{y} = \bar{y}\bar{x}$ (product in $\mathcal{S}(M)$) for all $x, y \in M$. Hence $\mathcal{S}(M)$ is commutative. Furthermore, writing $J_n := J \cap T_n(M)$, it can be shown that $J = \oplus_{\mathbb{N}} J_n$ and hence

$$\mathcal{S}(M) = \bigoplus_{\mathbb{N}} T_n(M)/J_n,$$

which makes $\mathcal{S}(M)$ an $\mathbb{N}$-graded algebra. The other assertions follow from the fact that $T_0(M) \cap J = 0$ and $T_1(M) \cap J = 0$. □

Remark. Since every permutation of n elements can be generated by transpositions, we know, for $n \geq 1$, that J_n is the R-submodule of $T_n(M)$ generated by all elements of the form

$$x_1 \otimes x_2 \otimes \cdots \otimes x_n - x_{\sigma(1)} \otimes x_{\sigma(2)} \cdots \otimes x_{\sigma(n)},$$

where all $x_i \in M$ and the σ are permutations of n-elements.

Universal property of $\mathcal{S}(M)$. *Let B be a unital associative and commutative R-algebra and $f : M \to B$ an R-linear map. Then there exists a unique R-algebra morphism $g : \mathcal{S}(M) \to B$ such that $f = g|_M$. This yields a bijective correspondence*

$$\mathrm{Alg}_R(\mathcal{S}(M), B) \to \mathrm{Hom}_R(M, B).$$

Proof. By the universal property of $T(M)$ there exists an algebra map $\tilde{f} : T(M) \to B$ such that

$$\tilde{f}(x \otimes y) = g(x)g(y) = g(y)g(x) = \tilde{f}(y \otimes x).$$

Hence all $x \otimes y - y \otimes x$ are in $\mathrm{Ke}\,\tilde{f}$, implying $J \subset \mathrm{Ke}\,\tilde{f}$, and by factorisation we obtain the assertion. □

Proposition. *For any two R-modules M_1, M_2, there is an R-algebra isomorphism*

$$g : \mathcal{S}(M_1 \oplus M_2) \to \mathcal{S}(M_1) \otimes_R \mathcal{S}(M_2).$$

Proof. Put $M = M_1 \oplus M_2$. The canonical injections $e_i : M_i \to M, i = 1, 2$, yield R-algebra morphisms $s(e_i) : \mathcal{S}(M_i) \to \mathcal{S}(M)$, $i = 1, 2$. Since $\mathcal{S}(M)$ is commutative, there exists a unique R-algebra map $h : \mathcal{S}(M_1) \otimes \mathcal{S}(M_2) \to \mathcal{S}(M)$, such that $s(e_i) = h \circ f_i$, $i = 1, 2$, where $f_i : \mathcal{S}(M_i) \to \mathcal{S}(M_1) \otimes \mathcal{S}(M_2)$ are the canonical morphisms. They can be extended to an R-algebra morphism $g : \mathcal{S}(M) \to \mathcal{S}(M_1) \otimes \mathcal{S}(M_2)$, which is the inverse map for h. □

The last proposition is the key for constructing the coalgebra structure of $\mathcal{S}(M)$. The diagonal map $\vartheta : M \to M \oplus M$, $x \longmapsto (x, x)$, is R-linear and induces an R-algebra morphism $s(\vartheta) : \mathcal{S}(M) \to \mathcal{S}(M \oplus M)$. Combined with the isomorphism $g : \mathcal{S}(M \oplus M) \to \mathcal{S}(M) \otimes_R \mathcal{S}(M)$, it yields an R-linear map

$$\Delta := g \circ s(\vartheta) : \mathcal{S}(M) \to \mathcal{S}(M) \otimes_R \mathcal{S}(M),$$

which makes $\mathcal{S}(M)$ a coalgebra. Tracing back the definitions we find that Δ is coassociative and is an R-algebra morphism with

$$\Delta(x) = g(x, x) = x \otimes 1 + 1 \otimes x, \quad \text{for all } x \in M.$$

The counit for this cocommutative comultiplication is

$$\varepsilon : \mathcal{S}(M) \to R, \quad 1 \mapsto 1, \ z \mapsto 0, \ \text{for } z \in \mathcal{S}_n(M), \ n \geq 1.$$

The structure maps $\Delta : \mathcal{S}(M) \to \mathcal{S}(M) \otimes_R \mathcal{S}(M)$ and $\varepsilon : \mathcal{S}(M) \to R$ are algebra morphisms and hence $\mathcal{S}(M)$ is a bialgebra.

For the antipode S of $T(M)$ and $x, y \in M$, we observe

$$S(x \otimes y - y \otimes x) = S(y)S(x) - S(x)S(y) = (-y) \otimes (-x) - (-x) \otimes (-y),$$

that is, $S(J) \subset J$ for the defining ideal $J \subset T(M)$ of $\mathcal{S}(M)$. Hence S factors to a map $\mathcal{S}(M) \to \mathcal{S}(M)$ that is an antipode, making $\mathcal{S}(M)$ a cocommutative Hopf algebra.

15.14. Exterior algebra of a module. Let M be an R-module and $T(M)$ its tensor algebra. The *exterior algebra* of M, denoted by $\Lambda(M)$, is the factor algebra of $T(M)$ by the ideal K generated by the subset

$$\{x \otimes x \mid x \in M\} \subset T(M).$$

Putting $K_n := K \cap T_n(M)$ and $\Lambda_n(M) := T_n(M)/K_n$, we obtain

$$K = \bigoplus\nolimits_{\mathbb{N}} K_n \quad \text{and} \quad \Lambda(M) = \bigoplus\nolimits_{\mathbb{N}} \Lambda_n(M).$$

It is easy to check that this makes $\Lambda(M)$ an $\mathbb{N}$-graded algebra.

Similar to the symmetric case, $K_0 = K_1 = 0$, and hence we may assume $R \subset \Lambda(M)$ and $M \subset \Lambda(M)$. The product of two elements $u, v \in \Lambda(M)$ is usually written as $u \wedge v$ and is called the *exterior product* of u and v. By construction, the elements of $\Lambda_n(M), n \geq 2$, are sums of elements of the form $x_1 \wedge x_2 \wedge \cdots \wedge x_n$ with $x_i \in M$. It is an elementary computation to verify that this product is zero if any two of the x_i are equal (or, over a field, if the $x_1, \ldots, x_n$ are linearly dependent).

Universal property of $\Lambda(M)$. *Let B be an associative R-algebra and let $f : M \to B$ be an R-linear map such that $f(x)^2 = 0$, for all $x \in M$. Then there exists a unique R-algebra morphism $h : \Lambda(M) \to B$ such that $f = h|_M$.*

As for the symmetric algebra, for any R-modules M_1, M_2, there is an isomorphism $g : \Lambda(M_1 \oplus M_2) \simeq \Lambda(M_1) \otimes_R \Lambda(M_2)$. The diagonal map $\delta : M \to M \oplus M$, $x \mapsto (x, x)$, induces an R-algebra morphism

$$\Lambda(\delta) : \Lambda(M) \to \Lambda(M \oplus M).$$

Combined with the isomomorphism $g : \Lambda(M \oplus M) \to \Lambda(M) \otimes_R \Lambda(M)$, this gives an R-linear map

$$g \circ \Lambda(\delta) : \Lambda(M) \to \Lambda(M) \otimes_R \Lambda(M), \qquad x \mapsto x \otimes 1 + 1 \otimes x,$$

which makes $\Lambda(M)$ a coalgebra. The counit is

$$\varepsilon : \Lambda(M) \to R, \quad 1 \mapsto 1, \ z \mapsto 0, \text{ for } z \in \Lambda_n(M), \ n \geq 1.$$

For the antipode S of $T(M)$, obviously $S(K) \subset K$ for the defining ideal $K \subset T(M)$, and hence it factorises to an antipode $\Lambda(M) \to \Lambda(M)$.

15.15. Exercises

(1) Let H be a Hopf algebra with antipode S that is finitely generated and projective as an R-module. Show that H^* with the canonical structure maps is again a Hopf algebra.

(2) Prove that for a Hopf algebra H with antipode S, the following are equivalent:
   (a) for any $h \in H$, $\sum S(h_{\underline{2}})h_{\underline{1}} = \varepsilon(h)1_H$;
   (b) for any $h \in H$, $\sum h_{\underline{2}}S(h_{\underline{1}}) = \varepsilon(h)1_H$;
   (c) $S \circ S = I_H$.

(3) Let $H = R[X]/J$ be the Hopf algebra considered in 15.11. Show that every element in H is contained in an H-subcomodule that is finitely generated as an R-module.

(4) Let H, K be Hopf algebras with antipodes S_H, S_K, respectively. Prove that, for any bialgebra morphism $f : H \to K$, $S_K \circ f = f \circ S_H$.

(5) Let L be a Lie algebra over R and assume L to be free as an R-module. In the tensor algebra $T(L)$ (over the R-module L) consider the ideal J generated by the subset
$$\{x \otimes y - y \otimes x - xy \mid x, y \in L\}.$$
The factor algebra $U(L) := T(L)/J$ is called the *universal enveloping R-algebra* of L. Prove that $U(L)$ is a Hopf algebra.

(6) Let R be a Noetherian ring and B an R-bialgebra. Assume that the finite dual B° is a pure R-submodule of R^B. Prove that B° is again a bialgebra with the coproduct as given in 5.7, and the product induced by the convolution product on B^*. Prove also that, if B is a Hopf algebra, then B° is a Hopf algebra.

(7) Let G be a group and R a Noetherian ring. Prove that if if $R[G]^\circ$ is R-pure in $R[G]^*$, then $R[G]^\circ$ is a Hopf-algebra with antipode
$$R[G]^\circ \to R[G]^\circ, \quad \xi \mapsto [g \mapsto \xi(g^{-1}), \text{ for } g \in G].$$

(8) Let R be a Noetherian ring and $R[X]$ the polynomial ring. Prove:
   (i) The coalgebra structure Δ_1, ε_1 on $R[X]$ (see 1.8) induces an algebra structure on the finite dual $R[X]^\circ$ with product ($\xi, \zeta \in R[X]^\circ$)
$$(\xi \cdot \zeta)(x^i) = (\xi \otimes \zeta)\Delta_1(x^i) = \xi(x^i)\zeta(x^i), \; i \geq 0,$$
   and unit $u_1 : R \to R[x]^\circ$, $1 \mapsto [x^i \mapsto 1, i \geq 0]$, which is compatible with the coalgebra structure on $R[X]^\circ$ (see 5.13).
   (ii) The coalgebra structure Δ_2, ε_2 on $R[X]$ (see 1.8) induces an algebra structure on $R[X]^\circ$ with product
$$(\xi * \zeta)(x^i) = (\xi \otimes \zeta)\Delta_2(x^i) = \sum_{j=0}^{i} \binom{i}{j} \xi(x^j)\zeta(x^{i-j}), \; i \geq 0,$$
   and unit $u_2 : R \to R[x]^\circ$, $1 \mapsto [x^i \mapsto \delta_{i,0}, i \geq 0]$, which makes $R[x]^\circ$ a Hopf algebra with antipode
$$R[x]^\circ \to R[x]^\circ, \quad \xi \mapsto [x^i \mapsto (-1)^i \, \xi(x^i), \; i \geq 0].$$

References. Bourbaki [5]; Dăscălescu, Năstăsescu and Raianu [14]; Lomp [153]; Montgomery [37]; Nakajima [164]; Pareigis [174]; Sweedler [45]; Wisbauer [212].

16 Trace ideal and integrals for Hopf algebras

Sections 4, 7 and 8 displayed clearly the significance and usefulness of the dual algebra C^* of a coalgebra C for analysing the structure of C and of the category of its comodules $\mathbf{M}^C$. The relationship with C^* allows one to turn questions about the structure of $\mathbf{M}^C$ into questions about modules over C^*. Since a Hopf algebra H is a coalgebra with an additional structure, H^* is a (convolution) algebra, and the results in Sections 4, 7 and 8 can be applied in particular to comodules of a Hopf algebra. In this section we investigate the relationship between H and H^*. We concentrate on properties that are typical for Hopf algebras, in particular, on those that are related to Hopf modules. One of such properties is the fact that the antipode S of H induces a right H-module structure on H^* that makes the rational module $\mathrm{Rat}^H(H^*)$ (cf. 7.1) a right Hopf module.

16.1. Right H-module structure on H^*. *Let H be a Hopf R-algebra, locally projective as an R-module, and suppose that R is Noetherian. Let $T = \mathrm{Rat}^H(H^*)$ be the left trace ideal with the right H-coaction $\varrho^T : T \to T \otimes_R H$. Then H^* is a right H-module with the multiplication*

$$\rightharpoondown_s \,: H^* \otimes_R H \to H^*, \quad f \otimes a \mapsto S(a) \rightharpoondown f\,.$$

Explicitly, for each $c \in H$ (cf. 14.14 for the definition of $\rightharpoondown$),

$$[f \rightharpoondown_s a](c) = [S(a) \rightharpoondown f](c) = f(cS(a)).$$

(1) *For all $a \in H$ and $g, f \in H^*$,*

$$g * (f \rightharpoondown_s a) = \sum [(a_{\underline{2}} \rightharpoondown g) * f] \rightharpoondown_s a_{\underline{1}}\,.$$

(2) *With this structure T is a right H-submodule of H^* and ϱ^T is a Hopf module structure map.*

(3) *H_R is finitely generated if and only if H^* is a Hopf module.*

Proof. Since S is a ring anti-morphism, the definition clearly yields a right module structure on H^*.

(1) Evaluating both expressions at $x \in H$, we obtain

$$\begin{aligned}
\sum([(a_{\underline{2}} \rightharpoondown g) * f] \rightharpoondown_s a_{\underline{1}})(x) &= \sum((a_{\underline{2}} \rightharpoondown g) \otimes f)(x_{\underline{1}} S(a_{\underline{12}}) \otimes x_{\underline{2}} S(a_{\underline{11}})) \\
&= \sum g(x_{\underline{1}} S(a_{\underline{2}}) a_{\underline{3}}) f(x_{\underline{2}} S(a_{\underline{1}})) \\
&= \sum g(x_{\underline{1}} \varepsilon(a_{\underline{2}})) f(x_{\underline{2}} S(a_{\underline{1}})) \\
&= \sum g(x_{\underline{1}}) f(x_{\underline{2}} S(a)) \;=\; g * (f \rightharpoondown_s a)(x)\,.
\end{aligned}$$

(2) Let $f \in T$ and $\varrho^T(f) = \sum f_{\underline{0}} \otimes f_{\underline{1}}$. For any $g \in H^*$, $g * f = \sum g(f_{\underline{1}}) f_{\underline{0}}$, and for all $a \in H$ we obtain from (1),

$$g * (f \leftharpoonup a) = \sum [(a_{\underline{2}} \rightharpoonup g)(f_{\underline{1}}) f_{\underline{0}}] \leftharpoonup a_{\underline{1}} = \sum g(f_{\underline{1}} a_{\underline{2}})(f_{\underline{0}} \leftharpoonup a_{\underline{1}}) .$$

This shows that the left ideal generated by $f \leftharpoonup a$ in H^* is a submodule of the R-module generated by the finite family $f_{\underline{0}} \leftharpoonup a_{\underline{1}}$. Since R is Noetherian, $H^* * (f \leftharpoonup a)$ is a finitely presented R-module and hence $f \leftharpoonup a \in T$ by 7.5. This proves that T is a right H-submodule in H^*. Furthermore, since the identity in (1) holds for all $g \in H^*$, the local projectivity implies that

$$\varrho^T(f \leftharpoonup a) = \sum (f_{\underline{0}} \leftharpoonup a_{\underline{1}}) \otimes f_{\underline{1}} a_{\underline{2}} = \varrho^T(f) \Delta a ,$$

so that T is an H-Hopf module with the specified structure maps.

(3) If H_R is finitely generated, then $\mathrm{Rat}^H(H^*) = H^*$ is a Hopf module. On the other hand, if H^* is a Hopf module, then in particular it is an H-comodule and hence H_R is finitely generated (see 4.7). □

16.2. Coinvariants in H^*. *Let H be a Hopf R-algebra that is locally projective as an R-module, and let $T = \mathrm{Rat}^H(H^*)$. Then for all $t \in T$ the following statements are equivalent:*

(a) t is a left integral on H;

(b) $t \in T^{coH}$.

If R is Noetherian, then (a) and (b) are equivalent to:

(c) $\alpha : H \to H^*$, $b \mapsto t \leftharpoonup b$, is a left H^*-morphism;

(d) $\beta : H \times H \to R$, $(c, d) \mapsto (t \leftharpoonup d)(c)$, is an H-balanced bilinear form.

Proof. (a) ⇔ (b) By 13.6, (a) implies $t \in T$ and $\varrho^T(t) = t \otimes 1_H$. This is equivalent to $t \in T^{coH}$ (see 14.7).

Now assume that R is Noetherian.

(b) ⇒ (c) Let $t \in T^{coH}$. From 16.1 we know that $\alpha(t) \subset T$, so that there is a commutative diagram

$$\begin{array}{ccc} H & \xrightarrow{\alpha} & T \\ {\scriptstyle\Delta}\downarrow & & \downarrow{\scriptstyle\varrho^T} \\ H \otimes_R H & \xrightarrow{\alpha \otimes I_H} & T \otimes_R H \end{array} \qquad \begin{array}{ccc} b & \longmapsto & t \leftharpoonup b \\ \downarrow & & \downarrow \\ \Delta(b) & \longmapsto & \sum t \leftharpoonup b_{\underline{1}} \otimes b_{\underline{2}} , \end{array}$$

showing that α is a right H-comodule (hence a left H^*-module) morphism.

(c) ⇒ (a) Since α is a left H^*-module morphism, $\alpha(1_H) = t \in T$ and $t \otimes 1_H = \varrho^T(t)$ (see 13.6).

(c) $\Leftrightarrow$ (d) By definition, (c) is satisfied if and only if

$$t \leftharpoondown_s (f \rightharpoonup d) = f * (t \leftharpoondown_s d), \text{ for } d \in H, \, f \in H^*.$$

For any $c, d \in H$, $f \in H^*$,

$$\begin{aligned} \beta(c \leftharpoonup f, d) &= (t \leftharpoondown_s d)(c \leftharpoonup f) = [f * (t \leftharpoondown_s d)](c), \\ \beta(c, f \rightharpoonup d) &= [t \leftharpoondown_s (f \rightharpoonup d)](c). \end{aligned}$$

From this the assertion is clear. □

16.3. The trace ideal as projective generator. *Let H be a Hopf R-algebra that is a locally projective R-module, and let $T := Rat^H(H^*)$. Consider the following properties:*

(i) T is a projective generator in $\mathbf{M}^H$,

(ii) T is a faithfully projective R-module,

(iii) T^{coH} is a faithfully flat R-module,

(iv) H is cogenerated by H^* as left H^*-module.

Then there is a chain of implications (i) $\Rightarrow$ (ii) $\Rightarrow$ (iii) $\Rightarrow$ (iv).

(1) If (iii) holds, then S is injective.

(2) If R is a QF ring, then (iv) $\Rightarrow$ (i).

Proof. (i) $\Rightarrow$ (ii) By 3.22, a projective object T in $\mathbf{M}^H$ is projective in $\mathbf{M}_R$. Moreover, T generates H and hence also R. This shows that T is faithfully projective in $\mathbf{M}_R$.

(ii) $\Rightarrow$ (iii) By the Fundamental Theorem 15.5, $T \simeq T^{coH} \otimes_R H$. Since H and T are faithfully flat R-modules, we conclude that so is T^{coH}.

(iii) $\Rightarrow$ (iv) For any $t \in T^{coH}$, the map $H \to H^*$, $b \mapsto t \leftharpoondown_s b$, is a left H^*-morphism. Since $T^{coH} \otimes b \neq 0$, for any nonzero $b \in H$, the isomorphism $T^{coH} \otimes_R H \to T$, $t \otimes b \mapsto t \leftharpoondown_s b$, implies that there exists $t \in T^{coH}$ such that $t \leftharpoondown_s b \neq 0$. Hence H is cogenerated by H^*.

(1) To prove that S is injective take any $a \in H$ such that $S(a) = 0$. Then

$$T^{coH} \otimes a \simeq T^{coH} \leftharpoondown_s a = S(a) \rightharpoondown T^{coH} = 0.$$

This implies that $a = 0$ and hence S is injective.

(2) The implication (iv) $\Rightarrow$ (i) over QF rings will be shown in 16.8. □

16.4. Proposition. *Let H be a Hopf R-algebra that is a locally projective R-module, and let $T = Rat^H(H^*)$. For the conditions*

(i) there exists a generator P in $\mathbf{M}^H$ which is projective in ${}_{H^*}\mathbf{M}$,

(ii) $\mathbf{M}^H$ is closed under extensions in ${}_{H^*}\mathbf{M}$,

(iii) H is a generator in $\mathbf{M}^H$,

(iv) if R is Artinian then T and T^{coH} are faithfully projective as R-modules,

the implications (i) $\Rightarrow$ (ii) $\Rightarrow$ (iii) and (ii) $\Rightarrow$ (iv) hold true.

Proof. (i) $\Rightarrow$ (ii) is a special case of Corollary 42.18.

(ii) $\Rightarrow$ (iii) By 42.16, T is a generator in $\mathbf{M}^H$ and, by 15.5, H generates T (as a right Hopf module). Thus H is a generator in $\mathbf{M}^H$.

(ii) $\Rightarrow$ (iv) Since R is Artinian and H is projective as an R-module, H^* is also a projective R-module. By 7.11, H^*/T is flat as a right H^*-module and hence is a direct limit of projective H^*-modules, which are also projective R-modules. Therefore H^*/T is projective as an R-module and so is T. This also implies that T^{coB} is faithfully projective (see 16.3, (ii) $\Rightarrow$ (iii)). □

Although in general left semiperfect coalgebras need not be right semiperfect, the above proposition implies that for Hopf algebras over QF rings these two notions are equivalent.

16.5. Corollary. *Let H be a right semiperfect Hopf R-algebra with H_R locally projective, and let R be a QF ring. Then:*

(1) H is cogenerated by H^* as a left H^*-module.

(2) H is left semiperfect as coalgebra and $\mathrm{Rat}^H(H^*) = {}^H\mathrm{Rat}(H^*)$.

Proof. (1) follows from 16.3 (iii)$\Rightarrow$(iv).

(2) As shown in 9.13, (1) implies that H is left semiperfect. Now it follows by 9.9 that the left and right trace ideals coincide. □

Next we prove a *uniqueness theorem* for the coinvariants of Hopf algebras over QF rings.

16.6. Lemma. *Let H be a right semiperfect Hopf R-algebra with H_R locally projective, and let $T = \mathrm{Rat}^H(H^*)$. If R is a QF ring, then:*

(1) for every $M \in \mathbf{M}^H$ that is finitely generated as an R-module,

$$\mathrm{length}_R(\mathrm{Hom}_{H^*}(H, M)) \leq \mathrm{length}_R(M),$$

where $\mathrm{length}_R(M)$ *denotes the composition length of the R-module M.*

(2) In particular, $T^{coH} = R\chi \simeq R$, for some $\chi \in T^{coH}$.

(3) There exists $t \in T$ with $t(1_H) = 1_R$.

(4) For any $\chi \in T^{coH}$ with $T^{coH} = R\chi$, there exists some $a \in H$ such that $\chi \leftharpoonup a(1_H) = 1_R$.

Proof. (1) By 16.4(3), R is a direct summand of T^{coH}. By the Fundamental Theorem 15.5, this implies that $H \simeq R \otimes_R H$ is a direct summand of $T^{coH} \otimes_R H \simeq T$ in $\mathbf{M}_H^H$, and hence also in $\mathbf{M}^H$, yielding an epimorphism $\mathrm{Hom}_{H^*}(T, M) \to \mathrm{Hom}_{H^*}(H, M)$. Under the given conditions, 9.7 implies that

$$M \simeq \mathrm{Hom}_{H^*}(H^*, M) \simeq \mathrm{Hom}_{H^*}(T, M),$$

and from this the assertion follows.

(2) With the trivial coaction $R \to R \otimes_R H$, $r \mapsto r \otimes 1_H$, R is a right H-comodule. Now (1) implies that $\mathrm{length}_R(\mathrm{Hom}_{H^*}(H, R)) \leq \mathrm{length}_R(R)$. Since $T = \mathrm{Rat}^H(H^*)$ by 16.5, we know from 13.6 that $\mathrm{Hom}_{H^*}(H, R) = T^{coH}$, so that $\mathrm{length}_R(T^{coH}) \leq \mathrm{length}_R(R)$. Since R is a direct summand of T^{coH}, this implies that $R \simeq T^{coH}$.

(3) By 9.6 and 7.11, for any $N \in \mathbf{M}^H$, the canonical map $T \otimes_{H^*} N \to N$ is an isomorphism. In particular, the map $T \otimes_{H^*} R \to R$, $t \otimes r \mapsto t \rightharpoonup r = rt(1_H)$, is an isomorphism. Therefore there exist $t_1, \ldots, t_n \in T$ and $r_1, \ldots, r_n \in R$, such that

$$1_R = \sum_{i=1}^n t_i \rightharpoonup r_i = \sum_{i=1}^n r_i t_i(1_H) = [\sum_{i=1}^n r_i t_i](1_H).$$

Hence $t := \sum_{i=1}^n r_i t_i \in T$ and $t(1_H) = 1_R$.

(4) By (2), there exists $\chi \in T^{coH}$ such that $T^{coH} = R\chi \simeq R$. By the Fundamental Theorem 15.5, the map $T^{coH} \otimes_R H \to T$, $\chi \otimes h \mapsto \chi \leftharpoondown h$, is an isomorphism in $\mathbf{M}_H^H$. Thus there exists $a \in H$ such that $\chi \leftharpoondown a = t$. □

16.7. Bijective antipode. *Let H be a (right) semiperfect Hopf R-algebra that is locally projective as an R-module. If R is a QF ring, then the antipode S of H is bijective.*

Proof. Let $T := \mathrm{Rat}^H(H^*)$ and let $T^{coH} = R\chi$ (see 16.6). By 16.3, S is injective. Now, suppose that $S(H) \neq H$. Since $S(H)$ is a subcoalgebra, we may consider it as a left subcomodule of H. Then $0 \neq H/S(H) \in {}^H\mathbf{M}$, and hence there is a nonzero morphism $\omega : H/S(H) \to E(U)$ in ${}^H\mathbf{M}$, for some simple object U with injective hull $E(U)$ in ${}^H\mathbf{M}$. Since R is a cogenerator in $\mathbf{M}_R$, there is an R-morphism $\alpha : E(U) \to R$ with $\alpha \circ \omega \neq 0$. The composition of $\alpha \circ \omega$ with the canonical projection, $\pi : H \to H/S(H)$, gives a nonzero R-morphism $\lambda := \alpha \circ \omega \circ \pi : H \to R$. Note that $\mathrm{Ke}\,\lambda \supset N \supset S(H)$, where $\mathrm{Ke}\,\omega = N/S(H)$. By definition, $N \subset H$ is a left subcomodule and H/N is finitely R-generated (since $E(U)$ is). So, by 7.5, $\lambda \in T$ and there exists $b \in H$ such that $\lambda = \chi \leftharpoondown b$. By construction, $\lambda(S(H)) = \chi \leftharpoondown b(S(H)) = 0$. Therefore, for any $h \in H$,

$$0 = \chi \leftharpoondown b(S(h)) = \chi(S(h)S(b)) = \chi(S(bh)) = \chi \circ S(bh),$$

and we conclude that $\chi \circ S(bH) = 0$. It is straightforward to prove that, for the left integral χ, the composition $\chi \circ S$ is a right integral and hence $\chi \circ S(H^* \rightharpoonup bH) = 0$.

Since H is a progenerator in $\mathbf{M}_H^H$ (see 15.5), there exists an ideal $J \subset R$ such that the Hopf submodule $H^* \rightharpoonup bH \subset H$ is of the form JH, and

$$0 = \chi \circ S(H^* \rightharpoonup bH) = \chi \circ S(JH) = J\chi \circ S(H).$$

As we have seen in 16.6, there exists $a \in H$ with $\chi \circ S(a) = 1_R$. This implies $J = 0$ and $bH \subset JH = 0$, that is, $b = 0$, contradicting the fact that by construction $0 \neq \lambda = \chi \leftharpoondown b$. □

16.8. Semiperfect Hopf algebras over QF rings. *Let H be a Hopf R-algebra that is locally projective as an R-module, and let $T = \text{Rat}^H(H^*)$. If R is a QF ring, then the following are equivalent:*

(a) *H is a right semiperfect coalgebra;*

(b) *T is a projective generator in $\mathbf{M}^H$;*

(c) *T is a faithful and flat R-module;*

(d) *T^{coH} is a faithful and flat R-module;*

(e) *H is cogenerated by H^* as a left H^*-module (left QcF);*

(f) *H is projective in $\mathbf{M}^H$;*

(g) *H is a projective generator in $\mathbf{M}^H$;*

(h) *$T^{coH} = R\chi \simeq R$, for some $\chi \in T^{coH}$;*

(i) *T is a flat R-module and the injective hull of R in ${}^H\mathbf{M}$ is finitely generated as an R-module;*

(j) *there exists a left H^*-monomorphism $H \to H^*$, that is, H is a* left co-Frobenius Hopf algebra;

(k) *H is a left semiperfect coalgebra.*

The left-side versions of (b)–(j) are also equivalent to (a).

Proof. (a) ⇒ (c) ⇒ (d) ⇒ (e) We know from 9.6 that $\mathbf{M}^H$ has a generator that is projective in ${}_{H^*}\mathbf{M}$. So the assertions follows from 16.4.

(e) ⇒ (f) ⇒ (k) are clear by 16.5 and 9.13.

(a) ⇒ (b) Since (a) ⇒ (k), we obtain from 9.9 that T is a ring with enough idempotents. From 9.6 (and [46, 49.1]) we know that T is a projective generator in $\mathbf{M}^H (= {}_T\mathbf{M})$.

(a) ⇒ (g) The implications (a) ⇒ (f) and (a) ⇒ (k) imply that H is projective as a left and right comodule. So, by 9.13 and 9.15, H is a projective generator in $\mathbf{M}^H$ (and ${}^H\mathbf{M}$).

(b) ⇒ (a), (g) ⇒ (a), and (j) ⇒ (e) are trivial, while (a) ⇒ (h) is shown in 16.6(2).

(a) ⇒ (i) As mentioned before, 16.4 applies and so T_R is projective, and by 9.6, injective hulls of simple comodules in ${}^H\mathbf{M}$ are finitely generated.

(i) ⇒ (c) Let $E(R)$ denote the injective hull of R in ${}^H\mathbf{M}$. Assume it to be finitely generated as an R-module. Then $E(R)^*$ is projective and cogenerated by T. The inclusion $R \subset E(R)^*$ implies that R is cogenerated by T. Hence T is a faithful R-module.

(h) ⇒ (j) This follows from the fact that $H \to H^*$, $b \mapsto \chi \leftharpoondown b$, is a monomorphism.

(k) ⇒ (a) This follows from (a) ⇒ (k) by left-right symmetry. □

Over a field every nonzero vector space is faithfully flat, so 16.8 implies:

16.9. Corollary. *For a Hopf algebra H over a field F and $T = \mathrm{Rat}^H(H^*)$, the following are equivalent:*

(a) H is a right semiperfect coalgebra;

(b) $T \neq 0$;

(c) $T^{\mathrm{co}H} \neq 0$;

(d) $T^{\mathrm{co}H}$ is one-dimensional over F;

(e) the injective hull of F in ${}^H\mathbf{M}$ is finite dimensional;

(f) H is cogenerated by H^* as a left H^*-module;

(g) H is projective (and a generator) in $\mathbf{M}^H$;

(h) H is left co-Frobenius;

(i) H is a left semiperfect coalgebra.

The left side versions of (b)–(g) are also equivalent to (a).

If these conditions are satisfied, then the antipode of H is bijective.

Definitions. Let B be an R-bialgebra. A left integral $t \in B^*$ is called a *total left integral on B* if $t(1_B) = 1_R$ or, equivalently, $t \circ \iota = 1_R$.

A Hopf algebra H is said to be *left (H,R)-cosemisimple* if, for any $M \in {}^H\mathbf{M}$, subcomodules of M that are R-direct summands of M are also direct summands as comodules. H is said to be *left cosemisimple* if it is semisimple as a left comodule.

16.10. Total integrals on H. *For a Hopf R-algebra H, the following are equivalent:*

(a) there exists a total left integral t on H;

(b) there exists an R-linear map $\alpha : H \otimes_R H \to R$ satisfying

$$\alpha \circ \Delta = \varepsilon \quad \text{and} \quad (I_H \otimes \alpha) \circ (\Delta \otimes I_H) = (\alpha \otimes I_H) \circ (I_H \otimes \Delta);$$

(c) every left H-comodule is (H, R)-cosemisimple;

(d) H is (H, R)-cosemisimple as a left H-comodule.

Proof. (a) $\Rightarrow$ (b) Let $t : H \to R$ be a total left integral on H and consider the map

$$\alpha : H \otimes_R H \xrightarrow{I_H \otimes S} H \otimes_R H \xrightarrow{\mu} H \xrightarrow{t} R, \quad h \otimes h' \mapsto t(hS(h')).$$

The properties of the antipode immediately imply that

$$\alpha \circ \Delta = t \circ \mu \circ (I_H \otimes S) \circ \Delta = t \circ \iota \circ \varepsilon = \varepsilon.$$

Furthermore, using 13.6(1), we obtain for all $h, h' \in H$,

$$\begin{aligned}(\alpha \otimes I_H)(I_H \otimes \Delta)(h \otimes h') &= \sum t(hS(h'_{\underline{1}}))h'_{\underline{2}} \\ &= \sum (hS(h'_{\underline{1}}))_{\underline{1}} t((hS(h'_{\underline{1}}))_{\underline{2}})h'_{\underline{2}} \\ &= \sum h_{\underline{1}} S(h'_{\underline{2}})h'_{\underline{3}} t(h_{\underline{2}} S(h'_{\underline{1}})) \\ &= \sum h_{\underline{1}} t(h_{\underline{2}} S(h')) = (I_H \otimes \alpha)(\Delta \otimes I_H)(h \otimes h').\end{aligned}$$

(b) $\Rightarrow$ (c) Suppose there is an R-linear map $\alpha : H \otimes_R H \to R$ satisfying the conditions in (b). Let M be a left H-comodule with subcomodule $i : N \to M$, which is an R-direct summand, that is, there exists an R-linear $p : M \to N$ such that $p \circ i = I_N$. Define

$$\begin{aligned}\beta &:= (\alpha \otimes I_N) \circ (I_H \otimes {}^N\varrho) \circ (I_H \otimes p) \circ {}^M\varrho : M \to N, \\ & m \mapsto \sum \alpha(m_{\underline{-1}} \otimes p(m_{\underline{0}})_{\underline{-1}})p(m_{\underline{0}})_{\underline{0}}.\end{aligned}$$

We have to show that β is a left H-comodule morphism satisfying $\beta \circ i = I_N$. Compute

$$\begin{aligned}\beta \circ i &= (\alpha \otimes I_N) \circ (I_H \otimes {}^N\varrho) \circ (I_H \otimes p) \circ {}^M\varrho \circ i \\ &= (\alpha \otimes I_N) \circ (I_H \otimes {}^N\varrho) \circ (I_H \otimes p) \circ (I_H \otimes i) \circ {}^N\varrho \\ &= (\alpha \otimes I_N) \circ (I_H \otimes {}^N\varrho) \circ {}^N\varrho = (\alpha \otimes I_N) \circ (\Delta \otimes I_N) \circ {}^N\varrho \\ &= (\alpha \circ \Delta \otimes I_N) \circ {}^N\varrho = (\varepsilon \otimes I_N) \circ {}^N\varrho = I_N .\end{aligned}$$

To verify that β is H-colinear, we prove that $(I_H \otimes \beta) \circ {}^M\varrho = {}^N\varrho \circ \beta$. For this we use the properties of α and compute, for all $m \in M$,

$$\begin{aligned}(I_H \otimes \beta) \circ {}^M\varrho(m) &= \sum m_{\underline{-2}} \otimes \alpha(m_{\underline{-1}} \otimes p(m_{\underline{0}})_{\underline{-1}})p(m_{\underline{0}})_{\underline{0}} \\ &= \sum p(m_{\underline{0}})_{\underline{-1}} \alpha(m_{\underline{-1}} \otimes p(m_{\underline{0}})_{\underline{-2}})p(m_{\underline{0}})_{\underline{0}} \\ &= {}^N\varrho \circ \beta(m),\end{aligned}$$

as required.

(c) $\Rightarrow$ (d) is trivial.

(d) $\Rightarrow$ (a) The map $\iota : R \to H$ is a left H-comodule map that splits as an R-linear map. Hence there is a left H-comodule morphism $t : H \to R$ such that $t \circ \iota = 1_R$, that is, t is a total left integral. $\square$

16.11. Corollary. *Let H be a Hopf R-algebra with a total left integral over a semisimple ring R. Then H is a direct sum of simple left comodules, that is, it is a semisimple right H^*-module.*

For a Hopf algebra H there is an interesting relationship between integrals in H and the centre of $H \otimes_R H$ considered as an H-bimodule in the natural way. Let

$$Z(H \otimes_R H) = \{u \in H \otimes_R H \mid hu = uh \text{ for all } h \in H\} \simeq {}_H\mathrm{Hom}_H(H, H \otimes_R H).$$

16.12. Coinvariants and $Z(H \otimes_R H)$. *Define the R-linear maps*

$$\begin{aligned} \gamma: &\ {}^H H \to Z(H \otimes_R H), \quad h \mapsto (I_H \otimes S)\Delta(h), \\ \delta: &\ Z(H \otimes_R H) \to {}^H H, \quad \textstyle\sum_i a_i \otimes b_i \mapsto \sum a_i \varepsilon(b_i), \end{aligned}$$

where ${}^H H = \{h \in H \mid$ for all $h' \in H$, $h'h = \varepsilon(h')h\}$ is the R-module of left invariants (cf. 14.12). Then $\delta \circ \gamma = I_{{}^H H}$.

Proof. First we need to show that the image of γ lies in $Z(H \otimes_R H)$. Consider the following equalities for all $h \in {}^H H$ and $a \in H$,

$$\Delta(h) \otimes a = \sum \Delta(\varepsilon(a_{\underline{1}})h) \otimes a_{\underline{2}} = \sum \Delta(a_{\underline{1}}h) \otimes a_{\underline{2}} = \sum a_{\underline{1}}h_{\underline{1}} \otimes a_{\underline{2}}h_{\underline{2}} \otimes a_{\underline{3}}.$$

Now apply $I_H \otimes S$ and $I_H \otimes \mu$ to obtain

$$\begin{aligned} (I_H \otimes S)\Delta(h)\, a = \textstyle\sum h_{\underline{1}} \otimes S(h_{\underline{2}})a &= \textstyle\sum a_{\underline{1}}h_{\underline{1}} \otimes S(a_{\underline{2}}h_{\underline{2}})a_{\underline{3}} \\ &= \textstyle\sum a_{\underline{1}}h_{\underline{1}} \otimes S(h_{\underline{2}})S(a_{\underline{2}})a_{\underline{3}} \\ &= \textstyle\sum ah_{\underline{1}} \otimes S(h_{\underline{2}}) \;=\; a\,(I_H \otimes S)\Delta(h). \end{aligned}$$

Thus the image of γ is in $Z(H \otimes_R H)$, as required. The multiplicativity of ε ensures that the image of δ is in ${}^H H$. Finally, the fact that γ is a section of δ follows immediately from the counit properties and from the counitality of the antipode. This completes the proof. □

Notice that similar maps exist for the right-hand versions.

16.13. Separable Hopf algebras. *For any Hopf R-algebra H, the following are equivalent:*

(a) *H is a separable R-algebra;*

(b) *H is left (right) (H, R)-semisimple;*

(c) *R is projective as a left (right) H-module;*

(d) *there exists a left (right) integral h in H with $\varepsilon(h) = 1_R$.*

Proof. (a) $\Rightarrow$ (b) $\Rightarrow$ (c) are well known in classical ring theory.

(c) $\Rightarrow$ (d) Since R is a projective left H-module, the counit $\varepsilon : H \to R$ is split by a left H-module morphism $\alpha : R \to H$. Then $\alpha(1_R) \in {}^H H$ and $\varepsilon(\alpha(1_R)) = 1_R$, that is, $\alpha(1_R)$ is a normalised left integral in H (see 14.12). Similar arguments apply for the right-hand case.

(d) $\Rightarrow$ (a) Take any $h \in {}^H H$ with $\varepsilon(h) = 1_R$. Then $\gamma(h) = (I_H \otimes S)\Delta(h) \in Z(H \otimes_R H)$ (see 16.12) and

$$\mu(\gamma(h)) = \sum h_{\underline{1}} S(h_{\underline{2}}) = \varepsilon(h) = 1_R.$$

Therefore $\gamma(h)$ is a separability idempotent. □

Recall that a separable R-algebra that is projective as an R-module is a finitely generated R-module. In particular, if R is a field, we obtain

16.14. Separable Hopf algebras over fields. *For any Hopf algebra H over a field F, the following are equivalent:*

(a) *H is a separable F-algebra;*

(b) *H is semisimple as a left (or right) H-module;*

(c) *F is projective as a left (or right) H-module;*

(d) *there exists a left (right) integral $h \in H$ with $\varepsilon(h) \neq 0$.*

Remarks. The assertions of 16.10 were proved in [148, Theorem]. The uniqueness of integrals of Hopf algebras over fields (see 16.6) was shown in [191]. Our proof adapts techniques of the proof given in [60, Theorem 3.3].

It was shown in [177, Proposition 2] that for semiperfect Hopf algebras over fields the antipode is bijective. The original proof was simplified in [88]. We essentially followed these ideas to prove the corresponding result for Hopf algebras over QF rings in 16.7.

Some of the equivalences given in 16.9 appear in [152, Theorem 3]. The characterisation of these algebras by (g) is given in [191, Theorem 1] and for *affine group schemes* it is shown in [109]. The characterisation of Hopf algebras in 16.13 is taken from [153].

16.15. Exercises

Let H be a Hopf R-algebra that is finitely generated and projective as an R-module. Prove:

(i) The antipode S of H is bijective.

(ii) The right coinvariants $(H^*)^{coH}$ of H^* form a finitely generated projective R-module of rank 1.

(iii) If $(H^*)^{coH} \simeq R$, then $H \simeq H^*$ as left H-modules (that is, H is a Frobenius algebra).

References. Beattie, Dăscălescu, Grünenfelder and Năstăsescu [60]; Dăscălescu, Năstăsescu and Raianu [14]; Larson [148]; Lomp [153]; Menini, Torrecillas and Wisbauer [157]; Pareigis [173]; Sullivan [191].

Chapter 3

Corings and comodules

Our adventure with corings starts here. Corings should be seen as one of the most fundamental algebraic structures that include rings and coalgebras as special cases. They appear naturally in the following chain of generalisations: coalgebras over fields; coalgebras over commutative rings; and coalgebras over noncommutative rings (= corings). The scope of applications of corings is truly amazing, and their importance can be explained at least on two levels. First, corings are a "mild" generalisation of coalgebras, in the sense that several properties of coalgebras over commutative rings carry over to corings. In particular, various general properties of corings can be proven by using the same techniques as for coalgebras over rings. From this point of view the step from coalgebras over fields to coalgebras over commutative rings is much bolder and adventurous than that from coalgebras over rings to corings. Second, the range of problems that can be described with the use of corings is much wider than the problems that could ever be addressed by coalgebras. In situations such as ring extensions, even if a commutative ring is extended, one will always require corings. In addition to all this, the theory of corings also has an extremely useful unifying power. We shall soon see that several results about the structure of Hopf modules (cf. Section 14), including the Fundamental Theorem of Hopf algebras 15.5, and their generalisations to different classes of Hopf-type modules, are simply special cases of structure theorems for corings.

In this chapter we outline the scenery for the theory of corings. We stress the relationship between corings and coalgebras over rings, and thus the pattern of our presentation follows that of Chapter 1. We often use the same or very similar techniques as in Chapter 1 to prove more general results about corings. At the same time we indicate the differences and pay particular attention to several fine points arising from the noncommutativity of the base algebra. Section 24 has the full coring flavour: we discuss the effects of the change of base for the corings and their comodules.

In the present chapter we concentrate on the theory of corings and their comodules. Examples can be found in Chapter 4 and, specifically, in Chapters 5 and 6. Throughout, R denotes a commutative and associative ring with a unit and A an associative R-algebra with a unit.

17 Corings and their morphisms

Extending the notion of an R-coalgebra to noncommutative base rings leads to *corings over an algebra A*. They are based on bimodules over A (instead of R-modules). The formalism related to the basic notions is very much the same as for coalgebras, except that we have to pay attention to the left-sided and right-sided properties of the bimodules.

The notion of a coring can be viewed as an example of a coalgebra in a monoidal category also called a *tensor category* (see 38.31). For this the category of (A, A)-bimodules ${}_A\mathbf{M}_A$ is taken with the usual tensor product over A, and the neutral object is A itself. Thus an A-coring is simply a coalgebra in the monoidal category $({}_A\mathbf{M}_A, \otimes_A, A)$. More explicitly, we have the following definition.

17.1. Corings. An *A-coring* is an (A, A)-bimodule $\mathcal{C}$ with (A, A)-bilinear maps

$$\underline{\Delta} : \mathcal{C} \to \mathcal{C} \otimes_A \mathcal{C} \quad \text{and} \quad \underline{\varepsilon} : \mathcal{C} \to A,$$

called *(coassociative) coproduct* and *counit*, with the properties

$$(I_\mathcal{C} \otimes \underline{\Delta}) \circ \underline{\Delta} = (\underline{\Delta} \otimes I_\mathcal{C}) \circ \underline{\Delta}, \text{ and } (I_\mathcal{C} \otimes \underline{\varepsilon}) \circ \underline{\Delta} = I_\mathcal{C} = (\underline{\varepsilon} \otimes I_\mathcal{C}) \circ \underline{\Delta}.$$

These can be expressed by the commutativity of the diagrams

$$\begin{array}{ccc} \mathcal{C} & \xrightarrow{\underline{\Delta}} & \mathcal{C} \otimes_A \mathcal{C} \\ {\scriptstyle \underline{\Delta}}\downarrow & & \downarrow{\scriptstyle I_\mathcal{C} \otimes \underline{\Delta}} \\ \mathcal{C} \otimes_A \mathcal{C} & \xrightarrow{\underline{\Delta} \otimes I_\mathcal{C}} & \mathcal{C} \otimes_A \mathcal{C} \otimes_A \mathcal{C}, \end{array} \qquad \begin{array}{ccc} \mathcal{C} & \xrightarrow{\underline{\Delta}} & \mathcal{C} \otimes_A \mathcal{C} \\ {\scriptstyle \underline{\Delta}}\downarrow & \searrow^{I_\mathcal{C}} & \downarrow{\scriptstyle \underline{\varepsilon} \otimes I_\mathcal{C}} \\ \mathcal{C} \otimes_A \mathcal{C} & \xrightarrow{I_\mathcal{C} \otimes \underline{\varepsilon}} & \mathcal{C}. \end{array}$$

As for coalgebras, we will use the *Σ-notation*, writing for $c \in \mathcal{C}$,

$$\underline{\Delta}(c) = \sum_{i=1}^{k} c_i \otimes \tilde{c}_i = \sum c_{\underline{1}} \otimes c_{\underline{2}},$$

and – again as for coalgebras – the coassociativity of $\underline{\Delta}$ is expressed by the formulae

$$\begin{aligned} \sum \underline{\Delta}(c_{\underline{1}}) \otimes c_{\underline{2}} &= \sum c_{\underline{11}} \otimes c_{\underline{12}} \otimes c_{\underline{2}} = \sum c_{\underline{1}} \otimes c_{\underline{2}} \otimes c_{\underline{3}} \\ &= \sum c_{\underline{1}} \otimes c_{\underline{21}} \otimes c_{\underline{22}} = \sum c_{\underline{1}} \otimes \underline{\Delta}(c_{\underline{2}}), \end{aligned}$$

and the conditions on the counit are

$$\sum \underline{\varepsilon}(c_{\underline{1}}) c_{\underline{2}} = c = \sum c_{\underline{1}} \underline{\varepsilon}(c_{\underline{2}}).$$

17.2. Base ring extension. Given an R-algebra morphism $\phi : A \to B$, any B-module has a natural A-module structure, and for an A-coring $\mathcal{C}$, the map

$$\vartheta : \mathcal{C} \to \mathcal{C} \otimes_A \mathcal{C} \to \mathcal{C} \otimes_A B \otimes_A \mathcal{C}, \quad c \mapsto \sum c_{\underline{1}} \otimes c_{\underline{2}} \mapsto \sum c_{\underline{1}} \otimes 1_B \otimes c_{\underline{2}},$$

induces a comultiplication and a counit,

$$\underline{\Delta}_{BCB} = I_B \otimes \vartheta \otimes I_B : B \otimes_A \mathcal{C} \otimes_A B \to B \otimes_A \mathcal{C} \otimes_A B \otimes_A \mathcal{C} \otimes_A B,$$

$$\underline{\varepsilon} : B \otimes_A \mathcal{C} \otimes_A B \to B, \quad b \otimes c \otimes b' \mapsto b\underline{\varepsilon}(c)b',$$

which makes $BCB = B \otimes_A \mathcal{C} \otimes_A B$ a B-coring, called a *base ring extension of* $\mathcal{C}$ (compare 1.4 for the commutative case).

Let us consider some examples of corings.

17.3. The trivial A-coring. The algebra A is a coring with the canonical isomorphism $A \to A \otimes_A A$ as a coproduct and the identity map $A \to A$ as a counit. This coring is known as a *trivial* A-coring.

17.4. Coring defined by preordered sets. Let G be a set and $Q \subset G \times G$ a relation on G that is reflexive, transitive and locally finite; that is, for any $g, h \in G$,

$$\omega(g,h) = \{f \in G \mid (g,f) \in Q \text{ and } (f,h) \in Q\} \text{ is a finite set.}$$

For any ring A, let $\mathcal{C} = A^{(Q)}$ be the free left A-module with basis Q and consider $\mathcal{C}$ as an A-bimodule by

$$a(\sum a_\lambda(g_\lambda, h_\lambda)) = \sum aa_\lambda(g_\lambda, h_\lambda), \quad (\sum a_\lambda(g_\lambda, h_\lambda))b = \sum a_\lambda b(g_\lambda, h_\lambda),$$

where $(g_\lambda, h_\lambda) \in Q$ and all a_λ and a, b are in A.

Define comultiplication and counit by the (A, A)-bilinear maps

$$\begin{aligned} \underline{\Delta} : \mathcal{C} \to \mathcal{C} \otimes_A \mathcal{C}, &\quad (g,h) \mapsto \textstyle\sum_{f \in \omega(g,h)} (g,f) \otimes (f,h), \\ \underline{\varepsilon} : \mathcal{C} \to A, &\quad (g,h) \mapsto \delta_{gh}, \qquad \text{for } (g,h) \in Q, \end{aligned}$$

where δ_{gh} is the Kronecker symbol. Now, transitivity of Q implies that $\underline{\Delta}$ is coassociative, and reflexivity of Q implies that $\underline{\varepsilon}$ is a counit. Hence $\mathcal{C}$ is an A-coring. Notice that $\mathcal{C}$ is generated by A as an (A, A)-bimodule.

Putting $Q = \{(g,g) \mid g \in G\}$ in the above example, we obtain the

17.5. Grouplike coring on a set G. For any set G, $\mathcal{C} = A^{(G)}$ is an A-coring by the comultiplication and counit

$$\begin{aligned} \underline{\Delta} : \mathcal{C} \to \mathcal{C} \otimes_A \mathcal{C}, &\quad g \mapsto g \otimes g, \\ \underline{\varepsilon} : \mathcal{C} \to A, &\quad g \mapsto 1_A, \qquad \text{for } g \in G. \end{aligned}$$

We have seen in 1.9 that finitely generated projective R-modules lead to interesting examples of coalgebras. For bimodules, a similar construction provides important corings.

17.6. Coring of a projective module. For R-algebras A, B, let P be a (B, A)-bimodule that is finitely generated and projective as a right A-module. Let $p_1, \ldots, p_n \in P$ and $\pi_1, \ldots, \pi_n \in P^* = \mathrm{Hom}_A(P, A)$ be a dual basis for P_A. Then there is a (B, B)-bimodule isomorphism

$$P \otimes_A P^* \to \mathrm{End}_A(P), \quad p \otimes f \mapsto [q \mapsto pf(q)].$$

Furthermore, the (A, A)-bimodule $P^* \otimes_B P$ is an A-coring with coproduct and counit defined by

$$\underline{\Delta}: P^* \otimes_B P \to (P^* \otimes_B P) \otimes_A (P^* \otimes_B P), \quad f \otimes p \mapsto \sum f \otimes p_i \otimes \pi_i \otimes p,$$
$$\underline{\varepsilon}: P^* \otimes_B P \to A, \quad f \otimes p \mapsto f(p).$$

Note that $\underline{\Delta}$ is well defined. Indeed, denote the endomorphism ring $\mathrm{End}_A(P)$ by S. Using the above isomorphism, we can identify $(P^* \otimes_B P) \otimes_A (P^* \otimes_B P)$ with $P^* \otimes_B S \otimes_B P$ and the map $\underline{\Delta}$ with $f \otimes p \mapsto f \otimes 1_S \otimes p$. The latter is well defined since the canonical map $B \to S$ (provided by left multiplication) is an algebra map. This also shows that $\underline{\Delta}$ is independent of the choice of the dual basis for P_A. Clearly, $\underline{\Delta}$ is coassociative. The dual basis property implies that $\underline{\varepsilon}$ is a counit.

As a special case we may consider the (A, A)-bimodule $P = A^n$, for $n \in \mathbb{N}$. Then $P^* \otimes_A P$ can be identified with the ring $M_n(A)$ of all $n \times n$ matrices with entries from A, and this leads to the matrix coring. Notice that this can also be derived from the construction in 17.4.

17.7. Matrix coring. Let $\{e_{ij}\}_{1 \leq i,j \leq n}$ be the canonical A-basis for $M_n(A)$ and define the coproduct by

$$\underline{\Delta}: M_n(A) \to M_n(A) \otimes_A M_n(A): e_{ij} \mapsto \sum\nolimits_k e_{ik} \otimes e_{kj}$$

and the counit by $\underline{\varepsilon}: M_n(A) \to A: e_{ij} \mapsto \delta_{ij}$. The resulting coring is called the (n, n)-*matrix coring over* A and is denoted by $M_n^c(A)$.

Notation. From now on A will always denote an associative R-algebra with unit (A, μ, ι) and $\mathcal{C}$ will stand for a coassociative A-coring $(\mathcal{C}, \underline{\Delta}, \underline{\varepsilon})$. $Z(S)$ denotes the centre of any ring S. The A-linear maps $\mathcal{C} \to A$ have ring structures that allow one to describe properties of the coring itself, and we put

$$\mathcal{C}^* = \mathrm{Hom}_A(\mathcal{C}, A), \quad {}^*\mathcal{C} = {}_A\mathrm{Hom}(\mathcal{C}, A), \quad {}^*\mathcal{C}^* = {}_A\mathrm{Hom}_A(\mathcal{C}, A) = {}^*\mathcal{C} \cap \mathcal{C}^*.$$

17.8. The dual rings.

(1) $\mathcal{C}^*$ is a ring with unit $\underline{\varepsilon}$ by the product (for $f, g \in \mathcal{C}^*$, $c \in \mathcal{C}$)

$$f *^r g : \mathcal{C} \xrightarrow{\underline{\Delta}} \mathcal{C} \otimes_A \mathcal{C} \xrightarrow{f \otimes I_{\mathcal{C}}} \mathcal{C} \xrightarrow{g} A, \quad f *^r g(c) = \sum g(f(c_{\underline{1}})c_{\underline{2}}),$$

and there is a ring anti-morphism $i_R : A \to \mathcal{C}^*$, $a \mapsto \underline{\varepsilon}(a-)$.

(2) ${}^*\mathcal{C}$ is a ring with unit $\underline{\varepsilon}$ by the product (for $f, g \in {}^*\mathcal{C}$, $c \in \mathcal{C}$)

$$f *^l g : \mathcal{C} \xrightarrow{\underline{\Delta}} \mathcal{C} \otimes_A \mathcal{C} \xrightarrow{I_{\mathcal{C}} \otimes g} \mathcal{C} \xrightarrow{f} A, \quad f *^l g(c) = \sum f(c_{\underline{1}} g(c_{\underline{2}})),$$

and there is a ring anti-morphism $i_L : A \to {}^*\mathcal{C}$, $a \mapsto \underline{\varepsilon}(-a)$.

(3) ${}^*\mathcal{C}^*$ is a ring with unit $\underline{\varepsilon}$ by the product (for $f, g \in {}^*\mathcal{C}^*$, $c \in \mathcal{C}$)

$$f * g(c) = \sum f(c_{\underline{1}}) g(c_{\underline{2}}),$$

and there is a ring morphism $Z(A) \to Z({}^*\mathcal{C}^*)$, $a \mapsto \underline{\varepsilon}(a-) = \underline{\varepsilon}(-a)$.

(4) There are inclusions $Z({}^*\mathcal{C}) \subset Z({}^*\mathcal{C}^*)$ and $Z(\mathcal{C}^*) \subset Z({}^*\mathcal{C}^*)$.

Proof. (1) For every $c \in \mathcal{C}$ and $f \in \mathcal{C}^*$,

$$f *^r \underline{\varepsilon}(c) = \sum \underline{\varepsilon}(f(c_{\underline{1}})c_{\underline{2}}) = f(c), \quad \underline{\varepsilon} *^r f(c) = \sum f(\underline{\varepsilon}(c_{\underline{1}})c_{\underline{2}}) = f(c),$$

showing that $\underline{\varepsilon}$ is the unit in $\mathcal{C}^*$. Associativity of the product follows from the equalities, for $f, g, h \in \mathcal{C}^*$, $c \in \mathcal{C}$,

$$\begin{aligned} f *^r (g *^r h)(c) &= \sum h(g *^r f(c_{\underline{1}})c_{\underline{2}}) = \sum h(g(f(c_{\underline{1}})c_{\underline{2}})c_{\underline{3}}) \\ &= \sum g *^r h(f(c_{\underline{1}})c_{\underline{2}}) = (f *^r g) *^r h(c). \end{aligned}$$

Finally, for all $a, a' \in A$ and $c \in \mathcal{C}$ we compute

$$\begin{aligned} (i_R(a) *^r i_R(a'))(c) &= \sum i_R(a')(i_R(a)(c_{\underline{1}})c_{\underline{2}}) = \sum \underline{\varepsilon}(a' \underline{\varepsilon}(a c_{\underline{1}}) c_{\underline{2}}) \\ &= \underline{\varepsilon}(a'ac) = i_R(a'a)(c). \end{aligned}$$

This proves that i_R is an anti-algebra map.

Similar computations show that ${}^*\mathcal{C}$ is an associative algebra and that i_L is an algebra anti-morphism, thus proving (2) and (3).

(4) For $f \in Z({}^*\mathcal{C})$, $a \in A$, and $c \in \mathcal{C}$,

$$f(ca) = f *^l \underline{\varepsilon}(-a)(c) = \underline{\varepsilon}(-a) *^l f(c) = f(c)a,$$

which shows $f \in \mathcal{C}^*$ and hence $Z({}^*\mathcal{C}) \subset {}^*\mathcal{C}^*$. This obviously implies $Z({}^*\mathcal{C}) \subset Z({}^*\mathcal{C}^*)$. Similarly we get $Z(\mathcal{C}^*) \subset Z({}^*\mathcal{C}^*)$. □

Of course, if $A = R$, then $\mathcal{C}^* = {}^*\mathcal{C} = {}^*\mathcal{C}^*$, and 17.8 reduces to 1.3.

17.9. Dual rings of a coring associated to a projective module. Consider the A-coring $P^* \otimes_B P$ of Example 17.6. Its left dual algebra is isomorphic to ${}_B\mathrm{End}(P)$ by the bijective maps

$$^*(P^* \otimes_B P) = {}_A\mathrm{Hom}(P^* \otimes_B P, A) \simeq {}_B\mathrm{Hom}(P, {}^*(P^*)) \simeq {}_B\mathrm{End}(P),$$

which yield a ring isomorphism provided the product in ${}_B\mathrm{End}(P)$ is given by the composition $\varphi\varphi' = \varphi \circ \varphi'$. Explicitly, the isomorphisms are given by

$${}_B\mathrm{End}(P) \to {}^*(P^* \otimes_B P), \quad \varphi \mapsto [f \otimes p \mapsto f(\varphi(p))],$$

and the inverse $\xi \mapsto [p \mapsto \sum_i p_i \xi(\pi_i \otimes p)]$, with $\{p_i, \pi_i\}$ a dual basis of P_A.

For the right dual algebra there are the isomorphisms

$$(P^* \otimes_B P)^* = \mathrm{Hom}_A(P^* \otimes_B P, A) \simeq \mathrm{End}_B(P^*).$$

It is interesting to observe how a dual of a ring extension can be made into a coring provided suitable finitely generated projective module type assumptions are made. This is well known and straightforward to see for the dual of algebras but needs some more care over noncommutative rings. The construction goes under the name of the *dual coring theorem*. First we consider the following situation.

17.10. Ring anti-morphism. Let $\phi : A \to S$ be a ring anti-morphism and as usual view S as an (A, A)-bimodule with left and right multiplication of $s \in S$ by $a, b \in A$,

$$as = s\phi(a), \quad sb = \phi(b)s.$$

Note that on the right-hand sides the product in S is used. Since S is an (A, A)-bimodule, one can consider two types of dual bimodules, the right dual with left and right A-action

$$S^* = \mathrm{Hom}_A(S, A), \quad (a\sigma b)(s) = a\sigma(bs) = a\sigma(s\phi(b)), \text{ for } \sigma \in S^*,$$

and the left dual with multiplications

$${}^*S = {}_A\mathrm{Hom}(S, A), \quad (a\sigma b)(s) = \sigma(sa)b = \sigma(\phi(a)s)b, \text{ for } \sigma \in {}^*S.$$

Given the above situation, S^* and ${}^*S$ are corings under suitable conditions.

17.11. Dual coring theorem. *Let $\phi : A \to S$ be an algebra anti-morphism.*

(1) If S is a finitely generated projective right A-module, then there is a unique A-coring structure on S^* whereby the evaluation map

$$\beta : S \to {}^*(S^*) = {}_A\mathrm{Hom}(\mathrm{Hom}_A(S, A), A), \quad s \mapsto [\sigma \mapsto \sigma(s)],$$

is an algebra isomorphism. Here $^*(S^*)$ *has an algebra structure as in 17.8, and the following diagram, where* i_L *is as in 17.8(2), commutes:*

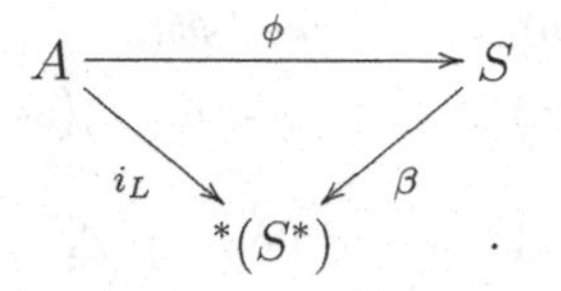

(2) *If* S *is a finitely generated projective left* A*-module, then there is a unique* A*-coring structure on* $^*S$ *whereby the evaluation map*

$$\beta : S \to (^*S)^* = {}_A\mathrm{Hom}(\mathrm{Hom}_A(S, A), A), \quad s \mapsto [\sigma \mapsto \sigma(s)],$$

is an algebra isomorphism. Here $(^*S)^*$ *has an algebra structure as in 17.8, and the following diagram, where* i_R *is as in 17.8(1), commutes:*

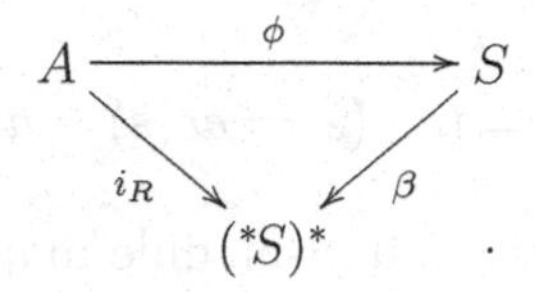

Proof. (1) Write $\mathcal{C} = S^*$ and define $\underline{\varepsilon} : \mathcal{C} \to A$ by $c \mapsto c(1_S)$. Then, for $a, b \in A$ and $c \in \mathcal{C}$,

$$\underline{\varepsilon}(acb) = acb(1_S) = ac(\phi(b)) = ac(1_S b) = ac(1_S)b = a\underline{\varepsilon}(c)b,$$

where we used that c is a right A-module map $c : S \to A$. Thus $\underline{\varepsilon}$ is an (A, A)-bimodule morphism. Since S is a finitely generated projective right A-module, the natural map

$$\theta : S^* \otimes_A S^* \to \mathrm{Hom}_A(S \otimes_A S, A), \quad c \otimes_A c' \mapsto [s \otimes_A s' \mapsto c(c'(s)s')],$$

is an (A, A)-bimodule isomorphism with the inverse map

$$\theta^{-1} : \mathrm{Hom}_A(S \otimes_A S, A) \to S^* \otimes_A S^*, \quad \sigma \mapsto \sigma(s_i \otimes_A -) \otimes_A c^i,$$

where $\{s_i\}_{i \in I} \subset S$, $\{c^i\}_{i \in I} \subset S^* = \mathcal{C}$ is a dual basis in S. Consider the map

$$\kappa : \mathcal{C} \to \mathrm{Hom}_A(S \otimes_A S, A), \quad c \mapsto [s \otimes_A s' \mapsto c(s's)].$$

First note that κ is well defined, that is, $\kappa(c)$ is a right A-module map for any $c \in \mathcal{C}$. Indeed,

$$\kappa(c)(s \otimes_A s'a) = c((s'a)s) = c(\phi(a)s's) = c((s's)a) = c(s's)a = \kappa(c)(s \otimes_A s')a,$$

for all $s, s' \in S$ and $a \in A$. Then note that κ is an (A, A)-bimodule map, since, for all $a, b \in A$, $s, s' \in S$ and $c \in \mathcal{C}$,

$$\begin{aligned} \kappa(acb)(s \otimes_A s') &= (acb)(s's) = ac(s's\phi(a)), \quad \text{and} \\ (a\kappa(c)b)(s \otimes_A s') &= a\kappa(c)(bs \otimes_A s') = a\kappa(c)(s\phi(b) \otimes_A s') = ac(s's\phi(a)). \end{aligned}$$

Thus we can define an (A, A)-bimodule map $\underline{\Delta} = \theta^{-1} \circ \kappa : \mathcal{C} \to \mathcal{C} \otimes_A \mathcal{C}$.

Evaluated at $s \otimes s' \in S \otimes_A S$, for $c \in \mathcal{C}$, the map $\underline{\Delta}(c) \in \mathcal{C} \otimes_A \mathcal{C}$ reads,

$$\underline{\Delta}(c)(s \otimes s') = c(s's).$$

The associativity of the product in S implies the coassociativity of $\underline{\Delta}$. Similarly, the fact that 1_S is a unit in S implies that $\underline{\varepsilon}$ is a counit for $\underline{\Delta}$. In this way a coring structure on $\mathcal{C} = S^*$ has been constructed.

Now we look at the map β, which is well defined since, for all $c \in S^*$, $a \in A$ and $s \in S$,

$$\beta(s)(ac) = (ac)(s) = ac(s) = a\beta(s)(c),$$

that is, for all $s \in S$, $\beta(s)$ is a left A-module map $S^* \to A$. It is a bijection since S_A is a finitely generated and projective module. We need to show that β is an algebra map, where ${}^*\mathcal{C} = {}^*(S^*)$ has the algebra structure given in 17.8(2). Recall that the product in ${}^*\mathcal{C}$ is given by

$$(f *^l g)(c) = \sum f(c_{\underline{1}}g(c_{\underline{2}})), \quad \text{for all } f, g \in {}^*\mathcal{C},\ c \in \mathcal{C}.$$

In particular, using the explicit form of θ^{-1} given in terms of a dual basis in S, for all $s, s' \in S$, we obtain

$$\begin{aligned} (\beta(s) *^l \beta(s'))(c) &= \sum \beta(s)(c_{\underline{1}}\beta(s')(c_{\underline{2}})) = \sum (c_{\underline{1}}\beta(s')(c_{\underline{2}}))(s) \\ &= \sum (c_{\underline{1}}c_{\underline{2}}(s'))(s) = \sum_i c(ss_i)c^i(s') \\ &= \sum_i c(ss_ic^i(s')) = c(ss') = \beta(ss')(c). \end{aligned}$$

Therefore β is a ring morphism, as claimed.

The uniqueness of $\underline{\Delta}$ and $\underline{\varepsilon}$ follows directly from their construction. Finally, recall that, for all $c \in \mathcal{C}$ and $a \in A$, $i_L(a)(c) = \underline{\varepsilon}(c)a$, but note also that

$$\beta(\phi(a))(c) = c(\phi(a)) = c(1_S a) = c(1_S)a = \underline{\varepsilon}(c)a,$$

so that $\beta \circ \phi = i_L$, as required. This completes the proof of the first part of the theorem.

(2) This is proven in a way similar to the proof of part (1). □

17.12. Coring morphism. Given two A-corings $\mathcal{C}, \mathcal{C}'$, an (A,A)-bilinear map $f : \mathcal{C} \to \mathcal{C}'$ is said to be a *coring morphism* provided the diagrams

$$\begin{array}{ccc} \mathcal{C} & \xrightarrow{f} & \mathcal{C}' \\ {\scriptstyle \underline{\Delta}}\downarrow & & \downarrow{\scriptstyle \underline{\Delta}'} \\ \mathcal{C}\otimes_A\mathcal{C} & \xrightarrow{f\otimes f} & \mathcal{C}'\otimes_A\mathcal{C}' \end{array} \qquad \begin{array}{ccc} \mathcal{C} & \xrightarrow{f} & \mathcal{C}' \\ & {\scriptstyle \underline{\varepsilon}}\searrow & \downarrow{\scriptstyle \underline{\varepsilon}'} \\ & & A \end{array}$$

are commutative. Explicitly, we require that

$$\underline{\Delta}' \circ f = (f \otimes f) \circ \underline{\Delta} \quad \text{and} \quad \underline{\varepsilon}' \circ f = \underline{\varepsilon},$$

that is, for all $c \in \mathcal{C}$,

$$\sum f(c_{\underline{1}}) \otimes f(c_{\underline{2}}) = \sum f(c)_{\underline{1}} \otimes f(c)_{\underline{2}}, \quad \text{and} \quad \underline{\varepsilon}'(f(c)) = \underline{\varepsilon}(c).$$

Notice that for a coring morphism $f : \mathcal{C} \to \mathcal{C}'$ that is bijective, the inverse map $f^{-1} : \mathcal{C}' \to \mathcal{C}$ is again a coring morphism (see Exercise 2.15(2)).

As shown in 17.8, the contravariant functors $\mathrm{Hom}_A(-,A)$ and ${}_A\mathrm{Hom}(-,A)$ turn corings into rings. They also turn coring morphisms into ring morphisms.

17.13. Duals of coring morphisms. *Let $f : \mathcal{C} \to \mathcal{C}'$ be an A-coring morphism. Then the following maps are ring morphisms:*

$$\begin{array}{rll} \mathrm{Hom}_A(f,A): & \mathcal{C}'^* \to \mathcal{C}^*, & \\ {}_A\mathrm{Hom}(f,A): & {}^*\mathcal{C}' \to {}^*\mathcal{C}, & g \mapsto g \circ f, \\ {}_A\mathrm{Hom}_A(f,A): & {}^*\mathcal{C}'^* \to {}^*\mathcal{C}^*. & \end{array}$$

Proof. For $g, h \in \mathcal{C}^*$ and $c \in \mathcal{C}$,

$$\begin{aligned} (g *^r h) \circ f(c) &= \sum h(g(f(c_{\underline{1}}))f(c_{\underline{2}})) \\ &= \sum h \circ f(g \circ f(c_{\underline{1}})c_{\underline{2}}) = (g \circ f) *^r (h \circ f)(c), \end{aligned}$$

where we used the fact that f is an (A,A)-bimodule map to derive the second equality. The other assertions are shown by similar computations. □

Coideals of an A-coring $\mathcal{C}$ are defined as the kernels of surjective A-coring morphisms $\mathcal{C} \to \mathcal{C}'$.

17.14. Coideals. *For an (A,A)-sub-bimodule $K \subset \mathcal{C}$ and the canonical projection $p : \mathcal{C} \to \mathcal{C}/K$, the following are equivalent:*

(a) K is a coideal;

(b) $\mathcal{C}/K$ has a coring structure such that p is a coring morphism;

(c) $\underline{\Delta}(K) \subset \mathrm{Ke}\,(p \otimes p)$ and $\underline{\varepsilon}(K) = 0$.

If $K \subset \mathcal{C}$ is left and right $\mathcal{C}$-pure as an A-module, then (c) is equivalent to:

(d) $\underline{\Delta}(K) \subset \mathcal{C} \otimes_A K + K \otimes_A \mathcal{C}$ *and* $\underline{\varepsilon}(K) = 0$.

Proof. In the proof of 2.4, the commutativity of the base ring is of no relevance, and hence this proof can be transferred to ${}_A\mathbf{M}_A$. □

17.15. Factorisation theorem. *Let $f : \mathcal{C} \to \mathcal{C}'$ be a morphism of A-corings. If $K \subset \mathcal{C}$ is a coideal and $K \subset \operatorname{Ke} f$, then there is a commutative diagram of coring morphisms*

$$\begin{array}{ccc} \mathcal{C} & \xrightarrow{p} & \mathcal{C}/K \\ & \searrow^{f} & \downarrow \bar{f} \\ & & \mathcal{C}' . \end{array}$$

To show this we refer the reader to the proof of 2.5, and for our next observation the proof of 2.6 applies.

17.16. The counit as coring morphism. *For any A-coring $\mathcal{C}$,*

(1) $\underline{\varepsilon}$ *is a coring morphism.*

(2) *If $\underline{\varepsilon}$ is surjective, then $\operatorname{Ke} \underline{\varepsilon}$ is a coideal.*

17.17. Subcorings. We call an (A, A)-sub-bimodule $\mathcal{D} \subset \mathcal{C}$ a *subcoring*, provided $\mathcal{D}$ has a coring structure such that the inclusion map is a coring morphism. The image of any A-coring map $f : \mathcal{C} \to \mathcal{C}'$ is a subcoring of $\mathcal{C}'$. To characterise sub-bimodules as subcorings, purity conditions on both sides are needed.

An (A, A)-sub-bimodule $\mathcal{D} \subset \mathcal{C}$ that is pure both as a left and as a right A-module is a subcoring provided that $\underline{\Delta}_{\mathcal{D}}(\mathcal{D}) \subset \mathcal{D} \otimes_A \mathcal{D} \subset \mathcal{C} \otimes_A \mathcal{C}$ and $\underline{\varepsilon}|_{\mathcal{D}} : \mathcal{D} \to A$ is a counit for $\mathcal{D}$.

Since the coproduct of bimodules is also based on the coproduct of Abelian groups, it is straightforward to obtain the

17.18. Coproduct of corings. For a family $\{\mathcal{C}_\lambda\}_\Lambda$ of A-corings, consider $\mathcal{C} = \bigoplus_\Lambda \mathcal{C}_\lambda$, $i_\lambda : \mathcal{C}_\lambda \to \mathcal{C}$ the canonical inclusions and the (A, A)-bilinear maps

$$\underline{\Delta}_\lambda : \mathcal{C}_\lambda \longrightarrow \mathcal{C}_\lambda \otimes \mathcal{C}_\lambda \subset \mathcal{C} \otimes \mathcal{C}, \quad \underline{\varepsilon}_\lambda : \mathcal{C}_\lambda \to A.$$

By properties of coproducts of (A, A)-bimodules there exist unique (A, A)-bilinear maps

$$\underline{\Delta} : \mathcal{C} \to \mathcal{C} \otimes_A \mathcal{C} \text{ with } \underline{\Delta} \circ i_\lambda = \underline{\Delta}_\lambda, \quad \underline{\varepsilon} : \mathcal{C} \to A \text{ with } \underline{\varepsilon} \circ i_\lambda = \underline{\varepsilon}_\lambda.$$

$(\mathcal{C}, \underline{\Delta}, \underline{\varepsilon})$ is the *coproduct* of the corings $\mathcal{C}_\lambda$ in the category of A-corings.

Similarly, the *direct limit* of A-corings can be obtained by the corresponding construction for (A, A)-bimodules (compare 2.11).

Remarks. The notion of a coring was introduced by Sweedler in [193] but can be traced back to the work of Jonah [22] on the cohomology of coalgebras.

The coring associated to a finitely generated projective module as in 17.6 was introduced by El Kaoutit and Gómez-Torrecillas in [111] and was termed *comatrix coring* there.

The construction in 17.11 was first considered by Sweedler in [193].

References. Brzeziński [73]; El Kaoutit and Gómez-Torrecillas [111]; Guzman [126]; Sweedler [193]; Wisbauer [212].

18 Comodules over corings

Following the pattern for comodules over R-coalgebras C, we introduce and study right (left) comodules over any A-coring $\mathcal{C}$. The structure of the category of right (left) $\mathcal{C}$-comodules is very similar to the structure of comodules over an R-coalgebra C; in particular, it depends on the A-module properties of $\mathcal{C}$ in a significant way. Thus the discussion of this dependence in the case of R-coalgebras presented in Section 3 comes in very handy here. Similarly as for comodules of coalgebras, it will turn out in Section 19 that comodules of corings $\mathcal{C}$ are closely related to modules over the dual algebras. In the case of corings there are two dual algebras (cf. 17.8), ${}^*\mathcal{C}$ and $\mathcal{C}^*$, and the right $\mathcal{C}$-comodules are left modules of the former while the left comodules are right modules of the latter.

Throughout, $(\mathcal{C}, \underline{\Delta}, \underline{\varepsilon})$ denotes an A-coring.

18.1. Right $\mathcal{C}$-comodules. Let M be a right A-module. An A-linear map $\varrho^M : M \to M \otimes_A \mathcal{C}$ is called a *right coaction* of $\mathcal{C}$ on M and is said to be *coassociative* and *counital*, provided the diagrams

$$\begin{array}{ccc} M & \xrightarrow{\varrho^M} & M \otimes_A \mathcal{C} \\ {\scriptstyle \varrho^M}\downarrow & & \downarrow{\scriptstyle I_M \otimes \underline{\Delta}} \\ M \otimes_A \mathcal{C} & \xrightarrow{\varrho^M \otimes I_{\mathcal{C}}} & M \otimes_A \mathcal{C} \otimes_A \mathcal{C} \end{array} \qquad \begin{array}{ccc} M & \xrightarrow{\varrho^M} & M \otimes_A \mathcal{C} \\ & \searrow^{=} & \downarrow{\scriptstyle I_M \otimes \underline{\varepsilon}} \\ & & M \end{array}$$

are commutative. Adopting the notation from comodules over coalgebras (see 3.1) we write, for all $m \in M$, $\varrho^M(m) = \sum m_{\underline{0}} \otimes m_{\underline{1}}$. Then the commutativity of the above diagrams means explicitly

$$\sum \varrho^M(m_{\underline{0}}) \otimes m_{\underline{1}} = \sum m_{\underline{0}} \otimes \underline{\Delta}(m_{\underline{1}}) = \sum m_{\underline{0}} \otimes m_{\underline{1}} \otimes m_{\underline{2}},$$

where the last expression is a notation and $m = \sum m_{\underline{0}} \underline{\varepsilon}(m_{\underline{1}})$. An A-module with a coassociative counital right coaction is called a *right $\mathcal{C}$-comodule*.

18.2. Comodule morphisms. A *comodule morphism* $f : M \to N$ between right $\mathcal{C}$-comodules is an A-linear map f inducing a commutative diagram

$$\begin{array}{ccc} M & \xrightarrow{f} & N \\ {\scriptstyle \varrho^M}\downarrow & & \downarrow{\scriptstyle \varrho^N} \\ M \otimes_A \mathcal{C} & \xrightarrow{f \otimes I_{\mathcal{C}}} & N \otimes_A \mathcal{C}, \end{array}$$

which means $\varrho^N \circ f = (f \otimes I_{\mathcal{C}}) \circ \varrho^M$ and, for any $m \in M$ (as in 3.3),

$$\sum f(m)_{\underline{0}} \otimes f(m)_{\underline{1}} = \sum f(m_{\underline{0}}) \otimes m_{\underline{1}}.$$

Instead of *comodule morphism* we will also say $\mathcal{C}$*-morphism* or *($\mathcal{C}$-)colinear map*. The set $\mathrm{Hom}^{\mathcal{C}}(M,N)$ of $\mathcal{C}$-morphisms from M to N is an Abelian group, and it follows from the definition that it is determined by the exact sequence in $\mathbf{M}_R$,

$$0 \to \mathrm{Hom}^{\mathcal{C}}(M,N) \to \mathrm{Hom}_A(M,N) \xrightarrow{\gamma} \mathrm{Hom}_A(M, N\otimes_A \mathcal{C}),$$

where $\gamma(f) = \varrho^N \circ f - (f\otimes I_{\mathcal{C}})\circ \varrho^M$, or, equivalently, by the pullback diagram

$$\begin{array}{ccc} \mathrm{Hom}^{\mathcal{C}}(M,N) & \longrightarrow & \mathrm{Hom}_A(M,N) \\ \downarrow & & \downarrow {\scriptstyle \varrho^N \circ -} \\ \mathrm{Hom}_A(M,N) & \xrightarrow{(-\otimes I_{\mathcal{C}})\circ \varrho^M} & \mathrm{Hom}_A(M, N\otimes_A \mathcal{C}). \end{array}$$

The category of right $\mathcal{C}$-comodules and comodule morphisms is denoted by $\mathbf{M}^{\mathcal{C}}$. Since $\mathrm{Hom}^{\mathcal{C}}(M,N)$ is an Abelian group for any right comodules M, N, $\mathbf{M}^{\mathcal{C}}$ is a preadditive category.

18.3. Left $\mathcal{C}$-comodules. Symmetrically, a *left $\mathcal{C}$-comodule* is defined as a left A-module M, with a coassociative and counital *left $\mathcal{C}$-coaction*, that is, an A-linear map ${}^M\!\varrho : M \to \mathcal{C}\otimes_A M$ for which

$$(\underline{\Delta}\otimes I_M)\circ {}^M\!\varrho = (I_{\mathcal{C}}\otimes {}^M\!\varrho)\circ {}^M\!\varrho, \quad (\underline{\varepsilon}\otimes I_M)\circ {}^M\!\varrho = I_M.$$

For $m \in M$ we write ${}^M\!\varrho(m) = \sum m_{\underline{-1}}\otimes m_{\underline{0}}$, and the above conditions are expressed by $m = \sum \underline{\varepsilon}(m_{\underline{-1}})m_{\underline{0}}$ and

$$\sum m_{\underline{-1}}\otimes {}^M\!\varrho(m_{\underline{0}}) = \sum \underline{\Delta}(m_{\underline{-1}})\otimes m_{\underline{0}} = \sum m_{\underline{-2}}\otimes m_{\underline{-1}}\otimes m_{\underline{0}}.$$

$\mathcal{C}$-morphisms between left $\mathcal{C}$-comodules M, N are defined in an obvious way, and the R-module of all such $\mathcal{C}$-morphisms is denoted by ${}^{\mathcal{C}}\mathrm{Hom}\,(M,N)$. Left $\mathcal{C}$-comodules and their morphisms again form a preadditive category that is denoted by ${}^{\mathcal{C}}\mathbf{M}$.

18.4. $(\mathcal{C},\mathcal{C})$-bicomodules. An (A,A)-bimodule M that is a right $\mathcal{C}$-comodule and left $\mathcal{C}$-comodule by $\varrho^M : M \to M\otimes_A \mathcal{C}$ and ${}^M\!\varrho : M \to \mathcal{C}\otimes_A M$, respectively, is called a $(\mathcal{C},\mathcal{C})$*-bicomodule* provided it satifies the compatibility condition

$$(I_{\mathcal{C}}\otimes \varrho^M)\circ {}^M\!\varrho = ({}^M\!\varrho\otimes I_{\mathcal{C}})\circ \varrho^M.$$

Morphisms $f : M \to N$ of bicomodules are maps that are both left and right $\mathcal{C}$-colinear, and the R-module of all these maps is denoted by ${}^{\mathcal{C}}\mathrm{Hom}^{\mathcal{C}}(M,N)$. With these morphisms the $(\mathcal{C},\mathcal{C})$-bimodules form a preadditive category that we denote by ${}^{\mathcal{C}}\mathbf{M}^{\mathcal{C}}$.

Clearly $\mathcal{C}$ itself is a $(\mathcal{C},\mathcal{C})$-bicomodule by the structure map $\underline{\Delta}$. More generally, bicomodules over different corings are considered in Section 22.

18.5. Comodules of a trivial coring. View A as a trivial A-coring as in 17.3. Let M be a right A-module. By the canonical isomorphism $M \otimes_A A \simeq M$, right A-coactions in M are identified with right A-module endomorphisms of M. Any such endomorphism $\varrho^M \in \mathrm{End}_A(M)$ is coassociative, while the counit property requires that $\varrho^M = I_M$. Thus the category of right A-comodules $\mathbf{M}^A$ is the same as the category of right A-modules $\mathbf{M}_A$. Similarly, the category of left A-comodules can be identified with ${}_A\mathbf{M}$ and the category of (A, A)-bicomodules is just ${}_A\mathbf{M}_A$.

As for corings, several constructions for $\mathcal{C}$-comodules are built upon the corresponding constructions for A-modules. Recalling the diagrams in 3.5 we obtain

18.6. Kernels and cokernels in $\mathbf{M}^{\mathcal{C}}$. For any $f : M \to N$ in $\mathbf{M}^{\mathcal{C}}$, the factor module $N/\mathrm{Im}\, f$ in $\mathbf{M}_A$ has a comodule structure such that the projection $g : N \to N/\mathrm{Im}\, f$ is a cokernel of f in $\mathbf{M}^{\mathcal{C}}$. This means that $\mathbf{M}^{\mathcal{C}}$ has cokernels.

A similar construction yields *kernels* for morphisms $f : M \to N$ in $\mathbf{M}^{\mathcal{C}}$ provided that f is $\mathcal{C}$-pure as a right A-morphism. This shows in particular that $\mathbf{M}^{\mathcal{C}}$ is an Abelian (in fact, Grothendieck) category provided $\mathcal{C}$ is flat as a left A-module.

18.7. $\mathcal{C}$-subcomodules. Let M be a right $\mathcal{C}$-comodule. An A-submodule $K \subset M$ is called a *$\mathcal{C}$-subcomodule of M* provided that K has a right comodule structure such that the inclusion is a comodule morphism. A $\mathcal{C}$-pure A-submodule $K \subset M$ is a subcomodule of M provided $\varrho^K(K) \subset K \otimes_A \mathcal{C} \subset N \otimes_A \mathcal{C}$. Notice that the kernel K in $\mathbf{M}_A$ of a comodule morphism $f : M \to N$ need not be a subcomodule of M unless it is a $\mathcal{C}$-pure A-submodule (compare the discussion in 3.6).

The coproduct of A-modules also provides the coproduct for $\mathcal{C}$-comodules (compare 3.7).

18.8. Coproducts in $\mathbf{M}^{\mathcal{C}}$. Let $\{M_\lambda\}_\Lambda$ be a family of $\mathcal{C}$-comodules. For $M = \bigoplus_\Lambda M_\lambda$ in $\mathbf{M}_A$, the canonical maps

$$\varrho^M_\lambda : M_\lambda \longrightarrow M_\lambda \otimes_A \mathcal{C} \subset M \otimes_A \mathcal{C}$$

yield a unique right A-module morphism $\varrho^M : M \to M \otimes_A \mathcal{C}$ which makes M a right $\mathcal{C}$-comodule such that the inclusions $M_\lambda \to M$ are $\mathcal{C}$-morphisms. This is obviously the coproduct of $\{M_\lambda\}_\Lambda$ in $\mathbf{M}^{\mathcal{C}}$.

As for comodules over coalgebras, an important class of $\mathcal{C}$-comodules is obtained by tensoring A-modules with $\mathcal{C}$. More precisely, the arguments from 3.8 yield the following.

18.9. Comodules and tensor products. *Let $M \in \mathbf{M}^{\mathcal{C}}$ and $f : X \to X'$ in $\mathbf{M}_A$. Then:*

(1) *$X \otimes_A \mathcal{C}$ is a right $\mathcal{C}$-comodule by*

$$I_X \otimes \underline{\Delta} : X \otimes_A \mathcal{C} \longrightarrow X \otimes_A \mathcal{C} \otimes_A \mathcal{C},$$

and the map $f \otimes I_{\mathcal{C}} : X \otimes_A \mathcal{C} \to X' \otimes_A \mathcal{C}$ is a $\mathcal{C}$-morphism.

(2) *For any index set Λ, $A^{(\Lambda)} \otimes_A \mathcal{C} \simeq \mathcal{C}^{(\Lambda)}$, and there exists a surjective $\mathcal{C}$-morphism*

$$\mathcal{C}^{(\Lambda')} \to M \otimes_A \mathcal{C}, \quad \text{for some } \Lambda'.$$

(3) *The structure map $\varrho^M : M \to M \otimes_A \mathcal{C}$ is a comodule morphism, and hence M is a subcomodule of a $\mathcal{C}$-generated comodule.*

(4) *Let B be an R-algebra such that M is a (B, A)-bimodule and ϱ^M is (B, A)-linear. Then, for any $Y \in \mathbf{M}_B$, the right A-module $Y \otimes_B M$ is a right $\mathcal{C}$-comodule by*

$$I_Y \otimes \varrho^M : Y \otimes_B M \to Y \otimes_B M \otimes_A \mathcal{C}.$$

As for coalgebras, relations between the Hom and tensor functors are of fundamental importance for corings.

18.10. Hom-tensor relations for right $\mathcal{C}$-comodules. *Let A, B be R-algebras, $M, N \in \mathbf{M}^{\mathcal{C}}$, and $X \in \mathbf{M}_A$.*

(1) *There is a bijective R-linear map*

$$\varphi : \mathrm{Hom}^{\mathcal{C}}(M, X \otimes_A \mathcal{C}) \to \mathrm{Hom}_A(M, X), \ f \mapsto (I_X \otimes \underline{\varepsilon}) \circ f,$$

with inverse map $h \mapsto (h \otimes I_{\mathcal{C}}) \circ \varrho^M$.

(2) *Suppose that M is a (B, A)-bimodule such that ϱ^M is also B-linear, and view $\mathrm{Hom}^{\mathcal{C}}(M, N)$ as a right B-module via $(hb)(m) = h(bm)$. Then, for any $Y \in \mathbf{M}_B$, there is a bijective R-linear map*

$$\psi : \mathrm{Hom}^{\mathcal{C}}(Y \otimes_B M, N) \to \mathrm{Hom}_B(Y, \mathrm{Hom}^{\mathcal{C}}(M, N)), \ g \mapsto [y \mapsto g(y \otimes -)],$$

with inverse map $h \mapsto [y \otimes m \mapsto h(y)(m)]$.

Proof. (1) The proof of 3.9(1) can be transferred to noncommutative base rings.

(2) By the Hom-tensor relations for modules, there is an isomorphism of R-modules

$$\psi : \mathrm{Hom}_A(Y \otimes_B M, N) \to \mathrm{Hom}_B(Y, \mathrm{Hom}_A(M, N)), \ g \mapsto [y \mapsto g(y \otimes -)].$$

For all $y \in Y$, the commutativity of the diagram

$$\begin{array}{ccc} M & \xrightarrow{y\otimes -} & Y \otimes_B M \\ {\scriptstyle \varrho^M}\downarrow & & \downarrow{\scriptstyle I_Y \otimes \varrho^M} \\ M \otimes_A \mathcal{C} & \xrightarrow{(y\otimes -)\otimes I_{\mathcal{C}}} & Y \otimes_B M \otimes_A \mathcal{C} \end{array} \qquad \begin{array}{ccc} m & \longmapsto & y \otimes m \\ \downarrow & & \downarrow \\ \varrho^M(m) & \longmapsto & y \otimes \varrho^M(m) \end{array}$$

implies that the map $y \otimes -$ is a $\mathcal{C}$-morphism. So, for $g \in \mathrm{Hom}^{\mathcal{C}}(Y \otimes_B M, N)$, the composition $g \circ (y \otimes -)$ is a $\mathcal{C}$-morphism, too.

On the other hand, for any $h \in \mathrm{Hom}_B(Y, \mathrm{Hom}^{\mathcal{C}}(M, N))$ there is the commutative diagram

$$\begin{array}{ccc} Y \otimes_B M & \longrightarrow & N \\ {\scriptstyle I_Y\otimes\varrho^M}\downarrow & & \downarrow{\scriptstyle \varrho^N} \\ Y \otimes_B M \otimes_A \mathcal{C} & \longrightarrow & N \otimes_A \mathcal{C} \end{array} \qquad \begin{array}{ccc} y \otimes m & \longmapsto & h(y)(m) \\ \downarrow & & \downarrow \\ y \otimes \varrho^M(m) & \longmapsto & (h(y) \otimes I_{\mathcal{C}}) \circ \varrho^M(m)\,. \end{array}$$

This shows that $\psi^{-1}(h)$ lies in $\mathrm{Hom}^{\mathcal{C}}(Y \otimes_B M, N)$ and implies that ψ induces a bijection $\mathrm{Hom}^{\mathcal{C}}(Y \otimes_B M, N) \to \mathrm{Hom}_B(Y, \mathrm{Hom}^{\mathcal{C}}(M, N))$, as required. □

We also state the left-hand version of these relations for completeness.

18.11. Hom-tensor relations for left $\mathcal{C}$-comodules. *Let A, B be R-algebras, $M, N \in {}^{\mathcal{C}}\mathbf{M}$, and $X \in {}_A\mathbf{M}$.*

(1) There is an R-isomorphism

$$\varphi' : {}^{\mathcal{C}}\mathrm{Hom}\,(M, \mathcal{C} \otimes_A X) \to {}_A\mathrm{Hom}(M, X), \ f \mapsto (\underline{\varepsilon} \otimes I_X) \circ f\,,$$

with inverse map $h \mapsto (I_{\mathcal{C}} \otimes_A h) \circ {}^M\varrho$.

(2) Suppose that M is an (A, B)-bimodule such that ${}^M\varrho$ is also B-linear. Then, for any $Y \in {}_B\mathbf{M}$, there is an R-isomorphism

$$\psi' : {}^{\mathcal{C}}\mathrm{Hom}(M \otimes_B Y, N) \to {}_B\mathrm{Hom}(Y, {}^{\mathcal{C}}\mathrm{Hom}(M, N)), \ g \mapsto [x \mapsto g(-\otimes x)],$$

with inverse map $h \mapsto [m \otimes x \mapsto h(x)(m)]$.

Putting $X = A$ and $M = \mathcal{C}$ in 18.10 or 18.11, the isomorphisms φ and φ' describe comodule endomorphisms of $\mathcal{C}$.

18.12. Comodule endomorphisms of $\mathcal{C}$.

(1) There is an algebra anti-isomorphism $\varphi : \mathrm{End}^{\mathcal{C}}(\mathcal{C}) \to \mathcal{C}^*, \ f \mapsto \underline{\varepsilon} \circ f$, *with inverse map* $h \mapsto (h \otimes I_{\mathcal{C}}) \circ \underline{\Delta}$.

(2) There is an algebra isomorphism $\varphi' : {}^{\mathcal{C}}\mathrm{End}(\mathcal{C}) \to {}^*\mathcal{C}, \ f \mapsto \underline{\varepsilon} \circ f$, *with inverse map* $h \mapsto (I_{\mathcal{C}} \otimes h) \circ \underline{\Delta}$.

(3) φ and φ' are homeomorphism with respect to the finite topologies.

Proof. (1) Definition and bijectivity of the maps follow by 18.10.

Let $f, g \in \mathrm{End}^{\mathcal{C}}(\mathcal{C})$. Recall that $(f \otimes I_{\mathcal{C}}) \circ \underline{\Delta} = \underline{\Delta} \circ f$ and consider the product, for $c \in \mathcal{C}$,

$$\begin{aligned}(\underline{\varepsilon} \circ f) *^r (\underline{\varepsilon} \circ g)(c) &= \sum \underline{\varepsilon} \circ g(\underline{\varepsilon} \circ f(c_{\underline{1}})c_{\underline{2}}) \\ &= \underline{\varepsilon} \circ g\,[(\underline{\varepsilon} \otimes I_{\mathcal{C}}) \circ (f \otimes I_{\mathcal{C}}) \circ \underline{\Delta}(c)] \\ &= \underline{\varepsilon} \circ g\,[(\underline{\varepsilon} \otimes I_{\mathcal{C}}) \circ \underline{\Delta} \circ f(c)] \\ &= \underline{\varepsilon} \circ (g \circ f)(c)\,.\end{aligned}$$

Clearly the unit of $\mathrm{End}^{\mathcal{C}}(\mathcal{C})$ is mapped to the unit of $\mathcal{C}^*$, and this shows that φ is an algebra anti-isomorphism.

Part (2) is symmetric to (1), and for (3) adapt the proof of 3.12(3). □

Notice that in 18.12 the morphisms are written on the left side of the argument. Writing morphisms of right comodules on the right side yields an isomorphism between $\mathcal{C}^*$ and $\mathrm{End}^{\mathcal{C}}(\mathcal{C})$. We summarise the preceding observations on the category of comodules.

18.13. The category $\mathbf{M}^{\mathcal{C}}$.

(1) *The category $\mathbf{M}^{\mathcal{C}}$ has direct sums and cokernels, and $\mathcal{C}$ is a subgenerator. $\mathbf{M}^{\mathcal{C}}$ has kernels provided $\mathcal{C}$ is a flat left A-module.*

(2) *The functor $-\otimes_A \mathcal{C} : \mathbf{M}_A \to \mathbf{M}^{\mathcal{C}}$ is right adjoint to the forgetful functor $(-)_A : \mathbf{M}^{\mathcal{C}} \to \mathbf{M}_A$.*

(3) *For any monomorphism $f : K \to L$ in $\mathbf{M}_A$,*

$$f \otimes I_{\mathcal{C}} : K \otimes_A \mathcal{C} \to L \otimes_A \mathcal{C}$$

is a monomorphism in $\mathbf{M}^{\mathcal{C}}$.

(4) *For any family $\{M_\lambda\}_\Lambda$ of right A-modules, $(\prod_\Lambda M_\lambda) \otimes_A \mathcal{C}$ is the product of the $M_\lambda \otimes_A \mathcal{C}$ in $\mathbf{M}^{\mathcal{C}}$.*

Proof. In view of the preceding observations on $\mathcal{C}$-comodules, the proof of 3.13 can be suitably modified. □

As indicated in 18.13(1), the category of right $\mathcal{C}$-comodules has particularly nice properties when $\mathcal{C}$ is a flat left A-module.

18.14. $\mathcal{C}$ as a flat A-module. *The following are equivalent:*

(a) *$\mathcal{C}$ is flat as a left A-module;*

(b) *every monomorphism in $\mathbf{M}^{\mathcal{C}}$ is injective;*

(c) *every monomorphism $U \to \mathcal{C}$ in $\mathbf{M}^{\mathcal{C}}$ is injective;*

(d) the forgetful functor $(-)_A : \mathbf{M}^{\mathcal{C}} \to \mathbf{M}_A$ respects monomorphisms.

If these conditions hold, then $\mathbf{M}^{\mathcal{C}}$ is a Grothendieck category.

Proof. As shown in 18.13(3), the functor $- \otimes_A \mathcal{C}$ preserves monomorphisms. Now the proof of 3.14 can be transferred. □

18.15. $- \otimes_A \mathcal{C}$ as left adjoint functor. *If the functor $- \otimes_A \mathcal{C} : \mathbf{M}_A \to \mathbf{M}^{\mathcal{C}}$ is left adjoint to the forgetful functor $(-)_A : \mathbf{M}^{\mathcal{C}} \to \mathbf{M}_A$, then $\mathcal{C}$ is finitely generated and projective as a left A-module.*

Proof. By 18.13 and 18.14, the proof for coalgebras 3.15 applies. □

The proof of 3.16 can be repeated to give the following finiteness property.

18.16. Finiteness Theorem (1). *Assume the coring $\mathcal{C}$ to be flat as a left A-module and let $M \in \mathbf{M}^{\mathcal{C}}$. Then every finite subset of M is contained in a subcomodule of M, which is contained in a finitely generated A-submodule.*

If ${}_A\mathcal{C}$ is flat, we can work with short exact sequences in $\mathbf{M}^{\mathcal{C}}$. Referring to the characterising sequence of the morphisms in 18.2, we obtain

18.17. Exactness of the $\mathrm{Hom}^{\mathcal{C}}$-functor. *Let ${}_A\mathcal{C}$ be flat and $M \in \mathbf{M}^{\mathcal{C}}$. Then:*

(1) $\mathrm{Hom}^{\mathcal{C}}(-, M) : \mathbf{M}^{\mathcal{C}} \to \mathbf{M}_R$ is a left exact functor.

(2) $\mathrm{Hom}^{\mathcal{C}}(M, -) : \mathbf{M}^{\mathcal{C}} \to \mathbf{M}_R$ is a left exact functor.

Proof. The diagrams in the proof of 3.19 can be constructed in this more general situation and the same arguments apply. □

As for coalgebras, a comodule $M \in \mathbf{M}^{\mathcal{C}}$ is termed an *A-relative injective comodule* or a *$(\mathcal{C}, A)$-injective comodule* provided that, for every $\mathcal{C}$-comodule map $i : N \to L$ that is a coretraction in $\mathbf{M}_A$, every diagram

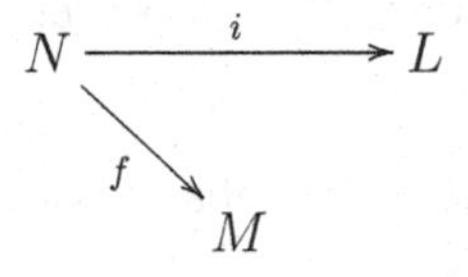

in $\mathbf{M}^{\mathcal{C}}$ can be completed commutatively by some $g : L \to M$ in $\mathbf{M}^{\mathcal{C}}$.

18.18. $(\mathcal{C}, A)$-injectivity.

(1) For any $M \in \mathbf{M}^{\mathcal{C}}$, the following are equivalent:

(a) M is $(\mathcal{C}, A)$-injective;

(b) any $\mathcal{C}$-comodule map $i : M \to L$ that is a coretraction in $\mathbf{M}_A$ is also a coretraction in $\mathbf{M}^{\mathcal{C}}$;

(c) $\varrho^M : M \to M \otimes_A \mathcal{C}$ *is a coretraction in* $\mathbf{M}^{\mathcal{C}}$.

(2) *For any* $X \in \mathbf{M}_A$, $X \otimes_A \mathcal{C}$ *is* $(\mathcal{C}, A)$*-injective.*

(3) *If* $M \in \mathbf{M}^{\mathcal{C}}$ *is* $(\mathcal{C}, A)$*-injective, then, for any* $L \in \mathbf{M}^{\mathcal{C}}$, *the canonical sequence (see 18.2)*

$$0 \longrightarrow \mathrm{Hom}^{\mathcal{C}}(L, M) \xrightarrow{i} \mathrm{Hom}_A(L, M) \xrightarrow{\gamma} \mathrm{Hom}_A(L, M \otimes_A \mathcal{C})$$

splits in $\mathbf{M}_B$, *where* $B = \mathrm{End}^{\mathcal{C}}(L)$ *and* $\gamma(f) = \varrho^M \circ f - (f \otimes I_{\mathcal{C}}) \circ \varrho^L$.

Proof. In the proof of 3.18, R can be readily replaced by A. □

Let ${}_A\mathcal{C}$ be flat. Then a short exact sequence in $\mathbf{M}^{\mathcal{C}}$ is called $(\mathcal{C}, A)$*-exact* provided it splits in $\mathbf{M}_A$. A functor on $\mathbf{M}^{\mathcal{C}}$ is said to be *left (right)* $(\mathcal{C}, A)$*-exact* if it is left (right) exact on short $(\mathcal{C}, A)$*-exact* sequences. It is obvious that a comodule M is $(\mathcal{C}, A)$-injective if and only if the functor $\mathrm{Hom}^{\mathcal{C}}(-, M)$ is $(\mathcal{C}, A)$-exact.

An object $Q \in \mathbf{M}^{\mathcal{C}}$ is *injective in* $\mathbf{M}^{\mathcal{C}}$ if, for any monomorphism $f : M \to N$ in $\mathbf{M}^{\mathcal{C}}$, the canonical map $\mathrm{Hom}^{\mathcal{C}}(N, Q) \to \mathrm{Hom}^{\mathcal{C}}(M, Q)$ is surjective. If ${}_A\mathcal{C}$ is flat, injectives $Q \in \mathbf{M}^{\mathcal{C}}$ are characterised by the exactness of $\mathrm{Hom}^{\mathcal{C}}(-, Q)$.

18.19. Injectives in $\mathbf{M}^{\mathcal{C}}$. *Assume* ${}_A\mathcal{C}$ *to be flat.*

(1) *If* X *is injective in* $\mathbf{M}_A$, *then* $X \otimes_A \mathcal{C}$ *is injective in* $\mathbf{M}^{\mathcal{C}}$.

(2) *If* M *is* $(\mathcal{C}, A)$*-injective and injective in* $\mathbf{M}_A$, *then* M *is injective in* $\mathbf{M}^{\mathcal{C}}$.

(3) *If* A *is injective in* $\mathbf{M}_A$, *then* $\mathcal{C}$ *is injective in* $\mathbf{M}^{\mathcal{C}}$.

Proof. (1) This follows from the isomorphism in 18.10(1).

(2) Let M be injective in $\mathbf{M}_A$. Then, by (1), $M \otimes_A \mathcal{C}$ is injective in $\mathbf{M}^{\mathcal{C}}$, and, by 18.18, M is a direct summand of $M \otimes_A \mathcal{C}$ as a comodule and hence it is also injective in $\mathbf{M}^{\mathcal{C}}$.

(3) This is a special case of (1). □

An object $P \in \mathbf{M}^{\mathcal{C}}$ is *projective in* $\mathbf{M}^{\mathcal{C}}$ if, for any epimorphism $M \to N$ in $\mathbf{M}^{\mathcal{C}}$, the canonical map $\mathrm{Hom}^{\mathcal{C}}(P, M) \to \mathrm{Hom}^{\mathcal{C}}(P, N)$ is surjective.

18.20. Projectives in $\mathbf{M}^{\mathcal{C}}$. *Consider any* $P \in \mathbf{M}^{\mathcal{C}}$.

(1) *If* P *is projective in* $\mathbf{M}^{\mathcal{C}}$, *then* P *is projective in* $\mathbf{M}_A$.

(2) *If* ${}_A\mathcal{C}$ *is flat, the following are equivalent:*

(a) P *is projective in* $\mathbf{M}^{\mathcal{C}}$;

(b) $\mathrm{Hom}^{\mathcal{C}}(P, -) : \mathbf{M}^{\mathcal{C}} \to \mathbf{M}_R$ *is exact.*

Proof. (1) Recalling the functorial isomorphism in 18.10(1) and taking care of sides, we can follow the same steps as in the proof of 3.22.

(2) By 18.17, the functor $\mathrm{Hom}^{\mathcal{C}}(P, -)$ is left exact. Hence it is exact if and only if it preserves epimorphisms (surjective morphisms). □

Although $\mathbf{M}^{\mathcal{C}}$ need not be a full subcategory of some module category, the Hom and tensor functors have essentially the same properties.

18.21. Hom and tensor functors. *For any $M \in \mathbf{M}^{\mathcal{C}}$ and $S = \mathrm{End}^{\mathcal{C}}(M)$, writing morphisms on the left side, there is an adjoint pair of functors*

$$- \otimes_S M : \mathbf{M}_S \to \mathbf{M}^{\mathcal{C}}, \quad \mathrm{Hom}^{\mathcal{C}}(M, -) : \mathbf{M}^{\mathcal{C}} \to \mathbf{M}_S,$$

that is, there are canonical isomorphisms, for $N \in \mathbf{M}^{\mathcal{C}}$ and $X \in \mathbf{M}_S$,

$$\mathrm{Hom}^{\mathcal{C}}(X \otimes_S M, N) \to \mathrm{Hom}_S(X, \mathrm{Hom}^{\mathcal{C}}(M, N)), \ \delta \mapsto [x \mapsto \delta(x \otimes -)],$$

with inverse map $\varphi \mapsto [x \otimes m \mapsto \varphi(x)(m)]$.

Explicitly, the counit and unit of adjunction are

$$\begin{aligned} \mu_N &: \mathrm{Hom}^{\mathcal{C}}(M, N) \otimes_S M \to N, \quad f \otimes m \mapsto f(m), \\ \nu_X &: X \to \mathrm{Hom}^{\mathcal{C}}(M, X \otimes_S M), \quad x \mapsto [m \mapsto x \otimes m]. \end{aligned}$$

Proof. In the standard Hom-tensor relation (see 43.9) is it easy to verify that restriction to colinear maps yields the given mapping. □

The classes of modules for which the unit and counit of adjunction in 18.21 are isomorphisms are of particular interest.

18.22. Static and adstatic comodules. Given a right $\mathcal{C}$-comodule M, let $S = \mathrm{End}^{\mathcal{C}}(M)$. A right $\mathcal{C}$-comodule N is said to be *M-static* if μ_N (in 18.21) is an isomorphism. The class of all M-static comodules is denoted by $\mathrm{Stat}(M)$. A right S-module X is called *M-adstatic* if ν_X (in 18.21) is an isomorphism.

It is easy to see that, for every M-static comodule N, $\mathrm{Hom}^{\mathcal{C}}(M, N)$ is M-adstatic, and for each M-adstatic module X_S, $X \otimes_S M$ is M-static.

18.23. Generators in $\mathbf{M}^{\mathcal{C}}$. *Let ${}_A\mathcal{C}$ be flat, $M \in \mathbf{M}^{\mathcal{C}}$, and $S = \mathrm{End}^{\mathcal{C}}(M)$. The following are equivalent:*

(a) M is a generator in $\mathbf{M}^{\mathcal{C}}$;

(b) the functor $\mathrm{Hom}^{\mathcal{C}}(M, -) : \mathbf{M}^{\mathcal{C}} \to \mathbf{M}_S$ is faithful;

(c) M generates every subcomodule of $\mathcal{C}^{(\mathbb{N})}$;

(d) for every subcomodule $K \subset \mathcal{C}^{(\mathbb{N})}$, μ_K (see 18.21) is surjective;

(e) ${}_SM$ is flat and every (finitely) $\mathcal{C}$-generated comodule is M-static;

(f) ${}_SM$ is flat and every injective comodule in $\mathbf{M}^{\mathcal{C}}$ is M-static;

(g) $\mathrm{Stat}(M) = \mathbf{M}^{\mathcal{C}}$.

Proof. (a) $\Leftrightarrow$ (b) holds in any category, and (b) $\Leftrightarrow$ (c) $\Leftrightarrow$ (d) is clear because the subcomodules of $\mathcal{C}^{(\mathbb{N})}$ form a set of generators in $\mathbf{M}^{\mathcal{C}}$.

(a) $\Leftrightarrow$ (g) The proof of 43.12, (a) $\Leftrightarrow$ (h), can be applied.

(e) $\Leftrightarrow$ (f) $\Leftrightarrow$ (g) The proof of [46, 15.9] shows that any generator in $\mathbf{M}^{\mathcal{C}}$ is flat over its endomorphism ring. Now the corresponding proofs of 43.12 hold. □

In case monomorphisms are injective in $\mathbf{M}^{C}$, the properties of projective generators can be transferred from module theory and 43.13 leads to the following characterisations.

18.24. Projective generators in $\mathbf{M}^{\mathcal{C}}$. *Assume ${}_A\mathcal{C}$ to be flat. Let $M \in \mathbf{M}^{\mathcal{C}}$ with M_A finitely generated and $S = \mathrm{End}^{\mathcal{C}}(M)$. The following are equivalent:*

(a) *M is a projective generator in $\mathbf{M}^{\mathcal{C}}$;*

(b) *M is a generator in $\mathbf{M}^{\mathcal{C}}$ and ${}_SM$ is faithfully flat;*

(c) *$\mathrm{Hom}^{\mathcal{C}}(M, -) : \mathbf{M}^{\mathcal{C}} \to \mathbf{M}_S$ induces an equivalence of categories.*

In the situation described in 18.23, in particular the coalgebra $\mathcal{C}$ itself is M-static. This property is of more general interest and leads to the following definition.

18.25. Galois comodules. Let M be a right $\mathcal{C}$-comodule, and let $S = \mathrm{End}^{\mathcal{C}}(M)$. M is termed a *Galois comodule* if M_A is finitely generated and projective, and the evaluation map

$$\mathrm{Hom}^{\mathcal{C}}(M, \mathcal{C}) \otimes_S M \to \mathcal{C}, \;\; f \otimes m \mapsto f(m),$$

is an isomorphism of right $\mathcal{C}$-comodules.

For any finitely generated projective $P \in \mathbf{M}_A$ and $S = \mathrm{End}_A(P)$, $P^* \otimes_S P$ has a coring structure (see 17.6), and, by the isomorphism

$$\mathrm{Hom}^{P^*\otimes_S P}(P, P^* \otimes_S P) \otimes_S P \simeq \mathrm{Hom}_A(P, A) \otimes_S P = P^* \otimes_S P,$$

P is a Galois comodule for $P^* \otimes_S P$.

Notice that, by the identification $P^* \otimes_S P \otimes_A P^* \otimes_S P \simeq P^* \otimes_S S \otimes_S P \simeq P^* \otimes_S P$, this coring structure is very close to the *trivial coproduct* $\underline{\Delta}'(f \otimes_S p) = f \otimes_S p$. It follows from the next theorem that this is the prototype of corings that have Galois comodules.

18.26. Characterisation of Galois comodules. *Let $M \in \mathbf{M}^{\mathcal{C}}$ with M_A finitely generated and projective and $S = \mathrm{End}^{\mathcal{C}}(M)$. Consider $M^* \otimes_S M$ as an A-coring (via 17.6). Then the following are equivalent:*

(a) *M is a Galois comodule;*

(b) *there is a (coring) isomorphism*

$$\mathsf{can}_M : M^* \otimes_S M \to \mathcal{C}, \quad \xi \otimes m \mapsto \sum \xi(m_{\underline{0}}) m_{\underline{1}};$$

(c) *for every $(\mathcal{C}, A)$-injective comodule $N \in \mathbf{M}^{\mathcal{C}}$, the evaluation*

$$\varphi_N : \mathrm{Hom}^{\mathcal{C}}(M, N) \otimes_S M \to N, \quad f \otimes m \mapsto f(m),$$

is a (comodule) isomorphism;

(d) *for every right A-module X, the map*

$$\varphi_N : \mathrm{Hom}_A(M, X) \otimes_S M \to X \otimes_A \mathcal{C}, \quad g \otimes m \mapsto \sum g(m_{\underline{0}}) \otimes m_{\underline{1}},$$

is a (comodule) isomorphism.

Proof. By the Hom-tensor relations 18.10, for any $X \in \mathbf{M}_A$ and $M \in \mathbf{M}^{\mathcal{C}}$, there is a commutative diagram of right $\mathcal{C}$-comodule maps ($\otimes$ is $\otimes_A$),

$$\begin{array}{ccccccc}
\mathrm{Hom}^{\mathcal{C}}(M, X \otimes \mathcal{C}) \otimes_S M & \longrightarrow & X \otimes \mathcal{C} & & f \otimes m & \longmapsto & f(m) \\
\downarrow \simeq & & \downarrow = & & \downarrow & & \downarrow = \\
\mathrm{Hom}_A(M, X) \otimes_S M & \longrightarrow & X \otimes \mathcal{C} & & (I \otimes \underline{\varepsilon}) \circ f \otimes m & \longmapsto & \sum (I \otimes \underline{\varepsilon})(f(m_{\underline{0}})) \otimes m_{\underline{1}}.
\end{array}$$

Since $X \otimes_A \mathcal{C}$ is a $(\mathcal{C}, A)$-injective comodule (cf. 18.18(2)), the above diagram implies that (c) $\Rightarrow$ (d). The equivalence of (a) and (b) follows from the diagram by putting $X = A$. Since $\mathcal{C}$ is $(\mathcal{C}, A)$-injective, the implication (c) $\Rightarrow$ (a) is trivial.

To prove that can_M is a coring map, we first need to show the commutativity of the diagram ($\otimes$ is $\otimes_A$)

$$\begin{array}{ccccccc}
M^* \otimes_S M & \xrightarrow{\delta} & \mathcal{C} & & \xi \otimes m & \longmapsto & \sum \xi(m_{\underline{0}}) m_{\underline{1}} \\
\downarrow \underline{\Delta}_{M^* \otimes_S M} & & \downarrow \underline{\Delta}_{\mathcal{C}} & & \downarrow & & \downarrow \\
(M^* \otimes_S M) \otimes (M^* \otimes_S M) & \xrightarrow{\delta \otimes \delta} & \mathcal{C} \otimes \mathcal{C} & & \sum_i \xi \otimes p_i \otimes \pi_i \otimes m & \longmapsto & \sum \xi(m_{\underline{0}}) m_{\underline{1}} \otimes m_{\underline{2}},
\end{array}$$

where $\delta = \mathsf{can}_M$, and $\pi_i \in M^*$, $p_i \in M$ denote a dual basis for M_A. Take any $m \in M$, $\xi \in M^*$ and compute

$$\begin{aligned}
\textstyle\sum_i \xi(p_{i\underline{0}}) p_{i\underline{1}} \otimes \pi_i(m_{\underline{0}}) m_{\underline{1}} &= (\xi \otimes I_P \otimes I_{\mathcal{C}})(\textstyle\sum_i \varrho^M(p_i \pi_i(m_{\underline{0}})) \otimes m_{\underline{1}}) \\
&= (\xi \otimes I_P \otimes I_{\mathcal{C}})(\textstyle\sum \varrho^M(m_{\underline{0}}) \otimes m_{\underline{1}}) \\
&= \textstyle\sum \xi(m_{\underline{0}}) m_{\underline{1}} \otimes m_{\underline{2}}.
\end{aligned}$$

Furthermore,

$$\underline{\varepsilon}_{\mathcal{C}} \circ \mathsf{can}_M(\xi \otimes m) = \xi(m) = \underline{\varepsilon}_{M^* \otimes_S M}(\xi \otimes m).$$

This completes the proof that can_M is a coring morphism.

(b) $\Rightarrow$ (d) Consider any $X \in \mathbf{M}_A$. Since M_A is finitely generated and projective and can_M is bijective, there are isomorphisms

$$\mathrm{Hom}_A(M,X)\otimes_S M \simeq X\otimes_A M^*\otimes_S M \simeq X\otimes_A \mathcal{C}.$$

(d) $\Rightarrow$ (c) Assume that $N \in \mathbf{M}^{\mathcal{C}}$ is $(\mathcal{C},A)$-injective. Then, by 18.18, the canonical sequence

$$0 \longrightarrow \mathrm{Hom}^{\mathcal{C}}(M,N) \xrightarrow{\ i\ } \mathrm{Hom}_A(M,N) \xrightarrow{\ \gamma\ } \mathrm{Hom}_A(M,N\otimes_A \mathcal{C})$$

is (split and hence) pure in $\mathbf{M}_S$, where $\gamma(f) = \varrho^N\circ f - (f\otimes I_{\mathcal{C}})\circ \varrho^M$. Hence tensoring with ${}_SM$ yields the commutative diagram with exact rows ($\otimes$ is $\otimes_S$):

$$\begin{array}{ccccccc}
0 \longrightarrow & \mathrm{Hom}^{\mathcal{C}}(M,N)\otimes M & \longrightarrow & \mathrm{Hom}_A(M,N)\otimes M & \longrightarrow & \mathrm{Hom}_A(M,N\otimes_A\mathcal{C})\otimes M \\
 & \downarrow{\scriptstyle \varphi_N} & & \downarrow{\scriptstyle \simeq} & & \downarrow{\scriptstyle \simeq} \\
0 \longrightarrow & N & \longrightarrow & N\otimes_A \mathcal{C} & \longrightarrow & N\otimes_A\mathcal{C}\otimes_A\mathcal{C}\,,
\end{array}$$

where (d) yields the vertical isomorphisms. From this the bijectivity of φ_N follows. □

The next theorem shows which additional condition on a Galois comodule M is sufficient to make it a (projective) generator in $\mathbf{M}^{\mathcal{C}}$.

18.27. The Galois comodule structure theorem. *Let $M \in \mathbf{M}^{\mathcal{C}}$ such that M_A is finitely generated and projective, and put $S = \mathrm{End}^{\mathcal{C}}(M)$.*

(1) The following are equivalent:

(a) M is a Galois comodule and ${}_SM$ is flat;

(b) ${}_A\mathcal{C}$ is flat and M is a generator in $\mathbf{M}^{\mathcal{C}}$.

(2) The following are equivalent:

(a) M is a Galois comodule and ${}_SM$ is faithfully flat;

(b) ${}_A\mathcal{C}$ is flat and M is a projective generator in $\mathbf{M}^{\mathcal{C}}$;

(c) ${}_A\mathcal{C}$ is flat and $\mathrm{Hom}^{\mathcal{C}}(M,-) : \mathbf{M}^{\mathcal{C}} \to \mathbf{M}_S$ is an equivalence with the inverse $-\otimes_S M : \mathbf{M}_S \to \mathbf{M}^{\mathcal{C}}$.

Proof. (1) (a) $\Rightarrow$ (b) Assume M to be a Galois comodule. Then, in the diagram of the proof 18.26, (b) $\Rightarrow$ (c), the top row is exact by the flatness of ${}_SM$ without any condition on $N \in \mathbf{M}^{\mathcal{C}}$. So $\mathrm{Hom}^{\mathcal{C}}(M,N)\otimes_S M \to N$ is surjective (bijective), showing that M is a generator. Since M^* is projective in ${}_A\mathbf{M}$, the isomorphism $-\otimes_A \mathcal{C} \simeq -\otimes_A (M^*\otimes_S M)$ implies that ${}_A\mathcal{C}$ is flat.

(b) ⇒ (a) If ${}_A\mathcal{C}$ is flat, it follows from 18.23 that the generator M in $\mathbf{M}^{\mathcal{C}}$ is flat over its endomorphism ring S, and $\mathrm{Hom}^{\mathcal{C}}(M,N)\otimes_S M \simeq N$, for all $N \in \mathbf{M}^{\mathcal{C}}$.

(2) This is clear by 18.24. □

Notice that any A-coring $\mathcal{C}$ with $\mathcal{C}_A$ finitely generated and projective is a Galois comodule in $\mathbf{M}^{\mathcal{C}}$ (by the isomorphism $(\mathcal{C}^*)^{\mathrm{op}} \otimes_{(\mathcal{C}^*)^{\mathrm{op}}} \mathcal{C} \simeq \mathcal{C}$), which need not be flat over its endomorphism ring $\mathcal{C}^*$.

At the end of this section we relate the category of comodules with more general constructions from category theory. The composite of the functor $-\otimes_A \mathcal{C} : \mathbf{M}_A \to \mathbf{M}^{\mathcal{C}}$ with the forgetful functor $\mathbf{M}^{\mathcal{C}} \to \mathbf{M}_A$ provides one with the category theory or universal (co)algebra interpretation of corings. We refer to 38.26 and 38.29 for the discussion of comonads and their coalgebras.

18.28. Corings as comonads. *Let $\mathcal{C}$ be an (A,A)-bimodule.*

(1) The following are equivalent:

(a) $\mathcal{C}$ is an A-coring;

(b) the functor $F = -\otimes_A \mathcal{C} : \mathbf{M}_A \to \mathbf{M}_A$ is a comonad;

(c) the functor $\bar{F} = \mathcal{C}\otimes_A - : {}_A\mathbf{M} \to {}_A\mathbf{M}$ is a comonad.

(2) If $\mathcal{C}$ is an A-coring, then the category of coalgebras of the comonad F (resp. $\bar{F}$) is isomorphic to the category of right (resp. left) $\mathcal{C}$-comodules.

Proof. (1) (a) ⇒ (b) If $\mathcal{C}$ is a coring, then, by 18.13, F is a composite of a pair of adjoint functors and thus it is a comonad by 38.29. Note that, for all $M \in \mathbf{M}_A$, the coproduct is given by $\delta_M = I_M \otimes \underline{\Delta} : M\otimes_A \mathcal{C} \to M\otimes_A\mathcal{C}\otimes_A\mathcal{C}$, and the counit is $\psi_M = I_M\otimes\underline{\varepsilon} : M\otimes_A\mathcal{C}\to M$.

(b) ⇒ (a) Now suppose that $F = -\otimes_A\mathcal{C} : \mathbf{M}_A \to \mathbf{M}_A$ is a comonad with coproduct δ and counit ψ. Define $\underline{\Delta} = \delta_A : \mathcal{C} \to \mathcal{C}\otimes_A\mathcal{C}$ and $\underline{\varepsilon} = \psi_A : \mathcal{C}\to A$. To any element a in A associate a morphism in $\mathbf{M}_A$, $\ell_a : A \to A$, $a' \mapsto aa'$. Since δ is a natural map, we have $\delta_A\circ(\ell_a\otimes I_{\mathcal{C}}) = (\ell_a\otimes I_{\mathcal{C}}\otimes I_{\mathcal{C}})\circ\delta_A$. Evaluating this equality at $1_A\otimes c$ for all $c\in\mathcal{C}$, we obtain $\underline{\Delta}(ac) = a\underline{\Delta}(c)$, that is, $\underline{\Delta}$ is a left A-linear map. Since δ_A is a morphism in $\mathbf{M}_A$, $\underline{\Delta}$ is also a right A-module map, and we conclude that it is an (A,A)-bimodule map. Similarly, the naturality of ψ evaluated at ℓ_a implies that $\underline{\varepsilon}$ is a left A-module map and thus an (A,A)-bimodule morphism. The coassociativity of $\underline{\Delta}$ and the counit property of $\underline{\varepsilon}$ follow then from the axioms of a comonad. More precisely, first note that $F(\delta_A) = \delta_A\otimes I_{\mathcal{C}}$. Next, for all $c\in\mathcal{C}$ consider a morphism $\ell^c : A\to\mathcal{C}$ in $\mathbf{M}_A$, given by $\ell^c : a\mapsto ca$. Since the coproduct δ is a natural transformation, $\delta_{\mathcal{C}}\circ(\ell^c\otimes I_{\mathcal{C}}) = (\ell^c\otimes I_{\mathcal{C}}\otimes I_{\mathcal{C}})\circ\delta_A$. Evaluating this equality at $1_A\otimes c'$, with $c'\in\mathcal{C}$, we thus obtain $\delta_{\mathcal{C}}(c\otimes c') = c\otimes\delta_A(c')$, that is, $\delta_{\mathcal{C}} = I_{\mathcal{C}}\otimes\delta_A$. Thus

$$\delta_{F(A)} = \delta_{A\otimes_A\mathcal{C}} = \delta_{\mathcal{C}} = I_{\mathcal{C}}\otimes\delta_A.$$

The fact that δ is a coproduct for a comonad F implies that

$$(I_{\mathcal{C}} \otimes \delta_A) \circ \delta_A = (\delta_A \otimes I_{\mathcal{C}}) \circ \delta_A,$$

so that $\underline{\Delta} = \delta_A$ is a coassociative coproduct for $\mathcal{C}$. Similarly, note that $F(\psi_A) = \psi_A \otimes I_{\mathcal{C}}$ and $\psi_{F(A)} = I_{\mathcal{C}} \otimes \psi_A$. Whence the fact that ψ is a counit for a comonad F means in particular that

$$(\psi_A \otimes I_{\mathcal{C}}) \circ \delta_A = (I_{\mathcal{C}} \otimes \psi_A) \circ \delta_A = I_{\mathcal{C}},$$

so that $\underline{\varepsilon} = \psi_A$ is a counit for the coproduct $\underline{\Delta} = \delta_A$.

(a) $\Leftrightarrow$ (c) This is proven along the same lines as the equivalence (a) $\Leftrightarrow$ (b).

(2) Note that, for all $M \in \mathbf{M}_A$, $F(M) = M \otimes_A \mathcal{C}$; hence the structure map of a coalgebra (M, ϱ^M) of the comonad $\mathbb{F} = (F, I_{\mathbf{M}_A} \otimes \underline{\Delta}, I_{\mathbf{M}_A} \otimes \underline{\varepsilon})$ comes out as $\varrho^M : M \to M \otimes_A \mathcal{C}$. The diagrams defining an $\mathbb{F}$-coalgebra with the underlying object M and the structure map ϱ^M are precisely the same as the diagrams required for M to be a right $\mathcal{C}$-comodule with coaction ϱ^M. A symmetric argument shows the second isomorphism of categories. □

In addition to the forgetful functor $(-)_A$, one can also study the forgetful functor $(-)_R : \mathbf{M}^{\mathcal{C}} \to \mathbf{M}_R$, which turns out to provide a categorical interpretation of dual rings defined in 17.8.

18.29. Dual rings as natural endomorphism rings.

(1) The ring $\mathrm{Nat}((-)_R, (-)_R)$ *of natural endomorphisms of the forgetful functor* $(-)_R : \mathbf{M}^{\mathcal{C}} \to \mathbf{M}_R$, *with the product given by* $\phi\phi' = \phi \circ \phi'$, *is isomorphic to the ring* ${}^*\mathcal{C} = {}_A\mathrm{Hom}(\mathcal{C}, A)$ *with product* $*^l$.

(2) The ring $\mathrm{Nat}({}_R(-), {}_R(-))$ *of natural endomorphisms of the forgetful functor* ${}_R(-) : {}^{\mathcal{C}}\mathbf{M} \to {}_R\mathbf{M}$ *is isomorphic to the ring* $\mathcal{C}^* = \mathrm{Hom}_A(\mathcal{C}, A)$ *with product* $*^r$.

Proof. We only prove (1), since (2) will follow by the left-right symmetry. If ϕ is an endomorphism of $F = (-)_R$ and $f : M \to N$ a morphism in $M^{\mathcal{C}}$, then $F(f) \circ \phi_M = \phi_N \circ F(f)$. For any $V \in \mathbf{M}_A$ and $v \in V$, define a morphism in $\mathbf{M}^{\mathcal{C}}$, $f_v : \mathcal{C} \to V \otimes_A \mathcal{C}$, by $c \to v \otimes c$. Then, for all $v \in V$, $c \in \mathcal{C}$,

$$\phi_{V \otimes_A \mathcal{C}}(v \otimes c) = v \otimes \phi_{\mathcal{C}}(c), \quad \text{thatis,} \quad \phi_{V \otimes_A \mathcal{C}} = I_V \otimes \phi_{\mathcal{C}}. \qquad (*)$$

Similarly, since $\varrho^M : M \to M \otimes_A \mathcal{C}$ is a morphism in $\mathbf{M}^{\mathcal{C}}$,

$$\phi_{M \otimes_A \mathcal{C}} \circ \varrho^M = \varrho^M \circ \phi_M. \qquad (**)$$

Applying $(*)$ to $V = A$ and identifying $A \otimes_A \mathcal{C}$ with $\mathcal{C}$, one obtains that $\phi_{\mathcal{C}}$ is left A-linear. Since $\underline{\varepsilon}$ is an (A, A)-bimodule map, the composite $\underline{\varepsilon} \circ \phi_{\mathcal{C}}$ is an element of ${}^*\mathcal{C} = {}_A\mathrm{Hom}(\mathcal{C}, A)$, and thus there is a map

$$\Pi : \mathrm{Nat}(F, F) \to {}_A\mathrm{Hom}(\mathcal{C}, A), \quad \phi \mapsto \underline{\varepsilon} \circ \phi_{\mathcal{C}}.$$

Conversely, for any $\xi \in {}^*\mathcal{C}$ and $M \in \mathbf{M}^{\mathcal{C}}$, define

$$\widetilde{\Pi}(\xi)_M = (I_M \otimes \xi) \circ \varrho^M : M \to M, \quad m \mapsto \sum m_{\underline{0}}\,\xi(m_{\underline{1}}).$$

For any $f \in \mathrm{Hom}^{\mathcal{C}}(M, N)$, $\xi \in {}^*\mathcal{C}$, and $m \in M$,

$$\widetilde{\Pi}(\xi)_N(f(m)) = \sum f(m)_{\underline{0}}\,\xi(f(m)_{\underline{1}}) = \sum f(m_{\underline{0}})\xi(m_{\underline{1}}) = f(\widetilde{\Pi}(\xi)_M(m)),$$

where we used the fact that f is a morphism of right $\mathcal{C}$-comodules. This proves that $\widetilde{\Pi}(\xi)_M$ is natural in M, and hence is an endomorphism of the forgetful functor F inducing a map

$$\widetilde{\Pi} : {}^*\mathcal{C} \to \mathrm{Nat}(F, F), \ \xi \mapsto \widetilde{\Pi}(\xi).$$

Notice that $\widetilde{\Pi}$ is a ring map, since for any $\xi, \xi' \in {}^*\mathcal{C}$, $M \in \mathbf{M}^{\mathcal{C}}$, $m \in M$,

$$\begin{aligned}(\widetilde{\Pi}(\xi) \circ \widetilde{\Pi}(\xi'))_M(m) &= \sum \widetilde{\Pi}(\xi)_M(m_{\underline{0}}\xi'(m_{\underline{1}})) = \sum m_{\underline{0}}\xi(m_{\underline{1}}\xi'(m_{\underline{2}})) \\ &= \sum m_{\underline{0}}(\xi *^l \xi')(m_{\underline{1}}) = \widetilde{\Pi}(\xi *^l \xi')_M(m).\end{aligned}$$

It remains to prove that $\widetilde{\Pi}$ is the inverse of Π. First take any $\xi \in {}^*\mathcal{C}$ and compute

$$\Pi(\widetilde{\Pi}(\xi)) = \underline{\varepsilon} \circ \widetilde{\Pi}(\xi)_{\mathcal{C}} = \underline{\varepsilon} \circ (I_{\mathcal{C}} \otimes \xi) \circ \underline{\Delta} = \xi,$$

by definition of $\underline{\varepsilon}$. Conversely, for an endomorphism ϕ of F, and any right $\mathcal{C}$-comodule M,

$$\begin{aligned}\widetilde{\Pi}(\Pi(\phi))_M &= (I_M \otimes \Pi(\phi)) \circ \varrho^M = (I_M \otimes \underline{\varepsilon}) \circ (I_M \otimes \phi_{\mathcal{C}}) \circ \varrho^M \\ &= (I_M \otimes \underline{\varepsilon}) \circ \phi_{M \otimes_A \mathcal{C}} \circ \varrho^M \\ &= (I_M \otimes \underline{\varepsilon}) \circ \varrho^M \circ \phi_M \ = \ \phi_M,\end{aligned}$$

where equation (*) was used in the derivation of the third equality, while the fourth equality follows from equation (**). This completes the proof of the theorem. □

References. Brzeziński [73]; El Kaoutit and Gómez-Torrecillas [111]; Guzman [126]; Sweedler [193]; Wisbauer [212].

19 $\mathcal{C}$-comodules and $\mathcal{C}^*$-modules

Again $\mathcal{C}$ denotes an A-coring. As noticed in 17.8, a left dual ring ${}^*\mathcal{C}$ (and a right dual ring $\mathcal{C}^*$) is associated to $\mathcal{C}$. This section starts with the key observation in 19.1 that any right $\mathcal{C}$-comodule is also a left ${}^*\mathcal{C}$-module, and hence there is a faithful functor from $\mathbf{M}^{\mathcal{C}}$ to ${}_{^*\mathcal{C}}\mathbf{M}$. One then would like to determine when $\mathbf{M}^{\mathcal{C}}$ is a full subcategory of ${}_{^*\mathcal{C}}\mathbf{M}$ or when these two categories are isomorphic to each other. These are the topics covered in the present section, which brings the corresponding problems for coalgebras considered in Section 4 to a more general, less symmetric level. In particular, we introduce the α-condition for corings and prove an important Finiteness Theorem.

19.1. $\mathcal{C}$-comodules and ${}^*\mathcal{C}$-modules. *Any $M \in \mathbf{M}^{\mathcal{C}}$ is a left ${}^*\mathcal{C}$-module by*

$$\rightharpoonup : {}^*\mathcal{C} \otimes_R M \to M, \quad f \otimes m \mapsto (I_M \otimes f) \circ \varrho^M(m).$$

Any morphism $h : M \to N$ in $\mathbf{M}^{\mathcal{C}}$ is a left ${}^*\mathcal{C}$-module morphism, so

$$\mathrm{Hom}^{\mathcal{C}}(M,N) \subset {}_{^*\mathcal{C}}\mathrm{Hom}\,(M,N)$$

and there is a faithful functor $\mathbf{M}^{\mathcal{C}} \to \sigma[{}_{^*\mathcal{C}}\mathcal{C}] \subset {}_{^*\mathcal{C}}\mathbf{M}$.

Proof. By definition, for $f, g \in {}^*\mathcal{C}$ and $m \in M$, the actions $f\rightharpoonup(g\rightharpoonup m)$ and $(f *^l g)\rightharpoonup m$ are the compositions of the upper and lower maps in the diagram

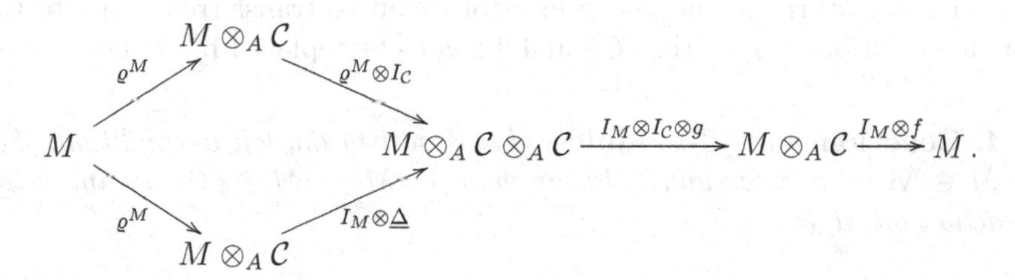

Clearly $\underline{\varepsilon}\rightharpoonup m = m$ for any $m \in M$, and so M is a left ${}^*\mathcal{C}$-module. For the remaining assertions the proofs of 4.1 apply. □

To assure that $\mathbf{M}^{\mathcal{C}}$ is a full subcategory of ${}_{^*\mathcal{C}}\mathbf{M}$ we need

19.2. The left α-condition. $\mathcal{C}$ is said to *satisfy the left α-condition* if the map

$$\alpha_N : N \otimes_A \mathcal{C} \to \mathrm{Hom}_A({}^*\mathcal{C}, N), \quad n \otimes c \mapsto [f \mapsto nf(c)],$$

is injective, for every $N \in \mathbf{M}_A$. By 42.10, the following are equivalent:

(a) *$\mathcal{C}$ satisfies the left α-condition;*

(b) *for* $N \in \mathbf{M}_A$ *and* $u \in N \otimes_A \mathcal{C}$, $(I_N \otimes f)(u) = 0$ *for all* $f \in {}^*\mathcal{C}$, *implies* $u = 0$;

(c) $\mathcal{C}$ *is locally projective as a left* A*-module.*

The left α-condition enforces $\mathcal{C}$ to be flat and cogenerated by A as a left A-module. Symmetrically, the *right α-condition* for $\mathcal{C}$ is defined and induces corresponding (left-right symmetric) properties.

19.3. $\mathbf{M}^\mathcal{C}$ as full subcategory of ${}_{*\mathcal{C}}\mathbf{M}$. *For $\mathcal{C}$ the following are equivalent:*

(a) $\mathbf{M}^\mathcal{C} = \sigma[{}_{*\mathcal{C}}\mathcal{C}]$;

(b) $\mathbf{M}^\mathcal{C}$ *is a full subcategory of* ${}_{*\mathcal{C}}\mathbf{M}$;

(c) *for all* $M, N \in \mathbf{M}^\mathcal{C}$, $\mathrm{Hom}^\mathcal{C}(M,N) = {}_{*\mathcal{C}}\mathrm{Hom}(M,N)$;

(d) $\mathcal{C}$ *satisfies the left α-condition;*

(e) *every left* ${}^*\mathcal{C}$*-submodule of* $\mathcal{C}^n$, $n \in \mathbb{N}$, *is a subcomodule of* $\mathcal{C}^n$.

If these conditions are satisfied, the inclusion functor $\mathbf{M}^\mathcal{C} \to {}_{*\mathcal{C}}\mathbf{M}$ *has a right adjoint, and for any family* $\{M_\lambda\}_\Lambda$ *of* A*-modules,*

$$(\prod\nolimits_\Lambda M_\lambda) \otimes_A \mathcal{C} \simeq \prod\nolimits_\Lambda^{\mathcal{C}} (M_\lambda \otimes_A \mathcal{C}) \subset \prod\nolimits_\Lambda (M_\lambda \otimes_A \mathcal{C}),$$

where $\prod^\mathcal{C}$ *denotes the product in* $\mathbf{M}^\mathcal{C}$.

Proof. The proof of 4.3 is based on 42.10, which holds for general noncommutative rings, and hence most of it can be transferred easily to the present situation. Notice that C^* in 4.3 has to be replaced by ${}^*\mathcal{C}$ here. □

19.4. Coaction and ${}^*\mathcal{C}$-modules. *Let $\mathcal{C}$ satisfy the left α-condition. For an* $M \in \mathbf{M}_A$, *consider any* A*-linear map* $\rho : M \to M \otimes_A \mathcal{C}$. *Define a left* ${}^*\mathcal{C}$*-action on* M *by*

$$\rightharpoonup : {}^*\mathcal{C} \otimes_R M \to M, \quad f \otimes m \mapsto (I_M \otimes f) \circ \rho(m).$$

Then the following are equivalent:

(a) ρ *is coassociative and counital;*

(b) M *is a (unital)* ${}^*\mathcal{C}$*-module by* $\rightharpoonup$.

Proof. (a) $\Rightarrow$ (b) is shown in 19.2.
(b) $\Rightarrow$ (a) With obvious modifications the proof of 4.4 applies. □

By symmetry we have the corresponding relationships between left $\mathcal{C}$-comodules and right $\mathcal{C}^*$-modules, which we formulate for completeness.

19.5. Left $\mathcal{C}$-comodules and right $\mathcal{C}^*$-modules.

(1) Any $M \in {}^{\mathcal{C}}\mathbf{M}$ is a right $\mathcal{C}^*$-module by

$$\leftharpoonup : M \otimes_R \mathcal{C}^* \to M, \quad m \otimes f \mapsto (f \otimes I_M) \circ {}^{M}\varrho(m).$$

(2) Any morphism $h : M \to N$ in ${}^{\mathcal{C}}\mathbf{M}$ is a right $\mathcal{C}^*$-module morphism, so

$${}^{\mathcal{C}}\mathrm{Hom}\,(M, N) \subset \mathrm{Hom}_{\mathcal{C}^*}(M, N)$$

and there is a faithful functor ${}^{\mathcal{C}}\mathbf{M} \to \sigma[\mathcal{C}_{\mathcal{C}^*}]$.

(3) $\mathcal{C}$ satisfies the right α-condition if and only if ${}^{\mathcal{C}}\mathbf{M} = \sigma[\mathcal{C}_{\mathcal{C}^*}]$.

The answer to the question when all ${}^*\mathcal{C}$-modules are $\mathcal{C}$-comodules is similar to the coalgebra case.

19.6. When is $\mathbf{M}^{\mathcal{C}} = {}_{^*\mathcal{C}}\mathbf{M}$? *For $\mathcal{C}$ the following are equivalent:*

(a) $\mathbf{M}^{\mathcal{C}} = {}_{^*\mathcal{C}}\mathbf{M}$;

(b) the functor $-\otimes_A \mathcal{C} : \mathbf{M}_A \to {}_{^*\mathcal{C}}\mathbf{M}$ has a left adjoint;

(c) ${}_A\mathcal{C}$ is finitely generated and projective;

(d) ${}_A\mathcal{C}$ is locally projective and $\mathcal{C}$ is finitely generated as a right $\mathcal{C}^*$-module.

Proof. (a) $\Rightarrow$ (b) By 18.13, the forgetful functor is left adjoint to $-\otimes_A \mathcal{C}$.

(b) $\Rightarrow$ (c) The condition implies that the functor $-\otimes_A \mathcal{C}$ commutes with products and hence ${}_A\mathcal{C}$ is finitely generated and projective.

(c) $\Rightarrow$ (d) Since there is a ring anti-morphism $A \to \mathcal{C}^*$ (see 17.8), (c) implies that $\mathcal{C}$ is finitely generated as a right $\mathcal{C}^*$-module.

(d) $\Rightarrow$ (a) Since $\mathcal{C}$ satisfies the α-condition, $\mathcal{C}^*$ is the endomorphism ring of the faithful left ${}^*\mathcal{C}$-module $\mathcal{C}$. So ${}_{^*\mathcal{C}}\mathcal{C}$ is a faithful module that is finitely generated as a module over its endomorphism ring, and, by 41.7(3), this implies $\mathbf{M}^{\mathcal{C}} = \sigma[{}_{^*\mathcal{C}}\mathcal{C}] = {}_{^*\mathcal{C}}\mathbf{M}$. $\square$

Notice how both left and right duals of $\mathcal{C}$ are used in 19.6. This is an important difference between the coring and coalgebra cases.

19.7. The category $\mathbf{M}^{P^*\otimes_B P}$. For R-algebras A, B, let $P \in {}_B\mathbf{M}_A$ be such that P_A is finitely generated and projective with dual basis $p_1, \ldots, p_n \in P$ and $\pi_1, \ldots, \pi_n \in P^* = \mathrm{Hom}_A(P, A)$. Consider $P^* \otimes_B P$ as an A-coring as in 17.6. Recall that ${}^*(P^* \otimes_B P) \simeq {}_B\mathrm{End}(P)$ (cf. 17.9). It is straightforward to verify (compare 4.9) that P is a right $P^* \otimes_B P$-comodule by

$$\varrho^P : P \to P \otimes_A P^* \otimes_B P, \quad p \mapsto \textstyle\sum_i p_i \otimes \pi_i \otimes p,$$

and P generates $P^* \otimes_B P$ as a right comodule. So, by 19.1 and 17.9, P is a right module over ${}_B\mathrm{End}(P)$ (with product $\phi'\phi = \phi \circ \phi'$, in which case the isomorphism discussed in 17.9 is an anti-algebra map) and there is a faithful functor $\mathbf{M}^{P^*\otimes_B P} \to \sigma[P_{_B\mathrm{End}(P)}]$. Properties of the right comodule category depend on properties of the left A-module $P^* \otimes_B P$ which in general is neither projective nor finitely generated. These properties depend on the B-module structure of P. More precisely:

(1) If ${}_BP$ is flat, then $\mathbf{M}^{P^*\otimes_B P}$ is an Abelian category.

(2) If ${}_BP$ is locally projective, then $\mathbf{M}^{P^*\otimes_B P} = \sigma[P_{_B\mathrm{End}(P)}]$.

(3) If ${}_BP$ is finitely generated and projective, then $\mathbf{M}^{P^*\otimes_B P} = \mathbf{M}_{_B\mathrm{End}(P)}$.

Proof. (1) Since ${}_AP^*$ is projective, the flatness of ${}_BP$ implies the flatness of $P^* \otimes_B P$ as a left A-module. Now the assertion follows by 18.14.

(2) It is enough to show that, for all $N \in \mathbf{M}_A$, the canonical map

$$\alpha : N \otimes_A (P^* \otimes_B P) \to \mathrm{Hom}_A({}^*(P^* \otimes_B P), N)$$

(see 42.9) is injective. With the canonical map

$$\delta : {}^*P \otimes_B P \to {}_B\mathrm{End}(P), \quad h \otimes p \mapsto [q \mapsto h(q)p],$$

and the isomorphism ${}^*(P^* \otimes_B P) \simeq {}_B\mathrm{End}(P)$, we obtain the commutative diagram

$$\begin{array}{ccccc} N \otimes_A (P^* \otimes_B P) & \xrightarrow{\alpha} & \mathrm{Hom}_A({}^*(P^* \otimes_B P), N) & \xrightarrow{\simeq} & \mathrm{Hom}_A({}_B\mathrm{End}(P), N) \\ \downarrow{\scriptstyle\simeq} & & & & \downarrow{\scriptstyle\mathrm{Hom}(\delta,N)} \\ \mathrm{Hom}_A(P,N) \otimes_B P & \longrightarrow & \mathrm{Hom}_B({}^*P, \mathrm{Hom}_A(P,N)) & \xrightarrow{\simeq} & \mathrm{Hom}_A({}^*P \otimes_B P, N), \end{array}$$

where the first bottom map is injective by the local projectivity of ${}_BP$. This implies that α is injective, hence $P^* \otimes_B P$ is locally projective as a left A-module. By 19.3, this means that $\mathbf{M}^{P^*\otimes_B P}$ is a full subcategory of $\mathbf{M}_{_B\mathrm{End}(P)}$, and, since P is a subgenerator in $\mathbf{M}^{P^*\otimes_B P}$, this category is equal to $\sigma[P_{_B\mathrm{End}(P)}]$.

(3) If ${}_BP$ is finitely generated and projective, then ${}^*P \otimes_B P \simeq {}_B\mathrm{End}(P)$, and hence $\mathbf{M}^{P^*\otimes_B P} = \mathbf{M}_{_B\mathrm{End}(P)}$. □

Under the α-condition, projectives and injectives in $\mathbf{M}^{\mathcal{C}}$ have similar characterisations as for comodules over coalgebras. Again, both types of duals of $\mathcal{C}$ play an important role. Since we can identify $\mathbf{M}^{\mathcal{C}}$ with $\sigma[{}_{^*\mathcal{C}}\mathcal{C}]$, we may formulate 41.4 in the following way.

19.8. Injectives in $\mathbf{M}^{\mathcal{C}}$. *Let $\mathcal{C}$ satisfy the left α-condition. For $Q \in \mathbf{M}^{\mathcal{C}}$ the following are equivalent:*

(a) Q is injective in $\mathbf{M}^{\mathcal{C}}$;

(b) the functor $\mathrm{Hom}^{\mathcal{C}}(-, Q) : \mathbf{M}^{\mathcal{C}} \to \mathbf{M}_R$ is exact;

(c) *Q is $\mathcal{C}$-injective (as left $\mathcal{C}^*$-module);*

(d) *Q is N-injective for every (finitely generated) subcomodule $N \subset \mathcal{C}$;*

(e) *every exact sequence $0 \to Q \to N \to L \to 0$ in $\mathbf{M}^{\mathcal{C}}$ splits.*

Moreover, injectives in $\mathbf{M}^{\mathcal{C}}$ are $\mathcal{C}$-generated and every comodule has an injective hull in $\mathbf{M}^{\mathcal{C}}$.

As for injectives, we also derive characterisations for projective comodules from the module case (see 41.6):

19.9. Projectives in $\mathbf{M}^{\mathcal{C}}$. *Let $\mathcal{C}$ satisfy the left α-condition. For $P \in \mathbf{M}^{\mathcal{C}}$ the following are equivalent:*

(a) *P is projective in $\mathbf{M}^{\mathcal{C}}$;*

(b) *the functor $\mathrm{Hom}^{\mathcal{C}}(P,-) : \mathbf{M}^{\mathcal{C}} \to \mathbf{M}_R$ is exact;*

(c) *P is $\mathcal{C}^{(\Lambda)}$-projective, for any set Λ;*

(d) *every exact sequence $0 \to K \to N \to P \to 0$ in $\mathbf{M}^{\mathcal{C}}$ splits.*

If P is finitely generated in ${}_{*\mathcal{C}}\mathbf{M}$ (or $\mathbf{M}_A$), then (a)–(d) are equivalent to:

(e) *P is $\mathcal{C}$-projective as ${}^*\mathcal{C}$-module;*

(f) *every exact sequence $0 \to K' \to N \to P \to 0$ in $\mathbf{M}^{\mathcal{C}}$ with $K' \subset \mathcal{C}$ splits.*

Notice that projectives need not exist in $\mathbf{M}^{\mathcal{C}}$. It is shown in 18.20 that projective objects in $\mathbf{M}^{\mathcal{C}}$ are also projective in $\mathbf{M}_A$.

$\mathcal{C}$ is a left and right comodule; thus we can consider

19.10. $\mathcal{C}$ as $({}^*\mathcal{C},\mathcal{C}^*)$-bimodule. *$\mathcal{C}$ is a $({}^*\mathcal{C},\mathcal{C}^*)$-bimodule by*

$$\rightharpoonup : {}^*\mathcal{C} \otimes_R \mathcal{C} \to \mathcal{C}, \quad f \otimes c \mapsto f\rightharpoonup c = (I_{\mathcal{C}} \otimes f) \circ \underline{\Delta}(c) = \sum c_{\underline{1}} f(c_{\underline{2}}),$$
$$\leftharpoonup : \mathcal{C} \otimes_R \mathcal{C}^* \to \mathcal{C}, \quad c \otimes g \mapsto c\leftharpoonup g = (g \otimes I_{\mathcal{C}}) \circ \underline{\Delta}(c) = \sum g(c_{\underline{1}}) c_{\underline{2}}.$$

(1) *For any $f \in {}^*\mathcal{C}$, $g \in \mathcal{C}^*$, and $c \in \mathcal{C}$,*

$$(f\rightharpoonup c)\leftharpoonup g = f\rightharpoonup(c\leftharpoonup g), \text{ and } g(f\rightharpoonup c) = f(c\leftharpoonup g).$$

(2) *$\mathcal{C}$ is faithful as a left ${}^*\mathcal{C}$- and as a right $\mathcal{C}^*$-module.*

(3) *If $\mathcal{C}$ is cogenerated by A as a left A-module, then, for all $f \in Z({}^*\mathcal{C})$, $c \in \mathcal{C}$, $f\rightharpoonup c = c\leftharpoonup f$, and $Z({}^*\mathcal{C}) \subset Z(\mathcal{C}^*)$.*

(4) *If $\mathcal{C}$ is cogenerated by A as an (A,A)-bimodule, then, for all $f \in Z({}^*\mathcal{C}^*)$, $c \in \mathcal{C}$, $f\rightharpoonup c = c\leftharpoonup f$, and $Z({}^*\mathcal{C}) = Z({}^*\mathcal{C}^*) = Z(\mathcal{C}^*)$.*

(5) *If $\mathcal{C}$ satisfies the left and right α-condition, then $\mathcal{C}$ is a balanced $({}^*\mathcal{C},\mathcal{C}^*)$-bimodule, that is,*

$${}_{*\mathcal{C}}\mathrm{End}(\mathcal{C}) = \mathrm{End}^{\mathcal{C}}(\mathcal{C}) \simeq \mathcal{C}^*, \quad \mathrm{End}_{\mathcal{C}^*}(\mathcal{C}) = {}^{\mathcal{C}}\mathrm{End}(\mathcal{C}) \simeq {}^*\mathcal{C}, \text{ and}$$

$$ {}_{^*\mathcal{C}}\mathrm{End}_{\mathcal{C}^*}(\mathcal{C}) = {}^{\mathcal{C}}\mathrm{End}^{\mathcal{C}}(\mathcal{C}) \simeq Z(\mathcal{C}^*) = Z({}^*\mathcal{C}), $$

where morphisms are written opposite to scalars. In this case a left and right pure (A,A)-sub-bimodule $D \subset \mathcal{C}$ is a subcoring if and only if D is a $({}^*\mathcal{C}, \mathcal{C}^*)$-sub-bimodule.

Proof. (1) These identities are easily verified.

(2) Assume $f \rightharpoonup c = 0$ for all $c \in C$. Then $0 = \underline{\varepsilon}(f \rightharpoonup c) = f(c \leftharpoonup \underline{\varepsilon}) = f(c)$, and hence f is the zero map.

(3) Let $f \in Z({}^*\mathcal{C})$. By 17.8(4), $f \in {}^*\mathcal{C}^*$ and, for any $g \in {}^*\mathcal{C}$ and $c \in \mathcal{C}$,

$$ \begin{aligned} g(f \rightharpoonup c) &= \sum g(c_{\underline{1}} f(c_{\underline{2}})) = g *^l f(c) = f *^l g(c) \\ &= \sum g(f(c_{\underline{1}}) c_{\underline{2}}) = g(c \leftharpoonup f), \end{aligned} $$

and the cogenerating condition implies $f \rightharpoonup c = c \leftharpoonup f$. Now, for any $h \in \mathcal{C}^*$,

$$ \begin{aligned} h *^r f(c) &= \sum f(h(c_{\underline{1}}) c_{\underline{2}}) = \sum h(c_{\underline{1}}) f(c_{\underline{2}}) = h(f \rightharpoonup c) \\ &= h(c \leftharpoonup f) = \sum h(f(c_{\underline{1}}) c_{\underline{2}}) = f *^r h(c), \end{aligned} $$

showing $f \in Z(\mathcal{C}^*)$ and hence $Z({}^*\mathcal{C}) \subset Z(\mathcal{C}^*)$.

(4) By 17.8(4), $Z({}^*\mathcal{C}) \subset Z({}^*\mathcal{C}^*)$. Conversely, for $f \in Z({}^*\mathcal{C}^*)$, $c \in \mathcal{C}$, and all $g \in {}^*\mathcal{C}^*$, the equalities in (1) imply

$$ g(c \leftharpoonup f) = f * g(c) = g * f(c) = g(f \rightharpoonup c). $$

If $\mathcal{C}$ is cogenerated by A as an (A,A)-bimodule, we conclude that $c \leftharpoonup f = f \rightharpoonup c$. As shown in the proof of (3), this implies $f \in Z({}^*\mathcal{C})$ and so $Z({}^*\mathcal{C}^*) \subset Z({}^*\mathcal{C})$. We derive $Z({}^*\mathcal{C}^*) \subset Z(\mathcal{C}^*)$, similarly.

(5) Notice that the left (right) α-condition implies that $\mathcal{C}$ is cogenerated by A as a left (right) A-module. Hence (3) applies on the left and the right side. The isomorphisms follow from 18.12 and 19.3.

Let $D \subset \mathcal{C}$ be a left and right pure (A,A)-sub-bimodule. If D is a subcoring, then it is a right and left subcomodule and hence a $({}^*\mathcal{C}, \mathcal{C}^*)$-sub-bimodule. Conversely, assume that D is a $({}^*\mathcal{C}, \mathcal{C}^*)$-sub-bimodule. Then the restriction of $\underline{\Delta}$ yields a left and right $\mathcal{C}$-coaction on D and, by 40.16,

$$ \underline{\Delta}(D) \subset D \otimes_A \mathcal{C} \, \cap \, \mathcal{C} \otimes_A D = D \otimes_A D, $$

proving that D is a subcoring. □

19.11. Factor corings. *Let $\mathcal{C}$ be cogenerated by A as a left A-module. Then, for any idempotent $e \in Z({}^*\mathcal{C})$, $e \rightharpoonup \mathcal{C}$ is an A-coring and both $e \rightharpoonup \, : \mathcal{C} \to e \rightharpoonup \mathcal{C}$ and the inclusion $e \rightharpoonup \mathcal{C} \to \mathcal{C}$ are coring morphisms.*

Proof. As shown in 19.10, e is also in $Z(\mathcal{C}^*)$ and $e \rightharpoonup \mathcal{C} \leftharpoonup e = \mathcal{C} \leftharpoonup e = e \rightharpoonup \mathcal{C}$. For all $c \in \mathcal{C}$, a coaction on $e \rightharpoonup \mathcal{C}$ is defined by

$$e \rightharpoonup c \mapsto \sum c_{\underline{1}} \otimes e \rightharpoonup c_{\underline{2}} = \sum e \rightharpoonup c_{\underline{1}} \otimes c_{\underline{2}}.$$

It is routine to verify coassociativity, and e is the counit since

$$\begin{aligned}(e \otimes I_{\mathcal{C}})(\sum c_{\underline{1}} \otimes e \rightharpoonup c_{\underline{2}}) &= \sum e(c_{\underline{1}}) e \rightharpoonup c_{\underline{2}} = \sum e(c_{\underline{1}}) c_{\underline{2}} e(c_{\underline{3}}) \\ &= e \rightharpoonup (c \leftharpoonup e) = e \rightharpoonup (e \rightharpoonup c) = e \rightharpoonup c,\end{aligned}$$

where we used 19.10(3) to derive the penultimate equality. A similar computation applies for $I_{\mathcal{C}} \otimes e$. □

Even if $\mathcal{C}$ is not finitely generated as an A-module, it is $(\mathcal{C}^*, A)$-finite as defined in 41.22 provided it satisfies the α-condition.

19.12. Finiteness Theorem (2).

(1) *Let $\mathcal{C}$ satisfy the left α-condition and $M \in \mathbf{M}^{\mathcal{C}}$. Then every finite subset of M is contained in a subcomodule of M that is finitely generated as a right A-module. In particular, minimal ${}^*\mathcal{C}$-submodules are finitely generated as right A-modules.*

(2) *Let $\mathcal{C}$ satisfy the left and right α-condition. Then any finite subset of $\mathcal{C}$ is contained in a $({}^*\mathcal{C}, \mathcal{C}^*)$-sub-bimodule that is finitely generated as an (A, A)-bimodule. In particular, minimal $({}^*\mathcal{C}, \mathcal{C}^*)$-sub-bimodules of $\mathcal{C}$ are finitely generated as (A, A)-bimodules.*

Proof. (1) The proof from 4.12 applies: we show that, for each $m \in M$, the sucobmodule ${}^*\mathcal{C} \rightharpoonup m$ is finitely generated as an A-module. Write $\varrho^M(m) = \sum_{i=1}^k m_i \otimes c_i$, where $m_i \in {}^*\mathcal{C} \rightharpoonup m$, $c_i \in \mathcal{C}$. Then, for any $f \in {}^*\mathcal{C}$,

$$f \rightharpoonup m = (I_M \otimes f) \circ \varrho^M(m) = \sum_{i=1}^k m_i\, f(c_i)\,.$$

So, as a right A-module, ${}^*\mathcal{C} \rightharpoonup m$ is (finitely) generated by $m_1, \ldots, m_k$.

(2) It is enough to prove the assertion for single elements $c \in \mathcal{C}$. By (1), ${}^*\mathcal{C} \rightharpoonup c$ is generated as a right A-module by some $c_1, \ldots, c_k \in \mathcal{C}$. By symmetry, each $c_i \leftharpoonup \mathcal{C}^*$ is a finitely generated left A-module. Hence ${}^*\mathcal{C} \rightharpoonup c \leftharpoonup \mathcal{C}^*$ is a finitely generated (A, A)-bimodule. □

By the definitions in 38.9, a right $\mathcal{C}$-comodule N is *semisimple* (in $\mathbf{M}^{\mathcal{C}}$) if every $\mathcal{C}$-monomorphism $U \to N$ is a coretraction, and N is *simple* if all these monomorphisms are isomorphisms. The semisimplicity of N is equivalent to the fact that every right $\mathcal{C}$-comodule is N-injective. (Semi)simple left $\mathcal{C}$-comodules and $(\mathcal{C}, \mathcal{C})$-bicomodules are defined similarly. A coring $\mathcal{C}$ is said to be *left (right) semisimple* if it is semisimple as a left (right) comodule. $\mathcal{C}$ is called a *simple coring* if it is simple as a $(\mathcal{C}, \mathcal{C})$-bicomodule.

19.13. Semisimple comodules. *Let $\mathcal{C}$ be flat as a left A-module.*

(1) Any $N \in \mathbf{M}^{\mathcal{C}}$ is simple if and only if N has no nontrivial subcomodules.

(2) For $N \in \mathbf{M}^{\mathcal{C}}$ the following are equivalent:

(a) N is semisimple (in $\mathbf{M}^{\mathcal{C}}$, as defined above);

(b) every subcomodule of N is a direct summand;

(c) N is a sum of simple subcomodules;

(d) N is a direct sum of simple subcomodules.

Proof. By 18.14, all monomorphisms in $\mathbf{M}^{\mathcal{C}}$ are injective maps and the intersection of any two subcomodules is again a subcomodule. Hence the proof of 4.13 can be followed. □

We are now able to characterise an important class of corings.

19.14. Right semisimple corings. *For $\mathcal{C}$ the following are equivalent:*

(a) $\mathcal{C}$ is a semisimple right $\mathcal{C}$-comodule;

(b) ${}_A\mathcal{C}$ is flat and every right subcomodule of $\mathcal{C}$ is a direct summand;

(c) ${}_A\mathcal{C}$ is flat and $\mathcal{C}$ is a (direct) sum of simple right comodules;

(d) ${}_A\mathcal{C}$ is flat and every comodule in $\mathbf{M}^{\mathcal{C}}$ is semisimple;

(e) ${}_A\mathcal{C}$ is flat and every short exact sequence in $\mathbf{M}^{\mathcal{C}}$ splits;

(f) ${}_A\mathcal{C}$ is flat and every comodule in $\mathbf{M}^{\mathcal{C}}$ is projective;

(g) ${}_A\mathcal{C}$ is projective and $\mathcal{C}$ is a semisimple left ${}^*\mathcal{C}$-module;

(h) every comodule in $\mathbf{M}^{\mathcal{C}}$ is ($\mathcal{C}$-)injective;

(i) $\mathcal{C}$ is a direct sum of simple corings that are right (left) semisimple;

(j) $\mathcal{C}_A$ is projective and $\mathcal{C}$ is a semisimple right $\mathcal{C}^*$-module;

(k) $\mathcal{C}$ is a semisimple left $\mathcal{C}$-comodule.

Proof. (a) ⇒ (b) ⇒ (c) ⇒ (d) ⇒ (e) ⇒ (f) If $\mathcal{C}$ is right semisimple, then every monomorphism $U \to \mathcal{C}$ in $\mathbf{M}^{\mathcal{C}}$ is a coretraction and hence injective. By 18.14, this implies that $\mathcal{C}$ is flat in ${}_A\mathbf{M}$. Now we can apply the arguments used in the proof of 4.14.

(a) ⇔ (h) This follows from 38.13. Notice that (as a consequence of (g)) a right comodule is $\mathcal{C}$-injective if and only if it is injective in $\mathbf{M}^{\mathcal{C}}$.

(f) ⇒ (j) In particular, $\mathcal{C}$ is projective in $\mathbf{M}^{\mathcal{C}}$, and hence, by 18.20, it is projective in $\mathbf{M}_A$. By a slight refinement of the proof of 41.8 we show that $\mathcal{C}$ is a semisimple right module over $\mathcal{C}^* = \mathrm{End}^{\mathcal{C}}(\mathcal{C})$: let $K \subset \mathcal{C}$ be a simple right subcomodule. We show that, for any $k \in K$, $k \leftharpoonup \mathcal{C}^* \subset \mathcal{C}$ is a simple $\mathcal{C}^*$-submodule: for any $t \in \mathcal{C}^*$ with $k \leftharpoonup t \neq 0$, $K \simeq K \leftharpoonup t$. Since these are direct summands in $\mathcal{C}$, there exists some $f \in \mathcal{C}^*$ with the property $k \leftharpoonup t \leftharpoonup f = k$ and hence $k \leftharpoonup \mathcal{C}^* = k \leftharpoonup t \leftharpoonup \mathcal{C}^*$, implying that $k \leftharpoonup \mathcal{C}^*$ has no nontrivial

$\mathcal{C}^*$-submodules. Notice that the right subcomodule K is a left ${}^*\mathcal{C}$-module and hence ${}^*\mathcal{C}\rightharpoonup K = K$.

As a right semisimple comodule, $\mathcal{C} = \sum_\Lambda K_\lambda$, where the K_λ are simple right subcomodules. Now, $\mathcal{C} = {}^*\mathcal{C}\rightharpoonup(\sum_\Lambda K_\lambda \leftharpoonup \mathcal{C}^*)$, showing that $\mathcal{C}$ is a sum of simple right $\mathcal{C}^*$-modules $f\rightharpoonup k_\lambda \leftharpoonup \mathcal{C}^*$, where $f \in {}^*\mathcal{C}$ and $k_\lambda \in K_\lambda$.

(j) $\Rightarrow$ (k) Since $\mathcal{C}_A$ is projective, ${}^{\mathcal{C}}\mathbf{M} = \sigma[\mathcal{C}_{\mathcal{C}^*}]$ and the assertion is obvious.

(k) $\Rightarrow$ (g) By symmetry this can be shown with the proof (a) $\Rightarrow$ (j).

(g) $\Rightarrow$ (a) This is clear since ${}_A\mathcal{C}$ projective implies $\mathbf{M}^{\mathcal{C}} = \sigma[{}_{\mathcal{C}^*}\mathcal{C}]$.

(k) $\Rightarrow$ (i) From what we have shown so far, (k) implies that $\mathcal{C}$ is projective in ${}_A\mathbf{M}$ and $\mathbf{M}_A$. Hence, by 19.10(5), the direct summands as $({}^*\mathcal{C}, \mathcal{C}^*)$-subbimodules are subcorings, and the decomposition is a particular case of the fully invariant decomposition of the semisimple left ${}^*\mathcal{C}$-module $\mathcal{C}$ (see 41.8).

(i) $\Rightarrow$ (g) We know that the simple semisimple subcorings of $\mathcal{C}$ are projective as left (and right) A-modules. From this the assertion is clear. □

The above observations also imply characterisations of simple corings:

19.15. Simple corings. *For $\mathcal{C}$ the following are equivalent:*

(a) $\mathcal{C}$ is a simple coring that is right semisimple;

(b) $\mathcal{C}$ is projective in ${}_A\mathbf{M}$ and is a simple $({}^*\mathcal{C}, \mathcal{C}^*)$-bimodule with a minimal right $\mathcal{C}^*$-submodule;

(c) $\mathcal{C}$ is right semisimple and all simple comodules are isomorphic;

(d) $\mathcal{C}$ is a simple coring that is left semisimple;

(e) there is a Galois comodule $M \in \mathbf{M}^{\mathcal{C}}$ such that $\mathrm{End}^{\mathcal{C}}(M)$ is a division algebra;

(f) there is a division R-algebra T and a (T, A)-bimodule P such that P_A is finitely generated and $P^* \otimes_T P \simeq \mathcal{C}$ as corings.

Proof. (a) $\Leftrightarrow$ (b) This follows by 19.14.

(b) $\Leftrightarrow$ (c) Obviously all simple subcomodules of $\mathcal{C}$ are isomorphic, and all simple comodules are isomorphic to subcomodules of $\mathcal{C}$.

(a) $\Leftrightarrow$ (d) The left-right symmetry is shown in 19.14.

(a) $\Rightarrow$ (e) Let M be any simple left subcomodule of $\mathcal{C}$. Then M is a finitely generated projective generator in $\mathbf{M}^{\mathcal{C}}$, and, by Schur's Lemma, $\mathrm{End}^{\mathcal{C}}(M)$ is a division algebra. Furthermore, by 19.12 and 18.20, M_A is finitely generated and projective, and so M is Galois comodule by 28.19.

(e) $\Rightarrow$ (f) For the Galois module M and $S = \mathrm{End}^{\mathcal{C}}(M)$ there is a coring isomorphism $M^* \otimes_S M \simeq \mathcal{C}$ (see 18.26). Putting $P = M_A$ and $S = T$, the (T, A)-bimodule P has the properties required.

(f) $\Rightarrow$ (c) Under the given conditions $P^* \otimes_T P$ is a coring, the dual algebra ${}^*(P^* \otimes_T P)$ is isomorphic to ${}_T\mathrm{End}(P)$, and $\mathbf{M}^{P^* \otimes_T P}$ is equivalent to $\sigma[P_{{}_T\mathrm{End}(P)}]$ (see 19.7). Since P is a simple module over ${}_T\mathrm{End}(P)$, all modules in $\sigma[P_{T\mathrm{End}(P)}]$ are semisimple and it contains only one type of simple modules. Hence all comodules in $\mathbf{M}^{P^* \otimes_T P}$ are semisimple and there is only one simple comodule (up to isomorphisms). □

By the Finiteness Theorem 19.12 and the Hom relations 18.10, properties of the base ring A have a strong influence on the comodule properties of $\mathcal{C}$.

19.16. Corings over special rings. *Let $\mathcal{C}$ satisfy the left α-condition.*

(1) If A is right Noetherian, then $\mathcal{C}$ is a locally Noetherian right comodule, and direct sums of injectives in $\mathbf{M}^{\mathcal{C}}$ are injective.

(2) If A is left perfect, then every module in $\mathbf{M}^{\mathcal{C}}$ satisfies the descending chain condition on finitely generated subcomodules.

(3) If A is right Artinian, then every finitely generated module in $\mathbf{M}^{\mathcal{C}}$ has finite length.

Proof. (1) Let A be right Noetherian and $N \subset \mathcal{C}$ a finitely generated left ${}^*\mathcal{C}$-submodule. Then N is finitely generated – and hence Noetherian – as a right A-module. This obviously implies the ascending chain condition for subcomodules in N, and so $\mathcal{C}$ is locally Noetherian.

(2) For a left perfect ring A, any finitely generated right A-module satisfies the descending chain condition on cyclic (and finitely generated) A-submodules (cf. 41.17). By 19.12, this implies the descending chain condition for finitely generated subcomodules for any $M \in \mathbf{M}^{\mathcal{C}}$.

(3) Over a right Artinian ring A, finitely generated right A-modules have finite length. This implies finite length for finitely generated comodules. □

Recall that, over left Artinian (left perfect) rings A, the left α-condition on $\mathcal{C}$ is equivalent to ${}_A\mathcal{C}$ being projective (see 42.11). QF rings A are Artinian and are injective and cogenerators in ${}_A\mathbf{M}$ and $\mathbf{M}_A$ (see 43.6).

19.17. Corings over QF rings. *Let ${}_A\mathcal{C}$ be projective and A a QF ring.*

(1) $\mathcal{C}$ is a (big) injective cogenerator in $\mathbf{M}^{\mathcal{C}}$.

(2) Every comodule in $\mathbf{M}^{\mathcal{C}}$ is a subcomodule of some direct sum $\mathcal{C}^{(\Lambda)}$.

(3) $\mathcal{C}^*$ is an f-semiperfect ring.

(4) $K := \mathrm{Soc}_{{}^*\mathcal{C}}\mathcal{C} \trianglelefteq \mathcal{C}$ and $\mathrm{Jac}(\mathcal{C}^*) = \mathrm{Hom}^{\mathcal{C}}(\mathcal{C}/K, \mathcal{C}) \simeq \mathrm{Hom}_A(\mathcal{C}/K, A)$.

Proof. (1),(2) Over a QF ring A, every right A-module is contained in a free A-module $A^{(\Lambda)}$. This implies for any right $\mathcal{C}$-comodule M the injection

$$M \xrightarrow{\varrho^M} M \otimes_A \mathcal{C} \subset A^{(\Lambda)} \otimes_A \mathcal{C} \simeq \mathcal{C}^{(\Lambda)}.$$

(3) By 41.19, the endomorphism ring $\mathrm{End}^{\mathcal{C}}(\mathcal{C}) \simeq \mathcal{C}^*$ of the self-injective module ${}_{{}^*\mathcal{C}}\mathcal{C}$ is f-semiperfect.

(4) It follows from 19.16(3) that ${}_{^*\mathcal{C}}\mathcal{C}$ has an essential socle, and hence the assertion follows by 41.19. □

Notice that – in contrast to the situation for coalgebras considered in 9.1 – we do not have a left-right symmetry here since ${}_A\mathcal{C}$ projective need not imply that $\mathcal{C}_A$ is also projective.

Writing morphisms of right comodules on the right side, the image of the functor $\mathrm{Hom}^{\mathcal{C}}(-,\mathcal{C})$ lies in the category of right modules over $\mathrm{End}^{\mathcal{C}}(\mathcal{C}) \simeq \mathrm{Hom}_A(\mathcal{C},A) = \mathcal{C}^*$. This induces a connection between $\mathbf{M}^{\mathcal{C}}$ and $\mathbf{M}_{\mathcal{C}^*}$.

19.18. The functors $\mathrm{Hom}^{\mathcal{C}}(-,\mathcal{C})$ and $\mathrm{Hom}^{\mathcal{C}}(\mathcal{C},-)$.

(1) For any $M \in \mathbf{M}^{\mathcal{C}}$ the R-module isomorphism (see 18.10)

$$\varphi : \mathrm{Hom}^{\mathcal{C}}(M,\mathcal{C}) \to \mathrm{Hom}_A(M,A) = M^*$$

induces a right $\mathcal{C}^*$-module structure on M^*,

$$M^* \otimes_R \mathcal{C}^* \to M^*, \quad g \otimes f \mapsto [m \mapsto \sum f(g(m_{\underline{0}})m_{\underline{1}})],$$

and so φ is a morphism in $\mathbf{M}_{\mathcal{C}^*}$ and the contravariant functor

$$\mathrm{Hom}^{\mathcal{C}}(-,\mathcal{C}) \simeq \mathrm{Hom}_A(-,A) : \mathbf{M}^{\mathcal{C}} \to \mathbf{M}_{\mathcal{C}^*}$$

is left exact provided that ${}_A\mathcal{C}$ is flat.

(2) The covariant functor $\mathrm{Hom}^{\mathcal{C}}(\mathcal{C},-) : \mathbf{M}^{\mathcal{C}} \to {}_{\mathcal{C}^*}\mathbf{M}$ is left exact provided that ${}_A\mathcal{C}$ is flat.

Proof. We get the right action by $f \in \mathcal{C}^*$ on $g \in M^*$ for $m \in M$ by

$$\begin{aligned} g \cdot f(m) &= \varphi(\varphi^{-1}(g)(m) \leftharpoonup f) = \sum \varphi(g(m_{\underline{0}})m_{\underline{1}} \leftharpoonup f) \\ &= \sum f(g(m_{\underline{0}})m_{\underline{1}})\underline{\varepsilon}(m_{\underline{2}}) \;=\; \sum f(g(m_{\underline{0}})m_{\underline{1}}). \end{aligned}$$

The remaining assertions are clear from module theory and 18.17. □

Under certain finiteness conditions we can pass from left to right $\mathcal{C}$-comodules. This works in particular over QF rings. Recall from 40.12 that, for a finitely presented right A-module M and a flat right A-module $\mathcal{C}$, there is an isomorphism

$$\nu_M : \mathcal{C} \otimes_A \mathrm{Hom}_A(M,A) \to \mathrm{Hom}_A(M,\mathcal{C}), \quad c \otimes h \mapsto c \otimes h(-).$$

19.19. Comodules finitely presented as A-modules. *Let $M \in \mathbf{M}^{\mathcal{C}}$ and assume that M_A is finitely generated and projective, or that $\mathcal{C}_A$ is flat and M_A is finitely presented.*

(1) $M^* = \mathrm{Hom}_A(M, A)$ *is a left* $\mathcal{C}$*-comodule by the structure map*

$$\bar{\varrho} : M^* \to \mathrm{Hom}_A(M, \mathcal{C}) \simeq \mathcal{C} \otimes_A M^*, \quad g \mapsto (g \otimes I_{\mathcal{C}}) \circ \varrho^M.$$

The resulting right $\mathcal{C}^*$*-module structure on* M^* *is the map from 19.18,*

$$M^* \otimes_R \mathcal{C}^* \to M^*, \quad g \otimes f \mapsto [m \mapsto \sum f(g(m_{\underline{0}})m_{\underline{1}})].$$

(2) Assume also that ${}_A\mathcal{C}$ *is flat. If* M *is injective as a right* $\mathcal{C}$*-comodule and is contained in a free* A*-module, then* M^* *is projective in* $\mathbf{M}_{\mathcal{C}^*}$*.*

Proof. (1) The coassociativity of $\bar{\varrho}$ is shown with an obviously modified diagram from the proof of 3.11. Let $\beta : A^n \to M$ be an epimorphism in $\mathbf{M}_A$. Consider the commutative diagram

$$\begin{array}{ccccc}
 & & M^* & & \\
 & & \downarrow \bar{\varrho} & & \\
0 \longrightarrow & & \mathrm{Hom}_A(M, \mathcal{C}) & \xrightarrow{\mathrm{Hom}(\beta, \mathcal{C})} & \mathrm{Hom}_A(A^n, \mathcal{C}) \\
 & & \downarrow \nu_M^{-1} & & \downarrow \nu_{A^n}^{-1} \\
0 \longrightarrow & & \mathcal{C} \otimes_A M^* & \xrightarrow{I_{\mathcal{C}} \otimes \beta^*} & \mathcal{C} \otimes_A (A^n)^*,
\end{array}$$

where, for a dual basis $\{e_i, e^i \mid i = 1, \ldots, n\}$ of A^n,

$$\nu_{A^n}^{-1} : \mathrm{Hom}_A(A^n, \mathcal{C}) \to \mathcal{C} \otimes_A (A^n)^*, \quad \varphi \mapsto \sum_i \varphi(e_i) \otimes e^i.$$

For $g \in M^*$ we obtain

$$\nu_M^{-1}((g \otimes I_{\mathcal{C}}) \circ \varrho^M) = \nu_{A^n}^{-1}((g \otimes I_{\mathcal{C}}) \circ \varrho^M \circ \beta) = \sum_i (g \otimes I_{\mathcal{C}}) \circ \varrho^M \circ \beta(e_i) \otimes e^i.$$

The canonical right action of $\mathcal{C}^*$ on the left $\mathcal{C}$-comodule M^* is defined by the composition of the structure map $M^* \to \mathcal{C} \otimes_A M^*$ and

$$\mathcal{C} \otimes_A M^* \otimes_R \mathcal{C}^* \to M^*, \quad c \otimes g \otimes f \mapsto f(c)g.$$

So we obtain

$$g \otimes f \mapsto \sum_i f((g \otimes I_{\mathcal{C}}) \circ \varrho^M \circ \beta(e_i)) \otimes e^i,$$

and for $m \in M$ and $a \in A^n$ with $\beta(a) = m$, the right-hand side maps a to

$$\sum_i f((g \otimes I_{\mathcal{C}}) \circ \varrho^M \circ \beta(e_i)) \otimes e^i(a) = f((g \otimes I_{\mathcal{C}}) \circ \varrho^M \circ \beta(a)) = \sum f(g(m_{\underline{0}})m_{\underline{1}}).$$

(2) A monomorphism $M \to A^n$ in $\mathbf{M}_A$ induces a monomorphism

$$M \to M \otimes_A \mathcal{C} \to A^n \otimes_A \mathcal{C} \simeq \mathcal{C}^n$$

in $\mathbf{M}^{\mathcal{C}}$ that splits by assumption. This yields a commutative diagram

$$\begin{array}{ccccc} \mathrm{Hom}^{\mathcal{C}}(\mathcal{C}^n,\mathcal{C}) & \longrightarrow & \mathrm{Hom}^{\mathcal{C}}(M,\mathcal{C}) & \longrightarrow & 0 \\ \simeq\downarrow & & \downarrow\simeq & & \\ \mathrm{Hom}_A(\mathcal{C}^n,A) & \longrightarrow & \mathrm{Hom}_A(M,A) & \longrightarrow & 0 \end{array}$$

in which the upper – and hence the lower – row splits in $\mathbf{M}_{\mathcal{C}^*}$, and so M^* is a direct summand of $\mathcal{C}^{*n}$. □

With the module structures just described, the functors on $\mathbf{M}^{\mathcal{C}}$ considered in 19.18 can be extended to functors on ${}_{^*\mathcal{C}}\mathbf{M}$.

Let $\mathbf{M}[f]^{\mathcal{C}}$ denote the full subcategory of $\mathbf{M}^{\mathcal{C}}$ whose objects are finitely presented as right A-modules. This is clearly an Abelian category provided that A is right Noetherian. A similar notation is used for left comodules.

19.20. The functors $\mathrm{Hom}(-,\mathcal{C})$**.** *Suppose that $\mathcal{C}$ satisfies the left and right α-conditions. Then the $({}^*\mathcal{C},\mathcal{C}^*)$-bimodule $\mathcal{C}$ defines the right adjoint pair of contravariant functors*

$$\begin{aligned} {}_{^*\mathcal{C}}\mathrm{Hom}(-,\mathcal{C}) &: {}_{^*\mathcal{C}}\mathbf{M} \to \mathbf{M}_{\mathcal{C}^*}, \quad M \mapsto {}_{^*\mathcal{C}}\mathrm{Hom}(M,\mathcal{C}), \\ \mathrm{Hom}_{\mathcal{C}^*}(-,\mathcal{C}) &: \mathbf{M}_{\mathcal{C}^*} \to {}_{^*\mathcal{C}}\mathbf{M}, \quad X \mapsto \mathrm{Hom}_{\mathcal{C}^*}(X,\mathcal{C}). \end{aligned}$$

Restricted to the subcategory $\mathbf{M}^{\mathcal{C}}$, ${}_{^*\mathcal{C}}\mathrm{Hom}(-,\mathcal{C}) \simeq \mathrm{Hom}^{\mathcal{C}}(-,\mathcal{C})$.

If A is Noetherian, then there is a left exact functor

$$\mathrm{Hom}^{\mathcal{C}}(-,\mathcal{C}) : \mathbf{M}[f]^{\mathcal{C}} \to {}^{\mathcal{C}}\mathbf{M}[f], \quad M \mapsto \mathrm{Hom}^{\mathcal{C}}(M,\mathcal{C}),$$

which induces a duality provided that A is a QF ring.

Proof. The first assertions follow from general module theory (see 40.23). As shown in 19.19, for any $N \in \mathbf{M}^{\mathcal{C}}$ that is finitely presented as an A-module, $\mathrm{Hom}^{\mathcal{C}}(N,\mathcal{C})$ is a left $\mathcal{C}$-comodule that is finitely generated as a left A-module. If A is QF, then both N and $\mathrm{Hom}^{\mathcal{C}}(N,\mathcal{C})$ are $\mathcal{C}$-reflexive. □

19.21. The functor ${}_{^*\mathcal{C}}\mathrm{Hom}(\mathcal{C},-)$**.** *Let $\mathcal{C}$ satisfy the left α-condition. The $({}^*\mathcal{C},\mathcal{C}^*)$-bimodule $\mathcal{C}$ defines an adjoint pair of covariant functors*

$$\mathcal{C}\otimes_{\mathcal{C}^*} - : {}_{\mathcal{C}^*}\mathbf{M} \to \mathbf{M}^{\mathcal{C}}, \quad {}_{^*\mathcal{C}}\mathrm{Hom}(\mathcal{C},-) : \mathbf{M}^{\mathcal{C}} \to {}_{\mathcal{C}^*}\mathbf{M}.$$

For $X \in {}_{\mathcal{C}^*}\mathbf{M}$, the right A-module structure of $\mathcal{C}\otimes_{\mathcal{C}^*} X$ is derived from the left ${}^*\mathcal{C}$-module structure of $\mathcal{C}$ via the ring anti-morphism $i_L : A \to {}^*\mathcal{C}$ (cf. 17.8(2)). $\mathcal{C}\otimes_{\mathcal{C}^*} X$ is a left ${}^*\mathcal{C}$-module and, in fact, a right $\mathcal{C}$-comodule by

$$\mathcal{C}\otimes_{\mathcal{C}^*} X \to (\mathcal{C}\otimes_{\mathcal{C}^*} X)\otimes_A \mathcal{C}, \quad c\otimes x \mapsto \sum c_{\underline{1}}\otimes x\otimes c_{\underline{2}}.$$

Proof. Again the first assertions follow from module theory (43.9). To show that the coaction is well defined we have to verify that, for any $f \in \mathcal{C}^*$, $c \in \mathcal{C}$, and $x \in X$, the images of $c \otimes f \cdot x$ and $(f \otimes I_{\mathcal{C}}) \circ \underline{\Delta}(c) \otimes x$ are the same. This follows from the equality

$$\sum c_{\underline{1}} \otimes f \cdot x \otimes c_{\underline{2}} = \sum f(c_{\underline{1}}) c_{\underline{2}} \otimes x \otimes c_{\underline{3}}.$$

Furthermore, for any $a \in A$, $(c \otimes x)a = i_L(a)c \otimes x = \sum c_{\underline{1}} \varepsilon(c_{\underline{2}}a) \otimes x = ca \otimes x$, and this is mapped to

$$\sum c_{\underline{1}} \otimes x \otimes c_{\underline{2}} \varepsilon(c_{\underline{3}}a) = \sum c_{\underline{1}} \otimes x \otimes c_{\underline{2}}a,$$

showing that the coaction is A-linear. □

Over QF rings, for some modules injectivity and projectivity in $\mathbf{M}^{\mathcal{C}}$ extend to injectivity, resp. projectivity, in ${}_{\mathcal{C}^*}\mathbf{M}$.

19.22. Injectives – projectives. *Let ${}_A\mathcal{C}$ be projective, A a QF ring, and $M \in \mathbf{M}^{\mathcal{C}}$.*

(1) If M is projective in $\mathbf{M}^{\mathcal{C}}$, then M^* is $\mathcal{C}$-injective as a right $\mathcal{C}^*$-module and ${}^{\mathcal{C}}\mathrm{Rat}(M^*)$ is injective in ${}^{\mathcal{C}}\mathbf{M}$.

(2) If M is finitely presented as a left A-module, then:

(i) M is injective in $\mathbf{M}^{\mathcal{C}}$ if and only if M is injective in ${}_{*\mathcal{C}}\mathbf{M}$.

(ii) M is projective in $\mathbf{M}^{\mathcal{C}}$ if and only if M is projective in ${}_{*\mathcal{C}}\mathbf{M}$.

Proof. (1) We slightly modify the proof of 9.5. Consider any diagram with exact row in ${}^{\mathcal{C}}\mathbf{M}$,

$$\begin{array}{ccccc} 0 & \longrightarrow & K & \longrightarrow & N \\ & & \downarrow f & & \\ & & M^*, & & \end{array}$$

where N is finitely generated as a left A-module (right $\mathcal{C}^*$-module). Denoting ${}^*(-) = {}_A\mathrm{Hom}(-, A)$ and $(-)^* = \mathrm{Hom}_A(-, A)$, we obtain – with the canonical map $\Phi_M : M \to {}^*(M^*)$ – the diagram in ${}_{*\mathcal{C}}\mathbf{M}$

$$\begin{array}{ccccc} & & M & \xrightarrow{\Phi_M} & {}^*(M^*) & & \\ & & & & \downarrow {}^*f & & \\ & & {}^*N & \longrightarrow & {}^*K & \longrightarrow & 0, \end{array}$$

where the bottom row is in $\mathbf{M}^{\mathcal{C}}$ (by 19.19) and hence can be extended commutatively by some right comodule morphism $g : M \to {}^*N$. Again applying $(-)^*$ – and recalling that the composition $M^* \xrightarrow{\Phi_{M^*}} ({}^*(M^*))^* \xrightarrow{(\Phi_M)^*} M^*$ yields the identity (by 40.23) – we see that g^* extends f to N. This proves that M^* is N-injective for all comodules $N \in {}^{\mathcal{C}}\mathbf{M}$ that are finitely presented as A-modules and hence is $\mathcal{C}$-injective. A standard argument shows that ${}^{\mathcal{C}}\mathrm{Rat}(M^*)$ is injective in ${}^{\mathcal{C}}\mathbf{M}$.

(2) Since 3.11 was extended to corings in 19.19, we can follow the proof of 9.5(2). □

19.23. Cogenerator properties of $\mathcal{C}$. *Let $\mathcal{C}$ satisfy the α-condition and assume $\mathcal{C}$ to cogenerate all finitely $\mathcal{C}$-generated comodules. Then the following are equivalent:*

(a) ${}_{^*\mathcal{C}}\mathcal{C}$ is linearly compact;

(b) $\mathcal{C}_{\mathcal{C}^*}$ is $\mathcal{C}^*$-injective.

If A is right perfect, (a),(b) are equivalent to:

(c) ${}_{^*\mathcal{C}}\mathcal{C}$ is Artinian.

Proof. The equivalence of (a) and (b) follows by 43.2(4). If A is left perfect, then $\mathcal{C}$ is right semi-Artinian by 19.16, and hence, by 41.13, (a) implies that ${}_{^*\mathcal{C}}\mathcal{C}$ is Artinian. □

Over a (left and right) Noetherian ring A, $\mathcal{C}$ is left and right locally Noetherian as a $\mathcal{C}^*$-module (by 19.16), and therefore we can apply 43.4 to obtain:

19.24. $\mathcal{C}$ as injective cogenerator in $\mathbf{M}^{\mathcal{C}}$. *Let A be Noetherian and let $\mathcal{C}$ satisfy the left and right α-conditions. Then the following are equivalent:*

(a) $\mathcal{C}$ is an injective cogenerator in $\mathbf{M}^{\mathcal{C}}$;

(b) $\mathcal{C}$ is an injective cogenerator in ${}^{\mathcal{C}}\mathbf{M}$;

(c) $\mathcal{C}$ is a cogenerator both in $\mathbf{M}^{\mathcal{C}}$ and ${}^{\mathcal{C}}\mathbf{M}$.

Restricting to Artinian rings, we have interesting characterisations of $\mathcal{C}$ as an injective cogenerator not only in $\mathbf{M}^{\mathcal{C}}$ but also in $\mathbf{M}_{\mathcal{C}^*}$.

19.25. $\mathcal{C}$ as injective cogenerator in $\mathbf{M}_{\mathcal{C}^*}$. *Let A be Artinian and ${}_A\mathcal{C}$ and $\mathcal{C}_A$ projective. The following are equivalent:*

(a) $\mathcal{C}$ is an injective cogenerator in $\mathbf{M}_{\mathcal{C}^*}$;

(b) ${}_{^*\mathcal{C}}\mathcal{C}$ is Artinian and an injective cogenerator in $\mathbf{M}^{\mathcal{C}}$;

(c) ${}_{^*\mathcal{C}}\mathcal{C}$ is an injective cogenerator in $\mathbf{M}^{\mathcal{C}}$ and $\mathcal{C}^*$ is right Noetherian.

If these conditions hold, then $\mathcal{C}^*$ is a semiperfect ring and every right $\mathcal{C}^*$-module that is finitely generated as an A-module belongs to ${}^{\mathcal{C}}\mathbf{M}$.

Proof. Since A is Artinian, $\mathcal{C}$ has locally finite length as a left ${}^*\mathcal{C}$- and right $\mathcal{C}^*$-module.

(a) $\Rightarrow$ (b) Assume $\mathcal{C}$ to be an injective cogenerator in $\mathbf{M}_{\mathcal{C}^*}$. Then, by 19.24, $\mathcal{C}$ is an injective cogenerator in $\mathbf{M}^{\mathcal{C}}$. Now 43.8 implies that ${}_{^*\mathcal{C}}\mathcal{C}$ is Artinian.

(b) $\Rightarrow$ (a) and (b) $\Leftrightarrow$ (c) also follow from 43.8.

Assume the conditions hold. Then $\mathcal{C}^*$ is f-semiperfect (as the endomorphism ring of a self-injective module; cf. 41.19). So $\mathcal{C}^*/\mathrm{Jac}(\mathcal{C}^*)$ is von Neumann regular and right Noetherian, and hence right (and left) semisimple. This implies that $\mathcal{C}^*$ is semiperfect.

Let $L \in \mathbf{M}_{\mathcal{C}^*}$ be finitely generated as an A-module. Then L is finitely cogenerated as a $\mathcal{C}^*$-module, and hence it is finitely cogenerated by $\mathcal{C}$. This implies $L \in {}^{\mathcal{C}}\mathbf{M}$. □

Since over a QF ring any coring is an injective cogenerator for its comodules, the results from 19.25 simplify to the

19.26. Corollary. *If A is QF and ${}_A\mathcal{C}$ and $\mathcal{C}_A$ are projective, the following are equivalent:*

(a) $\mathcal{C}$ is injective in $\mathbf{M}_{\mathcal{C}^*}$;

(b) $\mathcal{C}$ is an injective cogenerator in $\mathbf{M}_{\mathcal{C}^*}$;

(c) ${}_{^*\mathcal{C}}\mathcal{C}$ is Artinian;

(d) $\mathcal{C}^*$ is a right Noetherian ring.

Proof. Since A is QF, $\mathcal{C}$ is an injective cogenerator in $\mathbf{M}^{\mathcal{C}}$ and ${}^{\mathcal{C}}\mathbf{M}$ (by 19.17). So the equivalence of (b), (c) and (d) follows from 19.25, and (a) $\Rightarrow$ (c) is a consequence of 19.23. □

Now we interpret the topological observations about modules in 42.3 in the case of corings.

19.27. The $\mathcal{C}$-adic topology in ${}^*\mathcal{C}$. Let $\mathcal{C}$ satisfy the left α-condition. Then the finite topology in ${}_A\mathrm{End}(\mathcal{C})$ induces the $\mathcal{C}$-adic topology on ${}^*\mathcal{C}$ and the open left ideals determine the right $\mathcal{C}$-comodules.

Open left ideals. A filter basis for the open left ideals of ${}^*\mathcal{C}$ is given by

$$\mathcal{B}_{\mathcal{C}} = \{\mathrm{An}_{^*\mathcal{C}}(E) \mid E \text{ a finite subset of } \mathcal{C}\},$$

where $\mathrm{An}_{^*\mathcal{C}}(E) = \{f \in {}^*\mathcal{C} \mid f \rightharpoonup E = 0\}$, and the filter of all open left ideals is

$$\mathcal{F}_{\mathcal{C}} = \{I \subset {}^*\mathcal{C} \mid I \text{ is a left ideal and } {}^*\mathcal{C}/I \in \mathbf{M}^{\mathcal{C}}\}.$$

Thus a generator in $\mathbf{M}^{\mathcal{C}}$ is given by

$$G = \bigoplus \{{}^*\mathcal{C}/I \mid I \in \mathcal{B}_{\mathcal{C}}\}.$$

Closed left ideals. *For a left ideal $I \subset {}^*\mathcal{C}$, the following are equivalent:*

(a) I is closed in the $\mathcal{C}$-adic topology;

(b) $I = \mathrm{An}_{^*\mathcal{C}}(W)$ for some $W \in \mathbf{M}^{\mathcal{C}}$;

(c) $^*\mathcal{C}/I$ is cogenerated (in $_{^*\mathcal{C}}\mathbf{M}$) by some (minimal) cogenerator of $\mathbf{M}^{\mathcal{C}}$;

(d) $I = \bigcap_\Lambda I_\lambda$, where $^*\mathcal{C}/I_\lambda \in \mathbf{M}^{\mathcal{C}}$ and is finitely cogenerated (cocyclic).

Over QF rings. *Let A be a QF ring and $_A\mathcal{C}$ and $\mathcal{C}_A$ projective.*

(1) Any finitely generated right ideal in $\mathcal{C}^*$ (left ideal in $^*\mathcal{C}$) is closed in the $\mathcal{C}$-adic topology.

(2) A left ideal $I \subset {}^*\mathcal{C}$ is open if and only if it is closed and $^*\mathcal{C}/I$ is finitely A-generated (finitely A-cogenerated).

Proof. Since A is a QF ring, $\mathcal{C}$ is injective in $\mathbf{M}^{\mathcal{C}}$ and $^{\mathcal{C}}\mathbf{M}$ (by 19.17), and hence finitely generated left, resp. right, ideals in the endomorphism rings are closed in the $\mathcal{C}$-adic topology (see 42.3). □

References. Brzeziński [73]; Cuadra and Gómez-Torrecillas [102]; El Kaoutit, Gómez-Torrecillas and Lobillo [112]; El Kaoutit and Gómez-Torrecillas [111]; Guzman [126, 127]; Wisbauer [210, 212].

20 The rational functor for corings

In Section 19 we revealed a relationship between comodules of an A-coring $\mathcal{C}$ and modules of a dual algebra. The main idea of this section is to use this relationship more fully, and to derive properties of comodules from the properties of corresponding ${}^*\mathcal{C}$-modules. Although every right $\mathcal{C}$-comodule is a left ${}^*\mathcal{C}$-module, not every left ${}^*\mathcal{C}$-module is a right $\mathcal{C}$-comodule. In this section we study a functor that carves out the part of a left ${}^*\mathcal{C}$-module on which a right $\mathcal{C}$-coaction can be defined. Similarly as for coalgebras, this functor is known as the *rational functor*.

To realise the program outlined above, one first needs to restrict oneself to the case when $\mathbf{M}^{\mathcal{C}}$ is a full subcategory of left ${}^*\mathcal{C}$-modules. Thus, for the whole of the section we assume that $\mathcal{C}$ is an A-coring that is locally projective as a left A-module, that is, it satisfies the left α-condition 19.2. Then the inclusion of $\mathbf{M}^{\mathcal{C}} = \sigma[{}_{^*\mathcal{C}}\mathcal{C}]$ into ${}_{^*\mathcal{C}}\mathbf{M}$ has a right adjoint (see 41.1), namely, the

20.1. Rational functor. For any left ${}^*\mathcal{C}$-module M, the *rational submodule* is defined by

$$\mathrm{Rat}^{\mathcal{C}}(M) = \sum \{\mathrm{Im}\, f \mid f \in {}_{^*\mathcal{C}}\mathrm{Hom}(U, M),\, U \in \mathbf{M}^{\mathcal{C}}\}.$$

Thus $\mathrm{Rat}^{\mathcal{C}}(M)$ is the largest submodule of M that is subgenerated by $\mathcal{C}$, and hence it is a right $\mathcal{C}$-comodule. The induced functor, a subfunctor of the identity, is called the *rational functor*:

$$\begin{aligned} \mathrm{Rat}^{\mathcal{C}} : {}_{^*\mathcal{C}}\mathbf{M} \to \mathbf{M}^{\mathcal{C}}, \qquad & M \mapsto \mathrm{Rat}^{\mathcal{C}}(M), \\ & f : M \to N \mapsto f|_{\mathrm{Rat}^{\mathcal{C}}(M)} : \mathrm{Rat}^{\mathcal{C}}(M) \to \mathrm{Rat}^{\mathcal{C}}(N). \end{aligned}$$

As in the case of coalgebras, $\mathrm{Rat}^{\mathcal{C}}(M) = M$ for $M \in {}_{^*\mathcal{C}}\mathbf{M}$ if and only if $M \in \mathbf{M}^{\mathcal{C}}$. The equality $\mathrm{Rat}^{\mathcal{C}}(M) = M$ holds for all left ${}^*\mathcal{C}$-modules M if and only if $\mathcal{C}$ is finitely generated as a left A-module (see 19.6 and 42.11(3)).

Let M be a left ${}^*\mathcal{C}$-module. Any $k \in M$ is called a *rational element* if there exists an element $\sum_i m_i \otimes c_i \in M \otimes_R \mathcal{C}$, such that (see 19.2)

$$fk = \sum_i m_i f(c_i), \text{ for all } f \in {}^*\mathcal{C}.$$

Since $\alpha_M : M \otimes_A \mathcal{C} \to \mathrm{Hom}_A({}^*\mathcal{C}, M)$ in 19.2 is assumed to be injective, such an element $\sum_i m_i \otimes c_i$ is uniquely determined.

20.2. Rational submodule. *Let M be a left ${}^*\mathcal{C}$-module.*

(1) *An element $k \in M$ is rational if and only if ${}^*\mathcal{C}k$ is a right $\mathcal{C}$-comodule with $fk = f \rightharpoonup k$, for all $f \in {}^*\mathcal{C}$.*

(2) $\mathrm{Rat}^{\mathcal{C}}(M) = \{k \in M \mid k$ *is rational*$\}$.

Proof. Replace C^* by ${}^*\mathcal{C}$ in the proof of 7.3. □

The rational submodule $\mathrm{Rat}^{\mathcal{C}}({}^*\mathcal{C})$ is a two-sided ideal in ${}^*\mathcal{C}$ and is called a *left trace ideal.* Clearly $\mathrm{Rat}^{\mathcal{C}}({}^*\mathcal{C}) = {}^*\mathcal{C}$ if and only if ${}_A\mathcal{C}$ is finitely generated (and hence projective by 42.11(3)).

Symmetrically, if $\mathcal{C}$ satifies the right α-condition, *right rational* $\mathcal{C}^*$*-modules* are defined, yielding the *right trace ideal* ${}^{\mathcal{C}}\mathrm{Rat}(\mathcal{C}^*)$.

Although some properties survive the generalisation from coalgebras to corings, not all characterisations of the trace ideal are preserved.

20.3. Properties of the left trace ideal. *Let* $T = \mathrm{Rat}^{\mathcal{C}}({}^*\mathcal{C})$.

(1) Let $f \in {}^*\mathcal{C}$ *and assume that* $f \rightharpoonup \mathcal{C}$ *is a finitely presented left A-module. Then* $f \in T$.

(2) For any $f \in T$*, the right comodule* ${}^*\mathcal{C} *^l f$ *is finitely generated as an A-module.*

Proof. (1) Assume the rational right $\mathcal{C}^*$-module $f \rightharpoonup \mathcal{C}$ to be a finitely presented left A-module. Then, by 19.19, ${}^*(f \rightharpoonup \mathcal{C})$ is a rational left ${}^*\mathcal{C}$-module. Since $\underline{\varepsilon}(f \rightharpoonup c) = f(c)$ for all $c \in \mathcal{C}$, $f \in {}^*(f \rightharpoonup \mathcal{C}) \subset {}^*\mathcal{C}$ and hence $f \in T$.

(2) This follows from the Finiteness Theorem 19.12. □

20.4. $\mathbf{M}^{\mathcal{C}}$ closed under extensions. *For* $\mathcal{C}$ *the following are equivalent:*

(a) $\mathbf{M}^{\mathcal{C}}$ *is closed under extensions in* ${}_{*\mathcal{C}}\mathbf{M}$*;*

(b) for every $X \in {}_{*\mathcal{C}}\mathbf{M}$, $\mathrm{Rat}^{\mathcal{C}}(X/\mathrm{Rat}^{\mathcal{C}}(X)) = 0$*;*

(c) there exists a ${}^*\mathcal{C}$*-injective* $Q \in {}_{*\mathcal{C}}\mathbf{M}$ *such that*

$$\mathbf{M}^{\mathcal{C}} = \{N \in {}_{*\mathcal{C}}\mathbf{M} \mid {}_{*\mathcal{C}}\mathrm{Hom}(N, Q) = 0\}.$$

Proof. The assertions follow from general module theory (see 42.14). □

20.5. $\mathbf{M}^{\mathcal{C}}$ closed under essential extensions. *For* $\mathcal{C}$ *the following assertions are equivalent:*

(a) $\mathbf{M}^{\mathcal{C}}$ *is closed under essential extensions in* ${}_{*\mathcal{C}}\mathbf{M}$*;*

(b) $\mathbf{M}^{\mathcal{C}}$ *is closed under injective hulls in* ${}_{*\mathcal{C}}\mathbf{M}$*;*

(c) every $\mathcal{C}$*-injective module in* $\mathbf{M}^{\mathcal{C}}$ *is* ${}^*\mathcal{C}$*-injective;*

(d) for every injective ${}^*\mathcal{C}$*-module* Q, $\mathrm{Rat}^{\mathcal{C}}(Q)$ *is a direct summand in* Q*;*

(e) for every injective ${}^*\mathcal{C}$*-module* Q, $\mathrm{Rat}^{\mathcal{C}}(Q)$ *is* ${}^*\mathcal{C}$*-injective.*

If $\mathbf{M}^{\mathcal{C}}$ *is closed under essential extensions, then* $\mathrm{Rat}^{\mathcal{C}}$ *is exact.*

Proof. This is an application of 42.20. □

For the next two propositions the α-condition on $\mathcal{C}$ is not required a priori.

20.6. Density in $^*\mathcal{C}$. *For a right A-submodule $U \subset {}^*\mathcal{C}$, the following statements are equivalent:*

(a) U is dense in $^*\mathcal{C}$ in the finite topology (of $A^{\mathcal{C}}$);

(b) U is a $\mathcal{C}$-dense subset of $^*\mathcal{C}$ (in the finite topology of $\mathrm{End}_R(\mathcal{C})$).

If ${}_A\mathcal{C}$ is cogenerated by A, then (a), (b) imply:

(c) $\mathrm{Ke}\, U = \{x \in \mathcal{C} \mid u(x) = 0 \text{ for all } u \in U\} = 0$.

If A is a cogenerator in ${}_A\mathbf{M}$ and $\mathbf{M}_A$, then (c) $\Rightarrow$ (b).

Proof. (a) $\Leftrightarrow$ (b) By 18.12, the finite topologies in $^*\mathcal{C}$ and $^{\mathcal{C}}\mathrm{End}(\mathcal{C})$ can be identified.

(a) $\Leftrightarrow$ (c) In view of the properties of $^*\mathcal{C}$-actions on $\mathcal{C}$ shown in 19.10(1), the proof of 7.9 can be adopted. □

20.7. Dense subalgebras of $^*\mathcal{C}$. *For a subring $T \subset {}^*\mathcal{C}$, the following are equivalent:*

(a) T is dense in $^*\mathcal{C}$ and $\mathcal{C}$ satisfies the left α-condition;

(b) $\mathbf{M}^{\mathcal{C}} = \sigma[{}_T\mathcal{C}]$.

If T is an ideal in $^*\mathcal{C}$, then (a) and (b) are equivalent to:

(c) $\mathcal{C}$ is an s-unital T-module and $\mathcal{C}$ satisfies the α-condition.

Proof. (a) $\Leftrightarrow$ (b) The inclusions $\mathbf{M}^{\mathcal{C}} \subset \sigma[{}_{^*\mathcal{C}}\mathcal{C}] \subset \sigma[{}_T\mathcal{C}]$ always hold. $\mathbf{M}^{\mathcal{C}} = \sigma[{}_{^*\mathcal{C}}\mathcal{C}]$ is equivalent to the α-condition (by 19.3), while $\sigma[{}_{^*\mathcal{C}}\mathcal{C}] = \sigma[{}_T\mathcal{C}]$ corresponds to the density property (see 42.2).

(a) $\Leftrightarrow$ (c) By 42.6, for an ideal $T \subset {}^*\mathcal{C}$ the density property is equivalent to the s-unitality of the T-module $\mathcal{C}$ (see 42.2). □

Now again the α-condition is assumed for $\mathcal{C}$. The properties of the trace functor observed in 42.16 imply:

20.8. The rational functor exact. *Let $T = \mathrm{Rat}^{\mathcal{C}}({}^*\mathcal{C})$. The following are equivalent for $\mathcal{C}$:*

(a) the functor $\mathrm{Rat}^{\mathcal{C}} : {}_{^*\mathcal{C}}\mathbf{M} \to \mathbf{M}^{\mathcal{C}}$ is exact;

(b) the category $\mathbf{M}^{\mathcal{C}}$ is closed under extensions in ${}_{^*\mathcal{C}}\mathbf{M}$ and the (torsionfree) class $\{X \in {}_{^*\mathcal{C}}\mathbf{M} \mid \mathrm{Rat}^{\mathcal{C}}(X) = 0\}$ is closed under factor modules;

(c) for every $N \in \mathbf{M}^{\mathcal{C}}$ (with $N \subset \mathcal{C}$), $TN = N$;

(d) for every $N \in \mathbf{M}^{\mathcal{C}}$, the canonical map $T \otimes_{^*\mathcal{C}} N \to N$ is an isomorphism;

(e) $\mathcal{C}$ is an s-unital left T-module;

(f) $T^2 = T$ and T is a generator in $\mathbf{M}^{\mathcal{C}}$;

(g) $T\mathcal{C} = \mathcal{C}$ and $^*\mathcal{C}/T$ is flat as a right $^*\mathcal{C}$-module;

(h) T is a $\mathcal{C}$-dense subring of $^*\mathcal{C}$.

Notice some consequences of the exactness of the Rat-functor from 42.17:

20.9. Corollary. *Assume* $\mathrm{Rat}^{\mathcal{C}}$ *to be exact and* $P \in \mathbf{M}^{\mathcal{C}}$. *Then:*

(1) $\mathbf{M}^{\mathcal{C}}$ *is closed under small epimorphisms in* ${}_{*\mathcal{C}}\mathbf{M}$.

(2) *If* P *is finitely presented in* $\mathbf{M}^{\mathcal{C}}$, *then* P *is finitely presented in* ${}_{*\mathcal{C}}\mathbf{M}$.

(3) *If* P *is projective in* $\mathbf{M}^{\mathcal{C}}$, *then* P *is projective in* ${}_{*\mathcal{C}}\mathbf{M}$.

20.10. Enough projectives in $\mathbf{M}^{\mathcal{C}}$. *Let* A *be right Noetherian and assume* ${}^*\mathcal{C}$ *to be f-semiperfect. Then the following are equivalent:*

(a) *the functor* $\mathrm{Rat}^{\mathcal{C}}$ *is exact;*

(b) $\mathbf{M}^{\mathcal{C}}$ *has a generator that is (locally) projective in* ${}_{*\mathcal{C}}\mathbf{M}$;

(c) *there are idempotents* $\{e_\lambda\}_\Lambda$ *in* ${}^*\mathcal{C}$ *such that the* ${}^*\mathcal{C} *^l e_\lambda$ *are in* $\mathbf{M}^{\mathcal{C}}$ *and form a generating set of* $\mathbf{M}^{\mathcal{C}}$.

Proof. (Compare 42.19) (a) $\Rightarrow$ (c) Let $M \in \mathbf{M}^{\mathcal{C}}$ be any simple comodule. M is finitely presented in $\mathbf{M}_A$ and hence in $\mathbf{M}^{\mathcal{C}}$. By 20.9(1), M is finitely presented in ${}_{*\mathcal{C}}\mathbf{M}$ and – since ${}^*\mathcal{C}$ is f-semiperfect – it has a projective cover P in ${}_{*\mathcal{C}}\mathbf{M}$ (see 41.18). By 20.9(1), $P \in \mathbf{M}^{\mathcal{C}}$ and clearly $P \simeq {}^*\mathcal{C} *^l e$ for some idempotent $e \in {}^*\mathcal{C}$. Now a representing set of simple comodules yields the family of idempotents required.

(c) $\Rightarrow$ (b) is obvious, and (b) $\Rightarrow$ (a) follows from 42.18. □

Notice that, by 19.17, over a QF ring A, for any A-coring $\mathcal{C}$ with ${}_A\mathcal{C}$ and $\mathcal{C}_A$ projective, the dual rings $\mathcal{C}^*$ and ${}^*\mathcal{C}$ are f-semiperfect. Hence combining 20.8 and 20.10, we obtain

20.11. Corollary. *Let* ${}_A\mathcal{C}$ *and* $\mathcal{C}_A$ *be projective and assume* A *to be a QF ring. Then the following are equivalent:*

(a) *the functor* $\mathrm{Rat}^{\mathcal{C}}$ *is exact;*

(b) *the left trace ideal* $\mathrm{Rat}^{\mathcal{C}}({}^*\mathcal{C})$ *is dense in* ${}^*\mathcal{C}$;

(c) *every simple comodule has a projective cover in* $\mathbf{M}^{\mathcal{C}}$;

(d) $\mathbf{M}^{\mathcal{C}}$ *has a generating set of finitely generated projectives.*

20.12. Trace ideal and decompositions. *Let* $\mathcal{C}$ *satisfy the left and right* α*-conditions and put* $T = \mathrm{Rat}^{\mathcal{C}}({}^*\mathcal{C})$. *If* $\mathcal{C}$ *is a direct sum of finitely generated right* $\mathcal{C}^*$*-modules (left* $\mathcal{C}$*-comodules), then there exists a set of orthogonal idempotents* $\{e_\lambda\}_\Lambda$ *in* T *such that*

$$\mathcal{C} = \bigoplus\nolimits_\Lambda e_\lambda \rightharpoonup \mathcal{C}.$$

In this case T *is* $\mathcal{C}$*-dense in* ${}^*\mathcal{C}$.

Proof. Any finitely generated direct summand $U \subset \mathcal{C}$ is equal to $e\rightharpoonup\mathcal{C}$ as a left comodule, for some idempotent e in ${}^{\mathcal{C}}\mathrm{End}(\mathcal{C}) = {}^*\mathcal{C}$. From 20.3 we deduce that $e \in T$. Now the assertion follows by the usual module decomposition of $\mathcal{C}$ as a $\mathcal{C}^*$-module.

Any $d \in \mathcal{C}$ is contained in a finite partial sum U of the decomposition, and U is finitely generated as a left A-module. Hence there exists some idempotent $e \in T$ with $e\rightharpoonup d = d$, showing that $\mathcal{C}$ is an s-unital T-module and hence 20.8 applies. □

20.13. Two-sided decompositions. *Assume $\mathcal{C}$ to satisfy the left and right α-conditions and put $T = \mathrm{Rat}^{\mathcal{C}}({}^*\mathcal{C})$.*

If $\mathcal{C}$ is a direct sum of $({}^*\mathcal{C}, \mathcal{C}^*)$-bimodules (subcorings) that are finitely generated as left A-modules, then

$$\mathcal{C} = \bigoplus_{\Lambda} e_\lambda \rightharpoonup \mathcal{C},$$

where $\{e_\lambda\}_\Lambda$ is a family of orthogonal central idempotents in T.

Proof. By 19.10, the given $({}^*\mathcal{C}, \mathcal{C}^*)$-bimodule decomposition of $\mathcal{C}$ can be described by central idempotents $e_\lambda \in {}^*\mathcal{C}$ and the $e_\lambda\rightharpoonup\mathcal{C}$ are subcorings (see 19.11). Moreover, the $e_\lambda\rightharpoonup\mathcal{C}$ are finitely generated and projective left A-modules, and hence $e_\lambda \in T$ by 20.3. □

References. Cuadra and Gómez-Torrecillas [102]; Gómez-Torrecillas and Năstăsescu [123]; Wisbauer [210].

21 Cotensor product over corings

In this section the cotensor product for comodules over coalgebras is extended to the cotensor product of comodules over corings. The corresponding properties related to this cotensor product of corings, such as the tensor-cotensor relations, coflatness, purity, and so on, are studied. It turns out that the cotensor product can be used to describe equivalences between comodule categories over corings, and thus we are led to the Morita-Takeuchi theory for corings in Section 23. The techniques are very similar to the coalgebra case but, obviously, some left-right symmetry is lost. Throughout, $\mathcal{C}$ denotes a coring over an R-algebra A.

21.1. Cotensor product of comodules. For $M \in \mathbf{M}^{\mathcal{C}}$ and $N \in {}^{\mathcal{C}}\mathbf{M}$, the *cotensor product* $M\Box_{\mathcal{C}}N$ is defined as the equaliser in $\mathbf{M}_R$,

$$M\Box_{\mathcal{C}}N \longrightarrow M \otimes_A N \underset{I_M\otimes {}^N\varrho}{\overset{\varrho^M\otimes I_N}{\rightrightarrows}} M \otimes_A \mathcal{C} \otimes_R N\,,$$

or, equivalently, by the following exact sequence of R-modules:

$$0 \longrightarrow M\Box_{\mathcal{C}}N \longrightarrow M \otimes_A N \xrightarrow{\omega_{M,N}} M \otimes_A \mathcal{C} \otimes_A N,$$

where $\omega_{M,N} = \varrho^M \otimes I_N - I_M \otimes {}^N\varrho$. It can also be characterised by the pullback diagram

$$\begin{array}{ccc} M\Box_{\mathcal{C}}N & \longrightarrow & M \otimes_A N \\ \downarrow & & \downarrow {\scriptstyle \varrho^M\otimes I_N} \\ M \otimes_A N & \xrightarrow{I_M\otimes {}^N\varrho} & M \otimes_A \mathcal{C} \otimes_A N\,. \end{array}$$

In particular, for the comodule $\mathcal{C}$, there are A-module isomorphisms

$$M\Box_{\mathcal{C}}\mathcal{C} = \varrho^M(M) \simeq M, \quad \mathcal{C}\Box_{\mathcal{C}}N = {}^N\varrho(N) \simeq N.$$

Proof. Apply the arguments of the proof of 10.1. It will be shown in 22.4 that these are in fact isomorphisms of comodules. □

The proof of 10.2 also yields the cotensor product of morphisms.

21.2. Cotensor product of comodule morphisms. *Consider morphisms $f : M \to M'$ in $\mathbf{M}^{\mathcal{C}}$ and $g : N \to N'$ in ${}^{\mathcal{C}}\mathbf{M}$. There exists a unique R-linear map*

$$f\Box g: \; M\Box_{\mathcal{C}}N \longrightarrow M'\Box_{\mathcal{C}}N',$$

yielding a commutative diagram (ω as in 21.1)

$$\begin{array}{ccccccc} 0 \longrightarrow & M\Box_{\mathcal{C}}N & \longrightarrow & M \otimes_A N & \xrightarrow{\omega_{M,N}} & M \otimes_A \mathcal{C} \otimes_A N \\ & \downarrow {\scriptstyle f\Box g} & & \downarrow {\scriptstyle f\otimes g} & & \downarrow {\scriptstyle f\otimes I_{\mathcal{C}}\otimes g} \\ 0 \longrightarrow & M'\Box_{\mathcal{C}}N' & \longrightarrow & M' \otimes_A N' & \xrightarrow{\omega_{M',N'}} & M' \otimes_A \mathcal{C} \otimes_A N'\,. \end{array}$$

As for coalgebras, the cotensor product over corings induces functors between comodule categories and $\mathbf{M}_R$. Thus, similarly to 10.3, 10.4 and 10.5, one can consider

21.3. The cotensor functor. *Any $M \in \mathbf{M}^{\mathcal{C}}$ induces a covariant functor*

$$M\Box_{\mathcal{C}}- : {}^{\mathcal{C}}\mathbf{M} \to \mathbf{M}_R, \qquad \begin{array}{rcl} N & \mapsto & M\Box_{\mathcal{C}}N, \\ f: N \to N' & \mapsto & I_M\Box f : M\Box_{\mathcal{C}}N \to M\Box_{\mathcal{C}}N'. \end{array}$$

(1) *Let ${}_A\mathcal{C}$ be flat, let $0 \longrightarrow N' \xrightarrow{f} N \xrightarrow{g} N''$ be an exact sequence in ${}^{\mathcal{C}}\mathbf{M}$, and assume*

(i) *M is flat as an A-module, or*

(ii) *the sequence is M-pure (in ${}_A\mathbf{M}$), or*

(iii) *the sequence is $(\mathcal{C}, A)$-exact.*

Then cotensoring with M yields an exact sequence of R-modules

$$0 \to M\Box_{\mathcal{C}}N' \xrightarrow{I_M\Box f} M\Box_{\mathcal{C}}N \xrightarrow{I_M\Box g} M\Box_{\mathcal{C}}N''.$$

(2) *The cotensor functor $M\Box_{\mathcal{C}}- : {}^{\mathcal{C}}\mathbf{M} \to {}_R\mathbf{M}$ is left $(\mathcal{C}, A)$-exact, and it is left exact provided that M is flat as a right A-module.*

(3) *For any direct family $\{N_\lambda\}_\Lambda$ in ${}^{\mathcal{C}}\mathbf{M}$,*

$$\varinjlim(M\Box_{\mathcal{C}}N_\lambda) \simeq M\Box_{\mathcal{C}} \varinjlim N_\lambda.$$

As for coalgebras, associativity properties between tensor and cotensor products are of fundamental importance. Here we have to distinguish between properties of left and right A-modules.

21.4. Tensor-cotensor relations. *Let A, S, T be R-algebras.*

(1) *Let $M \in {}_S\mathbf{M}^{\mathcal{C}}$, that is, M is an (S, A)-bimodule and the coaction $\varrho^M : M \to M \otimes_A \mathcal{C}$ is S-linear, and let $N \in {}^{\mathcal{C}}\mathbf{M}$. For any $W \in \mathbf{M}_S$, $W \otimes_S M$ has a canonical right $\mathcal{C}$-comodule structure (see 18.9(4)) and there exists a canonical R-linear map*

$$\tau_W : \quad W \otimes_S (M\Box_{\mathcal{C}}N) \to (W \otimes_S M)\Box_{\mathcal{C}}N.$$

The following are equivalent:

(a) *$\omega_{M,N} : M \otimes_A N \to M \otimes_A \mathcal{C} \otimes_A N$ (as in 21.1) is W-pure (in ${}_S\mathbf{M}$);*

(b) *τ_W is an isomorphism.*

(2) Let $M \in \mathbf{M}^{\mathcal{C}}$ and $N \in {}^{\mathcal{C}}\mathbf{M}_T$, that is, N is an (A,T)-bimodule and the left coaction ${}^N\varrho : N \to \mathcal{C} \otimes_A N$ is right T-linear. For any $V \in {}_T\mathbf{M}$, there exists a canonical R-linear map

$$\tau'_V : \ (M \Box_{\mathcal{C}} N) \otimes_T V \to M \Box_{\mathcal{C}} (N \otimes_T V).$$

The following are equivalent:

(a) $\omega_{M,N} : M \otimes_A N \to M \otimes_A \mathcal{C} \otimes_A N$ (as in 21.1) is V-pure (in $\mathbf{M}_T$);

(b) τ'_V is an isomorphism.

Proof. (1) With obvious maps there is the commutative diagram

$$\begin{array}{ccccccc} 0 \longrightarrow & W \otimes_S (M \Box_{\mathcal{C}} N) & \longrightarrow & W \otimes_S (M \otimes_A N) & \xrightarrow{I_W \otimes \omega_{M,N}} & W \otimes_S (M \otimes_A \mathcal{C} \otimes_A N) \\ & \downarrow \tau_W & & \downarrow \simeq & & \downarrow \simeq \\ 0 \longrightarrow & (W \otimes_S M) \Box_{\mathcal{C}} N & \longrightarrow & (W \otimes_S M) \otimes_A N & \xrightarrow{\omega_{W \otimes M, N}} & (W \otimes_S M) \otimes_A \mathcal{C} \otimes_A N, \end{array}$$

where the bottom row is exact (by definition). If $\omega_{M,N}$ is a W-pure morphism, then the top row is exact, implying that τ_W is an isomorphism. On the other hand, if τ_W is an isomorphism, then the exactness of the bottom row implies the exactness of the top row, showing that $\omega_{M,N}$ is a W-pure morphism.

(2) The arguments used in (1) also apply to τ'_V. □

We consider cases where the conditions in 21.4 are satisfied.

21.5. Purity conditions over corings. *Let A, S, T be R-algebras.*

(1) Let $M \in \mathbf{M}^{\mathcal{C}}$ and $N \in {}^{\mathcal{C}}\mathbf{M}_T$. If the functor $M \Box_{\mathcal{C}} -$ is right exact, then $\omega_{M,N}$ is a pure morphism in $\mathbf{M}_T$.

(2) Let $M \in {}_S\mathbf{M}^{\mathcal{C}}$ and $N \in {}^{\mathcal{C}}\mathbf{M}$. If the functor $- \Box_{\mathcal{C}} N$ is right exact, then $\omega_{M,N}$ is a pure morphism in ${}_S\mathbf{M}$.

(3) If $M \in \mathbf{M}^{\mathcal{C}}$ is $(\mathcal{C}, A)$-injective and $N \in {}^{\mathcal{C}}\mathbf{M}_T$, then the exact sequence

$$0 \longrightarrow M \Box_{\mathcal{C}} N \longrightarrow M \otimes_A N \xrightarrow{\omega_{M,N}} M \otimes_A \mathcal{C} \otimes_A N$$

splits in $\mathbf{M}_T$ (and hence is pure).

(4) If $N \in {}^{\mathcal{C}}\mathbf{M}$ is $(\mathcal{C}, A)$-injective and $M \in {}_S\mathbf{M}^{\mathcal{C}}$, then the exact sequence

$$0 \longrightarrow M \Box_{\mathcal{C}} N \longrightarrow M \otimes_A N \xrightarrow{\omega_{M,N}} M \otimes_A \mathcal{C} \otimes_A N$$

splits in ${}_S\mathbf{M}$ (and hence is pure).

Proof. (1) Let $M\Box_{\mathcal{C}}- : {}^{\mathcal{C}}\mathbf{M} \to \mathbf{M}_R$ be right exact and $V \in {}_T\mathbf{M}$. From an exact sequence $F_2 \to F_1 \to V \to 0$ with free left T-modules F_1, F_2, we obtain a commutative diagram

$$\begin{array}{ccccccc} (M\Box_{\mathcal{C}}N)\otimes_T F_2 & \longrightarrow & (M\Box_{\mathcal{C}}N)\otimes_T F_1 & \longrightarrow & (M\Box_{\mathcal{C}}N)\otimes_T V & \longrightarrow & 0 \\ \downarrow\simeq & & \downarrow\simeq & & \downarrow\tau'_V & & \\ M\Box_{\mathcal{C}}(N\otimes_T F_2) & \longrightarrow & M\Box_{\mathcal{C}}(N\otimes_T F_1) & \longrightarrow & M\Box_{\mathcal{C}}(N\otimes_T V) & \longrightarrow & 0, \end{array}$$

where both sequences are exact. The first two vertical maps are isomorphisms since tensor and cotensor functors commute with direct sums. This implies that τ'_V is an isomorphism and the assertion follows by 21.4.

(2) A similar proof applies for the case in which $-_{\mathcal{C}}\Box N$ is right exact.

(3) The proof is similar to the proof of (4) below.

(4) If N is $(\mathcal{C}, A)$-injective, the structure map ${}^N\varrho : N \to \mathcal{C}\otimes_A N$ is split by a left $\mathcal{C}$-comodule morphism $\lambda : \mathcal{C}\otimes_A N \to N$. Then – as in the proof of 10.7(2) – we see that the map

$$\beta = (I_M \otimes \lambda)\circ(\varrho^M \otimes I_N) : M\otimes_A N \to M\Box_{\mathcal{C}}N$$

is a retraction. Since $M \in {}_S\mathbf{M}^{\mathcal{C}}$, obviously β is S-linear and hence $M\Box_{\mathcal{C}}N$ is an S-direct summand of $M\otimes_A N$. For the remaining assertions we can also follow the proof of 10.7(2). □

21.6. Coflat comodules over corings. Let ${}_A\mathcal{C}$ be flat. A comodule $M \in \mathbf{M}^{\mathcal{C}}$ is said to be *coflat* if the functor $M\Box_{\mathcal{C}}- : {}^{\mathcal{C}}\mathbf{M} \to \mathbf{M}_R$ is exact. As in 10.8, it is easy to see that, for any coflat $M \in \mathbf{M}^{\mathcal{C}}$, M is flat as an A-module and that direct sums and direct limits of coflat $\mathcal{C}$-comodules are again coflat.

M is said to be *faithfully coflat* provided the functor $M\Box_{\mathcal{C}}- : {}^{\mathcal{C}}\mathbf{M} \to \mathbf{M}_R$ is exact and faithful. Faithfulness is equivalent to the requirement that the canonical map

$${}^{\mathcal{C}}\mathrm{Hom}(L, N) \to \mathrm{Hom}_R(M\Box_{\mathcal{C}}L, M\Box_{\mathcal{C}}N)$$

is injective, for any $L, N \in {}^{\mathcal{C}}\mathbf{M}$. The same arguments as in the proof of 10.9 can be used to derive the properties of

21.7. Faithfully coflat comodules over corings. *Let $\mathcal{C}$ be an A-coring with ${}_A\mathcal{C}$ flat. Then, for $M \in \mathbf{M}^{\mathcal{C}}$ the following are equivalent:*

(a) *M is faithfully coflat;*

(b) *$M\Box_{\mathcal{C}}- : {}^{\mathcal{C}}\mathbf{M} \to \mathbf{M}_R$ is exact and reflects exact sequences (zero morphisms);*

(c) *M is coflat and $M\Box_{\mathcal{C}}N \neq 0$, for any nonzero $N \in {}^{\mathcal{C}}\mathbf{M}$.*

Recall from 19.19 that, for $\mathcal{C}_A$ flat and any $N \in M^{\mathcal{C}}$ that is finitely presented as an A-module, $N^* = \mathrm{Hom}_A(N, A)$ is a left $\mathcal{C}$-comodule. Then an adaption of the proof of 10.11 yields

21.8. Hom-cotensor relation over corings. *Let $\mathcal{C}_A$ be flat and $M, L \in \mathbf{M}^{\mathcal{C}}$, such that M_A is flat and L_A is finitely presented. Then there exists a functorial isomorphism (natural in L)*

$$M\Box_{\mathcal{C}}L^* \xrightarrow{\simeq} \mathrm{Hom}^{\mathcal{C}}(L, M).$$

As for coalgebras, there is an interesting connection between

21.9. Coflatness and injectivity over corings. *For an A-coring $\mathcal{C}$, let $M \in \mathbf{M}^{\mathcal{C}}$ and assume that both M and $\mathcal{C}$ are flat as right A-modules.*

(1) Let $0 \to L_1 \to L_2 \to L_3 \to 0$ be an exact sequence in $\mathbf{M}^{\mathcal{C}}$, where each of the L_i is finitely presented in $\mathbf{M}_A$. If A is right injective, or the sequence is $(\mathcal{C}, A)$-exact, there is a commutative diagram

$$\begin{array}{ccccccccc} 0 & \longrightarrow & M\Box_{\mathcal{C}}L_3^* & \longrightarrow & M\Box_{\mathcal{C}}L_2^* & \longrightarrow & M\Box_{\mathcal{C}}L_1^* & \longrightarrow & 0 \\ & & \downarrow \simeq & & \downarrow \simeq & & \downarrow \simeq & & \\ 0 & \longrightarrow & \mathrm{Hom}^{\mathcal{C}}(L_3, M) & \longrightarrow & \mathrm{Hom}^{\mathcal{C}}(L_2, M) & \longrightarrow & \mathrm{Hom}^{\mathcal{C}}(L_1, M) & \longrightarrow & 0. \end{array}$$

So the upper sequence is exact if and only if the lower sequence is exact.

(2) Let A be a QF ring. Then

(i) M is coflat if and only if M is $\mathcal{C}$-injective;

(ii) M is faithfully coflat if and only if M is an injective cogenerator in $\mathbf{M}^{\mathcal{C}}$.

Proof. Since the L_i are finitely A-presented, thc isomorphisms are provided by 21.8. Now, the proof of 10.12, slightly modified, can be used. □

21.10. Exercises

(1) Consider a two-sided version of the canonical map in 19.2 as follows. For any $M \in {}_A\mathbf{M}_A$, $K \in \mathbf{M}_A$, and $N \in \mathbf{M}_A$, define an R-linear map

$$\alpha_{N,M,K} : N \otimes_A M \otimes_A K \to \mathrm{Hom}_R({}^*M^*, N \otimes_A K),\ n \otimes m \otimes k \mapsto [f \mapsto nf(m) \otimes k].$$

Prove:

(i) $\alpha_{N,M,K}$ is injective if and only if for any $u \in N \otimes_A M \otimes_A K$, the property $(I_N \otimes f \otimes I_K)(u) = 0$ for all $f \in {}^*M^*$, implies $u = 0$.

(ii) Assume that $M \in {}_A\mathbf{M}_A$ is a direct summand of some $A^{(\Lambda)}$ as an (A, A)-bimodule. Then $\alpha_{N,M,K}$ is injective, for any $N \in \mathbf{M}_A$ and $K \in {}_A\mathbf{M}$.

(2) Let $\mathcal{C}$ be an A-coring such that ${}_A\mathcal{C}_A$ is a direct summand of some $A^{(\Lambda)}$ as an (A,A)-bimodule. For any $M \in \mathbf{M}^{\mathcal{C}}$ and $N \in {}^{\mathcal{C}}\mathbf{M}$, consider $M \otimes_A N$ as a $({}^*\mathcal{C}^*, {}^*\mathcal{C}^*)$-bimodule. Prove that

$$M \Box_{\mathcal{C}} N \simeq {}_{^*\mathcal{C}^*}\mathrm{Hom}_{^*\mathcal{C}^*}({}^*\mathcal{C}^*, M \otimes_A N).$$

References. Al-Takhman [51]; Guzman [126, 127].

22 Bicomodules over corings

Similarly to coalgebras (cf. 11.1), given two A-corings one can study their bicomodules, that is, left comodules of one of the corings that are also right comodules of the other. In the case of corings, however, one allows for more freedom, by considering corings over different algebras.

Let A and B be R-algebras, $\mathcal{C}$ an A-coring, and $\mathcal{D}$ a B-coring.

22.1. Bicomodules. A (B,A)-bimodule M is called a $(\mathcal{D},\mathcal{C})$-*bicomodule* if M is a right $\mathcal{C}$-comodule and left $\mathcal{D}$-comodule with coactions

$$\varrho^M : M \to M \otimes_A \mathcal{C}, \quad {}^M\!\varrho : M \to \mathcal{D} \otimes_B M,$$

such that the diagram

$$\begin{array}{ccc} M & \xrightarrow{\varrho^M} & M \otimes_A \mathcal{C} \\ {\scriptstyle {}^M\!\varrho}\downarrow & & \downarrow{\scriptstyle {}^M\!\varrho \otimes I_{\mathcal{C}}} \\ \mathcal{D} \otimes_B M & \xrightarrow{I_{\mathcal{D}} \otimes \varrho^M} & \mathcal{D} \otimes_B M \otimes_A \mathcal{C} \end{array}$$

is commutative, that is, ϱ^M is a left $\mathcal{D}$-comodule morphism or, equivalently, ${}^M\!\varrho$ is a right $\mathcal{C}$-comodule morphism.

A morphism between two $(\mathcal{D},\mathcal{C})$-bicomodules $f : M \to N$ is an R-linear map that is both left $\mathcal{D}$-colinear and right $\mathcal{C}$-colinear. The category of $(\mathcal{D},\mathcal{C})$-bicomodules is denoted by ${}^{\mathcal{D}}\mathbf{M}^{\mathcal{C}}$.

22.2. Bicomodules and tensor products. For an R-algebra S, let ${}_S\mathbf{M}^{\mathcal{C}}$ denote the category whose objects are those right $\mathcal{C}$-comodules that are (S,A)-bimodules such that the coaction is an (S,A)-bimodule map. Morphisms are left S-linear right $\mathcal{C}$-comodule maps (see 39.2). Similarly, the category ${}^{\mathcal{D}}\mathbf{M}_S$ consists of those left $\mathcal{D}$-comodules that are (B,S)-bimodules such that the coaction is a (B,S)-bimodule map. Morphisms are right S-linear left $\mathcal{D}$-comodule maps. One easily proves the following extension of 18.9.

Propostion. *For any $M \in {}^{\mathcal{D}}\mathbf{M}_S$ and $N \in {}_S\mathbf{M}^{\mathcal{C}}$, $M \otimes_S N$ is a $(\mathcal{D},\mathcal{C})$-bicomodule with left coaction ${}^M\!\varrho \otimes I_N$ and right coaction $I_M \otimes \varrho^N$.*

Taking in turn $S = A$ and $S = B$, one concludes, in particular, that for all $M \in {}^{\mathcal{D}}\mathbf{M}_A$ and $N \in {}_B\mathbf{M}^{\mathcal{C}}$, $M \otimes_A \mathcal{C}$ and $\mathcal{D} \otimes_B N$ are $(\mathcal{D},\mathcal{C})$-bicomodules. In this context the morphisms between $(\mathcal{D},\mathcal{C})$-bicomodules can be viewed as defined by either of the exact sequences

$$0 \longrightarrow {}^{\mathcal{D}}\mathrm{Hom}^{\mathcal{C}}(L,M) \xrightarrow{\;i\;} {}_B\mathrm{Hom}^{\mathcal{C}}(L,M) \xrightarrow{\;\gamma_L\;} {}_B\mathrm{Hom}^{\mathcal{C}}(L,\mathcal{D} \otimes_B M),$$

where $\gamma_L(f) = {}^M\!\varrho \circ f - (I_{\mathcal{D}} \otimes f) \circ {}^L\!\varrho$, or

$$0 \longrightarrow {}^{\mathcal{D}}\mathrm{Hom}^{\mathcal{C}}(L,M) \xrightarrow{\;i\;} {}^{\mathcal{D}}\mathrm{Hom}_A(L,M) \xrightarrow{\;\gamma_R\;} {}^{\mathcal{D}}\mathrm{Hom}_A(L,M \otimes_A \mathcal{C}),$$

where $\gamma_R(g) = \varrho^M \circ g - (g \otimes I_{\mathcal{C}}) \circ \varrho^L$ (cf. 18.2).

We know that right $\mathcal{C}$-comodules are left ${}^*\mathcal{C} = {}_A\mathrm{Hom}(\mathcal{C}, A)$-modules, and left $\mathcal{D}$-comodules are right $\mathcal{D}^* = \mathrm{Hom}_B(\mathcal{D}, B)$-modules canonically. Hence any $(\mathcal{D}, \mathcal{C})$-bicomodule M is a left ${}^*\mathcal{C}$-module and a right $\mathcal{D}^*$-module, and the compatibility condition for bicomodules implies that M is in fact a $({}^*\mathcal{C}, \mathcal{D}^*)$-bimodule (cf. 11.1). Therefore there is a faithful functor ${}^{\mathcal{D}}\mathbf{M}^{\mathcal{C}} \to {}_{^*\mathcal{C}}\mathbf{M}_{\mathcal{D}^*}$.

In general, for $M \in \mathbf{M}^{\mathcal{C}}$ and $N \in {}^{\mathcal{C}}\mathbf{M}$, $M \Box_{\mathcal{C}} N$ is just an R-module. If M is a $(\mathcal{D}, \mathcal{C})$-bicomodule, the map

$$\omega_{M,N} : \varrho^M \otimes I_N - I_M \otimes {}^N\!\varrho : M \otimes_A N \to M \otimes_A \mathcal{C} \otimes_A N$$

is obviously a left $\mathcal{D}$-comodule morphism, and hence its kernel $M \Box_{\mathcal{C}} N$ is a $\mathcal{D}$-subcomodule of $M \otimes_A N$, provided $\omega_{M,N}$ is a $\mathcal{D}$-pure morphism in ${}_B\mathbf{M}$ (see 18.7, 40.13, 40.14). This implies:

22.3. Cotensor product of bicomodules over corings. *Let M be a $(\mathcal{D}, \mathcal{C})$-bicomodule, $L \in \mathbf{M}^{\mathcal{D}}$, and $N \in {}^{\mathcal{C}}\mathbf{M}$.*

(1) $M \Box_{\mathcal{C}} N$ is a left $\mathcal{D}$-comodule, provided that $\omega_{M,N}$ is $\mathcal{D}$-pure in ${}_B\mathbf{M}$.

(2) $L \Box_{\mathcal{D}} M$ is a right $\mathcal{C}$-comodule, provided that $\omega_{L,M}$ is $\mathcal{C}$-pure in $\mathbf{M}_A$.

(3) If N is a $(\mathcal{C}, \mathcal{D}')$-bicomodule for a B'-coring $\mathcal{D}'$, then $M \Box_{\mathcal{C}} N$ is a $(\mathcal{D}, \mathcal{D}')$-bicomodule, provided that $\omega_{M,N}$ is $\mathcal{D}$-pure in ${}_B\mathbf{M}$ and $\mathcal{D}'$-pure in $\mathbf{M}_{B'}$.

Notice that these conditions are in particular satisfied when ${}_A\mathcal{C}$, $\mathcal{D}_B$ and ${}_{B'}\mathcal{D}'$ are flat modules. Since $\mathcal{C}$ is $(\mathcal{C}, A)$-injective, the purity conditions are always satisfied for the $(\mathcal{C}, \mathcal{C})$-bicomodule $\mathcal{C}$ (see 21.5), thus yielding the following corollary of 22.3.

22.4. Cotensor product with $\mathcal{C}$. *For any $M \in \mathbf{M}^{\mathcal{C}}$ and $N \in {}^{\mathcal{C}}\mathbf{M}$, there are $\mathcal{C}$-comodule isomorphisms*

$$M \simeq M \Box_{\mathcal{C}} \mathcal{C}, \qquad N \simeq \mathcal{C} \Box_{\mathcal{C}} N.$$

22.5. Associativity of the cotensor product over corings. *Consider $M \in {}^{\mathcal{D}}\mathbf{M}^{\mathcal{C}}$, $L \in \mathbf{M}^{\mathcal{D}}$, and $N \in {}^{\mathcal{C}}\mathbf{M}$ such that the canonical maps yield the isomorphisms*

$$(L \Box_{\mathcal{D}} M) \otimes_A N' \simeq L \Box_{\mathcal{D}} (M \otimes_A N'), \quad \textit{for } N' = N, \mathcal{C}, \textit{ and } \mathcal{C} \otimes_A N,$$
$$L' \otimes_B (M \Box_{\mathcal{C}} N) \simeq (L' \otimes_B M) \Box_{\mathcal{C}} N, \quad \textit{for } L' = L, \mathcal{D}, \textit{ and } L \otimes_B \mathcal{D}.$$

Then $L \Box_{\mathcal{D}} M \in \mathbf{M}^{\mathcal{C}}$, $M \Box_{\mathcal{C}} N \in {}^{\mathcal{D}}\mathbf{M}$ and

$$(L \Box_{\mathcal{D}} M) \Box_{\mathcal{C}} N \simeq L \Box_{\mathcal{D}} (M \Box_{\mathcal{C}} N).$$

Proof. By 22.3, the conditions required imply that $L\Box_{\mathcal{D}}M$ and $M\Box_{\mathcal{C}}N$ are comodules. In the commutative diagram

$$\begin{array}{ccccccc}
0 \longrightarrow & (L\Box_{\mathcal{D}}M)\Box_{\mathcal{C}}N & \longrightarrow & (L\otimes_B M)\Box_{\mathcal{C}}N & \longrightarrow & (L\otimes_B D\otimes_B M)\Box_{\mathcal{C}}N \\
& \downarrow \psi_1 & & \downarrow \psi_2 & & \downarrow \psi_3 \\
0 \longrightarrow & L\Box_{\mathcal{D}}(M\Box_{\mathcal{C}}N) & \longrightarrow & L\otimes_B (M\Box_{\mathcal{C}}N) & \longrightarrow & L\otimes_B D\otimes_B (M\Box_{\mathcal{C}}N),
\end{array}$$

the top row is exact since $\omega_{L,M}$ is N-, $\mathcal{C}$-, and $\mathcal{C}\otimes_A N$-pure (see 21.4), and the bottom row is exact by definition of the cotensor product. The L- and $L\otimes_B\mathcal{D}$-purity of $\omega_{M,N}$ imply that ψ_2 and ψ_3 are isomorphisms, and so ψ_1 is an isomorphism. □

22.6. Proposition. *In the setup of 22.5, the canonical maps are isomorphisms provided that*

(i) L_B, $\mathcal{D}_B$, ${}_AN$ *and* ${}_A\mathcal{C}$ *are flat modules; or*

(ii) L *is coflat in* $\mathbf{M}^{\mathcal{D}}$ *and* $\mathcal{D}_B$ *is flat; or*

(iii) N *is coflat in* ${}^{\mathcal{C}}\mathbf{M}$ *and* ${}_A\mathcal{C}$ *is flat; or*

(iv) L *is* $(\mathcal{D},B)$*-injective and* N *is* $(\mathcal{C},A)$*-injective.*

Proof. It follows from 21.4 and 21.5 that each of the given sets of conditions implies the necessary isomorphisms. □

22.7. Cotensor product of coflat comodules. *Let* ${}_A\mathcal{C}$, $\mathcal{D}_B$ *be flat,* $L\in\mathbf{M}^{\mathcal{D}}$ *and* M *be a* $(\mathcal{D},\mathcal{C})$*-bicomodule. If* L *is* $\mathcal{D}$*-coflat and* M *is* $\mathcal{C}$*-coflat, then* $L\Box_{\mathcal{D}}M$ *is a coflat right* $\mathcal{C}$*-comodule.*

Proof. By the flatness conditions, $L\Box_{\mathcal{D}}M$ is a right $\mathcal{C}$-comodule and $M\Box_{\mathcal{C}}K$ is a left $\mathcal{D}$-comodule, for any $K\in{}^{\mathcal{C}}\mathbf{M}$ (see 22.3). Now, in view of 22.5, the proof of 11.7 can be used. □

For $(\mathcal{D},\mathcal{C})$-bicomodules one can study their properties relative to categories ${}_B\mathbf{M}^{\mathcal{C}}$ and ${}^{\mathcal{D}}\mathbf{M}_A$ (cf. 22.2), which are of significance for the cohomology of corings (cf. 30.3). In particular, a $(\mathcal{D},\mathcal{C})$-bicomodule M is called a $(B,\mathcal{C})$-*relative injective bicomodule* (resp. $(\mathcal{D},A)$-*relative injective bicomodule*) if, for every $(\mathcal{D},\mathcal{C})$-bicomodule map $i:N\to L$ that is a coretraction in ${}_B\mathbf{M}^{\mathcal{C}}$ (resp. in ${}^{\mathcal{D}}\mathbf{M}_A$), every diagram

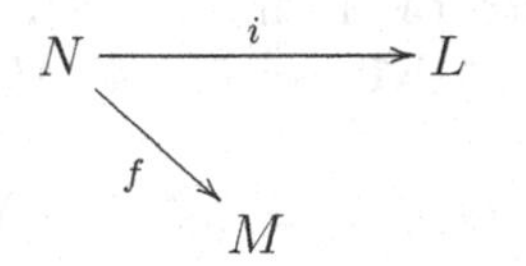

in ${}^{\mathcal{D}}\mathbf{M}^{\mathcal{C}}$ can be completed commutatively by some $g:L\to M$ in ${}^{\mathcal{D}}\mathbf{M}^{\mathcal{C}}$.

22.8. Relative injective bicomodules. *Take any $M, L \in {}^{\mathcal{D}}\mathbf{M}^{\mathcal{C}}$ and let $S = {}^{\mathcal{D}}\mathrm{End}^{\mathcal{C}}(L)$.*

(1) *The following are equivalent:*

(a) *M is a $(B, \mathcal{C})$-relative injective bicomodule;*

(b) *any morphism $i : M \to L$ in ${}^{\mathcal{D}}\mathbf{M}^{\mathcal{C}}$ that is a coretraction in ${}_B\mathbf{M}^{\mathcal{C}}$ is also a coretraction in ${}^{\mathcal{D}}\mathbf{M}^{\mathcal{C}}$;*

(c) *${}^{M}\varrho : M \to \mathcal{D} \otimes_B M$ is a coretraction in ${}^{\mathcal{D}}\mathbf{M}^{\mathcal{C}}$.*

If this holds, then the canonical sequence (cf. 22.2)

$$0 \longrightarrow {}^{\mathcal{D}}\mathrm{Hom}^{\mathcal{C}}(L, M) \xrightarrow{\ i\ } {}_B\mathrm{Hom}^{\mathcal{C}}(L, M) \xrightarrow{\ \gamma_L\ } {}_B\mathrm{Hom}^{\mathcal{C}}(L, \mathcal{D} \otimes_B M),$$

splits in $\mathbf{M}_S$.

(2) *The following are equivalent:*

(a) *M is a $(\mathcal{D}, A)$-relative injective bicomodule;*

(b) *any morphism $i : M \to L$ in ${}^{\mathcal{D}}\mathbf{M}^{\mathcal{C}}$ that is a coretraction in ${}^{\mathcal{D}}\mathbf{M}_A$ is also a coretraction in ${}^{\mathcal{D}}\mathbf{M}^{\mathcal{C}}$;*

(c) *$\varrho^M : M \to M \otimes_A \mathcal{C}$ is a coretraction in ${}^{\mathcal{D}}\mathbf{M}^{\mathcal{C}}$.*

If this holds, then the canonical sequence (cf. 22.2)

$$0 \longrightarrow {}^{\mathcal{D}}\mathrm{Hom}^{\mathcal{C}}(L, M) \xrightarrow{\ i\ } {}^{\mathcal{D}}\mathrm{Hom}_A(L, M) \xrightarrow{\ \gamma_R\ } {}^{\mathcal{D}}\mathrm{Hom}_A(L, M \otimes_A \mathcal{C})),$$

splits in $\mathbf{M}_S$.

(3) *For any $X \in {}_B\mathbf{M}^{\mathcal{C}}$ and $Y \in {}^{\mathcal{D}}\mathbf{M}_A$, $\mathcal{D} \otimes_B X$ is a $(B, \mathcal{C})$-relative injective bicomodule and $Y \otimes_A \mathcal{C}$ is a $(\mathcal{D}, A)$-relative injective bicomodule.*

Proof. This is proven by the same techniques as in 18.18 (cf. 3.18). More precisely, in the proof of the left (resp. right) comodule version of 18.18, all A-module maps can be replaced by morphisms in ${}_B\mathbf{M}^{\mathcal{C}}$ (resp. in ${}^{\mathcal{D}}\mathbf{M}_A$). □

For the remainder of this section let both $\mathcal{C}$ and $\mathcal{D}$ be A-corings. Take any $M \in {}^{\mathcal{D}}\mathbf{M}^{\mathcal{C}}$, $N \in \mathbf{M}^{\mathcal{C}}$ and consider the R-module $\mathrm{Hom}^{\mathcal{C}}(N, M)$. In case M is finitely A-presented, there is a Hom-cotensor relation $M \Box_{\mathcal{C}} N^* \simeq \mathrm{Hom}^{\mathcal{C}}(N, M)$ (see 21.8) and the left side is a left $\mathcal{D}$-comodule provided $\mathcal{D}_A$ is flat (see 22.3).

22.9. Comodule structure on $\mathrm{Hom}^{\mathcal{C}}(M, N)$. *For A-corings $\mathcal{C}$, $\mathcal{D}$, let $\mathcal{C}_A$ be flat, $M \in {}^{\mathcal{D}}\mathbf{M}^{\mathcal{C}}$ and $N \in \mathbf{M}^{\mathcal{C}}$ such that M_A is flat and N_A is finitely presented.*

If M is $(\mathcal{C}, A)$-injective, or if M is coflat in $\mathbf{M}^{\mathcal{C}}$, or if $\mathcal{D}_A$ is flat, then

$$M \Box_{\mathcal{C}} N^* \simeq \mathrm{Hom}^{\mathcal{C}}(N, M) \quad \text{in} \quad {}^{\mathcal{D}}\mathbf{M}.$$

Proof. As shown in 40.12, there are isomorphisms

$$\mathrm{Hom}_A(N, M) \simeq M \otimes_A N^* \text{ and } \mathrm{Hom}_A(N, M \otimes_A \mathcal{C}) \simeq M \otimes_A \mathcal{C} \otimes_A N^*,$$

where both modules are left $\mathcal{D}$-comodules (induced by M). With the help of 21.8 one can construct a commutative exact diagram as in the proof of 11.13 and, by 21.5, follow the arguments there. □

Special classes of bicomodules are derived from coring morphisms. We first consider the case of corings over the same algebra A. The general case, that is, that of corings over different algebras, is described in Section 24.

22.10. A-coring morphisms and comodules. Let $\gamma : \mathcal{C} \to \mathcal{D}$ be a morphism of A-corings, that is, there are commutative diagrams

$$\begin{array}{ccc} \mathcal{C} & \xrightarrow{\gamma} & \mathcal{D} \\ {\scriptstyle \Delta_\mathcal{C}}\downarrow & & \downarrow{\scriptstyle \Delta_\mathcal{D}} \\ \mathcal{C} \otimes_A \mathcal{C} & \xrightarrow{\gamma \otimes \gamma} & \mathcal{D} \otimes_A \mathcal{D}, \end{array} \qquad \begin{array}{ccc} \mathcal{C} & \xrightarrow{\gamma} & \mathcal{D} \\ {\scriptstyle \varepsilon_\mathcal{C}}\searrow & & \swarrow{\scriptstyle \varepsilon_\mathcal{D}} \\ & A. & \end{array}$$

Every right $\mathcal{C}$-comodule N is a right $\mathcal{D}$-comodule by

$$\varrho^N_\gamma = (I_N \otimes \gamma) \circ \varrho^N : N \to N \otimes_A \mathcal{C} \to N \otimes_A \mathcal{D},$$

and morphisms of right $\mathcal{C}$-comodules $f : N \to M$ are clearly morphisms of the induced $\mathcal{D}$-comodules. By symmetry, every left $\mathcal{C}$-comodule has a left $\mathcal{D}$-comodule structure, and $\mathcal{C}$ itself is a left and right $\mathcal{D}$-comodule and γ is a left and right $\mathcal{D}$-comodule morphism.

As in 11.8, we obtain that the composition of $\mathcal{D}$-colinear maps

$$N \xrightarrow{\varrho^N} N \Box_\mathcal{D} \mathcal{C} \xrightarrow{I_N \Box \gamma} N \Box_\mathcal{D} \mathcal{D} \simeq N$$

is the identity, and hence N is a direct summand of $N \Box_\mathcal{D} \mathcal{C}$ as a $\mathcal{D}$-comodule.

Coring morphisms induce functors between the comodule categories.

22.11. Coinduction and corestriction. Let $\gamma : \mathcal{C} \to \mathcal{D}$ be an A-coring morphism. The *corestriction functor* is defined by

$$(\ )_\gamma : \mathbf{M}^\mathcal{C} \to \mathbf{M}^\mathcal{D}, \quad (M, \varrho^M) \mapsto (M, (I_M \otimes \gamma) \circ \varrho^M)$$

(usually written as $(M)_\gamma = M$). Considering $\mathcal{C}$ as a $(\mathcal{C}, \mathcal{D})$-bicomodule (as above), $M \Box_\mathcal{C} \mathcal{C}$ is a right $\mathcal{D}$-comodule for any $M \in \mathbf{M}^\mathcal{C}$ (see 22.3) and the corestriction functor is isomorphic to $-\Box_\mathcal{C} \mathcal{C} : \mathbf{M}^\mathcal{C} \to \mathbf{M}^\mathcal{D}$. So it is left exact provided ${}_A\mathcal{C}$ is flat.

Symmetrically $\mathcal{C}$ is a $(\mathcal{D},\mathcal{C})$-bicomodule and for any $N \in \mathbf{M}^{\mathcal{D}}$ we can form $N\square_{\mathcal{D}}\mathcal{C}$. This is a right $\mathcal{C}$-comodule provided some pureness conditions hold (see 22.3). In this case there is a *coinduction functor*

$$-\square_{\mathcal{D}}\mathcal{C} : \mathbf{M}^{\mathcal{D}} \to \mathbf{M}^{\mathcal{C}}, \quad N \mapsto N\square_{\mathcal{D}}\mathcal{C},$$

where $N\square_{\mathcal{D}}\mathcal{C}$ is said to be *induced* by N. This functor is right exact provided $\mathcal{C}$ is coflat as a left $\mathcal{D}$-comodule.

The pureness conditions required are satified if, for example, ${}_A\mathcal{C}$ is flat. In this case the proof of 11.10 for coalgebras yields the adjointness of corestriction and coinduction.

22.12. Hom-cotensor relation for corings. *Let $\gamma : \mathcal{C} \to \mathcal{D}$ be an A-coring morphism and assume ${}_A\mathcal{C}$ to be flat. Then, for all $N \in \mathbf{M}^{\mathcal{C}}$ and $L \in {}^{\mathcal{D}}\mathbf{M}$, the map*

$$\mathrm{Hom}^{\mathcal{C}}(N, L\square_{\mathcal{D}}\mathcal{C}) \to \mathrm{Hom}^{\mathcal{D}}(N, L), \quad f \mapsto (I_L\square\gamma)\circ f,$$

is a functorial R-module isomorphism with inverse map $g \mapsto (g \otimes I_{\mathcal{C}})\circ \varrho^N$.

This is a special case of more general purity conditions to be considered in 24.8 and of an extended version of the adjointness isomorphism in 24.11.

Given a morphism of A-corings $\gamma : \mathcal{C} \to \mathcal{D}$, a short exact sequence in $\mathbf{M}^{\mathcal{C}}$ is called $(\mathcal{C},\mathcal{D})$-*exact* if it is splitting in $\mathbf{M}^{\mathcal{D}}$. A right $\mathcal{C}$-comodule N is called $(\mathcal{C},\mathcal{D})$-*injective* (resp. $(\mathcal{C},\mathcal{D})$-*projective*) if $\mathrm{Hom}^{\mathcal{C}}(-,N)$ (resp. $\mathrm{Hom}^{\mathcal{C}}(N,-)$) is exact with respect to $(\mathcal{C},\mathcal{D})$-exact sequences.

Notice that $\underline{\varepsilon} : \mathcal{C} \to A$ is an A-coring morphism, provided A is viewed as a trivial A-coring as in 17.3. Since the category of right A-comodules is isomorphic to $\mathbf{M}_A$ (cf. 18.5), the above definitions in particular yield $(\mathcal{C},A)$-exact sequences and $(\mathcal{C},A)$-injective comodules as considered in 18.18. The properties given there can be generalised by applying the isomorphism in the Hom-cotensor relation 22.12.

22.13. $(\mathcal{C},\mathcal{D})$-injectivity. *Let ${}_A\mathcal{C}$ be flat, $N \in \mathbf{M}^{\mathcal{C}}$, and $\gamma : \mathcal{C} \to \mathcal{D}$ an A-coring morphism.*

(1) If N is injective in $\mathbf{M}^{\mathcal{D}}$, then $N\square_{\mathcal{D}}\mathcal{C}$ is injective in $\mathbf{M}^{\mathcal{C}}$.

(2) The following are equivalent:

(a) N is $(\mathcal{C},\mathcal{D})$-injective;

(b) every $(\mathcal{C},\mathcal{D})$-exact sequence splits in $\mathbf{M}^{\mathcal{C}}$;

(c) the map $N \xrightarrow{\varrho^N} N\square_{\mathcal{D}}\mathcal{C}$ splits in $\mathbf{M}^{\mathcal{C}}$.

(3) If N is injective in $\mathbf{M}^{\mathcal{D}}$ and $(\mathcal{C},\mathcal{D})$-injective, then N is injective in $\mathbf{M}^{\mathcal{C}}$.

Proof. With the formalism in 22.10, the proof of 3.18 can be followed. □

$(\mathcal{C}, \mathcal{D})$-injectivity plays an important role in the description of coseparable corings. This is discussed in more detail in Section 26.

References. Gómez-Torrecillas [122]; Guzman [126, 127].

23 Functors between comodule categories

Given two corings over two different rings, it is natural to study relationships between the corresponding categories of comodules. For example, one would like to know when such categories are equivalent to each other, and what properties of corings can be derived from the equivalence of categories of their comodules. This leads, in particular, to the Morita-Takeuchi theory for corings.

For R-algebras A, B, let $\mathcal{C}$ be an A-coring and $\mathcal{D}$ a B-coring.

23.1. Functors between comodule categories over corings. *Consider an additive functor $F : \mathbf{M}^{\mathcal{C}} \to \mathbf{M}^{\mathcal{D}}$ that preserves colimits. Let ${}_B\mathbf{M}^{\mathcal{C}}$ denote the category of (B, A)-bimodules that are also right $\mathcal{C}$-comodules such that the coaction is left B-linear.*

(1) $F(\mathcal{C})$ is a $(\mathcal{C}, \mathcal{D})$-bicomodule and there exists a functorial isomorphism $\nu : -\Box_{\mathcal{C}} F(\mathcal{C}) \to F$.

(2) For any $W \in \mathbf{M}_B$ and $N \in {}_B\mathbf{M}^{\mathcal{C}}$,

$$W \otimes_B (N \Box_{\mathcal{C}} F(\mathcal{C})) \simeq (W \otimes_B N) \Box_{\mathcal{C}} F(\mathcal{C}).$$

(3) Let ${}_A\mathcal{C}$ and ${}_A\mathcal{D}$ be flat and assume F to preserve kernels. Then $F(\mathcal{C})$ is coflat as a left $\mathcal{C}$-comodule, and for all $W \in \mathbf{M}^{\mathcal{D}}$ and $N \in {}^{\mathcal{D}}\mathbf{M}^{\mathcal{C}}$,

$$W \Box_{\mathcal{D}} (N \Box_{\mathcal{C}} F(\mathcal{C})) \simeq (W \Box_{\mathcal{D}} N) \Box_{\mathcal{C}} F(\mathcal{C}).$$

Proof. (1) By 39.3, there is a functorial isomorphism

$$\Psi : - \otimes_B F(-) \to F(- \otimes_B -) \quad \text{of functors} \quad \mathbf{M}_B \times {}_B\mathbf{M}^{\mathcal{C}} \to \mathbf{M}^{\mathcal{D}}.$$

Moreover, by 39.7, for the $(\mathcal{C}, \mathcal{C})$-bicomodule $\mathcal{C}$, $F(\mathcal{C})$ is a left $\mathcal{C}$-comodule by the coaction

$${}^{F(\mathcal{C})}\varrho \;:\; F(\mathcal{C}) \xrightarrow{F(\Delta_{\mathcal{C}})} F(\mathcal{C} \otimes_A \mathcal{C}) \xrightarrow{\Psi^{-1}_{\mathcal{C},\mathcal{C}}} \mathcal{C} \otimes_A F(\mathcal{C}).$$

For any $M \in \mathbf{M}^{\mathcal{C}}$, the defining equalisers for the cotensor product give the following commutative diagram:

$$\begin{array}{ccccc}
M \Box_{\mathcal{C}} F(\mathcal{C}) & \longrightarrow & M \otimes_A F(\mathcal{C}) & \overset{\varrho^M \otimes I_{F(\mathcal{C})}}{\underset{I_M \otimes {}^{F(\mathcal{C})}\varrho}{\rightrightarrows}} & M \otimes_A \mathcal{C} \otimes_A F(\mathcal{C}) \\
 & & \downarrow{\scriptstyle \Psi_{M,\mathcal{C}}} & & \downarrow{\scriptstyle \Psi_{M \otimes \mathcal{C},\mathcal{C}}} \\
F(M) & \longrightarrow & F(M \otimes_A \mathcal{C}) & \overset{F(\varrho^M \otimes I_{\mathcal{C}})}{\underset{F(I_M \otimes \underline{\Delta})}{\rightrightarrows}} & F(M \otimes_A \mathcal{C} \otimes_A \mathcal{C}),
\end{array}$$

where the top sequence is an equaliser by the definition of a cotensor product, and the bottom sequence is an equaliser since F preserves kernels. From this we derive the functorial isomorphism $\nu_M : M\Box_{\mathcal{C}}F(C) \to F(M)$.

(2) This follows from the isomorphisms (tensors over B)

$$(W\otimes M)\Box_{\mathcal{C}}F(\mathcal{C}) \xrightarrow{\nu_{W\otimes M}} F(W\otimes M) \xrightarrow{\Psi^{-1}_{W,M}} W\otimes F(M) \xrightarrow{I\otimes\nu_M^{-1}} W\otimes(M\Box_{\mathcal{C}}F(\mathcal{C}))\,.$$

(3) Both $-\otimes_A \mathcal{C}$ and F are left exact functors, and thus so is their composition $F(-\otimes_A \mathcal{C}) \simeq -\otimes_A F(\mathcal{C})$, that is, $F(\mathcal{C})$ is a flat left A-module. Since F preserves epimorphisms, so does $-\Box_{\mathcal{C}}F(\mathcal{C})$, and hence $F(\mathcal{C})$ is coflat as a left $\mathcal{C}$-comodule and associativity of the cotensor products follows from 22.5 $\square$

23.2. Adjoint functors between comodule categories. *Let ${}_A\mathcal{C}$ and ${}_B\mathcal{D}$ be flat. Let (F,G) be an adjoint pair of additive functors, $F : \mathbf{M}^{\mathcal{C}} \to \mathbf{M}^{\mathcal{D}}$ and $G : \mathbf{M}^{\mathcal{D}} \to \mathbf{M}^{\mathcal{C}}$, with unit $\eta : I_{\mathbf{M}^{\mathcal{C}}} \to GF$ and counit $\psi : FG \to I_{\mathbf{M}^{\mathcal{D}}}$. Suppose that F preserves kernels and G preserves colimits. Then:*

(1) $F(\mathcal{C})$ is a $(\mathcal{C},\mathcal{D})$-bicomodule and there exists a functorial isomorphism $\nu : -\Box_{\mathcal{C}}F(\mathcal{C}) \to F$.

(2) $G(\mathcal{D})$ is a $(\mathcal{D},\mathcal{C})$-bicomodule and there exists a functorial isomorphism $\mu : -\Box_{\mathcal{D}}G(\mathcal{D}) \to G$.

(3) For any $M \in \mathbf{M}^{\mathcal{C}}$ and $N \in \mathbf{M}^{\mathcal{D}}$,

$$(M\Box_{\mathcal{C}}F(\mathcal{C}))\Box_{\mathcal{D}}G(\mathcal{D}) \simeq M\Box_{\mathcal{C}}(F(\mathcal{C})\Box_{\mathcal{D}}G(\mathcal{D})) \text{ and}$$
$$(N\Box_{\mathcal{D}}G(\mathcal{D}))\Box_{\mathcal{C}}F(\mathcal{C})) \simeq N\Box_{\mathcal{D}}(G(\mathcal{D})\Box_{\mathcal{C}}F(\mathcal{C})).$$

(4) There exist $(\mathcal{C},\mathcal{C})$-, resp. $(\mathcal{D},\mathcal{D})$-, bicomodule morphisms

$$\eta_{\mathcal{C}} : \mathcal{C} \to F(\mathcal{C})\Box_{\mathcal{D}}G(\mathcal{D}), \quad \psi_{\mathcal{D}} : G(\mathcal{D})\Box_{\mathcal{C}}F(\mathcal{C}) \to \mathcal{D},$$

such that the following compositions yield identities:

$$F(\mathcal{C}) \simeq \mathcal{C}\Box_{\mathcal{C}}F(\mathcal{C}) \xrightarrow{\eta_{\mathcal{C}}\Box I} F(\mathcal{C})\Box_{\mathcal{D}}G(\mathcal{D})\Box_{\mathcal{C}}F(\mathcal{C}) \xrightarrow{\psi_{F(\mathcal{C})}} F(\mathcal{C}),$$
$$G(\mathcal{D}) \xrightarrow{\eta_{G(\mathcal{D})}} G(\mathcal{D})\Box_{\mathcal{C}}F(\mathcal{C})\Box_{\mathcal{D}}G(\mathcal{D}) \xrightarrow{\psi_{\mathcal{D}}\Box I} D\Box_{\mathcal{D}}G(\mathcal{D}) \simeq G(\mathcal{D})\,.$$

(5) There is an adjoint pair of functors (G',F') with

$$G' = G(\mathcal{D})\Box_{\mathcal{C}}- : {}^{\mathcal{C}}\mathbf{M} \to {}^{\mathcal{D}}\mathbf{M} \quad \text{and } F' = F(\mathcal{C})\Box_{\mathcal{D}}- : {}^{\mathcal{D}}\mathbf{M} \to {}^{\mathcal{C}}\mathbf{M}\,.$$

Proof. (1),(2) Under the given conditions both F and G preserve kernels and colimits, and hence the assertions follow from 23.1.

(3) The two isomorphisms follow from 23.1.

(4) Notice that $\mathcal{C}$ is a $(\mathcal{C},\mathcal{C})$-bicomodule and GF preserves colimits. Under these conditions, 39.7 implies that $\eta_{\mathcal{C}}$ is a $(\mathcal{C},\mathcal{C})$-bicomodule morphism. Similar arguments apply to $\mathcal{D}$ and $\psi_{\mathcal{D}}$. The properties of the compositions are given in 38.21.

(5) This follows from the isomorphisms in (4). □

Notice that the conditions on the pair of functors (F, G) in 23.2 hold in particular when they form a Frobenius pair or if they induce an equivalence.

23.3. Equivalence between comodule categories over corings (1). *If ${}_A\mathcal{C}$ and ${}_B\mathcal{D}$ are flat, the following are equivalent:*

(a) there are functors $F : \mathbf{M}^{\mathcal{C}} \to \mathbf{M}^{\mathcal{D}}$ and $G : \mathbf{M}^{\mathcal{D}} \to \mathbf{M}^{\mathcal{C}}$ that establish an equivalence;

(b) there exist a $(\mathcal{C},\mathcal{D})$-bicomodule X, a $(\mathcal{D},\mathcal{C})$-bicomodule Y, and bicomodule isomorphisms $\gamma : \mathcal{D} \to Y\Box_{\mathcal{C}}X$ and $\delta : \mathcal{C} \to X\Box_{\mathcal{D}}Y$, such that

$$(I_Y\Box_{\mathcal{C}}\delta)\circ\varrho^Y = (\gamma\Box_{\mathcal{D}}I_Y)\circ{}^Y\varrho,\quad (\delta\Box_{\mathcal{D}}I_X)\circ{}^X\varrho = (I_X\Box_{\mathcal{D}}\gamma)\circ\varrho^X,$$

and (i) X is a coflat left $\mathcal{C}$-comodule, and Y is a coflat left $\mathcal{D}$-comodule, or (ii) ${}_AX$ and ${}_BY$ are flat modules, and the following pairs of morphisms are pure in ${}_B\mathbf{M}$ and ${}_A\mathbf{M}$, respectively,

$$Y\otimes_A X \underset{I_Y\otimes{}^X\varrho}{\overset{\varrho^Y\otimes I_X}{\rightrightarrows}} Y\otimes_A\mathcal{C}\otimes_A X,\quad X\otimes_B Y \underset{I_X\otimes{}^Y\varrho}{\overset{\varrho^X\otimes I_Y}{\rightrightarrows}} X\otimes_B D\otimes_B Y.$$

Proof. (a) ⇒ (b) By 23.1 and 23.2, the comodules $X = F(\mathcal{C})$ and $Y = G(\mathcal{D})$ have the properties required, and hence maps with the desired properties exist.

(b) ⇒ (a) The given conditions imply that $Y\Box_{\mathcal{C}}X$ is a $(\mathcal{C},\mathcal{D})$-bicomodule and $X\Box_{\mathcal{D}}Y$ is a $(\mathcal{D},\mathcal{C})$-bicomodule, and so there are functors (see 22.3)

$$-\Box_{\mathcal{C}}X : \mathbf{M}^{\mathcal{C}} \to \mathbf{M}^{\mathcal{D}}, \quad \text{and} \quad -\Box_{\mathcal{D}}Y : \mathbf{M}^{\mathcal{D}} \to \mathbf{M}^{\mathcal{C}}.$$

Furthermore, for any $M \in \mathbf{M}^{\mathcal{C}}$, $N \in \mathbf{M}^{\mathcal{D}}$, there are associativity relations

$$(M\Box_{\mathcal{C}}X)\Box_{\mathcal{D}}Y \simeq M\Box_{\mathcal{C}}(X\Box_{\mathcal{D}}Y),\quad (N\Box_{\mathcal{D}}Y)\Box_{\mathcal{C}}X \simeq N\Box_{\mathcal{D}}(Y\Box_{\mathcal{C}}X),$$

which imply functorial isomorphisms,

$$M\Box_{\mathcal{C}}X\Box_{\mathcal{D}}Y \simeq M\Box_{\mathcal{C}}\mathcal{C} \simeq M \ \text{ and } \ N\Box_{\mathcal{D}}Y\Box_{\mathcal{C}}X \simeq N\Box_{\mathcal{D}}D \simeq N,$$

thus proving that X and Y induce an equivalence. □

From 23.3 we know that functors describing equivalences between comodule categories are essentially cotensor functors, and we would like to know which properties of the comodules involved characterise equivalences.

23.4. Quasi-finite comodules. For an R-algebra B, let ${}_B\mathbf{M}^{\mathcal{C}}$ denote the category of (B, A)-bimodules that are also right $\mathcal{C}$-comodules such that the coaction is left B-linear. A comodule $Y \in {}_B\mathbf{M}^{\mathcal{C}}$ is called *$(B, \mathcal{C})$-quasi-finite* if the tensor functor $- \otimes_B Y : \mathbf{M}_B \to \mathbf{M}^{\mathcal{C}}$ has a left adjoint. This left adjoint is called the *Cohom functor* and is denoted by $h_{\mathcal{C}}(Y, -) : \mathbf{M}^{\mathcal{C}} \to \mathbf{M}_B$. Explicitly, this means that, for all $M \in \mathbf{M}^{\mathcal{C}}$ and $W \in \mathbf{M}_B$, there exists a functorial isomorphism

$$\Phi_{M,W} : \mathrm{Hom}_B(h_{\mathcal{C}}(Y, M), W) \to \mathrm{Hom}^{\mathcal{C}}(M, W \otimes_B Y).$$

If ${}_A\mathcal{C}$ is flat, then any quasi-finite comodule $Y \in {}_B\mathbf{M}^{\mathcal{C}}$ is flat as a left B-module (right adjoints respect monomorphisms). Since $h_{\mathcal{C}}(Y, -)$ is a left adjoint functor, it respects colimits (see 38.21), and, by 39.3, this implies a functorial isomorphism,

$$\Psi_{W,M} : W \otimes_B h_{\mathcal{C}}(Y, M) \simeq h_{\mathcal{C}}(Y, W \otimes_B M).$$

Moreover, if M is a $(\mathcal{D}, \mathcal{C})$-bicomodule, then $h_{\mathcal{C}}(Y, M)$ has a left $\mathcal{D}$-comodule structure,

$${}^{h_{\mathcal{C}}(Y,M)}\varrho = h_{\mathcal{C}}(Y, {}^M\varrho) : h_{\mathcal{C}}(Y, M) \to h_{\mathcal{C}}(Y, \mathcal{D} \otimes_B M) \simeq \mathcal{D} \otimes_B h_{\mathcal{C}}(Y, M),$$

such that the unit of the adjunction $\eta_M : M \to h_{\mathcal{C}}(Y, M) \otimes_B Y$ is a $(\mathcal{D}, \mathcal{C})$-bicomodule morphism (see 39.7).

By the Hom-tensor relations 18.10, $- \otimes_A \mathcal{C} : \mathbf{M}_A \to \mathbf{M}^{\mathcal{C}}$ is the right adjoint to the forgetful functor $\mathbf{M}^{\mathcal{C}} \to \mathbf{M}_A$, and hence $\mathcal{C}$ is an $(A, \mathcal{C})$-quasi-finite comodule and $h_{\mathcal{C}}(\mathcal{C}, -)$ is simply the forgetful functor. For any $(B, \mathcal{C})$-quasi-finite comodule Y and $n \in \mathbb{N}$, an isomorphism $\mathrm{Hom}_B(h_{\mathcal{C}}(Y, -), -) \to \mathrm{Hom}^{\mathcal{C}}(-, - \otimes_B Y)$ implies an isomorphism,

$$\mathrm{Hom}_R(h_{\mathcal{C}}(Y^n, -), -) \to \mathrm{Hom}^{\mathcal{C}}(-, - \otimes_B Y^n),$$

showing that Y^n is again a quasi-finite $\mathcal{C}$-comodule. Similarly it can be shown that $(B, \mathcal{C})$-direct summands of Y are again $(B, \mathcal{C})$-quasi-finite.

23.5. Quasi-finite bicomodules. *Let $Y \in {}^{\mathcal{D}}\mathbf{M}^{\mathcal{C}}$ be $(B, \mathcal{C})$-quasi-finite and denote by $\eta : I_{\mathbf{M}_B} \to h_{\mathcal{C}}(Y, -) \otimes_B Y$ the unit of the adjunction. There exists a unique $\mathcal{D}$-comodule structure map $\varrho^{h_{\mathcal{C}}(Y,M)} : h_{\mathcal{C}}(Y, M) \to h_{\mathcal{C}}(Y, M) \otimes_B \mathcal{D}$, with $(I_{h_{\mathcal{C}}(Y,M)} \otimes {}^Y\varrho) \circ \eta_M = (\varrho^{h_{\mathcal{C}}(Y,M)} \otimes I_Y) \circ \eta_M$ and $\mathrm{Im}\, \eta_M \subset h_{\mathcal{C}}(Y, M) \Box_{\mathcal{D}} Y$. This yields a functor,*

$$h_{\mathcal{C}}(Y, -) : \mathbf{M}^{\mathcal{C}} \longrightarrow \mathbf{M}^{\mathcal{D}}.$$

Proof. To shorten notation write I_h for $I_{h_{\mathcal{C}}(Y,M)}$. For the $\mathcal{C}$-colinear map

$$(I_h \otimes {}^Y\varrho) \circ \eta_M : M \longrightarrow h_{\mathcal{C}}(Y, M) \otimes_B \mathcal{D} \otimes_B Y,$$

there exists a unique right B-linear map $\varrho^{h_C(Y,M)} : h_C(Y,M) \to h_C(Y,M) \otimes_B \mathcal{D}$ such that $(I_h \otimes {}^Y\varrho) \circ \eta_M = (\varrho^{h_C(Y,M)} \otimes I_Y) \circ \eta_M$ (preimage of the given map under Φ). It is straightforward to prove that this coaction makes $h_C(Y,M)$ a right $\mathcal{D}$-comodule. The last equality means that Im η_M lies in the equaliser

$$h_C(Y,M) \otimes_B Y \overset{\varrho^{h_C(Y,M)} \otimes I_Y}{\underset{I_h \otimes {}^Y\varrho}{\rightrightarrows}} h_C(Y,M) \otimes_B \mathcal{D} \otimes_B Y,$$

that is, Im $\eta_M \subset h_C(Y,M) \Box_{\mathcal{D}} Y$. As in the proof of 12.6, assign to any morphism $f : M \to M'$ in $\mathbf{M}^C$ a $\mathcal{D}$-colinear map $h_C(Y,f) : h_C(Y,M) \to h_C(Y,M')$ to obtain the required functor. □

23.6. Cotensors with left adjoints. *Let ${}_A\mathcal{C}$ be flat. Then, for $Y \in {}^{\mathcal{D}}\mathbf{M}^{\mathcal{C}}$, the following are equivalent:*

(a) $- \otimes_B Y : \mathbf{M}_B \to \mathbf{M}^{\mathcal{C}}$ has a left adjoint (that is, Y is quasi-finite);

(b) $-\Box_{\mathcal{D}} Y : \mathbf{M}^{\mathcal{D}} \to \mathbf{M}^{\mathcal{C}}$ has a left adjoint.

Proof. (b) ⇒ (a) By the isomorphism $- \otimes_B Y \simeq (- \otimes_B D)\Box_{\mathcal{D}} Y$, the functor $- \otimes_B Y$ is composed by the functors $- \otimes_B \mathcal{D} : \mathbf{M}_B \to \mathbf{M}^{\mathcal{D}}$ and $-\Box_{\mathcal{D}} Y : \mathbf{M}^{\mathcal{D}} \to \mathbf{M}^{\mathcal{C}}$. We know that $-\otimes_B \mathcal{D}$ always has a left adjoint (forgetful functor). Hence our assertion follows from the fact that the composition of functors with left adjoints also has a left adjoint.

(a) ⇒ (b) Let Y be $(B,\mathcal{C})$-quasi-finite with left adjoint $h_C(Y,-)$ and denote by $\eta : I_{\mathbf{M}_B} \to h_C(Y,-) \otimes_B Y$ the unit of the adjunction. By 23.5, there is a functor $h_C(Y,-) : \mathbf{M}^{\mathcal{C}} \to \mathbf{M}^{\mathcal{D}}$, and, for any $M \in \mathbf{M}^{\mathcal{C}}$, Im $\eta_M \subset h_C(Y,M)\Box_{\mathcal{D}} Y$. Similarly as in the proof of 12.7, we can see that this functor is left adjoint to $-\Box_{\mathcal{D}} Y$. □

23.7. Exactness of the Cohom functor. *Let ${}_A\mathcal{C}$ and ${}_B\mathcal{D}$ be flat, and consider some $Y \in {}^{\mathcal{D}}\mathbf{M}^{\mathcal{C}}$ that is $(B,\mathcal{C})$-quasi-finite.*

(1) The following are equivalent:

(a) $h_C(Y,-) : \mathbf{M}^{\mathcal{C}} \to \mathbf{M}_B$ is exact;

(b) $W \otimes_B Y$ is injective in $\mathbf{M}^{\mathcal{C}}$, for every injective right B-module W;

(c) $h_C(Y,-) : \mathbf{M}^{\mathcal{C}} \to \mathbf{M}^{\mathcal{D}}$ is exact;

(d) $N\Box_{\mathcal{D}} Y$ is injective in $\mathbf{M}^{\mathcal{C}}$, for every injective comodule $N \in \mathbf{M}^{\mathcal{D}}$.

If this holds, then $h_C(Y,-) \simeq -\Box_{\mathcal{C}} h_C(Y,C)$ as functors $\mathbf{M}^{\mathcal{C}} \to \mathbf{M}_B$ (or $\mathbf{M}^{\mathcal{C}} \to \mathbf{M}^{\mathcal{D}}$) and hence $h_C(Y,C)$ is a coflat left $\mathcal{C}$-comodule.

(2) If $h_C(Y,-)$ is exact, the following are equivalent:

(a) $h_C(Y,-)$ is faithful;

(b) $h_C(Y,\mathcal{C})$ is a faithfully coflat left $\mathcal{C}$-comodule.

Proof. (1) (a) $\Rightarrow$ (b) and (c) $\Rightarrow$ (d) follow from the adjointness isomorphisms

$$\mathrm{Hom}_B(h_{\mathcal{C}}(Y,-),W) \simeq \mathrm{Hom}^{\mathcal{C}}(-,W \otimes_B Y),$$
$$\mathrm{Hom}^{\mathcal{D}}(h_{\mathcal{C}}(Y,-),N) \simeq \mathrm{Hom}^{\mathcal{C}}(-,N \Box_{\mathcal{D}} Y),$$

since the exactness of $h_{\mathcal{C}}(Y,-)$ and $\mathrm{Hom}_B(-,W)$ (resp. $\mathrm{Hom}^{\mathcal{D}}(-,N)$) implies the exactness of their composition, which means the injectivity of $W \otimes_B Y$ (resp. $N \Box_{\mathcal{D}} Y$) in $\mathbf{M}^{\mathcal{C}}$.

(a) $\Leftrightarrow$ (c) Any sequence of morphisms in $\mathbf{M}^{\mathcal{D}}$ is exact if and only if it is exact in $\mathbf{M}_B$.

(b) $\Rightarrow$ (a), (d) $\Rightarrow$ (c) We can transfer the corresponding proofs of 12.8 from $\mathbf{M}_R$ to $\mathbf{M}_B$.

(2) This is clear by the isomorphism $h_{\mathcal{C}}(Y,M) \simeq M \Box_{\mathcal{C}} h_{\mathcal{C}}(Y,\mathcal{C})$, for any $M \in \mathbf{M}^{\mathcal{C}}$. □

Definition. A comodule $Y \in {}_B\mathbf{M}^{\mathcal{C}}$ is called a $(B,\mathcal{C})$-*injector* provided that the functor $- \otimes_B Y : \mathbf{M}_B \to \mathbf{M}^{\mathcal{C}}$ respects injective objects. Left comodule injectors are defined similarly. We know from the Hom-tensor relations 18.10 that a coring $\mathcal{C}$ is both an $(A,\mathcal{C})$-injector and a $(\mathcal{C},A)$-injector.

Similar to the situation for coalgebras, one can prove that, for any $Y \in {}_B\mathbf{M}^{\mathcal{C}}$ that is $(B,\mathcal{C})$-quasi-finite, there is a B-coring $\mathcal{D}$ (the dual of the algebra of $\mathcal{C}$-colinear endomorphisms of Y) that makes Y a $(\mathcal{D},\mathcal{C})$-comodule.

Set $e_{\mathcal{C}}(Y) = h_{\mathcal{C}}(Y,Y)$. The unit of adjunction induces a morphism in $\mathbf{M}^{\mathcal{C}}$,

$$(I_{e_{\mathcal{C}}(Y)} \otimes \eta_Y) \circ \eta_Y : Y \longrightarrow e_{\mathcal{C}}(Y) \otimes_B e_{\mathcal{C}}(Y) \otimes_B Y .$$

By the adjunction isomorphism

$$\mathrm{Hom}_B(e_{\mathcal{C}}(Y), e_{\mathcal{C}}(Y) \otimes_B e_{\mathcal{C}}(Y)) \xrightarrow{\simeq} \mathrm{Hom}^{\mathcal{C}}(Y, e_{\mathcal{C}}(Y) \otimes_B e_{\mathcal{C}}(Y) \otimes_B Y),$$

there exists a unique (B,B)-bilinear map $\underline{\Delta}_e : e_{\mathcal{C}}(Y) \to e_{\mathcal{C}}(Y) \otimes_B e_{\mathcal{C}}(Y)$, such that $(I_{e_{\mathcal{C}}(Y)} \otimes \eta_Y) \circ \eta_Y = (\underline{\Delta}_e \otimes I) \circ \eta_Y$. Moreover, by the isomorphism $\Phi_{Y,B} : \mathrm{Hom}_B(e_{\mathcal{C}}(Y), B) \simeq \mathrm{Hom}^{\mathcal{C}}(Y,Y)$, there exists a unique (B,B)-linear map $\underline{\varepsilon}_e : e_{\mathcal{C}}(Y) \to B$, such that $(\underline{\varepsilon}_e \otimes I_Y) \circ \eta_Y = I_Y$.

23.8. Coendomorphism coring. *Let $Y \in {}_B\mathbf{M}^{\mathcal{C}}$ be $(B,\mathcal{C})$-quasi-finite and $e_{\mathcal{C}}(Y) = h_{\mathcal{C}}(Y,Y)$. Then the (B,B)-bilinear maps*

$$\underline{\Delta}_e : e_{\mathcal{C}}(Y) \to e_{\mathcal{C}}(Y) \otimes_B e_{\mathcal{C}}(Y) \quad \textit{and} \quad \underline{\varepsilon}_e : e_{\mathcal{C}}(Y) \to B ,$$

defined above, make $e_{\mathcal{C}}(Y)$ a B-coring. Furthermore, Y is an $(e_{\mathcal{C}}(Y),\mathcal{C})$-bicomodule by $\eta_Y : Y \to e_{\mathcal{C}}(Y) \otimes_B Y$, and there is a ring anti-isomorphism

$$e_{\mathcal{C}}(Y)^* = \mathrm{Hom}_B(e_{\mathcal{C}}(Y), B) \xrightarrow{\Phi_{Y,B}} \mathrm{End}^{\mathcal{C}}(Y).$$

Proof. Write E for $e_{\mathcal{C}}(Y)$ to shorten the notation. To show that $\underline{\varepsilon}_e$ is a counit for $\underline{\Delta}_e$, consider

$$\begin{aligned}
\Phi_{E,E}((I_E \otimes \underline{\varepsilon}_e) \circ \underline{\Delta}_e) &= ((I_E \otimes \underline{\varepsilon}_e) \circ \underline{\Delta}_e \otimes I_Y) \circ \eta_Y \\
&= (I_E \otimes \underline{\varepsilon}_e \otimes I_Y) \circ (\underline{\Delta}_e \otimes I_Y) \circ \eta_Y \\
&= (I_E \otimes \underline{\varepsilon}_e \otimes I_Y) \circ (I_E \otimes \eta_Y) \circ \eta_Y = \eta_Y, \\
\Phi_{E,E}((\underline{\varepsilon}_e \otimes I) \circ \underline{\Delta}_e) &= ((\underline{\varepsilon}_e \otimes I) \circ \underline{\Delta}_e \otimes I_Y) \circ \eta_Y \\
&= (\underline{\varepsilon}_e \otimes I_E \otimes I_Y) \circ (\underline{\Delta}_e \otimes I_Y) \circ \eta_Y \\
&= (\underline{\varepsilon}_e \otimes I_E \otimes I_Y) \circ (I_E \otimes \eta_Y) \circ \eta_Y = \eta_Y, \text{ and} \\
\Phi_{E,E}(I_E) &= \eta_Y .
\end{aligned}$$

Now the injectivity of Φ implies $(I \otimes \underline{\varepsilon}_e) \circ \underline{\Delta}_e = I_E = (\underline{\varepsilon}_e \otimes I) \circ \underline{\Delta}_e$.

To prove the coassociativity of $\underline{\Delta}_e$, we compute

$$\begin{aligned}
&((\underline{\Delta}_e \otimes I_E) \circ \underline{\Delta}_e \otimes I_Y) \circ \eta_Y = (\underline{\Delta}_e \otimes I_E \otimes I_Y) \circ (\underline{\Delta}_e \otimes I_Y) \circ \eta_Y \\
&= (\underline{\Delta}_e \otimes I_E \otimes I_Y) \circ (I_E \otimes \eta_Y) \circ \eta_Y = (I_E \otimes I_E \otimes \eta_Y) \circ (\underline{\Delta}_e \otimes I_Y) \circ \eta_Y \\
&= (I_E \otimes I_E \otimes \eta_Y) \circ (I_E \otimes \eta_Y) \circ \eta_Y = (I_E \otimes (I_E \otimes \eta_Y) \circ \eta_Y) \circ \eta_Y \\
&= (I_E \otimes (\underline{\Delta}_e \otimes I_Y) \circ \eta_Y) \circ \eta_Y = (I_E \otimes \underline{\Delta}_e \otimes I_Y) \circ (I_E \otimes \eta_Y) \circ \eta_Y \\
&= (I_E \otimes \underline{\Delta}_e \otimes I_Y) \circ (\underline{\Delta}_e \otimes I_Y) \circ \eta_Y = ((I_E \otimes \underline{\Delta}_e) \circ \underline{\Delta}_e \otimes I_Y) \circ \eta_Y,
\end{aligned}$$

and the adjointness isomorphism implies $(\underline{\Delta}_e \otimes I_E) \circ \underline{\Delta}_e = (I_E \otimes \underline{\Delta}_e) \circ \underline{\Delta}_e$.

By the definition of $\underline{\Delta}_e$ there is a commutative diagram,

$$\begin{array}{ccc}
Y & \xrightarrow{\eta_Y} & E \otimes_B Y \\
{\scriptstyle \eta_Y}\downarrow & & \downarrow{\scriptstyle \underline{\Delta} \otimes I_Y} \\
E \otimes_B Y & \xrightarrow{I_E \otimes \eta_Y} & E \otimes_B E \otimes_B Y ,
\end{array}$$

and the definition of $\underline{\varepsilon}_e$ shows that η_Y is a counital coaction. Since η_Y is $\mathcal{C}$-colinear, Y is an $(E, \mathcal{C})$-bicomodule. □

As an example we consider functors related to algebra extensions.

23.9. Functors induced by a base ring extension. Given an R-algebra morphism $\phi : A \to B$, an A-coring $\mathcal{C}$ induces a B-coring $B\mathcal{C}B = B \otimes_A \mathcal{C} \otimes_A B$ (see 17.2). For any $M \in \mathbf{M}$, $M \otimes_A B$ is a $B\mathcal{C}B$-comodule by

$$\varrho^{M \otimes_A B} : M \otimes_A B \to M \otimes_A B \otimes_B B\mathcal{C}B, \ m \otimes b \mapsto \sum m_{\underline{0}} \otimes 1_B \otimes m_{\underline{1}} \otimes b.$$

In particular, $\mathcal{C} \otimes_A B$ is a $(\mathcal{C}, B\mathcal{C}B)$-bicomodule, and, by symmetry, $B \otimes_A \mathcal{C}$ is a $(B\mathcal{C}B, \mathcal{C})$-bicomodule. Let ${}_A\mathcal{C}$ be flat. Then – since $\mathcal{C} \otimes_A B$ is left $(\mathcal{C}, A)$-injective (see 18.18) – the purity conditions are satisfied (see 21.5), thus yielding the functors (by 22.3)

$$-\Box_{\mathcal{C}}(\mathcal{C} \otimes_A B) : \mathbf{M}^{\mathcal{C}} \to \mathbf{M}^{B\mathcal{C}B}, \quad -\Box_{B\mathcal{C}B}(B \otimes_A \mathcal{C}) : \mathbf{M}^{B\mathcal{C}B} \to \mathbf{M}^{\mathcal{C}}.$$

The bicomodule maps

$$\gamma : (B \otimes_A \mathcal{C})\Box_{\mathcal{C}}(\mathcal{C} \otimes_A B) \xrightarrow{\simeq} B \otimes_A \mathcal{C} \otimes_A B,$$

$$\delta : \mathcal{C} \to (\mathcal{C} \otimes_A B)\Box_{BCB}(B \otimes_A \mathcal{C}), \quad c \mapsto \sum c_{\underline{1}} \otimes 1_B \otimes 1_B \otimes c_{\underline{2}},$$

induce the adjointness isomorphism ($\otimes$ means $\otimes_A$)

$$\mathrm{Hom}^{BCB}(L\Box_{\mathcal{C}}(\mathcal{C}\otimes B), N) \to \mathrm{Hom}^{\mathcal{C}}(L, N\Box_{BCB}(B\otimes\mathcal{C})),\ g \mapsto (g\Box I_{B\mathcal{C}})\circ(I_L\Box\delta),$$

with the inverse map $h \mapsto (I_N\Box\gamma)\circ(h\Box I_{\mathcal{C}B})$. The coendomorphism coring of the quasi-finite $\mathcal{C}$-comodule $B \otimes_A \mathcal{C}$ is then simply BCB.

We note that the isomorphism considered above is a special case of a more general adjointness theorem in 24.11.

23.10. Equivalence with comodules over $e_{\mathcal{C}}(Y)$. *Let ${}_A\mathcal{C}$ be flat and let $Y \in {}_B\mathbf{M}^{\mathcal{C}}$ be $(B,\mathcal{C})$-quasi-finite, faithfully coflat, and a $(B,\mathcal{C})$-injector in $\mathbf{M}^{\mathcal{C}}$. Denote by $e_{\mathcal{C}}(Y)$ the coendomorphism coring of Y. Then the functors*

$$-\Box_{e_{\mathcal{C}}(Y)}Y : \mathbf{M}^{e_{\mathcal{C}}(Y)} \to \mathbf{M}^{\mathcal{C}}, \quad h_{\mathcal{C}}(Y,-) : \mathbf{M}^{\mathcal{C}} \to \mathbf{M}^{e_{\mathcal{C}}(Y)},$$

where $h_{\mathcal{C}}(Y,-)$ is the left adjoint to $-\otimes_B Y$, are (inverse) equivalences.

Proof. We prove that the conditions of 23.3(b) are satisfied. By 23.8, Y is an $(e_{\mathcal{C}}(Y),\mathcal{C})$-bicomodule and the image of $h_{\mathcal{C}}(Y,-) : \mathbf{M}^{\mathcal{C}} \to \mathbf{M}_B$ lies in $\mathbf{M}^{e_{\mathcal{C}}(Y)}$ (see 23.5). Since Y is a $(B,\mathcal{C})$-injector, the functor $h_{\mathcal{C}}(Y,-)$ is exact (by 23.7), and hence $h_{\mathcal{C}}(Y,-) \simeq -\Box_{\mathcal{C}}h_{\mathcal{C}}(Y,\mathcal{C})$ (by 23.2) and so $h_{\mathcal{C}}(Y,\mathcal{C})$ is coflat as a left $\mathcal{C}$-comodule.

Take any $M \in \mathbf{M}^{\mathcal{C}}$ and consider the morphism

$$I_M\Box_{\mathcal{C}}\eta_{\mathcal{C}} : M \simeq M\Box_{\mathcal{C}}\mathcal{C} \to M\Box_{\mathcal{C}}(h_{\mathcal{C}}(Y,\mathcal{C}) \otimes_B Y) \simeq (M\Box_{\mathcal{C}}h_{\mathcal{C}}(Y,\mathcal{C})) \otimes_B Y,$$

where the last identification follows from the fact that ${}_BY$ is flat. There exists a unique right $e_{\mathcal{C}}(Y)$-colinear morphism $\delta_M : h_{\mathcal{C}}(Y,M) \to M\Box_{\mathcal{C}}h_{\mathcal{C}}(Y,\mathcal{C})$ such that $(\delta_M\Box_{e_{\mathcal{C}}(Y)}I)\circ\eta_M = I\Box_{\mathcal{C}}\eta_{\mathcal{C}}$. Note that δ_M is an isomorphism since $h_{\mathcal{C}}(Y,-)$ is exact (see 23.7). In particular, since Y is a left $e_{\mathcal{C}}(Y)$-comodule, the isomorphism

$$\delta_Y : e_{\mathcal{C}}(Y) \to Y\Box_{\mathcal{C}}h_{\mathcal{C}}(Y,C)$$

is in fact $(e_{\mathcal{C}}(Y), e_{\mathcal{C}}(Y))$-bicolinear by 39.7. Furthermore, Y is a faithfully coflat $\mathcal{C}$-comodule, and hence the bijectivity of $I_Y\Box_{\mathcal{C}}\eta_{\mathcal{C}}$ implies that the map $\eta_{\mathcal{C}} : C \to h_{\mathcal{C}}(Y,C)\Box_{e_{\mathcal{C}}(Y)}Y$ is an isomorphism in $\mathbf{M}^{\mathcal{C}}$. Since $\mathcal{C}$ is a bicomodule, this is a $(\mathcal{C},\mathcal{C})$-bicomodule morphism by 39.7.

To verify the purity conditions stated in 23.3(b) we have to show that the canonical maps

$$W \otimes_B (h_{\mathcal{C}}(Y,\mathcal{C})\Box_{e_{\mathcal{C}}(Y)}Y) \to (W \otimes_B h_{\mathcal{C}}(Y,\mathcal{C}))\Box_{e_{\mathcal{C}}(Y)}Y \text{ and}$$

$$W \otimes_B (Y\Box_{\mathcal{C}}h_{\mathcal{C}}(Y,\mathcal{C})) \to (W \otimes_B Y)\Box_{\mathcal{C}}h_{\mathcal{C}}(Y,\mathcal{C})$$

are isomorphisms for any $W \in \mathbf{M}_B$ (see 21.4). By 21.5, this follows from the coflatness of Y as a left $e_{\mathcal{C}}(Y)$-comodule and the coflatness of $h_{\mathcal{C}}(Y, C)$ as a left $\mathcal{C}$-comodule, respectively. □

Now we are in a position to describe equivalences between arbitrary comodule categories.

23.11. Equivalences and coendomorphism coring. *Let ${}_A\mathcal{C}$ and ${}_B\mathcal{D}$ be flat, and let $F : \mathbf{M}^{\mathcal{C}} \to \mathbf{M}^{\mathcal{D}}$ be an equivalence with inverse $G : \mathbf{M}^{\mathcal{D}} \to \mathbf{M}^{\mathcal{C}}$. Then:*

(1) $G(\mathcal{D})$ is $(B, \mathcal{C})$-quasi-finite, a $(B, \mathcal{C})$-injector, and faithfully coflat as a left $\mathcal{D}$-comodule.

(2) $F(\mathcal{C})$ is $(A, \mathcal{D})$-quasi-finite, an $(A, \mathcal{D})$-injector, and faithfully coflat as a left $\mathcal{C}$-comodule.

(3) There are coring isomorphisms

$$e_{\mathcal{C}}(G(\mathcal{D})) \simeq \mathcal{D} \simeq e_{\mathcal{C}}(F(\mathcal{C})) \quad \textit{and} \quad e_{\mathcal{D}}(G(\mathcal{D})) \simeq \mathcal{C} \simeq e_{\mathcal{D}}(F(\mathcal{C})).$$

Proof. (1) The proof is similar to the proof of (2).

(2) Since $F \simeq -\Box_{\mathcal{C}} F(\mathcal{C})$ is an equivalence, it has a left adjoint and is exact and faithful. Hence $F(\mathcal{C})$ is $(A, \mathcal{D})$-quasi-finite and faithfully flat as a left $\mathcal{C}$-comodule. For any injective $W \in \mathbf{M}_A$, $W \otimes_A \mathcal{C}$ is injective in $\mathbf{M}^{\mathcal{C}}$, by 18.10, and by properties of equivalences, $W \otimes_A F(\mathcal{C}) \simeq F(W \otimes_A \mathcal{C})$ is injective in $\mathbf{M}^{\mathcal{D}}$, that is, $F(\mathcal{C})$ is an $(A, \mathcal{D})$-injector.

(3) Since $G \simeq -\Box_{\mathcal{D}} G(\mathcal{D})$ is left adjoint to $F \simeq -\Box_{\mathcal{C}} F(\mathcal{C})$,

$$e_{\mathcal{D}}(F(\mathcal{C})) \simeq F(\mathcal{C}) \Box_{\mathcal{D}} G(\mathcal{D}) \simeq GF(\mathcal{C}) \simeq \mathcal{C}.$$

The other isomorphisms are obtained similarly. □

As for coalgebras (cf. Morita-Takeuchi Theorem 12.4) equivalences between comodule categories over corings can be described by the properties of a single bicomodule.

23.12. Equivalences between comodule categories over corings (2). *If ${}_A\mathcal{C}$ and ${}_B\mathcal{D}$ are flat, the following are equivalent:*

(a) the categories $\mathbf{M}^{\mathcal{C}}$ and $\mathbf{M}^{\mathcal{D}}$ are equivalent;

(b) there exists a $(\mathcal{D}, \mathcal{C})$-bicomodule Y that is $(B, \mathcal{C})$-quasi-finite, a $(B, \mathcal{C})$-injector, and faithfully coflat as a right $\mathcal{C}$-comodule, and $e_{\mathcal{C}}(Y) \simeq \mathcal{D}$ as corings;

(c) there is a $(\mathcal{C}, \mathcal{D})$-bicomodule X that is $(A, \mathcal{D})$-quasi-finite, an $(A, \mathcal{D})$-injector, and faithfully coflat as a right $\mathcal{D}$-comodule, and $e_{\mathcal{D}}(X) \simeq \mathcal{C}$ as corings.

Proof. (a) $\Rightarrow$ (b) Given an equivalence $F : \mathbf{M}^{\mathcal{C}} \to \mathbf{M}^{\mathcal{D}}$ with inverse $G : \mathbf{M}^{\mathcal{D}} \to \mathbf{M}^{\mathcal{C}}$, it was shown in 23.11 that $Y = G(\mathcal{D})$ satisfies the conditions stated.

(b) $\Rightarrow$ (a) This follows from 23.10.

(a) $\Leftrightarrow$ (c) The proof for this is symmetric to the proof (a) $\Leftrightarrow$ (b). □

23.13. Exercises

For an A-coring $\mathcal{C}$, let $P \in \mathbf{M}^{\mathcal{C}}$ be such that P_A is finitely generated and projective. Put $T = \mathrm{End}^{\mathcal{C}}(P)$, $P^* = \mathrm{Hom}_A(P, A)$, and consider T-modules as T-comodules canonically, that is, $\mathbf{M}^T = \mathbf{M}_T$. Prove ([111]):

(i) There is an adjoint pair of functors

$$- \otimes_T P : \mathbf{M}_T \to \mathbf{M}_A, \quad - \otimes_A P^* : \mathbf{M}_A \to \mathbf{M}_T.$$

(ii) P^* is a left $\mathcal{C}$-comodule and the functor $-\Box_{\mathcal{C}} P^* : \mathbf{M}^{\mathcal{C}} \to \mathbf{M}^T$ has a left adjoint.

(iii) The coendomorphism coring of P^* is isomorphic to $P^* \otimes_A P$ (with the structure from 17.6).

(iv) $-\Box_{\mathcal{C}} P^*$ induces an equivalence provided that P_A is a generator in $\mathbf{M}_A$.

References. Al-Takhman [51]; Gómez-Torrecillas [122]; El Kaoutit and Gómez-Torrecillas [111]; Guzman [126, 127].

24 The category of corings

To treat corings over different rings on the same footing, and thus to study relations between corings over different rings or *dynamical* properties of corings such as the change of the base ring, and so on, one needs to introduce the category of corings. In this section, in addition to the introduction of the category of corings, we also introduce the category of *representations* of corings. Such a category is defined for any pair of corings, and objects are morphisms between these corings. The fact that such a category can be defined is an intrinsic feature of the notion of a coring. We also study induction and coinduction functors, derive the general Hom-tensor relations, and associate an algebra to a pair of coring morphisms. This, in particular, gives an interpretation of a dual algebra of a coring. The category of corings also provides one with a nice setup for and the unification of the properties of corings discussed in the preceding sections.

24.1. Category of corings. Objects in the *category* **Crg** *of corings* are corings understood as pairs $(\mathcal{C} : A)$, where A is an R-algebra and $\mathcal{C}$ is an A-coring. A morphism between corings $(\mathcal{C} : A)$ and $(\mathcal{D} : B)$ is a pair of mappings $(\gamma : \alpha) : (\mathcal{C} : A) \to (\mathcal{D} : B)$ satisfying the following conditions:

(1) $\alpha : A \to B$ is an algebra map. Thus one can view $\mathcal{D}$ as an (A, A)-bimodule via α. Explicitly, $ada' = \alpha(a)d\alpha(a')$ for all $a, a' \in A$ and $d \in \mathcal{D}$.

(2) $\gamma : \mathcal{C} \to \mathcal{D}$ is an (A, A)-bimodule map such that

$$\chi \circ (\gamma \otimes_A \gamma) \circ \underline{\Delta}_{\mathcal{C}} = \underline{\Delta}_{\mathcal{D}} \circ \gamma, \quad \underline{\varepsilon}_{\mathcal{D}} \circ \gamma = \alpha \circ \underline{\varepsilon}_{\mathcal{C}},$$

where $\chi : \mathcal{D} \otimes_A \mathcal{D} \to \mathcal{D} \otimes_B \mathcal{D}$ is the canonical morphism of (A, A)-bimodules induced by α. Equivalently, we require that the induced map

$$I_B \otimes \gamma \otimes I_B : B \otimes_A \mathcal{C} \otimes_A B \to \mathcal{D}$$

be a morphism of B-corings, where $B \otimes_A \mathcal{C} \otimes_A B$ is the base ring extension of $\mathcal{C}$ (see 17.2).

Since any algebra A can be viewed as a trivial A-coring, the category **Crg** contains the category of R-algebras.

Crg is a monoidal category with the tensor product $(\mathcal{C} : A) \otimes (\mathcal{D} : B) = (\mathcal{C} \otimes_R \mathcal{D} : A \otimes_R B)$. Here $A \otimes_R B$ has the tensor product ring structure, the actions of $A \otimes_R B$ on $\mathcal{C} \otimes_R \mathcal{D}$ are given by $(a \otimes b)(c \otimes d)(a' \otimes b') = aca' \otimes bdb'$, and the coproduct and counit are

$$\underline{\Delta}_{\mathcal{C} \otimes_R \mathcal{D}}(c \otimes d) = \sum (c_{\underline{1}} \otimes d_{\underline{1}}) \otimes_{A \otimes_R B} (c_{\underline{2}} \otimes d_{\underline{2}}), \quad \underline{\varepsilon}_{\mathcal{C} \otimes_R \mathcal{D}} = \underline{\varepsilon}_{\mathcal{C}} \otimes \underline{\varepsilon}_{\mathcal{D}}.$$

The ring R viewed as a trivial R-coring is an identity object.

24.2. Examples of morphisms of corings. Any morphism between A-corings is a morphism in **Crg** (with $\alpha = I_A$). Of particular interest are the *identity morphism*, $(I_{\mathcal{C}} : I_A) : (\mathcal{C} : A) \to (\mathcal{C} : A)$, and the forgetful or *counit morphism*, $(\underline{\varepsilon} : I_A) : (\mathcal{C} : A) \to (A : A)$. In the latter case, A is viewed as a trivial A-coring. Note that the counit morphism satisfies the conditions in 24.1, since $\underline{\varepsilon}$ is the counit in $\mathcal{C}$ and so, for all $c \in \mathcal{C}$, $\sum \underline{\varepsilon}(c_{\underline{1}}) \otimes \underline{\varepsilon}(c_{\underline{2}}) = \underline{\varepsilon}(c)$.

24.3. The category of representations of corings. To any pair of corings $(\mathcal{C} : A)$ and $(\mathcal{D} : B)$ one can associate a category $\mathbf{Rep}(\mathcal{C} : A \mid \mathcal{D} : B)$ of *representations of a coring* $(\mathcal{C} : A)$ *in a coring* $(\mathcal{D} : B)$. The objects of $\mathbf{Rep}(\mathcal{C} : A \mid \mathcal{D} : B)$ are coring morphisms $(\gamma : \alpha) : (\mathcal{C} : A) \to (\mathcal{D} : B)$. For any pair of objects (γ_1, α_1) and (γ_2, α_2) in $\mathbf{Rep}(\mathcal{C} : A \mid \mathcal{D} : B)$, a morphism is an R-module map $f : \mathcal{C} \to B$ such that, for all $c \in \mathcal{C}$ and $a \in A$,

$$f(ca) = f(c)\alpha_2(a), \;\; f(ac) = \alpha_1(a)f(c), \;\; \sum f(c_{\underline{1}})\gamma_2(c_{\underline{2}}) = \sum \gamma_1(c_{\underline{1}})f(c_{\underline{2}}).$$

Composition of morphisms $f : (\gamma_1, \alpha_1) \to (\gamma_2, \alpha_2)$, $g : (\gamma_2, \alpha_2) \to (\gamma_3, \alpha_3)$ is defined by the convolution product, that is, for all $c \in \mathcal{C}$, $f * g(c) = \sum f(c_{\underline{1}})g(c_{\underline{2}})$. Note that $f * g$ is well defined since, for all $a \in A$ and $c, c' \in \mathcal{C}$,

$$f(ca)g(c') = f(c)\alpha_2(a)g(c') = f(c)g(ac').$$

One easily verifies that $f * g$ is a morphism $(\gamma_1, \alpha_1) \to (\gamma_3, \alpha_3)$ in the category $\mathbf{Rep}(\mathcal{C} : A \mid \mathcal{D} : B)$. Indeed, take any $c \in \mathcal{C}$ and $a \in A$ and compute

$$(f * g)(ca) = \sum f(c_{\underline{1}})g(c_{\underline{2}}a) = \sum f(c_{\underline{1}})g(c_{\underline{2}})\alpha_3(a) = (f * g)(c)\alpha_3(a),$$
$$(f * g)(ac) = \sum f(ac_{\underline{1}})g(c_{\underline{2}}) = \sum \alpha_1(a)f(c_{\underline{1}})g(c_{\underline{2}}) = \alpha_1(a)(f * g)(c),$$

$$\begin{aligned}\sum (f * g)(c_{\underline{1}})\gamma_3(c_{\underline{2}}) &= \sum f(c_{\underline{1}})g(c_{\underline{2}})\gamma_3(c_{\underline{3}}) = \sum f(c_{\underline{1}})\gamma_2(c_{\underline{2}})g(c_{\underline{3}}) \\ &= \sum \gamma_1(c_{\underline{1}})f(c_{\underline{2}})g(c_{\underline{3}}) = \sum \gamma_1(c_{\underline{1}})(f * g)(c_{\underline{2}}),\end{aligned}$$

where, at various places, the definitions of morphisms and the A-linearity of the coproduct in $\mathcal{C}$ were used.

24.4. Representations of a coring in an algebra. *Let $\mathcal{C}$ be an A-coring and view B as a trivial B-coring. Then objects in* $\mathbf{Rep}(\mathcal{C} : A \mid B : B)$ *are R-algebra maps $A \to B$. A morphism $f : \alpha_1 \to \alpha_2$ in* $\mathbf{Rep}(\mathcal{C} : A \mid B : B)$ *is an (A, A)-bimodule map $f : \mathcal{C} \to B$, where B is viewed as a left A-module via the map α_1 and as a right A-module via the map α_2.*

The category $\mathbf{Rep}(\mathcal{C} : A \mid B : B)$ is known as the *category of representation of a coring in an algebra.*

Proof. In this case any morphism of corings $(\gamma : \alpha) : (\mathcal{C} : A) \to (B : B)$ is fully determined by the algebra map $\alpha : A \to B$. Indeed, the condition

$\underline{\varepsilon}_B \circ \gamma = \alpha \circ \underline{\varepsilon}_{\mathcal{C}}$ implies that $\gamma = \alpha \circ \underline{\varepsilon}_{\mathcal{C}}$, for $\underline{\varepsilon}_B$ is the identity map on B. There are no further restrictions on α (or γ).

Clearly, a morphism $f : \alpha_1 \to \alpha_2$ in $\mathbf{Rep}(\mathcal{C} : A \mid B : B)$ must be an (A, A)-bimodule map, as stated. There are no further restrictions on f, since, for all $c \in \mathcal{C}$, $\sum f(c_{\underline{1}})\gamma_2(c_{\underline{2}}) = \sum f(c_{\underline{1}})\alpha_2(\underline{\varepsilon}_{\mathcal{C}}(c_{\underline{2}})) = f(c)$ and $\sum \gamma_1(c_{\underline{1}})f(c_{\underline{2}}) = \sum \alpha_1(\underline{\varepsilon}_{\mathcal{C}}(c_{\underline{1}}))f(c_{\underline{2}}) = f(c)$, and thus the relevant condition $\sum f(c_{\underline{1}})\gamma_2(c_{\underline{2}}) = \sum \gamma_1(c_{\underline{1}})f(c_{\underline{2}})$ is satisfied automatically. □

Dually, one can consider representations of a coalgebra in a coring.

24.5. Representations of a coalgebra in a coring. *For an R-coalgebra C and an A-coring $\mathcal{C}$, the objects in $\mathbf{Rep}(C : R \mid \mathcal{C} : A)$ are R-bilinear maps $\gamma : C \to \mathcal{C}$ that satisfy the following properties, for all $c \in C$:*

$$\underline{\varepsilon}_{\mathcal{C}}(\gamma(c)) = \varepsilon_C(c)1_A, \quad \sum \gamma(c_{\underline{1}}) \otimes_A \gamma(c_{\underline{2}}) = \underline{\Delta}_{\mathcal{C}}(\gamma(c)).$$

Morphisms $\gamma_1 \to \gamma_2$ are R-linear maps $f : C \to A$ such that, for all $c \in C$, $\sum \gamma_1(c_{\underline{1}})f(c_{\underline{2}}) = \sum f(c_{\underline{1}})\gamma_2(c_{\underline{2}})$.

Proof. In this case there is only one possible algebra map, $R \to A$ (the unit map), thus all objects in $\mathbf{Rep}(C : R \mid \mathcal{C} : A)$ are completely specified by maps $\gamma : C \to \mathcal{C}$. One easily checks that such maps, as well as morphisms between them, must satisfy the specified conditions. □

24.6. The induction functor. Take any $(\mathcal{C} : A), (\mathcal{D} : B) \in \mathbf{Crg}$. Given a morphism in $\mathbf{Crg}$, $(\gamma : \alpha) : (\mathcal{C} : A) \to (\mathcal{D} : B)$, define an *induction functor*

$$F : \mathbf{M}^{\mathcal{C}} \to \mathbf{M}^{\mathcal{D}}, \quad M \mapsto M \otimes_A B, \quad f \mapsto f \otimes I_B.$$

Here $F(M)$ is a right B-module via $(m \otimes b)b' = m \otimes bb'$, for all $m \in M$ and $b, b' \in B$. The right $\mathcal{D}$-coaction is given by

$$\varrho^{F(M)} : M \otimes_A B \to M \otimes_A B \otimes_B \mathcal{D} \simeq M \otimes_A \mathcal{D}, \; m \otimes b \mapsto \sum m_{\underline{0}} \otimes \gamma(m_{\underline{1}})b,$$

where $\sum m_{\underline{0}} \otimes m_{\underline{1}} = \varrho^M(m)$. Clearly $\varrho^{F(M)}$ is a right B-module map. It is a coaction, because, using the facts that ϱ^M is a coaction and that γ commutes with the coproduct, we can compute

$$\begin{aligned}(\varrho^{F(M)} \otimes_B I_{\mathcal{D}}) \circ \varrho^{F(M)}(m \otimes_A b) &= \sum \varrho^{F(M)}(m_{\underline{0}} \otimes_A 1_B) \otimes_B \gamma(m_{\underline{1}})b \\ &= \sum m_{\underline{0}} \otimes_A \gamma(m_{\underline{1}}) \otimes_B \gamma(m_{\underline{2}})b \\ &= \sum m_{\underline{0}} \otimes_A \gamma(m_{\underline{1}})_{\underline{1}} \otimes_B \gamma(m_{\underline{1}})_{\underline{2}}b \\ &= (I_{\mathcal{D}} \otimes_B \underline{\Delta}_{\mathcal{D}}) \circ \varrho^{F(M)}(m \otimes_A b),\end{aligned}$$

as required.

The induction functor associated to the identity morphism of $(\mathcal{C} : A)$, $(I_{\mathcal{C}} : I_A) : (\mathcal{C} : A) \to (\mathcal{C} : A)$, turns out to be the identity functor in $\mathbf{M}^{\mathcal{C}}$. Also, the forgetful functor $\mathbf{M}^{\mathcal{C}} \to \mathbf{M}_A$ can be viewed as an induction functor associated to the counit morphism $(\underline{\varepsilon}_{\mathcal{C}} : I_A) : (\mathcal{C} : A) \to (A : A)$.

Thus to any morphism of corings $(\gamma : \alpha) : (\mathcal{C} : A) \to (\mathcal{D} : B)$ one can associate a functor between the categories of comodules $\mathbf{M}^{\mathcal{C}} \to \mathbf{M}^{\mathcal{D}}$. It can be viewed as a composite of functors

$$-\Box_{\mathcal{C}}(\mathcal{C} \otimes_A B) : \mathbf{M}^{\mathcal{C}} \to \mathbf{M}^{B\mathcal{C}B}, \quad (\ \ )_{\tilde{\gamma}} : \mathbf{M}^{B\mathcal{C}B} \to \mathbf{M}^{\mathcal{D}},$$

where the first one is induced by the base ring extension $\alpha : A \to B$ (see 23.9) and the second one is the corestriction functor derived from the B-coring morphism $\tilde{\gamma} : B\mathcal{C}B \to \mathcal{D}$ (see 22.11). In certain circumstances one can also construct a functor in the opposite direction, $\mathbf{M}^{\mathcal{D}} \to \mathbf{M}^{\mathcal{C}}$. This construction is slightly more involved, and we explain it in a number of steps.

24.7. The coinduced module $G(N)$. For any morphism in the category **Crg**, $(\gamma : \alpha) : (\mathcal{C} : A) \to (\mathcal{D} : B)$, view the left B-module $B \otimes_A \mathcal{C}$ as a left $\mathcal{D}$-comodule via the coaction

$${}^{B\otimes_A\mathcal{C}}\varrho : B \otimes_A \mathcal{C} \to \mathcal{D} \otimes_B B \otimes_A \mathcal{C} \simeq \mathcal{D} \otimes_A \mathcal{C}, \quad b \otimes c \mapsto \sum b\gamma(c_{\underline{1}}) \otimes c_{\underline{2}}.$$

This is simply a left-handed version of the induction functor F considered in 24.6. Take any right $\mathcal{D}$-comodule N and consider a right A-module defined by the cotensor product

$$G(N) = N\Box_{\mathcal{D}}(B \otimes_A \mathcal{C}).$$

Recall from 21.1 that this means that $G(N)$ is an equaliser of right A-modules

$$G(N) \longrightarrow N \otimes_A \mathcal{C} \underset{b_N}{\overset{t_N}{\rightrightarrows}} N \otimes_B \mathcal{D} \otimes_A \mathcal{C},$$

where $t_N = \varrho^N \otimes I_{\mathcal{C}}$ and $b_N = (\chi \otimes I_{\mathcal{C}}) \circ (I_N \otimes \gamma \otimes I_{\mathcal{C}}) \circ (I_N \otimes \underline{\Delta}_{\mathcal{C}})$, with $\chi : N \otimes_A \mathcal{D} \to N \otimes_B \mathcal{D}$ the canonical map associated to $\alpha : A \to B$. Explicitly,

$$G(N) = \{\sum_i n^i \otimes c^i \in N \otimes_A \mathcal{C} \mid \sum_i n^i \otimes_B \gamma(c^i{}_{\underline{1}}) \otimes c^i{}_{\underline{2}} = \sum_i n^i{}_{\underline{0}} \otimes_B n^i{}_{\underline{1}} \otimes c^i\}.$$

24.8. Pure morphism of corings. A morphism $(\gamma : \alpha) : (\mathcal{C} : A) \to (\mathcal{D} : B)$ of corings is said to be *pure* if, for every right $\mathcal{D}$-comodule N, the right A-module morphism $\omega_{N,B\otimes_A\mathcal{C}} = t_N - b_N$ is $\mathcal{C}$-pure (see 40.13 for the definition of a $\mathcal{C}$-pure morphism). Note that $(\gamma : \alpha)$ is a pure morphism of corings if and only if, for every $\mathcal{D}$-comodule N, $G(N) \otimes_A \mathcal{C}$ is equal to the equaliser of $t_N \otimes I_{\mathcal{C}}$ and $b_N \otimes I_{\mathcal{C}}$, that is, $G(N)$ is a $\mathcal{C}$-pure equaliser in $\mathbf{M}_A$ (cf. 40.14). In particular, if $\mathcal{C}$ is a flat left A-module, then every morphism $(\mathcal{C} : A) \to (\mathcal{D} : B)$ is a pure morphism of corings.

As an example of a pure morphism of corings, take the counit morphism $(\underline{\varepsilon} : I_A) : (\mathcal{C} : A) \to (A : A)$ (cf. 24.2). In this case $\mathbf{M}^{\mathcal{D}} = \mathbf{M}^A = \mathbf{M}_A$, and for all $M \in \mathbf{M}_A$, $\omega_{M,\mathcal{C}}$ is a zero map $m \otimes c \mapsto 0$. Thus $\omega_{M,\mathcal{C}}$ is $\mathcal{C}$-pure. Note that in this case $G(M) = \ker \omega_{M,\mathcal{C}} = M \otimes_A \mathcal{C}$.

24.9. The coinduction functor. *Let $(\gamma : \alpha) : (\mathcal{C} : A) \to (\mathcal{D} : B)$ be a pure morphism of corings. Then, for all $N \in \mathbf{M}^{\mathcal{D}}$, $G(N)$ is a right $\mathcal{C}$-comodule with the coaction*

$$\varrho^{G(N)} : G(N) \to G(N) \otimes_A \mathcal{C}, \quad \sum_i n^i \otimes c^i \mapsto \sum_i n^i \otimes_B c^i{}_{\underline{1}} \otimes_A c^i{}_{\underline{2}},$$

provided $G(N)$ is viewed as a right A-module via $(\sum_i n^i \otimes c^i)\, a = \sum_i n^i \otimes c^i a$.
Furthermore, for any morphism $f : N \to N'$ in $\mathbf{M}^{\mathcal{D}}$,

$$G(f) : G(N) \to G(N'), \quad \sum_i n^i \otimes c^i \mapsto \sum_i f(n^i) \otimes c^i,$$

is a morphism in $\mathbf{M}^{\mathcal{C}}$.
Thus $G : \mathbf{M}^{\mathcal{D}} \to \mathbf{M}^{\mathcal{C}}$ is a covariant functor called a *coinduction functor.*

Proof. The fact that $G(N)$ is a right $\mathcal{C}$-comodule follows from 22.3(2). The definition of $G(f)$ makes sense because f is a morphism of right B-modules and hence of right A-modules (the A-action on N is given via $\alpha : A \to B$). Second, the image of $G(f)$ is in $G(N')$ since f is a right $\mathcal{D}$-comodule map. Explicitly,

$$\sum_i f(n^i)_{\underline{0}} \otimes_B f(n^i)_{\underline{1}} \otimes_A c^i = \sum_i f(n^i{}_{\underline{0}}) \otimes_B n^i{}_{\underline{1}} \otimes_A c^i = \sum_i f(n^i) \otimes_B \gamma(c^i{}_{\underline{1}}) \otimes_A c^i{}_{\underline{2}}.$$

Since $G(f)$ acts as identity on the part of $G(N)$ in $\mathcal{C}$, it is clear that $G(f)$ is a right $\mathcal{C}$-comodule map. Thus, with the above definitions, and provided that $(\gamma : \alpha)$ is a pure morphism of corings, G is a covariant functor, as required. □

To gain a better understanding of the functor G we compute it in a very simple, but important and general situation (cf. tensor-cotensor relations 21.4).

24.10. Action of the coinduction functor on induced comodules. *Let $(\gamma : \alpha) : (\mathcal{C} : A) \to (\mathcal{D} : B)$ be a pure morphism of corings and take any right B-module N. View $N \otimes_B \mathcal{D}$ as a right $\mathcal{D}$-comodule via the natural right B-action and the coaction $I_N \otimes_B \underline{\Delta}_{\mathcal{D}}$ (cf. 18.9). Then $G(N \otimes_B \mathcal{D}) \simeq N \otimes_A \mathcal{C}$ in $\mathbf{M}^{\mathcal{C}}$.*

Proof. The isomorphism reads

$$\theta : G(N \otimes_B \mathcal{D}) \to N \otimes_A \mathcal{C}, \quad \sum_i n^i \otimes_B d^i \otimes_A c^i \mapsto \sum_i n^i \underline{\varepsilon}_{\mathcal{D}}(d^i) \otimes_A c^i,$$

and its inverse is $\theta^{-1} : n \otimes_A c \mapsto \sum n \otimes_B \gamma(c_{\underline{1}}) \otimes_A c_{\underline{2}}$. One easily checks that $\mathrm{Im}\,(\theta^{-1}) \subseteq G(N \otimes_B \mathcal{D})$ and that θ^{-1} is the inverse of θ. Clearly, both θ and θ^{-1} are right $\mathcal{C}$-comodule maps, as required. □

24.11. Hom-tensor relations. *Let $(\gamma : \alpha) : (\mathcal{C} : A) \to (\mathcal{D} : B)$ be a pure morphism of corings, and let F and G be the induction and coinduction functors defined above. Then F is left adjoint to G. Consequently, for any pair of comodules $M \in \mathbf{M}^{\mathcal{C}}$, $N \in \mathbf{M}^{\mathcal{D}}$, there is an isomorphism of R-modules,*

$$\mathrm{Hom}^{\mathcal{C}}(M, G(N)) \simeq \mathrm{Hom}^{\mathcal{D}}(M \otimes_A B, N),$$

that is natural in M and N.

Proof. First we construct the unit of the adjunction. For any $M \in \mathbf{M}^{\mathcal{C}}$, consider an R-module map

$$\eta_M : M \to G(F(M)), \quad m \mapsto \sum m_{\underline{0}} \otimes 1_B \otimes m_{\underline{1}}.$$

The map η_M is well defined since, for all $m \in M$, $c \in \mathcal{C}$ and $a \in A$,

$$ma \otimes 1_B \otimes c = m \otimes \alpha(a) \otimes c = m \otimes 1_B \otimes ac.$$

Furthermore,

$$\begin{aligned} t_{F(M)} \circ \eta_M(m) &= t_{F(M)}(\textstyle\sum m_{\underline{0}} \otimes 1_B \otimes m_{\underline{1}}) \\ &= \textstyle\sum (m_{\underline{0}} \otimes 1_B)_{\underline{0}} \otimes_B (m_{\underline{0}} \otimes 1_B)_{\underline{1}} \otimes_A m_{\underline{1}} \\ &= \textstyle\sum m_{\underline{0}} \otimes \gamma(m_{\underline{1}}) \otimes m_{\underline{2}}, \quad \text{and} \\ b_{F(M)} \circ \eta_M(m) &= b_{F(M)}(\textstyle\sum m_{\underline{0}} \otimes 1_B \otimes m_{\underline{1}}) \\ &= \textstyle\sum m_{\underline{0}} \otimes 1_B \otimes_B \gamma(m_{\underline{1}}) \otimes m_{\underline{2}} \\ &= \textstyle\sum m_{\underline{0}} \otimes \gamma(m_{\underline{1}}) \otimes m_{\underline{2}}. \end{aligned}$$

Thus we conclude that the image of η_M is in the required equaliser. The way in which the definition of the map η_M depends upon the coaction ϱ^M ensures that η_M is a right A-module map (since ϱ^M is such a map). One easily checks that η_M is also a morphism in $\mathbf{M}^{\mathcal{C}}$. Next, take any $f : M \to M'$ in $\mathbf{M}^{\mathcal{C}}$ and compute for any $m \in M$,

$$\begin{aligned} \eta_{M'}(f(m)) &= \textstyle\sum f(m)_{\underline{0}} \otimes 1_B \otimes f(m)_{\underline{1}} \\ &= \textstyle\sum f(m_{\underline{0}}) \otimes 1_B \otimes m_{\underline{1}} \;=\; GF(f)(\eta_M(m)). \end{aligned}$$

The second equality follows since f is a morphism of right $\mathcal{C}$-comodules. Thus we have constructed a natural map $\eta : I_{\mathbf{M}^{\mathcal{C}}} \to GF$ that will be shown to be the unit of the adjunction.

Now, for any $N \in \mathbf{M}^{\mathcal{D}}$, consider a right B-module map

$$\psi_N : FG(N) \to N, \quad \sum_i n^i \otimes_A c^i \otimes_A b \mapsto \sum_i (n^i \alpha(\underline{\varepsilon}_{\mathcal{C}}(c^i))b.$$

The map ψ_N is a morphism in the category of right $\mathcal{D}$-comodules since, for all $n = \sum_i n^i \otimes_A c^i \otimes_A b \in FG(N)$,

$$\sum \psi_N(n_{\underline{0}}) \otimes_B n_{\underline{1}} = \sum_i \psi_N(n^i \otimes_A c^i{}_{\underline{1}} \otimes_A 1_B) \otimes_B \gamma(c^i{}_{\underline{2}})b = \sum_i n^i \otimes_A \gamma(c^i)b.$$

Since $\sum_i n^i \otimes_A c^i \in G(N)$, $\sum_i n^i{}_{\underline{0}} \otimes_B n^i{}_{\underline{1}} \otimes_A c^i = \sum^i n^i \otimes_B \gamma(c^i{}_{\underline{1}}) \otimes_A c^i{}_{\underline{2}}$, and thus the application of $I_N \otimes_B I_{\mathcal{D}} \otimes_A \underline{\varepsilon}_{\mathcal{C}}$ yields $\sum_i n^i{}_{\underline{0}} \otimes_B n^i{}_{\underline{1}}\alpha(\underline{\varepsilon}_{\mathcal{C}}(c^i)) = \sum_i n^i \otimes_B \gamma(c^i)$. Therefore we conclude

$$\sum \psi_N(n_{\underline{0}}) \otimes_B n_{\underline{1}} = \sum_i n^i{}_{\underline{0}} \otimes_B n^i{}_{\underline{1}}\alpha(\underline{\varepsilon}_{\mathcal{C}}(c^i))b = \sum \psi_N(n)_{\underline{0}} \otimes_B \psi_N(n)_{\underline{1}},$$

as required for a morphism in $\mathbf{M}^{\mathcal{D}}$. Thus for any right $\mathcal{D}$-comodule N we have constructed a morphism ψ_N in $\mathbf{M}^{\mathcal{D}}$. We need to show that the corresponding map $\psi : FG \to I_{\mathbf{M}^{\mathcal{D}}}$ is a morphism of functors. Take any $f : N \to N'$ in $\mathbf{M}^{\mathcal{D}}$ and $\sum_i n^i \otimes c^i \otimes b \in FG(N)$. Then, on the one hand,

$$f \circ \psi_N(\sum_i n^i \otimes c^i \otimes b) = \sum_i f(n^i \alpha(\underline{\varepsilon}_{\mathcal{C}}(c^{i)}b))) = \sum_i f(n^i)\alpha(\underline{\varepsilon}_{\mathcal{C}}(c^i))b,$$

since f is right B-linear. On the other hand,

$$\psi_{N'} \circ FG(f)(\sum_i n^i \otimes c^i \otimes b) = \psi_{N'}(\sum_i f(n^i) \otimes c^i \otimes b) = \sum_i f(n^i)\alpha(\underline{\varepsilon}_{\mathcal{C}}(c^i))b,$$

as required.

Finally, the fact that η and ψ are the unit and counit, respectively, can be verified by a simple calculation that uses the properties of the counit $\underline{\varepsilon}_{\mathcal{C}}$. □

Observe that if we take the counit morphism of corings 24.2, which is pure as explained in 24.8, then F is a forgetful functor by 24.6, while G is the induction functor $G = - \otimes_A \mathcal{C}$ of 18.9. The corresponding Hom-tensor relation is simply the Hom-tensor relation 18.10. Furthermore, if we take a morphism of A-corings (that is, the case $B = A$, $\alpha = I_A$) and assume that $\mathcal{C}$ is flat as a left A-module, then 24.11 reduces to 22.12. As yet another special case of the Hom-tensor relation described in 24.11, one obtains

24.12. Hom-tensor relation for a special comodule. *For a pure morphism of corings $(\gamma : \alpha) : (\mathcal{C} : A) \to (\mathcal{D} : B)$ and a comodule $M \in \mathbf{M}^{\mathcal{C}}$, there is an isomorphism of R-modules (natural in M)*

$$\mathrm{Hom}^{\mathcal{C}}(M, B \otimes_A \mathcal{C})) \simeq \mathrm{Hom}^{\mathcal{D}}(M \otimes_A B, \mathcal{D}).$$

Proof. Take $N = \mathcal{D}$ in 24.11 and apply 24.10. □

24.13. An algebra associated to a morphism of corings. *For a coring morphism* $(\gamma, \alpha) : (\mathcal{C} : A) \to (\mathcal{D} : B)$, *consider the R-module*

$$\mathfrak{A}(\alpha,\gamma) = \{a \in {}_A\mathrm{Hom}_A(\mathcal{C}, B) \mid \forall\, c \in \mathcal{C},\ \sum \gamma(c_{\underline{1}})a(c_{\underline{2}}) = \sum a(c_{\underline{1}})\gamma(c_{\underline{2}})\}.$$

Then $\mathfrak{A}(\alpha,\gamma)$ *is an R-algebra with product* $(a * a')(c) = \sum a(c_{\underline{1}})a'(c_{\underline{2}})$ *(convolution product) and unit* $1_{\mathfrak{A}(\alpha,\gamma)} = \alpha \circ \underline{\varepsilon}_{\mathcal{C}}$.

Proof. First note that the product is well defined since elements of $\mathfrak{A}(\alpha,\gamma)$ are (A,A)-bimodule maps. Furthermore, for all $c \in \mathcal{C}$,

$$\begin{aligned}\sum \gamma(c_{\underline{1}})(a * a')(c_{\underline{2}}) &= \sum \gamma(c_{\underline{1}})a(c_{\underline{2}})a'(c_{\underline{3}}) = \sum a(c_{\underline{1}})\gamma(c_{\underline{2}})a'(c_{\underline{3}})\\ &= \sum a(c_{\underline{1}})a'(c_{\underline{2}})\gamma(c_{\underline{3}}) = (a * a')(c_{\underline{1}})\gamma(c_{\underline{2}}),\end{aligned}$$

that is, $a * a' \in \mathfrak{A}(\alpha,\gamma)$. The coassociativity of $\underline{\Delta}_{\mathcal{C}}$ implies that the product in $\mathfrak{A}(\alpha,\gamma)$ is associative. Since the product in $\mathfrak{A}(\alpha,\gamma)$ is simply the convolution product, $\alpha \circ \underline{\varepsilon}_{\mathcal{C}}$ is the unit. Thus we need only to verify that $1_{\mathfrak{A}(\alpha,\gamma)}$ is in $\mathfrak{A}(\alpha,\gamma)$. For any $c \in \mathcal{C}$, by the axioms of the counit and the fact that γ is (A,A)-bilinear,

$$\begin{aligned}&\sum \gamma(c_{\underline{1}})1_{\mathfrak{A}(\alpha,\gamma)}(c_{\underline{2}}) = \sum \gamma(c_{\underline{1}})\alpha(\underline{\varepsilon}_{\mathcal{C}}(c_{\underline{2}})) = \sum \gamma(c_{\underline{1}}\varepsilon_{\mathcal{C}}(c_{\underline{2}})) = \gamma(c), \quad \text{and}\\ &\sum 1_{\mathfrak{A}(\alpha,\gamma)}(c_{\underline{1}})\gamma(c_{\underline{2}}) = \sum \alpha(\underline{\varepsilon}_{\mathcal{C}}(c_{\underline{1}}))\gamma(c_{\underline{2}}) = \sum \gamma(\alpha(\varepsilon_{\mathcal{C}}(c_{\underline{1}})c_{\underline{2}}) = \gamma(c).\end{aligned}$$

This shows that $1_{\mathfrak{A}(\alpha,\gamma)} \in \mathfrak{A}(\alpha,\gamma)$ and hence $\mathfrak{A}(\alpha,\gamma)$ is a unital, associative algebra as claimed. □

24.14. The endomorphism ring of an induction functor. *Given a morphism* $(\gamma : \alpha) : (\mathcal{C} : A) \to (\mathcal{D} : B)$ *in* **Crg**, *let* $F = -\otimes_A B : \mathbf{M}^{\mathcal{C}} \to \mathbf{M}^{\mathcal{D}}$ *be the corresponding induction functor. Then the algebra* $\mathfrak{A}(\alpha,\gamma)$ *is isomorphic to the R-algebra* $\mathrm{Nat}(F,F)$ *of natural transformations* $F \to F$ *(with respect to composition).*

Proof. Given a natural transformation $\phi : F \to F$, the corresponding morphism $\phi_M : M \otimes_A B \to M \otimes_A B$ in $M^{\mathcal{D}}$ is right B-linear and thus can be identified with a map $\tilde{\phi}_M : M \to M \otimes_A B$ via $\tilde{\phi}_M(m) = \phi_M(m \otimes 1_B)$ and $\phi_M(m \otimes b) = \tilde{\phi}_M(m)b$. The definition of $\tilde{\phi}_M$ immediately implies that for any other natural endomorphism $\phi' : F \to F$, and the corresponding morphism $\phi'_M : M \otimes_A B \to M \otimes_A B$, $\widetilde{(\phi \circ \phi')}_M = \phi_M \circ \tilde{\phi}'_M$. Take any right A-module M, and for any $m \in M$ consider a $\mathcal{C}$-comodule map $f_m : \mathcal{C} \to M \otimes_A \mathcal{C}$, $c \mapsto m \otimes c$. The naturality of ϕ then implies that

$$(f_m \otimes I_B) \circ \phi_{\mathcal{C}} = \phi_{M\otimes_A \mathcal{C}} \circ (f_m \otimes I_B).$$

This is equivalent to the following property of $\tilde{\phi}$:

$$\tilde{\phi}_{M\otimes_A\mathcal{C}}(m \otimes c) = m \otimes \tilde{\phi}_{\mathcal{C}}(c), \qquad (*)$$

for any $m \in M$ and $c \in \mathcal{C}$. On the other hand, for any $M \in \mathbf{M}^{\mathcal{C}}$, take $f = \varrho^M : M \to M \otimes_A \mathcal{C}$. This is a morphism in $\mathbf{M}^{\mathcal{C}}$, and hence the naturality of ϕ implies that $(\varrho^M \otimes I_B) \circ \phi_M = \phi_{M\otimes_A \mathcal{C}} \circ (\varrho^M \otimes I_B)$, or equivalently for $\tilde{\phi}$,

$$\tilde{\phi}_{M\otimes_A \mathcal{C}} \circ \varrho^M = (\varrho^M \otimes I_B) \circ \tilde{\phi}_M. \qquad (**)$$

Taking $M = A$ in equation $(*)$, one immediately concludes that $\tilde{\phi}_{\mathcal{C}}$ is left A-linear, and hence it is (A, A)-bilinear. Thus we can define an (A, A)-bimodule map, $a_\phi : \mathcal{C} \to B$, $a_\phi = (\underline{\varepsilon}_{\mathcal{C}} \otimes I_B) \circ \tilde{\phi}_{\mathcal{C}}$. Combining equations $(*)$ and $(**)$ for an arbitrary $M \in \mathbf{M}^{\mathcal{C}}$, one can obtain an explicit expression for $\tilde{\phi}_M$ in terms of a_ϕ. More precisely, equations $(*)$ and $(**)$ imply that, for all $m \in M$, $\sum m_{\underline{0}} \otimes \tilde{\phi}_{\mathcal{C}}(m_{\underline{1}}) = (\varrho^M \otimes I_B) \circ \tilde{\phi}_M(m)$. Thus, applying $I_M \otimes \underline{\varepsilon}_{\mathcal{C}} \otimes I_B$, one obtains that $\tilde{\phi}_M(m) = \sum m_{\underline{0}} \otimes a_\phi(m_{\underline{1}})$.

Since for any $M \in \mathbf{M}^{\mathcal{C}}$, ϕ_M is a morphism in $\mathbf{M}^{\mathcal{D}}$, one easily finds that $\sum \tilde{\phi}_M(m_{\underline{0}}) \otimes_B \gamma(m_{\underline{1}}) = \sum \tilde{\phi}_M(m)_{\underline{0}} \otimes_B \tilde{\phi}_M(m)_{\underline{1}}$. Thus in terms of the map a_ϕ we obtain

$$\sum m_{\underline{0}} \otimes a_\phi(m_{\underline{1}})\gamma(m_{\underline{2}}) = \sum m_{\underline{0}} \otimes \gamma(m_{\underline{1}})a_\phi(m_{\underline{2}}).$$

In particular, if we take $M = \mathcal{C}$ and apply $\underline{\varepsilon}_{\mathcal{C}} \otimes I_{\mathcal{D}}$, we obtain $\sum a_\phi(c_{\underline{1}})\gamma(c_{\underline{2}}) = \sum \gamma(c_{\underline{1}})a_\phi(c_{\underline{2}})$, that is, $a_\phi \in \mathfrak{A}(\alpha, \gamma)$. Thus the assignment $\phi \mapsto a_\phi$ defines an R-linear map $\mathrm{Nat}(F, F) \to \mathfrak{A}(\alpha, \gamma)$. To show that this is an algebra map, take any $\phi, \phi' \in \mathrm{Nat}(F, F)$, $c \in \mathcal{C}$, and compute

$$\begin{aligned} a_{\phi\circ\phi'}(c) &= (\underline{\varepsilon}_{\mathcal{C}} \otimes I_B) \circ (\widetilde{\phi \circ \phi'})_{\mathcal{C}}(c) \;=\; (\underline{\varepsilon}_{\mathcal{C}} \otimes I_B)(\phi_{\mathcal{C}}(\tilde{\phi}'_{\mathcal{C}}(c))) \\ &= \sum (\underline{\varepsilon}_{\mathcal{C}} \otimes I_B)(\phi_{\mathcal{C}}(c_{\underline{1}} \otimes a_{\phi'}(c_{\underline{2}}))) \\ &= \sum (\underline{\varepsilon}_{\mathcal{C}} \otimes I_B)(\phi_{\mathcal{C}}(c_{\underline{1}})a_{\phi'}(c_{\underline{2}})) \;=\; \sum a_\phi(c_{\underline{1}})a_{\phi'}(c_{\underline{2}}). \end{aligned}$$

Conversely, take any $a \in \mathfrak{A}(\alpha, \gamma)$ and, for any $M \in \mathbf{M}^{\mathcal{C}}$, define a right B-bimodule map

$$\phi_{a,M} : M \otimes_A B \to M \otimes_A B, \quad m \otimes b \mapsto \sum m_{\underline{0}} \otimes a(m_{\underline{1}})b.$$

Using the fact that $a \in \mathfrak{A}(\alpha, \gamma)$, one easily finds that $\phi_{a,M}$ is a right $\mathcal{D}$-comodule map. Next take any morphism $f : M \to N$ in $\mathbf{M}^{\mathcal{C}}$. Then, for any $m \in M$ and $b \in B$, $F(f) \circ \phi_{a,M}(m \otimes b) = \sum f(m_{\underline{0}}) \otimes a(m_{\underline{1}})b$, while

$$\phi_{a,N} \circ F(f)(m \otimes b) = \sum f(m)_{\underline{0}} \otimes a(f(m)_{\underline{1}})b = \sum f(m_{\underline{0}}) \otimes a(m_{\underline{1}})b.$$

Thus we conclude that the assignment $a \mapsto \phi_a$ defines a mapping $\mathfrak{A}(\alpha, \gamma) \to \mathrm{Nat}(F, F)$. It remains to show that this mapping is the inverse of the algebra map $\mathrm{Nat}(F,F) \to \mathfrak{A}(\alpha,\gamma)$, $\phi \mapsto a_\phi$, constructed previously. First take any $M \in \mathbf{M}^{\mathcal{C}}$, $m \in M$, and $b \in B$ and compute

$$\phi_{a_\phi,M}(m \otimes b) = \sum m_{\underline{0}} \otimes a_\phi(m_{\underline{1}})b = \tilde{\phi}_M(m)b = \phi_M(m \otimes b).$$

Second, for any $c \in \mathcal{C}$,

$$a_{\phi_a}(c) = (\underline{\varepsilon}_{\mathcal{C}} \otimes I_B) \circ \tilde{\phi}_{a,\mathcal{C}}(c) = \sum \underline{\varepsilon}_{\mathcal{C}}(c_{\underline{1}}) \otimes a(c_{\underline{2}}) = a(c).$$

This completes the proof of the theorem. □

24.15. The endomorphism ring of the coinduction functor. *Assume $(\gamma : \alpha) : (\mathcal{C} : A) \to (\mathcal{D} : B)$ to be a pure morphism of corings, and let G be the associated coinduction functor. Then there is an R-algebra isomorphism*

$$\mathrm{Nat}(G,G) \simeq \mathfrak{A}(\alpha,\gamma)^{op}.$$

Proof. This follows immediately from 24.14, since the endomorphism ring of a functor is isomorphic to the opposite endomorphism ring of its left or right adjoint functor (cf. [38, Section 2.1, Corollary 1]). Explicitly, the required isomorphisms $\mathrm{Nat}(G,G) \simeq \mathfrak{A}(\alpha,\gamma)^{op}$ are given as follows:

$$\mathrm{Nat}(G,G) \to \mathfrak{A}(\alpha,\gamma)^{op}, \quad \phi \mapsto (\underline{\varepsilon}_{\mathcal{D}} \otimes \underline{\varepsilon}_{\mathcal{C}}) \circ \tilde{\phi}_{\mathcal{D}} \circ (\gamma \otimes I_{\mathcal{C}}) \circ \underline{\Delta}_{\mathcal{C}},$$

where $\tilde{\phi}_{\mathcal{D}}$ is as in the proof of 24.14. In the converse direction, given $a \in \mathfrak{A}(\alpha,\gamma)^{op}$, one defines a natural endomorphism ϕ_a via

$$\phi_{a,N} : G(N) \to G(N), \quad \sum_i n^i \otimes c^i \mapsto \sum_i n^i a(c^i{}_{\underline{1}}) \otimes c^i{}_{\underline{2}},$$

for all $N \in \mathbf{M}^{\mathcal{D}}$. □

24.16. The endomorphism ring of the forgetful functor. Consider the counit morphism of corings $(\underline{\varepsilon} : I_A) : (\mathcal{C} : A) \to (A : A)$. Then

$$\begin{aligned}\mathfrak{A}(I_A,\underline{\varepsilon}) &= \{a \in {}_A\mathrm{Hom}_A(\mathcal{C},A) \mid \forall c \in \mathcal{C},\ \sum a(c_{\underline{1}})\underline{\varepsilon}(c_{\underline{2}}) = \sum \underline{\varepsilon}(c_{\underline{1}})a(c_{\underline{2}})\} \\ &= {}_A\mathrm{Hom}_A(\mathcal{C},A) = {}^*\mathcal{C}^*.\end{aligned}$$

Since in this case F is the forgetful functor $\mathbf{M}^{\mathcal{C}} \to \mathbf{M}_A$, we obtain an interpretation of the dual algebra ${}^*\mathcal{C}^*$ in 17.8(3) as an endomorphism ring of the forgetful functor $\mathbf{M}^{\mathcal{C}} \to \mathbf{M}_A$. This may be compared with the interpretation of the left-dual ring ${}^*\mathcal{C}$ as the endomorphism ring of the forgetful functor $\mathbf{M}^{\mathcal{C}} \to \mathbf{M}_R$ in 18.29.

The endomorphism rings of general induction and coinduction functors between categories $\mathbf{M}^{\mathcal{C}}$ and $\mathbf{M}^{\mathcal{D}}$ have a natural interpretation in terms of the category $\mathbf{Rep}(\mathcal{C} : A \mid \mathcal{D} : B)$ of representations of a coring in a coring.

24.17. Endomorphisms of the induction functor in the category of representations. *Let $F : \mathbf{M}^{\mathcal{C}} \to \mathbf{M}^{\mathcal{D}}$ be the induction functor associated to a morphism of corings $(\gamma : \alpha) : (\mathcal{C} : A) \to (\mathcal{D} : B)$. Then there is an R-algebra isomorphism*

$$\mathrm{Nat}(F, F) \simeq \mathrm{End}_{\mathbf{Rep}(\mathcal{C}:A|\mathcal{D}:B)}((\gamma : \alpha)).$$

Proof. By 24.14, the R-algebra of natural endomorphisms of F is isomorphic to

$$\mathfrak{A}(\alpha, \gamma) = \{a \in {}_A\mathrm{Hom}_A(\mathcal{C}, B) \mid \forall\, c \in \mathcal{C},\ \sum \gamma(c_{\underline{1}})a(c_{\underline{2}}) = \sum a(c_{\underline{1}})\gamma(c_{\underline{2}})\}.$$

View B as an A-bimodule via the algebra map α. Then an R-module map $f : \mathcal{C} \to B$ is in $\mathrm{End}_{\mathbf{Rep}(\mathcal{C}:A|\mathcal{D}:B)}((\gamma : \alpha))$ if and only if $f \in {}_A\mathrm{Hom}_A(\mathcal{C}, B)$ and for all $c \in \mathcal{C}$, $\sum f(c_{\underline{1}})\gamma(c_{\underline{2}}) = \sum \gamma(c_{\underline{1}})f(c_{\underline{2}})$, that is, $f \in \mathfrak{A}(\alpha, \gamma)$, as claimed.
□

References. Gómez-Torrecillas [122]; Rojter [182].

Chapter 4

Corings and extensions of rings

Corings appear naturally in the context of extensions of rings. In fact, they provide an alternative description of such extensions. This chapter is devoted to studies of corings related to extensions. Thus we start with the description of the canonical coring associated to an extension, its basic properties and its connection with noncommutative descent theory. We then study specific classes of corings that are closely related to separable, split and Frobenius extensions. Next we analyse corings that are characterised by the existence of a nonzero element, known as a *grouplike element*. The key property of such an element $g \in \mathcal{C}$ is that its coproduct has the simplest possible form, $\underline{\Delta}(g) = g \otimes g$. We study the basic properties of corings with a grouplike element and reveal that they exhibit a natural ring structure. Then we associate a cochain complex to a coring and a grouplike element. The constructed complex turns out to generalise the Amitsur complex familiar in the ring extension theory. Next we proceed to introduce a differential graded algebra associated to a coring and a grouplike element and study some elements of the theory of connections in this differential graded algebra. This leads to the equivalence between the notions of a comodule and a module admitting a flat connection. Next we define Galois corings. These are corings that are isomorphic to the canonical coring associated to a ring extension, and they provide one with a natural framework for studying Galois-type ring extensions.

In this chapter, A and B are algebras over a commutative ring R.

25 Canonical corings and ring extensions

In this section we introduce the fundamental example of a coring, which was the motivation for Sweedler's work on dualisation of the Bourbaki-Jacobson Theorem. Such a coring can be associated to any extension of rings. Due to its importance, we thus term it the *canonical coring.* We study comodules of this coring and show how they are related to a noncommutative descent theory.

25.1. The canonical coring. *Let $B \to A$ be an extension of R-algebras. Then $\mathcal{C} = A \otimes_B A$ is an A-coring with coproduct*

$$\underline{\Delta} : \mathcal{C} \to \mathcal{C} \otimes_A \mathcal{C} \simeq A \otimes_B A \otimes_B A, \quad a \otimes a' \mapsto a \otimes 1_A \otimes a',$$

and counit $\underline{\varepsilon}(a \otimes a') = aa'$. $\mathcal{C}$ *is called the* canonical *or* Sweedler A-coring *associated to a ring (algebra) extension* $B \to A$.

Proof. Note that A is a (B, A)-bimodule, which is finitely generated projective as a right A-module, $A^* = \mathrm{Hom}_A(A, A) \simeq A$, and the dual basis consists only of 1_A viewed as an element of A or A^*. Thus this is a special case of a coring associated to a finitely generated projective module in 17.6.

Alternatively, $\mathcal{C}$ can be seen as a base ring extension of the trivial B-coring B (cf. 17.2, 17.3). □

25.2. Dual algebras of a canonical coring. As special cases of 17.9 one easily computes the dual algebras of a canonical coring. First, the left dual ${}^*\mathcal{C} = {}_A\mathrm{Hom}(A \otimes_B A, A) \simeq {}_B\mathrm{End}(A)$, where the endomorphisms have the algebra structure provided by the composition of maps. Second, $\mathcal{C}^* = \mathrm{Hom}_A(A \otimes_B A, A) \simeq \mathrm{End}_B(A)$ is an anti-algebra isomorphism. Therefore $\mathcal{C}^* \simeq \mathrm{End}_B(A)^{op}$ as algebras. Finally,

$$ {}^*\mathcal{C}^* = {}_A\mathrm{Hom}_A(A \otimes_B A, A) \simeq A^B = \{a \in A \mid \text{ for all } b \in B,\ ab = ba\}. $$

25.3. Comodules of a canonical coring. *Let* $B \to A$ *be an algebra extension, and let* $\mathcal{C} = A \otimes_B A$ *be the canonical* A*-coring in 25.1. The objects in the category* $\mathbf{M}^{\mathcal{C}}$ *are pairs* (M, f)*, where* M *is a right* A*-module and* $f : M \to M \otimes_B A$ *is a right* A*-module morphism such that, writing* $f(m) = \sum_i m_i \otimes a_i$ *for any* $m \in M$,

(1) $\sum_i f(m_i) \otimes a_i = \sum_i m_i \otimes 1_A \otimes a_i$;

(2) $\sum_i m_i a_i = m$.

Proof. Let M be a right $\mathcal{C}$-comodule with a coaction ϱ^M. This induces a right A-module map $f : M \xrightarrow{\varrho^M} M \otimes_A A \otimes_B A \xrightarrow{\simeq} M \otimes_B A$. Explicitly, if we denote $\varrho^M(m) = \sum m_{\underline{0}} \otimes_A m_{\underline{1}} = \sum_i m_i \otimes_A \bar{a}_i \otimes_B a_i$, for $m \in M$, then $f(m) = \sum_i m_i \bar{a}_i \otimes_B a_i$. Since ϱ^M is a coaction of $\mathcal{C} = A \otimes_B A$ on M (compare the definition of the coproduct in $A \otimes_B A$ in 25.1),

$$ \textstyle\sum_{ij} m_{ij} \otimes_A \bar{a}_{ij} \otimes_B a_{ij} \otimes_A \bar{a}_i \otimes_B a_i = \sum_i m_i \otimes_A \bar{a}_i \otimes_B 1_A \otimes_A 1_A \otimes_B a_i, $$

where, for each i, we used the notation $\varrho^M(m_i) = \sum_j m_{ij} \otimes_A \bar{a}_{ij} \otimes_B a_{ij}$. Note that in this notation $f(m_i) = \sum_j m_{ij}\bar{a}_{ij} \otimes_B a_{ij}$. This in turn implies that

$$ \textstyle\sum_{i,j} m_{ij}\bar{a}_{ij} \otimes_B a_{ij}\bar{a}_i \otimes_B a_i = \sum_i m_i \bar{a}_i \otimes_B 1_A \otimes_B a_i. $$

The last equality is equivalent to condition (1) for the map f defined above. Now the fact that ϱ^M is counital means that $\sum_i m_i \bar{a}_i a_i = m$, which is precisely condition (2) for the above-defined f. Now let (M, f) be a pair satisfying conditions (1) and (2). Write $f(m) = \sum_i m_i \otimes a_i$ and define a right A-module map $\varrho^M : M \to M \otimes_A \mathcal{C}$, $m \mapsto \sum_i m_i \otimes_A 1_A \otimes_B a_i$. Condition (1) then implies

$$\begin{aligned} (\varrho^M \otimes_A I_\mathcal{C}) \circ \varrho^M(m) &= \textstyle\sum_i \varrho^M(m_i) \otimes_A 1_A \otimes_B a_i \\ &= \textstyle\sum_{i,j} m_{ij} \otimes_A 1_A \otimes_B a_{ij} \otimes_A 1_A \otimes_B a_i \\ &= \textstyle\sum_i m_i \otimes_A 1_A \otimes_B 1_A \otimes_A 1_A \otimes_B a_i \\ &= \textstyle\sum_i m_i \otimes_A \underline{\Delta}(1_A \otimes_B a_i) \\ &= (I_M \otimes_A \underline{\Delta}) \circ \varrho^M(m). \end{aligned}$$

Here we used the notation $f(m_i) = \sum_j m_{ij} \otimes_B a_{ij}$. On the other hand, condition (2) implies

$$(I_M \otimes_A \underline{\varepsilon}) \circ \varrho^M(m) = \textstyle\sum_i m_i \otimes_A \underline{\varepsilon}(1_A \otimes_B a_i) = \sum_i m_i a_i = m,$$

that is, M is a right $\mathcal{C}$-comodule with the coaction ϱ^M. □

25.4. Descent data. Given an algebra extension $B \to A$, the category consisting of pairs (M, f), where M is a right A-module and $f : M \to M \otimes_B A$ satisfies conditions (1) and (2) in 25.3, is known as the category of (right) *descent data* associated to a noncommutative algebra extension $B \to A$ [96]. The category of (right) descent data is denoted by $\mathbf{Desc}(A/B)$. A morphism $(M, f) \to (M', f')$ in $\mathbf{Desc}(A/B)$ is a right A-module map $\phi : M \to M'$ such that $f' \circ \phi = (\phi \otimes I_A) \circ f$. The category $\mathbf{Desc}(A/B)$ is a noncommutative generalisation of the category of descent data associated to an extension of commutative rings introduced by Knus and Ojanguren in [28] and forms a backbone of the noncommutative extension of the classical *descent theory* [17, 18]. Recall that the descent theory provides answers to the following types of questions.

(i) *Descent of modules*: Given an algebra extension $B \to A$ and a right A-module M, is there a right B-module N such that $M \simeq N \otimes_B A$ as right A-modules?

(ii) *Classification of A-forms*: Given a right B-module N, classify all right B-modules M such that $N \otimes_B A \simeq M \otimes_B A$ as right A-modules.

Thus the result of 25.3 can be equivalently stated as

Proposition. *Let $B \to A$ be an algebra map. Then the category of descent data associated to this extension is isomorphic to the category of right comodules of the canonical coring $A \otimes_B A$.*

Under this isomorphism of categories, the induction functor $- \otimes_A \mathcal{C} : \mathbf{M}_A \to \mathbf{M}^\mathcal{C}$ takes the form of the functor $- \otimes_B A : \mathbf{M}_A \to \mathbf{Desc}(A/B)$. An A-module M is sent to the descent datum $(M \otimes_B A, f)$, where

$$f : M \otimes_B A \to M \otimes_B A \otimes_B A, \quad m \otimes a \mapsto m \otimes 1_A \otimes a.$$

The theory of corings provides a natural framework for studying the descent of structures related to noncommutative rings. This is explored in Section 28.

25.5. The inclusion morphism. *Let $\mathcal{C} = A \otimes_B A$ be the canonical coring associated to an algebra inclusion $i : B \hookrightarrow A$. Identify B with its image $i(B)$. View the subalgebra B of A as a trivial B-coring as in 17.3. Let $\gamma : B \to A \otimes_B A$ be a (B, B)-bilinear map given by $\gamma : b \mapsto 1_A \otimes b = b \otimes 1_A$. Then $(\gamma : I_B) : (B : B) \to (A \otimes_B A : A)$ is a pure morphism of corings, known as an* inclusion morphism.

Proof. Obviously, I_B is an algebra map, and, by definition, γ is a (B, B)-bimodule map; therefore only conditions (2) in 24.1 need to be verified. Clearly, the second of these conditions holds, since the counit in a trivial coring is the identity map, and the counit in the canonical coring is given by the product in A. Finally, for all $b \in B$,

$$\begin{aligned}\chi \circ (\gamma \otimes_B \gamma) \circ \underline{\Delta}_B(b) &= \chi(\gamma(1_B) \otimes \gamma(b)) = \chi(1_A \otimes 1_A \otimes 1_A \otimes b) \\ &= 1_A \otimes 1_A \otimes b = \underline{\Delta}_{A \otimes_B A}(1_A \otimes b) = \underline{\Delta}_{A \otimes_B A}(\gamma(b)),\end{aligned}$$

as required. Thus $(\gamma : I_B)$ is a morphism of corings. It is a pure morphism of corings, since B is a flat left B-module. $\square$

25.6. Representations of the canonical coring in an algebra. *Let C be an R-algebra. The category* **Rep**$(A \otimes_B A : A \mid C : C)$ *of representations of the canonical A-coring $A \otimes_B A$ associated to an algebra map $i : B \to A$ in C has as objects the R-algebra maps $\alpha : A \to C$. Morphisms are given by*

$$\mathrm{Mor}_{\mathbf{Rep}(A \otimes_B A : A \mid C : C)}(\alpha_1, \alpha_2) = {}_{\alpha_1}C^B_{\alpha_2} \equiv \{c \in C \mid \forall\, b \in B,\ \alpha_1(i(b))c = c\alpha_2(i(b))\}.$$

Proof. This follows by 24.4 and the identification ${}_A\mathrm{Hom}_A(A \otimes_B A, C) \simeq {}_{\alpha_1}C^B_{\alpha_2}$. $\square$

25.7. Representations of a coalgebra in the canonical coring. *For an R-coalgebra C and a canonical coring $A \otimes_B A$, the objects of the category* **Rep**$(C : R \mid A \otimes_B A : A)$ *are R-linear maps $\gamma : C \to A \otimes_B A$ such that, writing $\gamma(c) = \sum \gamma(c)^{\tilde{1}} \otimes \gamma(c)^{\tilde{2}}$ for $c \in C$,*

(1) $\sum \gamma(c)^{\tilde{1}} \gamma(c)^{\tilde{2}} = \varepsilon(c) 1_A$;

(2) $\sum \gamma(c_{\underline{1}})^{\tilde{1}} \otimes \gamma(c_{\underline{1}})^{\tilde{2}} \gamma(c_{\underline{2}})^{\tilde{1}} \otimes \gamma(c_{\underline{2}})^{\tilde{2}} = \sum \gamma(c)^{\tilde{1}} \otimes 1_A \otimes \gamma(c)^{\tilde{2}}$.

Morphisms $\gamma_1 \to \gamma_2$ are R-linear maps $f : C \to A$ such that, for all $c \in C$, $\sum \gamma_1(c_{\underline{1}}) f(c_{\underline{2}}) = \sum f(c_{\underline{1}}) \gamma_2(c_{\underline{2}})$.

Proof. This follows immediately from 24.5 and the definition of the canonical coring. □

A concrete example of a representation of the type described in 25.7 is given in 34.5.

References. Brzeziński [73]; Cippola [96]; Grothendieck [17, 18]; Knus and Ojanguren [28]; Sweedler [193].

26 Coseparable and cosplit corings

Among all the types of extensions of rings, separable and split extensions form two important classes. The former generalise the notion of a separable algebra, and the latter is a complementary notion to the former. In this section we study two types of corings that are closely related to separable and split extensions of rings. These corings are respectively known as *coseparable* and *cosplit* corings, and their properties determine when the forgetful and induction functors are separable, in particular, we show that a coseparable coring has a structure of a nonunital ring whose (balanced) product has a section. Comodules of $\mathcal{C}$ can then be viewed as (firm) modules of the corresponding nonunital ring.

Definition. An A-coring $\mathcal{C}$ is said to be *coseparable* if the structure map $\underline{\Delta} : \mathcal{C} \to \mathcal{C} \otimes_A \mathcal{C}$ splits as a $(\mathcal{C}, \mathcal{C})$-bicomodule map, that is, if there exists an (A, A)-bimodule map $\pi : \mathcal{C} \otimes_A \mathcal{C} \to \mathcal{C}$ such that

$$(I_{\mathcal{C}} \otimes \pi) \circ (\underline{\Delta} \otimes I_{\mathcal{C}}) = \underline{\Delta} \circ \pi = (\pi \otimes I_{\mathcal{C}}) \circ (I_{\mathcal{C}} \otimes \underline{\Delta}) \quad \text{and} \quad \pi \circ \underline{\Delta} = I_{\mathcal{C}}.$$

This is obviously a left-right symmetric notion, which can be viewed as a dualisation of *separable extensions of algebras* (in which case the splitting of the product is required). The notion of a coseparable coring generalises the notion of a coseparable coalgebra introduced in 3.28, and, in fact, the characterisations of coseparable coalgebras in 3.29 can be transferred.

26.1. Coseparable corings. *For an A-coring $\mathcal{C}$ the following are equivalent:*

(a) $\mathcal{C}$ is coseparable;

(b) there exists an (A, A)-linear map $\delta : \mathcal{C} \otimes_A \mathcal{C} \to A$ satisfying

$$\delta \circ \underline{\Delta} = \underline{\varepsilon} \quad \text{and} \quad (I_{\mathcal{C}} \otimes \delta) \circ (\underline{\Delta} \otimes I_{\mathcal{C}}) = (\delta \otimes I_{\mathcal{C}}) \circ (I_{\mathcal{C}} \otimes \underline{\Delta});$$

(c) the forgetful functor $(-)_A : \mathbf{M}^{\mathcal{C}} \to \mathbf{M}_A$ is separable;

(d) the forgetful functor ${}_A(-) : {}^{\mathcal{C}}\mathbf{M} \to {}_A\mathbf{M}$ is separable;

(e) the forgetful functor ${}_A(-)_A : {}^{\mathcal{C}}\mathbf{M}^{\mathcal{C}} \to {}_A\mathbf{M}_A$ is separable;

(f) $\mathcal{C}$ is (A, A)-relative semisimple as a $(\mathcal{C}, \mathcal{C})$-bicomodule, that is, any monomorphism in ${}^{\mathcal{C}}\mathbf{M}^{\mathcal{C}}$ that splits as an (A, A)-morphism also splits in ${}^{\mathcal{C}}\mathbf{M}^{\mathcal{C}}$;

(g) $\mathcal{C}$ is (A, A)-relative injective as a $(\mathcal{C}, \mathcal{C})$-bicomodule;

(h) $\mathcal{C}$ is $(A, \mathcal{C})$-relative injective as a $(\mathcal{C}, \mathcal{C})$-bicomodule;

(i) $\mathcal{C}$ is $(\mathcal{C}, A)$-relative injective as a $(\mathcal{C}, \mathcal{C})$-bicomodule.

If these conditions are satisfied, then $\mathcal{C}$ is left and right $(\mathcal{C}, A)$-semisimple, that is, all comodules in $\mathbf{M}^{\mathcal{C}}$ and ${}^{\mathcal{C}}\mathbf{M}$ are $(\mathcal{C}, A)$-injective.

Proof. The proof of 3.29 can be adapted easily. Note that (g) implies both (h) and (i), while (h) and (i) imply (a) since $\underline{\Delta}$ is split by a left A-module right $\mathcal{C}$-comodule map $\underline{\varepsilon} \otimes I_{\mathcal{C}}$ as well as by a right A-module left $\mathcal{C}$-comodule map $I_{\mathcal{C}} \otimes \underline{\varepsilon}$ (cf. 22.8). □

26.2. Cointegrals and Maschke-type theorem. Any map $\delta : \mathcal{C} \otimes_A \mathcal{C} \to A$ satisfying conditions 26.1(b) is known as a *cointegral* in $\mathcal{C}$. Note that the relationship between a splitting map π and the corresponding cointegral is given by the formulae $\delta = \underline{\varepsilon} \circ \pi$ and

$$\pi(c \otimes c') = \sum \delta(c \otimes c'_{\underline{1}})c'_{\underline{2}} = \sum c_{\underline{1}}\delta(c_{\underline{2}} \otimes c'),$$

for all $c, c' \in \mathcal{C}$ (compare the proof of 3.29).

Note also that the final assertion in 26.1 involving the relative semisimplicity of $\mathcal{C}$ can be viewed as a Maschke-type theorem for corings.

Thus, over a coseparable coring any comodule is relative injective, and so the purity conditions 21.5 are satisfied. Hence, similarly as for coalgebras (see 11.5), 22.3 implies:

26.3. Cotensor product over coseparable corings. *Let $\mathcal{C}$ be an A-coring and $\mathcal{D}$ a B-coring. If $\mathcal{C}$ is a coseparable coring, then:*

(1) for any $M \in {}^{\mathcal{D}}\mathbf{M}^{\mathcal{C}}$ and $N \in {}^{\mathcal{C}}\mathbf{M}$, $M \Box_{\mathcal{C}} N$ is a left $\mathcal{D}$-comodule.

(2) For any $M \in \mathbf{M}^{\mathcal{C}}$ and $N \in {}^{\mathcal{C}}\mathbf{M}^{\mathcal{D}}$, $M \Box_{\mathcal{C}} N$ is a right $\mathcal{D}$-comodule.

(3) For any $M \in {}^{\mathcal{D}}\mathbf{M}^{\mathcal{C}}$ and $N \in {}^{\mathcal{C}}\mathbf{M}^{\mathcal{D}'}$, $M \Box_{\mathcal{C}} N$ is a $(\mathcal{D}, \mathcal{D}')$-bicomodule.

26.4. Dual rings and cosperable corings. *Let B be the opposite algebra of the left dual algebra ${}^*\mathcal{C}$ of an A-coring $\mathcal{C}$, and let $i_L : A \to B$, $a \mapsto \underline{\varepsilon}(-a)$, be the corresponding algebra map as in 17.8. If $\mathcal{C}$ is cogenerated by A as a left A-module and the extension $i_L : A \to B$ is separable, then $\mathcal{C}$ is a coseparable coring.*

Proof. Let $e = \sum_i f_i \otimes g_i \in (B \otimes_A B)^B$ be the separability idempotent. The B-centrality of e implies that, for all $c, c' \in \mathcal{C}$ and $f \in {}^*\mathcal{C}$,

$$\sum_i g_i(cf_i(c'_{\underline{1}}f(c'_{\underline{2}}))) = \sum_i f(c_{\underline{1}}g_i(c_{\underline{2}}f_i(c'))), \tag{$*$}$$

while the normalisation of e means that $\sum_i g_i(c_{\underline{1}}f_i(c_{\underline{2}})) = \underline{\varepsilon}(c)$. Define a left A-linear map $\delta : \mathcal{C} \otimes_A \mathcal{C} \to A$, $c \otimes c' \mapsto \sum_i g_i(cf_i(c'))$. For any $a \in A$, take the map $i_L(a)$ for f in $(*)$ to find that δ is also right A-linear. For all $f \in {}^*\mathcal{C}$ and $c, c' \in \mathcal{C}$, one can use $(*)$ and the (A, A)-bilinearity of δ to compute

$$\begin{aligned} \sum f(c_{\underline{1}}\delta(c_{\underline{2}} \otimes c')) &= \sum_i f(c_{\underline{1}}g_i(c_{\underline{2}}f_i(c'))) = \sum_i g_i(cf_i(c'_{\underline{1}}f(c'_{\underline{2}}))) \\ &= \sum \delta(c \otimes c'_{\underline{1}}f(c'_{\underline{2}})) = \sum f(\delta(c \otimes c'_{\underline{1}})c'_{\underline{2}}). \end{aligned}$$

Since ${}_A\mathcal{C}$ is cogenerated by A, this implies that $\sum c_{\underline{1}}\delta(c_{\underline{2}}\otimes c') = \sum \delta(c\otimes c'_{\underline{1}})c'_{\underline{2}}$. On the other hand, the unitality of e precisely means that $\delta\circ\underline{\Delta} = \underline{\varepsilon}$. Therefore δ is a cointegral, and hence $\mathcal{C}$ is a coseparable A-coring. □

Coseparable corings also turn out to be closely related to the following generalisation of separable extensions of rings.

26.5. Separable A-rings. Let A be an R-algebra with unit 1_A and let B be an R-algebra (possibly) without unit. B is said to be a *separable A-ring* if

(1) B is an (A, A)-bimodule;

(2) the product $\mu : B \otimes_R B \to B$ is an A-balanced (A, A)-bimodule map, that is, for all $a \in A$ and $b, b' \in B$, $\mu(ab \otimes b') = a\mu(b \otimes b')$, $\mu(b \otimes b'a) = \mu(b \otimes b')a$ and $\mu(ba \otimes b') = \mu(b \otimes ab')$;

(3) the induced (B, B)-bimodule map $\mu_{B/A} : B \otimes_A B \to B$, $b \otimes b \mapsto bb'$ has a section.

26.6. A coseparable A-coring is a separable A-ring. *If $\mathcal{C}$ is a coseparable A-coring, then $\mathcal{C}$ is a separable A-ring.*

Proof. Let $\pi : \mathcal{C} \otimes_A \mathcal{C} \to \mathcal{C}$ be a bicomodule retraction of the coproduct $\underline{\Delta}_{\mathcal{C}}$, and let $\delta = \underline{\varepsilon} \circ \pi$ be the corresponding cointegral. We claim that $\mathcal{C}$ is an associative R-algebra (without unit) with product $cc' = \pi(c \otimes c')$. Indeed, since alternative expressions for the product are $cc' = \sum \delta(c \otimes c'_{\underline{1}})c'_{\underline{2}} = \sum c_{\underline{1}}\delta(c_{\underline{2}} \otimes c')$ (cf. 26.2), for all $c, c', c'' \in \mathcal{C}$, use of the left A-linearity of δ and $\underline{\Delta}$ yields

$$(cc')c'' = \sum(\delta(c \otimes c'_{\underline{1}})c'_{\underline{2}})c'' = \sum \delta(c \otimes c'_{\underline{1}})\delta(c'_{\underline{2}} \otimes c''_{\underline{1}})c''_{\underline{2}}.$$

On the other hand, the colinearity of π, the right A-linearity of δ and the left A-linearity of $\underline{\Delta}$ imply

$$\begin{aligned} c(c'c'') &= \sum c(\delta(c' \otimes c''_{\underline{1}})c''_{\underline{2}}) = \sum \delta(c \otimes \delta(c' \otimes c''_{\underline{1}})c''_{\underline{2}})c''_{\underline{3}} \\ &= \sum \delta(c \otimes c'_{\underline{1}}\delta(c'_{\underline{2}} \otimes c''_{\underline{1}}))c''_{\underline{2}} = \sum \delta(c \otimes c'_{\underline{1}})\delta(c'_{\underline{2}} \otimes c''_{\underline{1}})c''_{\underline{2}}. \end{aligned}$$

This explicitly proves that the product in $\mathcal{C}$ is associative. Clearly this product is (A, A)-bilinear and A-balanced, and the induced map $\mu_{\mathcal{C}/A}$ is precisely π. Note that $\underline{\Delta}$ is a $(\mathcal{C}, \mathcal{C})$-bimodule map since

$$c\underline{\Delta}(c') = \sum cc'_{\underline{1}} \otimes c'_{\underline{2}} = \sum \pi(c \otimes c'_{\underline{1}}) \otimes c'_{\underline{2}} = \underline{\Delta} \circ \pi(c \otimes c') = \underline{\Delta}(cc'),$$

by right colinearity of π, and similarly for left $\mathcal{C}$-linearity. Finally, π is split by $\underline{\Delta}$ since π is a retraction of $\underline{\Delta}$. This proves that $\mathcal{C}$ is a separable A-ring. □

26.7. Comodules of a coseparable coring. *Let $\mathcal{C}$ be a coseparable coring with cointegral δ. Then any right $\mathcal{C}$-comodule M is a right $\mathcal{C}$-module with the product $mc = \sum m_{\underline{0}}\delta(m_{\underline{1}} \otimes c)$. This product is right A-linear and A-balanced. Furthermore M is* firm, *that is, $M \otimes_{\mathcal{C}} \mathcal{C} \simeq M$ as right $\mathcal{C}$-modules.*

Proof. Take any $m \in M$ and $c, c' \in \mathcal{C}$ and compute

$$\begin{aligned}(mc)c' &= \sum(m_{\underline{0}}\delta(m_{\underline{1}} \otimes c))c' = \sum m_{\underline{0}}\delta(m_{\underline{1}}\delta(m_{\underline{2}} \otimes c) \otimes c')\\ &= \sum m_{\underline{0}}\delta(\delta(m_{\underline{1}} \otimes c_{\underline{1}})c_{\underline{2}} \otimes c') = \sum m_{\underline{0}}\delta(m_{\underline{1}} \otimes c_{\underline{1}})\delta(c_{\underline{2}} \otimes c')\\ &= \sum m_{\underline{0}}\delta(m_{\underline{1}} \otimes c_{\underline{1}}\delta(c_{\underline{2}} \otimes c')) = \sum m_{\underline{0}}\delta(m_{\underline{1}} \otimes cc') = m(cc'),\end{aligned}$$

as required. We use the colinearity of δ to derive the third equality and (A, A)-bilinearity of δ to derive the fourth and fifth equalities. Obviously the multiplication is right A-linear. It is A-balanced since the coaction is right A-linear.

Note that $M \otimes_{\mathcal{C}} \mathcal{C}$ is defined as a cokernel of the right $\mathcal{C}$-linear map

$$\lambda : M \otimes_A \mathcal{C} \otimes_A \mathcal{C} \to M \otimes_A \mathcal{C}, \quad m \otimes c \otimes c' \mapsto mc \otimes c' - m \otimes cc'.$$

Consider now the following right $\mathcal{C}$-linear map:

$$\varrho_{M/A} : M \otimes_A \mathcal{C} \to M, \quad m \otimes c \mapsto mc = \sum m_{\underline{0}}\delta(m_{\underline{1}} \otimes c).$$

Since δ is a cointegral, $\varrho_{M/A}$ is a right $\mathcal{C}$-linear retraction of ϱ^M, and hence, in particular it is surjective and there is the following sequence of right $\mathcal{C}$-module maps:

$$M \otimes_A \mathcal{C} \otimes_A \mathcal{C} \xrightarrow{\lambda} M \otimes_A \mathcal{C} \xrightarrow{\varrho_{M/A}} M \longrightarrow 0.$$

We need to show that this sequence is exact. Clearly associativity of the product in M implies that $\varrho_{M/A} \circ \lambda = 0$, so that $\mathrm{Im}\,\lambda \subseteq \mathrm{Ke}\,\varrho_{M/A}$. Furthermore, for all $m \in M$ and $c \in \mathcal{C}$,

$$\begin{aligned}&(\varrho^M \circ \varrho_{M/A} - \lambda \circ (I_M \otimes \underline{\Delta}))(m \otimes c)\\ &= \sum m_{\underline{0}} \otimes m_{\underline{1}}\delta(m_{\underline{2}} \otimes c) - \sum mc_{\underline{1}} \otimes c_{\underline{2}} + \sum m \otimes \pi(c_{\underline{1}} \otimes c_{\underline{2}})\\ &= \sum m_{\underline{0}}\delta(m_{\underline{1}} \otimes c_{\underline{1}}) \otimes c_{\underline{2}} - \sum m_{\underline{0}}\delta(m_{\underline{1}} \otimes c_{\underline{1}}) \otimes c_{\underline{2}} + m \otimes c \;=\; m \otimes c,\end{aligned}$$

where we used the properties of a cointegral. This implies $\mathrm{Ke}\,\varrho_{M/A} \subseteq \mathrm{Im}\,\lambda$, that is, the above sequence is exact, as required. □

In fact 26.6 has the following (partly) converse.

26.8. Coassociative coproduct in a separable A-ring. *Let B be a separable A-ring. Then the (B, B)-bimodule map $\Delta : B \to B \otimes_A B$ splitting the product $\mu_{B/A}$ is coassociative.*

Proof. Since Δ is assumed to be a (B,B)-bimodule map, there is the following commutative diagram:

$$\begin{array}{ccccc} B\otimes_A B\otimes_A B & \xleftarrow{I_B\otimes\Delta} & B\otimes_A B & \xrightarrow{\Delta\otimes I_B} & B\otimes_A B\otimes_A B \\ \downarrow{\scriptstyle \mu_{B/A}\otimes I_B} & & \downarrow{\scriptstyle \mu_{B/A}} & & \downarrow{\scriptstyle I_B\otimes\mu_{B/A}} \\ B\otimes_A B & \xleftarrow[\Delta]{} & B & \xrightarrow[\Delta]{} & B\otimes_A B\,. \end{array}$$

For all $b\in B$ we write $(\Delta\otimes I_B)\circ\Delta(b)=\sum b_{\underline{11}}\otimes b_{\underline{12}}\otimes b_{\underline{2}}$ and $(I_B\otimes\Delta)\circ\Delta(b)=\sum b_{\underline{1}}\otimes b_{\underline{21}}\otimes b_{\underline{22}}$, and use the above diagram to obtain

$$\begin{aligned}\Delta(b) &= (\Delta\circ\mu_{B/A}\circ\Delta)(b) \;=\; (I_B\otimes\mu_{B/A})\circ(\Delta\otimes I_B)\circ\Delta(b)\\ &= \sum b_{\underline{11}}\otimes\mu_{B/A}(b_{\underline{12}}\otimes b_{\underline{2}}) \;=\; (\mu_{B/A}\otimes I_B)\circ(I_B\otimes\Delta)\circ\Delta(b)\\ &= \sum \mu_{B/A}(b_{\underline{1}}\otimes b_{\underline{21}})\otimes b_{\underline{22}}.\end{aligned}$$

Using these identities we compute

$$\begin{aligned}(I_B\otimes\Delta)\circ\Delta(b) &= \sum b_{\underline{11}}\otimes(\Delta\circ\mu_{B/A})(b_{\underline{12}}\otimes b_{\underline{2}})\\ &= \sum b_{\underline{11}}\otimes((\mu_{B/A}\otimes I_B)\circ(I_B\otimes\Delta))(b_{\underline{12}}\otimes b_{\underline{2}})\\ &= \sum b_{\underline{11}}\otimes\mu_{B/A}(b_{\underline{12}}\otimes b_{\underline{21}})\otimes b_{\underline{22}}\\ &= \sum((I_B\otimes\mu_{B/A})\circ(\Delta\otimes I_B))(b_{\underline{1}}\otimes b_{\underline{21}})\otimes b_{\underline{22}}\\ &= \sum(\Delta\circ\mu_{B/A})(b_{\underline{1}}\otimes b_{\underline{21}})\otimes b_{\underline{22}}\\ &= (\Delta\otimes I_B)\circ\Delta(b),\end{aligned}$$

that is, $(\Delta\otimes I_B)\circ\Delta=(I_B\otimes\Delta)\circ\Delta$. This completes the proof. □

Thus a separable A-ring B also can be viewed as an A-coring with coproduct Δ but without a counit (i.e., B is a *noncounital coring*). Note then that the diagram in the proof of 26.8 can be understood as a statement that $\mu_{B/A}$ is a (B,B)-bicomodule map. Since $\mu_{B/A}$ is a retraction for Δ, one can say that a separable A-ring is a *coseparable noncounital A-coring*.

The coseparability of the canonical coring associated to an algebra map $B\to A$ is related to the problem when this map defines a split extension. Recall that an extension of algebras $B\to A$ is called a *split extension* if there exists a (B,B)-bimodule map $E:A\to B$ such that $E(1_A)=1_B$.

26.9. Coseparability of Sweedler corings. *Let $\phi:B\to A$ be an algebra map, and let $\mathcal{C}=A\otimes_B A$ be the canonical A-coring in 25.1. If ϕ is a split extension, then $\mathcal{C}$ is a coseparable coring. Conversely, if $\mathcal{C}$ is coseparable and either ${}_BA$ or A_B is faithfully flat, then ϕ is a split extension.*

Proof. In the case of the canonical A-coring $A\otimes_B A$, the conditions required for the cointegral $\delta\in{}_A\mathrm{Hom}_A(A\otimes_B A\otimes_B A,A)$ in 26.1 read as

$$\delta(a \otimes 1_A \otimes a') = aa' \text{ and } a \otimes \delta(1_A \otimes a' \otimes a'') = \delta(a \otimes a' \otimes 1_A) \otimes a'',$$

for all $a, a', a'' \in A$. Since ${}_A\mathrm{Hom}_A(A \otimes_B A \otimes_B A, A) \simeq {}_B\mathrm{Hom}_B(A, A)$, the maps δ are in one-to-one correspondence with the maps $E \in {}_B\mathrm{Hom}_B(A, A)$ via $\delta(a \otimes a' \otimes a'') = aE(a')a''$. The first of the above conditions for δ is equivalent to the normalisation of E, $E(1_A) = 1_A$, while the second condition gives for all $a \in A$, $1_A \otimes_B E(a) = E(a) \otimes_B 1_A$. Thus, if the extension ϕ is split, such an E exists by definition. Conversely, if $\mathcal{C}$ is coseparable and either ${}_BA$ or A_B is faithfully flat, then ϕ is injective and the equation $1_A \otimes_B E(a) = E(a) \otimes_B 1_A$ implies that $E(a) \in B$, where B is identified with its image in A. □

The coseparability of $\mathcal{C}$ in relation to A as considered in 26.1 can also be studied in relation to other corings.

Definition. Given a morphism $\gamma : \mathcal{C} \to \mathcal{D}$ of A-corings, $\mathcal{C}$ is called $\mathcal{D}$-*coseparable* if $\underline{\Delta}_{\mathcal{C}} : \mathcal{C} \to \mathcal{C}\Box_{\mathcal{D}}\mathcal{C}$ splits as a $(\mathcal{C}, \mathcal{C})$-bicomodule map, that is, there exists a map $\pi : \mathcal{C}\Box_{\mathcal{D}}\mathcal{C} \to \mathcal{C}$ with the properties

$$(I_{\mathcal{C}}\Box_{\mathcal{D}}\pi) \circ (\underline{\Delta}_{\mathcal{C}}\Box_{\mathcal{D}}I_{\mathcal{C}}) = \underline{\Delta}_{\mathcal{C}} \circ \pi = (\pi\Box_{\mathcal{D}}I_{\mathcal{C}}) \circ (I_{\mathcal{C}}\Box_{\mathcal{D}}\underline{\Delta}_{\mathcal{C}}) \quad \text{and} \quad \pi \circ \underline{\Delta}_{\mathcal{C}} = I_{\mathcal{C}}.$$

$\mathcal{D}$-coseparable corings are closely related to the injectivity of a coring relative to another coring or $(\mathcal{C}, \mathcal{D})$-injectivity discussed in 22.13. In particular, the characterisation of coseparable corings in 26.1 extends to $\mathcal{D}$-coseparable corings in the following way.

26.10. $\mathcal{D}$-coseparable corings. *Let $\gamma : \mathcal{C} \to \mathcal{D}$ be a morphism of A-corings. Then the following are equivalent:*

(a) $\mathcal{C}$ is $\mathcal{D}$-coseparable;

(b) there exists a $(\mathcal{D}, \mathcal{D})$-bicolinear map $\delta : \mathcal{C}\Box_{\mathcal{D}}\mathcal{C} \to \mathcal{D}$ satisfying

$$\delta \circ \underline{\Delta}_{\mathcal{C}} = \gamma \quad \text{and} \quad (I_{\mathcal{C}}\Box_{\mathcal{D}}\delta) \circ (\underline{\Delta}_{\mathcal{C}}\Box_{\mathcal{D}}I_{\mathcal{C}}) = (\delta\Box_{\mathcal{D}}I_{\mathcal{C}}) \circ (I_{\mathcal{C}}\Box_{\mathcal{D}}\underline{\Delta}_{\mathcal{C}}).$$

If $\mathcal{C}$ satisfies the left and right α-conditions 19.2, (a) and (b) are also equivalent to:

(c) the corestriction functor $(-)_\gamma : \mathbf{M}^{\mathcal{C}} \to \mathbf{M}^{\mathcal{D}}$ is separable (cf. 22.11);

(d) the corestriction functor ${}_\gamma(-) : {}^{\mathcal{C}}\mathbf{M} \to {}^{\mathcal{D}}\mathbf{M}$ is separable;

(e) the corestriction functor ${}_\gamma(-)_\gamma : {}^{\mathcal{C}}\mathbf{M}^{\mathcal{C}} \to {}^{\mathcal{D}}\mathbf{M}^{\mathcal{D}}$ is separable;

(f) $\mathcal{C}$ is $\mathcal{D}$-relative semisimple as a $(\mathcal{C}, \mathcal{C})$-bicomodule;

(g) $\mathcal{C}$ is $\mathcal{D}$-relative injective as a $(\mathcal{C}, \mathcal{C})$-bicomodule.

If these conditions are satisfied, then all comodules in $\mathbf{M}^{\mathcal{C}}$ and ${}^{\mathcal{C}}\mathbf{M}$ are $(\mathcal{C}, \mathcal{D})$-injective.

Proof. Recall that the left and right α-conditions imply that the categories of $\mathcal{C}$-comodules are fully included in the categories of ${}^*\mathcal{C}$-modules or $\mathcal{C}^*$-modules. Since 39.5 holds for noncommmutative algebras, we can apply the proof of 11.12. □

The separability of the functor $- \otimes_A \mathcal{C}$ has a similar characterisation as for coalgebras.

26.11. Separability of $- \otimes_A \mathcal{C}$. *The following are equivalent for* $\mathcal{C}$:

(a) $- \otimes_A \mathcal{C} : \mathbf{M}_A \to \mathbf{M}^{\mathcal{C}}$ *is a separable functor;*

(b) $\mathcal{C} \otimes_A - : {}_A\mathbf{M} \to {}^{\mathcal{C}}\mathbf{M}$ *is a separable functor;*

(c) $\mathcal{C} \otimes_A - \otimes_A \mathcal{C} : {}_A\mathbf{M}_A \to {}^{\mathcal{C}}\mathbf{M}^{\mathcal{C}}$ *is a separable functor;*

(d) *there exists an invariant* $e \in \mathcal{C}^A$ *with* $\underline{\varepsilon}(e) = 1_A$.

Proof. (a) $\Rightarrow$ (d) Let ψ denote the counit of the adjoint pair of functors $((-)_A, - \otimes_A \mathcal{C})$. In particular, $\psi_A = I_A \otimes \underline{\varepsilon} : A \otimes_A \mathcal{C} \to A$, that is, $\psi_A = \underline{\varepsilon}$ by the obvious identification. By 38.24(2), ψ_A is split by a morphism $\nu_A : A \to \mathcal{C}$. Since A is also a left A-module, we know by 39.5 that ν_A is in fact left and right A-linear and hence $e = \nu_R(1_A)$ is in $\mathcal{C}^A$. By definition, $\underline{\varepsilon}(e) = \psi_A \circ \nu_A(1_A) = 1_A$.

(d) $\Rightarrow$ (a) Follow the proof of 3.30 and observe that $e \in \mathcal{C}^A$ implies that, for all $M \in \mathbf{M}_A$, the map $\nu_M : M \to M \otimes_A \mathcal{C}$, $m \mapsto m \otimes e$, is right A-linear and provides the required natural splitting of the counit of adjunction ψ.

The remaining arguments of the proof of 3.30 can be followed literally. □

26.12. Cosplit corings. Note that condition (d) in 26.11 simply means that there is an (A, A)-bimodule section of a counit. Thus it is natural to term an A-coring satisfying any of the equivalent conditions in 26.11 a *cosplit coring*. The corresponding invariant $e \in \mathcal{C}^A$ is called a *normalised integral in* $\mathcal{C}$.

The notion of a cosplit coring can be viewed as a dualisation of the notion of a split extension (hence the name). This can be formalised as the following observation.

26.13. Dual ring of a cosplit coring. *Let* $\mathcal{C}$ *be a cosplit* A*-coring, and let* $B = ({}^*\mathcal{C})^{op}$ *be the opposite left dual ring. Then the algebra extension* $i_L : A \to B$, $a \mapsto \underline{\varepsilon}(-a)$ *is a split extension.*

Proof. Let e be a normalised integral in $\mathcal{C}$, and consider an R-linear map $E : B \to A$, $b \mapsto b(e)$. Note that $E(1_B) = E(\underline{\varepsilon}) = \underline{\varepsilon}(e) = 1_A$ by the normalisation of e. Furthermore, for all $a \in A$ and $b \in B$ we can use the definition of $B = ({}^*\mathcal{C})^{op}$, the properties of the counit, and the fact that e is an invariant to compute

$$\begin{aligned} E(ab) &= (i_L(a)b)(e) = \sum b(e_{\underline{1}} i_L(a)(e_{\underline{2}})) = \sum b(e_{\underline{1}}\underline{\varepsilon}(e_{\underline{2}})a) \\ &= b(ea) = b(ae) = ab(e) = aE(b), \end{aligned}$$

showing that E is left A-linear. Similarly one shows that E is also right A-linear, as required for a split extension. $\square$

We noticed that a coseparable Sweedler coring corresponds to a split extension. Somewhat perversely, cosplit Sweedler corings correspond to separable extensions.

26.14. Cosplit Sweedler corings. *Let $B \to A$ be an algebra extension and let $\mathcal{C}$ be the canonical A-coring in 25.1. The extension $B \to A$ is separable if and only if $\mathcal{C}$ is a cosplit coring.*

Proof. Recall that an extension is separable if there exists an invariant $e \in (A \otimes_B A)^A$, such that $\mu_{A/B}(e) = 1_A$, where $\mu_{A/B} : A \otimes_B A \to A$ is the product. In view of this the assertion follows immediately from the characterisation of cosplit corings in 26.11. $\square$

Given a coring morphism $\mathcal{C} \to \mathcal{D}$, there is a functor $-\Box_{\mathcal{D}}\mathcal{C} : \mathbf{M}^{\mathcal{D}} \to \mathbf{M}^{\mathcal{C}}$, provided suitable pureness conditions are satisfied (cf. 22.3). The separability of the functor $-\Box_{\mathcal{D}}\mathcal{C}$ has similar characterisations as that of $- \otimes_A \mathcal{C}$.

26.15. Separability of $-\Box_{\mathcal{D}}\mathcal{C}$. *Let $\gamma : \mathcal{C} \to \mathcal{D}$ be a morphism of A-corings, assume ${}_A\mathcal{C}$ and $\mathcal{C}_A$ to be flat and $\mathcal{D}$ to satisfy the left and right α-condition. Then the following are equivalent:*

(a) $-\Box_{\mathcal{D}}\mathcal{C} : \mathbf{M}^{\mathcal{D}} \to \mathbf{M}^{\mathcal{C}}$ *is separable;*

(b) $\mathcal{C}\Box_{\mathcal{D}}- : {}^{\mathcal{D}}\mathbf{M} \to {}^{\mathcal{C}}\mathbf{M}$ *is separable;*

(c) $\mathcal{C}\Box_{\mathcal{D}} - \Box_{\mathcal{D}}\mathcal{C} : {}^{\mathcal{D}}\mathbf{M}^{\mathcal{D}} \to {}^{\mathcal{C}}\mathbf{M}^{\mathcal{C}}$ *is separable;*

(d) there exists a $(\mathcal{D},\mathcal{D})$-colinear map $\delta : \mathcal{D} \to \mathcal{C}$ with $\gamma \circ \delta = I_{\mathcal{D}}$.

Proof. (a) $\Rightarrow$ (d) Let ψ denote the counit of the adjoint functor pair $((-)_\gamma, -\Box_{\mathcal{D}}\mathcal{C})$. Then $\psi_{\mathcal{D}} = I_{\mathcal{D}}\Box\gamma : \mathcal{D}\Box_{\mathcal{D}}\mathcal{C} \to \mathcal{D}$, that is, $\psi_{\mathcal{D}} = \gamma$ by the obvious identification. By 38.24(2), $\psi_{\mathcal{D}}$ is split by a morphism $\nu_{\mathcal{D}} : \mathcal{D} \to \mathcal{C}$ in $\mathbf{M}^{\mathcal{D}}$. Since $\mathcal{D}$ is also a left $\mathcal{D}$-comodule – that is, a right $\mathcal{D}^*$-module – we know by 39.5 that $\nu_{\mathcal{D}}$ is in fact right $\mathcal{D}^*$-linear and hence it is $(\mathcal{D},\mathcal{D})$-colinear.

(d) $\Rightarrow$ (a) The map $I_M\Box\delta : M \simeq M\Box_{\mathcal{D}}\mathcal{D} \to M\Box_{\mathcal{D}}\mathcal{C}$ is right $\mathcal{D}$-colinear and also $(I_M\Box\gamma) \circ (I\Box\Delta) = I_M$. Hence it provides the required natural splitting of the counit of adjunction ψ.

The remaining implications follow by symmetry and the properties of separable functors (cf. 38.20). $\square$

References. Brzeziński [73]; Brzeziński, Kadison and Wisbauer [78]; Caenepeel [8]; Gómez-Torrecillas [122]; Guzman [126, 127]; Pierce [39].

27 Frobenius extensions and corings

Corings appear naturally in the context of Frobenius extensions of rings (cf. 40.21 for the definition). In this section we introduce a class of corings, termed *Frobenius corings*, that provide an equivalent description of such extensions. In particular, a Frobenius A-coring $\mathcal{C}$ is itself an R-algebra and a Frobenius extension of A. The equivalence of descriptions is formulated rigorously as the isomorphism of categories of Frobenius corings and Frobenius extensions. This in turn implies that, to any Frobenius coring or a Frobenius extension, one can associate a tower of Frobenius corings and Frobenius extensions.

27.1. Frobenius element and homomorphism. Recall that a ring extension $A \to B$ is called a *Frobenius extension* (of the first kind) if and only if B is a finitely generated projective right A-module and $B \simeq \mathrm{Hom}_A(B, A)$ as (A, B)-bimodules. Equivalently, B is a finitely generated projective left A-module and $B \simeq {}_A\mathrm{Hom}(B, A)$ as (B, A)-bimodules. The (A, B)-bimodule structure of $\mathrm{Hom}_A(B, A)$ is given by $(afb)(b') = af(bb')$, for $a \in A, b, b' \in B$ and $f \in \mathrm{Hom}_A(B, A)$. The statement that $A \to B$ is a Frobenius extension is equivalent to the existence of an (A, A)-bimodule map $E : B \to A$ and an element $\beta = \sum_i b_i \otimes \bar{b}^i \in B \otimes_A B$, such that, for all $b \in B$,

$$\sum_i E(bb_i)\bar{b}^i = \sum_i b_i E(\bar{b}^i b) = b.$$

E is called a *Frobenius homomorphism* and β is known as a *Frobenius element.* One can easily show that in fact a Frobenius element β is an invariant, that is, $\beta \in (B \otimes_A B)^B = \{m \in B \otimes_A B \mid \text{for all } b \in B,\ bm = mb\}$.

It turns out that, equivalently, one can describe Frobenius extensions as a certain type of coring.

27.2. The coring structure of a Frobenius extension. *A ring extension $A \to B$ is a Frobenius extension if and only if B is an A-coring such that the coproduct is a (B, B)-bimodule map.*

Proof. Suppose $A \to B$ is a Frobenius extension with a Frobenius homomorphism E and a Frobenius element $\beta = \sum_i b_i \otimes \bar{b}^i$. Define $\underline{\varepsilon} = E$, an (A, A)-bimodule map, and

$$\underline{\Delta} : B \to B \otimes_A B, \quad b \mapsto b\beta = \beta b,$$

which is clearly a (B, B)-bimodule map, and hence it is also an (A, A)-bimodule map. We need to check the coassociativity of $\underline{\Delta}$ and the counit properties of $\underline{\varepsilon}$. For all $b \in B$,

$$\begin{aligned}(\underline{\Delta} \otimes I_B) \circ \underline{\Delta}(b) &= \textstyle\sum_i \underline{\Delta}(b_i) \otimes \bar{b}^i b = \sum_{i,j} b_j \otimes \bar{b}^j b_i \otimes \bar{b}^i b \\ &= \textstyle\sum_j b_j \otimes \underline{\Delta}(\bar{b}^j b) = (I_B \otimes \underline{\Delta}) \circ \underline{\Delta}(b), \quad \text{and} \\ (\underline{\varepsilon} \otimes I_B) \circ \underline{\Delta}(b) &= \textstyle\sum_i E(bb_i)\bar{b}^i = b,\end{aligned}$$

and similarly for the second counit property. Thus we have proven that a Frobenius element and homomorphism for $A \to B$ produce an A-coring structure on B.

Conversely, suppose B is an A-coring with a (B,B)-bimodule comultiplication. Define $\beta = \underline{\Delta}(1_B) = \sum_i b_i \otimes \bar{b}^i$ and $E = \underline{\varepsilon}$. Since $\underline{\Delta}$ is a (B,B)-bimodule map, we obtain for all $b \in B$,

$$\sum b_{\underline{1}} \otimes b_{\underline{2}} = \underline{\Delta}(b) = \underline{\Delta}(b1_B) = b\Delta_B(1_B) = \sum_i bb_i \otimes \bar{b}^i,$$

and similarly $\sum b_{\underline{1}} \otimes b_{\underline{2}} = \sum_i b_i \otimes \bar{b}^i b$. Now, using the counit property of $\underline{\varepsilon}$ we obtain

$$\sum_i E(bb_i)\bar{b}^i = \sum \underline{\varepsilon}(b_{\underline{1}})b_{\underline{2}} = b = \sum b_{\underline{1}}\underline{\varepsilon}(b_{\underline{2}}) = \sum_i b_i E(\bar{b}^i b),$$

as required. $\square$

Note that 27.2 in fact establishes a bijective correspondence between Frobenius structures (i.e., elements and homomorphisms) for $A \to B$ and A-coring structures on B with a (B,B)-bimodule coproduct.

27.3. Diagrammatic definition of Frobenius extensions. Let $A \to B$ be a ring extension with a Frobenius homomorphism and element, E and β. Then β can be identified with a (B,B)-bimodule map $\underline{\Delta}$ defined in the proof of 27.2, and by this means it can be viewed as a map $\beta \in {}_B\mathrm{Hom}_B(B, B \otimes_A B)$. Then the defining properties of E and β described in 27.1 can be stated in terms of the following commutative diagram:

$$\begin{array}{ccc} B & \xrightarrow{\beta} & B \otimes_A B \\ {\scriptstyle \beta}\downarrow & \searrow^{I_B} & \downarrow{\scriptstyle I_B \otimes E} \\ B \otimes_A B & \xrightarrow{E \otimes I_B} & B\,. \end{array}$$

One can dualise the definition of a Frobenius extension by reversing the arrows in the above notions, thus obtaining the notions of Frobenius corings and Frobenius systems.

27.4. Frobenius corings. An A-coring $\mathcal{C}$ is called a *Frobenius coring* if there exist an (A,A)-bimodule map $\eta : A \to \mathcal{C}$ and a $(\mathcal{C},\mathcal{C})$-bicomodule map $\pi : \mathcal{C} \otimes_A \mathcal{C} \to \mathcal{C}$ yielding a commutative diagram

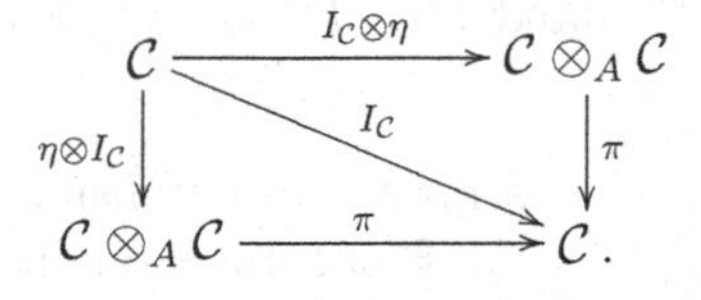

27.5. Frobenius systems. In view of the identification ${}_A\mathrm{Hom}_A(A, \mathcal{C}) \simeq \mathcal{C}^A$, an A-coring $\mathcal{C}$ is Frobenius if and only if there exist $e \in \mathcal{C}^A$ and a $(\mathcal{C}, \mathcal{C})$-bicomodule map $\pi : \mathcal{C} \otimes_A \mathcal{C} \to \mathcal{C}$ such that, for all $c \in \mathcal{C}$, $\pi(c \otimes e) = \pi(e \otimes c) = c$. Such a pair (π, e) is called a *Frobenius system* for $\mathcal{C}$. Furthermore, by the same argument as in 26.1 (cf. 3.29), a $(\mathcal{C}, \mathcal{C})$-bicomodule map $\pi : \mathcal{C} \otimes_A \mathcal{C} \to \mathcal{C}$ can be identified with a certain (A, A)-bimodule map $\delta : \mathcal{C} \otimes_A \mathcal{C} \to A$ via $\pi \mapsto \delta = \underline{\varepsilon} \circ \pi$ and $\delta \mapsto \pi = (\delta \otimes I_{\mathcal{C}}) \circ (I_{\mathcal{C}} \otimes \underline{\Delta}) = (I_{\mathcal{C}} \otimes \delta) \circ (\underline{\Delta} \otimes I_{\mathcal{C}})$. Using this identification one easily sees that $\mathcal{C}$ is Frobenius if and only if there exist $e \in \mathcal{C}^A$ and an (A, A)-bimodule map $\delta : \mathcal{C} \otimes_A \mathcal{C} \to A$ such that, for all $c, c' \in \mathcal{C}$,

$$\sum c_{\underline{1}} \delta(c_{\underline{2}} \otimes c') = \sum \delta(c \otimes c'_{\underline{1}}) c'_{\underline{2}}, \quad \delta(c \otimes e) = \delta(e \otimes c) = \underline{\varepsilon}(c).$$

The pair (δ, e) is called a *reduced Frobenius system* associated to a Frobenius system (π, e).

The following two observations explain the immediate relationship between Frobenius corings and Frobenius extensions, and thus can serve as a motivation for Definition 27.4.

27.6. Frobenius corings and Frobenius extensions. *Let $A \to B$ be a Frobenius extension with a Frobenius element β and a Frobenius homomorphism E. Then B is a Frobenius A-coring with a coproduct β (viewed as a (B, B)-bimodule map $B \to B \otimes_A B$), a counit E, and a Frobenius system $(\pi, 1_B)$, where $\pi : B \otimes_A B \to B$, $b \otimes_A b' \mapsto bb'$.*

Proof. By 27.2, B is an A-coring with the specified coproduct and counit. The fact that B is a Frobenius coring can be verified by direct calculations. We only note that if $\beta = \sum_i b_i \otimes \bar{b}^i$, then the fact that π is a bicomodule morphism means that, for all $b, b' \in B$,

$$\sum_i bb_i \otimes \pi(\bar{b}^i \otimes b') = \sum_i \pi(b \otimes b')b_i \otimes \bar{b}^i = \sum_i \pi(b \otimes b_i) \otimes \bar{b}^i b'.$$

This follows immediately from the definition of π (as a product) and from the fact that the Frobenius element β is B-central. □

In particular, 27.6 implies that the trivial A-coring is a Frobenius coring.

27.7. Sweedler corings and Frobenius extensions. *Let $\mathcal{C} = A \otimes_B A$ be the Sweedler coring associated to an extension $B \to A$. If $B \to A$ is a Frobenius extension, then $\mathcal{C}$ is a Frobenius coring. Conversely, if A is a faithfully flat left or right B-module and $\mathcal{C}$ is a Frobenius coring, then $B \to A$ is a Frobenius extension.*

Proof. Suppose $B \to A$ is a Frobenius extension with Frobenius element $\alpha = \sum_i a_i \otimes \bar{a}^i \in A \otimes_B A$ and Frobenius homomorphism $E : A \to B$.

Obviously $\alpha \in \mathcal{C}^A$. Let $\pi = I_A \otimes E \otimes I_A : A \otimes_B A \otimes_B A \simeq \mathcal{C} \otimes_A \mathcal{C} \to \mathcal{C}$ and $e = \alpha$. We claim that (π, e) is a Frobenius system for $\mathcal{C}$. Indeed, using the defining properties of Frobenius elements and homomorphisms, for all $a, a' \in A$,

$$\pi(e \otimes_A a \otimes_B a') = \sum_i \pi(a_i \otimes_B \bar{a}^i a \otimes_B a') = \sum_i a_i E(\bar{a}^i a) \otimes_B a' = a \otimes_B a'.$$

Similarly one shows that $\pi(a \otimes_B a' \otimes_A e) = a \otimes_B a'$. The map π is clearly (A, A)-bilinear. Its $(\mathcal{C}, \mathcal{C})$-bicolinearity can easily be checked by using the fact that the image of E is in B.

Conversely, suppose that ${}_BA$ or A_B is faithfully flat and that $\mathcal{C}$ is a Frobenius coring with a reduced Frobenius system $\delta : A \otimes_B A \otimes_B A \to A$ and $e = \sum_i a_i \otimes \bar{a}^i \in \mathcal{C}^A$. By the obvious identification ${}_A\mathrm{Hom}_A(A \otimes_B A \otimes_B A, A) \simeq {}_B\mathrm{End}_B(A)$, view δ as a (B, B)-bimodule map $E : A \to A$. Take any $a \in A$; then $a = \underline{\varepsilon}(1_A \otimes_B a) = \delta(1_A \otimes_B a \otimes_A e)$, that is, $a = \sum_i E(aa_i)\bar{a}^i$. Similarly one deduces that $a = \sum_i a_i E(\bar{a}^i a)$. Next, the properties of δ (see 27.5) imply, for all $a \in A$,

$$1_A \otimes_B E(a) = 1_A \otimes_B \delta(1_A \otimes_B a \otimes_B 1_A) = \delta(1_A \otimes_B a \otimes_B 1_A) \otimes_B 1_A = E(a) \otimes_B 1_A.$$

Since A is a faithfully flat left or right B-module, we conclude that $E(a) \in B$ and thus E is a Frobenius homomorphism and e is a Frobenius element. □

Recall that a functor is a *Frobenius functor* provided it has the same right and left adjoint (cf. 38.23).

27.8. Frobenius corings and Frobenius functors. *Let $\mathcal{C}$ be an A-coring. Then the following statements are equivalent:*

(a) $\mathcal{C}$ is a Frobenius coring;

(b) the forgetful functor $(-)_A : \mathbf{M}^{\mathcal{C}} \to \mathbf{M}_A$ is a Frobenius functor;

(c) the forgetful functor ${}_A(-) : {}^{\mathcal{C}}\mathbf{M} \to {}_A\mathbf{M}$ is a Frobenius functor.

Proof. (a) $\Leftrightarrow$ (b) We already know from 18.13 that the forgetful functor $F = (-)_A : \mathbf{M}^{\mathcal{C}} \to \mathbf{M}_A$ has a right adjoint $G = - \otimes_A \mathcal{C} : \mathbf{M}_A \to \mathbf{M}^{\mathcal{C}}$. Thus we need to show that $\mathcal{C}$ is a Frobenius coring if and only if the induction functor G is a left adjoint of the forgetful functor, that is, if and only if there exist natural maps $\eta : I_{\mathbf{M}_A} \to FG$ and $\psi : GF \to I_{\mathbf{M}^{\mathcal{C}}}$ that are a unit and a counit of the adjunction, respectively.

Suppose that $\mathcal{C}$ is a Frobenius coring with a reduced Frobenius system (δ, e). For all right A-modules M define an R-linear map

$$\eta_M : M \to FG(M) = M \otimes_A \mathcal{C}, \quad m \mapsto m \otimes e.$$

Since e is A-central, it is clear that η_M is a right A-module morphism. Take any $f : M \to N$ in $\mathbf{M}_A$ and compute for all $m \in M$,

$$(f \otimes I_{\mathcal{C}}) \circ \eta_M(m) = f(m) \otimes e = \eta_N(f(m)),$$

that is, there is a commutative diagram

$$\begin{array}{ccc} M & \xrightarrow{f} & N \\ {\scriptstyle \eta_M}\downarrow & & \downarrow{\scriptstyle \eta_N} \\ M \otimes_A \mathcal{C} & \xrightarrow{f \otimes I_{\mathcal{C}}} & N \otimes_A \mathcal{C}, \end{array}$$

thus proving that the assignment $M \mapsto \eta_M$ defines a natural transformation of functors, $\eta : I_{\mathbf{M}_A} \to FG$.

Next, for any $M \in \mathbf{M}^{\mathcal{C}}$, define a right A-module map

$$\psi_M : GF(M) = M \otimes_A \mathcal{C} \to M, \quad m \otimes c \mapsto \sum m_{\underline{0}} \delta(m_{\underline{1}} \otimes c),$$

and compute

$$\begin{aligned} \sum \psi_M(m \otimes c_{\underline{1}}) \otimes c_{\underline{2}} &= \sum m_{\underline{0}} \delta(m_{\underline{1}} \otimes c_{\underline{1}}) \otimes c_{\underline{2}} \\ &= \sum m_{\underline{0}} \otimes m_{\underline{1}} \delta(m_{\underline{2}} \otimes c) \;=\; \varrho^M(\psi_M(m \otimes c)), \end{aligned}$$

where we used the definition of δ to derive the second equality. This proves that ψ_M is a morphism of right $\mathcal{C}$-comodules. Take any morphism of right $\mathcal{C}$-comodules $f : M \to N$. Then, for all $m \in M$ and $c \in \mathcal{C}$,

$$\begin{aligned} \psi_M \circ (f \otimes I_{\mathcal{C}})(m \otimes c) &= \sum f(m)_{\underline{0}} \delta(f(m)_{\underline{1}} \otimes c) \;=\; \sum f(m_{\underline{0}}) \delta(m_{\underline{1}} \otimes c) \\ &= \sum f(m_{\underline{0}} \delta(m_{\underline{1}} \otimes c)) \;=\; f \circ \psi_M(m \otimes c), \end{aligned}$$

where the fact that f is a right $\mathcal{C}$-comodule map has been used repeatedly. Thus the diagram

$$\begin{array}{ccc} M \otimes_A \mathcal{C} & \xrightarrow{f \otimes I_{\mathcal{C}}} & N \otimes_A \mathcal{C} \\ {\scriptstyle \psi_M}\downarrow & & \downarrow{\scriptstyle \psi_N} \\ M & \xrightarrow{f} & N \end{array}$$

commutes, so that the assignment $M \mapsto \psi_M$ defines a natural transformation $\psi : GF \to I_{\mathbf{M}^{\mathcal{C}}}$.

It remains to prove that η and ψ are a unit and a counit of an adjunction. On one hand, for any $M \in \mathbf{M}^{\mathcal{C}}$ and any $m \in M$,

$$F(\psi_M) \circ \eta_{F(M)}(m) = \psi_M(m \otimes e) = \sum m_{\underline{0}} \delta(m_{\underline{1}} \otimes e) = \sum m_{\underline{0}} \underline{\varepsilon}(m_{\underline{1}}) = m,$$

while on the other hand, using the properties of δ, for any $M \in \mathbf{M}_A$, $m \in M$ and $c \in \mathcal{C}$,

$$\begin{aligned} \psi_{G(M)} \circ G(\eta_M)(m \otimes c) &= \psi_{M\otimes_A \mathcal{C}}(m \otimes e \otimes c) = \sum m \otimes e_{\underline{1}}\delta(e_{\underline{2}} \otimes c) \\ &= \sum m \otimes \delta(e \otimes c_{\underline{1}})c_{\underline{2}} = m \otimes c. \end{aligned}$$

This completes the proof that G is a left adjoint of F and, consequently, the forgetful functor F is a Frobenius functor.

Conversely, suppose now that F is a Frobenius functor with $\eta : I_{\mathbf{M}_A} \to FG$ and $\psi : GF \to I_{\mathbf{M}^{\mathcal{C}}}$ the corresponding unit and counit of the adjunction. Define

$$e = \eta_A(1_A) \in \mathcal{C}, \quad \pi = \psi_{\mathcal{C}} : \mathcal{C} \otimes_A \mathcal{C} \to \mathcal{C}.$$

We intend to show that (π, e) is a Frobenius system for $\mathcal{C}$. Since η_A is a right A-module map, for all $a \in A$, $\eta_A(a) = ea$. Using the fact that η is a natural transformation of functors in the context of the left multiplication morphism in $\mathbf{M}_A$, which for all $a \in A$ is given as $A \to A$, $a' \mapsto aa'$, one immediately obtains that η_A is a left A-module map as well. Thus for all $a \in A$, $ea = \eta_A(a) = a\eta_A(1_A) = ae$, that is, $e \in \mathcal{C}^A$, as needed.

Clearly, π is a right $\mathcal{C}$-comodule map. For any $a \in A$, consider a morphism of right $\mathcal{C}$-comodules $l_a : \mathcal{C} \to \mathcal{C}$, $c \mapsto ac$. In conjunction with this morphism, the fact that ψ is a natural transformation immediately yields that $\psi_{\mathcal{C}}$ is a left A-module map. Now, for any $c \in \mathcal{C}$, consider a morphism in $\mathbf{M}^{\mathcal{C}}$, $l^c : \mathcal{C} \to \mathcal{C} \otimes_A \mathcal{C}$, $c' \mapsto c \otimes c'$. Since ψ is a natural transformation, there is the commutative diagram

$$\begin{array}{ccc} \mathcal{C} \otimes_A \mathcal{C} & \xrightarrow{l^c \otimes I_{\mathcal{C}}} & \mathcal{C} \otimes_A \mathcal{C} \otimes_A \mathcal{C} \\ {\scriptstyle \psi_{\mathcal{C}}} \downarrow & & \downarrow {\scriptstyle \psi_{\mathcal{C}\otimes_A \mathcal{C}}} \\ \mathcal{C} & \xrightarrow{l^c} & \mathcal{C} \otimes_A \mathcal{C}, \end{array}$$

that is, for all $c, c', c'' \in \mathcal{C}$,

$$\psi_{\mathcal{C}\otimes_A \mathcal{C}}(c \otimes c' \otimes c'') = c \otimes \psi_{\mathcal{C}}(c' \otimes c'').$$

On the other hand, since the coproduct $\underline{\Delta} : \mathcal{C} \to \mathcal{C} \otimes_A \mathcal{C}$ is a morphism in $\mathbf{M}^{\mathcal{C}}$ and ψ is natural, we obtain for all $c, c' \in \mathcal{C}$,

$$\sum \psi_{\mathcal{C}\otimes_A \mathcal{C}}(c_{\underline{1}} \otimes c_{\underline{2}} \otimes c') = \sum \psi_{\mathcal{C}}(c \otimes c')_{\underline{1}} \otimes \psi_{\mathcal{C}}(c \otimes c')_{\underline{2}}.$$

Combining these two observations, one obtains for all $c, c' \in \mathcal{C}$,

$$\Delta_{\mathcal{C}}(\psi_{\mathcal{C}}(c \otimes c')) = \sum c_{\underline{1}} \otimes \psi_{\mathcal{C}}(c_{\underline{2}} \otimes c'),$$

that is, $\psi_{\mathcal{C}}$ is a left $\mathcal{C}$-comodule map. Therefore $\pi = \psi_{\mathcal{C}}$ is a $(\mathcal{C}, \mathcal{C})$-bicomodule morphism, as required.

Finally, since η and ψ are the unit and counit of an adjunction with F the right adjoint of G, first, for all $c \in \mathcal{C}$,

$$c = \psi_{\mathcal{C}}(\eta_A \otimes I_{\mathcal{C}})(c) = \psi_{\mathcal{C}}(e \otimes c),$$

that is, for all $c \in \mathcal{C}$, $\pi(e \otimes c) = c$. Second, for all $c \in \mathcal{C}$ consider a right A-module morphism $f_c : A \to \mathcal{C}$, $a \mapsto ca$. Since η is a natural transformation, there is a commutative diagram

$$\begin{array}{ccc} R & \xrightarrow{f_c} & \mathcal{C} \\ {\scriptstyle \eta_A}\downarrow & & \downarrow{\scriptstyle \eta_{\mathcal{C}}} \\ \mathcal{C} & \xrightarrow{f_c \otimes I_{\mathcal{C}}} & \mathcal{C} \otimes_A \mathcal{C}. \end{array}$$

Evaluation of this diagram at $a = 1_A$ yields $\eta_{\mathcal{C}}(c) = (f_c \otimes I_{\mathcal{C}})(e) = c \otimes e$, for all $c \in \mathcal{C}$. Since η is a unit and ψ is a counit, we obtain

$$c = \psi_{\mathcal{C}} \circ \eta_{\mathcal{C}}(c) = \psi_{\mathcal{C}}(c \otimes e),$$

that is, for all $c \in \mathcal{C}$, $\pi(c \otimes e) = c$. This completes the proof of the assertion that (π, e) is a Frobenius system for $\mathcal{C}$, and thus completes the proof of the first equivalence in the theorem.

(a) $\Leftrightarrow$ (c) Note that the definition of a Frobenius coring is left-right symmetric because it involves maps of $(\mathcal{C}, \mathcal{C})$-bicomodules and (A, A)-bimodules only. Thus a similar argument as in the proof of the first equivalence can be used to prove its left-handed version. This completes the proof of the theorem. □

27.9. The finiteness of a Frobenius coring. *If $\mathcal{C}$ is a Frobenius A-coring, then $\mathcal{C}$ is finitely generated and projective both as a right and a left A-module.*

Proof. This follows from 18.15 and 27.8. On the other hand, this can also be proven directly, by displaying explicitly the dual bases. Suppose $\mathcal{C}$ is a Frobenius A-coring with a reduced Frobenius system (δ, e). Write $\underline{\Delta}(e) = \sum_{i=1}^n e_i \otimes \bar{e}_i$. Taking $c' = e$ in the defining relations of the reduced Frobenius system in 27.5 we obtain $c = \sum_{i=1}^n \delta(c \otimes e_i)\bar{e}_i$. Similarly, taking $c = e$ we obtain $c' = \sum_{i=1}^n e_i \delta(\bar{e}_i \otimes c')$. Since δ is an (A, A)-bimodule map, for each $i \in \{1, 2, \ldots, n\}$, the map $\xi^i : \mathcal{C} \to A$, $c \mapsto \delta(c \otimes e_i)$, is left A-linear while the map $\bar{\xi}^i : \mathcal{C} \to A$, $c \mapsto \delta(\bar{e}_i \otimes c)$, is right A-linear. Hence $\{\xi^i, \bar{e}_i\}$ is a dual basis of ${}_A\mathcal{C}$, and $\{\bar{\xi}^i, e_i\}$ is a dual basis of $\mathcal{C}_A$. □

Recall that for every A-coring $\mathcal{C}$ there is an associated algebra extension $i_L : A \to ({}^*\mathcal{C})^{op}$, where $({}^*\mathcal{C})^{op}$ is the opposite algebra of the left-dual algebra to $\mathcal{C}$ in 17.8(2). It turns out that the notion of a Frobenius coring is closely related to the problem when this extension is a Frobenius extension.

27.10. Characterisation of Frobenius corings (I). *Let $\mathcal{C}$ be an A-coring and $S = ({}^*\mathcal{C})^{op}$. The following are equivalent:*

(a) *the forgetful functor $(-)_A : \mathbf{M}^{\mathcal{C}} \to \mathbf{M}_A$ is a Frobenius functor;*

(b) *$\mathcal{C}$ is a finitely generated projective left A-module and the ring extension $A \to S$ is Frobenius;*

(c) *$\mathcal{C}$ is a finitely generated projective left A-module and $\mathcal{C} \simeq S$ as (A, S)-bimodules, where $\mathcal{C}$ is a right S-module via $cs = \sum c_{\underline{1}} s(c_{\underline{2}})$, for all $c \in \mathcal{C}, s \in S$;*

(d) *$\mathcal{C}$ is a finitely generated projective left A-module and there exists $e \in \mathcal{C}^A$ such that the map $\phi_l : S \to \mathcal{C}$, $s \mapsto \sum e_{\underline{1}} s(e_{\underline{2}})$ is bijective.*

Proof. (a) $\Leftrightarrow$ (b) Since by 18.13 the functor $- \otimes_A \mathcal{C} : \mathbf{M}_A \to \mathbf{M}^{\mathcal{C}}$ is the right adjoint of the forgetful functor $(-)_A$, 27.10(a) is equivalent to the statement that $- \otimes_A \mathcal{C}$ is the left adjoint of $(-)_A$. By 27.9, (a) implies that $\mathcal{C}$ is a finitely generated projective left A-module. By 19.6, the category of right $\mathcal{C}$-comodules is isomorphic to the category of right S-modules, the forgetful functor is the restriction of scalars functor $\mathbf{M}_S \to \mathbf{M}_A$, and this functor has the right adjoint $- \otimes_A \mathcal{C}$. By 40.21, the restriction of scalars functor has the same left and right adjoint if and only if the extension $A \to S$ is Frobenius.

(b) $\Leftrightarrow$ (c) Since $\mathcal{C}$ is a finitely generated projective left A-module, the (A, S)-bimodule map

$$\alpha : \mathcal{C} \to \mathrm{Hom}_A(S, A) = \mathrm{Hom}_A({}_A\mathrm{Hom}(\mathcal{C}, A), A), \quad c \mapsto [s \mapsto s(c)],$$

is bijective. Thus $\mathcal{C} \simeq \mathrm{Hom}_A(S, A)$ as (A, S)-bimodules. The extension $A \to S$ is Frobenius if and only if $S \simeq \mathrm{Hom}_A(S, A)$, that is, if and only if $\mathcal{C} \simeq S$ as (A, S)-bimodules.

(c) $\Leftrightarrow$ (d) This follows from the bijective correspondence $\theta : \mathcal{C}^A \to {}_A\mathrm{Hom}_S(S, \mathcal{C})$, $\theta(e)(s) = es = \sum e_{\underline{1}} s(e_{\underline{2}})$, $\theta^{-1}(f) = f(\varepsilon_{\mathcal{C}})$. Note that the inverse of ϕ_l comes out explicitly as $\phi_l^{-1} : c \mapsto [c' \mapsto \delta(c \otimes c')]$. □

27.11. Endomorphism Ring Theorem. In the case of the canonical coring associated to a ring extension $B \to A$, 27.10 gives the criteria when the extension $A \to {}_B\mathrm{End}(A)$ is Frobenius. The ring structure on ${}_B\mathrm{End}(A)$ is given by the opposite composition of maps. For example, if $B \to A$ is itself a Frobenius extension with a Frobenius homomorphism $E \in {}_B\mathrm{Hom}_B(A, B)$ and a Frobenius element $\beta = \sum_i a_i \otimes \overline{a}^i$, then $A \to S$ is Frobenius by the Endomorphism Ring Theorem ([139, Section 2], [23, Theorem 2.5]). In this case the inverse of ϕ_l in 27.10(d) is given by $\phi_l^{-1}(a \otimes a')(a'') = E(a''a)a'$, for all $a, a', a'' \in A$.

By pulling back the algebra structure of S to $\mathcal{C}$ one obtains the following corollary of 27.10.

27.12. Ring structure of a Frobenius coring. *Let $\mathcal{C}$ be a Frobenius A-coring with a Frobenius system (π, e). Then*

(1) $\mathcal{C}$ is an algebra with product $cc' = \pi(c \otimes c')$ and unit $1_{\mathcal{C}} = e$.

(2) The extension $\iota_{\mathcal{C}} : A \to \mathcal{C}$, $a \mapsto ae = ea$, is Frobenius with Frobenius element $\underline{\Delta}(e)$ and Frobenius homomorphism $E = \underline{\varepsilon}$.

Proof. The algebra structure is induced from the algebra structure of $S = ({}^*\mathcal{C})^{op}$ via the maps ϕ_l and its inverse ϕ_l^{-1} described in 27.10. Note that the alternative expressions for the product are $cc' = \sum \delta(c \otimes c'_{\underline{1}})c'_{\underline{2}} = \sum c_{\underline{1}}\delta(c_{\underline{2}} \otimes c')$. When viewed through ϕ_l, the map $i_L : A \to S$ becomes the map $\iota_{\mathcal{C}} : A \to \mathcal{C}$ as described above. Since the extension $A \to S$ is Frobenius, so is the extension $\iota_{\mathcal{C}} : A \to \mathcal{C}$. One can verify explicitly that the Frobenius element and homomorphism have the form stated. By definition, the counit $\underline{\varepsilon}$ is an (A, A)-bimodule map. Note further that, for all $c \in \mathcal{C}$,

$$\begin{aligned} c\underline{\Delta}(e) &= \sum ce_{\underline{1}} \otimes e_{\underline{2}} = \sum c_{\underline{1}}\delta(c_{\underline{2}} \otimes e_{\underline{1}}) \otimes e_{\underline{2}} \\ &= \sum c_{\underline{1}} \otimes c_{\underline{2}}\delta(c_{\underline{3}} \otimes e) = \sum c_{\underline{1}} \otimes c_{\underline{2}}\underline{\varepsilon}(c_{\underline{3}}) = \sum c_{\underline{1}} \otimes c_{\underline{2}} = \underline{\Delta}(c). \end{aligned}$$

Similarly one shows that $\underline{\Delta}(c) = \underline{\Delta}(e)c$. Therefore $\underline{\Delta}(e) \in (\mathcal{C} \otimes_A \mathcal{C})^{\mathcal{C}}$, and

$$c = \sum \underline{\varepsilon}(c_{\underline{1}})c_{\underline{2}} = \sum \underline{\varepsilon}(ce_{\underline{1}})e_{\underline{2}}, \quad c = \sum c_{\underline{1}}\underline{\varepsilon}(c_{\underline{2}}) = \sum e_{\underline{1}}\underline{\varepsilon}(e_{\underline{2}}c),$$

thus proving that $\iota_{\mathcal{C}} : A \to \mathcal{C}$ is a Frobenius extension with Frobenius element $\underline{\Delta}(e)$ and Frobenius homomorphism $\underline{\varepsilon}$. □

Obviously, there is also a left-handed version of 27.10

27.13. Characterisation of Frobenius corings (II). *Let $\mathcal{C}$ be an A-coring and $T = \mathcal{C}^*$. The following are equivalent:*

(a) the forgetful functor $F : {}^{\mathcal{C}}\mathbf{M} \to {}_A\mathbf{M}$ is a Frobenius functor;

(b) $\mathcal{C}$ is a finitely generated projective right A-module and the ring extension $A^{op} \to T$ is Frobenius;

(c) $\mathcal{C}_A$ is finitely generated projective and $\mathcal{C} \simeq T$ as (A^{op}, T)-bimodules, where $\mathcal{C}$ is a right T-module via $ct = \sum t(c_{\underline{1}})c_{\underline{2}}$, for all $c \in \mathcal{C}, t \in T$;

(d) $\mathcal{C}_A$ is finitely generated projective and there exists $e \in \mathcal{C}^A$ such that the map $\phi_r : T \to \mathcal{C}$, $t \mapsto \sum t(e_{\underline{1}})e_{\underline{2}}$ is bijective.

We only point out here that the inverse of the map ϕ_r in 27.13(d) explicitly reads

$$\phi_r : \mathcal{C} \to \mathcal{C}^*, \quad \phi_r : c \mapsto [c' \mapsto \delta(c \otimes c')].$$

One can use this map to build an algebra structure on $\mathcal{C}$. The algebra structure obtained in this way coincides with the algebra structure constructed in 27.12.

Combining 27.10 with 27.13 and 27.8, one obtains the following

27.14. Characterisation of Frobenius corings (III). *Let $\mathcal{C}$ be an A-coring, $S = ({}^*\mathcal{C})^{op}$, and $T = \mathcal{C}^*$. Then the following are equivalent:*

(a) $\mathcal{C}$ is a Frobenius coring;

(b) $\mathcal{C}$ is a finitely generated projective left A-module and the ring extension $A \to S$ is Frobenius;

(c) $\mathcal{C}$ is a finitely generated projective right A-module and the ring extension $A^{op} \to T$ is Frobenius.

27.15. Co-Frobenius corings. The notion of a Frobenius coring should be compared with the notions of right and left co-Frobenius coalgebras introduced in [152]. The notions of right or left co-Frobenius coalgebras can be extended to the case of corings as follows. An A-coring $\mathcal{C}$ is said to be a *left co-Frobenius coring* when there is an injective morphism $\mathcal{C} \to {}^*\mathcal{C}$ of left ${}^*\mathcal{C}$-modules. $\mathcal{C}$ is a *right co-Frobenius coring* when there is an injective morphism $\mathcal{C} \to \mathcal{C}^*$ of right $\mathcal{C}^*$-modules.

By 27.10 and 27.13, a Frobenius A-coring $\mathcal{C}$ is isomorphic to ${}^*\mathcal{C}$ as a left ${}^*\mathcal{C}$-module via the map ϕ_l, and it is isomorphic to $\mathcal{C}^*$ as a right $\mathcal{C}^*$-module via the map ϕ_r. Thus, in particular, a Frobenius coring is left and right co-Frobenius.

The above discussion allows one to understand Frobenius corings as a different description of Frobenius extensions of rings.

27.16. Categories of Frobenius corings and Frobenius extensions. Define a *category* **Frob**(A) *of Frobenius extensions over an R-algebra A*, taking as objects quintuples $(M, \mu_{M/A}, \iota_M, E_M, \beta_M)$, where $(M, \mu_{M/A}, \iota_M)$ is a unital A-ring, that is, M is an R-algebra with multiplication $\mu_{M/A} : M \otimes_A M \to M$, $\iota_M : A \to M$ is an algebra map (cf. 26.5), and E_M, β_M are the Frobenius homomorphism and element for this extension. Thus objects in **Frob**(A) are Frobenius extensions over A. Morphisms

$$f : (M, \mu_{M/A}, \iota_M, E_M, \beta_M) \to (N, \mu_{N/A}, \iota_N, E_N, \beta_N)$$

are defined as R-linear maps $f : M \to N$ satisfying the following conditions:

(i) f is an A-ring map, that is, $f \circ \iota_M = \iota_N$ and $f \circ \mu_{M/A} = \mu_{N/A} \circ (f \otimes f)$;

(ii) $E_M = E_N \circ f$;

(iii) $(f \otimes f)(\beta_M) = \beta_N$.

Similarly, define the *category* **FrobCor**(A) *of Frobenius A-corings*, taking as objects quintuples $(\mathcal{C}, \underline{\Delta}_\mathcal{C}, \underline{\varepsilon}_\mathcal{C}, \pi_\mathcal{C}, e_\mathcal{C})$, where $(\mathcal{C}, \underline{\Delta}_\mathcal{C}, \underline{\varepsilon}_\mathcal{C})$ is an A-coring with a Frobenius system $(\pi_\mathcal{C}, e_\mathcal{C})$. Morphisms

$$f : (\mathcal{C}, \underline{\Delta}_\mathcal{C}, \underline{\varepsilon}_\mathcal{C}, \pi_\mathcal{C}, e_\mathcal{C}) \to (\mathcal{D}, \underline{\Delta}_\mathcal{D}, \underline{\varepsilon}_\mathcal{D}, \pi_\mathcal{D}, e_\mathcal{D})$$

are defined as A-coring maps $f : \mathcal{C} \to \mathcal{D}$ such that

(1) $e_{\mathcal{D}} = f(e_{\mathcal{C}})$;

(2) $\pi_{\mathcal{D}} \circ (f \otimes f) = f \circ \pi_{\mathcal{C}}$.

Theorem. *The functor* $F : \mathbf{FrobCor}(A) \to \mathbf{Frob}(A)$ *given by*

$$(\mathcal{C}, \underline{\Delta}, \underline{\varepsilon}, \pi, e) \mapsto (\mathcal{C}, \pi, \iota_{\mathcal{C}}, \underline{\varepsilon}, \underline{\Delta}(e)), \quad f \mapsto f,$$

is an isomorphism of categories. Here $\iota_{\mathcal{C}}$ *is an* (A, A)*-bimodule map induced by* e *as in 27.12(2). The inverse of* F *explicitly reads*

$$F^{-1} : (M, \mu_{M/A}, \iota_M, E_M, \beta_M) \mapsto (M, \beta_M, E_M, \mu_{M/A}, \iota_M(1_A) = 1_M).$$

Proof. The fact that F is well defined on objects follows immediately from 27.12. If $f : (\mathcal{C}, \underline{\Delta}_{\mathcal{C}}, \underline{\varepsilon}_{\mathcal{C}}, \pi_{\mathcal{C}}, e_{\mathcal{C}}) \to (\mathcal{D}, \underline{\Delta}_{\mathcal{D}}, \underline{\varepsilon}_{\mathcal{D}}, \pi_{\mathcal{D}}, e_{\mathcal{D}})$ is a morphism in $\mathbf{FrobCor}(A)$, then condition (2) above implies that f is a multiplicative map. Furthermore, since f is an (A, A)-bimodule map, for $a \in A$,

$$(f \circ \iota_{\mathcal{C}})(a) = f(ae_{\mathcal{C}}) = af(e_{\mathcal{C}}) = ae_{\mathcal{D}} = \iota_{\mathcal{D}}(a),$$

by condition (2). Thus f is an A-ring map. The fact that f is a coring morphism implies that conditions (ii) and (iii) are fulfilled.

Conversely, the functor F^{-1} is well defined on objects by 27.6. For morphisms, condition (ii) guarantees the compatibility of f with counits, while (iii) is responsible for the compatibility with coproducts. Furthermore, condition (i) implies that f satisfies condition (1), while (2) follows from the fact that f is an algebra map. Therefore f is a morphism of Frobenius A-corings and F^{-1} is a well-defined functor. Clearly F^{-1} is the inverse of F on morphisms. To check that this is also true on objects one only needs to observe that, for all $c \in \mathcal{C}$, $\underline{\Delta}(c) = \underline{\Delta}(e)c = c\underline{\Delta}(e)$ (compare the proof of 27.12) and that, for all $a \in A$,

$$\iota_{\mathcal{C}}(a)c = (ae)c = \pi(ae \otimes c) = a\pi(e \otimes c) = ac,$$

and similarly for the right A-multiplication. □

One of the most interesting features of Frobenius corings is that, given any such coring, one can construct a full tower of Frobenius corings.

27.17. Towers of Frobenius corings. Suppose that $\mathcal{C}$ is a Frobenius A-coring with a Frobenius system (π, e). Then, by 27.12, e viewed as a map $\iota_{\mathcal{C}} : A \to \mathcal{C}$ is a Frobenius extension with Frobenius element $\underline{\Delta}(e)$ and Frobenius homomorphism $\underline{\varepsilon}$. Now 27.7 implies that the Sweedler $\mathcal{C}$-coring $\mathcal{C} \otimes_A \mathcal{C}$ is Frobenius with the Frobenius system $(I_{\mathcal{C}} \otimes \underline{\varepsilon} \otimes I_{\mathcal{C}}, \Delta(e))$. Then $\mathcal{C} \otimes_A \mathcal{C}$ is a ring with unit $\underline{\Delta}(e)$ and product $(c \otimes c')(c'' \otimes c''') = c \otimes \delta(c' \otimes c'')c'''$, and the extension $\underline{\Delta} : \mathcal{C} \to \mathcal{C} \otimes_A \mathcal{C}$ is Frobenius by 27.12. The Frobenius element explicitly reads $\sum e_{\underline{1}} \otimes e \otimes e_{\underline{2}}$ and the Frobenius homomorphism is π. Apply 27.7 to deduce that Sweedler's $\mathcal{C}\otimes_A\mathcal{C}$-coring $(\mathcal{C}\otimes_A\mathcal{C})\otimes_{\mathcal{C}}(\mathcal{C}\otimes_A\mathcal{C}) \simeq \mathcal{C}\otimes_A\mathcal{C}\otimes_A\mathcal{C}$ is Frobenius. Iterating this procedure, we obtain the following

Theorem. *Let $\mathcal{C}$ be a Frobenius A-coring, and let $\mathcal{C}^k = \mathcal{C}^{\otimes_A k}$, $k = 1, 2, \ldots$, and $\mathcal{C}^0 = A$. Then there is a sequence of algebra maps*

$$\mathcal{C}^0 \xrightarrow{\iota_{\mathcal{C}}} \mathcal{C}^1 \xrightarrow{\underline{\Delta}} \mathcal{C}^2 \xrightarrow{I_{\mathcal{C}}\otimes\iota_{\mathcal{C}}\otimes I_{\mathcal{C}}} \mathcal{C}^3 \longrightarrow \ldots,$$

where, for all $k = 1, 2, \ldots$, $\mathcal{C}^{k-1} \to \mathcal{C}^k$ is a Frobenius extension and $\mathcal{C}^k$ is a Frobenius $\mathcal{C}^{k-1}$-coring.

This tower of corings bears very close resemblance to the tower of rings introduced by Jones [136] as means of classification of subfactors of von Neumann algebras.

References. Brzeziński [70, 75]; Caenepeel, DeGroot and Militaru [83]; Caenepeel, Ion and Militaru [84]; Jones [136]; Kadison [23]; Kasch [139]; Lin [152]; Menini and Năstăsescu [156]; Morita [162].

28 Corings with a grouplike element

Most closely related to ring extensions are those corings that have a *grouplike element* g. With the exception of Section 30, the remainder of this chapter is devoted to the studies of such corings. In this section we collect their basic properties, introduce the corresponding g-coinvariants functor and view it as a Hom-functor, and reveal that corings with a grouplike element exhibit a natural ring structure. We also introduce *Galois corings*, which are isomorphic to the Sweedler coring associated to a ring extension $B \to A$ induced by the existence of a grouplike element, and prove the Galois Coring Structure Theorem 28.19. This theorem determines when the g-coinvariants functor is an equivalence, and by this means generalises the faithfully flat (effective) descent theorem. It will be shown in Section 34 that the Fundamental Theorem of Hopf algebras 15.5 is a special case of the Galois Coring Structure Theorem 28.19 (via the structure theorem for Hopf-Galois extensions).

Throughout this section $\mathcal{C}$ denotes an A-coring.

28.1. Grouplike elements. An element $g \in \mathcal{C}$ is said to be *semi-grouplike* provided $\underline{\Delta}(g) = g \otimes g$, and g is called a *grouplike* element if $\underline{\Delta}(g) = g \otimes g$ and $\underline{\varepsilon}(g) = 1_A$.

Note that every coring has a semi-grouplike element (indeed, take $g = 0$). Note also that, for a semi-grouplike element g, $u = \underline{\varepsilon}(g) \in A$ is an idempotent in the centraliser of g in A, that is, $u^2 = u$ and $ug = gu$.

In the trivial A-coring A, 1_A is a grouplike element. Furthermore, if $B \to A$ is an algebra extension, then $g = 1_A \otimes 1_A$ is a grouplike element in the Sweedler A-coring $A \otimes_B A$. So corings with a grouplike element can be viewed as a generalisation of the Sweedler coring, and we keep the latter as our guiding example for general constructions.

28.2. Existence of grouplike elements. *$\mathcal{C}$ has a grouplike element if and only if A is a right or left $\mathcal{C}$-comodule.*

Proof. We prove the proposition in the right $\mathcal{C}$-comodule case. Let $g \in \mathcal{C}$ be a grouplike element. Define a right A-module map

$$\varrho^A : A \to A \otimes_A \mathcal{C} \simeq \mathcal{C}, \quad a \mapsto 1_A \otimes ga = ga.$$

By the fact that g is a grouplike element, for all $a \in A$,

$$(\varrho^A \otimes I_{\mathcal{C}}) \circ \varrho^A(a) = g \otimes 1_A \otimes ga = g \otimes ga = (I_A \otimes \underline{\Delta}) \circ \varrho^A(a),$$

and also $(I_A \otimes \underline{\varepsilon}) \circ \varrho^A(a) = \underline{\varepsilon}(ga) = \underline{\varepsilon}(g)a = a$. Therefore ϱ^A defines a right $\mathcal{C}$-comodule structure on A. Similarly, the left coaction of $\mathcal{C}$ on A is given by ${}^A\varrho(a) = ag$.

Conversely, let A be a right $\mathcal{C}$-comodule with a coaction $\varrho^A : A \to \mathcal{C}$, and put $g = \varrho^A(1_A) \in \mathcal{C}$. Since ϱ^A is a right coaction,

$$\underline{\Delta}(g) = (I_A \otimes \underline{\Delta})(\varrho^A(1_A)) = (\varrho^A \otimes I_{\mathcal{C}})(\varrho^A(1_A)) = \varrho^A(1_A) \otimes g = g \otimes g,$$

and also $\underline{\varepsilon}(g) = I \otimes \underline{\varepsilon}(\varrho^A(1_A) = 1_A$, showing that g is grouplike. □

Notation. If $g \in \mathcal{C}$ is grouplike, we write A_g or ${}_gA$ when we consider A with the right or left comodule structure induced by g.

In view of 25.3, 28.2 provides an algebra extension $B \to A$ with a descent datum (A, f), where $f : A \to A \otimes_B A$, $a \mapsto a \otimes 1_A$.

28.3. Kernel of the counit as a coideal. *If $\mathcal{C}$ has a grouplike element, then* $\mathrm{Ke}\,\underline{\varepsilon}$ *is a coideal in* $\mathcal{C}$.

Proof. Since $\mathcal{C}$ has a grouplike element, the map $\underline{\varepsilon}$ is surjective. Thus $\mathrm{Ke}\,\underline{\varepsilon}$ is a coideal by 17.16. □

28.4. Coinvariants. Given a grouplike element $g \in \mathcal{C}$ and $M \in \mathbf{M}^{\mathcal{C}}$, one defines *g-coinvariants* of M as the R-submodule

$$M_g^{co\mathcal{C}} = \{m \in M \mid \varrho^M(m) = m \otimes g\} = \mathrm{Ke}\,(\varrho^M - (- \otimes g)).$$

Proposition. *There is an R-module isomorphism*

$$\theta_M : \mathrm{Hom}^{\mathcal{C}}(A_g, M) \to M_g^{co\mathcal{C}}, \quad f \mapsto f(1_A).$$

Proof. For any $f \in \mathrm{Hom}^{\mathcal{C}}(A_g, M)$ there is the commutative diagram of right A-module maps

$$\begin{array}{ccc} A & \xrightarrow{\varrho^{\mathcal{C}}} & A \otimes_A \mathcal{C} \\ {\scriptstyle f}\downarrow & & \downarrow{\scriptstyle f \otimes I_{\mathcal{C}}} \\ M & \xrightarrow{\varrho^M} & M \otimes_A \mathcal{C} \end{array} \qquad \begin{array}{ccc} 1_A & \longmapsto & g \\ \downarrow & & \downarrow \\ f(1_A) & \longmapsto & \varrho^M(f(1_A)) = f(1_A) \otimes g, \end{array}$$

which shows that $f(1_A) \in M_g^{co\mathcal{C}}$.

Any $f \in \mathrm{Hom}^{\mathcal{C}}(A_g, M)$ is uniquely determined by $f(1_A)$, and for any $m \in M_g^{co\mathcal{C}}$ the map $h_m : A \to M, h(a) = ma$, is $\mathcal{C}$-colinear. One easily checks that the assignment $m \mapsto h_m$ is the inverse of θ_M. □

Similarly, g-coinvariants of any $N \in {}^{\mathcal{C}}\mathbf{M}$ are defined as

$${}^{co\mathcal{C}}_{\ \ g}N = \{n \in N \mid {}^N\varrho(n) = g \otimes n\} = \mathrm{Ke}\,({}^N\varrho - (g \otimes -)),$$

and there is an R-module isomorphism

$${}_N\theta : {}^{\mathcal{C}}\mathrm{Hom}({}_gA, N) \to {}^{co\mathcal{C}}_{\ \ g}N, \quad h \mapsto h(1_A).$$

Corings with a grouplike element lend themselves naturally to the extension theory. In fact, the existence of a grouplike element alone provides one with a pair of algebras connected by an algebra map.

28.5. Coinvariants of A and $\mathcal{C}$. *Let g be a grouplike element in $\mathcal{C}$ and let $B = \{b \in A \mid bg = gb\}$.*

(1) $A_g^{co\mathcal{C}} = {}^{co\mathcal{C}}_{\ g}A$ is a subalgebra of A equal to B.

(2) For any $M \in \mathbf{M}^{\mathcal{C}}$, $M_g^{co\mathcal{C}}$ is a right B-submodule of M, and for any $N \in {}^{\mathcal{C}}\mathbf{M}$, ${}^{co\mathcal{C}}_{\ g}N$ is a left B-submodule of N.

(3) For any $X \in \mathbf{M}_A$ and $Y \in {}_A\mathbf{M}$,

$$(X \otimes_A \mathcal{C})_g^{co\mathcal{C}} \simeq X \quad \text{in } \mathbf{M}_B, \qquad {}^{co\mathcal{C}}_{\ g}(\mathcal{C} \otimes_A Y) \simeq Y \quad \text{in } {}_B\mathbf{M}.$$

(4) In particular, $\mathcal{C}_g^{co\mathcal{C}} \simeq A$ as an (A, B)-bimodule, and ${}^{co\mathcal{C}}_{\ g}\mathcal{C} \simeq A$ as a (B, A)-bimodule.

Proof. (1) Since A is both a right and a left $\mathcal{C}$-comodule, the definition of $A_g^{co\mathcal{C}}$ or ${}^{co\mathcal{C}}_{\ g}A$ makes sense and

$$\mathrm{Hom}^{\mathcal{C}}(A_g, A_g) \simeq A_g^{co\mathcal{C}} = \{a \in A \,|\, ga = \varrho^A(a) = ag\} = B,$$

which is an (anti-)ring isomorphism, depending on which side the morphisms are written.

(2) Since B is the endomorphism ring $\mathrm{End}^{\mathcal{C}}(A_g)$, there is a natural right B-module structure on $\mathrm{Hom}^{\mathcal{C}}(A_g, M)$. In view of the isomorphism θ_M in 28.4, this B-module structure can be transported to $M_g^{co\mathcal{C}}$. One easily checks that the resulting B-multiplication in $M_g^{co\mathcal{C}}$ comes from the A-multiplication in M. The second statement follows from the isomorphism ${}_N\theta$ (see above) by a similar reasoning.

(3),(4) By the canonical Hom-tensor relation (see 18.10),

$$(X \otimes_A \mathcal{C})_g^{co\mathcal{C}} \simeq \mathrm{Hom}^{\mathcal{C}}(A_g, X \otimes_A \mathcal{C}) \simeq \mathrm{Hom}_A(A, X) \simeq X,$$

and for $X = A$,

$$\mathcal{C}_g^{co\mathcal{C}} \simeq \mathrm{Hom}^{\mathcal{C}}(A_g, \mathcal{C}) \simeq \mathrm{Hom}_A(A, A) \simeq A,$$

which is a left A- and right $\mathrm{End}^{\mathcal{C}}(A)$-morphism (when morphisms are written on the left). Similar isomorphisms exist for the left-hand side versions. □

28.6. Coinvariants of Sweedler corings. Let $A \otimes_Q A$ be the canonical Sweedler coring associated to a ring extension $Q \to A$. Take $g = 1_A \otimes 1_A$. Then $B = A_g^{co\mathcal{C}}$ is given by an equaliser,

$$B \longrightarrow A \rightrightarrows A \otimes_Q A,$$

where the maps are $a \mapsto a \otimes 1_A$ and $a \mapsto 1_A \otimes a$. In particular, if $Q \to A$ is faithfully flat, then $B = Q$.

28.7. The induction functor. Let $g \in \mathcal{C}$ be a grouplike element and $B = A_g^{co\mathcal{C}}$. Given any right B-module M, the tensor product $M \otimes_B A$ is a right $\mathcal{C}$-comodule via the coaction

$$\varrho^{M\otimes_B A} : M \otimes_B A \to M \otimes_B A \otimes_A \mathcal{C} \simeq M \otimes_B \mathcal{C}, \quad m \otimes a \mapsto m \otimes ga.$$

If $f : M \to N$ is a morphism in $\mathbf{M}_B$ then $f \otimes I_A : M \otimes_B A \to N \otimes_B A$ is a morphism in $\mathbf{M}^{\mathcal{C}}$ since, for all $a \in A$ and $m \in M$,

$$\begin{aligned}\varrho^{N\otimes_B A}(f(m) \otimes a) = f(m) \otimes ga &= (f \otimes I_{\mathcal{C}})(m \otimes ga) \\ &= (f \otimes I_{\mathcal{C}}) \circ \varrho^{M\otimes_B A}(m \otimes a).\end{aligned}$$

Hence the assignments $M \mapsto M \otimes_B A$ and $f \mapsto f \otimes I_A$ define a functor $- \otimes_B A : \mathbf{M}_B \to \mathbf{M}^{\mathcal{C}}$ known as an *induction functor.*

28.8. The g-coinvariants functor. *Let $g \in \mathcal{C}$ be a grouplike element and $B = A_g^{co\mathcal{C}}$. There is a pair of adjoint functors*

$$- \otimes_B A : \mathbf{M}_B \to \mathbf{M}^{\mathcal{C}}, \quad \mathrm{Hom}^{\mathcal{C}}(A_g, -) : \mathbf{M}^{\mathcal{C}} \to \mathbf{M}_B.$$

In view of the isomorphism θ_M in 28.4, the Hom-functor is isomorphic to the g-coinvariants functor,

$$G_g : \mathbf{M}^{\mathcal{C}} \to \mathbf{M}_B, \quad M \mapsto M_g^{co\mathcal{C}},$$

which acts on morphisms by restriction of the domain; that is, for $f : M \to N$ in $\mathbf{M}^{\mathcal{C}}$, $G_g(f) = f\,|_{M_g^{co\mathcal{C}}}$. For $N \in \mathbf{M}_B$, the unit of adjunction is given by

$$\eta_N : N \to (N \otimes_B A)_g^{co\mathcal{C}}, \quad n \mapsto n \otimes 1_A,$$

and for $M \in \mathbf{M}^{\mathcal{C}}$ the counit reads

$$\psi_M : M_g^{co\mathcal{C}} \otimes_B A \to M, \quad m \otimes a \mapsto ma.$$

Proof. Notice that $\varrho^{\mathcal{C}} : A \to \mathcal{C}$ is (B, A)-linear. Hence, by the Hom-tensor relation in 18.10(2), setting $M = A$ there, we obtain for all $M \in \mathbf{M}^{\mathcal{C}}$ and $N \in \mathbf{M}_B$, the functorial isomorphisms

$$\mathrm{Hom}^{\mathcal{C}}(N \otimes_B A, M) \simeq \mathrm{Hom}_B(N, \mathrm{Hom}^{\mathcal{C}}(A_g, M)) \simeq \mathrm{Hom}_B(N, M_g^{co\mathcal{C}}).$$

This proves the adjointness of the functors. The remaining assertions are easily verified. □

28.9. Proposition. *Let $g \in \mathcal{C}$ be a grouplike element and $B = A_g^{co\mathcal{C}}$. Then, for any right B-module N, there is a left A-module isomorphism*

$$\mathrm{Hom}^{\mathcal{C}}(N \otimes_B A, \mathcal{C}) \simeq \mathrm{Hom}_B(N, A).$$

Proof. By 28.5, $\mathcal{C}_g^{co\mathcal{C}} \simeq A$ as (A,B)-bimodules. Now 28.8 implies the isomorphism of R-modules,

$$\theta : \mathrm{Hom}^{\mathcal{C}}(N \otimes_B A, \mathcal{C}) \to \mathrm{Hom}_B(N, A), \quad f \mapsto \underline{\varepsilon}(f(- \otimes 1_A)),$$

with the inverse $h \mapsto [n \otimes a \mapsto h(n)ga]$.

$\mathrm{Hom}^{\mathcal{C}}(N \otimes_B A, \mathcal{C})$ is viewed as a left A-module via $(af)(n \otimes a') = af(n \otimes a')$ for all $f \in \mathrm{Hom}^{\mathcal{C}}(N \otimes_B A, \mathcal{C})$, $a, a' \in A$ and $n \in N$, while $\mathrm{Hom}_B(N, A)$ is a left A-module via $(ah)(n) = ah(n)$, for all $h \in \mathrm{Hom}_B(N, A)$, $n \in N$ and $a \in A$. Thus the isomorphism θ is clearly an isomorphism of left A-modules with specified A-multiplications. □

Another way of understanding the g-coinvariants functor is to view it as a coinduction functor associated to a pure morphism of corings.

28.10. The inclusion morphism and associated functors. *Let $g \in \mathcal{C}$ be a grouplike element and $B = A_g^{co\mathcal{C}}$ with inclusion $\alpha : B \to A$. Define a (B,B)-bimodule map $\gamma : B \to \mathcal{C}$, $b \mapsto gb = bg$, and view B as a trivial B-coring. Then $(\gamma : \alpha) : (B : B) \to (\mathcal{C} : B)$ is a pure morphism of corings known as a g-*inclusion morphism*. The associated induction functor has the form $F : \mathbf{M}_B \to \mathbf{M}^{\mathcal{C}}$, $M \mapsto M \otimes_B A$, as in 28.7. The associated coinduction functor $G : \mathbf{M}^{\mathcal{C}} \to \mathbf{M}_B$ is the g-coinvariants functor $N \mapsto N_g^{co\mathcal{C}}$.*

Proof. First note that γ is well defined since $B = \{b \in A \mid bg = gb\}$ by 28.5. The pair $(\gamma : \alpha)$ is a morphism indeed, since for all $b \in B$, $\underline{\varepsilon}(gb) = \underline{\varepsilon}(g)b = b = \underline{\varepsilon}_B(b)$ and, with the canonical map $\chi : \mathcal{C} \otimes_B \mathcal{C} \to \mathcal{C} \otimes_A \mathcal{C}$,

$$\chi \circ (\gamma \otimes_B \gamma) \circ \underline{\Delta}_B(b) = \chi(\gamma(1_B) \otimes \gamma(b)) = \chi(g \otimes gb) = g \otimes_A gb,$$

while on the other hand $\underline{\Delta} \circ \gamma(b) = \underline{\Delta}(gb) = g \otimes_A gb$. Thus all the conditions in 24.1 are satisfied as required. Since B is a flat B-module, this morphism is necessarily pure.

The form of the induction functor F follows immediately from the definition of $(\gamma : \alpha)$. Now, for any $N \in \mathbf{M}^{\mathcal{C}}$, $N \otimes_B B = N$ canonically and thus the equalising maps in the definition of $G(N)$ in 24.7 come out as $t_N, b_N : N \to N \otimes_A \mathcal{C}$, with $t_N = \varrho^N$ and $b_N : n \mapsto n \otimes \gamma(1_B) = n \otimes g$. Therefore $G(N) = N_g^{co\mathcal{C}} = \{n \in N \mid \varrho^N(n) = n \otimes g\}$, as claimed. □

In view of the general Hom-tensor relations in 24.11, 28.10 provides one with an alternative proof of the adjointness in 28.8.

28.11. Representations of an algebra in a coring. *Viewing an algebra B as a trivial B-coring, objects in the category $\mathbf{Rep}(B : B \mid \mathcal{C} : A)$ are pairs (g, α), where g is a grouplike element in $\mathcal{C}$ and α is an algebra map $B \to A_g^{co\mathcal{C}}$. The morphisms $(g_1, \alpha_1) \to (g_2, \alpha_2)$ are (B,B)-bimodule maps $f : B \to A$ such that $g_1 f(b) = f(b) g_2$ for all $b \in B$. Here A is a left (resp. right) B-module via the map α_1 (resp. α_2).*

Proof. If $(\gamma : \alpha)$ is an object in $\mathbf{Rep}(B : B \mid \mathcal{C} : A)$, then γ is a (B, B)-bimodule map, and hence it is fully determined by an element $g = \gamma(1_B)$. The fact that 1_B is a grouplike element in B implies that g is a grouplike element in $\mathcal{C}$. Furthermore, since γ is a (B, B)-bimodule map and $\mathcal{C}$ is viewed as a (B, B)-bimodule via α, we immediately obtain for all $b \in B$, $\alpha(b)g = \gamma(b) = g\alpha(b)$, that is, $\mathrm{Im}\,(\alpha) \in A_g^{co\mathcal{C}}$. Thus a morphism $(\gamma : \alpha)$ leads to a pair (g, α), as stated. In the converse direction, given any such pair, one defines $\gamma : B \to \mathcal{C}$ via $b \mapsto \alpha(b)g$ and easily checks that this satisfies all the requirements for a morphism of corings.

Since the maps γ_1, γ_2 are determined by grouplike elements g_1 and g_2, respectively, the definition of a morphism in $\mathbf{Rep}(B : B \mid \mathcal{C} : A)$ given in 24.3 immediately leads to a map f with the asserted properties. □

Another property characterising corings with a grouplike element is the fact that their dual rings are augmentation rings. From [10, p. 143] we know that a ring A is called a right (resp. left) *augmentation ring* if there exists an A-module M and a right (resp. left) A-module morphism $\pi : A \to M$. M is called an *augmentation module.*

28.12. Dual algebras as augmentation rings. *Let $g \in \mathcal{C}$ be a grouplike element. Then:*

(1) ${}^*\mathcal{C}$ is a left augmentation ring with an augmentation module A. The left action of ${}^*\mathcal{C}$ on A is provided by $\xi a = \xi(ga)$, for $a \in A$, $\xi \in {}^*\mathcal{C}$.

(2) $\mathcal{C}^*$ is a right augmentation ring with an augmentation module A. The right action of $\mathcal{C}^*$ on A is provided by $a\xi = \xi(ag)$, for $a \in A$, $\xi \in \mathcal{C}^*$.

Proof. We prove only part (1), since the second statement is proven in an analogous way. That the map given above defines a left action of ${}^*\mathcal{C}$ on A follows from the fact that A is a right $\mathcal{C}$-comodule, and hence a left ${}^*\mathcal{C}$-module by 19.1. The augmentation $\pi : {}^*\mathcal{C} \to A$ is given by $\xi \mapsto \xi(g)$. To show that π is a left ${}^*\mathcal{C}$-module map, take any $\xi, \xi' \in {}^*\mathcal{C}$ and obtain

$$\pi(\xi *^l \xi') = (\xi *^l \xi')(g) = \xi(g\xi'(g)) = \xi(g\pi(\xi')) = \xi\pi(\xi'),$$

as required. □

28.13. The splitting of the counit. *Let $g \in \mathcal{C}$ be a grouplike element and let $B = A_g^{co\mathcal{C}}$. Then the short sequence*

$$0 \longrightarrow \mathrm{Ke}\,\underline{\varepsilon} \xrightarrow{\ i\ } \mathcal{C} \xrightarrow{\ \underline{\varepsilon}\ } A \longrightarrow 0,$$

where i is the canonical inclusion, is split-exact as a sequence of (A, B)-bimodules and (B, A)-bimodules.

Proof. First note that the sequence under consideration is exact in the category of (A, A)-bimodules since $\underline{\varepsilon}$ is an (A, A)-bimodule map. Consider an R-linear map $i_R : A \to \mathcal{C}$, $a \mapsto ga$. Clearly, the map i_R is right A-linear. Furthermore, it is left B-linear since, for all $b \in B$, $i_R(ba) = gba = bga = bi_R(a)$, for B is the centraliser of g in A. Finally, for any $a \in A$, $\underline{\varepsilon}(i_R(a)) = \underline{\varepsilon}(ga) = a$. Therefore we conclude that i_R provides the required splitting of (B, A)-modules, and the sequence in the theorem is split-exact. The retraction of i is then given by $p_R : \mathcal{C} \to \mathrm{Ke}\,\underline{\varepsilon}$, $c \mapsto c - g\underline{\varepsilon}(c)$.

Similarly one proves that $i_L : A \to \mathcal{C}$, $a \mapsto ag$, and $p_L : \mathcal{C} \to \mathrm{Ke}\,\underline{\varepsilon}$, $c \mapsto c - \underline{\varepsilon}(c)g$, provide the splitting of the sequence of (A, B)-bimodules. □

28.14. Direct sum decompositions of $\mathcal{C}$. *Let $g \in \mathcal{C}$ be a grouplike element and $B = A_g^{co\mathcal{C}}$. Then:*

(1) $\mathcal{C} \simeq A \oplus \mathrm{Ke}\,\underline{\varepsilon}$ as (B, A)-bimodules with the isomorphism

$$u_R : \mathcal{C} \to A \oplus \mathrm{Ke}\,\underline{\varepsilon}, \quad c \mapsto (\underline{\varepsilon}(c), g\underline{\varepsilon}(c) - c).$$

(2) $\mathcal{C} \simeq A \oplus \mathrm{Ke}\,\underline{\varepsilon}$ as (A, B)-bimodules with the isomorphism

$$u_L : \mathcal{C} \to A \oplus \mathrm{Ke}\,\underline{\varepsilon}, \quad c \mapsto (\underline{\varepsilon}(c), c - \underline{\varepsilon}(c)g).$$

Proof. This follows immediately from 28.13. We only remark that the inverse u_R^{-1} of u_R explicitly reads, $u_R^{-1}(a, c) = ga - c$, while the inverse of u_L is given by $u_L^{-1}(a, c) = c + ag$. □

In view of 28.5 we obtain the

28.15. Corollary. *Let $g \in \mathcal{C}$ be a grouplike element and $B = A_g^{co\mathcal{C}}$. Then:*

(1) $\mathcal{C} \simeq {}^{co\mathcal{C}}_{g}\mathcal{C} \oplus \mathrm{Ke}\,\underline{\varepsilon}$ as (B, A)-bimodules.

(2) $\mathcal{C} \simeq \mathcal{C}_g^{co\mathcal{C}} \oplus \mathrm{Ke}\,\underline{\varepsilon}$ as (A, B)-bimodules.

Since $\mathrm{Ke}\,\underline{\varepsilon}$ is an (A, A)-bimodule (as a kernel of an (A, A)-bimodule map), $A \oplus \mathrm{Ke}\,\underline{\varepsilon}$ has a natural ring structure with the product given by $(a, c)(a', c') = (aa', ac' + ca')$ and the unit $(1, 0)$. Using the isomorphisms in 28.14, one can pull this ring structure back to $\mathcal{C}$. As a consequence one obtains

28.16. The algebra structure of $\mathcal{C}$. *Let $g \in \mathcal{C}$ be a grouplike element. Then:*

(1) $\mathcal{C}$ is an associative R-algebra with unit g and product

$$c \centerdot c' = \underline{\varepsilon}(c)c' + c\underline{\varepsilon}(c') - \underline{\varepsilon}(c)g\underline{\varepsilon}(c').$$

(2) The counit $\underline{\varepsilon} : \mathcal{C} \to A$ is an algebra map.

(3) *The R-linear maps $i_L, i_R : A \to \mathcal{C}$ given by $i_L : a \mapsto ag$, $i_R : a \mapsto ga$, are algebra morphisms splitting $\underline{\varepsilon}$. For all $a \in A$ and $c \in \mathcal{C}$,*

$$ac = i_L(a) \centerdot c, \quad ca = c \centerdot i_R(a).$$

Proof. (1) For any $c, c' \in \mathcal{C}$ define a product in $\mathcal{C}$ via the formula

$$c \centerdot c' = u_R^{-1}(u_R(c)u_R(c')),$$

where u_R is the isomorphism constructed in 28.14 and $u_R(c)u_R(c')$ is the natural product in $A \oplus \mathrm{Ke}\,\underline{\varepsilon}$ recalled in the preamble to the proposition. Then

$$\begin{aligned} c \centerdot c' &= u_R^{-1}((\underline{\varepsilon}(c), g\underline{\varepsilon}(c) - c)(\underline{\varepsilon}(c'), g\underline{\varepsilon}(c') - c')) \\ &= u_R^{-1}(\underline{\varepsilon}(c)\underline{\varepsilon}(c'), \underline{\varepsilon}(c)g\underline{\varepsilon}(c') - \underline{\varepsilon}(c)c' + g\underline{\varepsilon}(c)\underline{\varepsilon}(c') - c\underline{\varepsilon}(c')) \\ &= g\underline{\varepsilon}(c)\underline{\varepsilon}(c') - \underline{\varepsilon}(c)g\underline{\varepsilon}(c') + \underline{\varepsilon}(c)c' - g\underline{\varepsilon}(c)\underline{\varepsilon}(c') + c\underline{\varepsilon}(c') \\ &= \underline{\varepsilon}(c)c' + c\underline{\varepsilon}(c') - \underline{\varepsilon}(c)g\underline{\varepsilon}(c'). \end{aligned}$$

Furthermore, $u_R^{-1}(1, 0) = g$. Therefore $\mathcal{C}$ has an algebra structure, as claimed. Note that the same algebra structure can also be defined using the second isomorphism via $c \centerdot c' = u_L^{-1}(u_L(c)u_L(c'))$.

(2) Let $p : A \oplus \mathrm{Ke}\,\underline{\varepsilon} \to A$ be the canonical projection $p : (a, c) \mapsto a$. Then for all $c \in \mathcal{C}$ we obtain $p \circ u_R(c) = p(\underline{\varepsilon}(c), g\underline{\varepsilon}(c) - c) = \underline{\varepsilon}(c)$, and also $\underline{\varepsilon} = p \circ u_L$. Since p is an algebra morphism and both u_R and u_L are algebra maps by construction, so is $\underline{\varepsilon}$.

(3) Consider an algebra injection $i : A \to A \oplus \mathrm{Ke}\,\underline{\varepsilon}$, $a \mapsto (a, 0)$. For any $a \in A$, $u_R^{-1} \circ i(a) = u_R^{-1}(a, 0) = ga = i_R(a)$. Similarly, $i_L = u_L^{-1} \circ i$. Thus both i_R and i_L are algebra morphisms, and they are split by $\underline{\varepsilon}$ as i splits p (or directly from the split-exact sequence in 28.13). Furthermore,

$$i_L(a) \centerdot c = ac + i_L(a)\underline{\varepsilon}(c) - ag\underline{\varepsilon}(c) = ac + ag\underline{\varepsilon}(c) - ag\underline{\varepsilon}(c) = ac,$$

for all $a \in A$, $c \in \mathcal{C}$. Similarly one proves that $c \centerdot i_R(a) = ca$. □

28.17. Algebra structure of Sweedler corings. Let $\mathcal{C} = A \otimes_B A$ be the Sweedler coring associated to a ring extension $B \to A$. Then $\mathcal{C}$ is an algebra with unit $1_A \otimes 1_A$ and the product

$$(a_1 \otimes a_2) \centerdot (a_3 \otimes a_4) = a_1a_2a_3 \otimes a_4 + a_1 \otimes a_2a_3a_4 - a_1a_2 \otimes a_3a_4.$$

This product appears in [170, Section 1.2] in the context of a braiding related to the noncommutative descent theory.

In view of 28.2, an A-coring has a grouplike element if and only if A is a $\mathcal{C}$-comodule. Corings for which A is a Galois comodule (cf. 18.25) form an important class that is well worth distinguishing.

Definition. Let $g \in \mathcal{C}$ be a grouplike element and $B = A_g^{co\mathcal{C}}$. The pair $(\mathcal{C}, g)$ is said to be a *Galois coring* if A_g (equivalently ${}_gA$) is a Galois comodule (cf. 18.25), that is, if the canonical map

$$\varphi_{\mathcal{C}} : \mathrm{Hom}^{\mathcal{C}}(A_g, \mathcal{C}) \otimes_B A_g \to \mathcal{C}, \quad f \otimes a \mapsto f(a),$$

is an isomorphism of right $\mathcal{C}$-comodules.

28.18. Galois corings. *Let $g \in \mathcal{C}$ be a grouplike element and $B = A_g^{co\mathcal{C}}$. The following are equivalent:*

(a) $(\mathcal{C}, g)$ is a Galois coring;

(b) for every $(\mathcal{C}, A)$-injective comodule $N \in \mathbf{M}^{\mathcal{C}}$, the evaluation

$$\varphi_N : \mathrm{Hom}^{\mathcal{C}}(A_g, N) \otimes_B A_g \to N, \quad f \otimes a \mapsto f(a),$$

is an isomorphism of right $\mathcal{C}$-comodules;

(c) the (A, A)-bimodule map defined by

$$\mathsf{can}_A : A \otimes_B A \to \mathcal{C}, \quad a \otimes a' \mapsto aga',$$

is an isomorphism of A-corings.

The map can_A is called a *Galois isomorphism.*

Proof. This follows from 18.26 by setting $M = A_g$ and identifying B with $\mathrm{End}^{\mathcal{C}}(A_g)$ and $A^* = \mathrm{Hom}_A(A, A)$ with A. □

Note that 28.18 shows in particular that Galois corings are those corings $\mathcal{C}$ with a grouplike element g that are isomorphic to the canonical Sweedler coring associated to the algebra extension $A_g^{co\mathcal{C}} \to A$ (cf. [73]).

28.19. The Galois Coring Structure Theorem. *Let $g \in \mathcal{C}$ be a grouplike element, $B = A_g^{co\mathcal{C}}$, and let $G_g : \mathbf{M}^{\mathcal{C}} \to \mathbf{M}_B$, $M \mapsto M_g^{co\mathcal{C}}$, be the g-coinvariants functor.*

(1) The following are equivalent:

(a) $(\mathcal{C}, g)$ is a Galois coring and A is a flat left B-module;

(b) ${}_A\mathcal{C}$ is flat and A_g is a generator in $\mathbf{M}^{\mathcal{C}}$.

(2) The following also are equivalent:

(a) $(\mathcal{C}, g)$ is a Galois coring and ${}_BA$ is faithfully flat;

(b) ${}_A\mathcal{C}$ is flat and A_g is a projective generator in $\mathbf{M}^{\mathcal{C}}$;

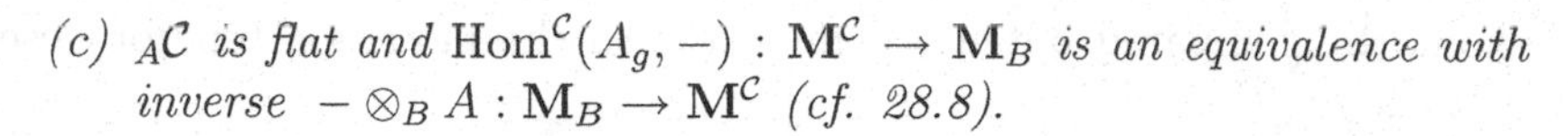

(c) ${}_A\mathcal{C}$ *is flat and* $\mathrm{Hom}^{\mathcal{C}}(A_g, -) : \mathbf{M}^{\mathcal{C}} \to \mathbf{M}_B$ *is an equivalence with inverse* $- \otimes_B A : \mathbf{M}_B \to \mathbf{M}^{\mathcal{C}}$ *(cf. 28.8).*

Proof. This follows from 18.27 by setting $M = A_g$. □

Note that not every canonical coring associated to an algebra extension $B \to A$ is a Galois coring with respect to a grouplike $1_A \otimes 1_A$. However, as noted in 28.6, if the extension $B \to A$ is faithfully flat, then $(A \otimes_B A, 1_A \otimes_B 1_A)$ is a Galois coring. As a particular example of this one can consider a Galois coring provided by

28.20. Sweedler's Fundamental Lemma. *Let A be a division ring and $g \in \mathcal{C}$ a grouplike element. Suppose that $\mathcal{C}$ is generated by g as an (A, A)-bimodule. Then $(\mathcal{C}, g)$ is a Galois coring.*

Proof. Under the given condition A is simple as a left $\mathcal{C}$-comodule, and it subgenerates $\mathcal{C}$ and hence $\mathbf{M}^{\mathcal{C}}$. This implies that $\mathcal{C}$ is a simple and right semisimple coring (see 19.15) and A_g is a projective generator in $\mathbf{M}^{\mathcal{C}}$. So $(\mathcal{C}, g)$ is a Galois coring by 28.19. □

More generally, we characterise simple corings with grouplike elements.

28.21. Simple corings. *Let $g \in \mathcal{C}$ be a grouplike element. Then the following are equivalent:*

(a) *$\mathcal{C}$ is a simple and left semisimple coring;*

(b) *$(\mathcal{C}, g)$ is a Galois coring and $\mathrm{End}^{\mathcal{C}}(A_g)$ is simple and left semisimple;*

(c) *$\mathsf{can}_A : A \otimes_B A \to \mathcal{C}$ is an isomorphism and B is a simple left semisimple subring of A;*

(d) *${}_A\mathcal{C}$ is flat, A_g is a projective generator in $\mathbf{M}^{\mathcal{C}}$, and $\mathrm{End}^{\mathcal{C}}(A_g)$ is simple and left semisimple;*

(e) *$\mathcal{C}_A$ is flat, ${}_gA$ is a projective generator in ${}^{\mathcal{C}}\mathbf{M}$, and $\mathrm{End}^{\mathcal{C}}({}_gA)$ is simple and left semisimple.*

Proof. If $\mathcal{C}$ is simple and semisimple, then every nonzero comodule is a projective generator in $\mathbf{M}^{\mathcal{C}}$ and $\mathrm{End}^{\mathcal{C}}(A_g)$ is simple and left (and right) semisimple. So the assertions follow by 28.18, 28.19 and 28.6. □

28.22. Comparison functor for a coring. In view of 25.1, the first part of Theorem 28.19 can be understood as a restatement of one of the main results in noncommutative descent theory (cf. [96, Theorem]). We refer to [3, Chapter 4] for a nice introduction to descent theory. Given an algebra extension $B \to A$, there is a *comparison functor* $- \otimes_B A : \mathbf{M}_B \to \mathbf{Desc}(A/B)$ that, to every right B-module M, assigns a descent datum $(M \otimes_B A, f)$ with $f : M \otimes_B A \to M \otimes_B A \otimes_B A$, $m \otimes a \mapsto m \otimes 1_A \otimes a$. The comparison functor should be compared but not confused with the induction functor described in 25.4.

An algebra morphism $B \to A$ is called an *effective descent morphism* if the comparison functor is an equivalence of the categories. Extending this terminology to corings, we propose the following:

Definition. Given a grouplike element $g \in \mathcal{C}$ and g-coinvariants B of A, we call the functor $- \otimes_B A : \mathbf{M}_B \to \mathbf{M}^{\mathcal{C}}$ a *comparison functor for a coring with a grouplike element* $(\mathcal{C}, g)$.

If $(\mathcal{C}, g)$ is a Galois coring, then the category of right $\mathcal{C}$-comodules is isomorphic to the category of descent data $\mathbf{Desc}(A/B)$, and thus the first part of 28.19 states that if $B \to A$ is faithfully flat, then it is an effective descent morphism. The second part of 28.19 can be viewed as a clarification of the idea of a Galois coring: Under the faithfully flat condition, Galois corings correspond to comparison functors that are equivalences.

28.23. Exercises

(1) Let $\mathcal{C}$ be an A-coring with a grouplike element g. Suppose that $\mathcal{C}$ is a coseparable coring with cointegral δ, and view it as an A-ring as in 26.6. Let $\mathbf{M}_{\mathcal{C}}$ denote the subcategory of $\mathbf{M}_A$ consisting of firm right $\mathcal{C}$-modules, that is, right A-modules M with associative A-linear $\mathcal{C}$-action such that $M \otimes_{\mathcal{C}} \mathcal{C} \simeq M$. Note that $\mathbf{M}^{\mathcal{C}}$ is a subcategory of $\mathbf{M}_{\mathcal{C}}$ by 26.7. For any $M \in \mathbf{M}_{\mathcal{C}}$ define

$$M^{\mathcal{C}}_{g,\delta} = \{m \in M \mid \text{for all } c \in \mathcal{C},\ mc = m\delta(g \otimes c)\}.$$

Prove ([78]):

(i) $B = A^{\mathcal{C}}_{g,\delta} = \{b \in A \mid \text{for all } c \in \mathcal{C},\ \delta(gb \otimes c) = b\delta(g \otimes c)\}$ is a subalgebra of A.

(ii) The assignment $(-)^{\mathcal{C}}_{g,\delta} : \mathbf{M}_{\mathcal{C}} \to \mathbf{M}_B$, $M \mapsto M^{\mathcal{C}}_{g,\delta}$, is a covariant functor that has a left adjoint $- \otimes_B A : \mathbf{M}_B \to \mathbf{M}_{\mathcal{C}}$.

(iii) $Q = \mathcal{C}^{\mathcal{C}}_{g,\delta}$ is a firm left ideal in $\mathcal{C}$ and hence a $(\mathcal{C}, B)$-bimodule.

(iv) For every $M \in \mathbf{M}_{\mathcal{C}}$, the additive map

$$\omega_M : M \otimes_{\mathcal{C}} Q \to M^{\mathcal{C}}_{g,\delta}, \qquad m \otimes q \mapsto mq,$$

is bijective.

(2) In the setup of Exercise (1), define two maps

$$\sigma : Q \otimes_B A \to \mathcal{C},\ q \otimes a \mapsto qa, \ \text{ and } \ \tau : A \otimes_{\mathcal{C}} Q \to B,\ a \otimes q \mapsto \delta(ga \otimes q).$$

Prove ([78]):

(i) σ is a $(\mathcal{C}, \mathcal{C})$-bilinear map.

(ii) τ is a (B, B)-bimodule isomorphism.

(iii) The maps σ and τ have the following *associativity property*: For all $a, a' \in A$ and $q, q' \in Q$,

$$\sigma(q \otimes a)q' = q\tau(a \otimes q'), \quad a\sigma(q \otimes a') = \tau(a \otimes q)a'.$$

In brief, this exercise shows that $(B, \mathcal{C}, A, Q, \tau, \sigma)$ is a *Morita context* (with a nonunital ring $\mathcal{C}$), in which τ is an isomorphism.

References. Borceux [3]; Brzeziński [71, 73, 74]; Brzeziński, Kadison and Wisbauer [78]; Cartan and Eilenberg [10]; Cipolla [96]; El Kaoutit, Gómez-Torrecillas and Lobillo [112]; Kleiner [140]; Nuss [170]; Sweedler [193].

29 Amitsur complex and connections

In this section we reveal a close relationship between corings with a grouplike element and noncommutative differential geometry. As a first step we show that to any coring with a grouplike element one can associate a differential graded algebra. This algebra can be viewed as a generalisation of the Amitsur complex associated to ring extensions, and hence it is termed an *Amitsur complex for a coring.* We study when this complex is acyclic. The Amitsur complex for a coring restricts to a graded differential algebra, termed *coring valued differential forms*, which can be interpreted as a generalisation of relative differential forms of noncommutative geometry. Motivated by noncommutative geometry, we study coring-valued connections. In particular, we show that the category of comodules of a coring with a grouplike element is isomorphic to the category of flat connections. This result provides us with a noncommutative geometric interpretation of comodules, in addition to giving a representation-theoretic point of view on noncommutative geometry.

29.1. Differential graded algebras. A *differential graded algebra* is an $\mathbb{N}\cup\{0\}$-graded R-algebra $\Omega = \bigoplus_{n=0}^{\infty}\Omega^n$ together with an R-linear degree-1 operation $d:\Omega^\bullet \to \Omega^{\bullet+1}$ such that $d(R1_A)=0$ and

(1) $d\circ d = 0$;

(2) d satisfies the graded Leibniz rule, that is, for all elements ω' and all degree-n elements ω,

$$d(\omega\omega') = d(\omega)\omega' + (-1)^n\omega d(\omega').$$

Condition (1) means that (Ω, d) is a *cochain complex* with a *coboundary operator* d, while (2) states that d is a *graded derivation* in Ω.

29.2. Differential algebra for a coring with a semi-grouplike element. *Let $g\in\mathcal{C}$ be a semi-grouplike element. Consider the tensor algebra*

$$\Omega(\mathcal{C}) = \bigoplus_{n=0}^{\infty}\Omega^n(\mathcal{C}),$$

where $\Omega^0(\mathcal{C}) = A$ and $\Omega^n(\mathcal{C}) = \mathcal{C}\otimes_A\mathcal{C}\otimes_A\cdots\otimes_A\mathcal{C}$ (n-times). Define a degree-1 linear map $d:\Omega(\mathcal{C})\to\Omega(\mathcal{C})$ via $d(a) = ga - ag$, for all $a\in A$, and

$$\begin{aligned} d(c^1\otimes\cdots\otimes c^n) &= g\otimes c^1\otimes\cdots\otimes c^n + (-1)^{n+1}c^1\otimes\cdots\otimes c^n\otimes g \\ &\quad + \sum_{i=1}^{n}(-1)^i c^1\otimes\cdots\otimes c^{i-1}\otimes\underline{\Delta}(c^i)\otimes c^{i+1}\otimes\cdots\otimes c^n. \end{aligned}$$

Then $\Omega(\mathcal{C})$ is a differential graded algebra.

Proof. Let $d^n = d|_{\Omega^n(\mathcal{C})}$. Since d^0 is given as a commutator, it satisfies the Leibniz rule. To show that $d^1 \circ d^0 = 0$, take any $a \in A$ and compute

$$d^1(d^0(a)) = d^1(ga - ag) = g \otimes ga - \underline{\Delta}(ga) + ga \otimes g - g \otimes ag + \underline{\Delta}(ag) - ag \otimes g = 0,$$

since g is a semi-grouplike element and $\underline{\Delta}$ is an (A, A)-bimodule map. Next we need to show that $d^{n+1} \circ d^n = 0$ for all $n \geq 1$. This can be done by a straightforward but lengthy calculation. The main points to notice here can be summarised as follows. The expression for $d^{n+1} \circ d^n$ involves sums of the following pairs of terms ($i \leq j$):

$$(-1)^{i+j} \ldots \otimes \underline{\Delta}(c^i) \otimes \ldots \otimes \underline{\Delta}(c^j) \otimes \ldots, \qquad \text{and}$$
$$(-1)^{i+j+1} \ldots \otimes \underline{\Delta}(c^i) \otimes \ldots \otimes \underline{\Delta}(c^j) \otimes \ldots.$$

The first term comes from the $i+1$-st term in the expansion of d^{n+1} applied to the $j+1$-st term in the expansion of d^n. The second is the $j+2$-nd term in the expansion of d^{n+1} applied to the $i+1$-st term in the expansion of d^n. Obviously such terms cancel each other. The sum of the $i+1$-st and $i+2$-nd terms in the expansion of d^{n+1} applied to the $i+1$-st term in the expansion of d^n reads

$$(-1)^{2i} \ldots \otimes (\underline{\Delta} \otimes I_{\mathcal{C}})\underline{\Delta}(c^i) \otimes \ldots + (-1)^{2i+1} \ldots \otimes (I_{\mathcal{C}} \otimes \underline{\Delta})\underline{\Delta}(c^i) \otimes \ldots$$

and vanishes because of the coassociativity of $\underline{\Delta}$.

Finally, d satisfies the graded Leibniz rule since

$$\begin{aligned}
&d^{m+n}(c^1 \otimes \cdots \otimes c^{m+n}) \\
&= g \otimes c^1 \otimes \cdots \otimes c^{m+n} + \sum_{i=1}^{m+n} (-1)^i c^1 \otimes \cdots \otimes c^{i-1} \otimes \underline{\Delta}(c^i) \otimes c^{i+1} \otimes \cdots \otimes c^{m+n} \\
&\quad + (-1)^{m+n+1} c^1 \otimes \cdots \otimes c^{m+n} \otimes g \\
&= g \otimes c^1 \otimes \cdots \otimes c^{m+n} + \sum_{i=1}^{m} (-1)^i c^1 \otimes \cdots \otimes c^{i-1} \otimes \underline{\Delta}(c^i) \otimes c^{i+1} \otimes \cdots \otimes c^{m+n} \\
&\quad + (-1)^{m+1} c^1 \otimes \cdots \otimes c^m \otimes g \otimes c^{m+1} \otimes \ldots \otimes c^{m+n} \\
&\quad + (-1)^{m} c^1 \otimes \cdots \otimes c^m \otimes g \otimes c^{m+1} \otimes \ldots \otimes c^{m+n} \\
&\quad + \sum_{i=m+1}^{m+n} (-1)^i c^1 \otimes \cdots \otimes c^{i-1} \otimes \underline{\Delta}(c^i) \otimes c^{i+1} \otimes \cdots \otimes c^{m+n} \\
&\quad + (-1)^{m+n+1} c^1 \otimes \cdots \otimes c^{m+n} \otimes g \\
&= d^m(c^1 \otimes \ldots \otimes c^m) \otimes c^{m+1} \otimes \ldots \otimes c^{m+n} \\
&\quad + (-1)^m c^1 \otimes \ldots \otimes c^m \otimes d^n(c^{m+1} \otimes \ldots \otimes c^{m+n}),
\end{aligned}$$

as required. Thus we conclude that $\Omega(\mathcal{C})$ is a differential graded algebra, as stated. □

Recall that, given an R-algebra B, an R-algebra A is called a *B-ring* or an *algebra over B* (with unit) if there is an algebra map $B \to A$. Thus the notion of a B-ring extends the notion of a B-algebra to the case in which B is neither commutative nor central in A. Extending this notion further to differential graded algebras, one says that Ω is a *B-relative differential graded algebra* or a *differential graded algebra over B* if there is an R-algebra map $B \to A = \Omega^0$ such that d is a (B, B)-bimodule map and $d(B) = 0$.

29.3. $\Omega(\mathcal{C})$ as a B-relative differential graded algebra. *Let $\mathcal{C}$ be an A-coring with a semi-grouplike element g, and $B = \{b \in A \mid bg = gb\}$. The differential graded algebra $\Omega(\mathcal{C})$ constructed in 29.2 is a B-relative differential graded algebra.*

Proof. This immediately follows from the facts that B is a centraliser of g in A (hence $d(B) = 0$) and that $\underline{\Delta}$ is a (B, B)-bimodule map. □

29.4. The Amitsur complex. Take an algebra extension $B \to A$, its Sweedler coring $\mathcal{C} = A \otimes_B A$, and the grouplike element $g = 1_A \otimes_B 1_A$. Then $\Omega^n(\mathcal{C}) = A^{\otimes_B n+1}$ and $d^n = \sum_{i=0}^{n+1} (-1)^i e_i^n : A^{\otimes_B n+1} \to A^{\otimes_B n+2}$, where

$$e_i^n : a_1 \otimes \ldots \otimes a_{n+1} \mapsto a_1 \otimes \ldots \otimes a_i \otimes 1_A \otimes a_{i+1} \otimes \ldots \otimes a_{n+1},$$

$i = 0, 1, \ldots, n+1$. This means that $\Omega(A \otimes_B A)$ is the Amitsur complex associated to an algebra extension $B \to A$ (cf. [55], [57]).

Motivated by 29.4, we call the cochain complex $\Omega(\mathcal{C})$ defined in 29.2 the *Amitsur complex* associated to a coring $\mathcal{C}$ and a semi-grouplike element g.

The faithfully flat descent theorem can be restated as a property of the Amitsur complex. If an algebra extension $B \to A$ is faithfully flat, then the corresponding Amitsur complex $\Omega(A/B)$ is acyclic (i.e., all the cohomology groups are trivial). Guided by the relationship between Galois corings and noncommutative descent theory, we can answer the following question.

29.5. When is the Amitsur complex acyclic? *The Amitsur complex of a Galois A-coring $(\mathcal{C}, g)$ is acyclic provided A is a faithfully flat left module over the subring B of its g-coinvariants.*

Proof. Since ${}_BA$ is faithfully flat and the Amitsur complex $(\Omega(\mathcal{C}), d)$ is a complex in the category of right B-modules, it suffices to show that the complex $(\Omega(\mathcal{C}) \otimes_B A, d \otimes I_A)$ is acyclic. Let $\mathsf{can}_A : A \otimes_B A \to \mathcal{C}$ be the Galois isomorphism of corings and write $\mathsf{can}_A^{-1}(c) = \sum c^{\tilde{1}} \otimes c^{\tilde{2}}$, for all $c \in \mathcal{C}$. First note that, for all $c \in \mathcal{C}$, $\sum c^{\tilde{1}} g c^{\tilde{2}} = c$, and then compute

$$\begin{aligned}(I_{\mathcal{C}} \otimes_A \mathsf{can}_A)(\textstyle\sum c^{\tilde{1}} g \otimes_A 1_A \otimes_B c^{\tilde{2}} - \sum c_{\underline{1}} \otimes_A {c_{\underline{2}}}^{\tilde{1}} \otimes_B {c_{\underline{2}}}^{\tilde{2}}) \\ = \textstyle\sum c^{\tilde{1}} g \otimes_A g c^{\tilde{2}} - \sum c_{\underline{1}} \otimes_A c_{\underline{2}} = \underline{\Delta}(\sum c^{\tilde{1}} g c^{\tilde{2}} - c) = 0.\end{aligned}$$

Since can_A is bijective, we conclude that for all $c \in \mathcal{C}$,

$$\sum c^{\tilde{1}} g \otimes c^{\tilde{2}} = \sum c_{\underline{1}} \mathsf{can}_A^{-1}(c_{\underline{2}}). \qquad (*)$$

Now, for any $n = 1, 2, \ldots$, consider an R-linear map

$$\begin{aligned} h^n : \Omega^n(\mathcal{C}) \otimes_B A &\longrightarrow \Omega^{n-1}(\mathcal{C}) \otimes_B A, \\ c^1 \otimes_A \ldots \otimes_A c^n \otimes_B a &\longmapsto (-1)^n c^1 \otimes_A \ldots \otimes_A c^{n-1} \mathsf{can}_A^{-1}(c^n a). \end{aligned}$$

We claim that the collection h of all such h^n is a contracting homotopy for $d \otimes I_A$. On the one hand, ($\otimes$ stands for $\otimes_A$)

$$\begin{aligned} h^{n+1}(d^n(c^1 \otimes \ldots \otimes c^n) \otimes_B a) &= (-1)^{n+1} g \otimes c^1 \otimes \ldots \otimes c^{n-1} \mathsf{can}_A^{-1}(c^n a) \\ &+ \sum_{i=1}^{n-1} (-1)^{n+i+1} c^1 \otimes \ldots \otimes \underline{\Delta}(c^i) \otimes \ldots \otimes c^{n-1} \mathsf{can}_A^{-1}(c^n a) \\ &- \sum c^1 \otimes \cdots \otimes c^n{}_{\underline{1}} \mathsf{can}_A^{-1}(c^n{}_{\underline{2}} a) + c^1 \otimes \cdots \otimes c^n \mathsf{can}_A^{-1}(ga) \\ &= (-1)^{n+1} g \otimes c^1 \otimes \ldots \otimes c^{n-1} \mathsf{can}_A^{-1}(c^n a) \\ &+ \sum_{i=1}^{n-1} (-1)^{n+i+1} c^1 \otimes \ldots \otimes \underline{\Delta}(c^i) \otimes \ldots \otimes c^{n-1} \mathsf{can}_A^{-1}(c^n a) \\ &- \sum c^1 \otimes \cdots \otimes c^{n\tilde{1}} g \otimes_B c^{n\tilde{2}} a + c^1 \otimes \cdots \otimes c^n \otimes_B a, \end{aligned}$$

where we have used equation $(*)$ and the fact that $\mathsf{can}_A^{-1}(g) = 1_A \otimes 1_A$. On the other hand,

$$\begin{aligned} d^{n-1}(h^n(c^1 \otimes \ldots \otimes c^n) \otimes_B a) &= (-1)^n g \otimes c^1 \otimes \ldots \otimes c^{n-1} \mathsf{can}_A^{-1}(c^n a) \\ &+ \sum_{i=1}^{n-1} (-1)^{n+i} c^1 \otimes \ldots \otimes \underline{\Delta}(c^i) \otimes \ldots \otimes c^{n-1} \mathsf{can}_A^{-1}(c^n a) \\ &+ \sum c^1 \otimes \cdots \otimes c^{n\tilde{1}} g \otimes_B c^{n\tilde{2}} a. \end{aligned}$$

Thus $h^{n+1} \circ d^n + d^{n-1} \circ h^n = I_{\Omega^n(\mathcal{C}) \otimes_B A}$. This means that $d \otimes I_A$ is homotopic to the identity, so the complex $(\Omega(\mathcal{C}) \otimes_B A, d \otimes I_A)$ is acyclic, and therefore the Amitsur complex is acyclic by virtue of the fact that the functor $- \otimes_B A$ reflects exact sequences (for A is a faithfully flat left B-module). □

In fact, since there is an algebra map $B \to A$, and hence a map $B \to \Omega(\mathcal{C})$, if $B \to A$ is faithfully flat and $(\mathcal{C}, g)$ is a Galois coring, then the associated Amitsur complex $\Omega(\mathcal{C})$ is a resolution of B.

The category of comodules of a coring with a grouplike element can be described naturally in terms of connections over an algebra.

29.6. Connections. Let $B \to A$ be an algebra extension, and let Ω be a B-relative differential graded algebra with $A = \Omega^0$. A *connection* in a right

A-module M is a right B-linear map $\nabla : M \otimes_A \Omega^\bullet \to M \otimes_A \Omega^{\bullet+1}$ such that, for all $\omega \in M \otimes_A \Omega^k$ and $\omega' \in \Omega$,

$$\nabla(\omega\omega') = \nabla(\omega)\omega' + (-1)^k \omega d(\omega'). \qquad (*)$$

A *curvature* of a connection ∇ is a right B-linear map

$$F_\nabla : M \to M \otimes_A \Omega^2,$$

defined as a restriction of $\nabla \circ \nabla$ to M, that is, $F_\nabla = \nabla \circ \nabla\,|_M$. A connection is said to be *flat* if its curvature is identically equal to 0.

It is important to note that a connection is fully determined by its restriction to the module M. Indeed, any element of $M \otimes_A \Omega$ is a sum of simple tensors $m \otimes \omega$ with $m \in M$ and $\omega \in \Omega$. Now, using the Leibniz rule $(*)$, the action of ∇ on $m \otimes \omega$ reads,

$$\nabla(m \otimes \omega) = \nabla(m)\omega + m \otimes d(\omega).$$

Here we concentrate on right connections (i.e., connections in right modules) and right comodules. Obviously parallel to this one can develop the left-handed version of the theory. To describe a relationship between connections and comodules of a coring, we first need to introduce an appropriate differential graded algebra.

29.7. Coring-valued differential forms. *Let $\mathcal{C}$ be an A-coring with a grouplike element $g \in \mathcal{C}$, and let B be the subalgebra of g-coinvariants of A. Then the associated Amitsur complex $(\Omega(\mathcal{C}), d)$ restricts to the B-relative differential graded algebra $(\Omega(\mathcal{C}/B), d)$ with $\Omega^0(\mathcal{C}/B) = B$ and*

$$\Omega^n(\mathcal{C}/B) = \mathrm{Ke}\,\underline{\varepsilon} \otimes_A \mathrm{Ke}\,\underline{\varepsilon} \otimes_A \ldots \otimes_A \mathrm{Ke}\,\underline{\varepsilon}$$

($\mathrm{Ke}\,\underline{\varepsilon}$ taken n-times). We term $\Omega(\mathcal{C}/B)$ the algebra of $\mathcal{C}$-valued differential forms on A.

Proof. The key observation here is that, first, for all $a \in A$, $\underline{\varepsilon}(d(a)) = \underline{\varepsilon}(ga) - \underline{\varepsilon}(ag) = a - a = 0$, and, second, for any $c^1, \ldots, c^n \in \mathrm{Ke}\,\underline{\varepsilon}$, the Amitsur coboundary operator d^n can be written equivalently as

$$d^n(c^1 \otimes \ldots \otimes c^n)$$
$$= \sum_{i=1}^{n} (-1)^i c^1 \otimes \ldots \otimes c^{i-1} \otimes (c^i{}_{\underline{1}} - g\underline{\varepsilon}(c^i{}_{\underline{1}})) \otimes (c^i{}_{\underline{2}} - \underline{\varepsilon}(c^i{}_{\underline{2}})g) \otimes c^{i+1} \otimes \ldots \otimes c^n.$$

This expression shows immediately that the image of d restricted to $(\mathrm{Ke}\,\underline{\varepsilon})^{\otimes_A n}$ is contained in $(\mathrm{Ke}\,\underline{\varepsilon})^{\otimes_A n+1}$, as required. Note how the canonical projections featuring in the description of $\mathcal{C}$ as a direct sum of A with $\mathrm{Ke}\,\underline{\varepsilon}$ discussed in 28.14 are used here. $\square$

29.8. Corings associated to differential graded algebras. The construction in 29.7 has an interesting converse. Let Ω be a differential graded algebra with $\Omega^0 = A$, and $\Omega^n = \Omega^1 \otimes_A \Omega^1 \otimes_A \ldots \otimes_A \Omega^1$ (n-times). Consider a left A-module $\mathcal{C} = A \oplus \Omega^1$. Let $g = (1, 0)$ and identify $(0, \omega)$ with ω so that $\mathcal{C}$ consists of elements of the form $ag + \omega$, where $a \in A$, $\omega \in \Omega^1$. One easily checks that $\mathcal{C}$ can be made into an (A, A)-bimodule with right multiplication $(ag + \omega)a' = aa'g + ad(a') + \omega a'$ for all $a, a' \in A$, $\omega \in \Omega^1$. Then $\mathcal{C}$ is an A-coring with coproduct and counit

$$\underline{\Delta}(ag) = ag \otimes g, \quad \underline{\Delta}(\omega) = g \otimes \omega + \omega \otimes g - d(\omega), \quad \underline{\varepsilon}(ag + \omega) = a,$$

for all $a \in A$ and $\omega \in \Omega^1$. The coassociativity of $\underline{\Delta}$ follows from the equality $\sum d(\omega^{\bar{1}}) \otimes \omega^{\bar{2}} + \omega^{\bar{1}} \otimes d(\omega^{\bar{2}}) = 0$, where $d(\omega) = \sum \omega^{\bar{1}} \otimes \omega^{\bar{2}}$ is a notation. This is a consequence of the Leibniz rule and the nilpotency of d. Clearly, g is a grouplike element and $\Omega = \Omega(\mathcal{C}/B)$, where $B = \ker(d : A \to \Omega^1)$. Note also that $d(a) = ga - ag$. In particular, some differential calculi of noncommutative geometry based on Dirac operators or Fredholm modules lead to corings with grouplike elements.

29.9. Relative differential forms. Given an algebra extension $B \to A$, there is an associated universal differential graded algebra over B generated by the B-ring A known as an *algebra of B-relative differential forms* $\Omega_B A$. Let A/B denote the cokernel of the algebra homomorphism $B \to A$ viewed as a map of (B, B)-bimodules, and let $\pi : A \to A/B$ be the canonical map of (B, B)-bimodules. Thus A/B is a (B, B)-bimodule, and for any $n \in \mathbb{N} \cup \{0\}$ one can consider an (A, B)-bimodule,

$$\Omega^n_B A = A \otimes_B (A/B)^{\otimes_B n} = A \otimes_B A/B \otimes_B A/B \otimes_B \cdots \otimes_B A/B,$$

and combine them into a direct sum $\Omega_B A = \bigoplus_{n=0}^{\infty} \Omega^n_B A$. $\Omega_B A$ is a cochain complex with a coboundary operator

$$d : \Omega^\bullet_B A \to \Omega^{\bullet+1}_B A, \quad a_0 \otimes a_1 \otimes \ldots \otimes a_n \mapsto 1_A \otimes \pi(a_0) \otimes a_1 \otimes \ldots \otimes a_n.$$

It is clear that $d \circ d = 0$ since $B \to A$ as an algebra homomorphism is a unit-preserving map so that $\pi(1_A) = 0$. Less trivial is the observation that $(\Omega_B A, d)$ is a B-relative differential graded algebra. The product in $\Omega_B A$ is given by the formula

$$(a_0, \ldots, a_n)(a_{n+1}, \ldots, a_m) = \sum_{i=0}^{n} (-1)^{n-i} (a_0, \ldots, a_{i-1}, a_i \cdot a_{i+1}, a_{i+2}, \ldots, a_m),$$

where we write $(a_0, \ldots, a_n)$ for $a_0 \otimes_B \ldots \otimes_B a_n$, and so on, to relieve the notation. Also, the notation $a_i \cdot a_{i+1}$ is a formal expression that is to be understood as follows. Take any $a'_i \in \pi^{-1}(a_i)$ and $a'_{i+1} \in \pi^{-1}(a_{i+1})$, $i = 1, \ldots n-1$. Then

$$a_i \cdot a_{i+1} = \begin{cases} a_i a'_{i+1} & \text{for } i = 0 \\ \pi(a'_i a'_{i+1}) & \text{for } 0 < i < n \\ \pi(a'_i a_{i+1}) & \text{for } i = n. \end{cases}$$

Although $a_i \cdot a_{i+1}$ depends on the choice of the a'_i, it can be easily shown that the product of cochains does not. It can also be shown that the above expression defines an associative product and that d satisfies the graded Leibniz rule. The details can be found in [103]. The algebra of B-relative differential forms has the following universality property. Given any graded differential algebra $\Omega = \bigoplus \Omega^n$ and an algebra homomorphism $u : A \to \Omega^0$ such that $d(u(b)) = 0$ for all $b \in B$, there exists a unique differential graded algebra homomorphism $u_* : \Omega_B A \to \Omega$ extending u. This means, in particular, that the identity map $A \to A$ extends to a map of B-relative differential graded rings $\Omega_B A \to \Omega(\mathcal{C}/B)$. This can be understood purely in terms of coring-valued differential forms.

29.10. Sweedler coring-valued differential forms. *Let $B \to A$ be an algebra extension. Take the Sweedler coring $\mathcal{C} = A \otimes_B A$ and a grouplike element $g = 1_A \otimes 1_A$. Then the differential graded algebra over B of $\mathcal{C}$-valued differential forms is isomorphic to the algebra of B-relative differential forms $\Omega_B A$.*

Proof. We first describe the structure of $\mathcal{C} = A \otimes_B A$-valued differential forms. Since the counit of the canonical coring coincides with the product map $\mu_{A/B} : A \otimes_B A \to A$, $a \otimes a' \mapsto aa'$, we have $\Omega^1(\mathcal{C}/B) = \mathrm{Ke}\,\underline{\varepsilon} = \mathrm{Ke}\,\mu_{A/B}$. Clearly, the Amitsur 0-differential $d : a \mapsto 1_A \otimes a - a \otimes 1_A$ has values restricted to $\mathrm{Ke}\,\mu_{A/B}$. Note that $\Omega^n(\mathcal{C}/B) = (\mathrm{Ke}\,\mu_{A/B})^{\otimes_A n}$. Thus, if we can show that $\mathrm{Ke}\,\mu_{A/B} \simeq A \otimes_B A/B$ as (A, A)-bimodules, then we will obtain the required form of $\Omega^n(\mathcal{C}/B)$. Indeed, the iteration

$$\begin{aligned} \Omega^n(\mathcal{C}/B) &= \Omega^{n-1}(\mathcal{C}/B) \otimes_A \mathrm{Ke}\,\mu_{A/B} \\ &\simeq \Omega^{n-1}(\mathcal{C}/B) \otimes_A A \otimes_B A/B \simeq \Omega^{n-1}(\mathcal{C}/B) \otimes_B A/B \end{aligned}$$

repeated n-times then yields the desired result.

Note that $\mathrm{Ke}\,\mu_{A/B} \simeq A \otimes_B A/B$ as (A, B)-bimodules via the map

$$\theta : \mathrm{Ke}\,\mu_{A/B} \to A \otimes_B A/B, \quad \textstyle\sum_i a_i \otimes a'_i \mapsto \sum_i a_i \otimes \pi(a'_i),$$

with the inverse $\theta^{-1} : a \otimes \pi(a') \mapsto a \otimes a' - a' \otimes 1_A$. Note also that the map θ^{-1} does not depend on the choice of a' in the inverse image of $\pi(a')$. Indeed, if $\pi(a') = 0$, then $a' = b1_A$ with $b \in B$ and hence $a \otimes a' - aa' \otimes 1_A = a \otimes b1_A - ab \otimes$

$1_A = 0$, as needed. The right A-module structure of $A \otimes_B A/B$ is derived from the product in $\Omega_B A$, that is, $(a_0 \otimes \pi(a_1))a = -a_0 a \otimes \pi(a_1) + a_0 \otimes \pi(a_1 a)$, and is well defined (does not depend on the choice of a_1) by a similar argument as above. Clearly θ and θ^{-1} are maps of right A-modules. The isomorphism θ extends to cochains of all degrees, and one can easily check that it provides an isomorphism of B-relative graded differential algebras. This isomorphism involves projections π in all bar the first tensorand and thus maps all the Amitsur operators e_i^n with $i > 0$ in 29.4 to 0. Thus the resulting differential d has the form $d : a_0 \otimes a_1 \otimes \ldots \otimes a_n \mapsto 1_A \otimes \pi(a_0) \otimes a_1 \otimes \ldots \otimes a_n$, as required. □

29.11. Existence of connections. *Let $\mathcal{C}$ be an A-coring with a grouplike element $g \in \mathcal{C}$, let $B = A_g^{co\mathcal{C}}$ be the subring of g-coinvariants of A, and let $\Omega(\mathcal{C}/B)$ be the algebra of $\mathcal{C}$-valued differential forms on A. A right A-module M admits an $\Omega(\mathcal{C}/B)$-valued connection if and only if $I_M \otimes_A \underline{\varepsilon}$ is a retraction in $\mathbf{M}_A$.*

Proof. Given a connection $\nabla : M \to M \otimes_A \Omega^1(\mathcal{C}/B)$, define an R-linear map

$$j_\nabla : M \to M \otimes_A \mathcal{C}, \quad m \mapsto \nabla(m) + m \otimes g.$$

Since $\mathrm{Im}(\nabla) \subseteq M \otimes_A \mathrm{Ke}\,\underline{\varepsilon}$ and $\underline{\varepsilon}(g) = 1_A$, the map j_∇ is an R-linear section of $I_M \otimes \underline{\varepsilon}$. Furthermore, for all $m \in M$ and $a \in A$,

$$\begin{aligned} j_\nabla(ma) &= \nabla(ma) + ma \otimes g = \nabla(m)a + m \otimes d(a) + m \otimes ag \\ &= \nabla(m)a + m \otimes ga - m \otimes ag + m \otimes ag = j_\nabla(m)a, \end{aligned}$$

where we used that ∇ is a connection to obtain the second equality. Thus j_∇ is a right A-linear section of $I_M \otimes \underline{\varepsilon}$.

Conversely, suppose $j : M \to M \otimes_A \mathcal{C}$ is a right A-linear section of $I_M \otimes \underline{\varepsilon}$, and define an R-linear map

$$\nabla_j : M \to M \otimes_A \Omega^1(\mathcal{C}/B) = M \otimes_A \mathrm{Ke}\,\underline{\varepsilon}, \quad m \mapsto j(m) - m \otimes g.$$

Note that ∇_j is well defined since the fact that j is a section implies that, for all $m \in M$, $\sum_i m^i \underline{\varepsilon}(c^i) = m$, where $\sum_i m^i \otimes c^i = j(m)$. Therefore

$$\nabla_j(m) = \textstyle\sum_i (m^i \otimes c^i - m^i \underline{\varepsilon}(c^i) \otimes g) = \sum_i m^i \otimes (c^i - \underline{\varepsilon}(c^i) g),$$

and each $c^i - \underline{\varepsilon}(c^i) \in \mathrm{Ke}\,\underline{\varepsilon}$. Finally, ∇_j is a connection since, for all $m \in M$ and $a \in A$,

$$\begin{aligned} \nabla_j(ma) &= j(ma) - ma \otimes g = j(m)a - m \otimes ag \\ &= j(m)a - m \otimes ga + m \otimes ga - m \otimes ag = \nabla_j(m)a + m \otimes d(a), \end{aligned}$$

as required. □

29.12. Corollary. *In the situation of 29.11, if a right A-module M admits an $\Omega(\mathcal{C}/B)$-valued connection, then M is a direct summand of a right A-module $M \otimes_A \mathcal{C}$.*

Proof. By 29.11, the map $M \xrightarrow{j_\nabla} M \otimes_A \mathcal{C}$ is split by $M \otimes_A \mathcal{C} \xrightarrow{I_M \otimes \underline{\varepsilon}} M$ in $\mathbf{M}_A$. □

29.13. Connections and cosplit corings. *If $\mathcal{C}$ is a cosplit A-coring with a grouplike element g and $B = A_g^{co\mathcal{C}}$, then any right A-module admits an $\Omega(\mathcal{C}/B)$-valued connection.*

Proof. If $\mathcal{C}$ is cosplit, then the counit $\underline{\varepsilon}$ has an (A,A)-bimodule section. Hence, for any right A-module M, the map $I_M \otimes_A \underline{\varepsilon}$ has a right A-module section, and thus M admits a connection by 29.11. Explicitly, given $e \in \mathcal{C}^A$ such that $\underline{\varepsilon}(e) = 1_A$, the connection ∇ is given by $\nabla : m \mapsto m \otimes (e - g)$. □

29.14. Comodules and flat connections. *Let $\mathcal{C}$ be an A-coring with a grouplike element g, $B = A_g^{co\mathcal{C}}$, and let $\Omega(\mathcal{C}/B)$ be the algebra of $\mathcal{C}$-valued differential forms on A. Then a right A-module M is a right $\mathcal{C}$-comodule if and only if it admits a flat connection $\nabla : M \to M \otimes_A \Omega(\mathcal{C}/B)$.*

Proof. Suppose M is a right $\mathcal{C}$-comodule with coaction ϱ^M and define

$$\nabla : M \to M \otimes_A \Omega^1(\mathcal{C}/B), \quad m \mapsto \varrho^M(m) - m \otimes g.$$

Since ϱ^M is a right A-module splitting of $I_M \otimes \underline{\varepsilon}$, the map ∇ is a connection by 29.11. We can now compute the curvature F_∇ of ∇. For any $m \in M$,

$$\begin{aligned} F_\nabla(m) &= \nabla(\textstyle\sum m_{\underline{0}} \otimes m_{\underline{1}} - m \otimes g) \\ &= \textstyle\sum \nabla(m_{\underline{0}}) \otimes m_{\underline{1}} + \sum m_{\underline{0}} \otimes dm_{\underline{1}} - \nabla(m) \otimes g + m \otimes dg = 0, \end{aligned}$$

by coassociativity of ϱ^M and the definition of d.

Conversely, suppose M is a right A-module with a flat connection $\nabla : M \to M \otimes_A \mathrm{Ke}\,\underline{\varepsilon}$. For any $m \in M$ write $\nabla(m) = \sum_i m^i \otimes c^i$. Then the flatness of ∇ means

$$0 = \nabla(\textstyle\sum_i m^i \otimes c^i) = \sum_i \nabla(m^i) \otimes c^i + \sum_i m^i \otimes d(c^i), \quad \text{i.e.,}$$

$$\textstyle\sum_i m^i \otimes c^i{}_{\underline{1}} \otimes c^i{}_{\underline{2}} = \sum_{i,j} m^{ij} \otimes \tilde{c}^{ij} \otimes c^i + \sum_i m^i \otimes g \otimes c^i + \sum_i m^i \otimes c^i \otimes g,$$

where $\nabla(m^i) = \sum_j m^{ij} \otimes \tilde{c}^{ij}$. Now define an R-linear map

$$\varrho^M : M \to M \otimes_A \mathcal{C}, \quad m \mapsto \nabla(m) + m \otimes g.$$

Note that ϱ^M coincides with the map j_∇ constructed in the proof of 29.11, and thus it is a right A-module section of $I_M \otimes \underline{\varepsilon}$, that is, $(I_M \otimes \underline{\varepsilon})\varrho^M(m) = m$.

So it only remains to show that ϱ^M is coassociative. Since explicitly $\varrho^M(m) = \sum_i m^i \otimes c^i + m \otimes g$, we can compute

$$\begin{aligned}(\varrho^M \otimes I_{\mathcal{C}})\varrho^M(m) &= \textstyle\sum_i \varrho^M(m^i) \otimes c^i + \varrho^M(m) \otimes g \\ &= \textstyle\sum_{ij} m^{ij} \otimes \tilde{c}^{ij} \otimes c^i + \sum_i m^i \otimes g \otimes c^i + \sum_i m^i \otimes c^i \otimes g + m \otimes g \otimes g \\ &= \textstyle\sum_i m^i \otimes c^i{}_{\underline{1}} \otimes c^i{}_{\underline{2}} + m \otimes g \otimes g \\ &= (I_M \otimes \underline{\Delta})\varrho^M(m),\end{aligned}$$

where we used the flatness of ∇ to derive the penultimate equality. This proves that (M, ϱ^M) is a right $\mathcal{C}$-comodule. □

In the proof of 29.14 we constructed two assignments. Given a right A-module M, to every right $\mathcal{C}$-coaction ϱ^M one assigns a flat connection by $\nabla_{\varrho^M} : m \mapsto \varrho^M(m) - m \otimes g$. Conversely, to any flat connection ∇ one assigns a right $\mathcal{C}$-coaction $\varrho^M_\nabla : m \mapsto \nabla(m) + m \otimes g$. Clearly these assignments are inverses of each other and hence establish an isomorphism of sets of flat connections and right $\mathcal{C}$-comodule structures. In fact, 29.14 describes an isomorphism of categories.

29.15. The category of connections. Consider an algebra extension $B \to A$ and a B-relative differential graded algebra Ω with $\Omega^0 = A$. The category of (right) connections with values in Ω, denoted by $\mathbf{Conn}(A/B, \Omega)$, consists of pairs (M, ∇), where M is a right A-module and $\nabla : M \otimes_A \Omega^\bullet \to M \otimes_A \Omega^{\bullet+1}$ is a connection. A morphism $(M, \nabla) \to (N, \nabla')$ in $\mathbf{Conn}(A/B, \Omega)$ is a right A-module map $f : M \to N$ inducing a commutative diagram,

$$\begin{array}{ccc} M & \xrightarrow{f} & N \\ {\scriptstyle\nabla}\downarrow & & \downarrow{\scriptstyle\nabla'} \\ M \otimes_A \Omega^1 & \xrightarrow{f \otimes I_\Omega} & N \otimes_A \Omega^1. \end{array}$$

Note that the Leibniz rule also implies that the diagram

$$\begin{array}{ccc} M \otimes_A \Omega^\bullet & \xrightarrow{f \otimes I_\Omega} & N \otimes_A \Omega^\bullet \\ {\scriptstyle\nabla}\downarrow & & \downarrow{\scriptstyle\nabla'} \\ M \otimes_A \Omega^{\bullet+1} & \xrightarrow{f \otimes I_\Omega} & N \otimes_A \Omega^{\bullet+1} \end{array}$$

is commutative. In particular, we can consider the diagram

$$\begin{array}{ccccc} M & \xrightarrow{\nabla} & M \otimes_A \Omega^1 & \xrightarrow{\nabla} & M \otimes_A \Omega^2 \\ {\scriptstyle f}\downarrow & & {\scriptstyle f\otimes I_\Omega}\downarrow & & \downarrow{\scriptstyle f\otimes I_\Omega} \\ N & \xrightarrow{\nabla'} & N \otimes_A \Omega^1 & \xrightarrow{\nabla'} & N \otimes_A \Omega^2, \end{array}$$

in which both left and right squares commute. This implies that the outer rectangle is commutative, and hence $F' \circ f = (f \otimes_A I_{\Omega^2}) \circ F$, where F is the curvature of ∇ and F' is the curvature of ∇'. This shows that $\mathbf{Conn}(A/B, \Omega)$ contains a full subcategory $\mathbf{Conn}_0(A/B, \Omega)$ of flat connections. Objects of $\mathbf{Conn}_0(A/B, \Omega)$ are pairs (M, ∇), where M is a right A-module and ∇ is a flat connection.

29.16. Isomorphism of categories of flat connections and comodules. *Let $\mathcal{C}$ be an A-coring with a grouplike element g, $B = A_g^{co\mathcal{C}}$, and let $\Omega(\mathcal{C}/B)$ be the algebra of $\mathcal{C}$-valued differential forms on A. Then $\mathbf{M}^{\mathcal{C}}$ is isomorphic to* $\mathbf{Conn}_0(A/B, \Omega(\mathcal{C}/B))$.

Proof. On objects, the isomorphism is provided by the assignement constructed in 29.14 while, on morphisms, $f : M \to N$, $f \leftrightarrow f$. Indeed, if f is a morphism of right $\mathcal{C}$-comodules, then, for all $m \in M$,

$$\begin{aligned} \nabla_{\varrho^N} \circ f(m) &= \varrho^N \circ f(m) - f(m) \otimes g \\ &= (f \otimes I_{\mathcal{C}})(\varrho^M(m) - m \otimes g) = (f \otimes I_{\mathcal{C}}) \circ \nabla_{\varrho^M}(m). \end{aligned}$$

If f is a morphism $(M, \nabla) \to (N, \nabla')$ in $\mathbf{Conn}_0(A/B, \Omega(\mathcal{C}/S))$, then

$$\begin{aligned} \varrho^N_{\nabla'} \circ f(m) &= \nabla'(f(m)) + f(m) \otimes g = (f \otimes I_{\mathcal{C}}) \circ \nabla(m) + f(m) \otimes g \\ &= (f \otimes I_{\mathcal{C}}) \circ (\nabla(m) + m \otimes g) = (f \otimes I_{\mathcal{C}}) \circ \varrho^M_{\nabla}. \end{aligned}$$

This completes the proof of the theorem. □

A similar isomorphism as in 29.16 can be established between the category of flat connections in left A-modules and the category of left $\mathcal{C}$-comodules.

References. Amitsur [55]; Artin [57]; Brzeziński [74]; Connes [101]; Cuntz and Quillen [103]; Nuss [170]; Rojter [182]; Sweedler [192].

30 Cartier and Hochschild cohomology

In this section we outline two cohomology theories of corings. Our main aim is to give a cohomological interpretation of coseparable and cosplit corings.

As before, A is an R-algebra and $\mathcal{C}$ is an A-coring.

30.1. The cobar complex. *Let* $\mathrm{Cob}(\mathcal{C}) = (\mathrm{Cob}(\mathcal{C})^\bullet, \delta)$ *be a complex given as* $\mathrm{Cob}(\mathcal{C})^n = \mathcal{C}^{\otimes_A n+2}$,

$$\delta^n = \sum_{k=0}^{n+1} (-1)^k I_{\mathcal{C}}^{\otimes_A k} \otimes \underline{\Delta} \otimes I_{\mathcal{C}}^{\otimes_A n-k+1} : \mathrm{Cob}(\mathcal{C})^n \to \mathrm{Cob}(\mathcal{C})^{n+1},$$

$n = 0, 1, 2\ldots$ *Then* $\mathrm{Cob}(\mathcal{C})$ *is a resolution of* $\mathcal{C}$ *in the category of* (A, A)*-bimodules and is called a* cobar complex *or a* cobar resolution *of* $\mathcal{C}$.

Proof. First we need to show that $\mathrm{Cob}(\mathcal{C})$ is a cochain complex, that is, for all $n \in \mathbb{N} \cup \{0\}$, $\delta^{n+1} \circ \delta^n = 0$. This follows directly from 29.2, by taking a semi-grouplike element $g = 0$ there.

Let M be an (A, A)-bimodule. Recall that a cochain complex of (A, A)-bimodules $X = (X^\bullet, \delta)$ is called a *resolution* of M in the category of (A, A)-bimodules if there exists an (A, A)-bimodule map $i : M \to X^0$ such that the sequence

$$0 \longrightarrow M \xrightarrow{\ i\ } X^0 \xrightarrow{\ \delta^0\ } X^1 \xrightarrow{\ \delta^1\ } X^2 \xrightarrow{\ \delta^2\ } \cdots$$

is exact. Next note that there is an injective map $\underline{\Delta} : \mathcal{C} \to \mathcal{C} \otimes_A \mathcal{C} = \mathrm{Cob}(\mathcal{C})^0$. We will construct a contracting homotopy for $\mathrm{Cob}(\mathcal{C})$, that is, a sequence of (A, A)-bimodule maps $(h_n)_{n \in \mathbb{N}\cup\{0\}}$, $h_n : \mathcal{C}^{\otimes_A n+2} \to \mathcal{C}^{\otimes_A n+1}$, with the property $h_{n+1} \circ \delta^n + \delta^{n-1} \circ h_n = I_{\mathcal{C}}^{\otimes n+2}$. Let $h_n = \underline{\varepsilon} \otimes I_{\mathcal{C}}^{\otimes n+1}$ for all $n \in \mathbb{N} \cup \{0\}$. Then

$$\begin{aligned} h_{n+1}\delta^n &= \sum_{k=0}^{n+1} (-1)^k (\underline{\varepsilon} \otimes I_{\mathcal{C}}^{\otimes n+1})(I_{\mathcal{C}}^{\otimes k} \otimes \underline{\Delta} \otimes I_{\mathcal{C}}^{\otimes n-k+1}) \\ &= I_{\mathcal{C}}^{\otimes n+2} - \sum_{k=0}^{n} (-1)^k (\underline{\varepsilon} \otimes I_{\mathcal{C}}^{\otimes k} \otimes \underline{\Delta} \otimes I_{\mathcal{C}}^{\otimes n-k+1}) = I_{\mathcal{C}}^{\otimes n+2} - \delta^{n-1} h_n, \end{aligned}$$

as required. The first equality follows from the counit property of $\underline{\varepsilon}$. Furthermore, note that

$$(h_1\delta^0 + \underline{\Delta} h_0) = (\underline{\varepsilon} \otimes I_{\mathcal{C}} \otimes I_{\mathcal{C}})(\underline{\Delta} \otimes I_{\mathcal{C}}) - (\underline{\varepsilon} \otimes I_{\mathcal{C}} \otimes I_{\mathcal{C}})(I_{\mathcal{C}} \otimes \underline{\Delta}) + \underline{\varepsilon} \otimes \underline{\Delta} = I_{\mathcal{C}} \otimes I_{\mathcal{C}}$$

by the counit property of $\underline{\varepsilon}$ again. By this equality $x \in \mathrm{Ke}\,\delta^0$ implies $x \in \mathrm{Im}\,\underline{\Delta}$. Combined with the existence of a contracting homotopy, we conclude that the sequence

$$0 \longrightarrow \mathcal{C} \xrightarrow{\ \underline{\Delta}\ } \mathcal{C} \otimes_A \mathcal{C} \xrightarrow{\ \delta^0\ } \mathcal{C}^{\otimes_A 3} \xrightarrow{\ \delta^1\ } \mathcal{C}^{\otimes_A 4} \xrightarrow{\ \delta^2\ } \cdots$$

is exact. Hence $\mathrm{Cob}(\mathcal{C})$ is a resolution of $\mathcal{C}$. □

30.2. The cobar resolution of a Sweedler coring. Consider an algebra extension $B \to A$, and let $\mathcal{C} = A \otimes_B A$ be the Sweedler A-coring. Then

$$\mathrm{Cob}(\mathcal{C}) = A^{\otimes_B n+3}, \qquad \delta^n = \sum_{i=0}^{n+1}(-1)^i e_{i+1}^{n+2} : A^{\otimes_B n+3} \to A^{\otimes_B n+4},$$

where $e_i^n : a_1 \otimes \ldots \otimes a_{n+1} \mapsto a_1 \otimes \ldots \otimes a_i \otimes 1_A \otimes a_{i+1} \otimes \ldots \otimes a_{n+1}$, for $i = 0, 1, \ldots, n+1$.

Note the similarity of this complex to the Amitsur complex $\Omega(A \otimes_B A)$ in 29.4. In fact, $\mathrm{Cob}(A \otimes_B A) = A \otimes_B \Omega(A \otimes_B A) \otimes_B A$. Thus it comes as no surprise that, although the Amitsur complex is not acyclic, the cobar complex is (cf. 29.5).

30.3. The Cartier complex. The cobar resolution $\mathrm{Cob}(\mathcal{C})$ of $\mathcal{C}$ can be viewed as a resolution in the category of $(\mathcal{C},\mathcal{C})$-bicomodules. The left coaction of $\mathcal{C}$ on $\mathrm{Cob}(\mathcal{C})^n = \mathcal{C}^{\otimes_A n+2}$ is given as $\underline{\Delta} \otimes I_{\mathcal{C}}^{\otimes n+1}$, while the right $\mathcal{C}$-coaction is $I_{\mathcal{C}}^{\otimes n+1} \otimes \underline{\Delta}$. Note that every $\mathrm{Cob}(\mathcal{C})^n$ is an $(A,\mathcal{C})$-relative injective $(\mathcal{C},\mathcal{C})$-bicomodule. This follows immediately from 22.8 since the left coactions have the $(\mathcal{C},\mathcal{C})$-bicomodule retractions $I_{\mathcal{C}} \otimes \underline{\varepsilon} \otimes I_{\mathcal{C}}^{\otimes n+1}$. The coassociativity of $\underline{\Delta}$ implies that all the coboundary operators δ^n are $(\mathcal{C},\mathcal{C})$-bicomodule maps. Furthermore, the contracting homotopy constructed in the proof of 30.1 consists of right $\mathcal{C}$-comodule left A-module maps. In other words we have

Corollary. $\mathrm{Cob}(\mathcal{C})$ *is an* $(A,\mathcal{C})$*-relative injective resolution of* $\mathcal{C}$ *in the category of* $(\mathcal{C},\mathcal{C})$*-bicomodules.*

Thus, for any $(\mathcal{C},\mathcal{C})$-bicomodule M, one can consider a cochain complex

$$C_{\mathrm{Ca}}(\mathcal{C}, M) = (C_{\mathrm{Ca}}(\mathcal{C}, M)^\bullet, d) = {}^{\mathcal{C}}\mathrm{Hom}^{\mathcal{C}}(M, \mathrm{Cob}(\mathcal{C})).$$

There is a Hom-tensor relation for $(\mathcal{C},\mathcal{C})$-bicomodules,

$$\theta : {}^{\mathcal{C}}\mathrm{Hom}^{\mathcal{C}}(M, \mathcal{C}^{\otimes_A n+2}) \xrightarrow{\simeq} {}_A\mathrm{Hom}_A(M, \mathcal{C}^{\otimes_A n}), \quad f \mapsto (\underline{\varepsilon} \otimes I_{\mathcal{C}}^{\otimes n} \otimes \underline{\varepsilon}) \circ f,$$

with the inverse $\theta^{-1}(g) = (I_{\mathcal{C}} \otimes g \otimes I_{\mathcal{C}}) \circ ({}^M\!\varrho \otimes I_{\mathcal{C}}) \circ \varrho^M$, where ${}^M\!\varrho$ is the left and ϱ^M is the right coaction of $\mathcal{C}$ on M. This Hom-tensor relation can be used to view the complex $C_{\mathrm{Ca}}(\mathcal{C}, M)$ as

$${}_A\mathrm{Hom}_A(M, A) \xrightarrow{d^0} {}_A\mathrm{Hom}_A(M, \mathcal{C}) \xrightarrow{d^1} {}_A\mathrm{Hom}_A(M, \mathcal{C} \otimes_A \mathcal{C}) \xrightarrow{d^2} \ldots,$$

where $d^n : {}_A\mathrm{Hom}_A(M, \mathcal{C}^{\otimes_A n}) \to {}_A\mathrm{Hom}_A(M, \mathcal{C}^{\otimes_A n+1})$ reads

$$d^n f = (I_{\mathcal{C}} \otimes f) \circ {}^M\!\varrho + \sum_{k=1}^{n}(-1)^k (I_{\mathcal{C}}^{\otimes k-1} \otimes \underline{\Delta} \otimes I_{\mathcal{C}}^{\otimes n-k}) \circ f + (-1)^{n+1}(f \otimes I_{\mathcal{C}}) \circ \varrho^M.$$

The complex $C_{\mathrm{Ca}}(\mathcal{C}, M)$ is called the *Cartier complex* of $\mathcal{C}$ with values in M. Its cohomology is called the *Cartier cohomology* of $\mathcal{C}$ with values in M and is denoted by $H_{\mathrm{Ca}}(\mathcal{C}, M)$.

The relative Hochschild cohomology detects when an extension of rings is separable. Similarly, the Cartier cohomology of a coring detects when the coring is coseparable.

30.4. The cohomological meaning of coseparability. *For an A-coring $\mathcal{C}$ the following statements are equivalent:*

(a) *$\mathcal{C}$ is coseparable;*

(b) *for all $(\mathcal{C},\mathcal{C})$-bicomodules M, $H^n_{\mathrm{Ca}}(\mathcal{C}, M) = 0$, $n \geq 1$;*

(c) *for all $(\mathcal{C},\mathcal{C})$-bicomodules M, $H^1_{\mathrm{Ca}}(\mathcal{C}, M) = 0$.*

Proof. The Cartier cohomology can be understood as a right-derived Ext-functor associated to an $(A,\mathcal{C})$-relative injective resolution of the $(\mathcal{C},\mathcal{C})$-bicomodule $\mathcal{C}$. Thus the assertion follows from standard arguments of relative homological algebra and from 26.1. The assertion of the theorem can also be proven directly as follows.

(a) $\Rightarrow$ (b) Let $\mathcal{C}$ be a coseparable coring with cointegral $\delta : \mathcal{C} \otimes_A \mathcal{C} \to A$ (cf. 26.2). For all $M \in {}^{\mathcal{C}}\mathbf{M}^{\mathcal{C}}$ and $n = 1, 2, \ldots$, the maps

$$h_n : C^n_{\mathrm{Ca}}(\mathcal{C}, M) \to C^{n-1}_{\mathrm{Ca}}(\mathcal{C}, M), \quad f \mapsto (\delta \otimes I_{\mathcal{C}}^{\otimes n-1}) \circ (I_{\mathcal{C}} \otimes f) \circ {}^M\varrho,$$

form a contracting homotopy for the Cartier complex. Hence, $H^n_{\mathrm{Ca}}(\mathcal{C}, M) = 0$ for all $n \geq 1$, as asserted.

(b) $\Rightarrow$ (c) is obvious.

(c) $\Rightarrow$ (a) Suppose that $H^1_{\mathrm{Ca}}(\mathcal{C}, M) = 0$ for all $M \in {}^{\mathcal{C}}\mathbf{M}^{\mathcal{C}}$. Since $\underline{\Delta} : \mathcal{C} \to \mathcal{C} \otimes_A \mathcal{C}$ is a $(\mathcal{C},\mathcal{C})$-bicomodule map, its cokernel $Q = \mathcal{C} \otimes_A \mathcal{C}/\underline{\Delta}(\mathcal{C})$ is a $(\mathcal{C},\mathcal{C})$-bicomodule (see 18.6). Let $p : \mathcal{C} \otimes_A \mathcal{C} \to Q$ be the canonical $(\mathcal{C},\mathcal{C})$-bicomodule epimorphism. The (A, A)-bimodule map $f : Q \to \mathcal{C}$ given by

$$f(p(c \otimes c')) = \underline{\varepsilon}(c)c' - c\underline{\varepsilon}(c'), \quad \text{for all } c, c' \in \mathcal{C},$$

is a one cocycle in the Cartier complex of $\mathcal{C}$ with values in Q. Since, by assumption, the first cohomology group is trivial, there exists $g \in {}_A\mathrm{Hom}_A(Q, A)$ such that $dg = f$. Then the (A, A)-bimodule map $\delta : \mathcal{C} \otimes_A \mathcal{C} \to A$, $\delta = g \circ p + \underline{\varepsilon} \otimes \underline{\varepsilon}$ is a cointegral in $\mathcal{C}$, and hence $\mathcal{C}$ is a coseparable coring by 26.1. □

In the classical paper [119] analysing the structure of the Hochschild cohomology of an associative ring, Gerstenhaber revealed a very rich algebraic structure that was initially called a *pre-Lie system* and later was renamed a *comp algebra*.

30.5. Comp algebras. A *(right) comp algebra* $(V^\bullet, \diamond, \pi)$ consists of a sequence of R-modules $V^0, V^1, V^2, \ldots$, an element $\pi \in V^2$, and R-linear operations

$$\diamond_i : V^m \otimes_R V^n \to V^{m+n-1} \text{ for } i \geq 0,$$

such that, for any $f \in V^m$, $g \in V^n$, $h \in V^p$,

(1) $f \diamond_i g = 0$ if $i > m-1$;

(2) $(f \diamond_i g) \diamond_j h = f \diamond_i (g \diamond_{j-i} h)$ if $i \leq j < n+i$;

(3) $(f \diamond_i g) \diamond_j h = (f \diamond_j h) \diamond_{i+p-1} g$ if $j < i$;

(4) $\pi \diamond_0 \pi = \pi \diamond_1 \pi$.

A comp algebra $(V^\bullet, \diamond, \pi)$ is said to be *strict* if there exists an (unique) element $u \in V^1$ such that, for all $f \in V^m$,

(5) $u \diamond_0 f = f \diamond_i u = f$ for all $i < m$.

A strict comp algebra is denoted by $(V^\bullet, \diamond, \pi, u)$. It is said to be *unital* if there is an (unique) element $1 \in V^0$ such that $\pi \diamond_0 1 = \pi \diamond_1 1 = u$. A unital comp algebra is denoted by $(V^\bullet, \diamond, \pi, u, 1)$. For any $f \in V^m$, the number m is called a *dimension* of f and $m-1$ is called a *degree* of f.

30.6. Remarks. It is an easy exercise to check that condition (3) of 30.5 implies also that

$$(f \diamond_i g) \diamond_j h = (f \diamond_{j-n+1} h) \diamond_i g \quad \text{if } j \geq n+i.$$

This equation is often built into the definition of a comp algebra. There is a close relationship between comp algebras and operads, as studied by May in [35] as a tool for the theory of iterated loop spaces. Nowadays, operads are a major tool in deformation-quantisation, including the Kontsevich theory of formal quantisation of Poisson structures [142], [143], [199] (cf. [27] and [29]). In fact, a system of R-modules $V^0, V^1, V^2, \ldots$, an element $u \in V^1$, and additive operations $\diamond_i : V^m \otimes_R V^n \to V^{m+n-1}$ for $i \geq 0$ satisfying conditions (1)–(3) and (5) of the above definition 30.5 is called a *preoperad* or a *composition system.* We refer the interested reader to [141], where several properties of preoperads are reviewed.

30.7. The structure of comp algebras. An interlude. The usefulness of comp algebras stems from the following basic facts (which we state without proofs). Let $(V^\bullet, \diamond, \pi)$ be a comp algebra over R.

The Lie structure. A comp algebra is a graded (by degree) Lie algebra with the bracket given by $[f,g] = f \diamond g - (-1)^{(m-1)(n-1)} g \diamond f$, for all $f \in V^m$ and $g \in V^n$. Here the R-linear operation $\diamond : V^m \otimes_R V^n \to V^{m+n-1}$ is defined by $f \diamond g = \sum_{i=0}^{m-1} (-1)^{i(n-1)} f \diamond_i g$, and is known as a *comp* or composition.

The cup product. $V = \bigoplus_{i=0} V^i$ is a nonunital graded algebra with the product defined for all $f \in V^m$, $g \in V^n$ by $f \cup g = (\pi \diamond_0 f) \diamond_m g = (\pi \diamond_1 g) \diamond_0 f$. The operation $\cup$ is known as a *cup product.* V has a unit if $(V^\bullet, \diamond, \pi)$ is unital.

Differential graded algebra structure and cohomology. A unital comp algebra is a differential graded algebra with the derivation $d : V^m \to V^{m+1}$, $df = -[\pi, f]$ and product $\cup$. The corresponding cohomology is denoted by $H(V)$. Since V is a differential graded algebra, the cup product descends to $H(V)$. Furthermore, for all $\alpha \in H^m(V)$ and $\beta \in H^n(V)$, $\alpha \cup \beta = (-1)^{mn} \beta \cup \alpha$. The Lie bracket also descends to $H(V)$, thus making $H(V)$ a graded (by degree) Lie algebra.

Gerstenhaber algebra structure. A *Gerstenhaber algebra* over R is a collection of R-modules $H^0, H^1, H^2, \ldots$ with a graded (by $\deg H^m = m-1$) Lie bracket $[-,-] : H^m \otimes_R H^n \to H^{m+n-1}$ and an associative unital graded (by $\dim H^m = m$) commutative product $\cup : H^m \otimes_R H^n \to H^{m+n}$, such that, for all $\alpha \in H^m$, $\beta \in H^n$ and $\gamma \in H$,

$$[\alpha, \beta \cup \gamma] = [\alpha, \beta] \cup \gamma + (-1)^{(m-1)n} \beta \cup [\alpha, \gamma].$$

The cohomology of a unital comp algebra is a Gerstenhaber algebra.

The relevance of this discussion of comp algebras to the Cartier cohomology of a coring is revealed in the following theorem.

30.8. Gerstenhaber algebra structure of the Cartier cohomology. *Let $C_{\mathrm{Ca}}(\mathcal{C}) = C_{\mathrm{Ca}}(\mathcal{C}, \mathcal{C})$ be the Cartier complex of $\mathcal{C}$ with values in $\mathcal{C}$. Then $C_{\mathrm{Ca}}(\mathcal{C})$ is a unital comp algebra with the compositions*

$$\diamond_i : C_{\mathrm{Ca}}(\mathcal{C})^m \otimes_R C_{\mathrm{Ca}}(\mathcal{C})^n \to C_{\mathrm{Ca}}(\mathcal{C})^{m+n-1}, \quad f \diamond_i g = (I_{\mathcal{C}}^{\otimes i} \otimes g \otimes I_{\mathcal{C}}^{\otimes m-i-1}) \circ f,$$

and distinguished elements $\pi = \underline{\Delta}$, $u = I_{\mathcal{C}}$, and $1 = \underline{\varepsilon}$. Consequently, the Cartier cohomology of $\mathcal{C}$ with values in $\mathcal{C}$ is a Gerstenhaber algebra.

Proof. This can be verified by lengthy but routine calculations. □

The Cartier cohomology of a coring is obtained by the dualisation of the relative Hochschild cohomology of a ring. In addition to the relative Hochschild complex, the Amitsur complex is associated to a ring extension. By dualising the Amitsur cochain complex one obtains a new kind of cohomology for corings, which we term the *Hochschild cohomology of corings*.

30.9. The Hochschild complex of a coring. *Let $X(\mathcal{C}) = (X(\mathcal{C})_\bullet, \delta)$ be a complex given by*

$$X(\mathcal{C})_n = \mathcal{C}^{\otimes_A n+1}, \quad \delta_n : \mathcal{C}^{\otimes_A n+2} \to \mathcal{C}^{\otimes_A n+1}, \quad \delta_n = \sum_{k=0}^{n+1} (-1)^k I_{\mathcal{C}}^{\otimes k} \otimes \underline{\varepsilon} \otimes I_{\mathcal{C}}^{\otimes n-k+1},$$

for all $n = 0, 1, 2, \ldots$. Then $X(\mathcal{C})$ is a chain complex.

Proof. We need to show that $\delta_{n-1}\circ\delta_n = 0$ for all $n = 1, 2, \ldots$. For all $k = 0, 1, \ldots, n$, define $\underline{\varepsilon}^{k,n} = I_{\mathcal{C}}^{\otimes k} \otimes \underline{\varepsilon} \otimes I_{\mathcal{C}}^{\otimes n-k} : \mathcal{C}^{\otimes_A n+1} \to \mathcal{C}^{\otimes_A n}$. Then $\delta_n = \sum_{k=0}^{n+1}(-1)^k \underline{\varepsilon}^{k,n+1}$, and therefore $\delta_{n-1}\circ\delta_n = \Gamma_1 + \Gamma_2$, where

$$\Gamma_1 = \sum_{l=0}^{n+1}\sum_{k=0}^{n}(-1)^{k+l}\underline{\varepsilon}^{k,n}\circ\underline{\varepsilon}^{l,n+1}, \quad \Gamma_2 = \sum_{k=0}^{n}(\underline{\varepsilon}^{k,n}\circ\underline{\varepsilon}^{k,n+1} - \underline{\varepsilon}^{k,n}\circ\underline{\varepsilon}^{k+1,n+1}).$$

There is a term-by-term cancellation in Γ_1 and also $\underline{\varepsilon}^{k,n}\circ\underline{\varepsilon}^{k,n+1} = \underline{\varepsilon}^{k,n}\circ\underline{\varepsilon}^{k+1,n+1}$, since $\underline{\varepsilon}$ is an (A, A)-bimodule map. This proves that $\Gamma_2 = 0$ as well, so that $\delta_{n-1}\circ\delta_n = 0$, as required. □

In general, the complex $(X(\mathcal{C}), \delta)$ is not acyclic; however, there are some instances in which its homology vanishes.

30.10. A sufficient condition for $X(\mathcal{C})$ to be acyclic. *If there exists an element $e \in \mathcal{C}$ such that $\underline{\varepsilon}(e) = 1_A$, then the complex $(X(\mathcal{C}), \delta)$ is acyclic.*

Proof. A contracting homotopy $h^\bullet$ for $(X(\mathcal{C}), \delta)$ can be constructed as $h^n : \mathcal{C}^{\otimes_A n+1} \to \mathcal{C}^{\otimes_A n+2}$, $x \mapsto e\otimes x$. Indeed,

$$\begin{aligned}\delta_n(h^n(c^0\otimes\ldots\otimes c^n)) &= \sum_{k=0}^{n+1}(-1)^k\underline{\varepsilon}^{k,n+1}(e\otimes c^0\otimes\ldots\otimes c^n)\\ &= c^0\otimes\ldots\otimes c^n - \sum_{k=0}^{n}(-1)^k e\otimes c^0\otimes\ldots\otimes c^k\underline{\varepsilon}(c^{k+1})\otimes\ldots\otimes c^n\\ &= c^0\otimes\ldots\otimes c^n - h^{n-1}(\delta_{n-1}(c^0\otimes\ldots\otimes c^n)).\end{aligned}$$

Therefore, $\delta_n\circ h^n + h^{n-1}\circ\delta_{n-1} = I_{\mathcal{C}}^{\otimes n+1}$, for all $n = 1, 2, \ldots$, that is, $h^\bullet$ is a contracting homotopy, as claimed. Thus the complex $(X(\mathcal{C})_\bullet, \delta)$ is acyclic. □

30.11. $X(\mathcal{C})$ for a coring with a grouplike element. *If $\mathcal{C}$ has a grouplike element, then the associated complex $(X(\mathcal{C})_\bullet, \delta)$ is acyclic.*

Proof. By definition, if $g \in \mathcal{C}$ is a grouplike element, then $\underline{\varepsilon}(g) = 1_A$ and the assertion follows immediately from 30.10. □

30.12. $X(\mathcal{C})$ of a Sweedler coring. *Consider an algebra extension $B \to A$ and the associated Sweedler A-coring $\mathcal{C} = A\otimes_B A$. Then the complex $(X(\mathcal{C})_\bullet, \delta)$ is the B-*relative bar resolution of *A, that is, $X(\mathcal{C})_n = A^{\otimes_B n+2}$ and*

$$\delta_n(a_0\otimes a_1\otimes\ldots\otimes a_{n+1}) = \sum_{k=0}^{n}(-1)^k a_0\otimes\ldots\otimes a_{k-1}\otimes a_k a_{k+1}\otimes\ldots\otimes a_{n+1},$$

for all $a_0\otimes a_1\otimes\ldots\otimes a_{n+1} \in A^{\otimes_B n+2}$.

Proof. This is clear from the natural identification $A^{\otimes_B n+2} \simeq \mathcal{C}^{\otimes_A n+1}$ and from the definition of the counit in this case. The complex $(X(\mathcal{C})_\bullet, \delta)$ is a resolution of A in the category of (B, B)-bimodules since, first, each of the δ_n is a (B, B)-bimodule map. Second, since $\underline{\varepsilon}(1_A \otimes 1_A) = 1_A$, we can take $e = 1_A \otimes 1_A$ in 30.10 and thus obtain a homotopy $h^\bullet$ with $h^n : A^{\otimes_B n+2} \to A^{\otimes_B n+3}$, $x \mapsto 1_A \otimes x$, $n = 0, 1, 2, \ldots$. Each of the h^n is a (B, B)-bimodule map. Thus there is the sequence

$$\ldots \xrightarrow{\delta_2} A^{\otimes_B 4} \xrightarrow{\delta_1} A^{\otimes_B 3} \xrightarrow{\delta_0} A^{\otimes_B 2} \xrightarrow{\mu_{A/B}} A \longrightarrow 0$$

of (B, B)-bimodule maps that is exact up to and including the map δ_0. Note that $\mu_{A/B} \circ \delta_0 = 0$ and consider a (B, B)-bimodule map $\sigma : A \to A \otimes_B A$, $a \mapsto 1_A \otimes a$. Then, for all $a, a' \in A$,

$$\begin{aligned}(\delta_0 \circ h^0 + \sigma \circ \mu_{A/B})(a \otimes a') &= \delta_0(1 \otimes a \otimes a') + 1_A \otimes aa' \\ &= a \otimes a' - 1_A \otimes aa' + 1_A \otimes aa' = a \otimes a'.\end{aligned}$$

This completes the proof for the example. □

30.13. $X(\mathcal{C})$ for a coring over a simple ring. Note that the hypothesis of 30.10 is satisfied if A is a simple ring, in particular a division ring. Indeed, there exists an element $\tilde{e} \in \mathcal{C}$ such that $\underline{\varepsilon}(\tilde{e}) = a \neq 0$. If A is a simple ring, there exist $a_i, b_i \in A$ such that $\sum a_i a b_i = 1_A$. Then, for $e = \sum a_i \tilde{e} b_i$, we obtain $\underline{\varepsilon}(e) = 1_A$, as required. Hence $X(\mathcal{C})$ is acyclic for any coring $\mathcal{C}$ over a simple (or division) ring.

30.14. $X(\mathcal{C})$ for a faithfully flat coring. *If $\mathcal{C}$ is faithfully flat as a right or left A-module, then the associated complex $(X(\mathcal{C})_\bullet, \delta)$ is acyclic.*

Proof. Suppose that $\mathcal{C}_A$ is faithfully flat. Since $(X(\mathcal{C})_\bullet, \delta)$ is a complex of (A, A)-bimodules, one can consider the derived complex $(\mathcal{C} \otimes_A X(\mathcal{C})_\bullet, I_\mathcal{C} \otimes \delta)$. Consider a collection of (A, A)-bimodule mappings $h^n : \mathcal{C}^{\otimes_A n+1} \to \mathcal{C}^{\otimes_A n+2}$, $h^n = \underline{\Delta} \otimes I_\mathcal{C}^{\otimes n}$. Then $h^n \circ (I_\mathcal{C} \otimes \delta_n) = \underline{\Delta} \otimes \delta_n$ and

$$\begin{aligned}(I_\mathcal{C} \otimes \delta_{n+1}) \circ h^{n+1} &= (I_\mathcal{C} \otimes \delta_{n+1}) \circ (\underline{\Delta} \otimes I_\mathcal{C}^{\otimes n+1}) \\ &= (I_\mathcal{C} \otimes \underline{\varepsilon} \otimes I_\mathcal{C}^{\otimes n+1}) \circ (\underline{\Delta} \otimes I_\mathcal{C}^{\otimes n+1}) - \sum_{k=0}^{n+1} (-1)^k \underline{\Delta} \otimes \underline{\varepsilon}^{k,n+1} \\ &= I_\mathcal{C}^{\otimes n+2} - \underline{\Delta} \otimes \delta_n.\end{aligned}$$

This proves that $h^n \circ (I_\mathcal{C} \otimes \delta_n) + (I_\mathcal{C} \otimes \delta_{n+1}) \circ h^{n+1} = I_\mathcal{C}^{\otimes n+2}$, which means that $h^\bullet$ is a contracting homotopy for $\mathcal{C} \otimes_A X(\mathcal{C})$. Therefore the derived complex $(\mathcal{C} \otimes_A X(\mathcal{C})_\bullet, I_\mathcal{C} \otimes \delta)$ is acyclic, and, since $\mathcal{C}$ is a faithfully flat right A-module, the complex $(X(\mathcal{C})_\bullet, \delta)$ is acyclic, too.

In case $\mathcal{C}$ is a faithfully flat left A module, consider the derived complex $(X(\mathcal{C})_\bullet \otimes_A \mathcal{C}, \delta \otimes I_\mathcal{C})$ and take $h^n = (-1)^n I_\mathcal{C}^{\otimes n} \otimes_A \underline{\Delta}$. □

30.15. The bimodule-valued Hochschild cohomology of a coring. Note that the complex $(X(\mathcal{C})_\bullet, \delta)$ is a complex of (A,A)-bimodules in which all the δ are (A,A)-bimodule maps. Thus, for any (A,A)-bimodule M, one can define a cochain complex $C_{\rm Ho}(\mathcal{C}, M)$ by applying the contravariant Hom-functor ${}_A\mathrm{Hom}_A(-,M)$ to $(X(\mathcal{C})_\bullet, \delta)$. Thus, explicitly, $C_{\rm Ho}(\mathcal{C},M) = (C^\bullet_{\rm Ho}(\mathcal{C},M), d^\bullet)$, where

$$C^n_{\rm Ho}(\mathcal{C},M) = {}_A\mathrm{Hom}_A(\mathcal{C}^{\otimes_A n+1}, M), \quad d^n(f) = \sum_{k=0}^{n+1}(-1)^k f\circ(I_\mathcal{C}^{\otimes k}\otimes\underline{\varepsilon}\otimes I_\mathcal{C}^{\otimes n+1-k}).$$

The complex $C_{\rm Ho}(\mathcal{C},M)$ is called the *Hochschild cochain complex* associated to $\mathcal{C}$ with values in M, and its cohomology is called the *Hochschild cohomology of $\mathcal{C}$ with values in M*. The Hochschild cohomology of $\mathcal{C}$ with values in M is denoted by $H_{\rm Ho}(\mathcal{C},M)$.

As explained in 30.4, the Cartier cohomology of a coring detects coseparability. Similarly, the Hochschild cohomology detects when a coring is cosplit (cf. 26.12 for the definition of a cosplit coring).

30.16. The cohomological interpretation of cosplit corings. *For an A-coring $\mathcal{C}$ the following statements are equivalent:*

(a) $\mathcal{C}$ is a cosplit coring;

(b) the counit of $\mathcal{C}$ is surjective and, for $n \geq 1$, $H^n_{\rm Ho}(\mathcal{C},M) = 0$, for all (A,A)-bimodules M;

(c) the counit of $\mathcal{C}$ is surjective and $H^1_{\rm Ho}(\mathcal{C},M) = 0$, for all (A,A)-bimodules M.

Proof. (a) ⇒ (b). Since $\mathcal{C}$ is a cosplit coring, there exists $e \in \mathcal{C}^A$ such that $\underline{\varepsilon}(e) = 1_A$ (cf. 26.11). The latter immediately implies that $\underline{\varepsilon}$ is surjective. Take any (A,A)-bimodule M, and for every nonnegative integer n define an R-module map

$$h_n : {}_A\mathrm{Hom}_A(\mathcal{C}^{\otimes_A n+1}, M) \to {}_A\mathrm{Hom}_A(\mathcal{C}^{\otimes_A n}, M), \quad f \mapsto [x \mapsto f(e\otimes x)].$$

One easily checks that the collection $(h_n)_{n\in\mathbb{N}}$ is a contracting homotopy. Thus $H^n_{\rm Ho}(\mathcal{C},M) = 0$, as claimed.

(b) ⇒ (c) is obvious.

(c) ⇒ (a) Suppose $H^1_{\rm Ho}(\mathcal{C},M) = 0$ for all (A,A)-bimodules M. In particular, take $M = \mathrm{Ke}\,\underline{\varepsilon}$ and consider $f = I_\mathcal{C}\otimes\underline{\varepsilon} - \underline{\varepsilon}\otimes I_\mathcal{C} \in {}_A\mathrm{Hom}_A(\mathcal{C}\otimes_A\mathcal{C}, \mathrm{Ke}\,\underline{\varepsilon})$. One easily checks that f is a 1-cocycle in $C_{\rm Ho}(\mathcal{C}, \mathrm{Ke}\,\underline{\varepsilon})$. By assumption, any 1-cocycle is a coboundary; thus there exists $h \in {}_A\mathrm{Hom}_A(\mathcal{C}, \mathrm{Ke}\,\underline{\varepsilon})$ such that $dh = f$, that is, such that, for all $c^1, c^2 \in \mathcal{C}$,

$$\underline{\varepsilon}(c^1)h(c^2) - h(c^1)\underline{\varepsilon}(c^2) = c^1\underline{\varepsilon}(c^2) - \underline{\varepsilon}(c^1)c^2. \qquad (*)$$

Since the counit $\underline{\varepsilon}$ is surjective, there exists $g \in \mathcal{C}$ such that $\underline{\varepsilon}(g) = 1_A$. Define $e = g + h(g)$. Since $h(g) \in \mathrm{Ke}\,\underline{\varepsilon}$, we immediately conclude that $\underline{\varepsilon}(e) = 1_A$. Furthermore, setting $c^1 = ga$ and $c^2 = g$ in $(*)$ we obtain $ah(g) - h(ga) = ga - ag$, for all $a \in A$. The right A-linearity of h now implies that the above equation can be transformed to $a(g + h(g)) = (g + h(g))a$, that is, $ae = ea$ for all $a \in A$. In view of 26.11, we conclude that $\mathcal{C}$ is a cosplit coring. □

30.17. Exercises

(1) Prove that condition (3) of 30.5 implies that

$$(f \diamond_i g) \diamond_j h = (f \diamond_{j-n+1} h) \diamond_i g \quad \text{if } j \geq n + i.$$

(2) A *cosimplicial object* in a category $\mathbf{A}$ is a collection of objects $X_0, X_1, \ldots$ in $\mathbf{A}$ and arrows $\delta_i^n : X_n \to X_{n+1}$, $i = 0, 1, \ldots, n$, and $\sigma_i^n : X_{n+1} \to X_n$, $i = 0, 1, \ldots, n-1$ such that

$$\delta_i^{n+1}\delta_j^n = \delta_{j+1}^{n+1}\delta_i^n, \qquad i \leq j,$$

$$\sigma_j^{n-1}\sigma_i^n = \sigma_i^{n-1}\sigma_{j+1}^n, \qquad i \leq j,$$

$$\sigma_j^n\delta_i^n = \begin{cases} \delta_i^{n-1}\sigma_{j-1}^{n-1} & \text{for } i < j \\ I_{X_n} & \text{for } i = j,\ i = j+1 \\ \delta_{i-1}^{n-1}\sigma_j^{n-1} & \text{for } i > j+1. \end{cases}$$

The morphisms δ_i^n are known as *coface operators* and the σ_j^n are known as *degeneracies*. Show:
Given an A-coring $\mathcal{C}$, there is a cosimplicial object in the category ${}_A\mathbf{M}_A$ of (A, A)-bimodules with $X_n = \mathcal{C}^{\otimes_A n+1}$, coface operators $\delta_i^n = I_{\mathcal{C}}^{\otimes_A i} \otimes_A \underline{\Delta} \otimes_A I_{\mathcal{C}}^{\otimes_A n-i}$, and degeneracies $\sigma_i^n = I_{\mathcal{C}}^{\otimes_A i+1} \otimes_A \underline{\varepsilon} \otimes_A I_{\mathcal{C}}^{\otimes_A n-i}$.

References. Cartier [90]; Gerstenhaber [119]; Gerstenhaber and Schack [120]; Guzman [126, 127]; Hochschild [132]; Kluge, Paal and Stasheff [141].

31 Bialgebroids

The idea of generalising bialgebras to the case of bimodules rather than modules goes back to Sweedler [192]. Since there is no obvious way in which a (part of) tensor product of bimodules can be equipped with an algebra structure, there is no obvious compatibility between algebra and coalgebra structures. This led to a number of different definitions of generalised bialgebras over noncommutative rings or *bialgebroids*. It is easy to overlook that those definitions are equivalent to each other. In this section we define bialgebroids following Lu [154], and then provide equivalent descriptions that appeared in the literature.

As before, R denotes a commutative ring and A is an R-algebra.

31.1. A-rings. Recall that a unital *A-ring* or an *algebra over A* is a pair (U, i), where U is an R-algebra and $i : A \to U$ is an algebra map. If (U, i) is an A-ring, then U is an (A, A)-bimodule with the structure provided by the map i, $aua' := i(a)ui(a')$. A map of A-rings $f : (U, i) \to (V, j)$ is an R-algebra map $f : U \to V$ such that $f \circ i = j$. Equivalently, a map of A-rings is an R-algebra map that is a left or right A-module map. Indeed, clearly, if $f : (U, i) \to (V, j)$ is a map of A-rings, it is an algebra and an A-bimodule map. Conversely, if f is a left A-linear algebra map, then, for all $a \in A$, $f(i(a)) = f(a1_U) = af(1_U) = j(a)$, and similarly in the right A-linear case.

31.2. Algebras over enveloping algebras: A^e-rings. Let $\bar{A} = A^{op}$ be the opposite algebra of A. For $a \in A$, $\bar{a} \in \bar{A}$ is the same a but now viewed as an element in $\bar{A}$, that is, $a \mapsto \bar{a}$ is an (obvious) anti-isomorphism of algebras. Let $A^e = A \otimes_R \bar{A}$ be the enveloping algebra of A. Note that a pair (H, i) is an A^e-ring if and only if there exist an algebra map $s : A \to H$ and an anti-algebra map $t : A \to H$, such that $s(a)t(b) = t(b)s(a)$, for all $a, b \in A$. Explicitly, $s(a) = i(a \otimes 1)$ and $t(a) = i(1 \otimes \bar{a})$, and, conversely, $i(a \otimes \bar{b}) = s(a)t(b)$.

In the sequel, the expression "let (H, s, t) be an A^e-ring" will be understood to mean an R-algebra H with algebra maps $s, t : A \to H$ as described above. A is called a *base algebra*, H a *total algebra*, s the *source* map and t the *target* map.

31.3. $\mathrm{End}_R(A)$ as an A^e-ring. An example of an A^e-ring is provided by $\mathrm{End}_R(A)$. In this case, $i : A \otimes_R \bar{A} \to \mathrm{End}_R(A)$, $i(a \otimes \bar{b})(x) = axb$. The source and the target come out as $s(a)(x) = ax$ and $t(b)(x) = xb$. It follows that $\mathrm{End}_R(A)$ is an A^e-bimodule via i. In particular, $\mathrm{End}_R(A)$ is a left A^e-module via

$$(af)(b) = af(b), \quad (\bar{a}f)(b) = f(b)a,$$

for all $a, b \in A$ and $f \in \mathrm{End}_R(A)$.

31.4. The $\times_A$-product. Let M and N be (A^e, A^e)-bimodules. Let

$$\int_a {}_{\bar{a}}M \otimes {}_aN := M \otimes_R N / < \{\bar{a}m \otimes n - m \otimes an \mid \text{ for all } a \in A\} >,$$

that is, $\int_a {}_{\bar{a}}M \otimes_a N = M \otimes_A N$, where the structure of M as a right A-module arises from that of M as a left $\bar{A}$-module. Let

$$\int^b M_{\bar{b}} \otimes N_b := \{\sum_i m_i \otimes n_i \in M \otimes_R N \mid \forall b \in A, \ \sum_i m_i\bar{b} \otimes n_i = \sum_i m_i \otimes n_i b\}.$$

Define the R-module

$$M \times_A N := \int^b \int_a {}_{\bar{a}}M_{\bar{b}} \otimes {}_aN_b.$$

Explicitly,

$$M \times_A N := \{\sum_i m_i \otimes n_i \in M \otimes_A N \mid \forall b \in A, \ \sum_i m_i\bar{b} \otimes n_i = \sum_i m_i \otimes n_i b\},$$

where again M is viewed as a right A-module through its left $\bar{A}$-module structure. The operation $- \times_A - : {}_{A^e}\mathbf{M}_{A^e} \times {}_{A^e}\mathbf{M}_{A^e} \to {}_{A^e}\mathbf{M}_{A^e}$ is a bifunctor. Here, for $M, N \in {}_{A^e}\mathbf{M}_{A^e}$, the product $M \times_A N$ is in ${}_{A^e}\mathbf{M}_{A^e}$ with the actions given by

$$(a' \otimes \bar{a})(\textstyle\sum_i m_i \otimes n_i)(b' \otimes \bar{b}) = \sum_i a' m_i b' \otimes \bar{a} n_i \bar{b}.$$

The importance of the notion of the $\times_A$-product stems from the following observation.

31.5. The $\times_A$-product of two A^e-rings. *For any pair of A^e-rings (U, i) and (V, j), the (A^e, A^e)-bimodule $U \times_A V$ is an A^e-ring with the algebra map $A \otimes_R \bar{A} \to U \times_A V$, $a \otimes \bar{b} \to i(a) \otimes j(\bar{b})$, the associative product*

$$(\textstyle\sum_i u^i \otimes v^i)(\sum_j \tilde{u}^j \otimes \tilde{v}^j) = \sum_{i,j} u^i\tilde{u}^j \otimes v^i\tilde{v}^j,$$

and the unit $1_U \otimes 1_V$.

Proof. The only nontrivial part is to check that the product is well defined, that is, it is independent of the choice of the representations. Since the operation clearly is linear, it suffices to show that the product yields zero provided one of the factors is zero.

First assume that $\sum_i u^i \otimes v^i = 0$. By the characterisation of 0 in tensor products, there exist finitely many $b_{ki} \in A$ and $y_k \in U$, such that $\sum_i b_{ki}v^i = 0$, for all k, and $u^i = \sum_k \bar{b}_{ki}y_k$ (e.g., [46, 12.10]). For any $\tilde{u} \otimes \tilde{v} \in U \otimes_R V$,

$$\begin{aligned}(\textstyle\sum_i u^i \otimes v^i)(\tilde{u} \otimes \tilde{v}) &= \textstyle\sum_i u^i\tilde{u} \otimes v^i\tilde{v} = \sum_{i,k} \bar{b}_{ki}y_k\tilde{u} \otimes v^i\tilde{v} \\ &= \textstyle\sum_k y_k\tilde{u} \otimes (\sum_i b_{ki}v^i)\tilde{v} = 0.\end{aligned}$$

Since no use is made of the fact that $\sum_i u^i \otimes v^i$ is an element of $U \times_A V$, this shows that $U \otimes_A V = \int_a {}_{\bar{a}}U \otimes_a V$ is a right module of the tensor algebra $U \otimes_R V$ and also implies that the formula for the product in $U \times_A V$ is independent of the representation of the left factors.

Now assume that $\sum_j \tilde{u}^j \otimes \tilde{v}^j = 0$. Then there exist finitely many $a_{kj} \in A$ and $x_k \in U$, such that $\sum_j a_{kj}\tilde{v}^j = 0$, for all k, and $\tilde{u}^j = \sum_k \bar{a}_{kj}x_k$. Using the definition of the $\times_A$-product, compute

$$
\begin{aligned}
(\textstyle\sum_i u^i \otimes v^i)(\sum_j \tilde{u}^j \otimes \tilde{v}^j) &= \textstyle\sum_{i,j} u^i\tilde{u}^j \otimes v^i\tilde{v}^j \\
&= \textstyle\sum_{i,j} u^i(\sum_k \bar{a}_{kj}x_k) \otimes v^i\tilde{v}^j \\
&= \textstyle\sum_{i,j,k}(u^i\bar{a}_{kj} \otimes v^i)(x_k \otimes \tilde{v}^j) \\
&= \textstyle\sum_{i,j,k}(u^i \otimes v^i a_{kj})(x_k \otimes \tilde{v}^j) \\
&= \textstyle\sum_{i,k} u^i x_k \otimes v^i(\sum_j a_{kj}\tilde{v}^j) \;=\; 0.
\end{aligned}
$$

Whence the product is also independent of the representation of right factors.

The fact that the right-hand side of the product formula is in $U \times_A V$ is proven by a similar straightforward calculation. □

With all these preliminary data at hand we can define the following generalisation of bialgebras introduced in [154].

31.6. Bialgebroids and Hopf algebroids. Let $(\mathcal{H}, s, t)$ be an A^e-ring. View $\mathcal{H}$ as an (A, A)-bimodule, with the left A-action given by the source map s, and the right A-action that descends from the left $\bar{A}$-action given by the target map t, that is,

$$ah = s(a)h, \quad ha = t(a)h, \qquad \text{for all } a \in A, h \in \mathcal{H}.$$

We say that $(\mathcal{H}, s, t, \underline{\Delta}, \underline{\varepsilon})$ is an *A-bialgebroid* if

(1) $(\mathcal{H}, \underline{\Delta}, \underline{\varepsilon})$ is an A-coring;

(2) $\mathrm{Im}(\underline{\Delta}) \subseteq \mathcal{H} \times_A \mathcal{H}$ and the corestriction of $\underline{\Delta}$ to $\underline{\Delta} : \mathcal{H} \to \mathcal{H} \times_A \mathcal{H}$ is an algebra map;

(3) $\underline{\varepsilon}(1_{\mathcal{H}}) = 1_A$, and, for all g, $h \in \mathcal{H}$,

$$\underline{\varepsilon}(gh) = \underline{\varepsilon}\left(gs(\underline{\varepsilon}(h))\right) = \underline{\varepsilon}\left(gt(\underline{\varepsilon}(h))\right).$$

An *antipode* for an A-bialgebroid $\mathcal{H}$ is an antialgebra map $\tau : \mathcal{H} \to \mathcal{H}$ such that

(i) $\tau \circ t = s$;

(ii) $\mu_{\mathcal{H}} \circ (\tau \otimes I_{\mathcal{H}}) \circ \underline{\Delta} = t \circ \underline{\varepsilon} \circ \tau$;

(iii) there exists a section $\gamma : \mathcal{H} \otimes_A \mathcal{H} \to \mathcal{H} \otimes_R \mathcal{H}$ of the natural projection $\mathcal{H} \otimes_R \mathcal{H} \to \mathcal{H} \otimes_A \mathcal{H}$ such that $\mu_{\mathcal{H}} \circ (I_{\mathcal{H}} \otimes \tau) \circ \gamma \circ \underline{\Delta} = s \circ \underline{\varepsilon}$.

An A-bialgebroid with an antipode is called a *Hopf algebroid*

31.7. Remarks about bialgebroids. (1) Observe that the counit property of $\underline{\varepsilon}$ in 31.6 explicitly means that, for all $h \in \mathcal{H}$,

$$\sum s\left(\underline{\varepsilon}(h_{\underline{1}})\right) h_{\underline{2}} = \sum t\left(\underline{\varepsilon}(h_{\underline{2}})\right) h_{\underline{1}} = h.$$

The first condition of 31.6(2) explicitly means $\sum h_{\underline{1}} t(a) \otimes h_{\underline{2}} = \sum h_{\underline{1}} \otimes h_{\underline{2}} s(a)$, for all $a \in A$ and $h \in \mathcal{H}$.

(2) The facts that $\underline{\varepsilon}$ preserves the unit and is an (A, A)-bimodule map imply that s and t are sections of $\underline{\varepsilon}$, that is, $\underline{\varepsilon}(s(a)) = \underline{\varepsilon}(t(a)) = a$ for all $a \in A$. Using this and 31.6(3), one easily finds $\underline{\varepsilon}(hs(a)) = \underline{\varepsilon}(ht(a))$, for all $a \in A$, $h \in \mathcal{H}$. Similarly, the facts that $\underline{\Delta}$ is a unital and an (A, A)-bimodule map imply

$$\underline{\Delta}(s(a)) = s(a) \otimes 1_{\mathcal{H}}, \quad \underline{\Delta}(t(a)) = 1_{\mathcal{H}} \otimes t(a).$$

(3) For an A^e-ring $(\mathcal{H}, s, t)$, let $F : {}_{\mathcal{H}}\mathbf{M} \to {}_A\mathbf{M}_A$ be the restriction of scalars functor. $\mathcal{H}$ is an (A, A)-bimodule as in 31.6. If $(\mathcal{H}, s, t, \underline{\Delta}, \underline{\varepsilon})$ is an A-bialgebroid, then ${}_{\mathcal{H}}\mathbf{M}$ has a monoidal structure such that F is a strict monoidal functor. For all $M, N \in {}_{\mathcal{H}}\mathbf{M}$, the tensor product $M \otimes_A N$ is in ${}_{\mathcal{H}}\mathbf{M}$ via $h(m \otimes n) = \sum h_{\underline{1}} m \otimes h_{\underline{2}} n$. Here M and N have the (A, A)-bimodule structures induced from their left $\mathcal{H}$-module structures via the source and target maps, that is, $ama' = s(a)t(a')m$, for all $a, a' \in A$ and $m \in M$ or $m \in N$. The right-hand side is well defined because $\mathrm{Im}(\underline{\Delta}) \subseteq \mathcal{H} \times_A \mathcal{H}$. A is the unit object, when viewed in ${}_{\mathcal{H}}\mathbf{M}$ via the action $h \triangleright a = \underline{\varepsilon}(hs(a)) = \underline{\varepsilon}(ht(a))$, for all $h \in \mathcal{H}$, $a \in A$. The fact that this is an action follows from 31.6(3). Note that the left $\mathcal{H}$-module structure on the tensor product of left $\mathcal{H}$-modules is an analog (and generalisation) of the module structure $- \otimes^b_R -$ defined in 13.4 for bialgebras. We thus denote the left $\mathcal{H}$-module $M \otimes_A N$ with the above left $\mathcal{H}$-module structure by $M \otimes^b_A N$. The corresponding left multiplication by $\mathcal{H}$ via the coproduct, $h!(m \otimes n) = \sum h_{\underline{1}} m \otimes h_{\underline{2}} n$ is known as a *left diagonal action* of $\mathcal{H}$ on $M \otimes^b_A N$.

(4) The notion of a bialgebroid can be understood as a dualisation (and generalisation) of the notion of a groupoid. Recall that a groupoid is defined as a small category in which all morphisms are isomorphisms. One can then consider the sets of points and arrows and maps from the latter to the former, which to each arrow associate its source and target. Intuitively, dualising the notion of a groupoid, thus in particular the base and total sets, and the source and target maps, one arrives at the notion of a bialgebroid. For this reason one often terms Hopf algebroids *quantum groupoids*. This connection between Hopf algebroids and groupoids can also be used to construct concrete examples of the former.

(5) The observation in item (3) indicates a marked difference between bialgebroids and bialgebras. Given an A-bialgebroid $\mathcal{H}$, ${}_{\mathcal{H}}\mathbf{M}$ is a monoidal

category, but it is not true that $\mathbf{M}_{\mathcal{H}}$ is a monoidal category. From this point of view bialgebroids defined in 31.6 are *one-sided* objects and often are termed *left bialgebroids*, since they allow only for left diagonal actions. Using the left-right symmetry one can easily define *right bialgebroids* as those that lead to a monoidal structure in $\mathbf{M}_{\mathcal{H}}$ via the right diagonal actions. We leave this to the reader.

An example of a bialgebroid is provided by the following construction.

31.8. The tensor product bialgebroid. *Let A be an R-algebra, and let B be an R-bialgebra with coproduct Δ and counit ε. Then $\mathcal{H} = A \otimes_R B \otimes_R \bar{A}$ is an A-bialgebroid with the natural tensor product algebra structure and the following structure maps:*

(i) *the source map $s : a \mapsto a \otimes 1_B \otimes 1_A$;*

(ii) *the target map $t : a \mapsto 1_A \otimes 1_B \otimes \bar{a}$;*

(iii) *the coproduct $\underline{\Delta} : a \otimes b \otimes \bar{a'} \mapsto \sum a \otimes b_{\underline{1}} \otimes 1_A \otimes 1_A \otimes b_{\underline{2}} \otimes \bar{a'}$;*

(iv) *the counit $\underline{\varepsilon} : a \otimes b \otimes \bar{a'} \mapsto \varepsilon(b)aa'$.*

Furthermore, if B is a Hopf algebra with antipode S, then $\mathcal{H}$ is a Hopf algebroid with antipode

$$\tau : a \otimes b \otimes \bar{a'} \mapsto a' \otimes S(b) \otimes \bar{a}.$$

In particular, A^e is a Hopf algebroid over A.

Proof. This is proven by routine checking of the axioms and is left to the reader as an exercise. □

A slightly more elaborate example of a bialgebroid termed an *Ehresmann-Schauenburg bialgebroid* or a *quantum gauge groupoid* is given in 34.14.

31.9. An anchor. *For an R-algebra A, view* $\mathrm{End}_R(A)$ *as an (A, A)-bimodule with the structure maps*

$$(af)(b) = af(b), \quad (fa)(b) = f(b)a,$$

for all a, $b \in A$ and $f \in \mathrm{End}_R(A)$. Let $(\mathcal{H}, s, t)$ be an A^e-ring, viewed as an (A, A)-bimodule as in 31.6. Suppose that

(1) *$\underline{\Delta} : \mathcal{H} \to \mathcal{H} \otimes_A \mathcal{H}$ is a coassociative (A, A)-bimodule map;*

(2) $\mathrm{Im}(\underline{\Delta}) \subseteq \mathcal{H} \times_A \mathcal{H}$ *and the corestriction of $\underline{\Delta}$ to $\underline{\Delta} : \mathcal{H} \to \mathcal{H} \times_A \mathcal{H}$ is an algebra map.*

Then $\mathcal{H}$ is an A-bialgebroid if and only if there exists an algebra and an (A, A)-bimodule map $\nu : \mathcal{H} \to \mathrm{End}_R(A)$ such that

(i) *$\sum s(h_{\underline{1}} \triangleright a)h_{\underline{2}} = hs(a)$;*

(ii) $\sum t(h_{\underline{2}} \triangleright a)h_{\underline{1}} = ht(a)$;
where $h \triangleright a := \nu(h)(a)$, *for all* $a \in A$ *and* $h \in \mathcal{H}$.

The map ν *is called an* anchor *for the bialgebroid* $\mathcal{H}$.

Proof. First, note that the left-hand sides of (i) and (ii) are well defined since ν is an (A, A)-bimodule map; that is, for all $a, b \in A$ and $h \in \mathcal{H}$,

$$\nu(s(a)h)(b) = a\nu(h)(b), \quad \nu(t(a)h)(b) = \nu(h)(b)a.$$

Suppose $(\mathcal{H}, s, t, \underline{\Delta}, \underline{\varepsilon})$ is an A-bialgebroid, and define

$$\nu = \nu_{\underline{\varepsilon}} : \mathcal{H} \to \mathrm{End}_R(A), \quad \nu(h)(a) = h \triangleright a := \underline{\varepsilon}(hs(a)) = \underline{\varepsilon}(ht(a)). \qquad (*)$$

The map ν is an algebra morphism since A is a left $\mathcal{H}$-module with the structure map $\triangleright$. The fact that ν is (A, A)-bilinear follows by an elementary calculation. Explicitly, for any $a, b \in A$, $h \in \mathcal{H}$,

$$\nu(ah)(b) = \nu(s(a)h)(b) = \underline{\varepsilon}(s(a)hs(b)) = a\underline{\varepsilon}(hs(b)) = a\nu(h)(b) = (a\nu(h))(b),$$

thus proving that ν is left A-linear. A similar calculation using the definition of the (A, A)-bimodule structure of $\mathrm{End}_R(A)$, proves the right A-linearity of ν. Next we prove that (i) and (ii) hold for ν. Using 31.6(2) and 31.7(2), we compute $\underline{\Delta}(hs(a)) = \underline{\Delta}(h)\underline{\Delta}(s(a)) = \sum h_{\underline{1}}s(a) \otimes h_{\underline{2}}$. Now, using the first part of the counit property 31.7(1) for $hs(a)$, we obtain $\sum s\left(\underline{\varepsilon}(h_{\underline{1}}s(a))\right) h_{\underline{2}} = hs(a)$, that is, (i) for ν. The condition (ii) follows from $h \triangleright a = \underline{\varepsilon}(ht(a))$ and the second part of the counit property 31.7(1) together with 31.6(2) and 31.7(2).

Conversely, suppose $(\mathcal{H}, s, t)$, $\underline{\Delta}$ and ν satisfy the hypothesis of the proposition, and let $\underline{\varepsilon} = \underline{\varepsilon}_\nu : \mathcal{H} \to A$, $h \mapsto \nu(h)(1_A)$. We claim that $\underline{\varepsilon}$ is a counit for $\underline{\Delta}$. Indeed, since $\underline{\varepsilon}$ is (A, A)-bilinear, $\underline{\varepsilon}(s(a)) = \underline{\varepsilon}(t(a)) = a$, and hence, in particular $\underline{\varepsilon}(1_{\mathcal{H}}) = 1_A$. Furthermore, note that for all $g, h \in \mathcal{H}$, $\underline{\varepsilon}(gh) = \nu(g)(\underline{\varepsilon}(h))$. Therefore $\underline{\varepsilon}(gh) = \nu(g)(\underline{\varepsilon}(h)) = \nu(g)(\underline{\varepsilon}(s(\underline{\varepsilon}(h)))) = \underline{\varepsilon}(hs(\underline{\varepsilon}(h)))$, and similarly for the target map t. This proves conditions 31.6(3), and we conclude that $(H, s, t, \Delta, \underline{\varepsilon}_\nu)$ is a bialgebroid, as required. □

The original formulation of bialgebroids in the works of Sweedler [192] and Takeuchi [196] makes more direct use of the $\times_A$-product. The main problem to overcome here is the fact that the $\times_A$-product is not associative.

31.10. $\times_A$-coalgebras. For M, N and $P \in {}_{A^e}\mathbf{M}_{A^e}$ define

$$M \times_A P \times_A N := \int^{s,u} \int_{r,t} {}_{\bar{r}}M_{\bar{s}} \otimes {}_{r,\bar{t}}P_{s,\bar{u}} \otimes {}_{t}N_{u}.$$

There exist obvious maps (identities on elements)

$$\alpha : (M \times_A P) \times_A N \to M \times_A P \times_A N, \quad \alpha' : M \times_A (P \times_A N) \to M \times_A P \times_A N.$$

The maps α, α' are not isomorphisms in general. Since $\mathrm{End}_R(A)$ is an A^e-ring by 31.3, it is an (A^e, A^e)-bimodule, so one can define the maps

$$\theta : M \times_A \mathrm{End}_R(A) \to M, \quad \theta(\sum_i m_i \otimes f_i) = \sum_i \overline{f_i(1)} m_i, \quad \text{and}$$
$$\theta' : \mathrm{End}_R(A) \times_A M \to M, \quad \theta'(\sum_i f_i \otimes m_i) = \sum_i f_i(1) m_i.$$

A triple (L, Δ, ν) is called a $\times_A$-*coalgebra* if L is an (A^e, A^e)-bimodule and

$$\Delta : L \to L \times_A L, \quad \nu : L \to \mathrm{End}_R(A),$$

are (A^e, A^e)-bimodule maps such that

$$\alpha \circ (\Delta \times_A I_L) \circ \Delta = \alpha' \circ (I_L \times_A \Delta) \circ \Delta, \quad \theta \circ (I_L \times_A \nu) \circ \Delta = I_L = \theta' \circ (\nu \times_A I_L) \circ \Delta.$$

Δ is called a *coproduct* and ν is called a *counit* of the $\times_A$-coalgebra L.

31.11. The coring structure of a $\times_A$-coalgebra. *Let L be an (A^e, A^e)-bimodule, and $\Delta : L \to L \times_A L$, $\nu : L \to \mathrm{End}_R(A)$ be (A^e, A^e)-bimodule maps. Let $i : L \times_A L \to L \otimes_A L$ be the canonical inclusion. Then (L, Δ, ν) is a $\times_A$-coalgebra if and only if $(L, \underline{\Delta}, \underline{\varepsilon}_\nu)$ is an A-coring, where $\underline{\Delta} = i \circ \Delta$ and $\underline{\varepsilon}_\nu(l) = \nu(l)(1_A)$.*

Proof. Clearly the coassociativity of Δ is equivalent to the coassociativity of $\underline{\Delta}$. The equivalence of the counit properties is checked by straightforward calculation. □

31.12. $\times_A$-bialgebras. *Let $(\mathcal{H}, s, t)$ be an A^e-ring. $\mathcal{H}$ is an A-bialgebroid if and only if there exists a $\times_A$-coalgebra structure on $\mathcal{H}$ such that both the coproduct Δ and the counit ν are algebra maps. In this case $(\mathcal{H}, \Delta, \nu)$ is called a $\times_A$-*bialgebra.

Proof. Let $(\mathcal{H}, s, t, \underline{\Delta}, \underline{\varepsilon})$ be an A-bialgebroid, and let ν be the corresponding anchor (cf. 31.9). Since ν and $\underline{\Delta}$ are (A, A)-bimodule maps, they are left A^e-module maps. Furthermore, both ν and the corestriction Δ of $\underline{\Delta}$ to $\mathcal{H} \times_A \mathcal{H}$ are R-algebra maps. Therefore, ν and Δ are maps of A^e-rings, and hence also maps of (A^e, A^e)-bimodules. Then 31.11 implies that $(\mathcal{H}, \Delta, \nu)$ is a $\times_A$-bialgebra.

In view of 31.11, the converse is obvious. □

A conceptual understanding of the notion of a bialgebroid is provided by the following equivalence obtained by Schauenburg [183].

31.13. The monoidal structure. *Let $(\mathcal{H}, s, t)$ be an A^e-ring. Then $\mathcal{H}$ is an A-bialgebroid if and only if ${}_{\mathcal{H}}\mathbf{M}$ is a monoidal category such that the forgetful functor $F : {}_{\mathcal{H}}\mathbf{M} \to {}_A\mathbf{M}_A$ is strict monoidal.*

Proof. If $\mathcal{H}$ is an A-bialgebroid, then ${}_{\mathcal{H}}\mathbf{M}$ is a monoidal category and F is a strict monoidal functor by 31.7(3).

Conversely, if ${}_{\mathcal{H}}\mathbf{M}$ is a monoidal category, then both $\mathcal{H}$ and $\mathcal{H} \otimes_A \mathcal{H}$ are left $\mathcal{H}$-modules. Define then a left $\mathcal{H}$-module map

$$\underline{\Delta} : \mathcal{H} \to \mathcal{H} \otimes_A \mathcal{H}, \quad h \mapsto h(1 \otimes 1) =: \sum h_{\underline{1}} \otimes h_{\underline{2}}.$$

Note that $\underline{\Delta}$ is an (A, A)-bimodule map, since H is an A^e-ring. Furthermore,

$$\begin{aligned}\sum h_{\underline{1}} \otimes h_{\underline{21}} \otimes h_{\underline{22}} &= \sum h_{\underline{1}} \otimes h_{\underline{2}}(1 \otimes 1) = h(1 \otimes (1 \otimes 1)) \\ &= h((1 \otimes 1) \otimes 1) = \sum h_{\underline{1}}(1 \otimes 1) \otimes h_{\underline{2}} = \sum h_{\underline{11}} \otimes h_{\underline{12}} \otimes h_{\underline{2}}.\end{aligned}$$

Hence $\underline{\Delta}$ is a coassociative coproduct. Next observe that $\mathrm{Im}(\underline{\Delta}) \subseteq \mathcal{H} \times_A \mathcal{H}$, since, for all $h \in \mathcal{H}$ and $a \in A$,

$$\sum h_{\underline{1}} \otimes h_{\underline{2}} s(a) = h(1 \otimes s(a)) = h(t(a) \otimes 1) = \sum h_{\underline{1}} t(a) \otimes h_{\underline{2}}.$$

Directly from the definition of $\underline{\Delta}$ it follows that $\underline{\Delta}$ is a multiplicative map. Since the forgetful functor F is a strict monoidal functor, A is a neutral object in ${}_{\mathcal{H}}\mathbf{M}$, and there is a left multiplication ${}_A\varrho : \mathcal{H} \otimes_R A \to A$ of $\mathcal{H}$ on A. Define $\underline{\varepsilon} : \mathcal{H} \to A$, $h \mapsto {}_A\varrho(h \otimes 1_A)$. One easily checks that $\underline{\varepsilon}$ is a counit for the bialgebroid $\mathcal{H}$ with coproduct $\underline{\Delta}$. This completes the proof. □

Examples of bialgebroids are provided by depth-2 ring extensions [138].

31.14. Depth-2 algebra extensions. An algebra extension $B \to D$ is said to be a *depth-2 extension* if

(i) the (B, D)-bimodule $D \otimes_B D$ is a direct summand of $\bigoplus^n D$ for some n;
(ii) the (D, B)-bimodule $D \otimes_B D$ is a direct summand of $\bigoplus^l D$ for some l.

Equivalently, $B \to D$ is a depth-2 extension if there exist $b_i, c_j \in (D \otimes_B D)^B$ and $\beta_i, \gamma_j \in {}_B\mathrm{Hom}_B(D, B)$, with i, j in finite index sets, such that, for all $d \in D$,

$$\textstyle\sum_i b_i \beta_i(d) = d \otimes 1, \qquad \sum_j \gamma_j(d) c_j = 1 \otimes d.$$

The system $\{b_i, \beta_i\}$ is known as a *left D2 quasibasis* while the system $\{c_j, \gamma_j\}$ is known as a *right D2 quasibasis* for the extension $B \to D$.

Examples of depth-2 algebra extensions are provided by *H-separable* extensions introduced in [130] (in this case one requires $D \otimes_B D$ to be a direct summand of $\bigoplus^n D$ for some n as a (D, D)-bimodule).

31.15. A bialgebroid associated to depth-2 algebra extensions. *Let $B \to D$ be a depth-2 algebra extension, $A = D^B = \{a \in D \mid$ for all $b \in B,\ ab = ba\}$, and view $\mathcal{H} = {}_B\mathrm{End}_B(D)$ as an algebra via the map composition. Then $\mathcal{H}$ is an A-bialgebroid with the following structure maps:*

(i) the source map $s : a \mapsto [d \mapsto ad]$;

(ii) the target map $t : a \mapsto [d \mapsto da]$;

(iii) the comultiplication $\underline{\Delta} : h \mapsto \sum_j \gamma_j \otimes c_j{}^{\tilde{1}} h(c_j{}^{\tilde{2}}-)$, *where the system* $\{c_j = \sum c_j{}^{\tilde{1}} \otimes c_j{}^{\tilde{2}}, \gamma_j\}$ *is a right D2 quasibasis;*

(iv) the counit $\underline{\varepsilon} : h \mapsto h(1_D)$.

Proof. Clearly s is an algebra map and t is an anti-algebra map. Note that, for all $a, a' \in A$, $h \in \mathcal{H}$ and $d \in D$,

$$(s(a) \circ t(a') \circ h)(d) = ah(d)a' = (t(a') \circ s(a) \circ h)(d),$$

so that $\mathcal{H}$ is an A^e-ring, and the (A, A)-bimodule structure of $\mathcal{H}$ induced by the source and target maps comes out as $aha' = ah(-)a'$. $\underline{\Delta}$ is obviously a right A-module map. Furthermore, using the left D2 quasibasis $\{b_i = \sum b_i{}^{\tilde{1}} \otimes b_i{}^{\tilde{2}}, \beta_i\}$, one finds the following alternative expression for $\underline{\Delta}$:

$$\begin{aligned}
\underline{\Delta}(h) &= \textstyle\sum_j \gamma_j \otimes c_j{}^{\tilde{1}} h(c_j{}^{\tilde{2}}-) = \sum_{i,j} \gamma_j \otimes c_j{}^{\tilde{1}} h(c_j{}^{\tilde{2}} b_i{}^{\tilde{1}}) b_i{}^{\tilde{2}} \beta_i(-) \\
&= \textstyle\sum_{i,j} \gamma_j(-) c_j{}^{\tilde{1}} h(c_j{}^{\tilde{2}} b_i{}^{\tilde{1}}) b_i{}^{\tilde{2}} \otimes \beta_i \\
&= \textstyle\sum_i h(-b_i{}^{\tilde{1}}) b_i{}^{\tilde{2}} \otimes \beta_i.
\end{aligned}$$

This immediately implies that $\underline{\Delta}$ is a left A-module map. The coassociativity of $\underline{\Delta}$ follows directly from the expressions of $\underline{\Delta}$ in terms of left and right quasibases. These expressions also imply that $\underline{\varepsilon}$ is the counit for $\underline{\Delta}$. Thus $\mathcal{H}$ is an A-coring.

Next, observe that for all $a \in A$, $b \in B$,

$$\textstyle\sum_i \beta_i(ab_i{}^{\tilde{1}}) b_i{}^{\tilde{2}} b = \sum_i \beta_i(abb_i{}^{\tilde{1}}) b_i{}^{\tilde{2}} = \sum_i b\beta_i(ab_i{}^{\tilde{1}}) b_i{}^{\tilde{2}},$$

so that $\sum_i \beta_i(ab_i{}^{\tilde{1}}) b_i{}^{\tilde{2}} \in A$. Using this observation one can easily prove that for all $h \in \mathcal{H}$ and $a \in A$,

$$\begin{aligned}
\underline{\Delta}(h) \circ (t(a) \otimes I_{\mathcal{H}}) &= \textstyle\sum_i h(-ab_i{}^{\tilde{1}}) b_i{}^{\tilde{2}} \otimes \beta_i \\
&= \textstyle\sum_{i,j} h(-b_j{}^{\tilde{1}}) b_j{}^{\tilde{2}} \beta_j(ab_i{}^{\tilde{1}}) b_i{}^{\tilde{2}} \otimes \beta_i \\
&= \textstyle\sum_i h(-b_i{}^{\tilde{1}}) b_i{}^{\tilde{2}} \otimes \beta_i(a-) = \underline{\Delta}(h) \circ (I_{\mathcal{H}} \otimes s(a)),
\end{aligned}$$

that is, $\underline{\Delta}(h) \in \mathcal{H} \times_A \mathcal{H}$. Using a similar type of arguments (and expressing $\underline{\Delta}$ in terms of left and right quasibases, as the need arises) one proves that $\underline{\Delta}$ is multiplicative. Obviously, $\underline{\Delta}(1) = 1 \otimes 1$ and $\underline{\varepsilon}(1) = 1$. Finally we compute for all $h, h' \in H$,

$$\underline{\varepsilon}(h \circ s(\underline{\varepsilon}(h'))) = h(h'(1)) = \underline{\varepsilon}(hh'),$$

and similarly for the target map t. Thus $\mathcal{H}$ is an A-bialgebroid. □

This intriguing connection between extensions and bialgebroids certainly deserves further investigation.

Similarly as for coalgebras, one can study modules associated to bialgebroids. However, since 31.6 defines left bialgebroids (cf. 31.7(5)), only left modules can be studied. The right-handed version requires the use of the notion of a right bialgebroid (which we do not develop here).

31.16. $\mathcal{H}$-bialgebroid modules. Given an A-bialgebroid $(\mathcal{H}, s, t, \underline{\Delta}, \underline{\varepsilon})$, a *left $\mathcal{H}$-bialgebroid module* is a left $\mathcal{H}$-module and a left (A-coring) $\mathcal{H}$-comodule M with the coaction ${}^M\varrho : M \to \mathcal{H} \otimes_A M$ such that

$$ {}^M\varrho(hm) = \sum h_{\underline{1}} m_{\underline{-1}} \otimes h_{\underline{2}} m_{\underline{0}}, $$

for all $m \in M$ and $h \in \mathcal{H}$. Here M is viewed as an (A, A)-bimodule via the source and target maps of $\mathcal{H}$, that is, $ama' = s(a)t(a')m$. Note that the right-hand side of the displayed equation is well defined since $\underline{\Delta}(\mathcal{H}) \subseteq \mathcal{H} \times_A \mathcal{H}$. A morphism of two $\mathcal{H}$-bialgebroid modules is a left $\mathcal{H}$-module left $\mathcal{H}$-comodule map. The category of left $\mathcal{H}$-bialgebroid modules is denoted by ${}^{\mathcal{H}}_{\mathcal{H}}\mathbf{M}$.

31.17. Trivial $\mathcal{H}$-bialgebroid modules. *Let $\mathcal{H}$ be an A-bialgebroid and let N be any left A-module.*

(1) $\mathcal{H} \otimes_A N$ is a left $\mathcal{H}$-bialgebroid module with the canonical structures

$$ \underline{\Delta} \otimes I_N : \mathcal{H} \otimes_A N \to \mathcal{H} \otimes_A (\mathcal{H} \otimes_A N), \quad \mu \otimes I_N : \mathcal{H} \otimes_R (\mathcal{H} \otimes_A N) \to \mathcal{H} \otimes_A N. $$

(2) For any $f : N \to N'$ in ${}_A\mathbf{M}$, the map $I_{\mathcal{H}} \otimes f : \mathcal{H} \otimes_A N \to \mathcal{H} \otimes_A N'$ is an $\mathcal{H}$-bialgebroid module morphism.

Proof. It is clear that $\mathcal{H} \otimes_A N$ is a left $\mathcal{H}$-module, and it is a comodule of the A-coring $\mathcal{H}$ by 18.9. The compatibility conditions follow immediately from the properties of a bialgebroid. Moreover, it is clear that $I_{\mathcal{H}} \otimes f$ is left $\mathcal{H}$-linear and it is left $\mathcal{H}$-colinear by 18.9. □

31.18. $\mathcal{H}$-modules and $\mathcal{H}$-bialgebroid modules. *For an A-bialgebroid $\mathcal{H}$, let $N \in {}_{\mathcal{H}}\mathbf{M}$.*

(1) The left $\mathcal{H}$-module $\mathcal{H} \otimes^b_A N$ is a left $\mathcal{H}$-bialgebroid module with the canonical comodule structure

$$ \underline{\Delta} \otimes I_N : \mathcal{H} \otimes^b_A N \to \mathcal{H} \otimes_A (\mathcal{H} \otimes^b_A N), \quad h \otimes n \mapsto \underline{\Delta}(h) \otimes n. $$

(2) For any $f : N \to N'$ in ${}_{\mathcal{H}}\mathbf{M}$, the map $I_{\mathcal{H}} \otimes f : \mathcal{H} \otimes^b_A N \to \mathcal{H} \otimes^b_A N'$ is an $\mathcal{H}$-bialgebroid module morphism.

(3) *There is a bialgebroid module morphism*

$$\gamma_N : \mathcal{H} \otimes_A N \to \mathcal{H} \otimes_A^b N, \quad h \otimes n \mapsto (1_{\mathcal{H}} \otimes n)\Delta(b).$$

Proof. The proof of 14.3 can be adapted to this situation. □

A number of properties of Hopf modules discussed in Section 14 can be easily generalised to the case of bialgebroid modules. Instead of repeating these properties we make an observation that allows us to deduce properties of bialgebroid modules from those of the comodules of a coring.

31.19. $\mathcal{H}$-bialgebroid modules as comodules of a coring. *Let $\mathcal{H}$ be an A-bialgebroid.*

(1) *View $\mathcal{C} = \mathcal{H} \otimes_A^b \mathcal{H}$ as an $(\mathcal{H}, \mathcal{H})$-bimodule with the left diagonal action and the right action $(h \otimes h')h'' = h \otimes h'h''$. Then $\mathcal{C}$ is an $\mathcal{H}$-coring with the coproduct*

$$\underline{\Delta}_{\mathcal{C}} : \mathcal{H} \otimes_A^b \mathcal{H} \to \mathcal{H} \otimes_A^b \mathcal{H} \otimes_{\mathcal{H}} \mathcal{H} \otimes_A^b \mathcal{H} \simeq \mathcal{H} \otimes_A^b \mathcal{H} \otimes_A^b \mathcal{H}, \quad \underline{\Delta}_{\mathcal{C}} = \underline{\Delta}_{\mathcal{H}} \otimes I_{\mathcal{H}},$$

and counit $\underline{\varepsilon}_{\mathcal{C}} = \underline{\varepsilon}_{\mathcal{H}} \otimes I_{\mathcal{H}}$.

(2) *The category ${}^{\mathcal{H}}_{\mathcal{H}}\mathbf{M}$ is isomorphic to the category of left $\mathcal{C}$-comodules ${}^{\mathcal{C}}\mathbf{M}$.*

Proof. (1) It is clear that $\mathcal{C}$ is an $(\mathcal{H}, \mathcal{H})$-bimodule, and it follows from 31.18 that $\underline{\Delta}_{\mathcal{C}}$ is an $(\mathcal{H}, \mathcal{H})$-bimodule map. The coassociativity of $\underline{\Delta}_{\mathcal{C}}$ follows from the coassociativity of $\underline{\Delta}_{\mathcal{H}}$. Clearly, $\underline{\varepsilon}_{\mathcal{C}}$ is a right $\mathcal{H}$-module map. The only nontrivial part is to show that it is left $\mathcal{H}$-linear as well. Take any $h, h', h'' \in \mathcal{H}$ and compute

$$\begin{aligned} \underline{\varepsilon}_{\mathcal{C}}(h(h' \otimes h'')) &= \sum \underline{\varepsilon}_{\mathcal{C}}(h_{\underline{1}}h' \otimes h_{\underline{2}}h'') = \sum \underline{\varepsilon}_{\mathcal{H}}(h_{\underline{1}}h')h_{\underline{2}}h'' \\ &= \underline{\varepsilon}_{\mathcal{H}}(h_{\underline{1}}t(\underline{\varepsilon}_{\mathcal{H}}(h')))h_{\underline{2}}h'' = \underline{\varepsilon}_{\mathcal{H}}(h_{\underline{1}})h_{\underline{2}}s(\underline{\varepsilon}_{\mathcal{H}}(h'))h'' \\ &= hs(\underline{\varepsilon}_{\mathcal{H}}(h'))h'' = h\underline{\varepsilon}_{\mathcal{C}}(h' \otimes h''), \end{aligned}$$

where we used 31.6(3) to derive the third equality and then the fact that $\underline{\Delta}_{\mathcal{H}}(h) \in \mathcal{H} \times_A \mathcal{H}$ to obtain the fourth equality. Finally, $\underline{\varepsilon}_{\mathcal{C}}$ is a counit for $\underline{\Delta}_{\mathcal{C}}$ since $\underline{\varepsilon}_{\mathcal{H}}$ is a counit for $\underline{\Delta}_{\mathcal{H}}$.

(2) For any left $\mathcal{H}$-module M, the canonical isomorphism $\mathcal{H} \otimes_A \mathcal{H} \otimes_{\mathcal{H}} M \simeq \mathcal{H} \otimes_A M$ allows one to view any $\mathcal{H}$-bialgebroid module as a left $\mathcal{C}$-comodule and vice versa. □

In view of 31.19 we immeditely deduce from (the left-handed versions of) 18.10, 18.13, 18.14 and 18.17 the following characterisation of $\mathcal{H}$-bialgebroid modules.

31.20. The category ${}^{\mathcal{H}}_{\mathcal{H}}\mathbf{M}$. *Let $\mathcal{H}$ be an A-bialgebroid. Then:*

(1) (i) *The left $\mathcal{H}$-bialgebroid module $\mathcal{H}\otimes^b_A\mathcal{H}$ is a subgenerator in ${}^{\mathcal{H}}_{\mathcal{H}}\mathbf{M}$.*

(ii) *For any $M\in{}^{\mathcal{H}}_{\mathcal{H}}\mathbf{M}$, $N\in{}_{\mathcal{H}}\mathbf{M}$,*

$${}^{\mathcal{H}}_{\mathcal{H}}\mathrm{Hom}(M,\mathcal{H}\otimes^b_A N)\to{}_{\mathcal{H}}\mathrm{Hom}(M,N),\quad f\mapsto(\underline{\varepsilon}\otimes I_N)\circ f,$$

is an A-module isomorphism with inverse map $g\mapsto(I_{\mathcal{H}}\otimes g)\circ{}^{M}\!\underline{\varrho}$.

(2) *If $\mathcal{H}$ is flat as a left A-module, then:*

(i) *${}^{\mathcal{H}}_{\mathcal{H}}\mathbf{M}$ is a Grothendieck category.*

(ii) *For all $M\in{}^{\mathcal{H}}_{\mathcal{H}}\mathbf{M}$, the functor ${}^{\mathcal{H}}_{\mathcal{H}}\mathrm{Hom}(M,-):{}^{\mathcal{H}}_{\mathcal{H}}\mathbf{M}\to{}_A\mathbf{M}$ is left exact.*

(iii) *For all $N\in{}^{\mathcal{H}}_{\mathcal{H}}\mathbf{M}$, the functor ${}^{\mathcal{H}}_{\mathcal{H}}\mathrm{Hom}(-,N):{}^{\mathcal{H}}_{\mathcal{H}}\mathbf{M}\to{}_A\mathbf{M}$ is left exact.*

Since bialgebroids have both an algebra and a coalgebra structure that are compatible with each other, it makes sense to consider not only modules and comodules but also module coalgebras (corings) and comodule algebras.

31.21. Module corings of a bialgebroid. Let $(\mathcal{H},s_{\mathcal{H}},t_{\mathcal{H}})$ be an A-bialgebroid. We say that a left $\mathcal{H}$-module $\mathcal{C}$ is a *left $\mathcal{H}$-module coring* if $\mathcal{C}$ is a coalgebra in the monoidal category $({}_{\mathcal{H}}\mathbf{M},\otimes_A)$ of left $\mathcal{H}$-modules (cf. 31.13 and 38.33).

Recall from 31.13 that ${}_{\mathcal{H}}\mathbf{M}$ has a monoidal structure defined as follows. For all $M,N\in{}_{\mathcal{H}}\mathbf{M}$, $M\otimes_A N\in{}_{\mathcal{H}}\mathbf{M}$ via $h(m\otimes n)=\sum h_{\underline{1}}m\otimes h_{\underline{2}}n$. A is the unit object, when viewed in ${}_{\mathcal{H}}\mathbf{M}$ via the action $h\triangleright a=\underline{\varepsilon}_{\mathcal{H}}(hs_{\mathcal{H}}(a))=\underline{\varepsilon}_{\mathcal{H}}(ht_{\mathcal{H}}(a))$. Thus, $\mathcal{C}$ is a left $\mathcal{H}$-module coring if and only if $\mathcal{C}$ is a left $\mathcal{H}$-module and $(\mathcal{C},\underline{\Delta}_{\mathcal{C}},\underline{\varepsilon}_{\mathcal{C}})$ is an A-coring, where $\mathcal{C}$ is viewed as an (A,A)-bimodule via $aca'=s_{\mathcal{H}}(a)t_{\mathcal{H}}(a')c$, such that $\underline{\Delta}_{\mathcal{C}}$, $\underline{\varepsilon}_{\mathcal{C}}$ are left $\mathcal{H}$-module maps, that is, for all $h\in\mathcal{H}$ and $c\in\mathcal{C}$,

$$\underline{\Delta}_{\mathcal{C}}(hc)=\sum h_{\underline{1}}c_{\underline{1}}\otimes h_{\underline{2}}c_{\underline{2}},\quad \underline{\varepsilon}_{\mathcal{C}}(hc)=h\triangleright\underline{\varepsilon}_{\mathcal{C}}(c)=\underline{\varepsilon}_{\mathcal{H}}(hs_{\mathcal{H}}(\underline{\varepsilon}_{\mathcal{C}}(c))).$$

A morphism between two $\mathcal{H}$-module corings is an $\mathcal{H}$-linear map of A-corings.

31.22. Examples of module corings. Let $\mathcal{H}$ be an A-bialgebroid.

(1) $(\mathcal{H},\underline{\Delta}_{\mathcal{H}},\underline{\varepsilon}_{\mathcal{H}})$ is a left $\mathcal{H}$-module coring with the left $\mathcal{H}$-action provided by the product. This is known as the *left regular module coring* of $\mathcal{H}$.

(2) View A as a trivial A-coring (cf. 17.3). Then $(A,\triangleright)$ is a left $\mathcal{H}$-module coring, known as a *trivial left $\mathcal{H}$-module coring* . Indeed, note that for all $a\in A$ and $h\in\mathcal{H}$ we know that $\sum s_{\mathcal{H}}(h_{\underline{1}}\triangleright a)h_{\underline{2}}=hs_{\mathcal{H}}(a)$, and then we apply the left A-module map $\underline{\varepsilon}_{\mathcal{H}}$ to obtain that $\underline{\varepsilon}_{\mathcal{H}}(hs_{\mathcal{H}}(a))=\sum\underline{\varepsilon}_{\mathcal{H}}(h_{\underline{1}}s_{\mathcal{H}}(a))\underline{\varepsilon}_{\mathcal{H}}(h_{\underline{2}})$. This is equivalent to the fact that $\underline{\Delta}_A=I_A$ is a left $\mathcal{H}$-linear map.

(3) $\mathcal{C} = A^e$ is an A-coring with the coproduct $\underline{\Delta}_{A^e}(a' \otimes \bar{a}) = a' \otimes 1_{\bar{A}} \otimes 1_A \otimes \bar{a}$ and the counit $\underline{\varepsilon}_{A^e}(a' \otimes \bar{a}) = a'a$, and it can be made into a left $\mathcal{H}$-module coring by the $\mathcal{H}$-action $h(a' \otimes \bar{a}) = \underline{\varepsilon}_{\mathcal{H}}(hs_{\mathcal{H}}(a')t_{\mathcal{H}}(\bar{a}))$.

31.23. Comodule algebras of a bialgebroid. Let $(\mathcal{H}, s_{\mathcal{H}}, t_{\mathcal{H}})$ be an A-bialgebroid. A *left $\mathcal{H}$-comodule algebra* is an R-algebra B together with an algebra map $s_B : A \to B$ and a left $\mathcal{H}$-module map ${}^B\varrho : B \to \mathcal{H} \otimes_A B$ such that

(1) B is a unital A-ring via the map s_B.

(2) B is a left comodule of the A-coring $\mathcal{H}$ with the coaction ${}^B\varrho$.

(3) $\mathrm{Im}({}^B\varrho) \subseteq \mathcal{H} \times_A B$ and its corestriction ${}^B\varrho : B \to \mathcal{H} \times_A B$ is an algebra map.

A morphism between two $\mathcal{H}$-comodule algebras is a left $\mathcal{H}$-colinear map that is also a morphism of A-rings (the latter defined in the obvious way).

31.24. Examples of comodule algebras of a bialgebroid. Let $\mathcal{H}$ be an A-bialgebroid.

(1) $\mathcal{H}$ is a left $\mathcal{H}$-comodule algebra with the map $s_{\mathcal{H}}$ given by the source map and with the regular coaction $\underline{\Delta}$. This is known as a *left regular comodule algebra* of $\mathcal{H}$.

(2) The algebra A is a left $\mathcal{H}$-comodule algebra, when viewed as an A-ring via the identity map $s_A = I_A$ and with the coaction

$$ {}^A\varrho : A \to \mathcal{H} \otimes_A A, \quad a \mapsto s_{\mathcal{H}}(a) \otimes 1_A. $$

This comodule algebra is known as the *trivial left $\mathcal{H}$-comodule algebra*.

(3) $B = A^e$ is a left $\mathcal{H}$-comodule algebra via $s_{B^e} = - \otimes 1_A$, and coaction ${}^{A^e}\varrho(a' \otimes \bar{a}) = s_{\mathcal{H}}(a') \otimes 1_A \otimes \bar{a}$.

(4) The above example can be generalised as follows. For an $\mathcal{H}$-comodule algebra B and an algebra B', $B \otimes_R B'$ is a left $\mathcal{H}$-comodule algebra with the structures arising from those of B. This defines a functor from the category of R-algebras to the category of $\mathcal{H}$-comodule algebras.

31.25. Strict comodule algebras. Note that the definition of a left $\mathcal{H}$-comodule algebra is not dual to that of a left $\mathcal{H}$-module coring. The reason is that, although the category of left $\mathcal{H}$-modules is monoidal, the category ${}^{\mathcal{H}}\mathbf{M}$ of left comodules of the A-coring $\mathcal{H}$ is not. Thus there is no way of defining a left $\mathcal{H}$-module algebra as an algebra in the category ${}^{\mathcal{H}}\mathbf{M}$. However, one can consider a more restrictive definition of a left comodule M of an A-bialgebroid $\mathcal{H}$ by requiring it to be an (A, A)-bimodule with an (A, A)-bimodule coaction ${}^M\varrho : M \to \mathcal{H} \otimes_A M$ such that $\mathrm{Im}({}^M\varrho) \subseteq \mathcal{H} \times_A M$, where

$$ \mathcal{H} \times_A M = \{\sum_i h^i \otimes m^i \in \mathcal{H} \otimes_A M \mid \forall a \in A, \sum_i h^i t_{\mathcal{H}}(a) \otimes m^i = \sum_i h^i \otimes m^i a\}. $$

Such an M is called a *strict left comodule* of an A-bialgebroid $\mathcal{H}$. The category of strict left $\mathcal{H}$-comodules is a subcategory of ${}^{\mathcal{H}}\mathbf{M}$, and it is monoidal. Explicitly, for all strict comodules M, N, the tensor product $M \otimes_A N$ is a strict $\mathcal{H}$-comodule via ${}^{M\otimes_A N}\varrho(m \otimes n) = \sum m_{\underline{-1}}n_{\underline{-1}} \otimes m_{\underline{0}} \otimes n_{\underline{0}}$. Note that the right-hand side is well defined because $\overline{\mathrm{Im}({}^M\varrho)} \subseteq \mathcal{H} \times_A M$. A is the unit object in the category of strict left comodules with the trivial coaction ${}^A\varrho(a) = s_{\mathcal{H}}(a) \otimes 1_A$. Furthermore, the forgetful functor from the category of strict left $\mathcal{H}$-comodules to the category of (A, A)-bimodules is strict monoidal. Now, one defines a *strict left $\mathcal{H}$-comodule algebra* as an algebra in the monoidal category of strict left $\mathcal{H}$-comodules. This definition, however, turns out to be too restrictive to cover the examples that we will encounter in Section 37.

31.26. Exercises

(1) Given an A-bialgebroid $\mathcal{H}$, use the fact that, $\mathrm{Im}(\underline{\Delta}) \subseteq \mathcal{H} \times_A \mathcal{H}$ to show that for any $M, N \in {}_{\mathcal{H}}\mathbf{M}$, the tensor product $M \otimes_A N$ is a left H-module with the product $h(m \otimes n) = \sum h_{\underline{1}}m \otimes h_{\underline{2}}n$. Here M and N have an (A, A)-bimodule structure induced from their left $\mathcal{H}$-module structures via the source and target maps, that is, $ama' = s(a)t(a')m$, for all $a, a' \in A$ and $m \in M$ or $m \in N$.

(2) Let $\mathcal{H}$ be an A-bialgebroid with source map s and target map t. Then the identity $1 \in \mathcal{H}$ is a grouplike element. Prove that

$$A_1^{co\mathcal{H}} = \{b \in A \mid s(b) = t(b)\}.$$

References. Brzeziński, Caenepeel and Militaru [76]; Brzeziński and Militaru [81]; Hirata [130]; Kadison [137]; Kadison and Szlachányi [138]; Lu [154]; Schauenburg [183, 185]; Sweedler [192]; Takeuchi [196]; Xu [215].

Chapter 5

Corings and entwining structures

In this chapter we introduce and analyse the main new class of examples of corings, that is, corings associated to entwining structures. An entwining structure can be understood as a generalisation of a bialgebra. In many applications, in particular in mathematical physics and noncommutative geometry, it can be viewed as a symmetry of a noncommutative manifold. From the Hopf algebra point of view, the introduction of an entwining structure leads to the unification of various categories of Hopf modules studied for over 30 years. Various properties of such modules can be then understood on a more fundamental level once they are formulated in terms of associated corings. Thus in this chapter we introduce the notion of an entwining structure, give numerous examples and study properties of associated corings and comodules. All aspects of the general theory of corings and comodules discussed in the previous chapters are thus illustrated and used in deriving properties of entwined modules.

Throughout this chapter, R is a commutative ring, A is an R-algebra and C is an R-coalgebra. The coproduct of C and any other coalgebra (including bialgebras and Hopf algebras) is denoted by Δ and its counit by ε. The product in A is denoted by μ, and the unit as a map is $\iota : R \to A$. If no confusion arises, we also write 1 for the element $1_A = \iota(1_R)$.

32 Entwining structures and corings

Entwining structures were introduced in [80] in order to recapture in noncommutative geometry symmetry properties of classical principal bundles. Although motivated by the geometry of noncommutative principal bundles, entwining structures have revealed a deep algebraic meaning leading far beyond their initial motivation. In this section we define entwining structures and entwined modules, and we prove an important theorem that will allow us to construct a coring associated to an entwining structure and then identify entwined modules with comodules of this coring. We also study the dual algebra of this coring, which turns out to be a *ψ-twisted convolution algebra*, and gather some general properties of entwining structures related to the notions introduced earlier for general corings. The latter include cointegrals (coseparability) and the Amitsur complex, as well as the ψ-equivariant cohomology obtained from the Cartier cohomology for the associated coring.

32.1. Entwining structures. A (right-right) *entwining structure* (over R) is a triple $(A, C)_\psi$ consisting of an R-algebra A, an R-coalgebra C and an R-module map $\psi : C \otimes_R A \to A \otimes_R C$ satisfying the following four conditions:

(1) $\psi \circ (I_C \otimes \mu) = (\mu \otimes I_C) \circ (I_A \otimes \psi) \circ (\psi \otimes I_A)$,

(2) $(I_A \otimes \Delta) \circ \psi = (\psi \otimes I_C) \circ (I_C \otimes \psi) \circ (\Delta \otimes I_A)$,

(3) $\psi \circ (I_C \otimes \iota) = \iota \otimes I_C$,

(4) $(I_A \otimes \varepsilon) \circ \psi = \varepsilon \otimes I_A$.

The map ψ is known as an *entwining map*, and C and A are said to be *entwined* by ψ.

32.2. The bow-tie diagram. The conditions of Definition 32.1 can be summarised in the following commutative *bow-tie diagram* (tensor over R):

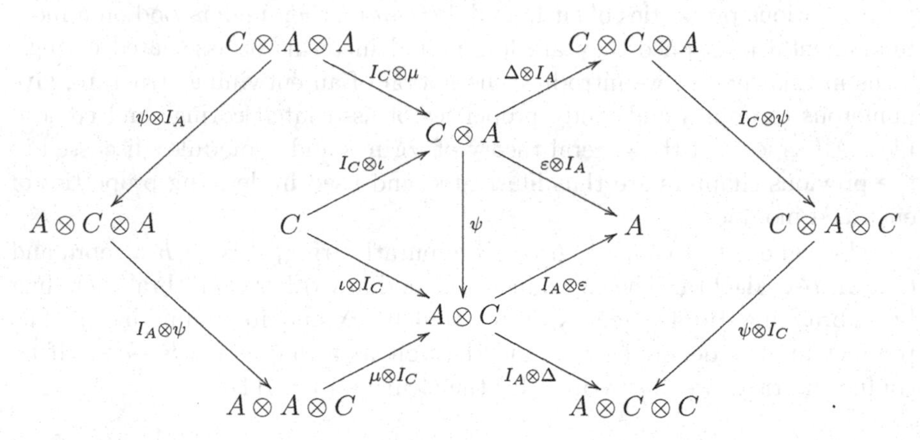

32.3. The α-notation. To denote the action of an entwining map ψ on elements we use the following α-*notation*:

$$\psi(c \otimes a) = \sum_\alpha a_\alpha \otimes c^\alpha, \quad (I_A \otimes \psi) \circ (\psi \otimes I_A)(c \otimes a \otimes a') = \sum_{\alpha,\beta} a_\alpha \otimes a'_\beta \otimes c^{\alpha\beta},$$

and so on, for all $a, a' \in A$, $c \in C$. This notation proves very useful in concrete computations involving ψ. The reader is advised to check that the bow-tie diagram is equivalent to the following relations, for all $a, a' \in A$, $c \in C$:

$$\begin{aligned}
\text{left pentagon:} &\quad \textstyle\sum_\alpha (aa')_\alpha \otimes c^\alpha = \sum_{\alpha,\beta} a_\alpha a'_\beta \otimes c^{\alpha\beta},\\
\text{left triangle:} &\quad \textstyle\sum_\alpha 1_\alpha \otimes c^\alpha = 1 \otimes c,\\
\text{right pentagon:} &\quad \textstyle\sum_\alpha a_\alpha \otimes c^\alpha{}_{\underline{1}} \otimes c^\alpha{}_{\underline{2}} = \sum_{\alpha,\beta} a_{\beta\alpha} \otimes c_{\underline{1}}{}^\alpha \otimes c_{\underline{2}}{}^\beta,\\
\text{right triangle:} &\quad \textstyle\sum_\alpha a_\alpha \varepsilon(c^\alpha) = a\varepsilon(c).
\end{aligned}$$

One can define right-left, left-right and left-left entwining structures by replacing the pair (A, C) with the pairs (A, C^{cop}), (A^{op}, C) and (A^{op}, C^{cop}), respectively, in the bow-tie diagram. All such combinations lead to equivalent theories; thus we concentrate on right-right entwining structures only.

32.4. Entwined modules. Associated to any entwining structure $(A, C)_\psi$ is the category of (right-right) $(A, C)_\psi$-*entwined modules* denoted by $\mathbf{M}_A^C(\psi)$. An object $M \in \mathbf{M}_A^C(\psi)$ is a right A-module with multiplication ϱ_M and a right C-comodule with coaction ϱ^M inducing a commutative diagram

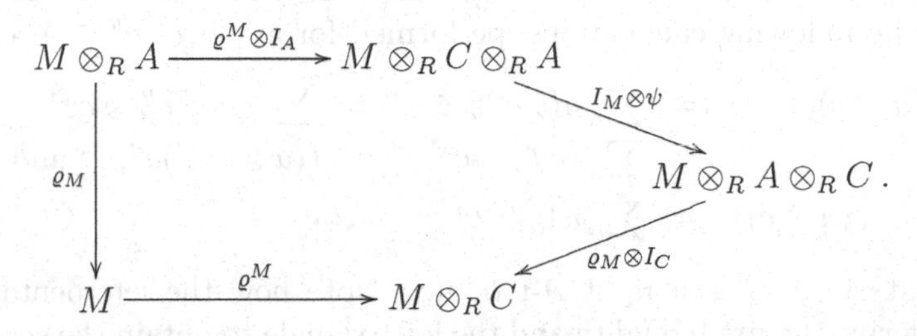

A morphism in $\mathbf{M}_A^C(\psi)$ is a right A-module map that is at the same time a right C-comodule map.

The category of entwined modules was introduced in [69], but special cases where studied intensively in the literature for many years, starting from the work by Sweedler on Hopf modules, through Doi's and Takeuchi's work on relative Hopf modules, up to Yetter-Drinfeld and Doi-Koppinen modules. These special examples of entwining structures and corresponding entwined modules are collected in the next section.

32.5. Self-duality of an entwining structure. One of the important properties of the notion of an entwining structure is its self-duality. The bow-tie diagram is invariant (bar a space rotation) under the operation consisting of interchanging A with C, μ with Δ, ι with ε, and reversing all the arrows. Note also that the notion of an entwined module is self-dual in the same sense. Explicitly, interchange C with A, ϱ^M with ϱ_M, and reverse all the arrows in the diagram in 32.4; then it remains unchanged.

32.6. Corings associated to entwining structures. *View $A \otimes_R C$ as a left A-module with the obvious left multiplication $a(a' \otimes c) = aa' \otimes c$, for all $a, a' \in A, c \in C$. Then:*

(1) For an entwining structure $(A, C)_\psi$, $\mathcal{C} = A \otimes_R C$ is an (A, A)-bimodule with right multiplication $(a' \otimes c)a = a'\psi(c \otimes a)$, and it is an A-coring with the coproduct and counit

$$\underline{\Delta} := I_A \otimes \Delta : \mathcal{C} \to A \otimes_R C \otimes_R C \simeq \mathcal{C} \otimes_A \mathcal{C}, \quad \underline{\varepsilon} := I_A \otimes \varepsilon : \mathcal{C} \to A.$$

(2) *If* $\mathcal{C} = A \otimes_R C$ *is an* A*-coring with coproduct* $\underline{\Delta} = I_A \otimes \Delta$ *and counit* $\underline{\varepsilon} = I_A \otimes \varepsilon$, *then* $(A, C)_\psi$ *is an entwining structure by*

$$\psi : C \otimes_R A \to A \otimes_R C, \quad c \otimes a \mapsto (1 \otimes c)a.$$

(3) *If* $\mathcal{C} = A \otimes_R C$ *is the* A*-coring associated to* $(A, C)_\psi$ *as in (1), then the category of* $(A, C)_\psi$*-entwined modules is isomorphic to the category of right* $\mathcal{C}$*-comodules.*

Proof. (1) It is obvious that $A \otimes_R C$ is a left A-module with the specified action. The following calculations, performed for any $a, a', a'' \in A$, $c \in C$,

$$\begin{aligned}(a \otimes c)(a'a'') &= \textstyle\sum_\alpha a(a'a'')_\alpha \otimes c^\alpha = \sum_{\alpha,\beta} aa'_\alpha a''_\beta \otimes c^{\alpha\beta} \\ &= \textstyle\sum_\alpha (aa'_\alpha \otimes c^\alpha)a'' = ((a \otimes c)a')a'', \quad \text{and} \\ (a \otimes c)1 &= \textstyle\sum_\alpha a1_\alpha \otimes c^\alpha = a \otimes c,\end{aligned}$$

prove that $A \otimes_R C$ is a right A-module. Note how the left pentagon was used to derive the first equality and the left triangle to obtain the second one. Thus $\mathcal{C}$ is an (A, A)-bimodule.

Next one has to check that $\underline{\varepsilon}$ and $\underline{\Delta}$ are (A, A)-bimodule maps. Clearly they are left A-linear. Take any $a, a' \in A$, $c \in C$ and compute

$$\underline{\varepsilon}((a \otimes c)a') = \textstyle\sum_\alpha \underline{\varepsilon}(aa'_\alpha \otimes c^\alpha) = \sum_\alpha aa'_\alpha \varepsilon(c^\alpha) = aa'\varepsilon(c) = \underline{\varepsilon}(a \otimes c)a',$$

where the right triangle was used for the penultimate equality. Furthermore,

$$\begin{aligned}\underline{\Delta}((a \otimes c)a') &= \textstyle\sum_\alpha aa'_\alpha \otimes c^\alpha{}_{\underline{1}} \otimes c^\alpha{}_{\underline{2}} = \sum_{\alpha,\beta} aa'_{\alpha\beta} \otimes c_{\underline{1}}{}^\beta \otimes c_{\underline{2}}{}^\alpha \\ &= \textstyle\sum_\alpha (a \otimes c_{\underline{1}})a'_\alpha \otimes c_{\underline{2}}{}^\alpha = \sum (a \otimes c_{\underline{1}}) \otimes_A (1 \otimes c_{\underline{2}})a' \\ &= \underline{\Delta}(a \otimes c)a'.\end{aligned}$$

Here the second equality follows from the right pentagon. Thus $\underline{\varepsilon}$ and $\underline{\Delta}$ are (A, A)-bimodule morphisms, as required. Now, the coassociativity of $\underline{\Delta}$ follows immediately from the coassociativity of Δ, while the counit property of $\underline{\varepsilon}$ is an immediate consequence of the fact that ε is a counit of C.

(2) Let $\mathcal{C} = A \otimes_R C$ be an A-coring with structure maps given in the hypothesis. Denote $\psi(c \otimes a) = (1 \otimes c)a = \sum_\alpha a_\alpha \otimes c^\alpha$. Since ψ is defined in terms of the right action, one finds $\psi(1 \otimes c) = c \otimes 1$, the left triangle, and

$$\begin{aligned}\psi(c \otimes aa') &= (1 \otimes c)aa' = ((1 \otimes c)a)a' = \textstyle\sum_\alpha (a_\alpha \otimes c^\alpha)a' \\ &= \textstyle\sum_\alpha a_\alpha(1 \otimes c^\alpha)a' = \sum_{\alpha,\beta} a_\alpha a'_\beta \otimes c^{\alpha\beta}\end{aligned}$$

the left pentagon. Furthermore, $\underline{\varepsilon}$ is right A-linear, and hence

$$\textstyle\sum_\alpha a_\alpha \varepsilon(c^\alpha) = \sum_\alpha \underline{\varepsilon}(a_\alpha \otimes c^\alpha) = \underline{\varepsilon}((1 \otimes c)a) = \underline{\varepsilon}(1 \otimes c)a = \varepsilon(c)a$$

the right triangle. Finally, $\underline{\Delta}$ is also right A-linear; thus

$$\begin{aligned}
\sum_{\alpha,\beta} a_{\alpha\beta} \otimes c_{\underline{1}}{}^{\beta} \otimes c_{\underline{2}}{}^{\alpha} &= \sum_{\alpha} (1 \otimes c_{\underline{1}}) a_{\alpha} \otimes c_{\underline{2}}{}^{\alpha} = \sum_{\alpha} (1 \otimes c_{\underline{1}}) \otimes_A a_{\alpha} \otimes c_{\underline{2}}{}^{\alpha} \\
&= \sum (1 \otimes c_{\underline{1}}) \otimes (1 \otimes c_{\underline{2}}) a \\
&= \underline{\Delta}(1 \otimes c) a = \underline{\Delta}((1 \otimes c) a) \\
&= \sum_{\alpha} \underline{\Delta}(a_{\alpha} \otimes c^{\alpha}) = \sum_{\alpha} a_{\alpha} \otimes c^{\alpha}{}_{\underline{1}} \otimes c^{\alpha}{}_{\underline{2}}
\end{aligned}$$

the right pentagon. Therefore ψ is an entwining map, as claimed.

(3) The key observation here is that if M is a right A-module, then $M \otimes_R C$ is a right A-module with the multiplication $(m \otimes c)a = \sum_{\alpha} m a_{\alpha} \otimes c^{\alpha}$. The statement "$M$ is an $(A, C)_{\psi}$-entwined module" is equivalent to the statement that ϱ^M is a right A-module map. By the canonical identification $M \otimes_R C \simeq M \otimes_A A \otimes_R C = M \otimes_A \mathcal{C}$, one can view a right C-coaction as a right A-module map $M \to M \otimes_A \mathcal{C}$, that is, as a right $\mathcal{C}$-coaction. Conversely, a right $\mathcal{C}$-coaction can be viewed as a right A-module map $\varrho^M : M \to M \otimes_R C$, thus providing a right $\mathcal{C}$-comodule with the structure of an $(A, C)_{\psi}$-entwined module. □

The isomorphism of categories in 32.6 allows one to derive various properties of entwined modules from general properties of the comodules of a coring. For example, 18.13(2) (cf. 18.10) implies the existence of the following

32.7. Induction functor for entwined modules. *Let $(A, C)_{\psi}$ be an entwining structure. The assignment $- \otimes_R C : M \mapsto M \otimes_R C$ for any right A-module M defines a covariant functor $- \otimes_R C : \mathbf{M}_A \to \mathbf{M}_A^C(\psi)$ that is the right adjoint of the forgetful functor $\mathbf{M}_A^C(\psi) \to \mathbf{M}_A$.*

Proof. This follows by identifications $- \otimes_A \mathcal{C} = - \otimes_A A \otimes_R C \simeq - \otimes_R C$. □

The self-duality of entwining structures explained in 32.5, allows one also to study the functor forgetting the A-action $(-)^C : \mathbf{M}_A^C(\psi) \to \mathbf{M}^C$. This functor can be understood as a coinduction functor associated to a particular morphism in the category of corings.

32.8. The functor forgetting action. *Let $\mathcal{C} = A \otimes_R C$ be the A-coring associated to an entwining structure $(A, C)_{\psi}$. Let $\alpha : R \to A$ be the unit map $\alpha = \iota$ and let $\gamma : C \to \mathcal{C}$ be an R-linear map given by $\gamma : c \mapsto 1 \otimes c$. Then:*

1. *$(\gamma : \alpha) : (C : R) \to (\mathcal{C} : A)$ is a pure morphism of corings.*
2. *For any $M \in \mathbf{M}^C$, $M \otimes_R A$ is an $(A, C)_{\psi}$-entwined module with the obvious right A-multiplication and the coaction*

$$\varrho^{M \otimes_R A} : M \otimes_R A \to M \otimes_R A \otimes_R C, \quad m \otimes a \mapsto \sum m_{\underline{0}} \otimes \psi(m_{\underline{1}} \otimes a).$$

(3) *The functor* $-\otimes_R A : \mathbf{M}^C \to \mathbf{M}^C_A(\psi)$ *is left adjoint to the forgetful functor* $(-)^C : \mathbf{M}^C_A(\psi) \to \mathbf{M}^C$.

Proof. (1) By definition, $\alpha = \iota$ is an algebra map and γ is an R-module map. Let $\chi : \mathcal{C} \otimes_R \mathcal{C} \to \mathcal{C} \otimes_A \mathcal{C}$ be the canonical projection. Then, for all $c \in C$,

$$\chi(\gamma \otimes_R \gamma)\Delta(c) = \sum (1 \otimes_R c_{\underline{1}}) \otimes_A (1 \otimes_R c_{\underline{2}}) = \sum 1 \otimes c_{\underline{1}} \otimes c_{\underline{2}} = \underline{\Delta}(1 \otimes c)$$

and $\underline{\varepsilon} \circ \gamma(c) = \iota\varepsilon(c) = \alpha \circ \varepsilon(c)$. This proves that $(\gamma : \alpha) : (C : R) \to (\mathcal{C} : A)$ is a morphism of corings.

Take an $(A, C)_\psi$-entwined module or, equivalently, a right $\mathcal{C}$-comodule M. Then the coinduced module $G(M)$ defined in 24.7 comes out as

$$G(M) = M \Box_C (A \otimes_R C) = M \Box_{\mathcal{C}} \mathcal{C} = \mathrm{Im}\,(\varrho^M),$$

where $\varrho^M : M \to M \otimes_R C$ is the C-coaction. Thus, to prove that $(\gamma : \alpha) : (C : R) \to (\mathcal{C} : A)$ is a pure morphism of corings, we need to show that $\mathrm{Im}\,(\varrho^M) \otimes_R C$ is equal to the equaliser of the maps $t_M \otimes I_C = \varrho^M \otimes I_C \otimes I_C$ and $b_M \otimes I_C = I_M \otimes \Delta \otimes I_C$ (cf. 24.8). The coassociativity of ϱ^M immediately implies that $\mathrm{Im}\,(\varrho^M) \otimes_R C$ is in the equaliser. Conversely, if the element $\sum_i m^i \otimes c^i \otimes \tilde{c}^i \in M \otimes_R C \otimes_R C$ is in this equaliser, then

$$\sum_i m^i{}_{\underline{0}} \otimes m^i{}_{\underline{1}} \otimes c^i \otimes \tilde{c}^i = \sum_i m^i \otimes c^i{}_{\underline{1}} \otimes c^i{}_{\underline{2}} \otimes \tilde{c}^i.$$

Applying $I_M \otimes I_C \otimes \varepsilon \otimes I_C$, we obtain

$$\sum_i m^i{}_{\underline{0}} \otimes m^i{}_{\underline{1}}\varepsilon(c^i) \otimes \tilde{c}^i = \sum_i m^i \otimes c^i \otimes \tilde{c}^i,$$

that is, $\sum_i m^i \otimes c^i \otimes \tilde{c}^i \in \mathrm{Im}\,(\varrho^M)$, as required.

(2) One immediately checks that the induction functor F associated to the coring morphism $(\gamma : \alpha)$ (cf. 24.6) comes out as $-\otimes_R A$.

(3) Note that $\mathrm{Im}\,(\varrho^M) \simeq M$ in $\mathbf{M}^C$, so that $G(M) = M$, and $G = (-)^C$. Therefore $(-)^C$ has the left adjoint $-\otimes_R A$ by 24.11. □

32.9. The dual ring of a coring associated to an entwining. *Let* $(A, C)_\psi$ *be an entwining structure and let* $\mathcal{C} = A \otimes_R C$ *be the associated* A*-coring. Then the* R*-algebra* ${}^*\mathcal{C}$ *is anti-isomorphic to the* ψ-twisted convolution algebra $\mathrm{Hom}_\psi(C, A)$. *The latter is isomorphic to* $\mathrm{Hom}_R(C, A)$ *as an* R*-module and has the product defined by*

$$(f *_\psi f')(c) = \sum\nolimits_\alpha f(c_{\underline{2}})_\alpha f'(c_{\underline{1}}{}^\alpha),$$

for all $f, f' \in \mathrm{Hom}_R(C, A)$ *and the unit* $\iota \circ \varepsilon$.

Proof. There is a canonical isomorphism of R-modules,

$$^*\mathcal{C} = {}_A\mathrm{Hom}(A \otimes_R C, A) \to \mathrm{Hom}_R(C, A), \quad \xi \mapsto [c \mapsto \xi(1 \otimes c)],$$

with the inverse $f \mapsto [a \otimes c \mapsto af(c)]$. Using this isomorphism, for any $\xi, \xi' \in {}^*\mathcal{C}$ define $f, f' \in \mathrm{Hom}_R(C, A)$ via

$$\xi(a \otimes c) = af(c), \quad \xi'(a \otimes c) = af'(c), \quad \text{for all } a \in A,\ c \in C,$$

and compute

$$\begin{aligned}(\xi *^l \xi')(a \otimes c) &= \sum \xi((a \otimes c_{\underline{1}})\xi'(1 \otimes c_{\underline{2}})) = \sum \xi((a \otimes c_{\underline{1}})f'(c_{\underline{2}})) \\ &= \sum_\alpha \xi(af'(c_{\underline{2}})_\alpha \otimes c_{\underline{1}}{}^\alpha) = \sum_\alpha af'(c_{\underline{2}})_\alpha f(c_{\underline{1}}{}^\alpha).\end{aligned}$$

Furthermore, $\underline{\varepsilon}(1 \otimes c) = 1\varepsilon(c)$. All this establishes that the isomorphism of R-modules extends to an anti-isomorphism of the corresponding algebras. □

The identification of a dual ring with the ψ-twisted convolution algebra allows one to derive various properties of the latter from general properties of the former. Thus, for example, using 19.3, 19.6 and 32.6, we obtain

32.10. Modules over ψ-twisted convolution algebras. *Let $(A, C)_\psi$ be an entwining structure and $A \otimes_R C$ the corresponding A-coring.*

(1) If C is a locally projective R-module, then the category $\mathbf{M}_A^C(\psi)$ is isomorphic to the category $\sigma[(A \otimes_R C)_{\mathrm{Hom}_\psi(C,A)}]$ of right $\mathrm{Hom}_\psi(C, A)$-modules subgenerated by $A \otimes_R C$.

(2) If C is a finitely generated projective R-module, then $\mathbf{M}_A^C(\psi)$ is isomorphic to the category $\mathbf{M}_{\mathrm{Hom}_\psi(C,A)}$ of right $\mathrm{Hom}_\psi(C, A)$-modules.

As yet another application of the results discussed in this section we obtain the following categorical interpretation of a twisted convolution algebra [131].

32.11. Twisted convolution algebra as endomorphism ring. *For an entwining structure $(A, C)_\psi$, let $F = (-)_R : \mathbf{M}_A^C \to \mathbf{M}_R$ be the forgetful functor. Then the R-algebra of endomorphisms of F, $\mathrm{Nat}(F, F)$ with the product given by $\phi\phi' = \phi' \circ \phi$ is isomorphic to the ψ-twisted convolution algebra $\mathrm{Hom}_\psi(C, A)$.*

Proof. This follows from 18.29 combined with 32.9 and 32.6. □

Next we derive a number of properties of entwining structures from general coring theory developed in preceding chapters. For the rest of this section $(A, C)_\psi$ denotes an entwining structure over R with the associated A-coring $\mathcal{C} = A \otimes_R C$.

32.12. Integral maps for entwining structures. *$\mathcal{C}$ is a coseparable A-coring if and only if there exists an R-linear map $\theta : C \otimes_R C \to A$ inducing the following three commutative diagrams:*

$$\begin{array}{ccccc} C\otimes_R C\otimes_R A & \xrightarrow{I_C\otimes\psi} & C\otimes_R A\otimes_R C & \xrightarrow{\psi\otimes I_C} & A\otimes_R C\otimes_R C \\ \downarrow{\scriptstyle \theta\otimes I_A} & & & & \downarrow{\scriptstyle I_A\otimes\theta} \\ A\otimes_R A & \xrightarrow{\mu} & A & \xleftarrow{\mu} & A\otimes_R A, \end{array}$$

$$\begin{array}{ccc} C & \xrightarrow{\Delta} & C\otimes_R C \\ \downarrow{\scriptstyle \varepsilon} & & \downarrow{\scriptstyle \theta} \\ R & \xrightarrow{\iota} & A, \end{array} \qquad \begin{array}{ccccc} C\otimes_R C & \xrightarrow{\Delta\otimes I_C} & C\otimes_R C\otimes_R C & \xrightarrow{I_C\otimes\theta} & C\otimes_R A \\ \downarrow{\scriptstyle I_C\otimes\Delta} & & & & \downarrow{\scriptstyle \psi} \\ C\otimes_R C\otimes_R C & & \xrightarrow{\theta\otimes I_C} & & A\otimes_R C, \end{array}$$

where μ is the product in A. Such a map θ is called a *normalised integral map in $(A, C)_\psi$.*

Proof. In this case, $\mathcal{C} \otimes_A \mathcal{C} \simeq A \otimes_R C \otimes_R C$. Using the natural isomorphism of R-modules, ${}_A\mathrm{Hom}(A \otimes_R C \otimes_R C, A) \simeq \mathrm{Hom}_R(C \otimes_R C, A)$, we identify the cointegral δ, which, in particular, is a left A-module map, with the mapping $\theta : C \otimes_R C \to A$. Thus $\theta(c \otimes c') = \delta(1 \otimes c \otimes c')$. Recall from 32.6 that the right multiplication of $\mathcal{C}$ by A is given by $(a \otimes c)a' = a\psi(c \otimes a')$, while the coproduct and counit are $I_A \otimes \Delta$ and $I_A \otimes \varepsilon$, respectively. In view of these definitions, the fact that δ is also a right A-module map is equivalent to the first diagram. Then the conditions required for a cointegral δ in 26.1(b) are equivalent to the second and third diagrams. □

32.13. Maschke-type theorem for entwined modules. *The forgetful functor $(-)_A : \mathbf{M}_A^C(\psi) \to \mathbf{M}_A$ is separable if and only if there exists a normalised integral map in $(A, C)_\psi$. In this case, any monomorphism of $(A, C)_\psi$-entwined modules that splits as a morphism of right A-modules also splits as a morphism of $(A, C)_\psi$-entwined modules.*

Proof. This follows from 32.12 combined with 26.1 and 32.6. □

32.14. Integrals in entwining structures. $\mathcal{C}$ is cosplit if and only if there exists $e = \sum_i a_i \otimes c_i \in A \otimes_R C$ such that

$$\textstyle\sum_i a_i\varepsilon(c_i) = 1 \quad \text{and} \quad \sum_i aa_i \otimes c_i = \sum_i a_i\psi(c_i \otimes a),$$

for all $a \in A$. Such an element e is called a *normalised integral in an entwining structure $(A, C)_\psi$.*

32.15. Twisted convolution algebra as a split extension. *If there exists an integral in an entwining structure* $(A, C)_\psi$, *then*

$$i_L : A \to \mathrm{Hom}_\psi(C, A), \quad a \mapsto [c \mapsto \varepsilon(c)a],$$

is a split extension.

Proof. This follows from 32.14 combined with 26.13 and 32.9. □

32.16. Characterisation of grouplike elements. *$\mathcal{C}$ has a grouplike element if and only if A is an $(A, C)_\psi$-entwined module. The grouplike element is $g = \varrho^A(1)$, where ϱ^A is the right coaction of C on A.*

Proof. By 32.6 there is a one-to-one correspondence between $(A, C)_\psi$-entwined modules and $\mathcal{C}$-comodules, so that the assertion follows immediately from 28.2. The proof of the latter also confirms the form of g. □

32.17. Characterisation of g-coinvariants. *Let $g = \varrho^A(1) = \sum 1_{\underline{0}} \otimes 1_{\underline{1}}$ be a grouplike element in $\mathcal{C}$. Then*

$$\begin{aligned} A_g^{co\mathcal{C}} &= \{b \in A \mid \varrho^A(b) = \textstyle\sum b1_{\underline{0}} \otimes 1_{\underline{1}}\} \\ &= \{b \in A \mid \textit{for all } a \in A,\ \varrho^A(ba) = b\varrho^A(a)\}. \end{aligned}$$

For $M \in \mathbf{M}_A^C(\psi)$, these g-coinvariants are denoted simply by M^{coC}, and the corresponding functor $\mathbf{M}_A^C(\psi) \to \mathbf{M}_{A^{coC}}$ is denoted by $(-)^{coC}$.

Proof. The first description of $A_g^{co\mathcal{C}}$ is simply the definition of g-coinvariants. Only the second description requires a justification. Suppose that $b \in A$ is such that, for all $a \in A$, $\varrho^A(ba) = b\varrho^A(a)$. Then, in particular, taking $a = 1$, $\varrho^A(b) = b\varrho(1) = bg$, that is, $b \in A_g^{co\mathcal{C}}$. Conversely, if $b \in A_g^{co\mathcal{C}}$, then, using the fact that A is an $(A, C)_\psi$-entwined module, for all $a \in A$,

$$\varrho^A(ba) = \sum b_{\underline{0}}\psi(b_{\underline{1}} \otimes a) = \sum b1_{\underline{0}}\psi(1_{\underline{1}} \otimes a) = b\varrho^A(1a) = b\varrho^A(a),$$

as required. □

32.18. Algebra structure. *Suppose that A is an $(A, C)_\psi$-entwined module. Then $A \otimes_R C$ is an R-algebra with the product given by the formula*

$$(a \otimes c) \bullet (a' \otimes c') = \varepsilon(c)aa' \otimes c' + \varepsilon(c')a\psi(c \otimes a') - \sum \varepsilon(c)\varepsilon(c')aa'_{\underline{0}} \otimes a'_{\underline{1}}$$

and the unit $\sum 1_{\underline{0}} \otimes 1_{\underline{1}} = \varrho^A(1)$.

Proof. This is a special case of 28.16. The first two terms in the product formula clearly follow from the definition of the left and right A-module structure of $A \otimes_R C$ in 32.6. To obtain the last term we note that $g = \sum 1_{\underline{0}} \otimes 1_{\underline{1}}$, and, since A is an $(A, C)_\psi$-entwined module, we find for all $a \in A$, $\varrho^A(a) = \varrho^A(1a) = \sum_\alpha 1_{\underline{0}} a_\alpha \otimes 1_{\underline{1}}{}^\alpha = \sum (1_{\underline{0}} \otimes 1_{\underline{1}})a$, as expected (cf. 32.16). □

32.19. The Amitsur complex. *Suppose that A is an $(A, C)_\psi$-entwined module. Then the Amitsur complex of $\mathcal{C}$ has the following explicit form: the n-cochains are $\Omega^n(\mathcal{C}) = A \otimes_R C^{\otimes_R n}$, and the coboundary is*

$$\begin{aligned} d^n(a \otimes c_1 \otimes \ldots \otimes c_n) &= \sum a_{\underline{0}} \otimes a_{\underline{1}} \otimes c_1 \otimes \ldots \otimes c_n \\ &\quad + \sum_{i=1}^{n} (-1)^i a \otimes c_1 \otimes \ldots \otimes \Delta(c_i) \otimes \ldots \otimes c_n \\ &\quad + (-1)^{n+1} \sum_{\alpha_1, \ldots, \alpha_n} a 1_{\underline{0}\alpha_n \ldots \alpha_1} \otimes c_1^{\alpha_1} \otimes \ldots \otimes c_n^{\alpha_n} \otimes 1_{\underline{1}}. \end{aligned}$$

Note that the product in $\Omega(\mathcal{C})$ reads

$$\begin{aligned} &(a \otimes c_1 \otimes \ldots \otimes c_m)(a' \otimes c_{m+1} \otimes \ldots \otimes c_{m+n}) \\ &= \sum_{\alpha_1, \ldots, \alpha_m} a a'_{\alpha_m \ldots \alpha_1} \otimes c_1^{\alpha_1} \otimes \ldots \otimes c_m^{\alpha_m} \otimes c_{m+1} \otimes \ldots \otimes c_{m+n}. \end{aligned}$$

Proof. Only the last term in the expression for d^n might require some explanation. This term is obtained by the following chain of identifications:

$$\begin{aligned} &\sum (a \otimes_R c_1) \otimes_A (1 \otimes_R c_2) \otimes_A \ldots \otimes_A (1 \otimes_A c_n) \otimes_A 1_{\underline{0}} \otimes_R 1_{\underline{1}} \\ &= \sum_{\alpha_n} (a \otimes_R c_1) \otimes_A (1 \otimes_R c_2) \otimes_A \ldots \otimes_A 1_{\underline{0}\alpha_n} \otimes_R c_n^{\alpha_n} \otimes_R 1_{\underline{1}} \\ &= \sum_{\alpha_{n-1}, \alpha_n} (a \otimes_R c_1) \otimes_A (1 \otimes_R c_2) \otimes_A \ldots \otimes_A 1_{\underline{0}\alpha_n \alpha_{n-1}} \otimes_R c_{n-1}^{\alpha_{n-1}} \otimes_R c_n^{\alpha_n} \otimes_R 1_{\underline{1}} \\ &\ldots \\ &= \sum_{\alpha_1, \ldots, \alpha_n} a 1_{\underline{0}\alpha_n \ldots \alpha_1} \otimes c_1^{\alpha_1} \otimes \ldots \otimes c_n^{\alpha_n} \otimes 1_{\underline{1}}. \end{aligned}$$

A similar chain leads to the product in $\Omega(\mathcal{C})$. □

Thus various properties of entwined modules can be obtained from the properties of the comodules of the corresponding coring. This allows a significant simplification of the theory of entwined modules, in addition to providing a better conceptual understanding of the latter. On the other hand, the knowledge of some specific properties of entwining structures might lead to new results in the theory of corings. In brief, there is a two-way interaction between corings and entwining structures. The following two examples serve as an illustration how, in view of 32.6, (obvious) $\mathcal{C}$-comodules can produce nontrivial entwined modules, and then how the self-duality property (cf. 32.5) applied to entwined modules obtained in this way can produce nontrivial comodules of a coring.

32.20. C-tensored entwined modules. *$M = A \otimes_R C^{\otimes_R n}$ is an $(A,C)_\psi$-entwined module with coaction $\varrho^M = I_A \otimes I_C^{\otimes n-1} \otimes \Delta$ and right multiplication*

$$(a' \otimes c_1 \otimes \ldots \otimes c_n)a = \sum_{\alpha_1,\ldots,\alpha_n} a' a_{\alpha_n \ldots \alpha_1} \otimes c_1^{\alpha_1} \otimes \ldots \otimes c_n^{\alpha_n}.$$

Proof. $M = \mathcal{C}^{\otimes_A n}$ is a right $\mathcal{C}$-comodule via $\varrho^M = I_{\mathcal{C}}^{\otimes_A n-1} \otimes_A \Delta_{\mathcal{C}}$. Taking $\mathcal{C} = A \otimes_R C$ and using the natural identification

$$\mathcal{C}^{\otimes_A n} = A \otimes_R C \otimes_A A \otimes C \otimes_A \ldots \otimes_A A \otimes_R C \simeq A \otimes_R C^{\otimes_R n},$$

one obtains the $(A,C)_\psi$-entwined module structure on $A \otimes_R C^{\otimes_R n}$, as stated. □

32.21. A-tensored entwined modules. *$M = C \otimes_R A^{\otimes_R n}$ is a right $\mathcal{C}$-comodule with the right A-multiplication $\varrho_M = I_C \otimes I_A^{\otimes n-1} \otimes \mu$ and the right C-coaction*

$$\begin{array}{rccl} \varrho^M : & C \otimes_R A^{\otimes_R n} & \longrightarrow & C \otimes_R A^{\otimes_R n} \otimes_A \mathcal{C} \simeq C \otimes_R A^{\otimes_R n} \otimes C, \\ & c \otimes a^1 \otimes \ldots a^n & \longmapsto & \sum_{\alpha_1,\ldots,\alpha_n} c_{\underline{1}} \otimes a^1_{\alpha_1} \otimes \ldots \otimes a^n_{\alpha_n} \otimes c_{\underline{2}}{}^{\alpha_1 \ldots \alpha_n}. \end{array}$$

Proof. This is an exercise in the self-duality of the notions of an entwining structure and an entwined module explained in 32.5. By 32.20 we know that $A \otimes_R C^{\otimes_R n}$ is an $(A,C)_\psi$-entwined module. By duality, $C \otimes_R A^{\otimes_R n}$ is also an $(A,C)_\psi$-module with the stated structure maps. This is most easily seen if one writes the right A-multiplication in 32.20 as a chain of maps and then reverses the order of composition and makes all the interchanges required for duality. In this way one obtains the map

$$(I_C \otimes I_A^{\otimes n-1} \otimes \psi) \circ (I_C \otimes I_A^{\otimes n-2} \otimes \psi \otimes I_A) \circ \ldots \circ (I_C \otimes \psi \otimes I_A^{\otimes n-1}) \circ (\Delta \otimes I_A^{\otimes n}),$$

which is precisely the asserted coaction. By 32.6, $C \otimes_R A^{\otimes_R n}$ is a right $\mathcal{C}$-comodule, as required. □

32.22. Flat connection in a C-tensored module. *Suppose that A is an $(A,C)_\psi$-entwined module. Then $A \otimes_R C^{\otimes_R n}$, $n = 1, 2, \ldots$, has a flat connection*

$$\nabla(a \otimes c_1 \otimes \ldots \otimes c_n) = a \otimes c_1 \otimes \ldots \otimes \Delta(c_n) - \sum_{\alpha_1,\ldots,\alpha_n} a 1_{\underline{0}\,\alpha_n \ldots \alpha_1} \otimes c_1^{\alpha_1} \otimes \ldots \otimes c_n^{\alpha_n} \otimes 1_{\underline{1}}.$$

Proof. This follows immediately from the fact that $A \otimes_R C^{\otimes_R n}$ is a right $\mathcal{C}$-comodule (or, equivalently, an $(A, C)_\psi$-entwined module) with the structure described in 32.20. □

32.23. Flat connection in an A-tensored module. *Suppose that A is an $(A, C)_\psi$-entwined module. Then $C \otimes_R A^{\otimes_R n}$, $n = 1, 2, \ldots$, has a flat connection*

$$\nabla(c \otimes a^1 \otimes \ldots \otimes a^n) = \sum_{\alpha_1, \ldots, \alpha_n} c_{\underline{1}} \otimes a^1_{\alpha_1} \otimes \ldots \otimes a^n_{\alpha_n} \otimes c_{\underline{2}}{}^{\alpha_1 \ldots \alpha_n} - \sum c \otimes a^1 \otimes \ldots \otimes a^n 1_{\underline{0}} \otimes 1_{\underline{1}}.$$

Proof. The assertion follows from the fact that $C \otimes_R A^{\otimes_R n}$ is a right $\mathcal{C}$-comodule (or, equivalently, an $(A, C)_\psi$-entwined module) with the structure described in 32.21. On the other hand, the result can be deduced from 32.22 by the duality arguments explained in 32.5. We encourage the reader to write out these arguments explicitly in this case. □

32.24. The ψ-equivariant cohomology of entwining structures. The ψ-equivariant cohomology $H_{\psi-e}(C)$ of an entwining structure $(A, C)_\psi$ is defined as the Cartier cohomology of the corresponding coring $\mathcal{C} = A \otimes_R C$ with values in $\mathcal{C}$, that is, $H^n_{\psi-e}(C) = H^n_{\mathrm{Ca}}(A \otimes_R C)$. Using the identifications of homomorphisms of (A, A)-bimodules with the R-submodules of homomorphisms of R-modules, we thus obtain that the n-cochains are given by

$$C^n_{\psi-e}(C) = \{f \in \mathrm{Hom}_R(C, A \otimes_R C^{\otimes_R n}) \mid \forall a \in A,\ c \in C,\ \sum_\alpha a_\alpha f(c^\alpha) = f(c)a\},$$

where the right action of A on the R-module $A \otimes_R C^{\otimes_R n}$ is given by 32.20. By 30.8, $C_{\psi-e}(C)$ is a comp algebra, and consequently $H_{\psi-e}(C)$ is a Gerstenhaber algebra. The composition comes out as

$$f \diamond_i g = \begin{cases} (\mu_i \otimes I_C^{\otimes m+n-i-1}) \circ (I_A \otimes I_C^{\otimes i} \otimes g \otimes I_C^{\otimes m-i-1}) \circ f & \text{if } 0 \leq i < m \\ 0 & \text{otherwise,} \end{cases}$$

for $f \in C^m_{\psi-e}(C)$, $g \in C^n_{\psi-e}(C)$. Here $\mu_i : A \otimes_R C^{\otimes_R i} \otimes_R A \to A \otimes_R C^{\otimes_R i}$ denotes the right A-multiplication described in 32.20. The distinguished elements are $\pi : c \mapsto 1 \otimes \Delta(c)$, $u : c \mapsto 1 \otimes c$ and $1 = \varepsilon$.

The ψ-equivariant cohomology of entwining structures was introduced in [72] (in the dual setup) without reference to corings. Its realisation as a cohomology of a coring leads to significant simplifications.

References. Brzeziński [69, 71, 72]; Brzeziński and Majid [80]; Caenepeel, Militaru and Zhu [9]; Hobst and Pareigis [131]; Takeuchi [195].

33 Entwinings and Hopf-type modules

In this section we collect examples of entwining structures and entwined modules (and thus of corings and comodules) that come from the theory of Hopf-type modules.

33.1. Bialgebra entwining and Hopf modules. *Let B be an R-algebra and an R-coalgebra. Consider an R-linear map*

$$\psi : B \otimes_R B \to B \otimes_R B, \quad b' \otimes b \mapsto \sum b_{\underline{1}} \otimes b' b_{\underline{2}}.$$

(1) *ψ entwines B with B if and only if B is a bialgebra. This entwining structure is known as a* bialgebra entwining.

(2) *An entwined module corresponding to a bialgebra entwining is a right B-module, right B-comodule M such that for all $m \in M$ and $b \in B$,*

$$\varrho^M(mb) = \sum m_{\underline{0}} b_{\underline{1}} \otimes m_{\underline{1}} b_{\underline{2}},$$

that is, the category of entwined modules $\mathbf{M}_B^B(\psi)$ is the same as the category $\mathbf{M}_B^B$ of Hopf modules studied in Section 14.

(3) *If B is a bialgebra, then $\mathcal{C} = B \otimes_R B$ is a B-coring with coproduct $\underline{\Delta} = I_B \otimes \Delta$, counit $\underline{\varepsilon} = I_B \otimes \varepsilon$, and the (B, B)-bimodule structure*

$$b(b' \otimes b'')c = \sum bb' c_{\underline{1}} \otimes b'' c_{\underline{2}},$$

for $b, b', b'', c \in B$; that is, as a right B-module, $\mathcal{C} = B \otimes_R^b B$ (cf. 13.4).

Proof. Suppose $(B, B)_\psi$ is an entwining structure. We need to show that the coproduct Δ and the counit ε of B are algebra maps. Evaluating the left pentagon identity at $1_B \otimes b \otimes b'$ for all $b, b' \in B$, we immediately conclude that Δ is a multiplicative map. Also, evaluating the left triangle at $1_B \otimes 1_B$, we obtain $\Delta(1_B) = 1_B \otimes 1_B$. Thus Δ is an algebra map, as required. Furthermore, the right triangle reads in this case $\sum b_{\underline{1}} \varepsilon(b' b_{\underline{2}}) = \varepsilon(b')b$. Now applying ε we deduce that the counit is a multiplicative map, as required.

Conversely, if B is a bialgebra, then B is a right B-module algebra via the coproduct. Thus the above entwining is a special case of the entwinings in 33.2 and 33.4 and follows from the latter.

The identification $\mathbf{M}_B^B(\psi) = \mathbf{M}_B^B$ and the description of the coring structure of $\mathcal{C}$ are immediate. □

The above example shows that, in a certain sense, an entwining structure is a generalisation of a bialgebra in which the coalgebra structure is separated from the algebra structure. Furthermore, in view of 32.6 and 33.1, the properties of Hopf modules described in Section 14 are simply special cases of general properties of comodules of corings described in Chapters 3 and 4. As an illustration of this look at the properties of comodules of a coring described in Section 18 to realise that 14.3 is a corollary of 18.9, 14.5 is a corollary of 18.13 and 18.10, and 14.6 is a corollary of 18.14 and 18.17. The reader is advised to search for other similar connections between Section 14 and Chapters 3 and 4 (for example, try and see how the theory of coinvariants of Hopf modules follows from the theory of coinvariants of corings with a grouplike element).

33.2. Relative entwining and relative modules. *Let B be an R-bialgebra and let A be a right B-comodule algebra, that is, an R-algebra and a right B-comodule such that the coaction $\varrho^A : A \to A \otimes_R B$ is an algebra map (when $A \otimes_R B$ is viewed as an ordinary tensor algebra). Define an R-linear map*

$$\psi : B \otimes_R A \to A \otimes_R B, \quad b \otimes a \mapsto \sum a_{\underline{0}} \otimes b a_{\underline{1}},$$

where $\sum a_{\underline{0}} \otimes a_{\underline{1}} = \varrho^A(a)$ as usual. Then:

(1) $(A, B)_\psi$ is an entwining structure.

(2) M is an $(A, B)_\psi$-entwined module if and only if it is a right A-module and a right B-comodule such that, for all $m \in M$ and $a \in A$,

$$\varrho^M(ma) = \sum m_{\underline{0}} a_{\underline{0}} \otimes m_{\underline{1}} a_{\underline{1}}.$$

Such a module is known as a relative Hopf module.

(3) $\mathcal{C} = A \otimes_R B$ is an A-coring with coproduct $\underline{\Delta} = I_A \otimes \Delta$, counit $\underline{\varepsilon} = I_A \otimes \varepsilon$, and the (A, A)-bimodule structure, for $a, a', a'' \in A$ and $b \in B$,

$$a''(a \otimes b)a' = \sum a'' a a'_{\underline{0}} \otimes b a'_{\underline{1}}.$$

Note that A itself is an $(A, B)_\psi$-entwined or relative Hopf module via the product and the B-coaction ϱ^A. The corresponding grouplike element in $\mathcal{C}$ comes out as $\varrho^A(1) = \sum 1_{\underline{0}} \otimes 1_{\underline{1}}$.

Proof. This example is a special case of the situation to be considered in 33.4 and hence follows from the latter. □

Using the natural left-right symmetry for modules, comodules, and so on, one can immediately associate a left-left entwining structure to a left comodule algebra of a bialgebra B. This is left for the reader as an exercise.

Using the self-duality of an entwining structure explained in 32.5, one obtains a dual version of 33.2 (and all other examples discussed later, for that matter).

33.3. A dual-relative entwining and $[C,A]$-Hopf modules. *Let A be an R-bialgebra and let C be a right A-module coalgebra, that is, an R-coalgebra and a right A-module such that the A-multiplication $\varrho_C : C \otimes_R A \to C$ is a coalgebra map (when $C \otimes_R A$ is viewed as an ordinary tensor coalgebra as in 2.12). Define an R-linear map*

$$\psi : C \otimes_R A \to A \otimes_R C, \quad c \otimes a \mapsto \sum a_{\underline{1}} \otimes c a_{\underline{2}}.$$

(1) $(A,C)_\psi$ is an entwining structure.

(2) M is an $(A,C)_\psi$-entwined module if and only if it is a right A-module and a right C-comodule such that, for all $m \in M$ and $a \in A$,

$$\varrho^M(ma) = \sum m_{\underline{0}} a_{\underline{1}} \otimes m_{\underline{1}} a_{\underline{2}}.$$

*Such a module is called a $[C,A]$-*Hopf module.

(3) $\mathcal{C} = A \otimes_R C$ is an A-coring with coproduct $\underline{\Delta} = I_A \otimes \Delta$, counit $\underline{\varepsilon} = I_A \otimes \varepsilon$, and the (A,A)-bimodule structure, for $a, a', a'' \in B$ and $c \in C$,

$$a''(a \otimes c)a' = \sum a'' a a'_{\underline{1}} \otimes c a'_{\underline{2}}.$$

Note that C itself is an $(A,C)_\psi$-entwined or $[C,A]$-Hopf module via the coproduct and the A-action.

The above examples are all special cases of the following construction.

33.4. Doi-Koppinen entwinings and Hopf modules. *Let B be an R-bialgebra. Let A be a right B-comodule algebra and let C be a right B-module coalgebra. Define an R-linear map*

$$\psi : C \otimes_R A \to A \otimes_R C, \quad c \otimes a \mapsto \sum a_{\underline{0}} \otimes c a_{\underline{1}},$$

where $\sum a_{\underline{0}} \otimes a_{\underline{1}} = \varrho^A(a) \in A \otimes_R B$ as usual. Then:

(1) $(A,C)_\psi$ is an entwining structure.

(2) M is an $(A,C)_\psi$-entwined module if and only if it is a right A-module and a right C-comodule such that, for all $m \in M$ and $a \in A$,

$$\varrho^M(ma) = \sum m_{\underline{0}} a_{\underline{0}} \otimes m_{\underline{1}} a_{\underline{1}}.$$

Modules satisfying the above condition are known as Doi-Koppinen *or* unifying *Hopf-modules.*

(3) $\mathcal{C} = A \otimes_R C$ is an A-coring with coproduct $\underline{\Delta} = I_A \otimes \Delta$, counit $\underline{\varepsilon} = I_A \otimes \varepsilon$, and the (A,A)-bimodule structure, for $a, a', b \in A$ and $c \in C$,

$$a'(a \otimes c)b = \sum a' a b_{\underline{0}} \otimes c b_{\underline{1}}.$$

(4) $\mathrm{Hom}_R(C, A)$ *is an associative algebra with the product* $*_\psi$ *given by*

$$(f *_\psi f')(c) = \sum f(c_{\underline{2}})_{\underline{0}} f'(c_{\underline{1}} f(c_{\underline{2}})_{\underline{1}}),$$

for all $c \in C$, *and unit* $\iota \circ \varepsilon$. *This algebra structure is also known as* Koppinen's smash product.

Proof. (1) This is shown by direct calculations. For the proof of the left pentagon take $a, a' \in A$ and $c \in C$ and compute

$$\begin{aligned} \textstyle\sum_\alpha (aa')_\alpha \otimes c^\alpha &= \textstyle\sum (aa')_{\underline{0}} \otimes (aa')_{\underline{1}} = \sum a_{\underline{0}} a'_{\underline{0}} \otimes c a_{\underline{1}} a'_{\underline{1}} \\ &= \textstyle\sum_\alpha a_\alpha a'_{\underline{0}} \otimes c^\alpha a'_{\underline{1}} = \sum_{\alpha,\beta} a_\alpha a'_\beta \otimes c^{\alpha\beta}, \end{aligned}$$

where we used the fact that the right coaction of B on A is an algebra map. Similarly, to prove that the left triangle commutes, we compute for all $c \in C$,

$$\textstyle\sum_\alpha 1_\alpha \otimes c^\alpha = \sum 1_{\underline{0}} \otimes c 1_{\underline{1}} = 1 \otimes c,$$

since the assumption that A is a right B-comodule algebra implies that $\varrho^A(1_A) = 1_A \otimes 1_B$. To prove the commutativity of the right pentagon, compute

$$\begin{aligned} \textstyle\sum_\alpha a_\alpha \otimes c^\alpha{}_{\underline{1}} \otimes c^\alpha{}_{\underline{2}} &= \textstyle\sum a_{\underline{0}} \otimes (c a_{\underline{1}})_{\underline{1}} \otimes (c a_{\underline{1}})_{\underline{2}} = \sum a_{\underline{0}} \otimes c_{\underline{1}} a_{\underline{1}} \otimes c_{\underline{2}} a_{\underline{2}} \\ &= \textstyle\sum a_{\underline{00}} \otimes c_{\underline{1}} a_{\underline{01}} \otimes c_{\underline{2}} a_{\underline{1}} = \sum_\alpha a_{\underline{0}\alpha} \otimes c_{\underline{1}}{}^\alpha \otimes c_{\underline{2}} a_{\underline{1}} \\ &= \textstyle\sum_{\alpha,\beta} a_{\beta\alpha} \otimes c_{\underline{1}}{}^\alpha \otimes c_{\underline{2}}{}^\beta, \end{aligned}$$

using the fact that the right multiplication of C by B is a coalgebra map. Finally,

$$\textstyle\sum_\alpha a_\alpha \varepsilon(c^\alpha) = \sum a_{\underline{0}} \varepsilon(c a_{\underline{1}}) = \sum a_{\underline{0}} \varepsilon(c) \varepsilon(a_{\underline{1}}) = a\varepsilon(c),$$

since the assumption that C is a right B-module coalgebra implies that $\varepsilon(cb) = \varepsilon(c)\varepsilon(b)$ for all $c \in C$ and $b \in B$. This shows that the right triangle commutes and thus completes the proof of (1).

Assertions (2), (3) and (4) can be seen immediately by explicitly writing out the condition for an entwined module, the structure of the corresponding coring from 32.6, and the dual algebra from 32.9. □

33.5. Doi-Koppinen datum. A triple (A, B, C) satisfying the conditions of 33.4 is known as a (right-right) *Doi-Koppinen datum* or a *Doi-Koppinen structure*, and the corresponding entwining structure is known as an *entwining structure associated to a Doi-Koppinen datum.*

The investigation of Doi-Koppinen structures and Doi-Koppinen Hopf modules was initiated independently in Doi [106] and Koppinen [144]. Doi-Koppinen data as a separate entity first appeared in [87], and then they were given a separate name in [85] (incidentally, they were called Doi-Hopf data). Various properties and applications of Doi-Koppinen structures are studied in the recent monograph [9]. Finally, the entwining structure associated to a Doi-Koppinen datum was first introduced in [69]. Now, in view of 32.6, numerous properties of Doi-Koppinen Hopf modules described in the literature on the subject (cf. [9] for a review and references) are consequences of the structure theorems for comodules of a coring in 33.4.

Another example of an entwining structure comes from the representation theory of quasi-triangular Hopf algebras (quantum groups).

33.6. Yetter-Drinfeld entwinings and crossed modules. *Let H be a Hopf algebra over R and consider an R-linear map*

$$\psi : H \otimes_R H \to H \otimes_R H, \quad h' \otimes h \mapsto \sum h_{\underline{2}} \otimes (Sh_{\underline{1}})h'h_{\underline{3}},$$

where S is the antipode in H. Then:

(1) *$(H, H)_\psi$ is an entwining structure.*

(2) *M is an $(H, H)_\psi$-entwined module if and only if it is a right H-module and a right H-comodule such that, for all $m \in M$ and $h \in H$,*

$$\varrho^M(mh) = \sum m_{\underline{0}}h_{\underline{2}} \otimes (Sh_{\underline{1}})m_{\underline{1}}h_{\underline{3}}.$$

Such modules M are known as Yetter-Drinfeld *or* crossed *modules (introduced in [217], [179]).*

(3) *$\mathcal{C} = H \otimes_R H$ is an H-coring with coproduct $\underline{\Delta} = I_H \otimes \Delta$, counit $\underline{\varepsilon} = I_H \otimes \varepsilon$, and the H-bimodule structure, for all $h, h', h'', k \in H$,*

$$k(h'' \otimes h')h = \sum kh''h_{\underline{2}} \otimes (Sh_{\underline{1}})h'h_{\underline{3}}.$$

Proof. This is a special case of an entwining structure associated to a Doi-Koppinen datum, as in 33.4. The relevant datum is given by the triple $(A = H, B = H^{op} \otimes H, C = A)$, where the right multiplication by $H^{op} \otimes H$ is given by $h(h' \otimes h'') = h'hh''$, and the right coaction of $H^{op} \otimes H$ on H is $\varrho^H(h) = \sum h_{\underline{2}} \otimes Sh_{\underline{1}} \otimes h_{\underline{3}}$. □

33.7. An alternative Doi-Koppinen entwining. *Let B be an R-bialgebra. Let A be a left B-module algebra, that is, an R-algebra and a left B-module such that the multiplication and unit in A are left B-module maps, where $B \otimes_R A$ is a B-module by the diagonal action (cf. 13.4) and R is a B-module via the counit. Explicitly, we require*

$$b(aa') = \sum (b_{\underline{1}}a)(b_{\underline{2}}a'), \quad b1_A = \varepsilon(b)1_A, \quad \textit{for all } a, a' \in A,\ b \in B.$$

Dually, let C be a left B-comodule coalgebra. Define an R-linear map

$$\psi : C \otimes_R A \to A \otimes_R C, \quad c \otimes a \mapsto \sum c_{\underline{-1}} a \otimes c_{\underline{0}},$$

where $\sum c_{\underline{-1}} \otimes c_{\underline{0}} = {}^C\varrho(c) \in B \otimes_R C$ is the left coaction of B on C. Then:

(1) $(A, C)_\psi$ is an entwining structure.

(2) M is an $(A, C)_\psi$-entwined module if and only if it is a right A-module and a right C-comodule such that, for all $m \in M$ and $a \in A$,

$$\varrho^M(ma) = \sum m_{\underline{0}}(m_{\underline{1}\underline{-1}} a) \otimes m_{\underline{1}\underline{0}}.$$

Note that the first Sweedler indices describe the right C-coaction on M, while the second correspond to the left B-coaction on C.

(3) $\mathcal{C} = A \otimes_R C$ is an A-coring with coproduct $\underline{\Delta} = I_A \otimes \Delta$, counit $\underline{\varepsilon} = I_A \otimes \varepsilon$, and the (A, A)-bimodule structure, for $a, a', a'' \in A$ and $c \in C$,

$$a''(a \otimes c)a' = \sum a'' a c_{\underline{-1}} a' \otimes c_{\underline{0}}.$$

Proof. This is proven by a direct calculation (cf. 33.4). □

A triple (A, B, C) satisfying the conditions of 33.7 is called an *alternative Doi-Koppinen datum.* Although Doi-Koppinen and alternative Doi-Koppinen data provide a rich source of entwining structures, they do not exhaust all possibilities. An example of an entwining structure that does not come from Doi-Koppinen data is described in Exercise 33.8.

33.8. Exercise.

([184]) In this exercise $R = F$ is a field. Take any entwining structure $(A, C)_\psi$ over F, and for $c \in C$ and $\xi \in C^*$ define a map $T_{c,\xi} : A \to A$, $a \mapsto \sum_\alpha a_\alpha \xi(c^\alpha)$.

(i) Let $(A, C)_\psi$ be a Doi-Koppinen entwining associated to a Doi-Koppinen datum (A, B, C) as in 33.4. Fix $c \in C$ and $\xi \in C^*$. Use the Finiteness Theorem 3.16 to show that any element of A is contained in a finite-dimensional $T_{c,\xi}$-invariant subspace of A.

(ii) Now let C be an F-coalgebra spanned as a vector space by e and t with the coproduct $\Delta(e) = e \otimes e$, $\Delta(t) = e \otimes t + t \otimes e$ and the counit $\varepsilon(e) = 1$, $\varepsilon(t) = 0$. Let A be a free algebra with generators x_i, $i \in \mathbb{Z}$. Define a linear map $\psi : C \otimes_R A \to A \otimes_R C$ by

$$\psi(e \otimes a) = a \otimes e, \qquad \psi(t \otimes x_{i_1} x_{i_2} \cdots x_{i_n}) = x_{i_1+1} x_{i_2+1} \cdots x_{i_n+1} \otimes t,$$

for all $a \in A$. Show that $(A, C)_\psi$ is an entwining structure.

(iii) In the setting of (ii), take $c = t$ and $\xi \in C^*$ such that $\xi(t) = 1$. Show that the $T_{t,\xi}$-invariant subspace of A generated by X_0 is infinite-dimensional, and deduce from (i) that $(A, C)_\psi$ is not a Doi-Koppinen entwining.

(iv) Modify (ii)–(iii) by taking a coalgebra spanned by a grouplike element e, and t_i with $\Delta(t_i) = e \otimes t_i + t_i \otimes e$, $i \in \mathbb{Z}$. Construct an entwining that is neither a Doi-Koppinen nor an alternative Doi-Koppinen entwining.

References. Brzeziński [69, 71, 72]; Brzeziński and Majid [80]; Caenepeel, Militaru and Zhu [9, 85]; Doi [105, 106]; Hobst and Pareigis [131]; Koppinen [144, 145]; Radford and Towber [179]; Schauenburg [184]; Takeuchi [195]; Yetter [217].

34 Entwinings and Galois-type extensions

Further examples of entwining structures, and thus of corings not related to Doi-Koppinen data in general, are provided by coalgebra-Galois extensions. In this section we describe such extensions as a generalisation of the Hopf-Galois theory (noncommutative affine schemes of a torsor or a principal fibre bundle). To any coalgebra-Galois extension an entwining structure, hence a coring, is associated. This coring turns out to be a Galois coring (cf. 28.18), and, in particular, the Galois Coring Structure Theorem 28.19 can be used to recover structure theorems for Hopf-Galois extensions and also the Fundamental Theorem of Hopf algebras 15.5. By this means the unifying and simplifying powers of the coring theory are illustrated. We also introduce the notion dual to a coalgebra-Galois extension, termed an *algebra-Galois coextension*, and describe the corresponding entwining structure and coring.

34.1. Coalgebra-Galois extensions. Let C be an R-coalgebra and A an R-algebra and a right C-comodule with coaction $\varrho^A : A \to A \otimes_R C$. Let B be the subalgebra of *coinvariants* of A, $B := \{b \in A \mid \text{for all } a \in A,\ \varrho^A(ba) = b\varrho^A(a)\}$. The extension $B \hookrightarrow A$ is called a *coalgebra-Galois extension* (or a *C-Galois extension*) if the canonical left A-module, right C-comodule map

$$\mathsf{can} : \ A \otimes_B A \to A \otimes C, \qquad a \otimes a' \mapsto a\varrho^A(a'),$$

is bijective. A C-Galois extension $B \hookrightarrow A$ is denoted by $A(B)^C$.

The notion of coalgebra-Galois extensions in the presented form was introduced in [77], following their appearance as generalised principal bundles in [80] (cf. [79] for the quantum group case) and an earlier appearance in the special case of quotients of Hopf algebras in [188]. The need for principal bundles with coalgebras playing the role of a structure group and coming from problems in noncommutative geometry based on quantum groups was a main geometric motivation for the introduction of the notion of a coalgebra-Galois extension. On an algebraic level, a coalgebra-Galois extension is a generalisation of the notion of a Hopf-Galois extension introduced by Kreimer and Takeuchi in [147] (cf. 34.7 below) and later intensively studied in particular by Doi, Takeuchi and Schneider in a series of papers [107], [108], [186], [187], [188] (cf. [37] and [13] for reviews). On the one hand, a Hopf-Galois extension brings the Galois theory in a realm of noncommutative rings. In fact, the introduction of Hopf-Galois extensions was motivated by the work of Chase and Sweedler on Galois theory [12]. On the other hand, a Hopf-Galois extension is a noncommutative generalisation of an affine group scheme corresponding to a torsor or a principal bundle.

A typical example of a coalgebra-Galois extension is provided by coideal subalgebras of Hopf algebras, also known as *quantum homogeneous spaces*.

34.2. Coideal subalgebras. *Let A be a Hopf R-algebra with coproduct Δ, counit ε, antipode S, and A_R flat. Let A_1 be a left A-coideal subalgebra, that is, a subalgebra of A such that $\Delta(A_1) \subset A \otimes_R A_1$ (an A-homogeneous quantum space). Let A_1^+ be the augmentation ideal, $A_1^+ = \mathrm{Ke}\,\varepsilon \cap A_1$. Define the quotient coalgebra $C = A/(A_1^+ A)$. There is a natural right coaction of C on A given as $\varrho^A = (I_A \otimes \pi) \circ \Delta$, where $\pi : A \to C$ is the canonical coalgebra surjection. Let $B = A^{coC} = \{b \in A \mid$ for all $a \in A,\ \varrho^A(ba) = b\varrho^A(a)\}$. Then $B \hookrightarrow A$ is a coalgebra-Galois extension, $A(B)^C$. Furthermore, if A is a faithfully flat left A_1-module, then $A_1 = B$.*

Proof. First observe that C is a quotient coalgebra, that is, $A_1^+ A$ is a coideal in A. Indeed, take any $a \in A_1^+$ and $a' \in A$, and compute

$$\begin{aligned}\Delta(aa') &= \sum (aa')_{\underline{1}} \otimes (aa')_{\underline{2}} = \sum a_{\underline{1}}a'_{\underline{1}} \otimes a_{\underline{2}}a'_{\underline{2}} \\ &= \sum a_{\underline{1}}a'_{\underline{1}} \otimes (a_{\underline{2}} - \varepsilon(a_{\underline{2}}))a'_{\underline{2}} + \sum aa'_{\underline{1}} \otimes a'_{\underline{2}}.\end{aligned}$$

Since A_1 is a left coideal subalgebra, the above calculation shows that

$$\Delta(A_1^+ A) \subset A_1^+ A \otimes_R A + A \otimes_R A_1^+ A,$$

that is, $A_1^+ A$ is a coideal, as required (see 2.4).

We need to construct the inverse for the canonical map can. This is given, for all $a \in A$ and $c \in C$, by

$$\mathsf{can}^{-1}(a \otimes c) = \sum aS(a'_{\underline{1}}) \otimes a'_{\underline{2}}, \quad a' \in \pi^{-1}(c).$$

To prove that this map is well defined (does not depend on the choice of a'), first note that A_1^+ is an ideal in A_1. This implies that $A_1^+ A$ is a left A_1-module, so that it immediately follows from its definition that ϱ^A is a left A_1-linear map. Therefore, $A_1 \subseteq B$. Take any $a \in A$ and $a' \in A_1^+$. Since A_1 is a left comodule subalgebra of A and S is an antialgebra map, we can compute

$$\begin{aligned}\sum S(a'a)_{\underline{1}} \otimes (a'a)_{\underline{2}} &= \sum S(a_{\underline{1}})S(a'_{\underline{1}}) \otimes a'_{\underline{2}}a_{\underline{2}} \\ &= \sum S(a_{\underline{1}})S(a'_{\underline{1}})a'_{\underline{2}} \otimes a_{\underline{2}} \\ &= \sum S(a_{\underline{1}})\varepsilon(a') \otimes a_{\underline{2}} = 0,\end{aligned}$$

where we used the fact that A_1 is a left A-coideal subalgebra of A to obtain the second equality. Since any element of $\pi^{-1}(0)$ is necessarily an R-linear combination of products $a'a$, with $a' \in A_1^+$, we conclude that the map can^{-1} is well defined. Furthermore, it is the inverse to can since, on one hand, for all $a, a' \in A$,

$$\begin{aligned}\mathsf{can}^{-1}(\mathsf{can}(a \otimes a')) &= \sum \mathsf{can}^{-1}(aa'_{\underline{1}} \otimes \pi(a'_{\underline{2}})) \\ &= \sum aa'_{\underline{1}}Sa'_{\underline{2}} \otimes a'_{\underline{3}} = a \otimes a',\end{aligned}$$

while on the other, for all $a \in A$ and $c = \pi(a')$,

$$\begin{aligned} \mathsf{can}(\mathsf{can}^{-1}(a \otimes c)) &= \sum \mathsf{can}(aSa'_{\underline{1}} \otimes a'_{\underline{2}}) = \sum aS(a'_{\underline{1}})a'_{\underline{2}} \otimes \pi(a'_{\underline{3}}) \\ &= a \otimes \pi(a') = a \otimes c. \end{aligned}$$

Clearly can^{-1} is a left A-module map. It is also a right C-comodule map since, for all $c \in C$, $a \in A$ and $a' \in \pi^{-1}(c)$,

$$\begin{aligned} \sum \mathsf{can}^{-1}(a \otimes c_{\underline{1}}) \otimes c_{\underline{2}} &= \sum \mathsf{can}^{-1}(a \otimes \pi(a')_{\underline{1}}) \otimes \pi(a')_{\underline{2}} \\ &= \sum \mathsf{can}^{-1}(a \otimes \pi(a'_{\underline{1}})) \otimes \pi(a'_{\underline{2}}) \\ &= \sum aS(a'_{\underline{1}}) \otimes a'_{\underline{2}} \otimes \pi(a'_{\underline{3}}) \\ &= (I_A \otimes \varrho^A)(\mathsf{can}^{-1}(a \otimes c)), \end{aligned}$$

where the second equality is a consequence of the fact that π is a coalgebra map. Therefore there is a coalgebra-Galois extension $A(B)^C$, as claimed.

We have already observed that $A_1 \subseteq B$. Thus, to prove the second statement, we need to show that this inclusion is in fact an equality provided A is a faithfully flat left A_1-module. This follows from the faithfully flat descent. Note that the coinvariants can also be described as

$$B = \{b \in A \mid b_{\underline{1}} \otimes \pi(b_{\underline{2}}) = b \otimes \pi(1)\}.$$

Indeed, if $b \in B$, then taking $a = 1$ in the definition of coinvariants one immediately obtains $\sum b_{\underline{1}} \otimes \pi(b_{\underline{2}}) = b \otimes \pi(1)$. Conversely, since $A_1^+ A$ is a right A-module, so is C. Thus π is a right A-module map, that is, $\pi(aa') = \pi(a)a'$. Therefore, if $b \in A$ satisfies $\sum b_{\underline{1}} \otimes \pi(b_{\underline{2}}) = b \otimes \pi(1)$, then for all $a \in A$,

$$\begin{aligned} \varrho^A(ba) &= \sum b_{\underline{1}}a_{\underline{1}} \otimes \pi(b_{\underline{2}}a_{\underline{2}}) = \sum b_{\underline{1}}a_{\underline{1}} \otimes \pi(b_{\underline{2}})a_{\underline{2}} \\ &= \sum ba_{\underline{1}} \otimes \pi(1)a_{\underline{2}} = \sum ba_{\underline{1}} \otimes \pi(a_{\underline{2}}) = b\varrho^A(a), \end{aligned}$$

that is, b is in the subalgebra of coinvariants. Now there is the following commutative diagram:

$$\begin{array}{ccccccc} 0 & \longrightarrow & A_1 & \longrightarrow & A & \longrightarrow & A \otimes_{A_1} A \\ & & \downarrow & & \downarrow {\scriptstyle =} & & \downarrow {\scriptstyle \mathsf{can}} \\ 0 & \longrightarrow & B & \longrightarrow & A & \longrightarrow & A \otimes_R C, \end{array}$$

where the maps in the top row are the obvious inclusion and the assignment $a \mapsto 1 \otimes_{A_1} a - a \otimes_{A_1} 1$, and the maps in the bottom row are the obvious inclusion and $a \mapsto \sum a_{\underline{1}} \otimes \pi(a_{\underline{2}}) - a \otimes \pi(1)$. The top row is exact by the faithfully flat descent, and the bottom row is exact by the above argument on the form of the coinvariants (it can be viewed as a defining sequence for coinvariants). Since can is bijective, we conclude that $A_1 = B$. □

Another construction of a similar kind that uses coideals in a dual of a Hopf algebra is presented in [163, Theorem 2.2].

34.3. The translation map. Consider a coalgebra-Galois extension $A(B)^C$. From the geometric point of view, the map

$$\tau : C \to A \otimes_B A, \quad c \mapsto \mathsf{can}^{-1}(1 \otimes c),$$

can be understood as a generalisation of the *translation function* [21], which is crucial for the definition of a principal fibre bundle (cf. [67] for the quantum principal bundle case). Thus τ is termed the *translation map* of a C-Galois extension $A(B)^C$. To denote the action of τ on elements of C we use the Sweedler-like notation

$$\tau(c) = \sum c^{\tilde{1}} \otimes c^{\tilde{2}} \in A \otimes_B A,$$

where summation is understood over a (undisplayed) finite index set. For future reference we gather basic properties of τ.

34.4. Translation Map Lemma. *Let $B \hookrightarrow A$ be a C-Galois extension, $A(B)^C$. The corresponding translation map τ has the following properties. For all $a \in A$, $c \in C$,*

(1) $\sum c^{\tilde{1}} c^{\tilde{2}}{}_{\underline{0}} \otimes c^{\tilde{2}}{}_{\underline{1}} = 1 \otimes c$.

(2) $\sum c^{\tilde{1}} c^{\tilde{2}} = \varepsilon(c)1$.

(3) $\sum a_{\underline{0}} a_{\underline{1}}{}^{\tilde{1}} \otimes a_{\underline{1}}{}^{\tilde{2}} = 1 \otimes a$.

(4) $\sum c^{\tilde{1}} \otimes c^{\tilde{2}}{}_{\underline{0}} \otimes c^{\tilde{2}}{}_{\underline{1}} = \sum c_{\underline{1}}{}^{\tilde{1}} \otimes c_{\underline{1}}{}^{\tilde{2}} \otimes c_{\underline{2}}$.

(5) $\sum c_{\underline{1}}{}^{\tilde{1}} \otimes c_{\underline{1}}{}^{\tilde{2}} c_{\underline{2}}{}^{\tilde{1}} \otimes c_{\underline{2}}{}^{\tilde{2}} = \sum c^{\tilde{1}} \otimes 1 \otimes c^{\tilde{2}}$.

Proof. Assertion (1) follows directly from the definition of the translation map. Assertion (2) follows from (1) by applying $I_A \otimes \varepsilon$. Again using (1), we obtain for all $a \in A$,

$$\begin{aligned} \sum \mathsf{can}(a_{\underline{0}} a_{\underline{1}}{}^{\tilde{1}} \otimes a_{\underline{1}}{}^{\tilde{2}}) &= \sum a_{\underline{0}} a_{\underline{1}}{}^{\tilde{1}} a_{\underline{1}}{}^{\tilde{2}}{}_{\underline{0}} \otimes a_{\underline{1}}{}^{\tilde{2}}{}_{\underline{1}} \\ &= \sum a_{\underline{0}} \otimes a_{\underline{1}} \;=\; \mathsf{can}(1 \otimes a). \end{aligned}$$

Thus, applying can^{-1}, we obtain assertion (3). Again making use of (1), we compute for all $c \in C$,

$$\begin{aligned} \sum \mathsf{can}(c^{\tilde{1}} \otimes c^{\tilde{2}}{}_{\underline{0}}) \otimes c^{\tilde{2}}{}_{\underline{1}} &= \sum c^{\tilde{1}} c^{\tilde{2}}{}_{\underline{0}} \otimes c^{\tilde{2}}{}_{\underline{1}} \otimes c^{\tilde{2}}{}_{\underline{2}} \\ &= \sum 1 \otimes c_{\underline{1}} \otimes c_{\underline{2}} \\ &= \sum c_{\underline{1}}{}^{\tilde{1}} c_{\underline{1}}{}^{\tilde{2}}{}_{\underline{0}} \otimes c_{\underline{1}}{}^{\tilde{2}}{}_{\underline{1}} \otimes c_{\underline{2}} \\ &= \sum \mathsf{can}(c_{\underline{1}}{}^{\tilde{1}} \otimes c_{\underline{1}}{}^{\tilde{2}}) \otimes c_{\underline{2}}, \end{aligned}$$

and hence (4) follows. Finally, applying $I_A \otimes \mathsf{can}^{-1}$ to (4) we derive for all $c \in C$,

$$\sum c^{\tilde{1}} \otimes c^{\tilde{2}}{}_{\underline{0}} c^{\tilde{2}}{}_{\underline{1}}{}^{\tilde{1}} \otimes c^{\tilde{2}}{}_{\underline{1}}{}^{\tilde{2}} = \sum c_{\underline{1}}{}^{\tilde{1}} \otimes c_{\underline{1}}{}^{\tilde{2}} c_{\underline{2}}{}^{\tilde{1}} \otimes c_{\underline{2}}{}^{\tilde{2}}.$$

Now, using (3) on the left-hand side, we immediately obtain assertion (5). □

34.5. Translation map as a morphism of corings. *The translation map of a coalgebra-Galois extension $A(B)^C$ is an object in the category of representations* $\mathbf{Rep}(C : R \mid A \otimes_B A : A)$.

Proof. This follows immediately from 25.7 and 34.4(2) and 34.4(5). □

34.6. Canonical entwining for a coalgebra-Galois extension. *Let A be a C-Galois extension of B. Then there exists a unique entwining map $\psi : C \otimes_R A \to A \otimes_R C$ such that $A \in \mathbf{M}_A^C(\psi)$ by the structure maps μ and ϱ^A. The map ψ is called the* canonical entwining map *associated to a C-Galois extension $B \hookrightarrow A$.*

Proof. Assume that $B \hookrightarrow A$ is a coalgebra-Galois extension and let $\tau : C \to A \otimes_B A$, $\tau(c) = \mathsf{can}^{-1}(1 \otimes c)$, be the corresponding translation map. We use the notation $\tau(c) = \sum c^{\tilde{1}} \otimes c^{\tilde{2}}$ (cf. the translation map lemma 34.4). Define a map $\psi : C \otimes_R A \to A \otimes_R C$ by

$$\psi = \mathsf{can} \circ (I_A \otimes \mu) \circ (\tau \otimes I_A), \quad \psi : c \otimes a \mapsto \sum c^{\tilde{1}}(c^{\tilde{2}}a)_{\underline{0}} \otimes (c^{\tilde{2}}a)_{\underline{1}}. \qquad (*)$$

We show that ψ entwines C and A. By the definition of the translation map (cf. 34.4(1)),

$$\psi(c \otimes 1) = \sum c^{\tilde{1}} c^{\tilde{2}}{}_{\underline{0}} \otimes c^{\tilde{2}}{}_{\underline{1}} = 1 \otimes c.$$

Thus the left triangle in the bow-tie diagram commutes. Furthermore, by property 34.4(2) of the translation map,

$$(I_A \otimes \varepsilon) \circ \psi(c \otimes a) = \sum c^{\tilde{1}}(c^{\tilde{2}}a)_{\underline{0}} \otimes \varepsilon((c^{\tilde{2}}a)_{\underline{1}}) = \sum c^{\tilde{1}} c^{\tilde{2}} a = \varepsilon(c)a,$$

that is, the right triangle in the bow-tie diagram commutes, too. Next we compute

$$\begin{aligned}
&(\mu \otimes I_C) \circ (I_A \otimes \psi) \circ (\psi \otimes I_A)(c \otimes a \otimes a') \\
&\quad = (\mu \otimes I_C) \circ (I_A \otimes \psi)(\sum c^{\tilde{1}}(c^{\tilde{2}}a)_{\underline{0}} \otimes (c^{\tilde{2}}a)_{\underline{1}} \otimes a') \\
&\quad = \sum c^{\tilde{1}}(c^{\tilde{2}}a)_{\underline{0}}(c^{\tilde{2}}a)_{\underline{1}}{}^{\tilde{1}}((c^{\tilde{2}}a)_{\underline{1}}{}^{\tilde{2}}a')_{\underline{0}} \otimes ((c^{\tilde{2}}a)_{\underline{1}}{}^{\tilde{2}}a')_{\underline{1}} \\
&\quad = \sum c^{\tilde{1}}(c^{\tilde{2}}aa')_{\underline{0}} \otimes (c^{\tilde{2}}aa')_{\underline{1}} = \psi \circ (I_C \otimes \mu)(c \otimes a \otimes a'),
\end{aligned}$$

where we used the property 34.4(3) of the translation map to derive the third equality. Hence the left pentagon in the bow-tie diagram commutes. Similarly for the right pentagon,

$$
\begin{aligned}
&(\psi\otimes I_C)\circ(I_C\otimes\psi)\circ(\Delta\otimes I_A)(c\otimes a)\\
&\quad = (\psi\otimes I_C)(\sum c_{\underline{1}}\otimes c_{\underline{2}}{}^{\tilde{1}}(c_{\underline{2}}{}^{\tilde{2}}a)_{\underline{0}}\otimes(c^{\tilde{2}}a)_{\underline{1}})\\
&\quad = \sum c_{\underline{1}}{}^{\tilde{1}}(c_{\underline{1}}{}^{\tilde{2}}c_{\underline{2}}{}^{\tilde{1}}(c_{\underline{2}}{}^{\tilde{2}}a)_{\underline{0}})_{\underline{0}}\otimes(c_{\underline{1}}{}^{\tilde{2}}c_{\underline{2}}{}^{\tilde{1}}(c_{\underline{2}}{}^{\tilde{2}}a)_{\underline{0}})_{\underline{1}}\otimes(c_{\underline{2}}{}^{\tilde{2}}a)_{\underline{1}}\\
&\quad = \sum c^{\tilde{1}}(c^{\tilde{2}}{}_{\underline{0}}c^{\tilde{2}}{}_{\underline{1}}{}^{\tilde{1}}(c^{\tilde{2}}{}_{\underline{1}}{}^{\tilde{2}}a)_{\underline{0}})_{\underline{0}}\otimes(c^{\tilde{2}}{}_{\underline{0}}c^{\tilde{2}}{}_{\underline{1}}{}^{\tilde{1}}(c^{\tilde{2}}{}_{\underline{1}}{}^{\tilde{2}}a)_{\underline{0}})_{\underline{1}}\otimes(c^{\tilde{2}}{}_{\underline{1}}{}^{\tilde{2}}a)_{\underline{1}}\\
&\quad = \sum c^{\tilde{1}}((c^{\tilde{2}}a)_{\underline{0}})_{\underline{0}}\otimes((c^{\tilde{2}}a)_{\underline{0}})_{\underline{1}}\otimes(c^{\tilde{2}}a)_{\underline{1}}\\
&\quad = \sum c^{\tilde{1}}(c^{\tilde{2}}a)_{\underline{0}}\otimes(c^{\tilde{2}}a)_{\underline{1}}\otimes(c^{\tilde{2}}a)_{\underline{2}} \;=\; (I_A\otimes\Delta)\circ\psi(c\otimes a).
\end{aligned}
$$

We used property 34.4(4) of the translation map to derive the third equality and then property 34.4(3) to derive the fourth one. Hence C and A are entwined by ψ, as required. Now, again by 34.4(3), for all $a, a' \in A$,

$$
\begin{aligned}
\sum a_{\underline{0}}\psi(a_{\underline{1}}\otimes a') &= \sum a_{\underline{0}}a_{\underline{1}}{}^{\tilde{1}}(a_{\underline{1}}{}^{\tilde{2}}a')_{\underline{0}}\otimes(a_{\underline{1}}{}^{\tilde{2}}a')_{\underline{1}}\\
&= \sum (aa')_{\underline{0}}\otimes(aa')_{\underline{1}} = \varrho^A(aa'),
\end{aligned}
$$

that is, A is an entwined $(A,C)_\psi$-module with structure maps μ and ϱ^A. It remains to prove the uniqueness of the entwining map ψ given by $(*)$. Suppose that there is an entwining map $\tilde{\psi}$ such that $A \in \mathbf{M}_A^C(\tilde{\psi})$ with structure maps μ and ϱ^A. Then, for all $a \in A$, $c \in C$,

$$
\psi(c\otimes a) = \sum c^{\tilde{1}}(c^{\tilde{2}}a)_{\underline{0}}\otimes(c^{\tilde{2}}a)_{\underline{1}} = \sum c^{\tilde{1}}c^{\tilde{2}}{}_{\underline{0}}\tilde{\psi}(c^{\tilde{2}}{}_{\underline{1}}\otimes a) = \tilde{\psi}(c\otimes a),
$$

using the definition of the translation map to obtain the last equality. □

34.7. Canonical entwining associated to a Hopf-Galois extension. Let H be a Hopf algebra and let $B \hookrightarrow A$ be a Hopf-Galois extension, that is, A is a right H-comodule algebra, $B = A^{coH} = \{b \in A \mid \varrho^A(b) = b \otimes 1_H\}$, and the canonical map

$$
\mathsf{can} : A\otimes_B A \to A\otimes_R H, \quad a\otimes a' \mapsto a\varrho^A(a'),
$$

is bijective. The canonical entwining structure $(A,H)_\psi$ associated to $B \hookrightarrow A$ is given by

$$
\psi : H\otimes_R A \to A\otimes_R H, \qquad h\otimes a \mapsto \sum a_{\underline{0}}\otimes ha_{\underline{1}}
$$

and thus coincides with the one described in 33.2. The category of right comodules of the corresponding A-coring $\mathcal{C} = A \otimes_R H$ is isomorphic to the category of (A,H)-relative Hopf modules in 33.2.

34.8. Canonical entwining associated to a coideal subalgebra. Let $A(B)^C$ be a coalgebra-Galois extension associated to a left coideal subalgebra of a Hopf algebra A with A_R flat as described in 34.2. Let $\pi : A \to C$ be the canonical surjection. The translation map in this case is $\tau(c) = \sum Sa'_{\underline{1}} \otimes a'_{\underline{2}}$, where $a' \in \pi^{-1}(c)$. Therefore the canonical entwining structure $(A, C)_\psi$ comes out as

$$\begin{aligned}\psi(c \otimes a) &= \sum S(a'_{\underline{1}})(a'_{\underline{2}}a)_{\underline{1}} \otimes \pi((a'_{\underline{2}}a)_{\underline{2}}) \\ &= \sum S(a'_{\underline{1}})a'_{\underline{2}}a_{\underline{1}} \otimes \pi(a'_{\underline{3}}a_{\underline{2}}) = \sum a_{\underline{1}} \otimes \pi(a'a_{\underline{2}}).\end{aligned}$$

Now note that A is a right A-comodule algebra via the coproduct. Furthermore, C is a right A-module with the multiplication $ca = \pi(a'a)$, where $a' \in \pi^{-1}(c)$. Thus we can write

$$\psi(c \otimes a) = \sum a_{\underline{1}} \otimes ca_{\underline{2}}.$$

Since π is a coalgebra map, C is a right A-module coalgebra. Thus there is a Doi-Koppinen datum (A, A, C), and the canonical entwining structure is an entwining associated to this Doi-Koppinen datum as in 33.4. Right comodules of the corresponding coring $A \otimes_R C$ are therefore Doi-Koppinen modules as considered in 33.4.

34.9. Canonical entwining corings vs. Sweedler corings. *Let $B \hookrightarrow A$ be a coalgebra-Galois extension $A(B)^C$, and let $\psi : C \otimes_R A \to A \otimes_R C$ be the canonical entwining map. View $A \otimes_B A$ as the Sweedler A-coring associated to the extension $B \hookrightarrow A$, and view $A \otimes_R C$ as an A-coring associated to the entwining structure $(A, C)_\psi$ as explained in 32.6. Then the canonical map $\mathsf{can} : A \otimes_B A \to A \otimes_R C$ is an isomorphism of A-corings.*

Proof. By the definition of a C-Galois extension, the canonical map is bijective. By construction, it is a left A-module map. Thus we need to show that can is also a right A-module map with respect to the prescribed right multiplications and that it respects coring structures. Take any $a, a', a'' \in A$, and use the fact that A is an $(A, C)_\psi$-entwined module to compute

$$\begin{aligned}\mathsf{can}(a \otimes a'a'') &= a\varrho^A(a'a'') = \sum\nolimits_\alpha aa'_{\underline{0}}a''_\alpha \otimes a'_{\underline{1}}{}^\alpha \\ &= \sum (aa'_{\underline{0}} \otimes a'_{\underline{1}})a'' = \mathsf{can}(a \otimes a')a'',\end{aligned}$$

where $\psi(c \otimes a) = \sum_\alpha a_\alpha \otimes c^\alpha$ is the α-notation for an entwining map. From this we conclude that can is a right A-module map. Next, since A is an $(A, C)_\psi$-entwined module,

$$\varrho^A(a) = \varrho^A(1a) = \sum 1_{\underline{0}}a_\alpha \otimes 1_{\underline{1}}{}^\alpha.$$

This formula then facilitates the following calculation for all $a, a' \in A$:

$$\begin{aligned}\mathsf{can}(a \otimes 1) \otimes_A \mathsf{can}(1 \otimes a') &= \sum (a1_{\underline{0}} \otimes 1_{\underline{1}}) \otimes_A a'_{\underline{0}} \otimes a'_{\underline{1}} \\ &= \sum (a1_{\underline{0}} \otimes 1_{\underline{1}}) a'_{\underline{0}} \otimes a'_{\underline{1}} \\ &= \sum\nolimits_{\alpha} a1_{\underline{0}} a'_{\underline{0}\alpha} \otimes 1_{\underline{1}}{}^{\alpha} \otimes a'_{\underline{1}} \\ &= \sum aa'_{\underline{0}} \otimes a'_{\underline{1}} \otimes_A 1 \otimes a'_{\underline{2}} \\ &= \underline{\Delta}_{A\otimes C} \circ \mathsf{can}(a \otimes_B a').\end{aligned}$$

Finally, $\underline{\varepsilon}_{A\otimes C} \circ \mathsf{can}(a \otimes a') = \sum aa'_{\underline{0}}\varepsilon(a'_{\underline{1}}) = aa' = \underline{\varepsilon}_{A\otimes_B A}(a \otimes a')$.

This completes the proof that can is a coring map. Since it is bijective, can^{-1} is also an A-coring morphism, thus proving the proposition. □

34.10. $A \otimes_R C$ as a Galois coring. *Let $(A, C)_\psi$ be an entwining structure over R, and let $\mathcal{C} = A \otimes_R C$ be the associated A-coring as in 32.6. Then $\mathcal{C}$ is a Galois A-coring if and only if $(A, C)_\psi$ is the canonical entwining structure of a C-Galois extension $B \hookrightarrow A$.*

Proof. If $\mathcal{C}$ corresponds to the canonical entwining structure of a C-Galois extension $B \hookrightarrow A$, then A is an $(A, C)_\psi$-entwined module, and thus $\mathcal{C}$ has a grouplike element $g = \varrho^A(1)$ by 32.16 and $B = A_g^{co\mathcal{C}}$ by 32.17. By 34.9, $A \otimes_B A \simeq A \otimes_R C$ as A-corings via the canonical map can. Note that $\mathsf{can}(1 \otimes 1) = \varrho^A(1) = g$. Hence $(\mathcal{C}, g)$ is a Galois coring. Conversely, if $(\mathcal{C}, g)$ is a Galois coring, then A is a right $\mathcal{C}$-comodule, and by the correspondence in 32.6 it is an $(A, C)_\psi$-module. The corresponding grouplike in $\mathcal{C}$ is $g = \varrho^A(1) = \sum 1_{\underline{0}} \otimes 1_{\underline{1}}$. Furthermore, $A_g^{co\mathcal{C}} = \{b \in A \mid \varrho^A(b) = \sum b1_{\underline{0}} \otimes 1_{\underline{1}}\} = B$, since $A \in \mathbf{M}_A^C(\psi)$ (cf. 32.17). For the same reason the A-coring isomorphism $\mathsf{can}_A : A \otimes_B A \to A \otimes_R C$ (see 28.18) explicitly reads

$$\mathsf{can}_A(a \otimes_B a') = a(\sum 1_{\underline{0}} \otimes 1_{\underline{1}})a' = \sum\nolimits_\alpha a1_{\underline{0}}a'_\alpha \otimes 1_{\underline{1}}{}^\alpha = \sum aa'_{\underline{0}} \otimes a'_{\underline{1}}$$

and thus coincides with the canonical map can. This proves that $B \hookrightarrow A$ is a C-Galois extension and, by the uniqueness of the canonical entwining structure (cf. 34.6), $(A, C)_\psi$ must be the canonical entwining structure associated to $B \hookrightarrow A$. □

34.11. Structure theorem for coalgebra-Galois extensions. *For an entwining structure $(A, C)_\psi$ over R, with C flat as an R-module, the following are equivalent:*

(a) $A(B)^C$ is a C-Galois extension with the canonical entwining map ψ and A is faithfully flat as a left B-module.

(b) $A \in \mathbf{M}_A^C(\psi)$ *and the induction functor* $- \otimes_B A : \mathbf{M}_B \to \mathbf{M}_A^C(\psi)$ *is an equivalence of categories.*

Proof. This follows immediately from 28.19 and 34.10 □

Theorem 34.11 shows that 28.19 is the origin of an important structure theorem in Hopf-Galois and coalgebra-Galois extensions theory. As a special case of 34.11 we obtain a

34.12. Structure theorem of Hopf-Galois extensions. *Let H be a Hopf algebra, and let $C = H/I$ be a quotient coalgebra and a right H-module (cf. Example 34.2). Let $\pi : H \to C$ denote the canonical surjection. Let A be a right H-comodule algebra with coaction $\varrho : A \to A \otimes_R H$ and view it as a right C-comodule via $\varrho^A = (I_A \otimes \pi) \circ \varrho$. Thus (A, H, C) is a Doi-Koppinen datum, and we denote by $\mathbf{M}_A^C(H)$ the corresponding category of Doi-Koppinen Hopf modules (cf. Example 33.4). Let $B = \{b \in A \mid \varrho^A(b) = b \otimes \pi(1)\}$ and suppose that C is R-flat. Then the following are equivalent:*

(a) *A is a faithfully flat left B-module and A is a coalgebra-Galois extension of B by C.*

(b) *The induction functor $- \otimes_B A : \mathbf{M}_B \to \mathbf{M}_A^C(H)$ is an equivalence of categories.*

To any coalgebra-Galois extension $A(B)^C$ one can associate a B-coring, provided certain flatness-type conditions are satisfied. This coring can be viewed as a generalisation of a *gauge* or *Ehresmann* groupoid that is associated to a principal fibre bundle (cf. [32]).

34.13. The Ehresmann coring. *Given a C-Galois extension $B \hookrightarrow A$ with translation map τ, consider a (B, B)-bimodule*

$$\mathcal{C} = \{\sum_i a^i \otimes \tilde{a}^i \in A \otimes_R A \mid \sum_i a^i{}_{\underline{0}} \otimes \tau(a^i{}_{\underline{1}})\tilde{a}^i = \sum_i a^i \otimes \tilde{a}^i \otimes_B 1\}.$$

If A is faithfully flat as a left B-module, then $\mathcal{C}$ is a B-coring with the coproduct and counit

$$\underline{\Delta}(\sum_i a^i \otimes \tilde{a}^i) = \sum_i a^i{}_{\underline{0}} \otimes \tau(a^i{}_{\underline{1}}) \otimes \tilde{a}^i, \quad \underline{\varepsilon}(\sum_i a^i \otimes \tilde{a}^i) = \sum_i a^i \tilde{a}^i.$$

The B-coring $\mathcal{C}$ is called the Ehresmann *or* gauge *coring and is denoted by $\mathcal{E}(A(B)^C)$.*

Proof. Clearly, $\mathcal{C}$ is a (B, B)-bimodule. Directly from the definitions one finds that $\underline{\Delta}$ and $\underline{\varepsilon}$ are both B-bilinear (in the case of $\underline{\Delta}$ one uses that ϱ^A is left B-linear by the construction of B). First we need to show that $\underline{\Delta}$ is well defined.

Let ψ be the canonical entwining map associated to $A(B)^C$. Then, for any (B,B)-bimodule M, one can view $M\otimes_B A\otimes_R A$ as an $(A,C)_\psi$-entwined module with right coaction $\varrho^{M\otimes_B A\otimes_R A}(m\otimes a\otimes a') = \sum m\otimes a_{\underline{0}}\otimes\psi(a_{\underline{1}}\otimes a')$ and A-multiplication $(m\otimes a\otimes a')a'' = m\otimes a\otimes a'a''$. By applying $I_A\otimes$ can to the defining relation of $\mathcal{C}$ and using the definition of ψ in terms of τ, one easily finds that

$$\begin{aligned}\mathcal{C} &= (A\otimes_R A)^{coC}\\ &= \{\textstyle\sum_i a^i\otimes\tilde{a}^i\in A\otimes_R A \mid \sum_i a^i{}_{\underline{0}}\otimes\psi(a^i{}_{\underline{1}}\otimes\tilde{a}^i) = \sum_i a^i\otimes\tilde{a}^i\varrho^A(1)\}.\end{aligned}$$

Since A is a faithfully flat left B-module, 28.19 and 34.10 imply that the functors $-\otimes_B A$, $(-)^{coC}$ are inverse equivalences. In particular, the morphism in $\mathbf{M}_A^C(\psi)$,

$$\theta_M : M\otimes_B A\otimes_R A\to(M\otimes_B A\otimes_R A)^{coC}\otimes_B A,\ m\otimes a\otimes a'\mapsto\sum m\otimes a_{\underline{0}}\otimes\tau(a_{\underline{1}})a',$$

is an isomorphism with the inverse $\sum_i m^i\otimes a^i\otimes\tilde{a}^i\otimes a\mapsto\sum_i m^i\otimes a^i\otimes\tilde{a}^i a$. Clearly this is also a left B-module isomorphism. Choosing $M = B$, we obtain the isomorphism $\mathcal{C}\otimes_B A\simeq A\otimes_R A$. The form of θ_B immediately confirms that for all $c\in\mathcal{C}$, $\underline{\Delta}(c)\in\mathcal{C}\otimes_B A\otimes_R A$. Next note that in fact $\underline{\Delta}(c)\in(\mathcal{C}\otimes_B A\otimes_R A)^{coC}$ since, for all $c = \sum_i a^i\otimes\tilde{a}^i$,

$$\begin{aligned}\textstyle\sum_i a^i{}_{\underline{0}}\otimes a^i{}_{\underline{1}}{}^{\tilde{1}}\otimes a^i{}_{\underline{1}}{}^{\tilde{2}}{}_{\underline{0}}\otimes\psi(a^i{}_{\underline{1}}{}^{\tilde{2}}{}_{\underline{1}}\otimes\tilde{a}^i) &= \textstyle\sum_i a^i{}_{\underline{0}}\otimes\tau(a^i{}_{\underline{1}})\otimes\psi(a^i{}_{\underline{2}}\otimes\tilde{a}^i)\\ &= \textstyle\sum_i a^i{}_{\underline{0}}\otimes\tau(a^i{}_{\underline{1}})\otimes\tilde{a}^i\varrho^A(1),\end{aligned}$$

where we used 34.4(4) to derive the second equality. Now take $M = \mathcal{C}$ and use $\theta_{\mathcal{C}}$ and θ_B to derive the following chain of isomorphisms in $\mathbf{M}_A^C(\psi)$:

$$(\mathcal{C}\otimes_B A\otimes_R A)^{coC}\otimes_B A\simeq\mathcal{C}\otimes_B A\otimes_R A\simeq\mathcal{C}\otimes_B\mathcal{C}\otimes_B A.$$

Since ${}_BA$ is a faithfully flat module we conclude that $(\mathcal{C}\otimes_B A\otimes_R A)^{coC}\simeq\mathcal{C}\otimes_B\mathcal{C}$, that is, $\underline{\Delta}(\mathcal{C})\subseteq\mathcal{C}\otimes_B\mathcal{C}$, as required.

Now it remains to be proven that $\underline{\Delta}$ is coassociative and that $\underline{\varepsilon}$ is a counit. For the former take $c = \sum_i a^i\otimes\tilde{a}^i\in\mathcal{C}$ as before and compute

$$\begin{aligned}(\underline{\Delta}\otimes I_{\mathcal{C}})\circ\underline{\Delta}(c) &= \textstyle\sum_i\underline{\Delta}(a^i{}_{\underline{0}}\otimes a^i{}_{\underline{1}}{}^{\tilde{1}})\otimes a^i{}_{\underline{1}}{}^{\tilde{2}}\otimes\tilde{a}^i\\ &= \textstyle\sum_i a^i{}_{\underline{0}}\otimes a^i{}_{\underline{1}}{}^{\tilde{1}}\otimes a^i{}_{\underline{1}}{}^{\tilde{2}}\otimes a^i{}_{\underline{2}}{}^{\tilde{1}}\otimes a^i{}_{\underline{2}}{}^{\tilde{2}}\tilde{a}^i\\ &= \textstyle\sum_i a^i{}_{\underline{0}}\otimes\tau(a^i{}_{\underline{1}})\otimes\tau(a^i{}_{\underline{2}})\otimes\tilde{a}^i.\end{aligned}$$

On the other hand,

$$\begin{aligned}(I_{\mathcal{C}}\otimes\underline{\Delta})\circ\underline{\Delta}(c) &= \textstyle\sum_i a^i{}_{\underline{0}}\otimes a^i{}_{\underline{1}}{}^{\tilde{1}}\otimes\underline{\Delta}(a^i{}_{\underline{1}}{}^{\tilde{2}}\otimes\tilde{a}^i)\\ &= \textstyle\sum_i a^i{}_{\underline{0}}\otimes a^i{}_{\underline{1}}{}^{\tilde{1}}\otimes a^i{}_{\underline{1}}{}^{\tilde{2}}{}_{\underline{0}}\otimes a^i{}_{\underline{1}}{}^{\tilde{2}}{}_{\underline{1}}{}^{\tilde{1}}\otimes a^i{}_{\underline{1}}{}^{\tilde{2}}{}_{\underline{1}}{}^{\tilde{2}}\otimes\tilde{a}^i\\ &= \textstyle\sum_i a^i{}_{\underline{0}}\otimes\tau(a^i{}_{\underline{1}})\otimes\tau(a^i{}_{\underline{2}})\otimes\tilde{a}^i,\end{aligned}$$

where 34.4(4) was used to derive the last equality. This proves the coassociativity of $\underline{\Delta}$. The counit property of $\underline{\varepsilon}$ is verified by the following simple calculations, which use 34.4(3) and 34.4(2), respectively:

$$(\underline{\varepsilon} \otimes I_{\mathcal{C}}) \circ \underline{\Delta}(c) = \sum_i a^i{}_{\underline{0}} a^i{}_{\underline{1}}{}^{\tilde{1}} \otimes a^i{}_{\underline{1}}{}^{\tilde{2}} \otimes \tilde{a}^i = \sum_i a^i \otimes \tilde{a}^i = c,$$

where we used that $c = \sum_i a^i \otimes \tilde{a}^i$ is in $\mathcal{C}$, and

$$(I_{\mathcal{C}} \otimes \varepsilon_{\mathcal{C}}) \circ \underline{\Delta}(c) = \sum_i a^i{}_{\underline{0}} \otimes a^i{}_{\underline{1}}{}^{\tilde{1}} a^i{}_{\underline{1}}{}^{\tilde{2}} \tilde{a}^i = \sum_i a^i{}_{\underline{0}} \varepsilon(a^i{}_{\underline{1}}) \otimes \tilde{a}^i = \sum_i a^i \otimes \tilde{a}^i = c.$$

This completes the proof of the theorem. □

In the case of a Hopf-Galois extension, the Ehresmann coring of 34.13 is in fact a bialgebroid (see 31.6 for the definition of a bialgebroid).

34.14. The Ehresmann-Schauenburg bialgebroid. *Let H be a Hopf algebra and let $B \hookrightarrow A$ be a Hopf-Galois extension (cf. 34.7). Suppose that A is a faithfully flat left B-module and let $\mathcal{E}(A(B)^H)$ be the associated Ehresmann coring (cf. 34.13). Then $\mathcal{E}(A(B)^H)$ is a subalgebra of A^e, and it is a B-bialgebroid with the source map $s : a \mapsto a \otimes 1$ and the target map $t : a \mapsto 1 \otimes \bar{a}$.*

Proof. In view of 34.7, the right coaction of H on $A \otimes_R A$ comes out as $\varrho^{A \otimes_R A}(a \otimes a') = \sum a_{\underline{0}} \otimes a'_{\underline{0}} \otimes a_{\underline{1}} a'_{\underline{1}}$. Since the right coaction of H on A is an algebra map, one easily checks that the coinvariants $(A \otimes_R A)^{coH}$ are a subalgebra of A^e. Then clearly s and t make $\mathcal{E}(A(B)^H)$ into an A^e-ring, and condition 31.6(3) is checked by a routine calculation. □

The notion of a coalgebra-Galois extension can be dualised. As a result, one obtains the notion of an algebra-Galois coextension to which one can associate a unique, canonical entwining and thus also a coring. We now describe this construction following [77]. In the rest of this section, $R = F$ is a (commutative) field and all algebras, coalgebras, and so on, are over F. Any unadorned tensor product of vector spaces is also over F.

34.15. Coalgebra coextensions. *Let A be an F-algebra, and let C be an F-coalgebra and a right A-module with A-multiplication $\varrho_C : C \otimes A \to C$. Denote $C^* = \mathrm{Hom}_F(C, F)$. Let J be a subspace of C defined as*

$$J = \mathrm{span}\{\sum (ca)_{\underline{1}} \xi((ca)_{\underline{2}}) - \sum c_{\underline{1}} \xi(c_{\underline{2}} a) \mid a \in A, c \in C, \xi \in C^*\}.$$

Let $B = C/J$ and $\pi : C \to B/J$ be the canonical surjection. Then:

(1) J is a coideal in C, and hence B is a coalgebra.

(2) View C as a left B-comodule with coaction ${}^C\varrho = (\pi \otimes I_C) \circ \Delta$ and view $C \otimes A$ as a left C comodule via ${}^C\varrho \otimes I_A$. Then ϱ_C is a left C-comodule map.

(3) $\operatorname{Im}((I_C \otimes \varrho_C) \circ (\Delta \otimes I_A)) \subseteq C\Box_B C$.

Proof. (1) By 6.3(4), it suffices to prove that $D := \{\zeta \in C^* \mid \zeta(J) = 0\}$ is a subalgebra of the dual algebra C^* (cf. 1.3 for the definition of the convolution product and dual algebra). Note that $\varepsilon \in D$. Furthermore,

$$D = \{\zeta \in C^* \mid (\zeta * \xi)(ca) = \sum \zeta(c_{\underline{1}})\xi(c_{\underline{2}}a), \text{ for all } a \in A,\, c \in C,\, \xi \in C^*\}.$$

If $\zeta, \zeta' \in D$, then for all $a \in A$, $c \in C$ and $\xi \in C^*$,

$$\begin{aligned} ((\zeta * \zeta') * \xi)(ca) &= (\zeta * (\zeta' * \xi))(ca) = \sum \zeta(c_{\underline{1}})(\zeta' * \xi)(c_{\underline{2}}a) \\ &= \sum \zeta(c_{\underline{1}})\zeta'(c_{\underline{2}})\xi(c_{\underline{3}}a) = \sum (\zeta * \zeta')(c_{\underline{1}})\xi(c_{\underline{2}}a). \end{aligned}$$

Hence $\zeta * \zeta' \in D$, and D is a subalgebra of C^*, as needed.

(2) Choose $a \in A$, $c \in C$. The fact that ϱ_C is a morphism of left B-comodules can be written as $\sum \pi((ca)_{\underline{1}}) \otimes (ca)_{\underline{2}} = \sum \pi(c_{\underline{1}}) \otimes c_{\underline{2}}a$, which is equivalent to the condition

$$\sum \pi((ca)_{\underline{1}})\xi((ca)_{\underline{2}}) = \sum \pi(c_{\underline{1}})\xi(c_{\underline{2}}a), \quad \text{for all } \xi \in C^*.$$

This proves the assertion.

(3) Observe that this assertion can be stated as

$$\sum c_{\underline{1}} \otimes \pi(c_{\underline{2}}) \otimes c_{\underline{3}}a = \sum c_{\underline{1}} \otimes \pi((c_{\underline{2}}a)_{\underline{1}}) \otimes (c_{\underline{2}}a)_{\underline{2}}.$$

This is true by the left B-colinearity argument in part (2) of the proposition applied to the last two tensorands. □

34.16. Algebra-Galois coextensions. Let A be an algebra, C a coalgebra and right A-module with the A-multiplication ϱ_C, and $B = C/J$, where J is the coideal of 34.15. We say that C is a (right) *algebra-Galois coextension* (or A-Galois coextension) of B if and only if the canonical left C-comodule right A-module map

$$\overline{\mathsf{can}} = (I_C \otimes \varrho_C) \circ (\Delta \otimes I_A) : C \otimes A \to C\Box_B C$$

is bijective. An A-Galois coextension $C \to B$ is denoted by $C(B)_A$.

Notice that the map $\overline{\mathsf{can}}$ is well defined by 34.15.

34.17. Entwining structures and algebra-Galois coextensions. *Let C be an A-Galois coextension of B, $C(B)_A$. Then there exists a unique map $\psi : C \otimes A \to A \otimes C$ entwining C with A, such that $C \in \mathbf{M}_A^C(\psi)$ with the structure maps Δ and ϱ_C. The map ψ is called the* canonical entwining map *associated to an A-Galois coextension $C \to B$.*

Proof. This is shown by dualising the proof of 34.6, details are left to the reader. We only mention that, for any A-Galois coextension $C \to B$, one can define a *cotranslation map* $\check{\tau} : C\Box_B C \to A$, $\check{\tau} = (\varepsilon \otimes I_A) \circ \overline{\mathsf{can}}^{-1}$. By dualising the properties of the translation map listed in the Translation Map Lemma 34.4 (or directly from the definition of $\check{\tau}$), one can establish the following properties of the cotranslation map:

(i) $\check{\tau} \circ \Delta = \iota \circ \varepsilon$.

(ii) $\varrho_C \circ (I_C \otimes \check{\tau}) \circ (\Delta \otimes I_C) = \varepsilon \otimes I_C$ on $C\Box_B C$.

(iii) $\check{\tau} \circ (I_C \otimes \varrho_C) = \mu \circ (\check{\tau} \otimes I_A)$ on $C\Box_B C \otimes A$.

These can be used to prove that a map $\psi : C \otimes A \to A \otimes C$ given by

$$\psi = (\check{\tau} \otimes I_C) \circ (I_C \otimes \Delta) \circ \overline{\mathsf{can}}, \quad \psi(c \otimes a) = \sum \check{\tau}(c_{\underline{1}}, (c_{\underline{2}}a)_{\underline{1}}) \otimes (c_{\underline{2}}a)_{\underline{2}},$$

is the required entwining. □

Thus, in view of 32.6 and 34.17, given an algebra-Galois coextension $C(B)_A$ there is an associated A-coring $\mathcal{C} = A \otimes C$. The (A, A)-bimodule structure in $\mathcal{C}$ is given by

$$a(a' \otimes c)a'' = \sum aa'\check{\tau}(c_{\underline{1}}, (c_{\underline{2}}a'')_{\underline{1}}) \otimes (c_{\underline{2}}a'')_{\underline{2}},$$

where $\check{\tau}$ is the cotranslation map. The coproduct and counit are $I_A \otimes \Delta$ and $I_A \otimes \varepsilon$, respectively.

34.18. Algebra-Galois coextension and coideal subalgebras. *Let C be a Hopf algebra with coproduct Δ, counit ε, and the antipode S. Let A be a right C-coideal subalgebra of C, that is, a subalgebra of C such that $\Delta(A) \subseteq A \otimes C$. View C as a right A-module via the product in C, and suppose that A is a faithfully flat right A-module. Consider the coideal $J \subset C$ as defined in 34.15 and the corresponding quotient coalgebra $B = C/J$. Then $C \to B$ is an A-Galois coextension, and the canonical entwining map is*

$$\psi(c \otimes a) = \sum a_{\underline{1}} \otimes ca_{\underline{2}}, \quad \textit{for all } a \in A,\ c \in C,$$

that is, it coincides with the entwining constructed in 33.2 (note that A is a right C-comodule algebra here, since Δ is an algebra map).

Proof. First note that $J = CA^+$, where $A^+ = A \cap \operatorname{Ke} \varepsilon$ is the augmentation ideal. Indeed, taking $\xi = \varepsilon$ in the definition of J, and using the fact that the coproduct and counit in C are algebra maps, one immediately concludes that $CA^+ \subseteq J$. Conversely, any element of J is an F-linear combination of elements $x = \sum (ca)_{\underline{1}}\xi((ca)_{\underline{2}}) - \sum c_{\underline{1}}\xi(c_{\underline{2}}a)$, for some $a \in A$, $c \in C$ and $\xi \in C^*$. By the counit property of ε, this x can also be written as

$$x = \sum (ca)_{\underline{1}}\xi((ca)_{\underline{2}}) - \sum c_{\underline{1}}\varepsilon((c_{\underline{2}}a)_{\underline{1}})\xi((c_{\underline{2}}a)_{\underline{2}}).$$

Since C is a Hopf algebra, both Δ and ε are multiplicative maps, so that

$$\begin{aligned} x &= \sum c_{\underline{1}}a_{\underline{1}}\xi(c_{\underline{2}}a_{\underline{2}}) - \sum c_{\underline{1}}\varepsilon(c_{\underline{2}}a_{\underline{1}})\xi(c_{\underline{3}}a_{\underline{2}}) \\ &= \sum c_{\underline{1}}a_{\underline{1}}\xi(c_{\underline{2}}a_{\underline{2}}) - \sum c_{\underline{1}}\varepsilon(c_{\underline{2}})\varepsilon(a_{\underline{1}})\xi(c_{\underline{3}}a_{\underline{2}}) \\ &= \sum c_{\underline{1}}(a_{\underline{1}} - \varepsilon(a_{\underline{1}}))\xi(c_{\underline{2}}a_{\underline{1}}). \end{aligned}$$

Since $\Delta(A) \subset A \otimes C$, we conclude that $x \in CA^+$, as needed. Thus $B = C/(CA^+)$. In particular, the canonical coalgebra surjection $\pi : C \to B$ is a left C-module map. If C is a faithfully flat right A-module, then by the same argument as in 34.2, we obtain the identification

$$A = {}^{coB}C = \{a \in C \mid \sum \pi(a_{\underline{1}}) \otimes a_{\underline{2}} = \pi(1_C) \otimes a\}.$$

We claim that the canonical map $\overline{\mathsf{can}}$, which in this case is of the form $\overline{\mathsf{can}}: c \otimes a \mapsto \sum c_{\underline{1}} \otimes c_{\underline{2}}a$, has the inverse

$$\overline{\mathsf{can}}^{-1} : C\Box_B C \to C \otimes A, \qquad \sum_i c^i \otimes \tilde{c}^i \mapsto \sum_i c^i{}_{\underline{1}} \otimes S(c^i{}_{\underline{2}})\tilde{c}^i.$$

First note that $\overline{\mathsf{can}}^{-1}$ is well defined since, for all $x = \sum_i c^i \otimes \tilde{c}^i \in C\Box_B C$,

$$\begin{aligned} (I_C \otimes \pi \otimes I_C)(I_C \otimes \Delta)\overline{\mathsf{can}}^{-1}(x) &= \textstyle\sum_i c^i{}_{\underline{1}} \otimes \pi((S(c^i{}_{\underline{2}})\tilde{c}^i)_{\underline{1}}) \otimes (S(c^i{}_{\underline{2}})\tilde{c}^i)_{\underline{2}} \\ &= \textstyle\sum_i c^i{}_{\underline{1}} \otimes \pi(S(c^i{}_{\underline{3}})\tilde{c}^i{}_{\underline{1}}) \otimes S(c^i{}_{\underline{2}})\tilde{c}^i{}_{\underline{2}} \\ &= \textstyle\sum_i c^i{}_{\underline{1}} \otimes S(c^i{}_{\underline{3}})\pi(\tilde{c}^i{}_{\underline{1}}) \otimes S(c^i{}_{\underline{2}})\tilde{c}^i{}_{\underline{2}} \\ &= \textstyle\sum_i c^i{}_{\underline{1}} \otimes S(c^i{}_{\underline{3}})\pi(c^i{}_{\underline{4}}) \otimes S(c^i{}_{\underline{2}})\tilde{c}^i \\ &= \textstyle\sum_i c^i{}_{\underline{1}} \otimes \pi(S(c^i{}_{\underline{3}})c^i{}_{\underline{4}}) \otimes S(c^i{}_{\underline{2}})\tilde{c}^i \\ &= \textstyle\sum_i c^i{}_{\underline{1}} \otimes \pi(1_C) \otimes S(c^i{}_{\underline{2}})\tilde{c}^i, \end{aligned}$$

where we used the fact that π is a left C-module map to derive the third and fifth equalities, and the fact that $x \in C\Box_B C$ to obtain the fourth equality. This means that

$$\overline{\mathsf{can}}^{-1}(C\Box_B C) \subseteq C \otimes {}^{coB}C = C \otimes A,$$

as required. Clearly $\overline{\mathsf{can}}^{-1}$ is a right A-module and a left C-comodule map. Furthermore, for all $a \in A$ and $c \in C$,

$$\overline{\mathsf{can}}^{-1}(\overline{\mathsf{can}}(c \otimes a)) = \overline{\mathsf{can}}^{-1}(\textstyle\sum c_{\underline{1}} \otimes c_{\underline{2}}a) = \sum c_{\underline{1}} \otimes S(c_{\underline{2}})c_{\underline{3}}a = c \otimes a,$$

and for all $x = \sum_i c^i \otimes \tilde{c}^i \in C\Box_B C$,

$$\overline{\mathsf{can}}(\overline{\mathsf{can}}^{-1}(x)) = \overline{\mathsf{can}}(\textstyle\sum_i c^i{}_{\underline{1}} \otimes S(c^i{}_{\underline{2}})\tilde{c}^i) = \sum_i c^i{}_{\underline{1}} \otimes c^i{}_{\underline{2}}S(c^i{}_{\underline{3}})\tilde{c}^i = x.$$

Therefore $\overline{\mathsf{can}}^{-1}$ is the inverse of the canonical map $\overline{\mathsf{can}}$, as claimed. Note that the cotranslation map reads

$$\check{\tau}(\sum_i c^i \otimes \tilde{c}^i) = \sum_i \varepsilon(c^i{}_{\underline{1}})S(c^i{}_{\underline{2}})\tilde{c}^i = \sum_i S(c^i)\tilde{c}^i.$$

Applying this formula to the definition of the canonical entwining map ψ in the proof of 34.17, we obtain, for all $a \in A$ and $c \in C$,

$$\psi(c \otimes a) = \sum \check{\tau}(c_{\underline{1}} \otimes c_{\underline{2}}a_{\underline{1}}) \otimes c_{\underline{3}}a_{\underline{2}} = \sum S(c_{\underline{1}})c_{\underline{2}}a_{\underline{1}} \otimes c_{\underline{3}}a_{\underline{2}} = \sum a_{\underline{1}} \otimes ca_{\underline{2}},$$

as asserted. □

Thus to any left coideal subalgebra of a Hopf algebra one can associate a coalgebra-Galois extension by 34.2, and to any right coideal subalgebra one can associate an algebra-Galois coextension by 34.18, provided some faithful flatness conditions are met. It has been observed in [68] that in many examples of particular interest in quantum geometry, such as quantum spheres, left and right coideal subalgebras are actually isomorphic one to the other. This results in the isomorphism of quotient coalgebras and thus gives one two geometric interpretations for a coideal subalgebra. One can interpret such a subalgebra either as a base manifold or else as a fibre of corresponding principal bundles with isomorphic total spaces.

34.19. Exercises

(1) Suppose that $R = F$ is a field, and let $A(B)^C$ be a coalgebra-Galois extension associated to a coideal subalgebra A_1 of a Hopf algebra A, as described in 34.2. Suppose that A is a faithfully flat left A_1-module (thus, in particular $A_1 = B = A^{coC}$). Show that the Ehresmann coring $\mathcal{E}(A(B)^C)$ (cf. 34.13) is isomorphic to the B-coring $\mathcal{C} = A \otimes_F B$. A (B, B)-bimodule structure on $A \otimes_F B$ is given by

$$b(a \otimes b')b'' = \sum b_{\underline{1}}a \otimes b_{\underline{2}}b'b'', \quad \text{for all } b, b', b'' \in B,\ a \in A,$$

and the coproduct and counit are

$$\underline{\Delta}(a \otimes b) = \sum a_{\underline{1}} \otimes a_{\underline{2}} \otimes b, \quad \underline{\varepsilon}(a \otimes b) = \varepsilon(a)b,$$

for all $a \in A$, $b \in B$. Here the natural identification $A \otimes_F B \otimes_B A \otimes_F B = A \otimes_F A \otimes_F B$ has been used. (Hint: the isomorphism is $\theta : \mathcal{C} \to \mathcal{E}(A(B)^C)$, $a \otimes b \mapsto \sum a_{\underline{1}} \otimes S(a_{\underline{2}})b$, and its inverse is $\theta^{-1} : \sum_i a^i \otimes \tilde{a}^i \mapsto \sum_i a^i{}_{\underline{1}} \otimes a^i{}_{\underline{2}}\tilde{a}^i$.)

(2) Use the properties of the cotranslation map to prove that the map ψ defined in the proof of 34.17 satisfies the bow-tie diagram conditions.

References. Brzeziński [67, 68, 73]; Brzeziński and Hajac [77]; Brzeziński and Majid [79, 80]; Chase and Sweedler [12]; Doi and Takeuchi [107, 108]; Husemoller [21]; Kreimer and Takeuchi [147]; Mackenzie [32]; Montgomery [37]; Müller and Schneider [163]; Schauenburg [183]; Schneider [186, 187, 188]; Takeuchi [198].

Chapter 6

Weak corings and entwinings

This chapter is devoted to a generalisation of corings in which it is not assumed that a coring is a unital (A, A)-bimodule. Such generalised corings are known as *weak corings*. Various parts of the theory of corings can be transferred to this more general situation. We discuss this transfer. Weak corings can be seen as a natural algebraic framework for a generalisation of bialgebras known as *weak bialgebras*. We derive weak bialgebras from weak corings and study their properties. Finally, we consider examples of weak corings coming from a modification of entwining structures.

35 Weak corings

Usually modules over associative algebras A are assumed to be unital. It turns out, however, that there are (A, A)-bimodules $\mathcal{C}$ that satisfy all the conditions for corings except unitality, that is, the multiplication with 1_A is no longer the identity map in $\mathcal{C}$. Such algebraic structures are called *weak A-corings*, and these together with their *weak comodules* are the subject of this section. In particular, we study when A itself is a weak comodule and define *Galois weak A-corings*. Applications to weak Hopf algebras and weak entwining structures are subjects of the following sections.

As before, A is an R-algebra with unit 1_A (often written as 1).

35.1. Nonunital modules. By $\widetilde{\mathbf{M}}_A$ (resp. ${}_A\widetilde{\mathbf{M}}$) we denote the category of all (not necessarily unital) right (left) A-modules with the usual homomorphisms. $\widetilde{\mathbf{M}}_A$ (resp. ${}_A\widetilde{\mathbf{M}}$) contains unital modules as a full subcategory.

For an R-algebra B, ${}_A\widetilde{\mathbf{M}}_B$ denotes the category of (A, B)-bimodules, which need not be unital either on the left or on the right; that is, for any $M \in {}_A\widetilde{\mathbf{M}}_B$ and $m \in M$, $a \in A$, $b \in B$, $(am)b = a(mb)$ but possibly $m1_B \neq m$ and $1_Am \neq m$. For $M, N \in {}_A\widetilde{\mathbf{M}}_B$, the set of bimodule morphisms $M \to N$ will be denoted by ${}_A\mathrm{Hom}_B(M, N)$. Again unital bimodules ${}_A\mathbf{M}_B$ form a full subcategory of ${}_A\widetilde{\mathbf{M}}_B$.

For any $M \in \widetilde{\mathbf{M}}_A$ there is a splitting A-epimorphism,

$$- \otimes 1_A : M \to M \otimes_A A, \quad m \mapsto m \otimes 1_A,$$

which is injective (bijective) if and only if M is a unital A-module. The canonical isomorphisms $M \otimes_A A \to MA$, $m \otimes a \mapsto ma$, and $\mathrm{Hom}_A(A, M) \to$

MA, $f \mapsto f(1_A)$, are used to identify $M \otimes_A A$ and $\mathrm{Hom}_A(A, M)$ with MA. In particular, $MA = M1_A$.

For any A-module morphism $f : M \to N$, the map $f \otimes I_A : M \otimes_A A \to N \otimes_A A$ can be identified with the restriction $f\,|_{MA}: MA \to NA$, which is also denoted by f. Since MA is a unital module, there is a functor

$$- \otimes_A A : \widetilde{\mathbf{M}}_A \to \mathbf{M}_A \subset \widetilde{\mathbf{M}}_A, \quad M \mapsto M \otimes_A A,\ f \mapsto f \otimes I_A,$$

which is left (right) adjoint to itself; that is, for any $M, N \in \widetilde{\mathbf{M}}_A$,

$$\mathrm{Hom}_A(M \otimes_A A, N) \simeq \mathrm{Hom}_A(M \otimes_A A, N \otimes_A A) \simeq \mathrm{Hom}_A(M, N \otimes_A A).$$

In particular, this implies $\mathrm{Hom}_A(M, A) \simeq \mathrm{Hom}_A(MA, A)$.

Similar constructions and properties hold for $A \otimes_A -$ and left A-modules. For any $M \in {}_A\widetilde{\mathbf{M}}_A$, this induces a splitting (A, A)-morphism,

$$1_A \otimes - \otimes 1_A : M \to A \otimes_A M \otimes_A A \simeq AMA, \quad m \mapsto 1_A \otimes m \otimes 1_A = 1_A m 1_A,$$

and isomorphisms ${}_A\mathrm{Hom}_A(M, A) \simeq {}_A\mathrm{Hom}_A(MA, A) \simeq {}_A\mathrm{Hom}_A(AMA, A)$.

35.2. Weak A-corings. A bimodule $\mathcal{C} \in {}_A\widetilde{\mathbf{M}}_A$ is called a *weak A-coring* provided there are two (A, A)-bilinear maps

$$\underline{\Delta} : \mathcal{C} \to \mathcal{C} \otimes_A A \otimes_A \mathcal{C}, \quad \underline{\varepsilon} : \mathcal{C} \to A,$$

the *weak coproduct* and *weak counit*, rendering commutative the diagrams

$$\begin{array}{ccc}
\mathcal{C} & \xrightarrow{\underline{\Delta}} & \mathcal{C} \otimes_A A \otimes_A \mathcal{C} \\
{\scriptstyle \underline{\Delta}}\downarrow & & \downarrow{\scriptstyle I_{\mathcal{C}} \otimes I_A \otimes \underline{\Delta}} \\
\mathcal{C} \otimes_A A \otimes_A \mathcal{C} & \xrightarrow{\underline{\Delta} \otimes I_A \otimes I_{\mathcal{C}}} & \mathcal{C} \otimes_A A \otimes_A \mathcal{C} \otimes_A A \otimes_A \mathcal{C},
\end{array}$$

and

$$\begin{array}{ccccc}
 & \swarrow^{\underline{\Delta}} & \mathcal{C} & ^{\underline{\Delta}}\searrow & \\
\mathcal{C} \otimes_A A \otimes_A \mathcal{C} & & {\scriptstyle 1_A \otimes -}\downarrow{\scriptstyle - \otimes 1_A} & & \mathcal{C} \otimes_A A \otimes_A \mathcal{C} \\
 & {}_{\underline{\varepsilon} \otimes I_{AC}}\searrow & \mathcal{C}. & \swarrow_{I_{CA} \otimes \underline{\varepsilon}} &
\end{array}$$

Writing $\underline{\Delta}(c) = \sum c_{\underline{1}} \otimes 1_A \otimes c_{\underline{2}}$ for $c \in \mathcal{C}$, these conditions are expressed by

$$\sum c_{\underline{11}} \otimes 1_A \otimes c_{\underline{12}} \otimes 1_A \otimes c_{\underline{2}} = \sum c_{\underline{1}} \otimes 1_A \otimes c_{\underline{21}} \otimes 1_A \otimes c_{\underline{22}},$$

$$\sum \underline{\varepsilon}(c_{\underline{1}}) c_{\underline{2}} = 1_A c 1_A = \sum c_{\underline{1}} \underline{\varepsilon}(c_{\underline{2}}).$$

A weak A-coring $\mathcal{C}$ is said to be *right (left) unital* provided $\mathcal{C}$ is unital as a right (left) A-module, and clearly $\mathcal{C}$ is an *A-coring* provided $\mathcal{C}$ is unital both as a left and right A-module. Notice that $\underline{\Delta}$ need not split as an (A,A)-bimodule morphism unless $\mathcal{C}$ is left and right unital.

For the rest of the section $(\mathcal{C}, \underline{\Delta}, \underline{\varepsilon})$ denotes a weak A-coring.

35.3. Corings and weak corings.

(1) $(\mathcal{C}A, \underline{\Delta}, \underline{\varepsilon})$ is a (right unital) weak A-coring;
(2) $(A\mathcal{C}, \underline{\Delta}, \underline{\varepsilon})$ is a (left unital) weak A-coring;
(3) $(A\mathcal{C}A, \underline{\Delta}, \underline{\varepsilon})$ is an A-coring.

There are various types of duals of $\mathcal{C}$, and we use 35.1 to derive the canonical isomorphisms

$$\begin{array}{rcll} \mathcal{C}^* & := & \mathrm{Hom}_A(\mathcal{C}, A) & \simeq \ \mathrm{Hom}_A(\mathcal{C}A, A), \\ (A\mathcal{C})^* & := & \mathrm{Hom}_A(A\mathcal{C}, A) & \simeq \ \mathrm{Hom}_A(A\mathcal{C}A, A), \\ {}^*\mathcal{C} & := & {}_A\mathrm{Hom}(\mathcal{C}, A) & \simeq \ {}_A\mathrm{Hom}(A\mathcal{C}, A), \\ {}^*(\mathcal{C}A) & := & {}_A\mathrm{Hom}(\mathcal{C}A, A) & \simeq \ {}_A\mathrm{Hom}(A\mathcal{C}A, A), \\ {}^*\mathcal{C}^* & := & {}_A\mathrm{Hom}_A(\mathcal{C}, A) & \simeq \ {}_A\mathrm{Hom}_A(A\mathcal{C}A, A) = {}^*\mathcal{C} \cap \mathcal{C}^*. \end{array}$$

35.4. Dual algebras.

(1) $\mathcal{C}^*$ is an algebra (without a unit) by the product $f *^r g(c) = \sum g(f(c_{\underline{1}})c_{\underline{2}})$, for $f, g \in \mathcal{C}^*$, $c \in \mathcal{C}$. The weak counit $\underline{\varepsilon}$ is a central idempotent in $\mathcal{C}^*$, and $(A\mathcal{C})^* = \underline{\varepsilon} *^r \mathcal{C}^*$.

(2) ${}^*\mathcal{C}$ is an algebra (without a unit) by the product $f *^l g(c) = \sum f(c_{\underline{1}}g(c_{\underline{2}}))$, for $f, g \in \mathcal{C}^*$, $c \in \mathcal{C}$. The weak counit $\underline{\varepsilon}$ is a central idempotent in ${}^*\mathcal{C}$, and ${}^*(\mathcal{C}A) \simeq {}^*\mathcal{C} *^l \underline{\varepsilon}$.

(3) ${}^*\mathcal{C}^*$ is an algebra by the product $f * g(c) = \sum f(c_{\underline{1}})g(c_{\underline{2}})$, for $f, g \in {}^*\mathcal{C}^*$, $c \in \mathcal{C}$, with unit $\underline{\varepsilon}$.

Proof. Most of the assertions have the same proofs as for corings. For example, to prove the centrality of $\underline{\varepsilon}$ in assertion (1), take any $f \in \mathcal{C}^*$, $c \in \mathcal{C}$ and compute

$$f *^r \underline{\varepsilon}(c) = \sum \underline{\varepsilon}(f(c_{\underline{1}})c_{\underline{2}}) = \sum f(c_{\underline{1}})\underline{\varepsilon}(c_{\underline{2}}) = \sum f(c_{\underline{1}}\underline{\varepsilon}(c_{\underline{2}})) = f(1c1), \text{ and}$$
$$\underline{\varepsilon} *^r f(c) = \sum f(\underline{\varepsilon}(c_{\underline{1}})c_{\underline{2}}) = f(1c1).$$ □

35.5. Weak comodules. A module $M \in \widetilde{\mathbf{M}}_A$ is called a *right weak $\mathcal{C}$-comodule* provided there is an A-linear map $\varrho^M : M \to M \otimes_A A \otimes_A \mathcal{C}$ rendering commutative the diagrams

$$\begin{array}{ccc} M & \xrightarrow{\varrho^M} & M \otimes_A A\mathcal{C} \\ {\scriptstyle \varrho^M}\downarrow & & \downarrow{\scriptstyle I\otimes\underline{\Delta}} \\ M \otimes_A A\mathcal{C} & \xrightarrow{\varrho^M \otimes I} & M \otimes_A A\mathcal{C} \otimes_A A\mathcal{C}, \end{array} \qquad \begin{array}{ccc} M & \xrightarrow{\varrho^M} & M \otimes_A A\mathcal{C} \\ & {\scriptstyle -\otimes 1}\searrow & \downarrow{\scriptstyle I\otimes\underline{\varepsilon}} \\ & & M \otimes_A A. \end{array}$$

With the notation $\varrho^M(m) = \sum m_{\underline{0}} \otimes 1 \otimes m_{\underline{1}}$, for all $m \in M$, the conditions stated above are equivalent to the equalities

$$\sum m_{\underline{0}} \otimes 1_A \otimes \underline{\Delta}(m_{\underline{1}}) = \sum \varrho^M(m_{\underline{0}}) \otimes 1_A \otimes m_{\underline{1}}, \quad m1_A = \sum m_{\underline{0}}\, \underline{\varepsilon}(m_{\underline{1}}).$$

Left weak $\mathcal{C}$-comodules are defined in a symmetric way.

Notice that any weak A-coring $\mathcal{C}$ has a left and a right coaction (by $\underline{\Delta}$) which, however, need not be weakly counital. On the other hand, it is easy to see that the obvious right $\mathcal{C}$-coaction on $A\mathcal{C}$ is weakly counital, that is, $A\mathcal{C}$ is a right weak $\mathcal{C}$-comodule. Similarly, $\mathcal{C}A$ is a left weak $\mathcal{C}$-comodule.

35.6. Weak comodules and comodules. *Let M be a right weak $\mathcal{C}$-comodule.*

(1) *MA is a weak comodule over $\mathcal{C}$.*

(2) *MA is a weak comodule over the (left unital) weak A-coring $A\mathcal{C}$.*

(3) *MA is a weak comodule over the (right unital) weak A-coring $\mathcal{C}A$.*

(4) *MA is a comodule over the A-coring $A\mathcal{C}A$.*

Notice that the structure map $\varrho^M : M \to M \otimes_A A \otimes_A \mathcal{C}$ of weak comodules need not be injective even if $\mathcal{C}$ is a coring. For example, considering A as an A-coring (by $\underline{\Delta} : A \simeq A \otimes_A A$, $\underline{\varepsilon} = I_A$), every right A-module M is a weak A-comodule by the map $- \otimes 1 : M \to M \otimes_A A$, which is not injective unless M is unital.

35.7. Morphisms. A *morphism* of weak comodules $f : M \to N$ is an A-linear map such that the diagram

$$\begin{array}{ccc} M & \xrightarrow{f} & N \\ {\scriptstyle \varrho^M}\downarrow & & \downarrow{\scriptstyle \varrho^N} \\ M \otimes_A A\mathcal{C} & \xrightarrow{f \otimes I_{A\mathcal{C}}} & N \otimes_A A\mathcal{C} \end{array}$$

commutes, which means $\varrho^N \circ f = (f \otimes I_{A\mathcal{C}}) \circ \varrho^M$. The set $\mathrm{Hom}^{\mathcal{C}}(M, N)$ of weak comodule morphisms is an Abelian group, and by definition it is determined by an exact sequence,

$$0 \to \mathrm{Hom}^{\mathcal{C}}(M, N) \to \mathrm{Hom}_A(M, N) \xrightarrow{\gamma} \mathrm{Hom}_A(M, N \otimes_A A\mathcal{C}),$$

where $\gamma(f) := \varrho^N \circ f - (f \otimes I_{A\mathcal{C}}) \circ \varrho^M$.

The following observations are easy to verify.

35.8. Weak coaction and tensor products. *Let X be any unital right A-module. Then $X \otimes_A \mathcal{C}$ is a right weak $\mathcal{C}$-comodule by*

$$I_X \otimes \underline{\Delta} : X \otimes_A \mathcal{C} \longrightarrow X \otimes_A \mathcal{C} \otimes_A A\mathcal{C}\,.$$

For any morphism $f : X \to Y$ in $\mathbf{M}_A$,

$$f \otimes I_{\mathcal{C}} : X \otimes_A \mathcal{C} \to Y \otimes_A \mathcal{C}$$

is a morphism of weak $\mathcal{C}$-comodules.

For any index set Λ, $A^{(\Lambda)} \otimes_A \mathcal{C} \simeq A\mathcal{C}^{(\Lambda)}$ as weak comodules.

35.9. Kernels and cokernels. Let $f : K \to M$ be a morphism of weak $\mathcal{C}$-comodules. There is a commutative diagram with exact rows in $\widetilde{\mathbf{M}}_A$,

$$\begin{array}{ccccccc}
K & \xrightarrow{f} & M & \xrightarrow{g} & N & \longrightarrow & 0 \\
\downarrow{\scriptstyle \varrho^K} & & \downarrow{\scriptstyle \varrho^M} & & & & \\
K \otimes_A A\mathcal{C} & \xrightarrow{f \otimes I_{A\mathcal{C}}} & M \otimes_A A\mathcal{C} & \xrightarrow{g \otimes I_{A\mathcal{C}}} & N \otimes_A A\mathcal{C} & \longrightarrow & 0.
\end{array}$$

By the cokernel property of N in $\widetilde{\mathbf{M}}_A$, this can be completed commutatively by some A-linear map $\varrho^N : N \to N \otimes_A A\mathcal{C}$. An easy check shows that this makes N a weak $\mathcal{C}$-comodule, and so f has a cokernel as a weak comodule morphism. The existence of a kernel of f can be shown in a similar way provided the functor $- \otimes_A A\mathcal{C}$ respects injective morphisms, that is, $A\mathcal{C}$ is flat as a left A-module. The arguments are the same as for comodules (compare 3.5, 18.6).

35.10. The category $\widetilde{\mathbf{M}}^{\mathcal{C}}$. *The right weak $\mathcal{C}$-comodules with their morphisms form an additive category denoted by $\widetilde{\mathbf{M}}^{\mathcal{C}}$.*

1. *The category $\widetilde{\mathbf{M}}^{\mathcal{C}}$ has direct sums and cokernels, and it has kernels provided $A\mathcal{C}$ is flat as a left A-module.*
2. *For any $M \in \widetilde{\mathbf{M}}^{\mathcal{C}}$, $X \in \mathbf{M}_A$, the map*

$$\mathrm{Hom}^{\mathcal{C}}(MA, X \otimes_A \mathcal{C}) \to \mathrm{Hom}_A(MA, X), \quad f \mapsto (I_X \otimes \underline{\varepsilon}) \circ f,$$

   is an isomorphism natural in M and X. Its inverse is $h \mapsto (h \otimes I_{\mathcal{C}}) \circ \varrho^M$.

3. *The functor $- \otimes_A \mathcal{C}A : \mathbf{M}_A \to \widetilde{\mathbf{M}}^{\mathcal{C}}$ is right adjoint to the functor $- \otimes_A A : \widetilde{\mathbf{M}}^{\mathcal{C}} \to \mathbf{M}_A$.*

Proof. (1) It is easy to check that coproducts in $\widetilde{\mathbf{M}}_A$ yield coproducts in $\widetilde{\mathbf{M}}^{\mathcal{C}}$ in an obvious way. The rest is clear by the preceding remarks.

(2) For all $h \in \mathrm{Hom}_A(MA, X)$, the composition

$$MA \xrightarrow{\varrho^M} MA \otimes_A \mathcal{C} \xrightarrow{h \otimes I_{\mathcal{C}}} X \otimes_A \mathcal{C} \xrightarrow{I_X \otimes \underline{\varepsilon}} X$$

gives back the map h. Let $f \in \mathrm{Hom}^{\mathcal{C}}(MA, X\otimes_A \mathcal{C})$ and put $h = (I_X \otimes \underline{\varepsilon}) \circ f$. Then the composition $MA \xrightarrow{\varrho^M} MA\otimes_A\mathcal{C} \xrightarrow{h\otimes I_{\mathcal{C}}} X\otimes_A\mathcal{C}$ yields the map f. Thus the given assignments are inverses to each other. Any morphism $M \to N$ in $\widetilde{\mathbf{M}}_A$ induces a morphism $MA \to NA$, and so it is easy to see that the isomorphism is natural in both arguments.

(3) This follows from (2) by $\mathrm{Hom}^{\mathcal{C}}(MA, X\otimes_A\mathcal{C}) \simeq \mathrm{Hom}^{\mathcal{C}}(M, X\otimes_A\mathcal{C}A)$. □

35.11. Corollary. *There are isomorphisms*

$$\mathrm{End}^{\mathcal{C}}(A\mathcal{C}A) \simeq (A\mathcal{C})^*, \quad {}^{\mathcal{C}}\mathrm{End}(A\mathcal{C}A) \simeq {}^*(\mathcal{C}A),$$

both given by $f \mapsto \underline{\varepsilon} \circ f$, which are anti-ring and ring morphisms, respectively.

Proof. Set $X = A$ and $M = A\mathcal{C}$ in 35.10 and perform computations as in the proof of 18.12. □

35.12. Exactness of the $\mathrm{Hom}^{\mathcal{C}}$-functors. *Let ${}_A\mathcal{C}$ be flat and $M, N \in \widetilde{\mathbf{M}}^{\mathcal{C}}$.*

(1) $\mathrm{Hom}^{\mathcal{C}}(-, N) : \widetilde{\mathbf{M}}^{\mathcal{C}} \to \mathbf{M}_R$ *is left exact.*

(2) $\mathrm{Hom}^{\mathcal{C}}(M, -) : \widetilde{\mathbf{M}}^{\mathcal{C}} \to \mathbf{M}_R$ *is left exact.*

(3) *If A is right A-injective, then* $\mathrm{Hom}^{\mathcal{C}}(-, A\mathcal{C}A) : \widetilde{\mathbf{M}}^{\mathcal{C}} \to \mathbf{M}_R$ *is exact.*

Proof. One can follow the proofs of 3.19 and 18.17. □

35.13. A as a weak comodule. *The following are equivalent:*

(a) *A is a right (or left) weak $\mathcal{C}$-comodule;*

(b) *A is a right (or left) $A\mathcal{C}A$-comodule;*

(c) *there exists a grouplike element $g \in A\mathcal{C}A$ (that is, $\underline{\Delta}(g) = g \otimes_A g$ and $\underline{\varepsilon}(g) = 1$).*

If this holds, we write A_g (or ${}_gA$) when A is considered as a weak right (or left) $\mathcal{C}$-comodule.

Proof. (a) ⇔ (b) Let A be a right $\mathcal{C}$-comodule by $\varrho^A : A \to A\otimes_A \mathcal{C}$. Then $\mathrm{Im}\, \varrho^A \subset A\mathcal{C}A$ and A is a right $A\mathcal{C}A$-comodule. The converse is obvious.

(b) ⇔ (c) Since $A\mathcal{C}A$ is an A-coring, the claim follows by 28.2. For a grouplike element $g \in \mathcal{C}$, $\varrho^A : A \to A\otimes_A\mathcal{C}$, $a \mapsto 1\otimes ga$, is a coaction on A. □

If $A, M \in \widetilde{\mathbf{M}}^{\mathcal{C}}$, any comodule morphism $f : A \to M$ is uniquely determined by the image of $1_A \in A$ and one defines the

35.14. Coinvariants. *Let $g \in A\mathcal{C}A$ be a grouplike element.*

(1) *The* coinvariants *of any $M \in \widetilde{\mathbf{M}}^{\mathcal{C}}$ are defined by*

$$M^{\mathrm{co}\mathcal{C}} = \{f(1_A) \mid f \in \mathrm{Hom}^{\mathcal{C}}(A_g, M)\} = \{m \in MA \mid \varrho^M(m) = m \otimes g\}.$$

(2) In particular, for $M = A$ there is a subalgebra

$$A^{co\mathcal{C}} = \{f(1_A) \mid f \in \mathrm{End}^{\mathcal{C}}(A_g)\} = \{a \in A \mid ga = ag\} \subset A.$$

(3) The map $\mathrm{End}^{\mathcal{C}}(A_g) \to A^{co\mathcal{C}}$, $f \mapsto f(1_A)$, is a ring isomorphism, and

$$\mathrm{Hom}^{\mathcal{C}}(A_g, M) \to M^{co\mathcal{C}}, \quad f \mapsto f(1_A),$$

is a right $A^{co\mathcal{C}}$-module isomorphism, for $M \in \widetilde{\mathbf{M}}^{\mathcal{C}}$.

(4) For any $N \in \widetilde{\mathbf{M}}_A$ there are isomorphisms

$$\begin{array}{rcccl} \mathrm{Hom}^{\mathcal{C}}(A_g, N \otimes_A A\mathcal{C}A) & \to & \mathrm{Hom}_A(A, NA) & \to & NA, \\ f & \mapsto & (I_N \otimes \underline{\varepsilon}) \circ f & \mapsto & (I_N \otimes \underline{\varepsilon}) \circ f(1_A). \end{array}$$

(5) $(A\mathcal{C})^{co\mathcal{C}} \simeq \mathrm{Hom}^{\mathcal{C}}(A_g, A\mathcal{C}A) \simeq \mathrm{Hom}_A(A, A) \simeq A$, with the maps

$$\varphi_A : \mathrm{Hom}^{\mathcal{C}}(A_g, A \otimes_A A\mathcal{C}A) \to \mathrm{Hom}_A(A, A) \to A, \quad f \mapsto \underline{\varepsilon} \circ f \mapsto \underline{\varepsilon}(f(1_A)).$$

Proof. With canonical morphisms in 35.1, the proofs of 28.5 apply. □

The standard Hom-tensor relation yields:

35.15. The coinvariants functor. *Let $g \in A\mathcal{C}A$ be a grouplike element and $B = A^{co\mathcal{C}}$. There is an isomorphism*

$$\mathrm{Hom}^{\mathcal{C}}(N \otimes_B A, M) \simeq \mathrm{Hom}_B(N, \mathrm{Hom}^{\mathcal{C}}(A_g, M))$$

that is natural in $N \in \mathbf{M}_B$ and $M \in \widetilde{\mathbf{M}}^{\mathcal{C}}$. Thus the functor

$$(-)^{co\mathcal{C}} \simeq \mathrm{Hom}^{\mathcal{C}}(A_g, -) : \widetilde{\mathbf{M}}^{\mathcal{C}} \to \mathbf{M}_B, \quad M \mapsto M^{co\mathcal{C}},$$

is right adjoint to the induction functor $- \otimes_B A : \mathbf{M}_B \to \widetilde{\mathbf{M}}^{\mathcal{C}}$, where $N \otimes_B A$ is a right $\mathcal{C}$-comodule with coaction $I_N \otimes \varrho^A$.

Furthermore, if ${}_A\mathcal{C}$ is flat, then $(-)^{co\mathcal{C}}$ is exact if and only if A is a projective object in $\widetilde{\mathbf{M}}^{\mathcal{C}}$.

35.16. Galois weak A-corings. Let $g \in A\mathcal{C}A$ be a grouplike element and $B = A^{co\mathcal{C}}$. Then $(\mathcal{C}, g)$ is said to be a *Galois weak coring* if A_g (equivalently ${}_gA$) is a Galois comodule over $A\mathcal{C}A$, that is, there is an isomorphism

$$\mathrm{Hom}^{\mathcal{C}}(A_g, A\mathcal{C}) \otimes_B A_g \to A\mathcal{C}A, \quad f \otimes a \mapsto f(a).$$

By the morphisms in 35.1, the arguments of 28.18 (resp. 18.26) prove:

If $g \in A\mathcal{C}A$ is a grouplike element, then the following are equivalent:

(a) $(\mathcal{C}, g)$ is a Galois weak A-coring;

(b) *there is an isomorphism* $\mathsf{can}_A : A \otimes_B A \to ACA$, $a \otimes b \mapsto a\varrho^A(1)b$;

(c) (ACA, g) *is a Galois A-coring.*

With the results observed so far the proof of 28.19 yields:

35.17. The Galois weak coring structure theorem. *Let* $g \in ACA$ *be a grouplike element and* $B = A^{coC}$.

(1) *The following are equivalent:*

(a) $(\mathcal{C}, g)$ *is Galois weak coring and A is flat as a left B-module;*

(b) AC *is flat as a left A-module and A is a generator in* $\mathbf{M}^{ACA}$;

(c) AC *is flat as a left A-module, and, for any* $M \in \mathbf{M}^{ACA}$, *there is an isomorphism*

$$M^{co\mathcal{C}} \otimes_B A \to M, \quad m \otimes a \mapsto ma.$$

(2) *The following are equivalent:*

(a) $(\mathcal{C}, g)$ *is a Galois weak coring and* ${}_BA$ *is faithfully flat;*

(b) AC *is flat as a left A-module and* A_g *is a projective generator in* $\mathbf{M}^{ACA}$, *that is,* $\mathrm{Hom}^{\mathcal{C}}(A_g, -) : \mathbf{M}^{ACA} \to \mathbf{M}_B$ *is an equivalence.*

Comodules over corings are closely related to modules over the dual algebras of corings. To a certain extent this can be transferred to weak corings.

35.18. α-condition for weak corings. We say that $\mathcal{C}$ *satifies the left α-condition* if the map

$$\alpha_{N,\mathcal{C}} : N \otimes_A AC \to \mathrm{Hom}_R({}^*\mathcal{C}, NA), \quad n \otimes c \mapsto [f \mapsto nf(c)],$$

is injective for every right A-module N. By 19.3, this is satisfied if and only if AC is locally projective as a left A-module.

The *right α-condition* is defined and characterised similarly.

35.19. Weak $\mathcal{C}$-comodules and ${}^*\mathcal{C}$-modules. *Let* $M, N \in \widetilde{\mathbf{M}}^{\mathcal{C}}$.

(1) *The left action*

$$\rightharpoonup : {}^*\mathcal{C} \otimes_R M \to M, \quad f \otimes m \mapsto (I_{MA} \otimes f) \circ \varrho^M(m),$$

defines a left ${}^*\mathcal{C}$-module structure on M.

(2) $\mathrm{Hom}^{\mathcal{C}}(M, N) \subseteq {}_{{}^*\mathcal{C}}\mathrm{Hom}(M, N)$, *and equality holds for all* $M, N \in \widetilde{\mathbf{M}}^{\mathcal{C}}$ *if and only if AC is locally projective as a left A-module.*

Proof. Modulo the canonical maps for nonunital modules, the proof of 19.3 (resp. 4.3) applies. □

Left weak comodules are related to right $\mathcal{C}^*$ modules in a similar way.

35.20. $^*\mathcal{C}$- and $\mathcal{C}^*$-actions on $\mathcal{C}$. *For any weak A-coring $\mathcal{C}$ there are actions*

$$\begin{aligned} \rightharpoonup : {}^*\mathcal{C} \otimes_R \mathcal{C} \to \mathcal{C}A, \quad & f \otimes c \mapsto (I_{\mathcal{C}A} \otimes f) \circ \underline{\Delta}(c), \\ \leftharpoonup : \mathcal{C} \otimes_R \mathcal{C}^* \to A\mathcal{C}, \quad & c \otimes g \mapsto (g \otimes I_{A\mathcal{C}}) \circ \underline{\Delta}(c). \end{aligned}$$

For any $c \in \mathcal{C}$, $c \leftharpoonup \underline{\varepsilon} = 1_A c 1_A = \underline{\varepsilon} \rightharpoonup c$. Furthermore,

(1) for all $f \in {}^*\mathcal{C}$, $g \in \mathcal{C}^*$, $(f \rightharpoonup c) \leftharpoonup g = f \rightharpoonup (c \leftharpoonup f)$;

(2) for all $f \in {}^*\mathcal{C}$, $h \in {}^*\mathcal{C}^*$, $f *^l h(c) = h(f \rightharpoonup c) = f(c \leftharpoonup h)$.

Proof. This is proved by direct computation. □

35.21. The category of weak comodules. *Let $\mathcal{C}$ satisfy the left α-condition.*

(1) $\widetilde{\mathbf{M}}^{\mathcal{C}}$ is a full subcategory of ${}_{*\mathcal{C}}\mathbf{M}$.

(2) For every $M \in \widetilde{\mathbf{M}}^{\mathcal{C}}$, $M \otimes_A A\mathcal{C}$ is generated, and MA is subgenerated, by the right $\mathcal{C}$-comodule $A\mathcal{C}$.

(3) For every $M \in \widetilde{\mathbf{M}}^{\mathcal{C}}$, finitely generated $^*\mathcal{C}$-submodules of MA are finitely generated as (right) A-modules.

(4) If $A\mathcal{C}A$ is finitely generated as a right $\mathcal{C}^*$-module (left A-module), then ${}^*(A\mathcal{C}A) \in \widetilde{\mathbf{M}}^{\mathcal{C}}$.

Proof. (1) This is clear by 35.19.

(2) There is an epimorphism $A^{(\Lambda)} \to M \otimes_A A$ of right A-modules. By 35.8, this yields an epimorphism in $\widetilde{\mathbf{M}}^{\mathcal{C}}$,

$$(A \otimes_A \mathcal{C})^{(\Lambda)} \simeq A^{(\Lambda)} \otimes_A \mathcal{C} \to M \otimes_A A\mathcal{C}.$$

Notice that ϱ^M is a comodule morphism but need not be injective. However, the restriction to $MA \subset M$ is injective and hence MA is a subcomodule of $M \otimes_A A\mathcal{C}$.

(3) For $k \in MA$, consider the cyclic submodule $K := {}^*\mathcal{C} \rightharpoonup k \subset MA$. By 35.19, there exists a weak coaction $\varrho^K : K \to K \otimes_A A\mathcal{C}$, and let $\varrho^K(k) = \sum_{i=1}^r k_i \otimes c_i$, where $k_i \in K$, $c_i \in \mathcal{C}$. Hence, for any $f \in {}^*\mathcal{C}$, $f \rightharpoonup k = \sum_{i=1}^r k_i f(c_i)$, that is, K is a finitely generated right A-module.

(4) Let $A\mathcal{C}A$ be a finitely generated right $\mathcal{C}^*$-module (or left A-module) with generators $a_1, \ldots, a_r \in A\mathcal{C}A$. Consider the map

$${}^*(A\mathcal{C}A) \to {}^*(A\mathcal{C}A)\,(a_1, \ldots, a_r) \subset (A\mathcal{C}A)^r \subset (A\mathcal{C})^r, \quad f \mapsto f \rightharpoonup (a_1, \ldots, a_r).$$

Since ${}^*(A\mathcal{C}A)$ acts faithfully on $A\mathcal{C}A$, this is a monomorphism of left ${}^*(A\mathcal{C}A)$-modules. So ${}^*(A\mathcal{C}A)$ is a submodule of the weak comodule $(A\mathcal{C})^r$ and hence is a right weak $\mathcal{C}$-subcomodule (by 35.19). □

Reference. Wisbauer [212].

36 Weak bialgebras

Weak bialgebras were introduced in mathematical physics as a means for studying some types of integrable Hamiltonian systems and as symmetries of certain quantum field theories. They can be viewed as a generalisation of bialgebras in which the coproduct is required to be a multiplicative but not necessarily unit-preserving map. In this section we describe the basic properties of and present a conceptual framework for weak bialgebras.

Throughout, (B, μ, Δ) denotes an R-module B that is an R-algebra with multiplication μ and unit 1 as well as a coalgebra with comultiplication Δ and counit ε, such that

$$\Delta(ab) = \Delta(a)\Delta(b), \qquad \text{for all } a, b \in B.$$

In explicit formulae we often need two copies of $\Delta(1)$, and hence we write $\sum 1_{\underline{1}} \otimes 1_{\underline{2}}$ and $\sum 1_{\underline{1}'} \otimes 1_{\underline{2}'}$ for $\Delta(1)$

With the twist map tw we can form another product $\mu^{\mathsf{tw}} := \mu \circ \mathsf{tw}$ and coproduct $\Delta^{\mathsf{tw}} := \mathsf{tw} \circ \Delta$ for B, and the resulting structures $(B, \mu^{\mathsf{tw}}, \Delta^{\mathsf{tw}})$, $(B, \mu^{\mathsf{tw}}, \Delta)$ and $(B, \mu, \Delta^{\mathsf{tw}})$ are again algebras and coalgebras with multiplicative coproducts. Based on any of these data, $B \otimes_R B$ is, canonically, an algebra with unit $1 \otimes 1$ as well as a coalgebra with counit $\varepsilon \otimes \varepsilon$ (cf. 2.12). The question we want to study is, when is $B \otimes_R B$ a weak B-coring?

36.1. Comultiplications on $B \otimes_R B$. *Given (B, μ, Δ), consider $B \otimes_R B$ as a (B, B)-bimodule with multiplications*

$$a'(a \otimes b) \cdot c = (a'a \otimes b)\Delta(c) = \sum a'ac_{\underline{1}} \otimes bc_{\underline{2}}, \quad \textit{for all } a, a', b, c \in B.$$

(1) For (B, μ, Δ) define the maps

$$\begin{array}{rcccl} \underline{\Delta} : B \otimes_R B & \to & (B \otimes_R B) \otimes_B (B \otimes_R B) & \simeq & (B \otimes_R B) \cdot 1 \otimes_R B, \\ a \otimes b & \mapsto & \sum (a \otimes b_{\underline{1}}) \otimes_B (1 \otimes b_{\underline{2}}) & \mapsto & \sum a1_{\underline{1}} \otimes b_{\underline{1}}1_{\underline{2}} \otimes b_{\underline{2}}, \end{array}$$

$$\begin{array}{rcccl} \underline{\varepsilon} : B \otimes_R B & \to & (B \otimes_R B) \cdot 1 & \xrightarrow{I_B \otimes \varepsilon} & B, \\ a \otimes b & \mapsto & (a \otimes b) \cdot 1 & \longmapsto & \sum a1_{\underline{1}}\varepsilon(b1_{\underline{2}}). \end{array}$$

(2) For $(B, \mu^{\mathsf{tw}}, \Delta^{\mathsf{tw}})$ consider the maps

$$\underline{\Delta}^{tw} : a \otimes b \mapsto \sum (a \otimes b_{\underline{2}}) \otimes_B (1 \otimes b_{\underline{1}}), \quad {}^{tw}\underline{\varepsilon}^{tw} : a \otimes b \mapsto \sum 1_{\underline{2}} a\varepsilon(1_{\underline{1}} b).$$

The module $B \otimes_R B$ with these maps is denoted by $B \otimes_R^o B$.

(3) For $(B, \mu^{\mathsf{tw}}, \Delta)$ consider the maps

$$\underline{\Delta} : a \otimes b \mapsto \sum (a \otimes b_{\underline{1}}) \otimes_B (1 \otimes b_{\underline{2}}), \quad \underline{\varepsilon}^{tw} : a \otimes b \mapsto \sum 1_{\underline{1}} a\varepsilon(1_{\underline{2}} b).$$

(4) For (B, μ, Δ^{tw}) *consider the maps*

$$\underline{\Delta}^{tw}: a \otimes b \mapsto \sum (a \otimes b_{\underline{2}}) \otimes_B (1 \otimes b_{\underline{1}}), \quad {}^{tw}\underline{\varepsilon}: a \otimes b \mapsto \sum a 1_{\underline{2}} \varepsilon(b 1_{\underline{1}}).$$

Then all the $\underline{\Delta}$ *are* (B, B)*-bilinear weak coproducts on* $B \otimes_R B$ *and all the* $\underline{\varepsilon}$ *are left* B*-linear with*

$$(a \otimes b) \cdot 1 = (I_B \otimes \underline{\varepsilon}) \circ \underline{\Delta}(a \otimes b), \quad \textit{for all } a, b \in B.$$

Proof. (1) Clearly $\underline{\Delta}$ is a left B-module morphism. For $b, c \in B$,

$$\begin{aligned} \underline{\Delta}((1 \otimes b) \cdot c) &= \sum (c_{\underline{1}} \otimes (bc_{\underline{2}})_{\underline{1}}) \otimes_B (1 \otimes (bc_{\underline{2}})_{\underline{2}}) \\ &= \sum c_{\underline{1}} 1_{\underline{1}} \otimes b_{\underline{1}} c_{\underline{21}} 1_{\underline{2}} \otimes b_{\underline{2}} c_{\underline{22}} = \sum c_{\underline{1}} \otimes b_{\underline{1}} c_{\underline{2}} \otimes b_{\underline{2}} c_{\underline{3}} \\ &= \sum (1 \otimes b_{\underline{1}}) \otimes_B (c_{\underline{1}} \otimes b_{\underline{2}} c_{\underline{2}}) \\ &= \sum (1 \otimes b_{\underline{1}}) \otimes_B (1 \otimes b_{\underline{2}}) \cdot c = \underline{\Delta}(1 \otimes b) \cdot c. \end{aligned}$$

This shows that $\underline{\Delta}$ is right B-linear. The coassociativity of $\underline{\Delta}$ follows easily from the coassociativity of Δ. Clearly $\underline{\varepsilon}$ is left B-linear. Moreover, for $a, b \in B$,

$$\begin{aligned} (I_B \otimes \underline{\varepsilon}) \circ \underline{\Delta}(a \otimes b) &= \sum (a \otimes b_{\underline{1}}) \otimes_B 1_{\underline{1}} \varepsilon(b_{\underline{2}} 1_{\underline{2}}) = \sum a 1_{\underline{1}} \otimes b_{\underline{1}} 1_{\underline{2}} \varepsilon(b_{\underline{2}} 1_{\underline{3}}) \\ &= \sum a 1_{\underline{1}} \otimes b 1_{\underline{2}} = (a \otimes b) \cdot 1. \end{aligned}$$

The proofs for (2), (3) and (4) follow the same pattern. □

In general, the properties of $\underline{\Delta}$ and $\underline{\varepsilon}$ observed in 36.1 are not sufficient to make $B \otimes_R B$ a coring, in particular, neither $\underline{\varepsilon}$ is right B-linear nor holds $(\underline{\varepsilon} \otimes I_B) \circ \underline{\Delta}(a \otimes b) = (a \otimes b) \cdot 1$. To ensure these properties we have to pose additional conditions on ε and Δ. We say that (B, μ, Δ) *induces a (weak) coring structure* on $B \otimes_R B$ if the latter is a (weak) B-coring with the maps defined in 36.1. Recall that (B, μ, Δ) is a *bialgebra* provided Δ and ε are unital algebra morphisms.

36.2. $B \otimes_R B$ as a B-coring. *The followig are equivalent:*

(a) (B, μ, Δ) *is a bialgebra;*

(b) (B, μ, Δ) *induces a coring structure on* $B \otimes_R B$*;*

(c) $(B, \mu^{tw}, \Delta^{tw})$ *induces a coring structure on* $B \otimes_R B$*;*

(d) (B, μ^{tw}, Δ) *induces a coring structure on* $B \otimes_R B$*;*

(e) (B, μ, Δ^{tw}) *induces a coring structure on* $B \otimes_R B$*.*

Proof. The equivalence of (a) and (b) follows from 33.1 via the correspondence between corings and entwinings. The remaining assertions follow by symmetry and are easy to verify. □

Part of the symmetry is lost in the case of weak corings.

36.3. $B \otimes_R B$ as a weak B-coring.

(1) The following are equivalent:

(a) (B, μ, Δ) *induces a weak coring structure on* $B \otimes_R B$*;*

(b) $(B, \mu^{tw}, \Delta^{tw})$ *induces a weak coring structure on* $B \otimes_R B$*;*

(c) (w.1) $\varepsilon(abc) = \sum \varepsilon(ab_{\underline{2}})\varepsilon(b_{\underline{1}}c)$, *for* $a, b, c \in B$*;*

(w.2) $(I_B \otimes \Delta) \circ \Delta(1) = (1 \otimes \Delta(1))(\Delta(1) \otimes 1)$
$(= \sum 1_{\underline{1}} \otimes 1_{\underline{1'}} 1_{\underline{2}} \otimes 1_{\underline{2'}})$.

(2) The following are equivalent:

(a) (B, μ, Δ^{tw}) *induces a weak coring structure on* $B \otimes_R B$*;*

(b) (B, μ^{tw}, Δ) *induces a weak coring structure on* $B \otimes_R B$*;*

(c) (w^{tw}.1) $\varepsilon(abc) = \sum \varepsilon(ab_{\underline{1}})\varepsilon(b_{\underline{2}}c)$, *for* $a, b, c \in B$*;*

(w^{tw}.2) $(I_B \otimes \Delta) \circ \Delta(1) = (\Delta(1) \otimes 1)(1 \otimes \Delta(1))$
$(= \sum 1_{\underline{1}} \otimes 1_{\underline{2}} 1_{\underline{1'}} \otimes 1_{\underline{2'}})$.

Proof. (1) (a) $\Rightarrow$ (c) If $B \otimes_R B$ is a weak B-coring, then $\underline{\varepsilon}$ is right B-linear and

$$\begin{aligned} \underline{\varepsilon}((1 \otimes a) \cdot b \cdot c) &= \underline{\varepsilon}((1 \otimes a) \cdot b)c \;=\; \sum b_{\underline{1}} \varepsilon(ab_{\underline{2}})c \\ &= \underline{\varepsilon}((1 \otimes a) \cdot (bc)) \;=\; \sum (bc)_{\underline{1}} \varepsilon(a(bc)_{\underline{2}}). \end{aligned}$$

Now apply ε to obtain

$$\sum \varepsilon(ab_{\underline{2}})\varepsilon(b_{\underline{1}}c) = \sum \varepsilon((bc)_{\underline{1}})\varepsilon(a(bc)_{\underline{2}}) = \sum \varepsilon(a\varepsilon((bc)_{\underline{1}})(bc)_{\underline{2}})) = \varepsilon(abc).$$

Since $\underline{\varepsilon}$ is a weak counit, we obtain

$$(1 \otimes a) \cdot 1 = \sum \underline{\varepsilon}(1 \otimes a_{\underline{1}}) \otimes a_{\underline{2}} = \sum 1_{\underline{1}} \varepsilon(a_{\underline{1}} 1_{\underline{2}}) \otimes a_{\underline{2}}.$$

Setting a to be either $1_{\underline{1'}}$ or 1 yields

$$\begin{aligned} (1 \otimes 1_{\underline{1'}})\Delta(1) &= \sum 1_{\underline{1}} \varepsilon(1_{\underline{1'}\underline{1}} 1_{\underline{2}}) \otimes 1_{\underline{1'}\underline{2}}, \quad \text{and} \\ \Delta(1) &= \sum 1_{\underline{1}} \varepsilon(1_{\underline{1'}} 1_{\underline{1'}\underline{2}}) \otimes 1_{\underline{2'}}. \end{aligned}$$

Finally, evaluation of $I_B \otimes \Delta$ at the second equality yields

$$\begin{aligned} (I_B \otimes \Delta) \circ \Delta(1) &= \sum 1_{\underline{1}} \varepsilon(1_{\underline{1'}} 1_{\underline{2}}) \otimes 1_{\underline{2'}\,\underline{1}} \otimes 1_{\underline{2'}\,\underline{2}} \\ &= \sum 1_{\underline{1}} \varepsilon(1_{\underline{1'}\underline{1}} 1_{\underline{2}}) \otimes 1_{\underline{1'}\underline{2}} \otimes 1_{\underline{2'}} \;=\; \sum 1_{\underline{1}} \otimes 1_{\underline{1'}} 1_{\underline{2}} \otimes 1_{\underline{2'}}. \end{aligned}$$

(c) $\Rightarrow$ (a) Suppose that (w.1) and (w.2) are satisfied. Then, for all $a, b \in B$,

$$\begin{aligned} \underline{\varepsilon}((1 \otimes a) \cdot 1 \cdot b) &= \sum (I_B \otimes \varepsilon)(1_{\underline{1}} b_{\underline{1}} \otimes a 1_{\underline{2}} b_{\underline{2}}) = \sum 1_{\underline{1}} b_{\underline{1}}\, \varepsilon(a 1_{\underline{2}} b_{\underline{2}}) \\ \overset{\text{(w.1)}}{=} &\; \sum 1_{\underline{1}} b_{\underline{1}}\, \varepsilon(a 1_{\underline{3}})\, \varepsilon(1_{\underline{2}} b_{\underline{2}}) \\ \overset{\text{(w.1)}}{=} &\; \sum 1_{\underline{1}} b_{\underline{1}}\, \varepsilon(a 1_{\underline{4}})\, \varepsilon(1_{\underline{3}})\, \varepsilon(1_{\underline{2}} b_{\underline{2}}) \\ &= \sum 1_{\underline{1}} b\, \varepsilon(a 1_{\underline{2}}) \;=\; \underline{\varepsilon}(1 \otimes a) b, \end{aligned}$$

that is, $\underline{\varepsilon}$ is right B-linear. By (w.2), for all $a \in B$,

$$\begin{aligned}
\sum \underline{\varepsilon}(1 \otimes a_{\underline{1}}) \otimes a_{\underline{2}} &= \sum \underline{\varepsilon}(1 \otimes (a1)_{\underline{1}}) \otimes (a1)_{\underline{2}} \\
&= \sum (I_B \otimes \varepsilon)(1_{\underline{1}} \otimes a_{\underline{1}} 1_{\underline{1}'} 1_{\underline{2}}) \otimes a_{\underline{2}} 1_{\underline{2}'} \\
&\overset{(w.2)}{=} \sum (I_B \otimes \varepsilon)(1_{\underline{1}} \otimes a_{\underline{1}} 1_{\underline{2}}) \otimes a_{\underline{2}} 1_{\underline{3}} \\
&= \sum 1_{\underline{1}} \varepsilon(a_{\underline{1}} 1_{\underline{2}}) \otimes a_{\underline{2}} 1_{\underline{3}} = \sum 1_{\underline{1}} \otimes (\varepsilon \otimes I_B)\Delta(a 1_{\underline{2}}) \\
&= \sum 1_{\underline{1}} \otimes a 1_{\underline{2}} = (1 \otimes a)\Delta(1) = (1 \otimes a) \cdot 1 .
\end{aligned}$$

This shows that $\underline{\varepsilon}$ is weakly counital.

(b) $\Leftrightarrow$ (c) is shown with a similar computation.

(2) The proof is similar to the proof of (1). □

36.4. Grouplike elements. *Assume that* (B, μ, Δ) *induces a weak coring structure on* $B \otimes_R B$. *Then* $\Delta(1)$ *and* $\Delta^{tw}(1)$ *are grouplike elements for* $B \otimes_R B$ *and* $B \otimes_R^o B$, *respectively.*

(1) B *is a right* $B \otimes_R B$*-comodule, and, for any* $M \in \widetilde{\mathbf{M}}^{B \otimes_R B}$, *the coinvariants are*

$$M^{co(B \otimes_R B)} = \{m \in MB \mid \varrho^M(m) = \sum m 1_{\underline{1}} \otimes 1_{\underline{2}})\}, \quad \textit{and}$$
$$B^{co(B \otimes_R B)} = \{a \in B \mid \Delta(a) = \sum a 1_{\underline{1}} \otimes 1_{\underline{2}}\}.$$

(2) B *is a right* $B \otimes_R^o B$*-comodule, and, for any* $N \in \widetilde{\mathbf{M}}^{B \otimes_R^o B}$, *the coinvariants are*

$$N^{co(B \otimes_R^o B)} = \{n \in NB \mid \varrho^N(n) = \sum n 1_{\underline{2}} \otimes 1_{\underline{1}}\}, \quad \textit{and}$$
$$B^{co(B \otimes_R^o B)} = \{a \in B \mid \Delta(a) = \sum 1_{\underline{2}} a \otimes 1_{\underline{1}}\}.$$

Proof. $\Delta(1)$ is a grouplike element for $B \otimes_R B$ since

$$\begin{aligned}
\underline{\Delta}(\Delta(1)) &= \sum (1_{\underline{1}} \otimes 1_{\underline{21}}) \otimes_B (1 \otimes 1_{\underline{22}}) = \sum (1_{\underline{11}} \otimes 1_{\underline{12}}) \otimes_B (1 \otimes 1_{\underline{2}}) \\
&= \sum \Delta(1) \otimes_B (1_{\underline{1}} \otimes 1_{\underline{2}}) = \Delta(1) \otimes_B \Delta(1), \quad \text{and} \\
\underline{\varepsilon}(\Delta(1)) &= (I_B \otimes \varepsilon)(\Delta(1) \cdot 1) = \sum 1_{\underline{1}} \varepsilon(1_{\underline{2}}) = 1.
\end{aligned}$$

Similarly one can show that $\Delta^{tw}(1)$ is a grouplike element for $B \otimes_R^o B$.

(1) By 35.13, B is a right $B \otimes_R B$-comodule and 35.14(1) yields the given characterisation of the coinvariants.

(2) The proof is analogous to the proof of part (1). □

An algebra and coalgebra B with a multiplicative coproduct is called a *weak R-bialgebra* provided (B, μ, Δ), $(B, \mu^{tw}, \Delta^{tw})$, (B, μ^{tw}, Δ) and (B, μ, Δ^{tw}) all induce weak coring structures on $B \otimes_R B$. This is the most symmetric definition of a weak bialgebra, which immediately implies that if B is a weak bialgebra, so are all possible twists B^{op}, B^{cop} and $(B^{op})^{cop}$. Some of the requirements listed above, however, are redundant, since from 36.3 we obtain:

36.5. Weak bialgebras. *Let B be an algebra and a coalgebra with a multiplicative coproduct. Then the following are equivalent:*

(a) *B is a weak bialgebra;*

(b) *(B, μ, Δ) and (B, μ, Δ^{tw}) induce weak coring structures on $B \otimes_R B$;*

(c) *$(B, \mu^{tw}, \Delta^{tw})$ and (B, μ^{tw}, Δ) induce weak coring structures on $B \otimes_R B$;*

(d) *the conditions* (w.1), (w.2), (w^{tw}.1) *and* (w^{tw}.2) *in 36.3 are satisfied.*

In case $(B \otimes_R B, \underline{\Delta}, \underline{\varepsilon})$ is a B-coring, the condition $b \otimes 1 = \Delta(b)$ implies $b = \varepsilon(b)1$, which means $B^{co(B \otimes_R B)} = R1_B$ and R is an R-direct summand in B. This is no longer true in the weak case, but some results in this direction still hold.

36.6. Coinvariants in weak bialgebras. *Let B be a weak bialgebra.*

(1) *For any $a \in B$ the following are equivalent:*

(a) $\Delta(a) = \sum a1_{\underline{1}} \otimes 1_{\underline{2}}$ *(that is, $a \in B^{co(B \otimes_R B)}$);*

(b) $\Delta(a) = \sum 1_{\underline{1}}a \otimes 1_{\underline{2}}$;

(c) $a = \sum \varepsilon(a1_{\underline{1}})1_{\underline{2}}$;

(d) $a = \sum \varepsilon(1_{\underline{1}}a)1_{\underline{2}}$.

(2) *For any $a \in B$ the following are equivalent:*

(a) $\Delta(a) = \sum 1_{\underline{1}} \otimes 1_{\underline{2}}a$ *(that is, $a \in B^{co(B \otimes_R^o B)}$);*

(b) $\Delta(a) = \sum 1_{\underline{1}} \otimes a1_{\underline{2}}$;

(c) $a = \sum 1_{\underline{1}}\varepsilon(1_{\underline{2}}a)$;

(d) $a = \sum 1_{\underline{1}}\varepsilon(a1_{\underline{2}})$.

Proof. (1) (a) $\Rightarrow$ (c), (b) $\Rightarrow$ (d) Apply $\varepsilon \otimes I_B$ to the equality in (a) and (b), respectively.

(c) $\Rightarrow$ (a), (b) Assume $a = \sum \varepsilon(a1_{\underline{1}})1_{\underline{2}}$. Then

$$\begin{aligned}\Delta(a) = \sum \varepsilon(a1_{\underline{1}})1_{\underline{2}} \otimes 1_{\underline{3}} &\overset{(w^{tw}.2)}{=} \sum \varepsilon(a1_{\underline{1}})1_{\underline{2}}1_{\underline{1'}} \otimes 1_{\underline{2'}} = \sum a1_{\underline{1}} \otimes 1_{\underline{2}}, \\ &\overset{(w.2)}{=} \sum \varepsilon(a1_{\underline{1}})1_{\underline{1'}}1_{\underline{2}} \otimes 1_{\underline{2'}} = \sum 1_{\underline{1}}a \otimes 1_{\underline{2}}.\end{aligned}$$

(d) $\Rightarrow$ (a) is shown similarly.

(2) The proof goes along the lines of the proof of (1). □

Recall that, for any R-module B that is an R-coalgebra and an R-algebra, the convolution product $f * g = \mu \circ (f \otimes g) \circ \Delta$, for all $f, g \in \mathrm{End}_R(B)$, makes $(\mathrm{End}_R(B), *)$ an associative R-algebra with unit $\iota \circ \varepsilon$, that is, $\iota \circ \varepsilon(b) = \varepsilon(b)1_B$, for any $b \in B$ (see 15.1). Besides the unit for $*$ there are two other maps that are of particular interest for weak bialgebras.

36.7. The counital source maps. *Assume that (B, μ, Δ) induces a weak coring structure on $B \otimes_R B$. Define the maps*

$$\sqcap^R : B \xrightarrow{1\otimes -} B \otimes_R B \xrightarrow{\varepsilon} B, \quad b \mapsto \sum 1_{\underline{1}}\varepsilon(b1_{\underline{2}}),$$
$$\sqcap^L : B \xrightarrow{1\otimes -} B \otimes_R B \xrightarrow{tw_{\varepsilon}tw} B, \quad b \mapsto \sum \varepsilon(1_{\underline{1}}b)1_{\underline{2}},$$

which obviously satisfy $\sqcap^L * I_B = I_B = I_B * \sqcap^R$. For all $a, b \in B$:

(1) *(i)* $\sum b_{\underline{1}} \otimes \sqcap^L(b_{\underline{2}}) = \sum 1_{\underline{1}}b \otimes 1_{\underline{2}}$;
(ii) $a\sqcap^L(b) = \sum \sqcap^L(a_{\underline{1}}b)a_{\underline{2}}$ $(= \sum \varepsilon(a_{\underline{1}}b)a_{\underline{2}})$;
(iii) $f * \sqcap^L(b) = \sum f(1_{\underline{1}}b)1_{\underline{2}}$, *for any* $f \in \mathrm{End}_R(B)$;
(iv) $\sqcap^L \circ \sqcap^L = \sqcap^L$;
(v) $\varepsilon(ab) = \varepsilon(a\sqcap^L(b))$ *and* $\sqcap^L(ab) = \sqcap^L(a\sqcap^L(b))$;
(vi) $\sqcap^L(a)\sqcap^L(b) = \sqcap^L(\sqcap^L(a)b)$.
So $B^L := \sqcap^L(B)$ is a subalgebra of B and $\sqcap^L$ is left B^L-linear.

(2) *(i)* $\sum \sqcap^R(b_{\underline{1}}) \otimes b_{\underline{2}} = \sum 1_{\underline{1}} \otimes b1_{\underline{2}}$;
(ii) $\sqcap^R(b)a = \sum a_{\underline{1}}\sqcap^R(ba_{\underline{2}})$ $(= \sum a_{\underline{1}}\varepsilon(ba_{\underline{2}}))$;
(iii) $\sqcap^R * g(b) = \sum 1_{\underline{1}}g(b1_{\underline{2}})$, *for any* $g \in \mathrm{End}_R(B)$;
(iv) $\sqcap^R \circ \sqcap^R = \sqcap^R$;
(v) $\varepsilon(ab) = \varepsilon(\sqcap^R(a)b)$ *and* $\sqcap^R(ab) = \sqcap^R(\sqcap^R(a)b)$;
(vi) $\sqcap^R(a)\sqcap^R(b) = \sqcap^R(a\sqcap^R(b))$.
So $B^R := \sqcap^R(B)$ is a subalgebra of B and $\sqcap^R$ is right B^R-linear.

The maps $\sqcap^L$, $\sqcap^R$ are known as *left*, resp. *right, counital source maps.*

Proof. (1) (i), (ii) follow directly from (w.1) and (w.2); (iii) is a consequence of (i).

(iv) and (v) follow from the computations

$$\begin{aligned} \sqcap^L(\sqcap^L(a)) &= \sum \varepsilon(\varepsilon(1_{\underline{1}}a)1_{\underline{1}'}1_{\underline{2}})1_{\underline{2}'} = \sum \varepsilon(1_{\underline{1}}a)\,\varepsilon(1_{\underline{1}'}1_{\underline{2}})1_{\underline{2}'} \\ &\overset{(w.1)}{=} \sum \varepsilon(1_{\underline{1}'}a)1_{\underline{2}'} = \sqcap^L(a), \quad \text{and} \\ \varepsilon(a\sqcap^L(b)) &= \sum \varepsilon(a\varepsilon(1_{\underline{1}}b)1_{\underline{2}}) \\ &\overset{(w.1)}{=} \sum \varepsilon(a1_{\underline{2}})\varepsilon(1_{\underline{1}}b) = \varepsilon(ab). \end{aligned}$$

(vi) $\Delta(\sqcap^L(a)) = \sum 1_{\underline{1}}\sqcap^L(a) \otimes 1_{\underline{2}}$, and hence by (ii),

$$\sqcap^L(\sqcap^L(a)\sqcap^L(b)) = \sum \varepsilon(1_{\underline{1}}\sqcap^L(a)b)1_{\underline{2}} = \sqcap^L(\sqcap^L(a)b).$$

(2) If (B, μ, Δ) induces a weak coring structure on $B \otimes_R B$, then so does $(B, \mu^{\mathrm{tw}}, \Delta^{\mathrm{tw}})$ (see 36.3), and the proof is similar to the proof of (1). □

36.8. The base algebra of a weak bialgebra. *For any weak bialgebra B,*

(1) $B^{co(B\otimes_R B)} = B^L$ and B^L is a direct summand of B as a left B^L-module;

(2) $B^{co(B\otimes^{o}_R B)} = B^R$ and B^R is a direct summand of B as a right B^R-module;

(3) B^R and B^L are commuting subalgebras of B;

(4) B^L and B^R are separable and Frobenius as R-algebras.

Proof. (1) and (2) follow by 36.4, 36.6 and 36.7.

(3) Let $x \in B^L$, $y \in B^R$. Then, by (w.2) and (w^{tw}.2),

$$\begin{aligned} xy &= \sum \varepsilon(1_{\underline{1}}x)1_{\underline{2}}1_{\underline{1}'}\varepsilon(y1_{\underline{2}'}) = \sum \varepsilon(1_{\underline{1}}x)1_{\underline{1}'}1_{\underline{2}}\varepsilon(y1_{\underline{2}'}) \\ &= \sum 1_{\underline{1}'}\varepsilon(y1_{\underline{2}'})1_{\underline{2}}\varepsilon(1_{\underline{1}}x)1_{\underline{2}} = yx. \end{aligned}$$

(4) We prove the assertion for $B^L =: A$ by showing that

$$e = \sum \sqcap^L(1_{\underline{1}}) \otimes 1_{\underline{2}} \in A \otimes_R A$$

is a separability idempotent and a Frobenius element, and that ε induces a Frobenius homomorphism $E := \varepsilon|_A$. By 36.7(1)(i), $\Delta(1) = \sum 1_{\underline{1}} \otimes \sqcap^L(1_{\underline{2}}) \in B \otimes_R A$, and, consequently, $e = \sum \sqcap^L(1_{\underline{1}}) \otimes \sqcap^L(1_{\underline{2}}) \in A \otimes_R A$. Explicitly,

$$e = \sum_i a_i \otimes \tilde{a}^i = \sum \varepsilon(1_{\underline{1}'}1_{\underline{1}})1_{\underline{2}'} \otimes 1_{\underline{2}}.$$

Now, take any $b \in B$ and compute

$$\begin{aligned} \sum_i E(\sqcap^L(b)a_i)\tilde{a}^i &= \sum \varepsilon(1_{\underline{1}'}1_{\underline{1}})\varepsilon(\sqcap^L(b)1_{\underline{2}'})1_{\underline{2}} = \sum \varepsilon(1_{\underline{1}}\sqcap^L(b))1_{\underline{2}} \\ &= \sum \varepsilon(1_{\underline{1}'}b)\varepsilon(1_{\underline{2}'}1_{\underline{1}})1_{\underline{2}} = \sum \varepsilon(1_{\underline{1}}b)\varepsilon(1_{\underline{2}})1_{\underline{3}} = \sqcap^L(b), \end{aligned}$$

where we used (w.1) to obtain the second equality and then (w^{tw}.2) to obtain the penultimate one. Similarly,

$$\begin{aligned} \sum_i a_i E(\tilde{a}^i \sqcap^L(b)) &= \sum \varepsilon(1_{\underline{1}'}1_{\underline{1}})\varepsilon(1_{\underline{2}}\sqcap^L(b))1_{\underline{2}'} \\ &= \sum \varepsilon(1_{\underline{1}}\sqcap^L(b))1_{\underline{2}} = \sqcap^L(\sqcap^L(b)) = \sqcap^L(b). \end{aligned}$$

Here the second equality follows from (w^{tw}.1), while the last equality is the consequence of (1)(v) in 36.7. This shows that A is a Frobenius algebra with a Frobenius element e and Frobenius homomorphism E. It also implies that for all $a \in A$, $ae = ea$. Furthermore, the multiplicativity of Δ and the definition of the counit immediately imply that $\sum_i a_i\tilde{a}^i = \sum \varepsilon(1_{\underline{1}}1_{\underline{1}'})1_{\underline{2}}1_{\underline{2}'} = 1$. So e is a separability idempotent and A is a separable Frobenius algebra. □

By the definition, given a weak bialgebra B, $B \otimes_R B$ is a weak B-coring. It turns out that B itself is a coring. Even more, B is a bialgebroid (cf. 31.6) with the subalgebra B^L as a base algebra.

36.9. Bialgebroid structure of weak bialgebras. *Let B be a weak bialgebra with coproduct Δ, counit ε, and let $A = \mathrm{Im}(\sqcap^L)$ as in 36.8. Then B is an A-bialgebroid with source and target $s, t : A \to B$ given by*

$$s(a) = a, \quad t(a) = \sum \varepsilon(1_{\underline{2}}a)1_{\underline{1}},$$

and comultiplication $\underline{\Delta} : B \to B \otimes_A B$ and counit $\underline{\varepsilon} : B \to A$ given by

$$\underline{\Delta}(b) = (\chi \circ \Delta)(b) = \sum b_{\underline{1}} \otimes b_{\underline{2}}, \quad \underline{\varepsilon}(b) = \sqcap^L(b),$$

where $\chi : B \otimes_R B \to B \otimes_A B$ is the canonical projection.

Proof. First note that t is the restriction to A of a more general map $t : B \to B$, $b \mapsto \sum 1_{\underline{1}}\varepsilon(1_{\underline{2}}b)$. Now, (w.2) and (w^{tw}.2) imply for all $b, b' \in B$,

$$\sqcap^L(b)t(b') = \sum \varepsilon(1_{\underline{1}}b)1_{\underline{2}}1_{\underline{1'}}\varepsilon(1_{\underline{2'}}b') = \sum \varepsilon(1_{\underline{1}}b)1_{\underline{1'}}1_{\underline{2}}\varepsilon(1_{\underline{2'}}b') = t(b')\sqcap^L(b), \ (*)$$

and hence, for all $a, a' \in A$, $s(a)t(a') = t(a')s(a)$, as required for the source and target maps.

Observe that, for a weak bialgebra B, its co-opposite B^{cop} (B with flipped comultiplication) is also a weak bialgebra, and t is a left counital source map for B^{cop}. Thus the corresponding statements 36.7(1)(i)–(vi) hold for t, in particular $t(bb') = t(bt(b'))$ and $t(b)t(b') = t(t(b)b')$ for all $b, b' \in B$. Using these relations as well as equation $(*)$, we obtain for all $b, b' \in B$,

$$t(\sqcap^L(b)b') = t(\sqcap^L(b)t(b')) = t(t(b')\sqcap^L(b)) = t(b')t(\sqcap^L(b)).$$

This immediately implies that t is an anti-algebra map.

Again, since t is a left counital source map for B^{cop}, there is a version of 36.7(1)(i) for t, in particular, $\sum t(1_{\underline{1}}) \otimes 1_{\underline{2}} = \sum 1_{\underline{1}} \otimes 1_{\underline{2}}$. Furthermore, note that $t \circ \sqcap^L = t$, since, by (w.1), for all $b \in B$,

$$t \circ \sqcap^L(b) = \sum 1_{\underline{1}}\varepsilon(1_{\underline{2}}\sqcap^L(b)) = \sum 1_{\underline{1}}\varepsilon(1_{\underline{2}}1_{\underline{2'}})\varepsilon(1_{\underline{1'}}b) = \sum 1_{\underline{1}}\varepsilon(1_{\underline{2}}b) = t(b),$$

as required. Since $\sum \sqcap^L(1_{\underline{1}}) \otimes 1_{\underline{2}}$ is a separability element for A,

$$a \sum \sqcap^L(1_{\underline{1}}) \otimes 1_{\underline{2}} = \sum \sqcap^L(1_{\underline{1}}) \otimes 1_{\underline{2}}a,$$

for all $a \in A$. Now, applying $t \otimes I_B$ to this equality and using the above results (including the fact that t is an anti-algebra map), we deduce that

$$\sum 1_{\underline{1}}t(a) \otimes 1_{\underline{2}} = \sum 1_{\underline{1}} \otimes 1_{\underline{2}}a, \quad \text{for all } a \in A.$$

Therefore, for all $b \in B$ and $a \in A$,

$$\sum b_{\underline{1}} \otimes b_{\underline{2}}s(a) = \sum b_{\underline{1}}1_{\underline{1}} \otimes b_{\underline{2}}1_{\underline{2}}a = \sum b_{\underline{1}}1_{\underline{1}}t(a) \otimes b_{\underline{2}}1_{\underline{2}} = \sum b_{\underline{1}}t(a) \otimes b_{\underline{2}},$$

that is, $\underline{\Delta}(b) \in B \times_A B$. Finally, using (w^{tw}.1), note that for all $b, b' \in B$,

$$\sqcap^L(bt(b')) = \sum \varepsilon(1_{\underline{1}}b1_{\underline{1}'})\varepsilon(1_{\underline{2}'}b')1_{\underline{2}} = \sum \varepsilon(1_{\underline{1}}bb')1_{\underline{2}} = \sqcap^L(bb'), \qquad (**)$$

so that

$$\begin{aligned} \underline{\varepsilon}(bb') = \sqcap^L(bb') &= \sqcap^L(b\sqcap^L(b')) = \sqcap^L(bs(\underline{\varepsilon}(b'))), \\ &= \sqcap^L(bt(b')) = \sqcap^L(bt(\underline{\varepsilon}(b'))), \end{aligned}$$

as required for a counit of a bialgebroid. Note that we used 36.7(1)(v) to derive the first conclusion, and (**) together with the fact that $t = t \circ \sqcap^L$ to obtain the second conclusion. This completes the proof. □

36.10. Antipodes. An element $S \in \mathrm{End}_R(B)$ is called a *left antipode* if $S * I_B = \sqcap^R$ and $S * \sqcap^L = S$. Explicitly, this means that for all $b \in B$,

$$\sum (Sb_{\underline{1}})\, b_{\underline{2}} = \sum 1_{\underline{1}}\varepsilon(b1_{\underline{2}}) \quad \text{and} \quad \sum S(1_{\underline{1}}b)1_{\underline{2}} = S(b).$$

S is called a *right antipode* provided $I_B * S = \sqcap^L$ and $\sqcap^R * S = S$, that is,

$$\sum b_{\underline{1}}(Sb_{\underline{2}}) = \sum \varepsilon(1_{\underline{1}}b)1_{\underline{2}} \quad \text{and} \quad \sum 1_{\underline{1}}S(b1_{\underline{2}}) = S(b).$$

S is called an *antipode* if it is both a left and a right antipode, that is, the following identities hold:

$$S * I_B = \sqcap^R, \quad S * I_B * S = S, \quad I_B * S = \sqcap^L.$$

A weak bialgebra B with an antipode is called a *weak Hopf algebra.*

It is straightforward to see that the antipode of a weak bialgebra has the usual properties of the antipode in case B is a bialgebra (then $\sqcap^L$ and $\sqcap^R$ coincide with $\iota \circ \varepsilon$). On the other hand, the essential properties of the antipodes can also be shown for the weak case.

36.11. Antipode and source maps. *Let B be a weak Hopf algebra with antipode S. Then, for any $a, b \in B$:*

(1) $\sum \varepsilon(S(b)1_{\underline{1}})1_{\underline{2}} = \sqcap^L(b) = \sum S(1_{\underline{1}})\varepsilon(1_{\underline{2}}b)$.
(2) $\sum 1_{\underline{1}}\varepsilon(1_{\underline{2}}S(b)) = \sqcap^R(b) = \sum \varepsilon(b1_{\underline{1}})S(1_{\underline{2}})$.
(3) $\sqcap^L \circ \sqcap^R = \sqcap^L \circ S = S \circ \sqcap^R$ *and* $\sqcap^R \circ \sqcap^L = \sqcap^R \circ S = S \circ \sqcap^L$.
(4) $\sqcap^L(a\sqcap^L(b)) = \sum a_{\underline{1}}\sqcap^L(b)S(a_{\underline{2}})$ *and* $\sqcap^R(\sqcap^R(a)b) = \sum S(b_{\underline{1}})\sqcap^R(a)b_{\underline{2}}$.
(5) $S(B^L) = B^R$ *and* $S(B^R) = B^L$.

Proof. Let $b \in B$. The first equality in (1) follows by the computation

$$\begin{aligned} \sqcap^L(b) &= \sum \varepsilon(1_{\underline{1}}\sqcap^L(b))1_{\underline{2}} \\ \overset{36.6}{} &= \sum \varepsilon(\sqcap^L(b)1_{\underline{1}})1_{\underline{2}} = \sum \varepsilon(b_{\underline{1}}S(b_{\underline{2}})1_{\underline{1}})1_{\underline{2}} \\ \overset{36.7(2)}{} &= \sum \varepsilon(\sqcap^R(b_{\underline{1}})S(b_{\underline{2}})1_{\underline{1}})1_{\underline{2}} \\ \overset{36.10}{} &= \sum \varepsilon(S(b)1_{\underline{1}})1_{\underline{2}}. \end{aligned}$$

The first equality in (2) results from an analogous calculation. For the first equality in (3) compute (again referring to 36.6, 36.7 and 36.10)

$$\begin{aligned}\sqcap^L \circ S(b) &= \sum \varepsilon(1_{\underline{1}} S(b)) 1_{\underline{2}} = \sum \varepsilon(1_{\underline{1}} S(b_{\underline{1}}) b_{\underline{2}} S(b_{\underline{3}})) 1_{\underline{2}} \\ &= \sum \varepsilon(1_{\underline{1}} S(b_{\underline{1}}) \sqcap^L(b_{\underline{2}})) 1_{\underline{2}} = \sum \varepsilon(1_{\underline{1}} S(b_{\underline{1}}) b_{\underline{2}}) 1_{\underline{2}} = \sqcap^L \circ \sqcap^R(b),\end{aligned}$$

and an analogous computation proves $\sqcap^R \circ S = \sqcap^R \circ \sqcap^L$.

As an intermediate result we show

$$\sum S(b)_{\underline{1}} \otimes S(b)_{\underline{2}} = \sum 1_{\underline{1}} S(b_{\underline{2}}) \otimes 1_{\underline{2}} S(b_{\underline{1}}) \qquad (*)$$

by the computation

$$\begin{aligned}\Delta \circ S(b) &= \sum \Delta(S(b_{\underline{1}}) b_{\underline{2}} S(b_{\underline{3}})) = \sum \Delta(S(b_{\underline{1}}) \sqcap^L(b_{\underline{2}})) \\ &= \sum S(b_{\underline{1}})_{\underline{1}} \sqcap^L(b_{\underline{2}}) \otimes S(b_{\underline{1}})_{\underline{2}} = \sum S(b_{\underline{1}})_{\underline{1}} b_{\underline{2}} S(b_{\underline{3}}) \otimes S(b_{\underline{1}})_{\underline{2}} \\ &= \sum S(b_{\underline{1}})_{\underline{1}} b_{\underline{2}}\, \varepsilon(1_{\underline{1}} b_{\underline{3}}) S(b_{\underline{4}}) \otimes S(b_{\underline{1}})_{\underline{2}} 1_{\underline{2}} \\ &= \sum S(b_{\underline{1}})_{\underline{1}} b_{\underline{2}} S(b_{\underline{4}}) \otimes S(b_{\underline{1}})_{\underline{2}} \sqcap^L(b_{\underline{3}}) \\ &= \sum S(b_{\underline{1}})_{\underline{1}} b_{\underline{2}} S(b_{\underline{5}}) \otimes S(b_{\underline{1}})_{\underline{2}} b_{\underline{3}} S(b_{\underline{4}}) \\ &= \sum [S(b_{\underline{1}}) b_{\underline{2}}]_{\underline{1}} S(b_{\underline{4}}) \otimes [S(b_{\underline{1}}) b_{\underline{2}}]_{\underline{2}} S(b_{\underline{3}}) \\ &= \sum 1_{\underline{1}} S(b_{\underline{3}}) \otimes 1_{\underline{2}} \sqcap^R(b_{\underline{1}}) S(b_{\underline{2}}) = \sum 1_{\underline{1}} S(b_{\underline{2}}) \otimes 1_{\underline{2}} S(b_{\underline{1}}).\end{aligned}$$

The equality $(*)$ is then used to show

$$\begin{aligned}\sqcap^L \circ S(b) &= \sum S(b)_{\underline{1}} S(S(b)_{\underline{2}}) = \sum 1_{\underline{1}} S(b_{\underline{2}}) S(1_{\underline{2}}(S(b_{\underline{1}})) \\ &= \sum 1_{\underline{1}}(S(1_{\underline{2}} \sqcap^R(b)) = 1_{\underline{1}}(S(\sqcap^R(b) 1_{\underline{2}}) = S \circ \sqcap^R(b); \\ \sqcap^R \circ S(b) &= \sum S(S(b)_{\underline{1}}) S(b)_{\underline{2}} = \sum S(1_{\underline{1}} S(b_{\underline{2}})) 1_{\underline{2}}(S(b_{\underline{1}}) \\ &= \sum S^2(b_{\underline{2}}) S(b_{\underline{1}}) = S \circ \sqcap^L(b).\end{aligned}$$

The preceding results allow one to prove the second equality in (1),

$$\begin{aligned}\sum \varepsilon(1_{\underline{2}} b) S(1_{\underline{1}}) &= \sum \varepsilon(1_{\underline{2}} b) S \circ \sqcap^R(1_{\underline{1}}) = \sum \varepsilon(1_{\underline{2}} b) \sqcap^L \circ S(1_{\underline{1}}) \\ &= \sum \varepsilon(1_{\underline{2}} b)\, \varepsilon(1_{\underline{1'}} S(1_{\underline{1}})) 1_{\underline{2'}} = \sum \varepsilon(1_{\underline{2}} b)\, \varepsilon(1_{\underline{1'}} 1_{\underline{1}}) 1_{\underline{2'}} \\ &= \sum \varepsilon(1_{\underline{1'}} b)_{\underline{2'}} = \sqcap^L(b),\end{aligned}$$

and the second equality in (2) is shown with a similar proof. The statements in (4) are easily derived from (1) and 36.6. The assertions in (5) follow immediately from the equalities in (1), (2) and (3). □

36.12. Properties of the antipode. *Let B be a weak Hopf algebra with antipode S. Then for any $a, b \in B$:*

(1) $S(ab) = S(b)S(a)$, that is, S is an algebra anti-morphism. Furthermore, $\Delta(S(b)) = S \otimes S(\Delta^{tw}(b))$, that is, S is a coalgebra anti-morphism.

(2) $S(1_B) = 1_B$ and $\varepsilon \circ S = \varepsilon$, that is, unit and counit are S-invariant.

(3) $\sum b_{\underline{1}} \otimes S(b_{\underline{2}})b_{\underline{3}} = \sum b1_{\underline{1}} \otimes S(1_{\underline{2}})$ and $\sum b_{\underline{1}}S(b_{\underline{2}}) \otimes b_{\underline{3}} = \sum S(1_{\underline{1}}) \otimes 1_{\underline{2}}b$.

Proof. To show that S is an anti-morphism of algebras we compute

$$\begin{aligned} S(ab) &= \sum S(a_{\underline{1}}b_{\underline{1}})a_{\underline{2}}b_{\underline{2}}S(a_{\underline{3}}b_{\underline{3}}) \;=\; \sum S(a_{\underline{1}}b_{\underline{1}})\sqcap^L(a_{\underline{2}}\sqcap^L(b_{\underline{2}})) \\ &= \sum S(a_{\underline{1}}b_{\underline{1}})a_{\underline{2}}\sqcap^L(b_{\underline{2}})S(a_{\underline{3}}) \;=\; \sum \sqcap^R(\sqcap^R(a_{\underline{1}})b_{\underline{1}})S(b_{\underline{2}})S(a_{\underline{2}}) \\ &= \sum S(b_{\underline{1}})\sqcap^R(a_{\underline{1}})b_{\underline{2}}S(b_{\underline{3}})S(a_{\underline{3}}) \\ &= \sum S(b_{\underline{1}})b_{\underline{2}}S(b_{\underline{3}})S(a_{\underline{1}})a_{\underline{2}}S(a_{\underline{3}}) \;=\; S(b)S(a). \end{aligned}$$

To show that S is an anti-morphism of coalgebras we first compute

$$\begin{aligned} \sum S(1_{\underline{1}}) \otimes S(1_{\underline{2}}) &= \sum 1_{\underline{2'}}\,\varepsilon(1_{\underline{1'}}S(1_{\underline{1}})) \otimes S(1_{\underline{2}}) \\ &= \sum 1_{\underline{2'}} \otimes S(1_{\underline{2}})\varepsilon(1_{\underline{1'}}1_{\underline{1}}) \\ &= \sum 1_{\underline{2'}} \otimes \sqcap^R(1_{\underline{1'}}) \;=\; \sum 1_{\underline{2}} \otimes 1_{\underline{1}}. \end{aligned}$$

Now equation $(*)$ in the proof of 36.11 yields

$$\begin{aligned} \sum S(b)_{\underline{1}} \otimes S(b)_{\underline{2}} &= \sum 1_{\underline{1}}S(b_{\underline{2}}) \otimes 1_{\underline{2}}S(b_{\underline{1}}) \\ &= \sum S(1_{\underline{1}})S(b_{\underline{2}}) \otimes S(1_{\underline{1}})S(b_{\underline{1}}) \;=\; \sum S(b_{\underline{2}}) \otimes S(b_{\underline{1}}). \end{aligned}$$

For the claims in (2) observe that $S(1) = S(\sqcap^R(1)) = \sqcap^L(\sqcap^R(1)) = 1$, and

$$\begin{aligned} \varepsilon(S(b)) &= \sum \varepsilon(S(b_{\underline{1}})b_{\underline{2}}S(b_{\underline{3}})) = \sum \varepsilon(S(b_{\underline{1}})\sqcap^L(b_{\underline{2}})) \\ &= \sum \varepsilon(S(b_{\underline{1}})b_{\underline{2}}) = \varepsilon(\sqcap^R(b)) = \varepsilon(b). \end{aligned}$$

(3) The assertions follow by the computations

$$\begin{aligned} \sum b_{\underline{1}} \otimes S(b_{\underline{2}})b_{\underline{3}} &= \sum b_{\underline{1}} \otimes \varepsilon(b_{\underline{2}}1_{\underline{1}})S(1_{\underline{2}}) \\ &= \sum b_{\underline{1}}1_{\underline{1'}} \otimes \varepsilon(b_{\underline{2}}1_{\underline{2'}}1_{\underline{1}})S(1_{\underline{2}}) \\ &= \sum b_{\underline{1}}1_{\underline{1}} \otimes \varepsilon(b_{\underline{2}}1_{\underline{2}})S(1_{\underline{3}}) \;=\; \sum b1_{\underline{1}} \otimes S(1_{\underline{2}}), \\ \sum b_{\underline{1}}S(b_{\underline{2}}) \otimes b_{\underline{3}} &= \sum S(1_{\underline{1}})\,\varepsilon(1_{\underline{2}}b_{\underline{1}}) \otimes b_{\underline{2}} \\ &= \sum S(1_{\underline{1}})\,\varepsilon(1_{\underline{2}}1_{\underline{1'}}b_{\underline{1}}) \otimes 1_{\underline{2'}}b_{\underline{2}} \\ &= \sum S(1_{\underline{1}}) \otimes \varepsilon(1_{\underline{2}}b_{\underline{1}})1_{\underline{3}}b_{\underline{2}} \;=\; \sum S(1_{\underline{1}}) \otimes 1_{\underline{2}}b. \end{aligned}$$

□

36.13. Galois corings. Let B be a weak bialgebra. Then the weak B-coring $B \otimes_R B$ is said to be *Galois* if there is an isomorphism (see 35.16)

$$\mathsf{can}_B : B \otimes_{B^L} B \to (B \otimes_R B) \cdot 1, \quad a \otimes b \mapsto (a \otimes 1)\Delta(b).$$

Obviously, can_B is a left B-module morphism.

36.14. Existence of antipodes. *Let B be a weak bialgebra. Then:*

(1) B has a right antipode if and only if can_B has a left inverse in ${}_B\mathbf{M}$.

(2) can_B is an isomorphism if and only if B has an antipode.

Proof. (1) ($\Leftarrow$) To simplify notation put $\mathsf{can}_B = \gamma$. If β is a left inverse of γ, then $1 \otimes_{B^L} b = \beta \circ \gamma(1 \otimes_{B^L} b) = \beta(\Delta b)$, and application of $I_B \otimes \sqcap^L$ yields $\sqcap^L(b) = (I_B \otimes \sqcap^L) \circ \beta(\Delta b)$. The composition

$$S : B \xrightarrow{1\otimes -} B \otimes_R B \xrightarrow{-\cdot 1} (B \otimes_R B) \cdot 1 \xrightarrow{\beta} B \otimes_{B^L} B \xrightarrow{I_B \otimes \sqcap^L} B$$

is a right antipode since

$$\begin{aligned} \mu \circ (id \otimes S) \circ \Delta(b) &= \sum b_{\underline{1}}((I_B \otimes \sqcap^L)\beta(1_{\underline{1}} \otimes b_{\underline{2}}1_{\underline{2}})) \\ &= (I_B \otimes \sqcap^L) \circ \beta(\Delta b) = \sqcap^L(b), \text{ and} \\ \sqcap^R * S(b) &= \sum 1_{\underline{1}} S(b1_{\underline{2}}) = \sum (I_B \otimes \sqcap^L) \circ \beta(1_{\underline{1}} \otimes b1_{\underline{2}}) = S(b). \end{aligned}$$

($\Rightarrow$) Let $S : B \to B$ be a right antipode and consider the map

$$\beta : B \otimes_R B \to B \otimes_{B^L} B, \quad a \otimes b \mapsto \sum aS(b_{\underline{1}}) \otimes_{B^L} b_{\underline{2}}.$$

By the property

$$\begin{aligned} \beta((a \otimes b)\Delta(1)) &= \sum a1_{\underline{1}} S(b_{\underline{1}}1_{\underline{21}}e) \otimes_{B^L} b_{\underline{2}}1_{\underline{22}} \\ \overset{(\text{w.2})}{=} &\sum a1_{\underline{1}} S(b_{\underline{1}}1_{\underline{1}'}1_{\underline{2}}) \otimes_{B^L} b_{\underline{2}}1_{\underline{2}'} \\ &= \sum aS(b_{\underline{1}}1_{\underline{1}'}) \otimes_{B^L} b_{\underline{2}}1_{\underline{2}'} = \beta(a \otimes b) \end{aligned}$$

it induces a map $\beta : (B \otimes_R B) \cdot 1 \to B \otimes_{B^L} B$, which is a left inverse of γ since, for any $b \in B$,

$$\begin{aligned} \beta \circ \gamma(1 \otimes_{B^L} b) = \beta(\Delta b) &= \sum b_{\underline{1}} S(b_{\underline{21}}) \otimes_{B^L} b_{\underline{22}} = \sum b_{\underline{11}} e S(b_{\underline{12}}) \otimes_{B^L} b_{\underline{2}} \\ &= \sum \sqcap^L(b_{\underline{1}}) \otimes_{B^L} b_{\underline{2}} = 1 \otimes_{B^L} b\,. \end{aligned}$$

(2) ($\Rightarrow$) Assume that γ is bijective. By (1), there exists a right antipode S, and hence $I_B * S * I_B = \sqcap^L * I_B = I_B$. Any element in $(B \otimes B) \cdot 1$ can be written as $\sum_i a_i \Delta c_i$, for some $a_i, c_i \in B$, and

$$\textstyle\sum_i \mu \circ (I_B \otimes (S * I_B - \varepsilon_B))(a_i \Delta c_i) = \sum_i a_i (I_B * S * I_B - I_B * \varepsilon_B)(c_i) = 0.$$

This implies for $(1\otimes b)\Delta(1)\in(B\otimes B)\cdot 1$, where $b\in B$,

$$\begin{aligned}\sqcap^R(b)=\sum 1_{\underline{1}}\varepsilon(b1_{\underline{2}}) &= \sum 1_{\underline{1}}S*I_B(b1_{\underline{2}}) = \sum 1_{\underline{1}}S(b_{\underline{1}}1_{\underline{21}})\,b_{\underline{2}}1_{\underline{22}}\\ \text{(w.2)} &= \sum 1_{\underline{1}}S(b_{\underline{1}}1_{\underline{1'}}1_{\underline{2}})\,b_{\underline{2}}1_{\underline{2'}}\\ &= \sum S(b_{\underline{1}}1_{\underline{1'}})\,b_{\underline{2}}1_{\underline{2'}} = S*I_B(b).\end{aligned}$$

Moreover, $\sqcap^R*S=S*I_B*S=S*\sqcap^L=S$, showing that S is an antipode.

($\Leftarrow$) We know for β defined in (1) that $\beta\circ\gamma$ is the identity map on $B\otimes_{B^L}B$. Furthermore, for any $a,b\in B$,

$$\begin{aligned}\gamma\circ\beta((a\otimes b)\cdot 1) &= \sum(aS(b_{\underline{1}})\otimes 1)\Delta(b_{\underline{2}}) = \sum a\,S(b_{\underline{1}})b_{\underline{21}}\otimes b_{\underline{22}}\\ &= \sum a\,S(b_{\underline{11}})b_{\underline{12}}\otimes b_{\underline{2}} = a\sum\sqcap^R(b_{\underline{1}})\otimes b_{\underline{2}}\\ \text{36.7(1)}(i) &= a(1\otimes b)\cdot 1 = (a\otimes b)\cdot 1,\end{aligned}$$

showing that $\gamma\circ\beta$ is the identity on $(B\otimes_R B)\cdot 1$. So γ is an isomorphism. □

36.15. Weak Hopf modules. Let B be a weak bialgebra. Then weak right $B\widetilde{\otimes_R}B$-comodules are called *weak Hopf modules* and their category is denoted by $\widetilde{\mathbf{M}}^{B\otimes_R B}$. They have the following properties.

(1) *B is a right (and left) weak Hopf module with a grouplike element $\Delta(1_B)=\sum 1_{\underline{1}}\otimes 1_{\underline{2}}$.*

(2) *The coinvariants of any $M\in\widetilde{\mathbf{M}}^{B\otimes_R B}$ are*

$$M^{co(B\otimes_R B)}=\{m\in M\mid \varrho^M(m)=m\otimes_B\Delta_B(1_B)\}.$$

(3) *For any $N\in\widetilde{\mathbf{M}}_B$, the coinvariants of $N\otimes_B(B\otimes_R B)$ are*

$$\mathrm{Hom}^{B\otimes_R B}(B,N\otimes_B(B\otimes_R B))\simeq NB.$$

In particular, $B^{co(B\otimes_R B)}=B^L$ and $(B\otimes_R B)^{co(B\otimes_R B)}\simeq B$.

(4) *For any $m\in M\in\widetilde{\mathbf{M}}^{B\otimes_R B}$, $m\cdot 1=\sum m_{\underline{0}}\sqcap^R(m_{\underline{1}})$.*

Proof. By 36.4, $\Delta(1_B)$ is a grouplike element and the assertions (2) and (3) follow from 35.14. The coinvariants of B are considered in 36.6. Finally, (4) follows from the counitality of M and the definition of $\sqcap^R$. □

36.16. Fundamental theorem for weak Hopf algebras. *For a weak R-bialgebra B, the following are equivalent:*

(a) *B is a weak Hopf algebra;*

(b) *$B\otimes_R B$ is a Galois weak coring;*

(c) *$\mathrm{Hom}^{B\otimes_R B}(B,-):\mathbf{M}^{(B\otimes_R B)\cdot 1}\to\mathbf{M}_{B^L}$ is an equivalence with inverse $-\otimes_{B^L}B$;*

(d) *for every* $M \in \mathbf{M}^{(B\otimes_R B)\cdot 1}$, $\varphi_M : M^{co(B\otimes_R B)} \otimes_{B^L} B \to M$, $n \otimes b \mapsto nb$, *is an isomorphism.*

Proof. (a) $\Leftrightarrow$ (b) is shown in 36.14.
(b) $\Rightarrow$ (d) First observe that, for any $m \in M$,

$$\begin{aligned} \varrho^M(m_{\underline{0}}S(m_{\underline{1}})) &= \sum(m_{\underline{00}} \otimes m_{\underline{01}})(S(m_{\underline{1}})_{\underline{1}} \otimes S(m_{\underline{1}})_{\underline{2}}) \\ &= \sum m_{\underline{0}}S(m_{\underline{3}}) \otimes m_{\underline{1}}S(m_{\underline{2}}) \\ {\scriptstyle 36.12(3)} \quad &= \sum m_{\underline{0}}S(1_{\underline{2}}m_{\underline{1}}) \otimes S(1_{\underline{1}}) = \sum m_{\underline{0}}S(m_{\underline{1}})1_{\underline{1}} \otimes 1_{\underline{2}}, \end{aligned}$$

that is, $\sum m_{\underline{0}}S(m_{\underline{1}}) \in M^{co(B\otimes_R B)}$. Now define a map

$$\beta : M \to M^{co(B\otimes_R B)} \otimes_{B^L} B, \quad m \mapsto \sum m_{\underline{0}}S(m_{\underline{1}}) \otimes m_{\underline{2}}.$$

This is the inverse to φ since, on the one hand, for any $m \in M$,

$$\varphi \circ \beta(m) = \sum \varphi(m_{\underline{0}}S(m_{\underline{1}}) \otimes m_{\underline{2}}) = \sum m_{\underline{0}}\sqcap^R(m_{\underline{1}}) = m,$$

and on the other hand, for any $n \in M^{co(B\otimes_R B)}$, $b \in B$,

$$\begin{aligned} \beta \circ \varphi(n \otimes b) &= \beta(nb) = \beta(n)b = \sum n1_{\underline{1}}S(1_{\underline{2}}) \otimes 1_{\underline{3}}b \\ &= \sum n \otimes 1_{\underline{1}}S(_{\underline{2}})1_{\underline{3}}b = n \otimes b. \end{aligned}$$

(c) $\Leftrightarrow$ (d) This follows from the commutative diagram

$$\begin{array}{ccc} \mathrm{Hom}^{B\otimes_R B}(B, M) \otimes_{B^L} B & \longrightarrow & M \\ \simeq \downarrow & & \downarrow = \\ M^{co(B\otimes_R B)} \otimes_{B^L} B & \longrightarrow & M. \end{array}$$

(d) $\Rightarrow$ (b) This follows from the observation that $\mathsf{can}_B = \varphi_{(B\otimes_R B)\cdot 1}$. □

Recall that the category of comodules over a weak coring $B \otimes_R B$ is a Grothendieck category provided $B \otimes_R B$ is flat as a left B-module (see 35.10). It follows from 36.7(3) that any weak bialgebra B has B^L as a direct summand, which means that B is flat as a left B^L-module if and only if it is faithfully flat. Hence the characterisation of a ring as a generator for related comodules in 35.17 immediately implies the

36.17. Structure theorem for weak Hopf algebras. *For any weak R-bialgebra B, the following are equivalent:*

(a) *B is a weak Hopf algebra, and B is flat as a left* B^L*-module;*

(b) $B \otimes_R B$ *is flat as a left B-module and*

  (i) *B is a weak Hopf algebra, or*

(ii) *B is a (projective) generator in* $\mathbf{M}^{(B\otimes_R B)\cdot 1}$, *or*

(iii) $\mathrm{Hom}^{B\otimes_R B}(B,-) : \mathbf{M}^{(B\otimes_R B)\cdot 1} \to \mathbf{M}_{B^L}$ *is an equivalence, or*

(iv) *for every* $M \in \mathbf{M}^{(B\otimes_R B)\cdot 1}$, $M^{coB} \otimes_{B^L} B \to M$, $m \otimes b \mapsto mb$, *is an isomorphism.*

Proof. First note that the equivalence of all the assertions in (b) follows from 36.16.

(a) $\Rightarrow$ (b.i) If B is flat as left B^L-module, then $B \otimes_{B^L} B \simeq B \otimes_R B$ is flat as left B-module.

(b.ii) $\Rightarrow$ (a) By the flatness condition, monomorphisms in $\mathbf{M}^{(B\otimes_R B)\cdot 1}$ are injective, and hence the module-theoretic proof (see 43.12) works to show that the generator B is flat over its endomorphism ring B^L. □

36.18. Remarks.

(1) The description of weak bialgebras in 36.5 shows that the definition we use here is equivalent to the original definition in [65], which is stated in terms of conditions (w.1), (w.2), (w^{tw}.1) and (w^{tw}.2). The only difference is that no assumptions on R or the dimension of B are made here. Most of the computations can be found in [65].

Since $B \otimes_R B$ is flat (projective) as a left B-module provided B is flat as an R-module, the conditions of 36.17 hold for any weak Hopf algebra over a field, and for this case the statements are shown in [65, Theorem 3.9].

The antipodes satify in particular $S * I_B * S = S$ and $I_B * S * I_B = I_B$, the conditions used by Fang Li in [114] to define his "weak Hopf algebras".

(2) The relationship between weak bialgebras and bialgebroids described in 36.9 was first established in [113], under the stronger assumption that a weak bialgebra is a weak Hopf algebra with a bijective antipode. The fact that it suffices to take a weak bialgebra to produce a bialgebroid was realised by Schauenburg in [185]. Furthermore, in [113] the following refinement of 36.9 is proven:

If H is a weak Hopf algebra, then the corresponding bialgebroid over A is a Hopf algebroid.

On the other hand, the following converse to 36.9 is proven in [138] (cf. [185]):

Over a field F, if A is a Frobenius-separable algebra (that is, a Frobenius and separable algebra for which a Frobenius element coincides with the separability idempotent), then any A-bialgebroid is a weak bialgebra over F.

We note that a Frobenius-separable algebra is termed a *Frobenius index-1* algebra in [138].

(3) There are several examples of weak Hopf algebras that come from mathematical physics. In particular, weak Hopf algebras appear as symmetries of partition functions of the IRF-type integrable models ([128]) and also as algebraic structures related to the dynamical Yang-Baxter equation (cf. [113]). Other examples of weak Hopf algebras include the generalised Kac algebras introduced in [216]. On the other hand, examples of weak Hopf algebras were obtained as a generalisation of Ocneanu's paragroup [172] introduced in the context of depth-2 subfactors of von Neumann algebras. In view of 36.18, as a corollary of 31.15 one obtains that, given a depth-2 Frobenius algebra extension $B \to D$, with the centraliser $A = D^B$, which is a Frobenius-separable algebra, the endomorphisms space $H = {}_B\mathrm{End}_B(D)$ is a weak Hopf algebra [138].

References. Böhm, Nill and Szlachányi [65]; Böhm and Szlachányi [66]; Brzeziński, Caenepeel and Militaru [76]; Brzeziński and Militaru [81]; Caenepeel and DeGroot [82]; Etingof and Nikshych [113]; Fang Li [114]; Hirata [130]; Kadison [137]; Kadison and Szlachányi [138]; Lu [154]; Nakajima [164]; Nill [169]; Schauenburg [183, 185]; Sweedler [192]; Takeuchi [196]; Wisbauer [212]; Xu [215]; Yamanouchi [216].

37 Weak entwining structures

In this section A denotes an R-algebra with product μ and unit (map) ι, and C an R-coalgebra with coproduct Δ and counit ε. We have seen in 32.6 that entwining structures are in one-to-one correspondence with A-coring structures on $A \otimes_R C$. Now a natural question arises. Suppose that $A \otimes_R C$ is a weak coring. What is the relationship between A and C? This leads to the introduction of

37.1. Weak entwining structures. A triple (A, C, ψ) is said to be a (right-right) *weak entwining structure* (over R) if $\psi : C \otimes_R A \to A \otimes_R C$ is an R-module map satisfying the following four conditions for all $a, b \in A$, $c \in C$, where the α-notation $\psi(c \otimes a) = \sum_\alpha a_\alpha \otimes c^\alpha$ from 32.3 is used:

$$\text{(we.1)} \quad \sum_\alpha (ab)_\alpha \otimes c^\alpha = \sum_\alpha a_\alpha b_\beta \otimes c^{\alpha\beta}.$$

$$\text{(we.2)} \quad \sum_\alpha a_\alpha \psi(c^\alpha_{\underline{1}} \otimes 1) \otimes c^\alpha_{\underline{2}} = \sum_{\alpha,\beta} a_{\alpha\beta} \otimes c_{\underline{1}}{}^\beta \otimes c_{\underline{2}}{}^\alpha.$$

$$\text{(we.3)} \quad \sum_\alpha a_\alpha \varepsilon(c^\alpha) = \sum_\alpha \varepsilon(c^\alpha) 1_\alpha a.$$

$$\text{(we.4)} \quad \sum_\alpha 1_\alpha \otimes c^\alpha = \sum_\alpha \varepsilon(c_{\underline{1}}{}^\alpha) 1_\alpha \otimes c_{\underline{2}}.$$

In comparison with the definition of an entwining structure in 32.1, one sees that this generalisation is obtained by keeping the left pentagon equation but weakening the right pentagon equation and replacing the triangle equalities by hexagon equalities. Note that the right pentagon equation implies (we.2).

Weak entwining structures are in bijective correpondence with canonical weak coring structures on $A \otimes_R C$.

37.2. Weak corings and weak entwining structure. *Consider $A \otimes_R C$ as a left A-module canonically.*

(1) If (A, C, ψ) is a weak entwining structure, then $A \otimes_R C$ is a right A-module by

$$(a \otimes c) \cdot b = a\psi(c \otimes b), \quad \textit{for } a, b, \in A,\ c \in C,$$

and $A \otimes_R C$ is a left unital weak A-coring with coproduct and counit

$$\begin{array}{rcccl} \underline{\Delta} : A \otimes_R C & \to & (A \otimes_R C) \otimes_A (A \otimes_R C) & \simeq & (A \otimes_R C) \cdot 1 \otimes_R C, \\ a \otimes c & \mapsto & \sum (a \otimes c_{\underline{1}}) \otimes_A (1 \otimes c_{\underline{2}}) & \mapsto & \sum (a \otimes c_{\underline{1}}) \cdot 1 \otimes c_{\underline{2}}, \end{array}$$

$$\begin{array}{rcccl} \underline{\varepsilon} : A \otimes_R C & \to & (A \otimes_R C) \cdot 1 & \to & A\,, \\ a \otimes c & \mapsto & (a \otimes c) \cdot 1 & \mapsto & (I \otimes \varepsilon)((a \otimes c) \cdot 1). \end{array}$$

(2) Assume that $A \otimes_R C$ is an A-coring with $\underline{\Delta}$ and $\underline{\varepsilon}$ as defined in (1). Then the R-linear map

$$C \otimes_R A \to A \otimes_R C, \quad c \otimes a \mapsto (1 \otimes c) \cdot a,$$

is a weak entwining map for A and C.

Proof. Clearly both $\underline{\Delta}$ and $\underline{\varepsilon}$ are left A-linear. The equalities

$$(1 \otimes c) \cdot ab = \sum (ab)_\alpha \otimes c^\alpha, \quad (1 \otimes c) \cdot a \cdot b = \sum a_\alpha b_\beta \otimes c^{\alpha\beta},$$

show that (we.1) implies that $A \otimes_R C$ is a (nonunital) right A-module. By definition,

$$\begin{array}{rcl} \underline{\Delta}(1 \otimes c) \cdot a &=& \sum (1 \otimes c_{\underline{1}}) \otimes_A (1 \otimes c_{\underline{2}}) \cdot a = \sum (1 \otimes c_{\underline{1}}) \otimes_A (\sum a_\alpha \otimes c_{\underline{2}}{}^\alpha) \\ &=& \sum a_{\alpha\beta} \otimes c_{\underline{1}}{}^\beta \otimes c_{\underline{2}}{}^\alpha, \text{ and} \\ \underline{\Delta}((1 \otimes c) \cdot a) &=& \underline{\Delta}(\sum a_\alpha \otimes c^\alpha) \;=\; \sum (a_\alpha \otimes c^\alpha_{\underline{1}}) \cdot 1 \otimes c^\alpha_{\underline{2}} \\ &=& \sum a_\alpha \psi(c^\alpha_{\underline{1}} \otimes 1) \otimes c^\alpha_{\underline{2}}, \end{array}$$

and hence (we.2) implies that $\underline{\Delta}$ is right A-linear.

The remaining conditions are related to properties of $\underline{\varepsilon}$. The equalities

$$\begin{array}{l} \underline{\varepsilon}((1 \otimes c) \cdot a) = I_A \otimes \varepsilon((1 \otimes c) \cdot a) = \sum a_\alpha \varepsilon(c^\alpha), \text{ and} \\ \underline{\varepsilon}((1 \otimes c) \cdot 1) \cdot a = (I_A \otimes \varepsilon((1 \otimes c) \cdot 1)) \cdot a = \sum \varepsilon(c^\alpha) 1_\alpha a \end{array}$$

show that $\underline{\varepsilon}$ is a right A-module morphism provided (we.3) holds, and by

$$\begin{array}{rcl} (1 \otimes c) \cdot 1 &=& \sum 1_\alpha \otimes c^\alpha, \text{ and} \\ (\underline{\varepsilon} \otimes I_C) \circ \underline{\Delta}(1 \otimes c) &=& I_A \otimes \varepsilon \otimes I_C(\sum 1_\alpha \otimes c_{\underline{1}}{}^\alpha \otimes c_{\underline{2}}) = \sum \varepsilon(c_{\underline{1}}{}^\alpha) 1_\alpha \otimes c_{\underline{2}}, \end{array}$$

we see that $\underline{\varepsilon}$ is weakly counitary provided that (we.4) is satisfied.

(2) If $A \otimes_R C$ is a weak A-coring with the given maps, the reverse conclusions show that the map defined by the right multiplication yields a weak entwining between A and C. □

Immediately from the definition of a weak bialgebra (cf. 36.5) and 37.2 one obtains the following example of a weak entwining structure (cf. 33.1):

37.3. Weak bialgebra entwining. *Given a weak bialgebra B, the triple (B, B, ψ) with*

$$\psi : B \otimes_R B \to B \otimes_R B, \quad b' \otimes b \mapsto \sum b_{\underline{1}} \otimes b' b_{\underline{2}}.$$

is a weak entwining structure.

As pointed out in 35.3, associated to any weak coring $\mathcal{C}$ there is a coring $A\mathcal{C}A$. This yields in our situation:

37.4. A coring associated to a weak entwining structure. *Let (A, C, ψ) be a weak entwining structure. With the right A-module structure defined by ψ (see 37.2), the product $\mathcal{C} = (A \otimes_R C) \cdot A$ is an A-coring with coproduct and counit*

$$\begin{array}{l} \underline{\Delta} : \mathcal{C} \to \mathcal{C} \otimes_A \mathcal{C}, \;\; (1 \otimes c) \cdot 1 \mapsto \sum_{\alpha,\beta} 1_{\alpha\beta} \otimes c_{\underline{1}}{}^\beta \otimes c_{\underline{2}}{}^\alpha, \\ \underline{\varepsilon} : \mathcal{C} \to A, \;\; (1 \otimes c) \cdot 1 \mapsto \sum_\alpha 1_\alpha \varepsilon(c^\alpha). \end{array}$$

Proof. This follows from 35.3 with explicit formulae taken from 37.2. □

37.5. Weak entwined modules. For a weak entwining structure (A, C, ψ), right weak comodules over the canonical weak coring $A \otimes_R C$ are called *weak (A, C, ψ)-entwined modules* . In this context, the category $\widetilde{\mathbf{M}}^{A\otimes_R C}$ is denoted by $\widetilde{\mathbf{M}}_A^C(\psi)$.

For $M \in \widetilde{\mathbf{M}}^{A\otimes_R C}$ and $m \in M$, write $\varrho^M(m) = \sum m_{\underline{0}} \otimes m_{\underline{1}}$, where $m_{\underline{0}} \in MA$ and $m_{\underline{1}} \in C$, and coassociativity is given by the commutative diagram

$$\begin{array}{ccc} M & \xrightarrow{\varrho^M} & M \otimes_A (A \otimes_R C) \\ {\scriptstyle \varrho^M}\downarrow & & \downarrow{\scriptstyle I_M \otimes \underline{\Delta}} \\ M \otimes_A (A \otimes_R C) & \xrightarrow{\varrho^M \otimes I_{A\otimes_R C}} & M \otimes_A (A \otimes_R C) \otimes_A (A \otimes_R C), \end{array}$$

corresponding to the identity

$$\sum m_{\underline{0}} \otimes_A (1 \otimes_R m_{\underline{11}}) \otimes_A (1 \otimes_R m_{\underline{12}}) = \sum m_{\underline{00}} \otimes_A (1 \otimes_R m_{\underline{01}}) \otimes_A (1 \otimes_R m_{\underline{1}}).$$

Expressing the right multiplication by A in terms of ψ, this gives

$$\sum m_{\underline{0}}\psi(m_{\underline{11}} \otimes_R 1) \otimes_R m_{\underline{12}} = \sum m_{\underline{00}}\psi(m_{\underline{01}} \otimes_R 1) \otimes_R m_{\underline{1}}$$

in $MA \otimes_R C \otimes_R C$. A-linearity of ϱ^M is expressed by the relation

$$\varrho^M(ma) = \sum m_{\underline{0}}\psi(m_{\underline{1}} \otimes_R a) \ \text{ in } MA \otimes_R C.$$

For example, any unital right A-module with a coassociative right C-coaction is a weak (A, C, ψ)-entwined module provided it satisfies the latter compatibility condition.

As in the case of a bialgebra entwining in 33.1, modules for a weak bialgebra entwining in 37.3 are simply the weak Hopf modules in 36.15.

37.6. Dual algebra and smash product. Let $A \otimes_R C$ be a weak A-coring (as in 37.2). Then the isomorphism ${}_A\mathrm{Hom}(A\otimes_R C, A) \simeq \mathrm{Hom}_R(C, A)$ induces an associative algebra structure on $\mathrm{Hom}_R(C, A)$ with the product

$$f *^l g(c) = \sum g(c_{\underline{2}})_\psi f(c_{\underline{1}}{}^\psi), \quad \text{for all } f, g \in \mathrm{Hom}_R(C, A),\ c \in C.$$

This algebra is called the *smash product* of A and C and is denoted by $\#(C, A)$. Although $\#(C, A)$ has no unit, it has a central idempotent e,

$$e(c) := \underline{\varepsilon}(1 \otimes c) = I_A \otimes \varepsilon((1 \otimes c) \cdot 1), \ \textit{for } c \in C.$$

If C is locally projective as an R-module, then:

(1) The category $\widetilde{\mathbf{M}}^{A\otimes_R C}$ of right weak $A\otimes_R C$-comodules is a full subcategory of $\mathbf{M}_{\#(C,A)}$.

(2) $A\otimes_R C$ subgenerates all weak right $A\otimes_R C$-comodules that are unital right A-modules.

(3) If C is finitely generated as R-module, then $\#(C,A) *_l e \in \widetilde{\mathbf{M}}^{A\otimes_R C}$.

Proof. For $\tilde{f},\tilde{g} \in {}_A\mathrm{Hom}(A\otimes_R C, A) = {}^*(A\otimes_R C)$ (cf. 35.4),

$$\begin{aligned}\tilde{f} *^l \tilde{g} &= \sum \tilde{f}((1\otimes c_{\underline{1}})\cdot \tilde{g}(1\otimes c_{\underline{2}}))\\ &= \sum \tilde{f}(\tilde{g}(1\otimes c_{\underline{2}})_\psi \otimes c_1^{\psi}) \;=\; \sum \tilde{g}(1\otimes c_{\underline{2}})_\psi\, \tilde{f}(1\otimes c_1^{\psi}),\end{aligned}$$

and this induces the multiplication stated. Since $\underline{\varepsilon}$ is a central idempotent in ${}^*(A\otimes_R C)$ (see 35.4) and e is the image of $\underline{\varepsilon}$ under the isomorphism ${}_A\mathrm{Hom}(A\otimes_R C, A) \to \mathrm{Hom}_R(C,A)$, e is a central idempotent in $\#(C,A)$. If C is locally projective as an R-module, then $A\otimes_R C$ is a locally projective A-module and hence it satisfies the α-condition. So (1) and (2) are special cases of 35.21. Moreover, if C is finitely generated as an R-module, then $A\otimes_R C$ is finitely generated as an A-module, and so is its homomorphic image $(A\otimes_R C)\cdot A$. Now 35.21(4) implies that ${}^*((A\otimes_R C)\cdot A) \simeq {}^*(A\otimes_R C) *^l \underline{\varepsilon}$ is in $\widetilde{\mathbf{M}}^{A\otimes_R C}$, and this ring is isomorphic to $\#(C,A) *^l e$. □

37.7. The existence of a grouplike element. *Let (A,C,ψ) be a weak entwining structure. Then the associated A-coring $A\otimes_R C$ (see 36.5) has a grouplike element $g \in (A\otimes_R C)\cdot A$ if and only if A is an (A,C,ψ)-weak entwined module. The grouplike element is $g = \varrho^A(1_A)$, where ϱ^A is the right coaction of C on A.*

Proof. This is a special case of 35.13. □

General examples of entwining structures come from Doi-Koppinen data. Recall that such a datum comprises an algebra, a coalgebra and a bialgebra. A natural question arises: is it possible to replace a bialgebra in a Doi-Koppinen datum by a more general object, for example, by a weak bialgebra? Does such a new weakened Doi-Koppinen datum lead to a weak entwining structure? These questions were considered in [64] and [82] and led to the introduction of respectively weak Doi-Koppinen data and a certain class of weak entwining structures that we term *self-dual weak entwining structures.*

37.8. Self-dual weak entwining structures. A weak entwining structure (A,C,ψ) is said to be *self-dual* if

$$(I_A\otimes\Delta)\circ\psi = (\psi\otimes I_C)\circ(I_C\otimes\psi)\circ(\Delta\otimes I_A).$$

Explicitly, we require for all $a \in A$ and $c \in C$,

$$\sum_\alpha a_\alpha \otimes c^\alpha{}_{\underline{1}} \otimes c^\alpha{}_{\underline{2}} = \sum ab_{\alpha,\beta} a_{\beta\alpha} \otimes c_{\underline{1}}{}^\alpha \otimes c_{\underline{2}}{}^\beta. \qquad (S)$$

In other words, a self-dual weak entwining structure is a weak entwining structure that satisfies the right pentagon condition in the bow-tie diagram. Since (we.1) in 37.1 combined with a right pentagon implies (we.2), we can equivalently define a self-dual weak entwining structure as a triple (A, C, ψ) in which the map ψ satisfies condition (S) and conditions (we.1), (we.3) and (we.4) in 37.1. This can be expressed as the following bow-tie diagram:

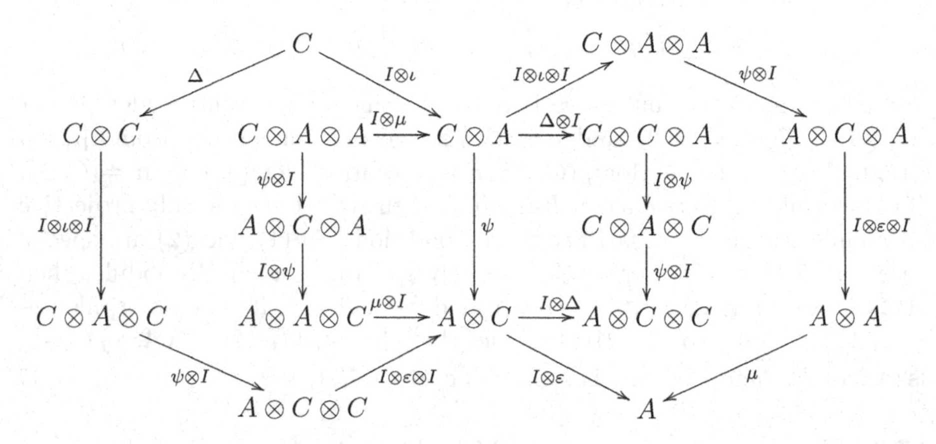

(tensor product over R). The above diagram is self-dual in the same sense as the bow-tie diagram defining an entwining structure (cf. 32.5). Interchanging C with A, ε with ι, Δ with μ, and reversing all the arrows, the above diagram stays invariant (only space rotated). This is the origin of the name of this particular class of weak entwining structures.

One easily checks that a weak bialgebra entwining in 37.3 is self-dual. More elaborate examples of self-dual weak entwining structures can be obtained by the following construction.

37.9. A weak coalgebra-Galois extension. *Let A be a right C-comodule with coaction ϱ^A, and let*

$$B = A^{coC} = \{b \in A \mid \textit{for all } a \in A,\ \varrho^A(ba) = b\varrho^A(a)\} \simeq \mathrm{End}_A^C(A), \quad \textit{and}$$
$$\mathsf{can} : A \otimes_B A \to A \otimes_R C, \quad a \otimes a' \mapsto a\varrho^A(a').$$

View $A \otimes_B A$ as a left A-module via $\mu \otimes I_A$ and a right C-comodule via $I_A \otimes \varrho^A$. View $A \otimes_R C$ as a left A-module via $\mu \otimes I_C$ and as a right C-comodule via $I_A \otimes \Delta$. Now suppose that can *is a split monomorphism in the category $\mathbf{M}_A^C$, that is, there exists a left A-module, right C-comodule map*

$\sigma : A \otimes_R C \to A \otimes_B A$ such that $\sigma \circ \mathsf{can} = I_{A\otimes_B A}$. Let $\tau : C \to A \otimes_B A$, $c \mapsto \sigma(1 \otimes c)$ and define

$$\psi : C \otimes_R A \to A \otimes_R C, \quad \psi = \mathsf{can} \circ (I_A \otimes \mu) \circ (\tau \otimes I_A).$$

Then (A, C, ψ) is a self-dual weak entwining structure. An extension of algebras $B \subset A$ satisfying the above conditions is called a weak coalgebra-Galois extension.

Proof. First we introduce a similar notation as for a translation map by writing $\tau(c) = \sum c^{\tilde{1}} \otimes c^{\tilde{2}}$, for all $c \in C$. Directly from the definition of τ one deduces the following two properties. First, from the fact that σ is a splitting of the canonical map can, it follows that for all $a \in A$,

$$\sum a_{\underline{0}} {a_{\underline{1}}}^{\tilde{1}} \otimes {a_{\underline{1}}}^{\tilde{2}} = 1 \otimes a. \tag{$*$}$$

Second, since σ is a right C-comodule morphism, for all $c \in C$,

$$\sum c^{\tilde{1}} \otimes {c^{\tilde{2}}}_{\underline{0}} \otimes {c^{\tilde{2}}}_{\underline{1}} = \sum {c_{\underline{1}}}^{\tilde{1}} \otimes {c_{\underline{1}}}^{\tilde{2}} \otimes c_{\underline{2}}. \tag{$**$}$$

Note that the equations $(*)$ and $(**)$ have exactly the same form as the conditions (3) and (4) in 34.4 for the translation map (thus it makes sense to term τ a *weak translation map*). These are the key properties needed to show that ψ satisfies the pentagon identities. Similar to the case of a coalgebra-Galois extension, for all $a, a' \in A$ and $c \in C$,

$$\begin{aligned}
&(\mu \otimes I_C) \circ (I_A \otimes \psi) \circ (\psi \otimes I_A)(c \otimes a \otimes a') \\
&= (\mu \otimes I_C) \circ (I_A \otimes \psi)\big(\sum c^{\tilde{1}} (c^{\tilde{2}} a)_{\underline{0}} \otimes (c^{\tilde{2}} a)_{\underline{1}} \otimes a'\big) \\
&= \sum c^{\tilde{1}} (c^{\tilde{2}} a)_{\underline{0}} {(c^{\tilde{2}} a)_{\underline{1}}}^{\tilde{1}} ({(c^{\tilde{2}} a)_{\underline{1}}}^{\tilde{2}} a')_{\underline{0}} \otimes ({(c^{\tilde{2}} a)_{\underline{1}}}^{\tilde{2}} a')_{\underline{1}} \\
&= \sum c^{\tilde{1}} (c^{\tilde{2}} a a')_{\underline{0}} \otimes (c^{\tilde{2}} a a')_{\underline{1}} \;=\; \psi \circ (I_C \otimes \mu)(c \otimes a \otimes a'),
\end{aligned}$$

where we used equation $(*)$ above to derive the third equality. Hence the left pentagon in the bow-tie diagram commutes. Similarly for the right pentagon,

$$\begin{aligned}
&(\psi \otimes I_C) \circ (I_C \otimes \psi) \circ (\Delta \otimes I_A)(c \otimes a) \\
&= (\psi \otimes I_C)\big(\sum c_{\underline{1}} \otimes {c_{\underline{2}}}^{\tilde{1}} ({c_{\underline{2}}}^{\tilde{2}} a)_{\underline{0}} \otimes ({c_{\underline{2}}}^{\tilde{2}} a)_{\underline{1}}\big) \\
&= \sum {c_{\underline{1}}}^{\tilde{1}} ({c_{\underline{1}}}^{\tilde{2}} {c_{\underline{2}}}^{\tilde{1}} ({c_{\underline{2}}}^{\tilde{2}} a)_{\underline{0}})_{\underline{0}} \otimes ({c_{\underline{1}}}^{\tilde{2}} {c_{\underline{2}}}^{\tilde{1}} ({c_{\underline{2}}}^{\tilde{2}} a)_{\underline{0}})_{\underline{1}} \otimes ({c_{\underline{2}}}^{\tilde{2}} a)_{\underline{1}} \\
&= \sum c^{\tilde{1}} ({c^{\tilde{2}}}_{\underline{0}} {{c^{\tilde{2}}}_{\underline{1}}}^{\tilde{1}} ({{c^{\tilde{2}}}_{\underline{1}}}^{\tilde{2}} a)_{\underline{0}})_{\underline{0}} \otimes ({c^{\tilde{2}}}_{\underline{0}} {{c^{\tilde{2}}}_{\underline{1}}}^{\tilde{1}} ({{c^{\tilde{2}}}_{\underline{1}}}^{\tilde{2}} a)_{\underline{0}})_{\underline{1}} \otimes ({{c^{\tilde{2}}}_{\underline{1}}}^{\tilde{2}} a)_{\underline{1}} \\
&= \sum c^{\tilde{1}} ((c^{\tilde{2}} a)_{\underline{0}})_{\underline{0}} \otimes ((c^{\tilde{2}} a)_{\underline{0}})_{\underline{1}} \otimes (c^{\tilde{2}} a)_{\underline{1}} \\
&= \sum c^{\tilde{1}} (c^{\tilde{2}} a)_{\underline{0}} \otimes (c^{\tilde{2}} a)_{\underline{1}} \otimes (c^{\tilde{2}} a)_{\underline{2}} \;=\; (I_A \otimes \Delta) \circ \psi(c \otimes a),
\end{aligned}$$

where we used equation $(*)$ to derive the third equality and then equation $(**)$ to derive the fourth one. It remains to show that (we.3) and (we.4) hold. On the one hand, for all $a \in A$ and $c \in C$,

$$\begin{aligned}\sum_\alpha a_\alpha \varepsilon(c^\alpha) &= \sum c^{\tilde{1}}(c^{\tilde{2}}a)_{\underline{0}}\,\varepsilon((c^{\tilde{1}}a)_{\underline{1}}) = \sum c^{\tilde{1}}c^{\tilde{2}}a \\ &= \sum c^{\tilde{1}}(c^{\tilde{2}}1)_{\underline{0}}\,a\,\varepsilon((c^{\tilde{1}}1)_{\underline{1}}) = \sum_\alpha 1_\alpha a\,\varepsilon(c^\alpha),\end{aligned}$$

while on the other hand, by a similar token and using equation $(**)$,

$$\begin{aligned}\sum_\alpha 1_\alpha \varepsilon(c_{\underline{1}}{}^\alpha) \otimes c_{\underline{2}} &= \sum c_{\underline{1}}{}^{\tilde{1}} c_{\underline{1}}{}^{\tilde{2}} \otimes c_{\underline{2}} = \sum c^{\tilde{1}} c^{\tilde{2}}{}_{\underline{0}} \otimes c^{\tilde{2}}{}_{\underline{1}} \\ &= \sum c^{\tilde{1}}(c^{\tilde{2}}1)_{\underline{0}} \otimes (c^{\tilde{2}}1)_{\underline{1}} = \sum_\alpha 1_\alpha \otimes c^\alpha,\end{aligned}$$

as required. □

Note that in 37.9, A is a weak (A, C, ψ)-entwined module with the multiplication given by the product μ of A and the C-coaction ϱ^A. Indeed, by using equation $(*)$ in the proof of 37.9, one can compute for all $a, a' \in A$,

$$\begin{aligned}\sum a_{\underline{0}}\psi(a_{\underline{1}} \otimes a') &= \sum a_{\underline{0}} a_{\underline{1}}{}^{\tilde{1}}(a_{\underline{1}}{}^{\tilde{2}}a')_{\underline{0}} \otimes (a_{\underline{1}}{}^{\tilde{2}}a')_{\underline{1}} \\ &= \sum (aa')_{\underline{0}} \otimes (aa')_{\underline{1}} = \varrho^A(aa').\end{aligned}$$

Similarly to the case of the canonical entwining structure associated to a coalgebra-Galois extension in 34.6, one shows that (A, C, ψ) constructed in 37.9 is the unique weak entwining structure for which A is a weak (A, C, ψ)-entwined module via ϱ^A and multiplication in A.

37.10. The Galois property. *Let (A, C, ψ) be a (self-dual) weak entwining structure corresponding to a weak C-Galois extension $B \hookrightarrow A$ as described in 37.9. Then the corresponding A-coring $(A \otimes_R C) \cdot A$ (given in 37.4) has a grouplike element $g = \varrho^A(1)$ and $((A \otimes_R C) \cdot A, g)$ is a Galois coring.*

Proof. It suffices to show that $\mathrm{Im}(\mathsf{can}) = \mathcal{C} := (A \otimes_R C) \cdot A$; then can will provide the required isomorphism of A-corings. Notice that, from the definition of ψ in 37.9, it follows that $\mathrm{Im}\psi \subseteq \mathrm{Im}(\mathsf{can})$. Since any element of $\mathcal{C}$ is an R-linear combination of typical elements of the form $\sum_\alpha a 1_\alpha \otimes c^\alpha$ and can is a left A-module map, $\sum_\alpha a 1_\alpha \otimes c^\alpha \in \mathrm{Im}(\mathsf{can})$. Therefore $\mathcal{C} \subseteq \mathrm{Im}(\mathsf{can})$. On the other hand, since A is a weak entwined module, for all $a \in A$,

$$\varrho^A(a) = \varrho^A(a1) = \sum_\alpha a_{\underline{0}} 1_\alpha \otimes a_{\underline{1}}{}^\alpha \in \mathcal{C}.$$

In view of the fact that $\mathsf{can}(a \otimes a') = a\varrho^A(a')$, this implies $\mathrm{Im}(\mathsf{can}) \subseteq \mathcal{C}$. □

A motivating example of a self-dual weak entwining structure comes from the notion of a weak Doi-Koppinen datum introduced in [64].

37.11. Right-right weak Doi-Koppinen modules. For a weak R-bialgebra H (cf. 36.5), a triple (H, A, C) is called a (right-right) *weak Doi-Koppinen datum* provided that

(1) (A, ϱ^A) is a right *weak H-comodule algebra*, that is, A is an R-algebra and a right H-comodule such that, for all $a, a' \in A$,
   (i) $\varrho^A(a)\varrho^A(a') = \varrho^A(aa')$;
   (ii) $(\varrho^A \otimes I_H) \circ \varrho^A(1_A) = \sum 1_{A\underline{0}} \otimes 1_{H\underline{1}} 1_{A\underline{1}} \otimes 1_{H\underline{2}}$.

(2) C is a right *weak H-module coalgebra*, that is, C is an R-coalgebra and a right H-module such that, for all $c \in C$, $h, g \in H$,
   (i) $\Delta_C(ch) = \sum c_{\underline{1}} h_{\underline{1}} \otimes c_{\underline{2}} h_{\underline{2}}$;
   (ii) $\varepsilon_C(c(gh)) = \sum \varepsilon_C(cg_{\underline{2}})\varepsilon_H(g_{\underline{1}}h)$.

A (right-right) *weak Doi-Koppinen module* associated to a weak Doi-Koppinen datum (H, A, C) is a right A-module and right C-comodule M with coaction ϱ^M, such that

$$\varrho^M(ma) = \sum m_{\underline{0}} a_{\underline{0}} \otimes m_{\underline{1}} a_{\underline{1}}, \quad \text{for all } a \in A,\ m \in M.$$

Note that $\varrho^A(a) = \sum a_{\underline{0}} \otimes a_{\underline{1}} \in A \otimes_R H$ and $\varrho^M(m) = \sum m_{\underline{0}} \otimes m_{\underline{1}} \in M \otimes_R C$. The category of weak Doi-Koppinen modules is denoted by $\widetilde{\mathbf{M}}(H)^C_A$.

37.12. Weak entwining associated to a weak Doi-Koppinen datum. *Let (H, A, C) be a weak Doi-Koppinen datum over a weak R-bialgebra H, and consider the R-linear map*

$$\psi : C \otimes_R A \to A \otimes_R A, \quad c \otimes a \mapsto \sum a_{\underline{0}} \otimes c a_{\underline{1}}.$$

Then (A, C, ψ) is a self-dual weak entwining structure and the category of weak entwined modules $\widetilde{\mathbf{M}}^C_A(\psi)$ is isomorphic to the category of weak Doi-Koppinen modules $\widetilde{\mathbf{M}}(H)^C_A$.

Proof. This is proven by direct calculations in a very similar fashion as 33.4 and is left to the reader as an exercise. □

In the context of weak Doi-Koppinen data, one can also consider a more general notion of a Doi-Koppinen datum over an algebra.

37.13. Doi-Koppinen datum over an algebra. Let $(\mathcal{H}, s_{\mathcal{H}}, t_{\mathcal{H}})$ be an A-bialgebroid (cf. 31.6). Then $(\mathcal{H}, B, \mathcal{C})_A$ is called a *(left-left) Doi-Koppinen datum over an algebra A* if B is a left $\mathcal{H}$-comodule algebra as in 31.23 and $\mathcal{C}$ is a left $\mathcal{H}$-module coring as in 31.21.

A *(left-left) Doi-Koppinen module* over A (associated to $(\mathcal{H}, B, \mathcal{C})_A$) is a left B-module and left $\mathcal{C}$-comodule M, such that, for all $b \in B$, $m \in M$,

$$^M\varrho(bm) = \sum b_{\underline{-1}} m_{\underline{-1}} \otimes b_{\underline{0}} m_{\underline{0}},$$

where ${}^M\!\varrho(m) = \sum m_{\underline{-1}} \otimes m_{\underline{0}}$ and ${}^B\!\varrho(b) = \sum b_{\underline{-1}} \otimes b_{\underline{0}}$.

Note that since M is a left B-module, it is also a left A-module via the source map $s_B : A \to B$. Note also that the right-hand side of the above equality is well defined since $\mathrm{Im}({}^B\!\varrho) \subseteq \mathcal{H} \times_A B$. The category of Doi-Koppinen modules associated to $(\mathcal{H}, B, \mathcal{C})_A$ is denoted by ${}^{\mathcal{C}}_B\mathbf{M}(\mathcal{H}; A)$.

There are various examples of special cases of the category ${}^{\mathcal{C}}_B\mathbf{M}(\mathcal{H}; A)$ obtained by setting $B = \mathcal{H}, A, A^e$ and $\mathcal{C} = \mathcal{H}, A, A^e$. In particular, the category of left $\mathcal{H}$-modules, the category of left $\mathcal{H}$-comodules or the category of left $\mathcal{H}$-bialgebroid modules in 31.16 are all special cases of the category of Doi-Koppinen modules over an algebra A.

In 37.11 we constructed right-right weak Doi-Koppinen data and modules as those built on a weak bialgebra. From 36.9 we know that a weak bialgebra can be viewed as a bialgebroid. It is therefore natural to ask for the relationship between weak Doi-Koppinen modules and Doi-Koppinen modules over an algebra. To consider this relationship we need to introduce left-left weak Doi-Koppinen data. This can be easily done, but to avoid any confusion we display the definition explicitly.

37.14. Left-left weak Doi-Koppinen modules. A (left-left) *weak Doi-Koppinen datum* is a triple (H, B, C), where H is a weak bialgebra over R (cf. 36.5) and

(1) $(B, {}^B\!\varrho)$ is a left *weak H-comodule algebra*, that is, B is an R-algebra and a left H-comodule such that, for all $b, b' \in B$,

  (i) ${}^B\!\varrho(b)\,{}^B\!\varrho(b') = {}^B\!\varrho(bb')$;

  (ii) $(I_H \otimes {}^B\!\varrho) \circ {}^B\!\varrho(1_B) = \sum 1_{H\underline{1}} \otimes 1_{B\underline{-1}} 1_{H\underline{2}} \otimes 1_{B\underline{0}}$.

(2) C is a left *weak H-module coalgebra*, that is, C is an R-coalgebra and a left H-module such that, for all $c \in C$, $h, g \in H$,

  (i) $\Delta_C(hc) = \sum h_{\underline{1}} c_{\underline{1}} \otimes h_{\underline{2}} c_{\underline{2}}$;

  (ii) $\varepsilon_C((gh)c) = \sum \varepsilon_H(h g_{\underline{2}}) \varepsilon_C(g_{\underline{1}} c)$.

A (left-left) *weak Doi-Koppinen module* associated to a weak Doi-Koppinen datum (H, B, C) is a left B-module and left C-comodule M with coaction ${}^M\!\varrho$, such that, for $b \in B$, $m \in M$, ${}^M\!\varrho(bm) = \sum b_{\underline{-1}} m_{\underline{-1}} \otimes b_{\underline{0}} m_{\underline{0}}$.

37.15. Doi-Koppinen data over A and weak Doi-Koppinen data. *Let H be a weak R-bialgebra with a counital source map $\sqcap^L$, and view it also as an A-bialgebroid, where $A = \mathrm{Im}(\sqcap^L)$ as in 36.9. Then there is a one-to-one correspondence between the weak Doi-Koppinen data with H and the Doi-Koppinen data over A with H. Furthermore, the corresponding categories of Doi-Koppinen modules are isomorphic.*

Proof. The proof relies on checking all axioms for weak comodule algebras and weak module coalgebras as well as for module coalgebras and comodule algebras over a bialgebroid. Details can be found in [76]; here we will only point out the correspondences.

If B is a left comodule algebra of a weak bialgebra H, then B is an A-ring with source map $s_B : A \to B$, $\sqcap^L(h) \mapsto \sum \varepsilon_H(1_{\underline{-1}}h)1_{\underline{0}}$. The A-coring comodule coaction for B is obtained by projecting its R-coalgebra coaction via the canonical map $H \otimes_R B \to H \otimes_A B$. The key observation here is that condition 37.14(1)(ii) implies for all $b \in B$, $\sum 1_{H\underline{1}}\varepsilon_H(b_{\underline{-1}}1_{H\underline{2}}) \otimes_R b_{\underline{0}} = \sum 1_{B\underline{-1}} \otimes_R b1_{\underline{0}}$. This equality in turn can be used to prove that, for all $a \in A$, $b \in B$,

$$\sum b_{\underline{-1}}t(a) \otimes_A b_{\underline{0}} = \sum b_{\underline{-1}} \otimes_A b_{\underline{0}}s_B(a),$$

where t is the target map $t : A \to H$, $a \mapsto \sum \varepsilon(1_{\underline{2}}a)1_{\underline{1}}$. Therefore, the A-coring coaction for B has its image in $H \times_A B$, as required.

Conversely, view H as an A-bialgebroid, and let B be a left H-comodule algebra. Since A is a separable Frobenius algebra by 36.8, the canonical map $H \otimes_R B \to H \otimes_A B$ has a section

$$\sigma : H \otimes_A B \to H \otimes_R B, \quad h \otimes b \mapsto \sum h\sqcap^L(1_{H\underline{1}}) \otimes 1_{H\underline{2}}b.$$

With the help of σ, the left A-coring coaction ${}^B\!\varrho : B \to H \otimes_A B$ can be lifted to the R-coalgebra coaction $\sigma \circ {}^B\!\varrho : B \to H \otimes_R B$.

If C is a left weak H-module coalgebra with counit ε_C and coproduct $\Delta_C(c) = \sum c_{\underline{1}} \otimes_R c_{\underline{2}}$, then C is an A-coring with coproduct $\underline{\Delta}_C(c) = \sum c_{\underline{1}} \otimes_A c_{\underline{2}}$ and counit $\underline{\varepsilon}_C(c) = \sum \varepsilon_C(c1_{\underline{1}})1_{\underline{2}}$.

Conversely, if $\mathcal{C}$ is a left module coring over an A-bialgebroid H, then its coproduct can be lifted to the coproduct of an R-coalgebra via the section σ above. Explicitly,

$$\Delta_{\mathcal{C}} : \mathcal{C} \to \mathcal{C} \otimes_R \mathcal{C}, \quad c \mapsto \sum c_{\underline{1}}\sqcap^L(1_{\underline{1}}) \otimes 1_{\underline{2}}c_{\underline{2}},$$

where $\underline{\Delta}_{\mathcal{C}}(c) = \sum c_{\underline{1}} \otimes c_{\underline{2}}$. □

37.16. Corings and Doi-Koppinen data over an algebra. *Let $(\mathcal{H}, B, \mathcal{C})_A$ be a Doi-Koppinen datum over A. Then $\mathcal{D} = \mathcal{C} \otimes_A B$ is a B-bimodule with right action given by the product in B and left action $b(c \otimes b') = \sum b_{\underline{-1}}c \otimes b_{\underline{0}}b'$, for all $b, b' \in B$, $c \in \mathcal{C}$. Furthermore, $\mathcal{D}$ is a B-coring with comultiplication $\underline{\Delta}_{\mathcal{D}} = \underline{\Delta}_{\mathcal{C}} \otimes I_B$ and counit $\underline{\varepsilon}_{\mathcal{D}} = \underline{\varepsilon}_{\mathcal{C}} \otimes I_B$, where $\underline{\Delta}_{\mathcal{C}}$, $\underline{\varepsilon}_{\mathcal{C}}$ are the coproduct and counit of the A-coring $\mathcal{C}$. In this case the categories of left $\mathcal{D}$-comodules and left-left Doi-Koppinen modules over A are isomorphic to each other.*

Proof. First note that the left action of B on $\mathcal{D}$ is well defined since the image of the left $\mathcal{H}$-coaction of B is required to be in $B \times_A \mathcal{H}$. The fact that it is an action indeed follows from the condition that B is a left $\mathcal{H}$-comodule algebra. Note also that in the definitions of $\underline{\Delta}_{\mathcal{D}}$ and $\underline{\varepsilon}_{\mathcal{D}}$ we use the natural isomorphisms $\mathcal{C}\otimes_A B\otimes_B\mathcal{C}\otimes_A B \simeq \mathcal{C}\otimes_A\mathcal{C}\otimes_A B$ and $A\otimes_A B \simeq B$, respectively. Clearly $\underline{\Delta}_{\mathcal{D}}$ is a right B-module map. To prove that it is left B-linear as well, take any $b, b' \in B$ and $c \in \mathcal{C}$, and compute

$$\underline{\Delta}_{\mathcal{D}}(b(c\otimes b')) = \sum (b_{\underline{-1}}c)_{\underline{1}} \otimes (b_{\underline{-1}}c)_{\underline{2}} \otimes b_{\underline{0}}b' = \sum b_{\underline{-11}}c_{\underline{1}} \otimes b_{\underline{-12}}c_{\underline{2}} \otimes b_{\underline{0}}b',$$

where we used the fact that $\mathcal{C}$ is a left $\mathcal{H}$-module coring. On the other hand,

$$\begin{aligned} b\underline{\Delta}_{\mathcal{D}}(c\otimes b') &= \sum b(c_{\underline{1}}\otimes 1)\otimes_B (c_{\underline{2}}\otimes b') = \sum b_{\underline{-1}}c_{\underline{1}}\otimes b_{\underline{0}}(c_{\underline{2}}\otimes b') \\ &= \sum b_{\underline{-11}}c_{\underline{1}}\otimes b_{\underline{-12}}c_{\underline{2}}\otimes b_{\underline{0}}b'. \end{aligned}$$

This proves that $\underline{\Delta}_{\mathcal{D}}$ is right B-linear, and hence it is a (B, B)-bimodule map, as required. The coassociativity of $\underline{\Delta}_{\mathcal{D}}$ follows directly from the definition.

Clearly $\underline{\varepsilon}_{\mathcal{D}}$ is right B-linear. To prove that it is also a left B-linear, take any $b, b' \in B$, $c \in \mathcal{C}$, and compute

$$\begin{aligned} \underline{\varepsilon}_{\mathcal{D}}(b(c\otimes b')) &= \sum \underline{\varepsilon}_{\mathcal{C}}(b_{\underline{-1}}c)(b_{\underline{0}}b') = \sum \underline{\varepsilon}_{\mathcal{H}}(b_{\underline{-1}}s_{\mathcal{H}}(\underline{\varepsilon}_{\mathcal{C}}(c)))(b_{\underline{0}}b') \\ &= \sum \underline{\varepsilon}_{\mathcal{H}}(b_{\underline{-1}}t_{\mathcal{H}}(\underline{\varepsilon}_{\mathcal{C}}(c)))(b_{\underline{0}}b') = \sum \underline{\varepsilon}_{\mathcal{H}}(b_{\underline{-1}})(b_{\underline{0}}s_B(\underline{\varepsilon}_{\mathcal{C}}(c))b') \\ &= b_{\underline{0}}s_B(\underline{\varepsilon}_{\mathcal{C}}(c))b' = b\,\underline{\varepsilon}_{\mathcal{D}}(c\otimes b'), \end{aligned}$$

where we used the fact that $\mathcal{C}$ is a left $\mathcal{H}$-module coring to obtain the second equality and then equation 31.6(3) to derive the third one, and the fact that the image of the left coaction of $\mathcal{H}$ on B is in $\mathcal{H}\times_A B$ to obtain the fourth equality. This proves that $\underline{\varepsilon}_{\mathcal{D}}$ is left B-linear, and hence it is (B, B)-bilinear. The fact that $\underline{\varepsilon}_{\mathcal{D}}$ is a counit of $\mathcal{D}$ follows directly from its definition. Thus we conclude that $\mathcal{D}$ is a B-coring, as stated.

To prove the isomorphism of the categories, take any left $\mathcal{D}$-comodule M and consider it as a Doi-Koppinen module via the same coaction ${}^M\!\varrho : M \to \mathcal{C}\otimes_A B\otimes_B M \simeq \mathcal{C}\otimes_A M$. Conversely, any Doi-Koppinen module M is also a left $\mathcal{D}$ comodule via ${}^M\!\varrho : M \to \mathcal{C}\otimes_A M \simeq \mathcal{C}\otimes_A B\otimes_B M = \mathcal{D}\otimes_B M$. □

Thus, instead of studying the structure of Doi-Koppinen Hopf modules over an algebra on there own, we can use the already developed coring theory.

37.17. Corings and weak Doi-Koppinen data. *Let (H, B, C) be a weak Doi-Koppinen datum as considered in 37.11. Define the corresponding B-coring $\mathcal{C} = \{\sum_i 1_{\underline{-1}}c^i \otimes 1_{\underline{0}}b^i \mid b^i \in B, c^i \in C\} \subseteq C\otimes_R B$ as in 37.4. View (H, B, C) as a Doi-Koppinen datum over the algebra $A = \mathrm{Im}\,\sqcap^L$ by 37.15, and let $\mathcal{D} = C\otimes_A B$ be the corresponding coring constructed in 37.16. Then $\mathcal{C} \simeq \mathcal{D}$ as B-corings.*

Proof. This is proven by direct computations, and the details are left to the reader. We only indicate that the isomorphism $\theta : \mathcal{D} \to \mathcal{C}$ and its inverse are given explicitly by

$$\theta : c \otimes_A b \mapsto \sum 1_{\underline{-1}} c \otimes_R 1_{\underline{0}} b, \qquad \theta^{-1} : \sum_i c^i \otimes_R b^i \mapsto \sum_i c^i \otimes_A b^i.$$

Note that the map θ is well defined by 37.18(ii). □

37.18. Exercise.

In the setting of 37.14:

(i) Show that the condition 37.14(1)(ii) implies that, for all $b \in B$,

$$\sum 1_{H\underline{1}} \varepsilon_H(b_{\underline{-1}} 1_{H\underline{2}}) \otimes_R b_{\underline{0}} = \sum 1_{B\underline{-1}} \otimes_R b 1_{B\underline{0}}.$$

(ii) Use the equality derived in (i) to prove that, for all $a \in A$, $b \in B$,

$$\sum b_{\underline{-1}} t(a) \otimes_A b_{\underline{0}} = \sum b_{\underline{-1}} \otimes_A b_{\underline{0}} s_B(a),$$

where t is the target map $t : A \to H$, $a \mapsto \sum \varepsilon(1_{\underline{2}} a) 1_{\underline{1}}$.

References. Böhm [64]; Brzeziński [73]; Brzeziński, Caenepeel and Militaru [76]; Caenepeel and DeGroot [82]; Wisbauer [212].

Appendix

38 Categories and functors

Basic definitions and theorems from general category theory are recalled here. The purpose of this exposition is to provide a convenient reference to the categorical notions used in coring and comodule theory. For more details we refer to [40], [42], [44] and [46].

38.1. Categories. A *category* $\mathbf{A}$ is defined as a class of objects $\mathrm{Obj}(\mathbf{A})$ and a class of morphism sets $\mathrm{Mor}(\mathbf{A})$, which satisfy the following axioms.

(i) For every ordered pair (A, B) of objects in $\mathbf{A}$ there is a set $\mathrm{Mor}_{\mathbf{A}}(A, B)$, the *morphisms* of A to B, such that

$$\mathrm{Mor}_{\mathbf{A}}(A, B) \cap \mathrm{Mor}_{\mathbf{A}}(A', B') = \emptyset \text{ for } (A, B) \neq (A', B').$$

(ii) For any $A, B, C \in \mathrm{Obj}(\mathbf{A})$ there is a map

$$\mathrm{Mor}_{\mathbf{A}}(A, B) \times \mathrm{Mor}_{\mathbf{A}}(B, C) \to \mathrm{Mor}_{\mathbf{A}}(A, C), \quad (f, g) \mapsto g \circ f,$$

called the *composition* of morphisms, which is associative (in an obvious sense). Often we write $g \circ f = gf$.

(iii) For every $A \in \mathrm{Obj}(\mathbf{A})$ there is a morphism $I_A \in \mathrm{Mor}_{\mathbf{A}}(A, A)$, such that $f \circ I_A = f$ and $I_A \circ g = g$, for any $f \in \mathrm{Mor}_{\mathbf{A}}(A, B)$, $g \in \mathrm{Mor}_{\mathbf{A}}(B, A)$ and $B \in \mathrm{Obj}(\mathbf{A})$. I_A is called the *identity morphism* of A.

For any category $\mathbf{A}$ the *dual category* $\mathbf{A}^{op}$ has the same class of objects but reversed morphisms, that is, $\mathrm{Mor}_{\mathbf{A}^{op}}(A, A') = \mathrm{Mor}_{\mathbf{A}}(A', A)$, for any objects $A, A' \in \mathrm{Obj}(\mathbf{A}^{op}) = \mathrm{Obj}(\mathbf{A})$.

Given two categories $\mathbf{A}$ and $\mathbf{B}$, the *product category* $\mathbf{A} \times \mathbf{B}$ has ordered pairs (A, B) with $A \in \mathbf{A}$, $B \in \mathbf{B}$ as objects and

$$\mathrm{Mor}_{\mathbf{A}\times\mathbf{B}}((A, B), (A', B')) = \mathrm{Mor}_{\mathbf{A}}(A, A') \times \mathrm{Mor}_{\mathbf{B}}(B, B')$$

as morphism sets.

For applications of categories (and to avoid the Russell paradox) it is essential that the objects be a class rather than a set. Categories in which the objects form a set are called *small categories*.

The category $\mathbf{A}$ is called *preadditive* if each set $\mathrm{Mor}_{\mathbf{A}}(A, A')$ is an Abelian group and the compositions $\mathrm{Mor}_{\mathbf{A}}(A, A') \times \mathrm{Mor}_{\mathbf{A}}(A', A'') \to \mathrm{Mor}_{\mathbf{A}}(A, A'')$ are bilinear maps.

38.2. Subcategory. A category $\mathbf{B}$ is called a *subcategory* of a category $\mathbf{A}$ if $\mathrm{Obj}(\mathbf{B}) \subset \mathrm{Obj}(\mathbf{A})$, $\mathrm{Mor}_{\mathbf{B}}(A, B) \subset \mathrm{Mor}_{\mathbf{A}}(A, B)$ for all $A, B \in \mathrm{Obj}(\mathbf{B})$, and the composition of morphisms in $\mathbf{B}$ is the restriction of the composition in $\mathbf{A}$. If $\mathrm{Mor}_{\mathbf{B}}(A, B) = \mathrm{Mor}_{\mathbf{A}}(A, B)$ for all $A, B \in \mathrm{Obj}(\mathbf{B})$, then $\mathbf{B}$ is called a *full subcategory* of $\mathbf{B}$. So a full subcategory of $\mathbf{A}$ is fully determined by its objects.

The prototype of a category is the category of sets, denoted by **Set**, whose objects are the class of sets and the morphisms are maps between sets. We write $\mathrm{Mor}_{\mathbf{Set}}(A, A') = \mathrm{Map}(A, A')$ for any sets A, A'. The motivating example of a preadditive category is the category **Ab** of Abelian groups whose objects are Abelian groups and morphisms are group homomorphisms. The usual notation is $\mathrm{Mor}_{\mathbf{Ab}}(A, A') = \mathrm{Hom}_{\mathbb{Z}}(A, A')$, for any Abelian groups A, A'.

38.3. Functors. A *covariant functor* $F : \mathbf{A} \to \mathbf{B}$ between categories consists of the assignments

$$\begin{array}{ll} \mathrm{Obj}(\mathbf{A}) \to \mathrm{Obj}(\mathbf{B}), & A \mapsto F(A), \\ \mathrm{Mor}(\mathbf{A}) \to \mathrm{Mor}(\mathbf{B}), & [f : A \to B] \mapsto [F(f) : F(A) \to F(B)], \end{array}$$

such that $F(I_A) = I_{F(A)}$ and $F(fg) = F(f)F(g)$ whenever fg is defined in $\mathbf{A}$. Dually, a *contravariant functor* $F : \mathbf{A} \to \mathbf{B}$ consists of the assignments

$$\begin{array}{ll} \mathrm{Obj}(\mathbf{A}) \to \mathrm{Obj}(\mathbf{B}), & A \mapsto F(A), \\ \mathrm{Mor}(\mathbf{A}) \to \mathrm{Mor}(\mathbf{B}), & [f : A \to B] \mapsto [F(f) : F(B) \to F(A)], \end{array}$$

such that $F(I_A) = I_{F(A)}$ and $F(fg) = F(g)F(f)$ when fg is defined in $\mathbf{A}$.

Clearly the composition of two functors is again a functor.

38.4. Properties of functors. Let $F : \mathbf{A} \to \mathbf{B}$ be a functor, $A \in \mathrm{Obj}(\mathbf{A})$ and $f \in \mathrm{Mor}(\mathbf{A})$. F is said to *preserve* a property of A (or f), if $F(A)$ (resp. $F(f)$) again has this property. F is said to *reflect* a property of A (resp. of f) if the fact that $F(A)$ (resp. $F(f)$) has this property implies that A (resp. f) has the same property.

By definition, *covariant* functors preserve identities and compositions of morphisms and commutative diagrams.

38.5. Mor-functors. For any morphism $f : B \to C$ in a category $\mathbf{A}$ and $A \in \mathrm{Obj}(\mathbf{A})$, composition yields maps between morphism sets:

$$\begin{array}{ll} \mathrm{Mor}(A, f) : \mathrm{Mor}_{\mathbf{A}}(A, B) \to \mathrm{Mor}_{\mathbf{A}}(A, C), & u \mapsto fu, \\ \mathrm{Mor}(f, A) : \mathrm{Mor}_{\mathbf{A}}(C, A) \to \mathrm{Mor}_{\mathbf{A}}(B, A), & v \mapsto vf. \end{array}$$

This defines a covariant and a contravariant functor,

$$\mathrm{Mor}(A, -) : \mathbf{A} \to \mathbf{Set}, \quad \mathrm{Mor}(-, A) : \mathbf{A} \to \mathbf{Set}.$$

If $\mathbf{A}$ is a preadditive category, then both of these functors have values in **Ab**.

38.6. Natural transformations. Given covariant functors $F, F' : \mathbf{A} \to \mathbf{B}$, a *natural transformation* or *functorial morphism* η: $F \to F'$ is determined by a class of morphisms $\eta_A : F(A) \to F'(A)$ in $\mathbf{B}$, $A \in \mathbf{A}$, such that every morphism $f : A \to B$ in $\mathbf{A}$ induces a commutative diagram,

$$\begin{array}{ccc} F(A) & \xrightarrow{F(f)} & F(B) \\ \downarrow{\scriptstyle \eta_A} & & \downarrow{\scriptstyle \eta_B} \\ F'(A) & \xrightarrow{F'(f)} & F'(B)\,. \end{array}$$

Natural transformations from F to F' are denoted by $\mathrm{Nat}(F, F')$.

Remarkably, any covariant functor $F : \mathbf{A} \to \mathbf{Set}$ is closely related to Mor-functors.

38.7. Yoneda Lemma. *Let $F : \mathbf{A} \to \mathbf{Set}$ be a covariant functor. For any $A \in \mathbf{A}$ there is a bijective map*

$$Y : \mathrm{Nat}(\mathrm{Mor}_{\mathbf{A}}(A, -), F) \to F(A), \quad \eta \mapsto \eta_A(I_A)\,.$$

Y is known as the Yoneda map.

Every covariant functor $F : \mathbf{A} \to \mathbf{B}$ to morphisms $A \to A'$ in $\mathbf{A}$ assigns morphisms $F(A) \to F(A')$ in $\mathbf{B}$, that is, for every pair A, A' in $\mathrm{Obj}(\mathbf{A})$ there is a (set) map

$$F_{A,A'} : \mathrm{Mor}_{\mathbf{A}}(A, A') \to \mathrm{Mor}_{\mathbf{B}}(F(A), F(A')).$$

If $\mathbf{A}$ and $\mathbf{B}$ are preadditive categories, then F is called *additive* provided all the $F_{A,A'}$ are homomorphisms (of Abelian groups).

In general one considers two bifunctors

$$\mathrm{Mor}_{\mathbf{A}}(-, -),\ \mathrm{Mor}_{\mathbf{B}}(F(-), F(-)) : \mathbf{A}^{op} \times \mathbf{A} \to \mathbf{Set},$$

and a functorial morphism

$$\mathcal{F} : \mathrm{Mor}_{\mathbf{A}}(-, -) \to \mathrm{Mor}_{\mathbf{B}}(F(-), F(-)), \quad f \mapsto F(f).$$

Their properties are significant for the properties of the functor F itself.

38.8. Special morphisms. A morphism $f : A \to B$ in $\mathbf{A}$ is called:

monomorphism	if, for $g, h \in \mathrm{Mor}_{\mathbf{A}}(C, A)$, $fg = fh$ implies $g = h$;
epimorphism	if, for $g, h \in \mathrm{Mor}_{\mathbf{A}}(B, D)$, $gf = hf$ implies $g = h$;
bimorphism	if f is both a mono- and an epimorphism;
retraction	if there exists $g \in \mathrm{Mor}_{\mathbf{A}}(B, A)$ with $fg = id_B$;
coretraction or *section*	if there exists $g \in \mathrm{Mor}_{\mathbf{A}}(B, A)$ with $gf = id_A$;
isomorphism	if f is both a retraction and a coretraction;
left zero morphism	if, for any $g, h \in \mathrm{Mor}_{\mathbf{A}}(D, A)$, $fg = fh$;
right zero morphism	if, for any $g, h \in \mathrm{Mor}_{\mathbf{A}}(B, C)$, $gf = hf$;
zero morphism	if f is both a left and right zero morphism.

In **Set**, monomorphisms are injective maps and epimorphisms are surjective maps. On the other hand, in any category $\mathbf{A}$, $f : A \to B$ is a monomorphism if and only if the map $\mathrm{Mor}(C, f) : \mathrm{Mor}_{\mathbf{A}}(C, A) \to \mathrm{Mor}_{\mathbf{A}}(C, B)$ is injective, for any $C \in \mathbf{A}$. Furthermore, f is an epimorphism if and only if $\mathrm{Mor}(f, D) : \mathrm{Mor}_{\mathbf{A}}(B, D) \to \mathrm{Mor}_{\mathbf{A}}(A, D)$ is injective, for any $D \in \mathbf{A}$.

38.9. Special objects. An object $A \in \mathbf{A}$ is called:

initial	if $\mathrm{Mor}_{\mathbf{A}}(A, B)$ has just one element, for any $B \in \mathbf{A}$;
terminal (final)	if $\mathrm{Mor}_{\mathbf{A}}(C, A)$ has just one element, for any $C \in \mathbf{A}$;
zero	if A is both an initial and a terminal object;
semisimple	if any monomorphism $B \to A$ in $\mathbf{A}$ is a coretraction;
simple	if any monomorphism $B \to A$ in $\mathbf{A}$ is an isomorphism.

38.10. Special functors. A covariant functor $F : \mathbf{A} \to \mathbf{B}$ is called:

faithful	if $F_{A,A'}$ is injective for all $A, A' \in \mathrm{Obj}(\mathbf{A})$;
full	if $F_{A,A'}$ is surjective for all $A, A' \in \mathrm{Obj}(\mathbf{A})$;
fully faithful	if F is full and faithful;
embedding	if the assignment $F : \mathrm{Mor}(\mathbf{A}) \to \mathrm{Mor}(\mathbf{B})$ is injective;
representative	if, for every $B \in \mathrm{Obj}(\mathbf{B})$, there exists an object A in $\mathbf{A}$ such that $F(A) \simeq B$.

Notice that a covariant faithful functor F reflects monomorphisms, epimorphisms, bimorphisms and commutative diagrams. If F is fully faithful, it also reflects retractions, coretractions and isomorphisms. If F is fully faithful and representative, it preserves and reflects mono-, epi- and bimorphisms (retractions, coretractions, isomorphisms and commutative diagrams).

38.11. Generators and cogenerators. An object A in $\mathbf{A}$ is said to be a

generator in $\mathbf{A}$	if $\mathrm{Mor}_{\mathbf{A}}(A, -) : \mathbf{A} \to \mathbf{Set}$ is faithful;
cogenerator in $\mathbf{A}$	if $\mathrm{Mor}_{\mathbf{A}}(-, A) : \mathbf{A} \to \mathbf{Set}$ is faithful.

Notice that A is a generator if and only if

$$\text{for any } f \neq g : B \to C \text{ in } \mathbf{A}, \text{ there exists } h : A \to B \text{ with } fh \neq gh,$$

and A is a cogenerator if and only if

$$\text{for any } f \neq g : B \to C \text{ in } \mathbf{A}, \text{ there exists } k : C \to A \text{ with } kf \neq kg.$$

38.12. Projectives and injectives. Let A, B be objects in $\mathbf{A}$. A is called:

B-projective	if, for any epimorphism $p : B \to C$ in $\mathbf{A}$, the mapping $\mathrm{Mor}(A, p) : \mathrm{Mor}_{\mathbf{A}}(A, B) \to \mathrm{Mor}_{\mathbf{A}}(A, C)$ is surjective;
projective (in $\mathbf{A}$)	if A is B-projective for all $B \in \mathbf{A}$;
B-injective	if, for any monomorphism $i : C \to B$, the mapping $\mathrm{Mor}(i, A) : \mathrm{Mor}_{\mathbf{A}}(C, A) \to \mathrm{Mor}_{\mathbf{A}}(B, A)$ is surjective;
injective (in $\mathbf{A}$)	if A is B-injective for all $B \in \mathbf{A}$.

38.13. Semisimple objects. *An object A in $\mathbf{A}$ is semisimple if and only if every object $C \in \mathbf{A}$ is A-injective.*

Proof. Consider the diagram in $\mathbf{A}$,

$$\begin{array}{ccc} B & \xrightarrow{i} & A \\ {\scriptstyle f}\downarrow & & \\ C, & & \end{array}$$

where i is a monomorphism. If A is semisimple, then there exists $g : A \to B$ with $gi = I_B$ and $fg : A \to C$ extends the diagram commutatively, that is, C is A-injective. Conversely, assume that all $C \in \mathbf{A}$ are A-injective. Putting $B = C$ and $f = I_B$, we get a morphism $h : A \to B$ with $hi = I_B$, thus showing that i is a coretraction. □

38.14. Equalisers and coequalisers. Let $f, g : A \to B$ be morphisms in the category $\mathbf{A}$.

A morphism $k : K \to A$ is called a *difference kernel* or an *equaliser* if $fk = gk$, and for every morphism $x : X \to A$ with $fx = gx$, there is a unique morphism $h : X \to K$ such that $x = kh$.

A morphism $c : B \to C$ is called a *difference cokernel* or a *coequaliser* if $cf = cg$, and for every morphism $y : B \to Y$ with $yf = yg$, there is a unique morphism $h : C \to Y$ such that $y = hc$.

Equalisers and coequalisers are denoted by the diagrams

$$K \xrightarrow{k} A \underset{g}{\overset{f}{\rightrightarrows}} B, \qquad A \underset{g}{\overset{f}{\rightrightarrows}} B \xrightarrow{c} C.$$

If $\mathbf{A}$ has zero morphisms, the equaliser of the pair $(f, 0)$ is called a *kernel of f*, and the coequaliser of $(f, 0)$ is the *cokernel of f*. Every equaliser – and hence every kernel – is a monomorphism, whereas coequalisers and cokernels are epimorphisms. If $\mathbf{A}$ is an additive category, the (co)equaliser of f, g can be characterised as the (co)kernel of $f - g$.

38.15. Products and coproducts. Let $\{A_\lambda\}_\Lambda$ be a family of objects in $\mathbf{A}$.

An object P in $\mathbf{A}$ with morphisms $\{\pi_\lambda : P \to A_\lambda\}_\Lambda$ is called a *product* of the A_λ if, for any family $\{f_\lambda : X \to A_\lambda\}_\Lambda$, there is a unique $f : X \to P$ with $f\pi_\lambda = f_\lambda$ for all $\lambda \in \Lambda$. For this object P we usually write $\prod_\Lambda A_\lambda$, and if all $A_\lambda = A$, we put $\prod_\Lambda A_\lambda = A^\Lambda$.

An object Q in $\mathbf{A}$ with morphisms $\{\epsilon_\lambda : A_\lambda \to Q\}_\Lambda$ is called the *coproduct* of the A_λ if, for any family $\{g_\lambda : A_\lambda \to Y\}_\Lambda$, there is a unique $g : Q \to Y$ with $g\epsilon_\lambda = g_\lambda$, for all $\lambda \in \Lambda$. For this object Q we usually write $\coprod_\Lambda A_\lambda$, and if all the $A_\lambda = A$, we put $\coprod_\Lambda A_\lambda = A^{(\Lambda)}$.

The product in **Set** is just the Cartesian product, and the universal property of the product in $\mathbf{A}$ can be expressed by the isomorphism (bijection) in

Set for any $X \in \mathbf{A}$,

$$\mathrm{Mor}_{\mathbf{A}}(X, \prod_{\Lambda} A_\lambda) \longrightarrow \prod_{\Lambda} \mathrm{Mor}_{\mathbf{A}}(X, A_\lambda), \quad f \mapsto \pi_\lambda f,$$

and dually the coproduct is characterised by the isomorphism, for any $Y \in \mathbf{A}$,

$$\mathrm{Mor}_{\mathbf{A}}(\coprod_{\Lambda} A_\lambda, Y) \longrightarrow \prod_{\Lambda} \mathrm{Mor}_{\mathbf{A}}(A_\lambda, Y), \quad g \mapsto g\epsilon_\lambda.$$

38.16. Colimits and limits. Let $\mathbf{\Lambda}$ be a small category and $L : \mathbf{\Lambda} \to \mathbf{A}$ a functor. A familiy of morphisms $g_\lambda : L(\lambda) \to Y$ in $\mathbf{A}$, $\lambda \in \mathbf{\Lambda}$, is said to be *compatible* if, for any morphism $h : \lambda \to \mu$ in $\mathbf{\Lambda}$, one has $g_\mu L(h) = g_\lambda$. A *colimit* or *inductive limit* for the functor L is an object $\varinjlim L$ in $\mathbf{A}$ with a compatible family $\epsilon_\lambda : L(\lambda) \to \varinjlim L$, such that, for any compatible family g_λ, there exists a unique morphism

$$g : \varinjlim L \to Y \text{ with } g_\lambda = g\epsilon_\lambda.$$

Dually *(projective) limits* are defined for compatible families $f_\lambda : X \to L(\lambda)$ of morphisms in $\mathbf{A}$.

Any quasi-ordered directed set $(\Lambda, \leq)$ can be considered as a category $\mathbf{\Lambda}$. Then the colimit of a functor $L : \mathbf{\Lambda} \to \mathbf{A}$ is called a *direct limit*, while the limit of a functor $L' : \mathbf{\Lambda}^{op} \to \mathbf{A}$ is called an *inverse limit*. For sets consisting of three elements, these constructions yield *pullbacks* and *pushouts* as special cases.

38.17. Complete, Abelian and Grothendieck categories. If, for any small category $\mathbf{\Lambda}$ and functor $L : \mathbf{\Lambda} \to \mathbf{A}$ the limit (colimit) exists, then $\mathbf{A}$ is called a *complete (cocomplete)* category. Notice that a preadditive category $\mathbf{A}$ is complete if and only if $\mathbf{A}$ has products and kernels, and $\mathbf{A}$ is cocomplete if and only if it has coproducts and cokernels.

A preadditive category $\mathbf{A}$ with finite products (and coproducts) is called *additive* and is called *Abelian* if every morphism has a kernel and a cokernel, and every morphism f has a factorisation $f = gh$, where h is a cokernel and g is a kernel.

In an Abelian category $\mathbf{A}$, a sequence of morphisms $A \xrightarrow{f} A' \xrightarrow{g} A''$ is called *exact at* A' provided $\mathrm{Im}\, f = \mathrm{Ke}\, g$ (as subobjects of A'), and any sequence of morphisms is called *exact* provided it is exact at each object. In particular, an exact sequence of the form

$$0 \longrightarrow A \longrightarrow A' \longrightarrow A'' \longrightarrow 0$$

is called a *short exact sequence.*

A cocomplete Abelian category $\mathbf{A}$ with a generator is called a *Grothendieck category* if the direct limits of short exact sequences are exact.

38.18. Separable functors. A covariant functor $F : \mathbf{A} \to \mathbf{B}$ is said to be a *separable functor* if

$$\mathcal{F} : \mathrm{Mor}_{\mathbf{A}}(-,-) \to \mathrm{Mor}_{\mathbf{B}}(F(-), F(-))$$

is a functorial coretraction, that is, there exists a functorial morphism

$$\Phi : \mathrm{Mor}_{\mathbf{B}}(F(-), F(-)) \to \mathrm{Mor}_{\mathbf{A}}(-,-),$$

with $\Phi \circ \mathcal{F} = I_{\mathrm{Mor}_{\mathbf{A}}}(-,-)$. Such a Φ is characterised by the properties

(1) for any $f : A \to A'$ in $\mathbf{A}$, $\Phi_{A,A'}(F(f)) = f$;

(2) for any $f : A \to A'$, $f_1 : A_1 \to A_1'$, any commutative diagram on the left induces the commutative diagram on the right:

$$\begin{array}{ccc} F(A) & \xrightarrow{h} & F(A_1) \\ {\scriptstyle F(f)}\downarrow & & \downarrow{\scriptstyle F(f_1)} \\ F(A') & \xrightarrow[h']{} & F(A_1'), \end{array} \qquad \begin{array}{ccc} A & \xrightarrow{\Phi_{A,A_1}(h)} & A_1 \\ {\scriptstyle f}\downarrow & & \downarrow{\scriptstyle f_1} \\ A' & \xrightarrow[\Phi_{A',A_1'}(h')]{} & A_1' . \end{array}$$

38.19. Properties of separable functors. *Let $F : \mathbf{A} \to \mathbf{B}$ be a separable functor with a splitting functorial morphism Φ.*

(1) Consider any diagram in $\mathbf{A}$ and its image in $\mathbf{B}$,

$$\begin{array}{ccc} A & \xrightarrow{f} & A' \\ {\scriptstyle g}\downarrow & & \\ M, & & \end{array} \qquad \begin{array}{ccc} F(A) & \xrightarrow{F(f)} & F(A') \\ {\scriptstyle F(g)}\downarrow & & \\ F(M). & & \end{array}$$

If the right-hand diagram can be completed commutatively by some $h : F(A') \to F(M)$, then $\Phi(h)$ completes the first diagram commutatively.

(2) Consider any diagram in $\mathbf{A}$ and its image in $\mathbf{B}$,

$$\begin{array}{ccc} & & M \\ & & \downarrow{\scriptstyle g} \\ A & \xrightarrow{f} & A', \end{array} \qquad \begin{array}{ccc} & & F(M) \\ & & \downarrow{\scriptstyle F(g)} \\ F(A) & \xrightarrow{F(f)} & F(A') . \end{array}$$

If the right-hand diagram can be completed commutatively by some $h : F(M) \to F(A)$, then $\Phi(h)$ completes the first diagram commutatively.

(3) F reflects retractions and coretractions.

(4) If F preserves epimorphisms, then F reflects projective objects.

(5) If F preserves monomorphisms, then F reflects injective objects.

38.20. Composition of separable functors. *Consider covariant functors $F : \mathbf{A} \to \mathbf{B}$ and $G : \mathbf{B} \to \mathbf{C}$.*

(1) If F and G are separable, then $GF : \mathbf{A} \to \mathbf{C}$ is separable.

(2) If GF is a separable functor, then so is F.

38.21. Adjoint functors. A pair (F, G) of covariant functors $F : \mathbf{A} \to \mathbf{B}$ and $G : \mathbf{B} \to \mathbf{A}$ is called an *adjoint pair* if there is a functorial isomorphism

$$\Omega : \mathrm{Mor}_{\mathbf{B}}(F(-), -) \to \mathrm{Mor}_{\mathbf{A}}(-, G(-))$$

of functors $\mathbf{A}^{op} \times \mathbf{B} \to \mathbf{Set}$; that is, for each pair of objects $A \in \mathbf{A}$ and $B \in \mathbf{B}$ there is an isomorphism $\Omega_{A,B} : \mathrm{Mor}_{\mathbf{B}}(F(A), B) \to \mathrm{Mor}_{\mathbf{A}}(A, G(B))$, which is natural in A and B. F is said to be *left adjoint* to G and G is *right adjoint* to F. An adjoint pair (F, G) is characterised by the existence of functorial morphisms

$$\begin{array}{lll} \eta : I_{\mathbf{A}} \to GF, & \text{defined by} \quad \eta_A = \Omega_{A,F(A)}(I_{F(A)}) : A \to GF(A), & A \in \mathbf{A}, \\ \psi : FG \to I_{\mathbf{B}}, & \text{defined by} \quad \psi_B = \Omega^{-1}_{G(B),B}(I_{G(B)}) : FG(B) \to B, & B \in \mathbf{B}, \end{array}$$

such that each of the following compositions yield the identity,

$$\begin{array}{l} F(A) \xrightarrow{F(\eta_A)} FGF(A) \xrightarrow{\psi_{F(A)}} F(A) \\ G(B) \xrightarrow{\eta_{G(B)}} GFG(B) \xrightarrow{G(\psi_B)} G(B). \end{array}$$

The transformation η is called a *unit* of adjunction and ψ is termed a *counit* of adjunction. In terms of unit and counit, the functorial isomorphism Ω and its inverse come out as

$$\Omega_{A,B} : \mathrm{Mor}_{\mathbf{B}}(F(A), B) \to \mathrm{Mor}_{\mathbf{A}}(A, G(B)), \quad g \mapsto G(g) \circ \eta_A,$$

$$\Omega^{-1}_{A,B} : \mathrm{Mor}_{\mathbf{A}}(A, G(B)) \to \mathrm{Mor}_{\mathbf{B}}(F(A), B), \quad f \mapsto \psi_B \circ F(f).$$

If (F, G) is an adjoint pair, then F preserves colimits (hence also epimorphisms and coproducts), while G preserves limits (monomorphisms and products). Furthermore, if G preserves epimorphisms, then F preserves projective objects, and if F preserves monomorphisms, then G preserves injective objects.

38.22. Equivalence of categories. An adjoint pair of covariant functors (F, G), $F : \mathbf{A} \to \mathbf{B}$ and $G : \mathbf{B} \to \mathbf{A}$, is called an *equivalence* if there are functorial isomorphisms $GF \simeq I_{\mathbf{A}}$ and $FG \simeq I_{\mathbf{B}}$. F and G are also called (inverse) *equivalences.*

Any functor is an equivalence if and only if it is full, faithful and representative.

38.23. Frobenius functors. An adjoint pair (F, G) of covariant functors $F : \mathbf{A} \to \mathbf{B}$ and $G : \mathbf{B} \to \mathbf{A}$ is called a *Frobenius pair* (and F, G *Frobenius functors*) if (G, F) also form an adjoint pair. Of course in this case F and G combine the properties of left and right adjoint functors (see 38.21) and so:

A Frobenius functor F preserves limits and colimits, projective and injective objects.

38.24. Adjoint pairs and separability. *Let (F, G) be an adjoint pair of covariant functors $F : \mathbf{A} \to \mathbf{B},\ G : \mathbf{B} \to \mathbf{A}$. Then:*

(1) *F is separable if and only if the unit $\eta : I_{\mathbf{A}} \to GF$ is a functorial coretraction; that is, for each object $A \in \mathbf{A}$, there exists a morphism $\nu_A : GF(A) \to A$ such that $\nu_A \circ \eta_A = I_A$, and any $f : A \to A'$ in $\mathbf{A}$ induces a commutative diagram,*

$$\begin{array}{ccc} GF(A) & \xrightarrow{GF(f)} & GF(A') \\ {\scriptstyle \nu_A}\downarrow & & \downarrow{\scriptstyle \nu_{A'}} \\ A & \xrightarrow[f]{} & A'. \end{array}$$

(2) *G is separable if and only if the counit $\psi : FG \to I_{\mathbf{B}}$ is a functorial retraction; that is, for each object $B \in \mathbf{B}$, there exists a morphism $\phi_B : B \to FG(B)$ such that $\psi_B \circ \phi_B = I_B$, and any $g \in B \to B'$ in $\mathbf{B}$ induces a commutative diagram,*

$$\begin{array}{ccc} B & \xrightarrow{g} & B' \\ {\scriptstyle \nu_B}\downarrow & & \downarrow{\scriptstyle \nu_{B'}} \\ FG(B) & \xrightarrow[FG(g)]{} & FG(B'). \end{array}$$

Proof. (1) "$\Rightarrow$" For any object $A \in \mathbf{A}$, the counit provides a morphism $\psi_{F(A)} : FGF(A) \to F(A)$. By the separability of F, there is a map

$$\Phi_{GF(A),A} : \mathrm{Mor}_{\mathbf{B}}(F(GF(A)), F(A)) \to \mathrm{Mor}_{\mathbf{A}}(GF(A), A),$$

which induces a morphism $\nu_A := \Phi_{GF(A),A}(\psi_{F(A)}) : GF(A) \to A$, functorial in A. From 38.21 we know that $\psi_{F(A)} \circ F(\eta_A) = I_{F(A)}$. Therefore

$$I_A = \Phi_{A,A}(I_{F(A)}) = \Phi_{GF(A),A}(\psi_{F(A)}) \circ \Phi_{A,GF(A)}(F(\eta_A)) = \nu_A \circ \eta_A.$$

"$\Leftarrow$" Assume there is a functorial morphism $\nu_A : GF(A) \to A$, such that $\nu_A \circ \eta_A = I_A$, where $A \in \mathbf{A}$. For any morphism $h : F(A) \to F(A_1)$ define

$$\Phi_{A,A_1}(h) \ :\ A \xrightarrow{\eta_A} GF(A) \xrightarrow{G(h)} GF(A_1) \xrightarrow{\nu_{A_1}} A_1.$$

The functoriality of ν_A implies that, for any $g : A \to A_1$, $\Phi_{A,A_1}(F(g)) = g$.

For morphisms $f : A \to A'$, $f_1 : A_1 \to A'_1$ in **A**, consider a commutative diagram in **B**,

$$\begin{array}{ccc} F(A) & \xrightarrow{h} & F(A_1) \\ {\scriptstyle F(f)}\downarrow & & \downarrow{\scriptstyle F(f_1)} \\ F(A') & \xrightarrow[h']{} & F(A'_1). \end{array}$$

Applying G we obtain the diagram

$$\begin{array}{ccccccc} A & \xrightarrow{\eta_A} & GF(A) & \xrightarrow{G(h)} & GF(A_1) & \xrightarrow{\nu_{A_1}} & A_1 \\ {\scriptstyle f}\downarrow & & {\scriptstyle GF(f)}\downarrow & & \downarrow{\scriptstyle GF(f_1)} & & \downarrow{\scriptstyle f_1} \\ A' & \xrightarrow[\eta_{A'}]{} & GF(A') & \xrightarrow[G(h')]{} & GF(A'_1) & \xrightarrow[\nu_{A'_1}]{} & A'_1, \end{array}$$

which obviously is commutative. The top sequence defines $\Phi_{A,A_1}(h)$ while the bottom sequence defines $\Phi_{A',A'_1}(h')$, and hence the conditions on the functor Φ stated in 38.18 are satisfied.

(2) The proof is similar to the proof of (1). □

38.25. F-coalgebras. For a category **A**, let $F : \mathbf{A} \to \mathbf{A}$ be a functor.

An *F-coalgebra* (N, ϱ^N) is an object N in **A** together with a morphism $\varrho^N : N \to F(N)$. A morphism $f : N \to N'$ in **A** between two F-coalgebras (N, ϱ^N) and $(N', \varrho^{N'})$ is called an *F-coalgebra morphism* provided it induces a commutative diagram,

$$\begin{array}{ccc} N & \xrightarrow{f} & N' \\ {\scriptstyle \varrho^N}\downarrow & & \downarrow{\scriptstyle \varrho^{N'}} \\ F(N) & \xrightarrow[F(f)]{} & F(N'). \end{array}$$

F-coalgebras together with F-coalgebra morphisms form a category that is denoted by **Coalg**(F).

Proposition. *Let $L : \Lambda \to$ **Coalg**(F) be any functor and Λ a small category.*

*(1) If $\varinjlim L$ exists in **A**, then it belongs to **Coalg**(F).*

*(2) Assume that F preserves limits. If $\varprojlim L$ exists in **A**, then it belongs to **Coalg**(F).*

Proof. (1) There is the diagram

$$\begin{array}{ccc} L(\lambda) & \xrightarrow{\epsilon_\lambda} & \varinjlim L \\ \downarrow{\scriptstyle \varrho^{L(\lambda)}} & & \\ F(L(\lambda)) & \xrightarrow{F(\epsilon_\lambda)} & F(\varinjlim L), \end{array}$$

where the $F(\epsilon_\lambda) \circ \varrho^{L(\lambda)}$ obviously form a compatible family of morphisms. By the universal property of the colimit we obtain a morphism $\varinjlim L \to F(\varinjlim L)$. Hence $\varinjlim L$ is an F-coalgebra.

(2) This is shown with a similar proof. □

As an example one can consider the functor $F = - \otimes_R C : \mathbf{M}_R \to \mathbf{M}_R$, where C is an (non-coassociative) R-coalgebra. In this case, F-coalgebras are R-modules with a right C-coaction. If C is coassociative, this functor F shows special properties that are axiomatised in the following notion. The resulting F-coalgebras will be right C-comodules.

38.26. Comonads and their coalgebras. A *comonad* or a *cotriple* is a triple $\mathbb{F} = (F, \delta, \psi)$, where $F : \mathbf{A} \to \mathbf{A}$ is a functor and $\delta : F \to F \circ F$, $\psi : F \to I_\mathbf{A}$ are natural transformations making the diagrams:

$$\begin{array}{ccc} F & \xrightarrow{\delta} & F \circ F \\ {\scriptstyle \delta}\downarrow & & \downarrow{\scriptstyle F\delta} \\ F \circ F & \xrightarrow[\delta_F]{} & F \circ F \circ F, \end{array} \qquad \begin{array}{ccc} F & \xrightarrow{\delta} & F \circ F \\ {\scriptstyle \delta}\downarrow & \searrow = & \downarrow{\scriptstyle \psi_F} \\ F \circ F & \xrightarrow[F\psi]{} & F \end{array}$$

commute. Explicitly, for each $N \in \mathbf{A}$, there exist $\delta_N : F(N) \to F(F(N))$ and $\psi_N : F(N) \to N$ such that $\delta_{F(N)} \circ \delta_N = F(\delta_N) \circ \delta_N$ and $\psi_{F(N)} \circ \delta_N = F(\psi_N) \circ \delta_N = I_{F(N)}$. The transformation δ is called a *coproduct* and ψ is called a *counit* of a comonad (F, δ, ψ).

An $\mathbb{F}$-*coalgebra* is a pair (N, ϱ^N), where $N \in \mathrm{Obj}(\mathbf{A})$ and $\varrho^N : N \to F(N)$ is a morphism in $\mathbf{A}$ such that the following diagrams:

$$\begin{array}{ccc} N & \xrightarrow{\varrho^N} & F(N) \\ {\scriptstyle \varrho^N}\downarrow & & \downarrow{\scriptstyle \delta_N} \\ F(N) & \xrightarrow[F\varrho^N]{} & F \circ F(N), \end{array} \qquad \begin{array}{ccc} N & \xrightarrow{\varrho^N} & F(N) \\ & \searrow^{I_N} & \downarrow{\scriptstyle \psi_N} \\ & & N \end{array}$$

commute. A morphism $f : N \to N'$ in $\mathbf{A}$ between two $\mathbb{F}$-coalgebras (N, ϱ^N) and $(N', \varrho^{N'})$ is called an $\mathbb{F}$-*coalgebra morphism* provided it induces a commutative diagram,

$$\begin{array}{ccc} N & \xrightarrow{f} & N' \\ {\scriptstyle \varrho^N}\downarrow & & \downarrow{\scriptstyle \varrho^{N'}} \\ F(N) & \xrightarrow[F(f)]{} & F(N'). \end{array}$$

$\mathbb{F}$-coalgebras together with $\mathbb{F}$-coalgebra morphisms form a category that is denoted by $\mathbf{Coalg}(\mathbb{F})$.

38.27. F-algebras. Let $F : \mathbf{A} \to \mathbf{A}$ be a functor. An *F-algebra* (M, ϱ_M) is an $M \in \mathrm{Obj}(\mathbf{A})$ together with a morphism $\varrho_M : F(M) \to M$. Morphisms between F-algebras are defined dually to F-coalgebra morphisms, and they yield the category $\mathbf{Alg}(\mathrm{F})$ of F-algebras. The behaviour towards limits and colimits is dual to coalgebras.

Proposition. *Let $L : \mathbf{\Lambda} \to \mathbf{Alg}(F)$ be any functor and $\mathbf{\Lambda}$ a small category.*

(1) If $\varprojlim L$ exists in $\mathbf{A}$, then it belongs to $\mathbf{Alg}(F)$.

(2) Assume that F preserves colimits. If $\varinjlim L$ exists in $\mathbf{A}$, then it belongs to $\mathbf{Alg}(F)$.

As an example consider the functor $F = A \otimes_R - : \mathbf{M}_R \to \mathbf{M}_R$, where A is some (nonassociative) R-algebra. If A is associative, this functor F shows special properties that are axiomatised in the following notion. In the case of the above example, the resulting F-algebras are simply left A-modules.

38.28. Monads and their algebras. A *monad* (or a *triple*) is a triple $\mathbb{F} = (F, \nu, \eta)$, where $F : \mathbf{A} \to \mathbf{A}$ is a functor and $\nu : F \circ F \to F$, $\eta : I_{\mathbf{A}} \to F$ are natural transformations making the following diagrams commute:

$$\begin{array}{ccc} F \circ F \circ F & \xrightarrow{\nu_F} & F \circ F \\ {\scriptstyle F\nu}\downarrow & & \downarrow{\scriptstyle \nu} \\ F \circ F & \xrightarrow[\nu]{} & F, \end{array} \qquad \begin{array}{ccc} F & \xrightarrow{\eta F} & F \circ F \\ {\scriptstyle F\eta}\downarrow & \searrow^{=} & \downarrow{\scriptstyle \nu} \\ F \circ F & \xrightarrow[\nu]{} & F. \end{array}$$

An $\mathbb{F}$-*algebra* is a pair (M, ϱ_M), where $M \in \mathrm{Obj}(\mathbf{A})$ and $\varrho_M : F(M) \to M$ is a morphism in M rendering commutative the following diagrams:

$$\begin{array}{ccc} F \circ F(M) & \xrightarrow{\nu_M} & F(M) \\ {\scriptstyle F\varrho_M}\downarrow & & \downarrow{\scriptstyle \varrho_M} \\ F(M) & \xrightarrow[\varrho_M]{} & M, \end{array} \qquad \begin{array}{ccc} M & \xrightarrow{\eta_M} & F(M) \\ & \searrow^{I_M} & \downarrow{\scriptstyle \varrho_M} \\ & & M. \end{array}$$

Morphisms between two $\mathbb{F}$-algebras are defined in an obvious way, and the category of $\mathbb{F}$-algebras is denoted by $\mathbf{Alg}(\mathbb{F})$.

General algebras and coalgebras of a functor have applications not only in mathematics but also in computer science. Monads and comonads are related to adjoint pairs.

38.29. (Co)monads and adjoint pairs. *Let $F : \mathbf{A} \to \mathbf{A}$ be a functor.*

(1) If $\mathbb{F} = (F, \nu, \eta)$ is a monad, then the forgetful functor $\mathbf{Alg}(\mathbb{F}) \to \mathbf{A}$ has the left adjoint

$$G : \mathbf{A} \to \mathbf{Alg}(\mathbb{F}), \quad M \mapsto (F(M),\ \nu_M : F \circ F(M) \to F(M)).$$

Explicitly, for $M \in \mathrm{Obj}(\mathbf{A})$, N *in* $\mathbf{Alg}(\mathbb{F})$, *the map*

$$\mathrm{Mor}_{\mathbf{Alg}(\mathbb{F})}(F(M), N) \longrightarrow \mathrm{Mor}_{\mathbf{A}}(M, N), \quad f \mapsto f \circ \eta_M,$$

is bijective with the inverse $h \mapsto \varrho_N \circ F(h)$.

(2) *If* $\mathbb{F} = (F, \delta, \psi)$ *is a comonad, then the forgetful functor* $\mathbf{Coalg}(\mathbb{F}) \to \mathbf{A}$ *has the right adjoint*

$$H : \mathbf{A} \to \mathbf{Coalg}(\mathbb{F}), \quad N \mapsto (F(N),\ \delta_N : F(N) \to F \circ F(N)).$$

Explicitly, for M *in* $\mathbf{Coalg}(\mathbb{F})$, $N \in \mathrm{Obj}(\mathbf{A})$, *the map*

$$\mathrm{Mor}_{\mathbf{Coalg}(\mathbb{F})}(M, F(N)) \to \mathrm{Mor}_{\mathbf{A}}(M, N), \quad f \mapsto \psi_N \circ f,$$

is bijective with the inverse $h \mapsto F(h) \circ \varrho^M$.

38.30. Adjoint pairs and (co)monads. *Let* $L : \mathbf{A} \to \mathbf{B}$ *and* $R : \mathbf{B} \to \mathbf{A}$ *be an adjoint pair of functors with unit* $\eta : I_{\mathbf{A}} \to RL$ *and counit* $\psi : LR \to I_{\mathbf{B}}$.

(1) $RL : \mathbf{A} \to \mathbf{A}$ *induces a monad with product* $\nu = R\psi_L : RLRL \to RL$ *and unit* $\eta : I_{\mathbf{A}} \to RL$.

(2) $LR : \mathbf{B} \to \mathbf{B}$ *induces a comonad with coproduct* $\delta = L\eta_R : LR \to LRLR$ *and counit* $\psi : LR \to I_{\mathbf{B}}$.

As already suggested by the motivating examples, monads and comonads arise most naturally in monoidal categories.

38.31. Monoidal category. A category $\mathbf{A}$ is called a *monoidal category* if there exist a bifunctor $- \otimes - : \mathbf{A} \times \mathbf{A} \to \mathbf{A}$, a distinguished neutral object E in $\mathbf{A}$ and natural isomorphisms

$$\alpha : (- \otimes -) \otimes - \to - \otimes (- \otimes -), \quad \lambda : E \otimes - \to I_{\mathbf{A}}, \quad \varrho : - \otimes E \to I_{\mathbf{A}}$$

such that, for all objects W, X, Y, Z in $\mathbf{A}$, the following two diagrams commute:

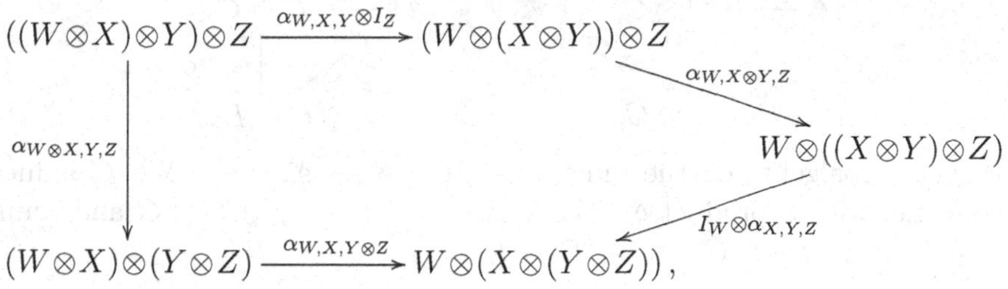

$$\begin{array}{ccc}
((W \otimes X) \otimes Y) \otimes Z & \xrightarrow{\alpha_{W,X,Y} \otimes I_Z} & (W \otimes (X \otimes Y)) \otimes Z \\
 & & \downarrow \alpha_{W,X \otimes Y,Z} \\
\downarrow \alpha_{W \otimes X,Y,Z} & & W \otimes ((X \otimes Y) \otimes Z) \\
 & & \downarrow I_W \otimes \alpha_{X,Y,Z} \\
(W \otimes X) \otimes (Y \otimes Z) & \xrightarrow{\alpha_{W,X,Y \otimes Z}} & W \otimes (X \otimes (Y \otimes Z)),
\end{array}$$

and

$$\begin{array}{ccc}
(X \otimes E) \otimes Y & \xrightarrow{\alpha_{X,E,Y}} & X \otimes (E \otimes Y) \\
{\scriptstyle \varrho_X \otimes I_Y} \searrow & & \swarrow {\scriptstyle I_X \otimes \lambda_Y} \\
 & X \otimes Y. &
\end{array}$$

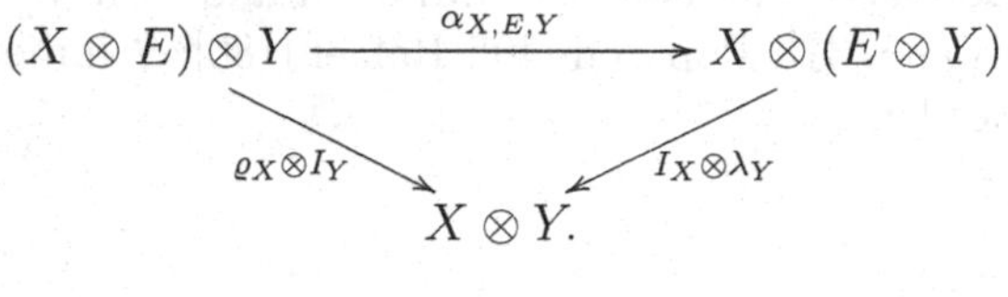

A monoidal category is denoted by $(\mathbf{A}, \otimes, E)$. It is said to be *strict* if the isomorphisms α, λ, ϱ are identity morphisms.

38.32. Algebras in monoidal categories. An (associative) *algebra* in a monoidal category $(\mathbf{A}, \otimes, E)$ is an object $A \in \mathbf{A}$ and a pair of morphisms $\mu_A : A \otimes A \to A$, $\iota_A : E \to A$, rendering commutative the following three diagrams:

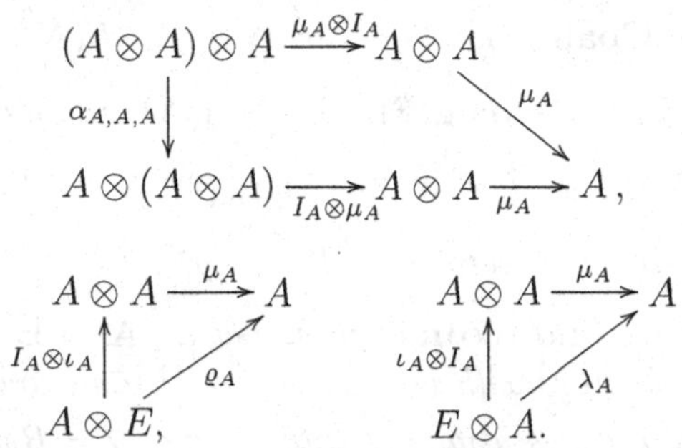

Given an algebra A, the functor $A \otimes - : \mathbf{A} \to \mathbf{A}$, $N \mapsto A \otimes N$, induces a monad, with product $\nu = \mu_A \otimes I_{\mathbf{A}} : A \otimes A \otimes - \to A \otimes -$ and unit $\eta = (\iota_A \otimes I_{\mathbf{A}})\lambda^{-1} : I_{\mathbf{A}} \to E \otimes - \to A \otimes -$.

38.33. Coalgebras in monoidal categories. A (coassociative) *coalgebra* in a monoidal category $(\mathbf{A}, \otimes, E)$ is an object $C \in \mathbf{A}$ and a pair of morphisms $\Delta_C : C \to C \otimes C$, $\varepsilon_C : C \to E$, inducing the following three commutative diagrams:

$$\begin{array}{ccccc}
C & \xrightarrow{\Delta_C} & C \otimes C & \xrightarrow{\Delta_C \otimes I_C} & (C \otimes C) \otimes C \\
& \searrow^{\Delta_C} & & & \downarrow \alpha_{C,C,C} \\
& & C \otimes C & \xrightarrow[I_C \otimes \Delta_C]{} & C \otimes (C \otimes C),
\end{array}$$

$$\begin{array}{ccc}
C & \xrightarrow{\Delta_C} & C \otimes C \\
& \searrow^{\lambda_C^{-1}} & \downarrow \varepsilon_C \otimes I_C \\
& & E \otimes C,
\end{array}
\qquad
\begin{array}{ccc}
C & \xrightarrow{\Delta_C} & C \otimes C \\
& \searrow^{\varrho_C^{-1}} & \downarrow I_C \otimes \varepsilon_C \\
& & C \otimes E.
\end{array}$$

For such a coalgebra C, the functor $- \otimes C : \mathbf{A} \to \mathbf{A}$, $N \mapsto N \otimes C$ induces a comonad with coproduct $\delta = I_{\mathbf{A}} \otimes \Delta_C : - \otimes C \to - \otimes C \otimes C$ and counit $\psi = \varrho(I_{\mathbf{A}} \otimes \varepsilon_C) : - \otimes C \to - \otimes E \to I_{\mathbf{A}}$.

References. Borceux [3]; Caenepeel, Militaru and Zhu [9]; Gumm [19]; Hughes [134]; Mac Lane [31]; Popescu [40]; Rafael [180]; Schubert [42]; Stenström [44]; Wisbauer [46].

39 Modules and Abelian categories

Many interesting preadditive categories have in fact an additional module structure over an associative ring. The following theorem of B. Mitchell shows that functorial morphisms defined on the ring can be extended to the whole module category.

39.1. Functors on $\mathbf{M}_T$. *Let $\mathbf{A}$ be a cocomplete Abelian category. For a ring T, denote by $\mathbf{M}_T$ the category of right T-modules, and let $F, F' : \mathbf{M}_T \to \mathbf{A}$ be additive functors. Suppose that F preserves colimits. Consider T as a subcategory (with one object) of $\mathbf{M}_T$ and assume that there exists a functorial morphism $\eta_T : F(T) \to F'(T)$ for the restricted functors $F, F' : T \to \mathbf{A}$. Then:*

(1) η_T can be extended to a functorial morphism $\eta : F \to F'$.

(2) If F' respects colimits, then η is a functorial isomorphism.

Proof. (1) For any free right T-module $T^{(\Lambda)}$, consider the morphism

$$\eta_{T^{(\Lambda)}} : F(T^{(\Lambda)}) \xrightarrow{\simeq} F(T)^{(\Lambda)} \xrightarrow{\eta_T^{(\Lambda)}} F'(T)^{(\Lambda)} \xrightarrow{\pi_\Lambda} F'(T^{(\Lambda)}),$$

where π_Λ is defined by the canonical injections $T \to T^{(\Lambda)}$ and the universal property of the coproduct $F'(T)^{(\Lambda)}$. For any $N \in \mathbf{M}_T$, consider a presentation $T^{(\Omega)} \to T^{(\Lambda)} \xrightarrow{p} N \to 0$. Then there is the following commutative diagram:

$$\begin{array}{ccccccc}
F(T^{(\Omega)}) & \longrightarrow & F(T^{(\Lambda)}) & \xrightarrow{F(p)} & F(N) & \longrightarrow & 0 \\
\downarrow{\scriptstyle \eta_{T^{(\Omega)}}} & & \downarrow{\scriptstyle \eta_{T^{(\Lambda)}}} & & & & \\
F'(T^{(\Omega)}) & \longrightarrow & F'(T^{(\Lambda)}) & \xrightarrow{F'(p)} & F'(N), & &
\end{array}$$

where the upper sequence is exact in $\mathbf{Ab}$ and the bottom is a zero sequence. By the cokernel property we obtain a morphism $\eta_N : F(N) \to F'(N)$ making the diagram commutative. To show that η_N is functorial in N, consider any morphism $f : N \to N'$ in $\mathbf{M}_T$ and the following diagram:

$$\begin{array}{ccccc}
T^{(\Lambda)} & \xrightarrow{p} & N & \longrightarrow & 0 \\
 & & \downarrow{\scriptstyle f} & & \\
T^{(\Lambda')} & \longrightarrow & N' & \longrightarrow & 0,
\end{array}$$

where the top and bottom sequences are used to define η. Choosing a map $g : T^{(\Lambda)} \to T^{(\Lambda')}$ that renders the diagram commutative, we obtain two commutative diagrams:

$$\begin{array}{ccc} F(T^{(\Lambda)}) & \xrightarrow{F(p)} & F(N) \\ {\scriptstyle F(g)}\downarrow & & \downarrow{\scriptstyle F(f)} \\ F(T^{(\Lambda')}) & \longrightarrow & F(N'), \end{array} \qquad \begin{array}{ccc} F'(T^{(\Lambda)}) & \xrightarrow{F'(p)} & F'(N) \\ {\scriptstyle F'(g)}\downarrow & & \downarrow{\scriptstyle F'(f)} \\ F'(T^{(\Lambda')}) & \longrightarrow & F'(N'), \end{array}$$

which are connected by η. In the resulting cube we find the square

$$\begin{array}{ccc} F(N) & \xrightarrow{F(f)} & F(N') \\ {\scriptstyle \eta_N}\downarrow & & \downarrow{\scriptstyle \eta_{N'}} \\ F'(N) & \xrightarrow{F'(f)} & F'(N'), \end{array}$$

and we derive the equality

$$F'(f) \circ \eta_N \circ F(p) = \eta_{N'} \circ F(f) \circ F(p).$$

Since $F(p)$ is surjective, this implies the commutativity of the previous diagram, showing that η_N is natural in N.

(2) If F' respects colimits, then $\eta_{T^{(\Omega)}}$ and $\eta_{T^{(\Lambda)}}$ in the diagram defining η_N (see proof of (1)) are isomorphisms, and we conclude that so is η_N. □

39.2. T-objects in a category. Let $\mathbf{A}$ be a cocomplete additive category of Abelian groups and T an associative ring. An object A in $\mathbf{A}$ is called a *T-object* if there is a ring morphism $\phi_A : T \to \mathrm{End}_{\mathbf{A}}(A)$. A morphism $f : A \to A'$ in $\mathbf{A}$ between T-objects is called *T-linear* if $\phi_{A'} \circ f = f \circ \phi_A$. The category consisting of T-objects in $\mathbf{A}$ and T-linear morphisms is denoted by ${}_T\mathbf{A}$. Any T-object A can be considered as a left T-module in a canonical way. This yields a T-linear isomorphism in $\mathbf{A}$, $T \otimes_T A \to A$, $t \otimes a \mapsto \phi_A(t)(a) =: ta$.

For any $N \in \mathbf{M}_T$ with the free presentation $T^{(\Omega)} \xrightarrow{h} T^{(\Lambda)} \longrightarrow N \longrightarrow 0$ and a T-object $A \in \mathbf{A}$, view $N \otimes_T A$ as an object in $\mathbf{A}$ by the commutative exact diagram,

$$\begin{array}{ccccccc} T^{(\Omega)} \otimes_T A & \xrightarrow{h \otimes I_A} & T^{(\Lambda)} \otimes_T A & \longrightarrow & N \otimes_T A & \longrightarrow & 0 \\ \simeq\downarrow & & \downarrow\simeq & & \downarrow\simeq & & \\ A^{(\Omega)} & \xrightarrow{\tilde{h}} & A^{(\Lambda)} & \longrightarrow & \mathrm{Coke}\,\tilde{h} & \longrightarrow & 0, \end{array}$$

and obviously for any morphism $g : N \to N'$ in $\mathbf{M}_T$ we obtain a morphism $g \otimes I_A : N \otimes_T A \to N' \otimes_T A$ in $\mathbf{A}$.

39.3. Functors and T-objects. *Let $\mathbf{A}, \mathbf{B}$ be cocomplete additive categories of Abelian groups, T any ring, and consider an additive covariant functor $F : \mathbf{A} \to \mathbf{B}$.*

(1) *For any* $A \in {}_T\mathbf{A}$, $F(A)$ *lies in* ${}_T\mathbf{B}$, *and there is a functorial morphism*

$$\Psi_{-,A} : - \otimes_T F(A) \to F(- \otimes_T A) \text{ of functors } \mathbf{M}_T \to \mathbf{B}.$$

$\Psi_{P,A}$ *is an isomorphism provided that* $P \in \mathbf{M}_T$ *is finitely generated and projective.*

(2) *If* F *preserves coproducts, then there is a functorial morphism*

$$\Psi : - \otimes_T F(-) \to F(- \otimes_T -) \text{ of functors } \mathbf{M}_T \times {}_T\mathbf{A} \to \mathbf{B}.$$

(3) *If* F *preserves colimits, then the* $\Psi_{N,A}$ *are isomorphisms, for any objects* $N \in \mathbf{M}_T$, $A \in {}_T\mathbf{A}$.

Proof. (1) $F(A)$ is a T-object by the composition of ring morphisms $T \to \text{End}_{\mathbf{A}}(A) \to \text{End}_{\mathbf{B}}(F(A))$. To construct the functorial morphism needed, $\Psi_{-,A} : - \otimes_T F(A) \to F(- \otimes_T A)$, use the canonical isomorphisms (see above) to define

$$\varphi_{T,A} : T \otimes_T F(A) \xrightarrow{\simeq} F(A) \xrightarrow{\simeq} F(T \otimes_T A).$$

For any $t \in T$ (considered as an element in $\text{End}_T(T)$) there is a diagram

$$\begin{array}{ccccc} T \otimes_T F(A) & \xrightarrow{\simeq} & F(A) & \xrightarrow{\simeq} & F(T \otimes_T A) \\ \downarrow{\scriptstyle t \otimes I_{F(A)}} & & \downarrow{\scriptstyle F(\phi_A(t))} & & \downarrow{\scriptstyle F(t \otimes I_A)} \\ T \otimes_T F(A) & \xrightarrow{\simeq} & F(A) & \xrightarrow{\simeq} & F(T \otimes_T A), \end{array}$$

where the left square is commutative by the definition of the module structure on $F(A)$, and the right square is the image under F of a commutative diagram in $\mathbf{A}$. So the outer diagram is commutative, which means that $\Psi_{-,A}$ is functorial with respect to internal morphisms $T \to T$. Now apply Mitchell's Theorem 39.1 to extend the functorial morphism to all of $\mathbf{M}_T$.

(2) Assume that F preserves coproducts. It remains to show that $\Psi_{N,-}$ is functorial for all $N \in \mathbf{M}_T$. Consider any $f : A \to A'$ in ${}_T\mathbf{A}$. For $N = T$ there is a diagram,

$$\begin{array}{ccccc} T \otimes_T F(A) & \xrightarrow{\simeq} & F(A) & \xrightarrow{\simeq} & F(T \otimes_T A) \\ \downarrow{\scriptstyle I_T \otimes F(f)} & & \downarrow{\scriptstyle F(f)} & & \downarrow{\scriptstyle F(I_T \otimes f)} \\ T \otimes_T F(A') & \xrightarrow{\simeq} & F(A') & \xrightarrow{\simeq} & F(T \otimes_T A'), \end{array}$$

where the left square is commutative by the definition of the T-module structure on $F(A)$ and $F(A')$ and functor properties of F, whereas the right-hand square is the image under F of a commutative diagram (characterising f as a T-morphism) in $\mathbf{A}$.

Since F preserves coproducts, $\Psi_{P,-}$ is functorial for any free module P in $\mathbf{M}_T$. For an arbitrary $N \in \mathbf{M}_T$, choose a free presentation $P \xrightarrow{p} N \to 0$. Then we obtain two commutative diagrams:

$$\begin{array}{ccc} P \otimes_T F(A) & \xrightarrow{I_P \otimes F(f)} & P \otimes_T F(A') \\ \downarrow{\scriptstyle p\otimes I_{F(A)}} & & \downarrow{\scriptstyle p\otimes I_{F(A')}} \\ N \otimes_T F(A) & \xrightarrow{I_N \otimes F(f)} & N \otimes_T F(A'), \end{array} \qquad \begin{array}{ccc} F(P \otimes_T A) & \xrightarrow{F(I_P \otimes f)} & F(P \otimes_T A') \\ \downarrow{\scriptstyle F(p\otimes I_A)} & & \downarrow{\scriptstyle F(p\otimes I_{A'})} \\ F(N \otimes_T A) & \xrightarrow{F(I_N \otimes f)} & F(N \otimes_T A'), \end{array}$$

which are connected by Ψ. In the resulting cube we find the diagram

$$\begin{array}{ccc} N \otimes_T F(A) & \xrightarrow{I_N \otimes F(f)} & N \otimes_T F(A') \\ \downarrow{\scriptstyle \Psi_{N,A}} & & \downarrow{\scriptstyle \Psi_{N,A'}} \\ F(N \otimes_T A) & \xrightarrow{F(I_N \otimes f)} & F(N \otimes_T A'), \end{array}$$

and we derive

$$F(I_N \otimes f) \circ \Psi_{N,A} \circ (p \otimes I_A) = \Psi_{N,A'} \circ (I_N \otimes F(f)) \circ (p \otimes I_A).$$

Since $p \otimes I$ is surjective, we conclude that the previous diagram is commutative, showing that $\Psi_{N,-}$ is a functorial morphism.

(3) If F respects colimits, then so does $F(- \otimes_T A)$ for any $A \in {}_T\mathbf{A}$ and the assertion follows from 39.1(2). □

As a corollary we obtain the important *Eilenberg-Watts Theorem*:

39.4. Functors from $\mathbf{M}_T$. *Let* $\mathbf{B}$ *be a cocomplete additive category of Abelian groups,* T *any ring, and* $F : \mathbf{M}_T \to \mathbf{B}$ *an additive covariant functor that preserves colimits. Then* $F(T)$ *is a left* T*-module and there is a functorial isomorphism*

$$F(-) \simeq - \otimes_T F(T).$$

Proof. Take $\mathbf{A} = \mathbf{M}_T$ and apply 39.3. □

39.5. Functorial morphisms and T-objects. *Let* $\mathbf{A}$, $\mathbf{B}$ *be cocomplete additive categories of Abelian groups and* T *any ring. Consider the additive covariant functors* $F, F' : \mathbf{A} \to \mathbf{B}$ *with a functorial morphism* $\eta : F \to F'$.

(1) *For any* $A \in {}_T\mathbf{A}$, *the map* $\eta_A : F(A) \to F'(A)$ *is* T*-linear.*

(2) *For* $N \in \mathbf{M}_T$ *and* $A \in {}_T\mathbf{A}$, *consider the diagram*

$$\begin{array}{ccc} N \otimes_T F(A) & \xrightarrow{I_N \otimes \eta_A} & N \otimes_T F'(A) \\ \downarrow{\scriptstyle \Psi_{N,A}} & & \downarrow{\scriptstyle \Psi'_{N,A}} \\ F(N \otimes_T A) & \xrightarrow{\eta_{N \otimes_T A}} & F'(N \otimes_T A), \end{array}$$

where Ψ and Ψ' denote the functorial morphisms induced by F and F', respectively (cf. 39.3). The diagram is commutative provided either N is finitely generated and projective, or both F and F' preserve colimits.

Proof. (1) With the canonical isomorphisms for T-objects (see 39.2) there is a commutative diagram,

$$\begin{array}{ccc} T \otimes_T F(A) & \xrightarrow{I_T \otimes \eta_A} & T \otimes_T F'(A) \\ \simeq \downarrow & & \downarrow \simeq \\ F(A) & \xrightarrow{\eta_A} & F'(A), \end{array}$$

which shows that η_A is T-linear.

(2) In the diagram

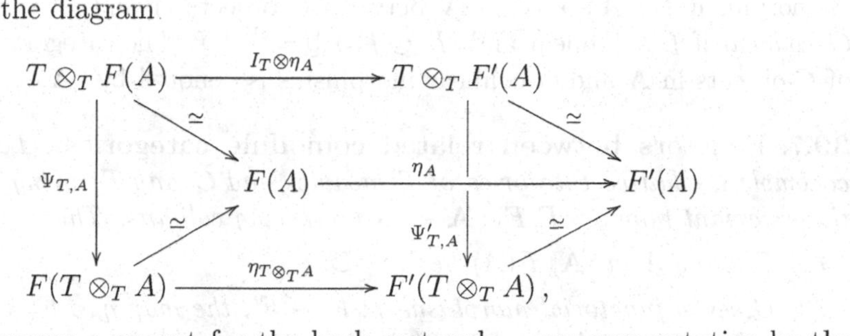

all subdiagrams – except for the back rectangle – are commutative by the definitions. This obviously implies that the back rectangle also commutes, and from this we derive the commutativity of the diagram in (2) under the given conditions. □

39.6. Separable functors between categories. *Let $F : \mathbf{A} \to \mathbf{B}$ be a separable additive functor between cocomplete additive categories whose objects are Abelian groups with a splitting morphism*

$$\Phi : \mathrm{Mor}_{\mathbf{B}}(F(-), F(-)) \to \mathrm{Mor}_{\mathbf{A}}(-,-).$$

Let $A, A' \in {}_T\mathbf{A}$ for some ring T. If $g : F(A) \to F(A')$ is a T-linear morphism in $\mathbf{B}$, then $\Phi_{A,A'}(g) : A \to A'$ is a T-linear morphism in $\mathbf{A}$.

Proof. If the action of T on A is given by the ring homomorphism $\phi_A : T \to \mathrm{End}_{\mathbf{A}}(A)$, the action on $F(A)$ is given by $F(\phi_A)$ (see 39.2). Given a T-morphism $g : F(A) \to F(A')$, for any $t \in T$, there is the commutative diagram on the left that – by Φ – induces the commutative diagram on the right:

$$\begin{array}{ccc} F(A) & \xrightarrow{F(\phi_A(t))} & F(A) \\ g \downarrow & & \downarrow g \\ F(A') & \xrightarrow[F(\phi_{A'}(t))]{} & F(A'), \end{array} \qquad \begin{array}{ccc} A & \xrightarrow{\phi_A(t)} & A \\ \Phi(g) \downarrow & & \downarrow \Phi(g) \\ A' & \xrightarrow[\phi_{A'}(t)]{} & A', \end{array}$$

thus showing that $\Phi(g)$ is T-linear. □

Let T be an associative ring, C a T-coring with coproduct Δ and counit ε, and let $\mathbf{A}$ be a cocomplete additive category. An object $A \in \mathbf{A}$ is called a *C-object* if A is a T-object and there is a morphism ${}^A\varrho : A \to C \otimes_T A$ inducing the commutative diagrams in $\mathbf{A}$,

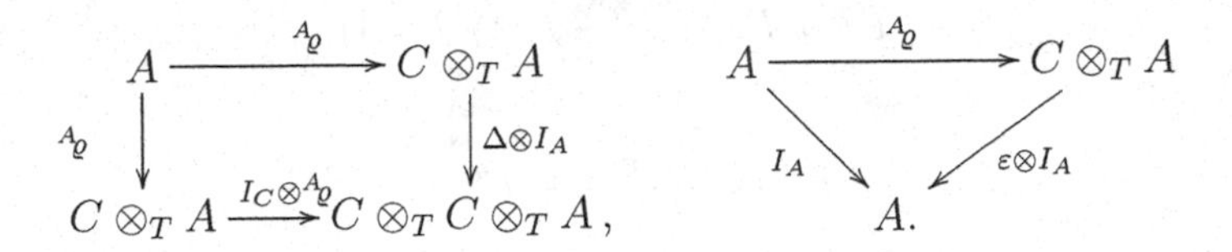

A morphism $f : A \to A'$ in $\mathbf{A}$ between C-objects A and A' is said to be *C-colinear* if f is T-linear and $(I_C \otimes f) \circ {}^A\varrho = {}^{A'}\varrho \circ f$. The category consisting of C-objects in $\mathbf{A}$ and C-colinear morphisms is denoted by ${}^C\mathbf{A}$.

39.7. Functors between related comodule categories. *Let $\mathbf{A}, \mathbf{B}$ be cocomplete additive categories of T-modules and C any T-coring. Consider the covariant functors $F, F' : \mathbf{A} \to \mathbf{B}$ preserving colimits. Then:*

(1) For any A in ${}^C\mathbf{A}$, $F(A)$ lies in ${}^C\mathbf{B}$.

(2) Given a functorial morphism $\eta : F \to F'$, the map $\eta_A : F(A) \to F'(A)$ is C-colinear, for any A in ${}^C\mathbf{A}$.

Proof. (1) By 39.3, $F(A) \in {}_T\mathbf{A}$ and there is a functorial morphism

$$\Psi_{-,A} : - \otimes_T F(A) \to F(- \otimes_T A) \quad \text{of functors} \quad \mathbf{M}_T \times {}_T\mathbf{A} \to \mathbf{B},$$

such that $\Psi_{N,A}$ is an isomorphism for each $N \in \mathbf{M}_T$. Define a comodule structure on $F(A)$ by the first row in the diagram

$$\begin{array}{ccccc}
F(A) & \xrightarrow{F({}^A\varrho)} & F(C \otimes_T A) & \xrightarrow{\Psi^{-1}_{C,A}} & C \otimes_T F(A) \\
\downarrow {\scriptstyle F({}^A\varrho)} & & \downarrow {\scriptstyle F(\Delta \otimes I_A)} & & \downarrow {\scriptstyle \Delta \otimes I_{F(A)}} \\
F(C \otimes_T A) & \xrightarrow{F(I_C \otimes {}^A\varrho)} & F(C \otimes_T C \otimes_T A) & \overset{\Psi^{-1}_{C\otimes C,A}}{\searrow} & \\
\downarrow {\scriptstyle \Psi^{-1}_{C,A}} & & \downarrow {\scriptstyle \Psi^{-1}_{C,C\otimes A}} & & \\
C \otimes_T F(A) & \xrightarrow[I_C \otimes F({}^A\varrho)]{} & C \otimes_T F(C \otimes_T A) & \xrightarrow[I_C \otimes \Psi^{-1}_{C,A}]{} & C \otimes_T C \otimes_T F(A).
\end{array}$$

Here the top left-hand square is commutative by the coassociativity of ${}^A\varrho$. By the functoriality of Ψ,

$$\Psi_{C,C,A} \circ (\Delta \otimes I_{F(A)}) = F(\Delta \otimes I_A) \circ \Psi_{C,A}.$$

Applying the corresponding inverse morphisms from the left and right, we obtain

$$(\Delta \otimes I_{F(A)}) \circ \Psi_{C,A}^{-1} = \Psi_{C\otimes_T C,A}^{-1} \circ F(\Delta \otimes I_A).$$

This shows that the top right-hand quadrangle in the diagram is commutative. Similarly, the commutativity of the bottom square is shown, and the triangle is commutative by the properties of Ψ. From this we see that the coaction on $F(A)$ is coassociative and the desired property for the counit can be seen from the commutative diagram

$$\begin{array}{ccccc} F(A) & \xrightarrow{F(^A\varrho)} & F(C\otimes_T A) & \xrightarrow{\Psi_{C,A}^{-1}} & C\otimes_T F(A) \\ & \searrow^{I_{F(A)}} & \downarrow{\scriptstyle F(\varepsilon\otimes I_A)} & \swarrow_{\varepsilon\otimes I_{F(A)}} & \\ & & F(A). & & \end{array}$$

(2) To prove that η_A is a C-colinear map, consider the diagram

$$\begin{array}{ccccc} F(A) & \xrightarrow{F(^A\varrho)} & F(C\otimes_T A) & \xrightarrow{\Psi_{C,A}^{-1}} & C\otimes_T F(A) \\ \downarrow{\scriptstyle \eta_A} & & \downarrow{\scriptstyle \eta_{C\otimes_T A}} & & \downarrow{\scriptstyle I_C\otimes\eta_A} \\ F'(A) & \xrightarrow[F(^A\varrho)]{} & F'(C\otimes_T A) & \xrightarrow[\Psi'^{-1}_{C,A}]{} & C\otimes_T F'(A)\,. \end{array}$$

Here the left-hand square is commutative by the properties of η and the right square is commutative by 39.5 and the fact that all the Ψ's are bijective. □

References. Gómez-Torrecillas [122]; Popescu [40]; Rafael [180].

40 Algebras over commutative rings

Although we assume the reader to be familiar with the basics of ring and module theory, and hence also with associative algebras, we gather here facts and some more nonstandard properties of algebras over a commutative ring. The aim of this presentation of the fundamentals of algebra is to provide the reader with an intuition in what sense coalgebras and corings are dual to algebras.

Throughout, R denotes an associative commutative ring with unit and $\mathbf{M}_R$ is the category of R-modules.

40.1. Tensor products. For any $M \in \mathbf{M}_R$, we will identify $M \otimes_R R$ with M by the isomorphism $\vartheta : R \otimes_R M \to M$, $r \otimes m \mapsto rm$. Let K, L, M, N be R-modules. We fix the notation for the twist isomorphisms,

$$\begin{array}{rlll} \mathsf{tw}: & M \otimes_R N & \to N \otimes_R M, & m \otimes n \mapsto n \otimes m, \\ \mathsf{tw}_{13}: & M \otimes_R N \otimes_R L & \to L \otimes_R N \otimes_R M, & (m \otimes n) \otimes l \mapsto l \otimes n \otimes m, \end{array}$$

and we write tw_{ij} for the permutation of position i with position j in a multiple tensor product.

Tensor product of morphisms. For R-morphisms $f : M \to N$, $g : K \to L$, there is a unique R-morphism,

$$f \underline{\otimes} g : M \otimes_R K \to N \otimes_R L, \quad m \otimes k \mapsto f(m) \otimes g(k),$$

called the *tensor product of f and g*. This induces an R-homomorphism,

$$\mathrm{Hom}_R(M, N) \otimes_R \mathrm{Hom}_R(K, L) \to \mathrm{Hom}_R(M \otimes_R K, N \otimes_R L), \quad f \otimes g \mapsto f \underline{\otimes} g.$$

In particular, for $N = L = R$, there is an R-homomorphism,

$$\mathrm{Hom}_R(M, R) \otimes_R \mathrm{Hom}_R(K, R) \to \mathrm{Hom}_R(M \otimes_R K, R),$$

which is an isomorphism if M (or K) is a finitely generated projective R-module. Usually we simply write $f \otimes g$ instead of $f \underline{\otimes} g$.

40.2. Algebras and their morphisms. An R-module A is an *associative R-algebra* (with unit) if there exist R-linear maps, $\mu : A \otimes_R A \to A$, called the *product* and $\iota : R \to A$ called the *unit*, such that

$$a(bc) = (ab)c \text{ and } a1_A = a = 1_A a, \quad \text{for all } a, b, c \in A,$$

where $\mu(a \otimes b) = ab$ and $1_A = \iota(1_R)$. A is called *commutative* if $ab = ba$, for all $a, b \in A$.

An R-linear map $f : A \to B$ between two R-algebras is a *(unital) algebra morphism*, if $f(ab) = f(a)f(b)$ for all $a, b \in A$, and $f(1_A) = 1_B$. The identity $I_A : A \to A$ is an algebra morphism, and the composition of algebra morphisms is again an algebra morphism. Hence the R-algebras and their morphisms form a category denoted by $\mathbf{Alg}(R)$.

40.3. Tensor product of algebras. The tensor product $A \otimes_R B$ of two R-algebras is an R-algebra defined by the product $\mu_{A\otimes B}$,

$$(A\otimes_R B)\otimes_R(A\otimes_R B) \xrightarrow{I_A\otimes \mathrm{tw}\otimes I_B} (A\otimes_R A)\otimes_R(B\otimes_R B) \xrightarrow{\mu_A\otimes\mu_B} A\otimes_R B\,,$$

and the unit $\iota_{A\otimes B}(1_R) = \iota_A(1_R)\otimes\iota_B(1_R) : R \to A\otimes_R B, 1_R \mapsto 1_A\otimes 1_B$. If B is commutative, then the map

$$\mu_{A\otimes B} : (A\otimes_R B)\otimes_B (A\otimes_R B) \to A\otimes_R B$$

is B-linear, and hence $A\otimes_R B$ is a B-algebra (*scalar extension* of A by B). $A\otimes_R B$ is commutative if both A and B are commutative algebras.

40.4. A-modules and homomorphisms. Let A be an R-algebra. An R-module M with an R-linear map

$$\varrho_M : A\otimes_R M \to M, \quad a\otimes m \mapsto am,$$

is called a *(unital) left A-module* if

$$\varrho_M\circ(\mu_A\otimes I_M) = \varrho_M\circ(I_M\otimes\varrho_M) \text{ and } \varrho_M\circ(\iota_A\otimes I_M) = I_M.$$

The second condition means that $1_A m = m$, for all $m\in M$. So there is an R-morphism $M \to A\otimes_R M$, $m\mapsto 1_A\otimes m$, and M is a direct summand of $A\otimes_R M$ as an R-module.

An R-linear map $g : M\to N$ between left A-modules is an *A-morphism* (*A-homomorphism, A-linear*) if $\varrho_M\circ(I_A\otimes g) = g\circ\varrho_N$, that is, $g(am) = ag(m)$, for all $a\in A$, $m\in M$. The set of A-morphisms $M\to N$ is denoted by ${}_A\mathrm{Hom}(M,N)$ and forms an Abelian group that can be characterised by the exact sequence

$$0 \longrightarrow {}_A\mathrm{Hom}(M,N) \longrightarrow \mathrm{Hom}_R(M,N) \xrightarrow{\beta} \mathrm{Hom}_R(A\otimes M,N),$$

where $\beta(f) = \varrho_N\circ(I_A\otimes f) - f\circ\varrho_M$. Left A-modules with the A-morphisms form an additive category that is denoted by ${}_A\mathbf{M}$. This is simply the category of F-algebras for the functor (monad) $A\otimes_R - : \mathbf{M}_R\to\mathbf{M}_R$ (see 38.28).

Right A-modules and related notions are defined in an obvious way, and the category of right A-modules is denoted by $\mathbf{M}_A$. For the morphisms between $M,N\in\mathbf{M}_A$ we write $\mathrm{Hom}_A(M,N)$. Of course A itself is a left and right A-module (defined by μ_A).

For notational convenience it is sometimes advantageous to write homomorphisms of left A-modules M on the right side of the argument. Then M is an (A,S)-bimodule, where S denotes ${}_A\mathrm{End}(M)$ acting from the right. Notice that the switch from the left action to the right action of endomorphisms changes the product to the opposite one. Symmetrically, morphisms of right modules are acting from the left.

40.5. Split exact sequences. A sequence $K \xrightarrow{f} L \xrightarrow{g} M$ of morphisms in ${}_A\mathbf{M}$ is called *exact* provided $\mathrm{Ke}\, g = \mathrm{Im}\, f$, and is said to be *split exact* if it is exact and the canonical morphism $L/\mathrm{Im}\, f \to M$ is a coretraction.

Sequences $K_1 \xrightarrow{f_1} K_2 \xrightarrow{f_2} K_3 \xrightarrow{f_3} \cdots$ of morphisms are called *(split) exact* if they are (split) exact at any K_i (where it makes sense).

40.6. Graded algebras. Let A be an R-algebra and G a monoid with a neutral element e. A is said to be a *G-graded algebra* if there exists an independent family of R-submodules $A_g \subset A, g \in G$, such that

$$A = \bigoplus\nolimits_G A_g \text{ and } A_g A_h \subset A_{gh}, \quad \text{for } g, h \in G.$$

This implies in particular that A_e is an R-subalgebra of A. If A has a unit 1_A and G is cancellable, then $1_A \in A_e$.

Let $A = \bigoplus_G A_g$ and $B = \bigoplus_G B_g$ be two G-graded R-algebras. A mapping $\phi : A \to B$ is called a *graded R-algebra morphism* if ϕ is an R-algebra morphism satisfying

$$\phi(A_g) \subset B_g \text{ for each } g \in G.$$

Graded modules. Let A be a G-graded algebra and M a left A-module. M is called *G-graded* if there exists a family of R-modules $\{M_g\}_G$ such that

$$M = \bigoplus\nolimits_G M_g \text{ and } A_g M_h \subset M_{gh}, \quad \text{for } g, h \in G.$$

G-graded *right A-modules* are defined symmetrically.

Obviously, any G-graded algebra A is a G-graded left and right module.

40.7. Product and coproduct. The product of a family of modules $\{M_\lambda\}_\Lambda$ in ${}_A\mathbf{M}$ is given by the Cartesian product $\prod_\Lambda M_\lambda$ with the canonical projections π_λ, where the module structure is defined by a componentwise operation. It is characterised by the group isomorphism, for any $N \in {}_A\mathbf{M}$,

$${}_A\mathrm{Hom}(N, \prod\nolimits_\Lambda M_\lambda) \to \prod\nolimits_\Lambda {}_A\mathrm{Hom}(N, M_\lambda), \quad f \mapsto (\pi_\lambda \circ f)_{\lambda \in \Lambda}.$$

The coproduct of $\{M_\lambda\}_\Lambda$ can be realised as a submodule of the product,

$$\bigoplus\nolimits_\Lambda M_\lambda = \{m \in \prod\nolimits_\Lambda M_\lambda \mid \pi_\lambda(m) \neq 0 \text{ only for finitely many } \lambda \in \Lambda\},$$

with the injections

$$\epsilon_\mu : M_\mu \to \bigoplus\nolimits_\Lambda M_\lambda, \quad m_\mu \mapsto (m_\mu \delta_{\mu\lambda})_{\lambda \in \Lambda}.$$

The defining property of the coproduct is the bijectivity of the map, $N \in {}_A\mathbf{M}$,

$${}_A\mathrm{Hom}(\bigoplus\nolimits_\Lambda M_\lambda, N) \to \prod\nolimits_\Lambda {}_A\mathrm{Hom}(M_\lambda, N), \quad g \mapsto (g \circ \epsilon_\lambda)_{\lambda \in \Lambda}.$$

40.8. Dual modules. For any $M \in {}_A\mathbf{M}$ and $X \in \mathbf{M}_R$, $\mathrm{Hom}_R(M, X)$ is a right A-module by the right A-action $fa(m) = f(am)$, for $f \in \mathrm{Hom}_R(M, X)$, $a \in A$ and $m \in M$. In particular, the R-dual $M^* = \mathrm{Hom}_R(M, R)$ is a right A-module by this multiplication.

Similar constructions apply to right A-modules and R-modules.

40.9. Bimodules. Let A, B be R-algebras. An R-module M is called an (A, B)*-bimodule* if it is a left A-module and a right B-module and $a(mb) = (am)b$, for all $a \in A$, $b \in B$ and $m \in M$. *Bimodule morphisms* between two (A, B)-bimodules M and N are maps that are both left A- and right B-linear, and we denote them by ${}_A\mathrm{Hom}_B(M, N)$. The category of all (A, B)-bimodules with these morphisms is denoted by ${}_A\mathbf{M}_B$. It can be identified with the category of left modules over the algebra $A \otimes_R B^{op}$.

40.10. Tensor product of A-modules. Let $N \in \mathbf{M}_A$ and $M \in {}_A\mathbf{M}$. The tensor product $N \otimes_A M$ is defined by the exact sequence of Abelian groups

$$N \otimes_R A \otimes_R M \xrightarrow{\delta} N \otimes_R M \longrightarrow N \otimes_A M \longrightarrow 0,$$

where $\delta = {}_N\varrho \otimes I_M - I_N \otimes \varrho_M$. For morphisms $f : N \to N'$ in $\mathbf{M}_A$ and $g \in M \to M'$ in ${}_A\mathbf{M}$, the tensor product $f \otimes g$ is defined by

$$f \otimes g : N \otimes_A M \to N' \otimes_A M', \quad n \otimes m \mapsto f(n) \otimes g(m).$$

40.11. Tensor functor. Any $L \in \mathbf{M}_A$ induces a covariant functor

$$L \otimes_A - : {}_A\mathbf{M} \to \mathbf{Ab}, \quad M \mapsto L \otimes_A M, \quad f \mapsto I_L \otimes f,$$

which is right exact and respects colimits. It is left exact if and only if L is a flat A-module.

40.12. Hom-tensor relations. *For R-algebras A, B, consider objects $M \in \mathbf{M}_A$, $N \in {}_B\mathbf{M}_A$ and $Q \in \mathbf{M}_B$. Then the canonical map*

$$\nu_M : Q \otimes_B \mathrm{Hom}_A(M, N) \to \mathrm{Hom}_A(M, Q \otimes_B N), \quad q \otimes h \mapsto q \otimes h(-),$$

is an isomorphism provided that (cf. [47, 15.7])

(1) Q is a flat B-module and M is a finitely presented A-module, or

(2) Q is a finitely generated and projective B-module, or

(3) M is a finitely generated and projective A-module.

40.13. Pure morphisms. Related to any morphism $f : M \to M'$ in ${}_A\mathbf{M}$, there is an exact sequence

$$0 \longrightarrow \mathrm{Ke}\, f \longrightarrow M \xrightarrow{f} M' \longrightarrow \mathrm{Coke}\, f \longrightarrow 0\,.$$

Given $L \in \mathbf{M}_A$, we say the morphism f is *L-pure* if tensoring this sequence with $L \otimes_A -$ yields an exact sequence (in $\mathbf{Ab}$). The morphism f is said to be *pure* if it is L-pure for every $L \in \mathbf{M}_A$. Since the tensor functor is right exact, the following are equivalent:

(a) f *is* L*-pure;*

(b) $0 \longrightarrow L \otimes_A \mathrm{Ke}\, f \longrightarrow L \otimes_A M \xrightarrow{I_L \otimes f} L \otimes_A M'$ *is exact;*

(c) $\mathrm{Ke}\, f \to M$ *and* $\mathrm{Im}\, f \to M'$ *are* L*-pure (mono) morphisms.*

For any inclusion $i : N \to M$, the image of the map

$$I_L \otimes i : L \otimes_A N \to L \otimes_A M$$

is called the *canonical image* of $L \otimes_A N$ in $L \otimes_A M$. If $I_L \otimes i$ is injective (i.e., i is an L-pure morphism), then N is said to be an *L-pure submodule* and we identify the canonical image of $I_L \otimes i$ with $L \otimes_A N$. Obviously, if L is a flat right A-module, then every morphism $f : M \to M'$ in ${}_A\mathbf{M}$ is L-pure.

More generally, a sequence $K_1 \xrightarrow{f_1} K_2 \xrightarrow{f_2} K_3 \xrightarrow{f_3} \cdots$ of morphisms in ${}_A\mathbf{M}$ is called *L-pure*, for $L \in \mathbf{M}_A$, if it remains an exact sequence under $L \otimes_A -$. If this holds for every $L \in \mathbf{M}_A$, the sequence is simply called *pure*. Split exact sequences (see 40.5) are of this type.

40.14. Pure equalisers. Let $f, g : M \to M'$ be two morphismsm in ${}_A\mathbf{M}$ and $L \in \mathbf{M}_A$. The equaliser

$$K \xrightarrow{k} M \underset{g}{\overset{f}{\rightrightarrows}} M'$$

is called *L-pure*, if it remains an equaliser after tensoring with $L \otimes_A -$, that is,

$$L \otimes_A K \xrightarrow{I_L \otimes k} L \otimes_A M \underset{I_L \otimes g}{\overset{I_L \otimes f}{\rightrightarrows}} L \otimes_A M'$$

is again an equaliser. An equaliser is called *pure* provided it is L-pure for all $L \in \mathbf{M}_A$. Since the equaliser of f, g can be characterised as the kernel of $f - g$, it is (L-)pure if and only if $f - g$ is an (L-)pure morphism.

40.15. Kernels of tensor products of maps. *If* $f : N \to N'$ *in* $\mathbf{M}_A$ *and* $g \in M \to M'$ *in* ${}_A\mathbf{M}$ *are surjective, then* $f \otimes g$ *is surjective and* $\mathrm{Ke}\,(f \otimes g)$ *is the sum of the canonical images of* $\mathrm{Ke}\, f \otimes_A M'$ *and* $N' \otimes_A \mathrm{Ke}\, g$.

In case $\mathrm{Ke}\, f \subset N$ *is* M*-pure and* $\mathrm{Ke}\, g \subset M$ *is* N*-pure,*

$$\mathrm{Ke}\,(f \otimes g) = \mathrm{Ke}\, f \otimes_A M + N \otimes_A \mathrm{Ke}\, g.$$

40.16. Intersection property. *Consider $N' \in \mathbf{M}_A$ and $M' \in {}_A\mathbf{M}$.*

(1) *Assume $N \subset N'$ in $\mathbf{M}_A$ and $M \subset M'$ in ${}_A\mathbf{M}$ to be pure submodules, or assume M' and M'/M to be (N'-) flat. Then*

$$N \otimes_A M = (N \otimes_A M') \cap (N' \otimes_A M).$$

(2) *Let $U, V \subset N'$ be submodules and assume M' to be flat. Then*

$$(U \otimes M') \cap (V \otimes M') = (U \cap V) \otimes M'.$$

Proof. (1) Under the given conditions we may identify $N \otimes_A M'$ and $N' \otimes_A M$ with their canonical images in $N' \otimes_A M'$ and obtain the exact commutative diagram in **Ab**,

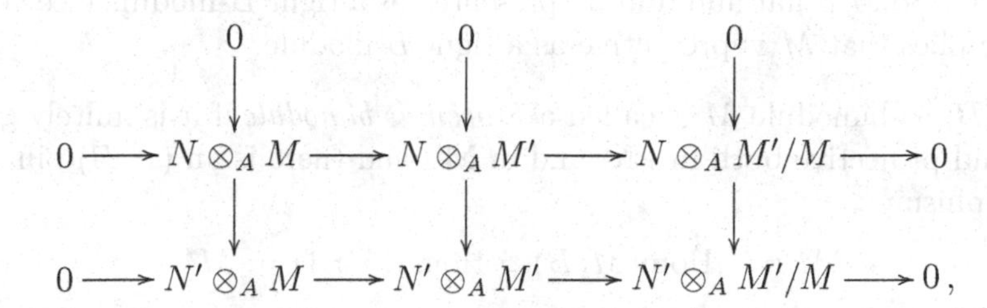

where the left square is a pullback (e.g., [46, 10.3]). This implies the identification stated.

(2) This is shown with a similar argument as in the proof of (1). □

It is well known that the tensor product respects direct sums (coproducts). The behaviour towards products is more complicated (e.g., [46, 12.9]).

40.17. Tensor product and products. For any family $\{M_\lambda\}_\Lambda$ of left A-modules and $U \in \mathbf{M}_A$, consider the map

$$\varphi_U : U \otimes_A \prod_\Lambda M_\lambda \to \prod_\Lambda U \otimes_A M_\lambda, \quad u \otimes (m_\lambda)_\Lambda \mapsto (u \otimes m_\lambda)_\Lambda.$$

(1) *φ_U is surjective for any family $\{M_\lambda\}_\Lambda$ if and only if U is a finitely generated A-module.*

(2) *φ_U is bijective for any family $\{M_\lambda\}_\Lambda$ if and only if U is a finitely presented A-module.*

U is called a *Mittag-Leffler module* if φ_U is injective for any family $\{M_\lambda\}_\Lambda$ in ${}_A\mathbf{M}$. Clearly, finitely presented and pure projective modules are Mittag-Leffler modules.

40.18. The functor $M \otimes_B -$. For $M \in {}_A\mathbf{M}_B$ and $X \in {}_B\mathbf{M}$, $M \otimes_B X$ is a left A-module by

$${}_M\varrho \otimes I_X : A \otimes_R M \otimes_B X \to M \otimes_B X,$$

and, for $N \in {}_A\mathbf{M}$, ${}_A\mathrm{Hom}(M,N)$ is a left B-module by $(bf)(m) = f(mb)$, for $b \in B$, $f \in {}_A\mathrm{Hom}(M,N)$ and $m \in M$. These observations lead to the functors

$$M \otimes_B - : {}_B\mathbf{M} \to {}_A\mathbf{M}, \qquad {}_A\mathrm{Hom}(M,-) : {}_A\mathbf{M} \to {}_B\mathbf{M},$$

which form an adjoint pair by the isomorphism

$${}_A\mathrm{Hom}(M \otimes_B X, N) \to {}_B\mathrm{Hom}(X, {}_A\mathrm{Hom}(M,N)), \quad g \mapsto [x \mapsto g(- \otimes x)],$$

with inverse map $h \mapsto [m \otimes x \mapsto h(x)(m)]$.

Proposition. *If $M \otimes_B -$ has a left adjoint, then M_B is finitely generated and projective.*

Proof. By 38.21, $M \otimes_B -$ preserves limits (hence monomorphisms) and products. So M is flat and finitely presented as a right B-module (see 40.17). This implies that M is projective as a right B-module. □

A (B,A)-bimodule M is called a *Frobenius bimodule* if it is finitely generated and projective both in ${}_B\mathbf{M}$ and in $\mathbf{M}_A$ and there is an (A,B)-bimodule isomorphism

$${}^*M := {}_B\mathrm{Hom}(M,B) \simeq \mathrm{Hom}_A(M,A) =: M^* .$$

40.19. Frobenius functors. *For additive covariant functors $F : {}_A\mathbf{M} \to {}_B\mathbf{M}$ and $G : {}_B\mathbf{M} \to {}_A\mathbf{M}$, the following are equivalent:*

(a) (F,G) is a Frobenius pair of functors;

(b) there exists a Frobenius bimodule ${}_BM_A$ such that

$$F \simeq M \otimes_A - \quad \text{and} \quad G \simeq {}^*M \otimes_B -;$$

(c) $F \simeq M \otimes_A -$ and $G \simeq N \otimes_B -$, for some bimodules ${}_BM_A$ and ${}_AN_B$ that are finitely generated and projective as A- and B-modules, and

$${}^*M \simeq {}_AN_B \quad \text{and} \quad N^* \simeq {}_BM_A \ \text{as bimodules.}$$

Proof. (a) ⇒ (c) By the Eilenberg-Watts Theorem 39.4, there exist bimodules ${}_BM_A = F(A)$ and ${}_AN_B = G(B)$ presenting the respective functors. By adjointness there are bimodule isomorphisms

$${}^*M \simeq {}_B\mathrm{Hom}(F(A),B) \simeq {}_A\mathrm{Hom}(A,G(B)) \simeq N.$$

It follows from 40.18 that M_A and N_B are finitely generated and projective, and the isomorphism implies that ${}_BM$ and ${}_AN$ are also finitely generated and projective.

(c) ⇒ (b) is obvious.

(b) ⇒ (a) Since ${}_BM$ is finitely generated and projective, ${}_B\mathrm{Hom}(M,-) \simeq {}^*M \otimes_B -$. Since ${}_B\mathrm{Hom}(M,-) : {}_B\mathbf{M} \to {}_A\mathbf{M}$ has a left adjoint, so has $G \simeq {}^*M \otimes_B -$. A similar argument shows that F has a right adjoint and so (F,G) is a Frobenius pair. □

40.20. Induction and coinduction functors. Consider a ring morphism $\phi : B \to A$. Then A is an (A, B)-bimodule and induces a covariant functor $A \otimes_B - : {}_B\mathbf{M} \to {}_A\mathbf{M}$, called an *induction functor.* This is left adjoint to the forgetful (restriction of scalars) functor ${}_B(-) : {}_A\mathbf{M} \to {}_B\mathbf{M}$ by the isomorphism

$$ {}_A\mathrm{Hom}(A \otimes_B N, M) \to {}_B\mathrm{Hom}(N, M), \quad f \mapsto f \circ \gamma, $$

for $N \in {}_B\mathbf{M}$, $M \in {}_A\mathbf{M}$, where $\gamma : N \to A \otimes_B N$, $n \mapsto 1_A \otimes n$.

Moreover, the *coinduction functor* ${}_B\mathrm{Hom}(A, -) : {}_B\mathbf{M} \to {}_A\mathbf{M}$ is right adjoint to ${}_B(-)$ by the isomorphism

$$ {}_A\mathrm{Hom}(M, {}_B\mathrm{Hom}(A, N)) \to {}_B\mathrm{Hom}(M, N), \quad g \mapsto g(m)(1), $$

with inverse map $h \mapsto [m \mapsto h(-m)]$, where $N \in {}_B\mathbf{M}$, $M \in {}_A\mathbf{M}$.

40.21. Frobenius extensions. *For a ring morphism $\phi : B \to A$ the following are equivalent:*

(a) *$(A \otimes_B -, {}_B(-))$ is a Frobenius pair;*

(b) *$({}_B(-), {}_B\mathrm{Hom}(A, -))$ is a Frobenius pair;*

(c) *${}_B\mathrm{Hom}(A, -) \simeq A \otimes_B -$;*

(d) *A is finitely generated and projective as left B-module and*

$$ {}_B\mathrm{Hom}(A, B) \simeq A \textit{ as } (A, B)\textit{-bimodule.} $$

If these conditions hold, then $\phi : B \to A$ is called a *Frobenius extension.*

Proof. By 40.20, the first three equivalences are obvious; (a) $\Leftrightarrow$ (d) follows from 40.19. □

40.22. Separability. *$A \otimes_B -$ is a separable functor if and only if B is a direct summand of A as a (B, B)-submodule.*

Proof. If $A \otimes_B -$ is separable, then the counit $\eta_N : N \to A \otimes_B N$ splits functorially in ${}_B\mathbf{M}$. In particular, $\eta_B : B \to A \otimes_B B \simeq A$ is split by some $\gamma : A \to B$ in ${}_B\mathbf{M}$. Since A is also a left B-module, γ is left B-linear by 39.5. Conversely, let $\gamma : A \to B$ be the splitting map in ${}_B\mathbf{M}_B$. Then $\gamma \otimes I_N : A \otimes_B N \to B \otimes_B N \simeq N$ is a functorial splitting of η_N. □

An adjoint pair of contravariant functors is given by (e.g., [46, 45.10])

40.23. The functors $\mathrm{Hom}(-, U)$. *Let U be an (A, B)-bimodule, $L \in \mathbf{M}_B$ and $N \in {}_A\mathbf{M}$. The functors*

$$ {}_A\mathrm{Hom}(-, U) : {}_A\mathbf{M} \to \mathbf{M}_B, \quad \mathrm{Hom}_B(-, U) : \mathbf{M}_B \to {}_A\mathbf{M}, $$

form a right adjoint pair by the functorial isomorphism

$$ \begin{aligned} \Phi_{L,N} : \mathrm{Hom}_B(L, {}_A\mathrm{Hom}(N, U)) &\longrightarrow {}_A\mathrm{Hom}(N, \mathrm{Hom}_B(L, U)), \\ f &\longmapsto [n \to [f(-)](n)]. \end{aligned} $$

(1) Associated with this are the (evaluation) homomorphisms

$$\Phi_N : \ N \to \mathrm{Hom}_B({}_A\mathrm{Hom}(N,U),U), \quad n \mapsto [\beta \mapsto (n)\beta],$$
$$\Phi_L : \ L \to {}_A\mathrm{Hom}(\mathrm{Hom}_B(L,U),U), \quad l \mapsto [\alpha \mapsto \alpha(l)].$$

(2) Φ_L *is injective if and only if* L *is cogenerated by* U_B*, and* Φ_N *is injective if and only if* N *is cogenerated by* ${}_AU$*.*

(3) Denoting ${}_A\mathrm{Hom}(-,U) = {}^*(\ )$ *and* $\mathrm{Hom}_B(-,U) = (\ )^*$,

$$^*(\Phi_L) \circ \Phi_{L^*} = I_{L^*} \ \text{in } {}_A\mathbf{M} \quad \text{and} \quad (\Phi_N)^* \circ \Phi_{^*N} = I_{^*N} \ \text{in } \mathbf{M}_B.$$

References. Bourbaki [5]; Caenepeel, Militaru and Zhu [9]; Wisbauer [46, 47].

41 The category $\sigma[M]$

Throughout A will denote an R-algebra and M a left A-module. The module structure of M is reflected by the smallest Grothendieck category of A-modules containing M, which we briefly describe in this and subsequent sections. For more details we refer the reader to [46].

An A-module N is called *M-generated* if there exists an epimorphism $M^{(\Lambda)} \to N$ for some set Λ. The class of all M-generated modules is denoted by $\mathrm{Gen}(M)$.

41.1. The category $\sigma[M]$. An A-module N is called *M-subgenerated* if it is (isomorphic to) a submodule of an M-generated module. By $\sigma[M]$ we denote the full subcategory of ${}_A\mathbf{M}$ whose objects are all M-subgenerated modules. This is the smallest full Grothendieck subcategory of ${}_A\mathbf{M}$ containing M. $\sigma[M]$ coincides with ${}_A\mathbf{M}$ if and only if A embeds into some (finite) coproduct of copies of M. This happens, for example, when M is a faithful A-module that is finitely generated as a module over its endomorphism ring (see [46, 15.4]).

The *trace functor* $\mathcal{T}^M : {}_A\mathbf{M} \to \sigma[M]$, which sends any $X \in {}_A\mathbf{M}$ to

$$\mathcal{T}^M(X) := \sum\{f(N) \mid N \in \sigma[M],\ f \in {}_A\mathrm{Hom}(N, X)\},$$

is right adjoint to the inclusion functor $\sigma[M] \to {}_A\mathbf{M}$ (e.g., [46, 45.11]). Hence, by 38.21, for any family $\{N_\lambda\}_\Lambda$ of modules in $\sigma[M]$, the product in $\sigma[M]$ is

$$\prod\nolimits^M_\Lambda N_\lambda = \mathcal{T}^M(\prod\nolimits_\Lambda N_\lambda),$$

where the unadorned $\prod$ denotes the usual (Cartesian) product of A-modules (see 40.7). Hence, for any $P \in \sigma[M]$,

$${}_A\mathrm{Hom}(P, \prod\nolimits^M_\Lambda N_\lambda) \simeq \prod\nolimits_\Lambda {}_A\mathrm{Hom}(P, N_\lambda).$$

Moreover, for any injective A-module Q, $\mathcal{T}^M(Q)$ is an injective object in the category $\sigma[M]$.

By definition, $\sigma[M]$ is closed under direct sums, factor modules and submodules in ${}_A\mathbf{M}$. Any subcategory that has these properties is said to be a *closed subcategory* (of ${}_A\mathbf{M}$ or $\sigma[M]$). It is straightforward to see that any closed subcategory is of type $\sigma[N]$ for some N in ${}_A\mathbf{M}$ or $\sigma[M]$, respectively.

$N \in \sigma[M]$ is said to be a *generator in $\sigma[M]$* if it generates all modules in $\sigma[M]$, and M is called a *self-generator* if it generates all its own submodules.

41.2. Properties of $\sigma[M]$. (Cf. [46, Section 15].)

(1) Let $\{N_\lambda\}_\Lambda$ be a family of modules in $\sigma[M]$. The coproduct (direct sum) $\bigoplus_\Lambda N_\lambda$ in ${}_A\mathbf{M}$ is also the coproduct in $\sigma[M]$.

(2) Finitely generated (cyclic) submodules of $M^{(\mathbb{N})}$ form a set of generators in $\sigma[M]$. The direct sum of these modules is a generator in $\sigma[M]$.

(3) Objects in $\sigma[M]$ are finitely (co)generated in $\sigma[M]$ if and only if they are finitely (co)generated in ${}_A\mathbf{M}$.

(4) Every simple module in $\sigma[M]$ is a subfactor of M.

41.3. Injective modules. Let U and M be A-modules. U is said to be *M-injective* if every diagram in ${}_A\mathbf{M}$ with exact row

$$\begin{array}{ccccc} 0 & \to & K & \to & M \\ & & \downarrow & & \\ & & U & & \end{array}$$

can be extended commutatively by some morphism $M \to U$. This holds if ${}_A\mathrm{Hom}(-, U)$ is exact with respect to all exact sequences of the form $0 \to K \to M \to N \to 0$ (in $\sigma[M]$). U is *injective in $\sigma[M]$ (in ${}_A\mathbf{M}$)* if it is N-injective, for every $N \in \sigma[M]$ ($N \in {}_A\mathbf{M}$, resp.).

U is said to be *weakly M-injective* if ${}_A\mathrm{Hom}(-, U)$ is exact on all exact sequences $0 \to K \to M^k \to N \to 0$, where $k \in \mathbb{N}$ and K is finitely generated. Weakly R-injective modules are also called *FP-injective*.

41.4. Injectives in $\sigma[M]$. (Cf. [46, 16.3, 16.11, 17.9].)

(1) For $Q \in \sigma[M]$ the following are equivalent:

(a) Q is injective in $\sigma[M]$;

(b) the functor ${}_A\mathrm{Hom}(-, Q) : \sigma[M] \to \mathbf{M}_R$ is exact;

(c) Q is M-injective;

(d) Q is N-injective for every (finitely generated) submodule $N \subset M$;

(e) every exact sequence $0 \to Q \to N \to L \to 0$ in $\sigma[M]$ splits.

(2) Every (weakly) M-injective object in $\sigma[M]$ is M-generated.

(3) Every object in $\sigma[M]$ has an injective hull.

41.5. Projectivity. Let M and P be A-modules. P is said to be *M-projective* if the functor ${}_A\mathrm{Hom}(P, -)$ is exact on all exact sequences of the form $0 \to K \to M \to N \to 0$ in ${}_A\mathbf{M}$. P is called *projective in $\sigma[M]$ (in ${}_A\mathbf{M}$)* if it is N-projective, for every $N \in \sigma[M]$ ($N \in {}_A\mathbf{M}$, repectively).

41.6. Projectives in $\sigma[M]$. (Cf. [46, 18.3].)
For $P \in \sigma[M]$ the following are equivalent:

(a) *P is projective in $\sigma[M]$;*

(b) *the functor ${}_A\mathrm{Hom}(P,-) : \sigma[M] \to \mathbf{M}_R$ is exact;*

(c) *P is $M^{(\Lambda)}$-projective, for any index set Λ;*

(d) *every exact sequence $0 \to K \to N \to P \to 0$ in $\sigma[M]$ splits.*

If P is finitely generated, then (a)–(d) are equivalent to:

(e) *P is M-projective;*

(f) *every exact sequence $0 \to K' \to N \to P \to 0$ in $\sigma[M]$ with $K' \subset M$ splits.*

A module $P \in \sigma[M]$ is called a *progenerator* in $\sigma[M]$ if it is finitely generated, projective and a generator in $\sigma[M]$. Notice that there may be no projective objects in $\sigma[M]$. A module $N \in \sigma[M]$ is a *subgenerator in $\sigma[M]$* if $\sigma[N] = \sigma[M]$.

41.7. Subgenerators. (Cf. [46, Section 15 and 16.3].)

(1) *For $N \in \sigma[M]$ the following are equivalent:*

  (a) *N is a subgenerator in $\sigma[M]$;*

  (b) *N generates all injective modules in $\sigma[M]$;*

  (c) *N generates the M-injective hull $\widehat{M}$ of M.*

  If $\sigma[M]$ has a progenerator G, then (a)–(c) are equivalent to:

  (d) *there exists a monomorphism $G \to N^k$, for some $k \in \mathbb{N}$.*

(2) *For an A-module M the following are equivalent:*

  (a) *M is a subgenerator in ${}_A\mathbf{M}$ (that is, $\sigma[M] = {}_A\mathbf{M}$);*

  (b) *M generates all injective modules in ${}_A\mathbf{M}$;*

  (c) *M generates the injective hull $E(A)$ of ${}_AA$;*

  (d) *there is a monomorphism $A \to M^k$, for some $k \in \mathbb{N}$.*

(3) *A faithful module ${}_AM$ is a subgenerator in ${}_A\mathbf{M}$ provided*

  (i) *${}_AM$ is finitely generated over $\mathrm{End}_A(M)$, or*

  (ii) *${}_AA$ is finitely cogenerated, or*

  (iii) *$\sigma[M]$ is closed under products in ${}_A\mathbf{M}$.*

41.8. Semisimple modules.

(1) *The following are equivalent:*

  (a) *M is a (direct) sum of simple modules;*

  (b) *every submodule of M is a direct summand;*

(c) *every module (in $\sigma[M]$) is M-projective (or M-injective);*

(d) *every short exact sequence in $\sigma[M]$ splits;*

(e) *every simple module (in $\sigma[M]$) is M-projective;*

(f) *every cyclic module (in $\sigma[M]$) is M-injective.*

Modules M with these properties are called *semisimple modules.*

(2) *Assume M to be semisimple.*

(i) *There exists a fully invariant decomposition*

$$M = \bigoplus \mathrm{Tr}(E_\lambda, M),$$

where $\{E_\lambda\}_\Lambda$ is a minimal representing set of simple submodules of M and the $\mathrm{Tr}(E_\lambda, M)$ are minimal fully invariant submodules.

(ii) *The ring $S = {}_A\mathrm{End}(M)$ is von Neumann regular and M is semisimple as a right S-module.*

(iii) *If all simple submodules of ${}_AM$ are isomorphic, then all simple submodules of M_S are isomorphic.*

Proof. The first parts are shown in [46, 20.2–20.6].

(2)(ii) Let $Am \subset M$ be a simple submodule. We show that $mS \subset M$ is a simple S-submodule. For any $t \in S$ with $mt \neq 0$, $Am \simeq Amt$. Since these are direct summands in M, there exists some $\phi \in S$ with $mt\phi = m$ and hence $mS = mtS$, implying that mS has no nontrivial S-submodules. As a semisimple module, $M = \sum_\Lambda Am_\lambda$ with Am_λ simple. Now $M = A(\sum_\Lambda m_\lambda S)$, showing that M is a sum of simple S-modules $am_\lambda S$, where $a \in A$.

(2)(iii) It is straightforward to show that, for any $m, n \in M$, $Am \simeq An$ implies $mS \simeq nS$. □

A finitely generated module $N \in \sigma[M]$ is said to be *finitely presented* in $\sigma[M]$ if, for any exact sequence $0 \to K \to L \to N \to 0$ in $\sigma[M]$, L finitely generated implies that K is finitely generated.

41.9. Finitely presented modules in $\sigma[M]$. (Cf. [46, 25.2].)
For a finitely generated $P \in \sigma[M]$ the following are equivalent:

(a) *P is finitely presented in $\sigma[M]$;*

(b) *${}_A\mathrm{Hom}(P, -)$ commutes with direct limits in $\sigma[M]$;*

(c) *${}_A\mathrm{Hom}(P, -)$ commutes with direct limits of M-generated modules;*

(d) *${}_A\mathrm{Hom}(P, -)$ commutes with direct limits of weakly M-injective modules.*

Definitions. A module M has *finite length* if it is Noetherian and Artinian. M is called locally Noetherian (Artinian, of finite length) if every finitely generated submodule of M is Noetherian (Artinian, of finite length). M is called *semi-Artinian* if every factor module of M has a nonzero socle.

41.10. Local finiteness conditions. (Cf. [46, 27.5, 32.5].)

(1) *The following are equivalent for a left A-module M:*

(a) *M is locally Noetherian;*

(b) *every finitely generated module in $\sigma[M]$ is Noetherian;*

(c) *every finitely generated module is finitely presented in $\sigma[M]$;*

(d) *every weakly M-injective module is M-injective;*

(e) *any direct sum of M-injective modules is M-injective;*

(f) *every injective module in $\sigma[M]$ is a direct sum of uniform modules.*

(2) *The following are equivalent for a left A-module M:*

(a) *M is locally of finite length;*

(b) *every finitely generated module in $\sigma[M]$ has finite length;*

(c) *every injective module in $\sigma[M]$ is a direct sum of M-injective hulls of simple modules.*

(3) *A module M is locally Artinian if and only if every finitely generated module in $\sigma[M]$ is Artinian.*

(4) *A module M is semi-Artinian if and only if every module in $\sigma[M]$ has a nonzero socle.*

Definitions. A submodule K of M is said to be *superfluous* or *small* in M if, for every submodule $L \subset M$, $K+L=M$ implies $L=M$. A small submodule is denoted by $K \ll M$. An epimorphism $\pi : P \to N$ with P projective in $\sigma[M]$ and $Ke\,\pi \ll P$ is said to be a *projective cover of N in $\sigma[M]$*. A module is called *local* if it has a largest proper submodule.

41.11. Local modules. (Cf. [46, 19.7].)
For a projective module $P \in \sigma[M]$, the following are equivalent:

(a) *P is local;*

(b) *P has a maximal submodule that is superfluous;*

(c) *P is cyclic and every proper submodule is superfluous;*

(d) *P is a projective cover of a simple module in $\sigma[M]$;*

(e) $\mathrm{End}({}_AP)$ *is a local ring.*

Definitions. Let U be a submodule of the A-module M. A submodule $V \subset M$ is called a *supplement* of U in M if V is minimal with the property $U+V=M$. It is easy to see that V is a supplement of U if and only if $U+V=M$ and $U \cap V \ll V$. Notice that supplements need not exist in general. M is said to be *supplemented* provided each of its submodules has a supplement. Examples of supplemented modules are provided by linearly compact modules.

41.12. Linearly compact modules. A left A-module M is called a *linearly compact module* if, for any family of cosets $\{x_i + M_i\}_\Lambda$ with the finite intersection property, where $x_i \in M$, and $M_i \subset M$ are submodules, $\bigcap_\Lambda (x_i + M_i) \neq \emptyset$. For further descriptions of these modules we refer the reader to [46, 29.7].

41.13. Properties of linearly compact modules. (Cf. [46, 29.8., 41.10].) *If M is a linearly compact left A-module, then:*

(1) *M is supplemented.*

(2) *$M/\text{Rad}(M)$ is finitely generated and semisimple.*

(3) *M is Noetherian if and only if* $\text{Rad}(U) \neq U$, *for all submodules $U \subset M$.*

(4) *M is Artinian if and only if M is semi-Artinian.*

Definitions. A module $P \in \sigma[M]$ is said to be *semiperfect in $\sigma[M]$* if every factor module of N has a projective cover in $\sigma[M]$. P is *perfect in $\sigma[M]$* if any direct sum $P^{(\Lambda)}$ is semiperfect in $\sigma[M]$.

41.14. Semiperfect modules. (Cf. [46, 42.5, 42.12].)
For a projective module P in $\sigma[M]$, the following are equivalent:

(a) *P is semiperfect in $\sigma[M]$;*

(b) *P is supplemented;*

(c) *every finitely P-generated module has a projective cover in $\sigma[M]$;*

(d) *(i) $P/Rad(P)$ is semisimple and* $\text{Rad}(P) \ll P$, *and*
(ii) decompositions of $P/\text{Rad}(P)$ *can be lifted to P;*

(e) *every proper submodule is contained in a maximal submodule of P, and every simple factor module of P has a projective cover in $\sigma[M]$;*

(f) *P is a direct sum of local modules and $Rad(P) \ll P$.*

41.15. Perfect modules. (Cf. [46, 43.2].)
For a projective module P in $\sigma[M]$, the following are equivalent:

(a) *P is perfect in $\sigma[M]$;*

(b) *P is semiperfect and, for any set Λ,* $\text{Rad}(P^{(\Lambda)}) \ll P^{(\Lambda)}$;

(c) *every P-generated module has a projective cover in $\sigma[M]$.*

Definition. We call $\sigma[M]$ a *(semi)perfect category* if every (simple) module in $\sigma[M]$ has a projective cover in $\sigma[M]$.

41.16. Semiperfect and perfect categories.

(1) *For an A-module M the following are equivalent:*

(a) *$\sigma[M]$ is semiperfect;*

(b) *$\sigma[M]$ has a generating set of local projective modules;*

(c) in $\sigma[M]$ every finitely generated module has a projective cover.

(2) For M the following are equivalent:

(a) $\sigma[M]$ is perfect;

(b) $\sigma[M]$ has a projective generator that is perfect in $\sigma[M]$.

Proof. (1) (a) $\Rightarrow$ (b) The projective covers of all simple objects in $\sigma[M]$ are local and form a generating set of $\sigma[M]$ (by [46, 18.5]). Notice that local modules are supplemented.

(b) $\Rightarrow$ (c) Any finite direct sum of supplemented modules is supplemented. Hence, for every finitely generated $N \in \sigma[M]$, there exists an epimorphism $P \to N$ with some supplemented projective module $P \in \sigma[M]$. By 41.14, every factor module of P has a projective cover in $\sigma[M]$, and so does N.

(c) $\Rightarrow$ (a) is trivial.

(2) (a) $\Rightarrow$ (b) Let P be the direct sum of projective covers of a representative set of the simple modules in $\sigma[M]$. Then P is a projective generator and every factor module of $P^{(\Lambda)}$ has a projective cover, and hence P is perfect.

(b) $\Rightarrow$ (a) is obvious. □

41.17. Left perfect rings. (Cf. [46, 43.9].)
For A the following are equivalent:

(a) A is a perfect module in ${}_A\mathbf{M}$;

(b) $A/\mathrm{Jac}(A)$ is left semisimple and $\mathrm{Jac}(A)$ is right t-nilpotent;

(c) every left A-module has a projective cover;

(d) every (indecomposable) flat left A-module is projective;

(e) A satisfies the descending chain condition (dcc) on cyclic right ideals;

(f) every right A-module has the dcc on cyclic (finitely generated) submodules.

41.18. (f-)semiperfect rings. A ring A is said to be *semiperfect* if A is semiperfect as a left A-module or – equivalently – as a right A-module. From 41.14(d) we see that A is semiperfect if and only if $A/\mathrm{Jac}(A)$ is left (and right) semisimple and idempotents lift modulo $\mathrm{Jac}(A)$. More generally, A is called *f-semiperfect* (or *semiregular*) if $A/\mathrm{Jac}(A)$ is von Neumann regular and idempotents lift modulo $\mathrm{Jac}(A)$. Note that A is semiperfect if and only if finitely generated left and right A-modules have projective covers, and A is f-semiperfect if and only if every finitely presented left (and right) A-module has a projective cover (see [46, 42.11]). From [46, 42.12, 22.1] we recall:

41.19. (f-)semiperfect endomorphism rings. *Put $S = {}_A\mathrm{End}(M)$.*

(1) Assume M to be projective in $\sigma[M]$. Then:

(i) *S is semiperfect if and only if M is finitely generated and semiperfect.*

(ii) *If M is semiperfect, then S is f-semiperfect.*

(iii) *If S is f-semiperfect, then* $\text{Rad}(M) \ll M$ *and M is a direct sum of cyclic modules.*

(2) *If M is self-injective, then S is f-semiperfect.*

(3) *If M is self-injective and* $\text{Soc}(M) \trianglelefteq M$, *then*

$$Jac(S) = {}_A\text{Hom}(M/\text{Soc}(M), M).$$

41.20. $\sum$-injective semiperfect modules. *Let P be semiperfect and projective in $\sigma[M]$. If P is $\sum$-injective in $\sigma[M]$, then P is perfect in $\sigma[M]$.*

Proof. Any coproduct $P^{(\Lambda)}$ is injective in $\sigma[M]$ and so ${}_A\text{End}(P^{(\Lambda)})$ is f-semiperfect. By 41.19, this implies $\text{Rad}(P^{(\Lambda)}) \ll P^{(\Lambda)}$, showing that P is perfect in $\sigma[M]$. □

Definitions. A module $P \in \sigma[M]$ is said to be *(semi)hereditary* in $\sigma[M]$ if every (finitely generated) submodule of P is projective in $\sigma[M]$. We call a module $P \in \sigma[M]$ *(semi)cohereditary* in $\sigma[M]$ if every factor module of P is (weakly) M-injective.

Any finite direct sum of cohereditary modules in $\sigma[M]$ is again cohereditary, and the corresponding statement for infinite direct sums holds provided that M is locally Noetherian (see [208, 6.3]).

Denote by Inj (M) the class of all injective modules in $\sigma[M]$.

41.21. Cohereditary modules. (Cf. [208, 6.4].)

(1) *If M is locally Noetherian, the following are equivalent:*

(a) *$\widehat{M}$ is cohereditary in $\sigma[M]$;*

(b) *every injective module is cohereditary in $\sigma[M]$;*

(c) *every indecomposable injective module is cohereditary in $\sigma[M]$.*

In case $\sigma[M]$ has a projective subgenerator L, (a)–(c) *are equivalent to:*

(d) *L is hereditary in $\sigma[M]$.*

(2) *For a subgenerator $P \in \sigma[M]$, the following are equivalent:*

(a) $\text{Gen}(P) = \text{Inj}\,(M)$;

(b) *P is locally Noetherian and cohereditary in $\sigma[M]$.*

In the remaining part of this section we consider the *relative properties* of A-modules related to a fixed ring morphism $\phi : B \to A$. In this case, any left A-module M is naturally a left B-module and there is an interplay between the properties of M as an A-module and those of M as a B-module.

41.22. (A, B)-finite modules. The module M is said to be *(A, B)-finite* if every finitely generated A-submodule of M is finitely generated as a B-module. $\sigma[M]$ is said to be *(A, B)-finite* if every module in $\sigma[M]$ is (A, B)-finite.

(1) The following are equivalent:

(a) $\sigma[M]$ is (A, B)-finite;

(b) $M^{(\mathbb{N})}$ is (A, B)-finite.

(2) If B is a left Noetherian ring and M is (A, B)-finite, then $\sigma[M]$ is (A, B)-finite.

(3) Let $\sigma[M]$ be (A, B)-finite.

(i) If B is a right perfect ring, then every module in $\sigma[M]$ has the dcc on finitely generated A-submodules.

(ii) If B is left Noetherian, then every module in $\sigma[M]$ is locally Noetherian.

(iii) If B is left Artinian, then every module in $\sigma[M]$ has locally finite length.

41.23. Socle of injectives. *Let A be an R-algebra and M an (A, R)-finite weakly M-injective A-module that is also (S, R)-finite, where $S = \text{End}(_AM)$. Assume that M has only finitely many nonisomorphic simple submodules. Then $\text{Soc}(_AM)$ is finitely generated as an R-module.*

Proof. Let $E_1, \ldots, E_k$ be a representative set of simple A-submodules of M. By injectivity, $\text{Soc}(_AM) = \sum_{i=1}^{k} E_iS$. Now (A, R)-finiteness implies that each E_i is finitely R-generated, and, by (S, R)-finiteness, each E_iS is finitely R-generated. □

The following observations essentially follow from [133], where the basic ideas were introduced, and from [47, Section 20], where these ideas were extended to modules over algebras.

41.24. Relative notions. An exact sequence $f_i : M_i \to M_{i+1}$ in $_A\mathbf{M}$, $i \in \mathbb{N}$, is called (A, B)-exact if Im f_i is a direct summand of M_{i+1} as a left B-module.

Let M, P, Q be left A-modules. P is called *(M, B)-projective* if $_A\text{Hom}(P, -)$ is exact with respect to all (A, B)-exact sequences in $\sigma[M]$. This is the case if and only if every (A, B)-exact sequence $L \to P \to 0$ in $\sigma[M]$ splits.

Q is called *(M, B)-injective* if $_A\text{Hom}(-, Q)$ is exact with respect to all (A, B)-exact sequences in $\sigma[M]$. This happens if and only if every (A, B)-exact sequence $0 \to Q \to L$ in $\sigma[M]$ splits.

Over a semisimple ring B, (M, B)-projective and (M, B)-injective are synonymous to projective and injective in $\sigma[M]$, respectively.

41.25. (A,B)-projectives and (A,B)-injectives.

(1) For any B-module X, $A \otimes_B X$ is (A,B)-projective.

(2) $P \in {}_A\mathbf{M}$ is (A,B)-projective if and only if the map $A \otimes_B P \to P$, $a \otimes p \mapsto ap$, splits in ${}_A\mathbf{M}$.

(3) For any B-module Y, $\mathrm{Hom}_B(A,Y)$ is (A,B)-injective.

(4) $Q \in {}_A\mathbf{M}$ is (A,B)-injective if and only if the map $Q \to \mathrm{Hom}_B(A,Q)$, $q \mapsto [a \mapsto aq]$, splits in ${}_A\mathbf{M}$.

The module M is called *(A,B)-semisimple* if every (A,B)-exact sequence in $\sigma[M]$ splits. The ring A is said to be *left (A,B)-semisimple* if A is (A,B)-semisimple as a left A-module.

41.26. (A,B)-semisimple modules.

(1) The following assertions are equivalent:

(a) M is (A,B)-semisimple;

(b) every A-module (in $\sigma[M]$) is (M,B)-projective;

(c) every A-module (in $\sigma[M]$) is (M,B)-injective.

(2) If M is (A,B)-semisimple, then:

(i) Every module in $\sigma[M]$ is (A,B)-semisimple.

(ii) For every ideal $J \subset B$, M/JM is $(A/JA, B/J)$-semisimple.

(iii) If B is hereditary and M is B-projective, then M is hereditary in $\sigma[M]$.

41.27. Left (A,B)-semisimple rings.

(1) For A the following are equivalent:

(a) A is left (A,B)-semisimple;

(b) every left A-module is (A,B)-projective;

(c) every left A-module is (A,B)-injective.

(2) Let A be left (A,B)-semisimple. Then:

(i) If M is a flat B-module, then M is a flat A-module.

(ii) If M is a projective B-module, then M is a projective A-module.

(iii) If M is an injective B-module, then M is an injective A-module.

(iv) If B is a left hereditary ring and A is a projective B-module, then A is a left hereditary ring.

References. Hochschild [133]; Wisbauer [46, 47, 207, 208].

42 Torsion-theoretic aspects

In the preceding section we dealt with the internal structure of the category $\sigma[M]$. Now we will look at the properties of $\sigma[M]$ as a class of modules in ${}_A\mathbf{M}$. This leads to some toplogical and torsion-theoretic considerations.

42.1. Finite topology. For sets X, Y we identify the set $\text{Map}(X,Y)$ of all maps $X \to Y$ with the Cartesian product Y^X. The *finite topology* on $\text{Map}(X,Y)$ is the product topology where Y is considered as a discrete space. For $f \in \text{Map}(X,Y)$, a basis of open neighbourhoods is given by the sets

$$\{g \in \text{Map}(X,Y) \,|\, g(x_i) = f(x_i) \quad \text{for } i = 1, \ldots, n\},$$

where $\{x_1, \ldots, x_n\}$ ranges over the finite subsets of X.

For subsets $U \subset V$ of $\text{Map}(X,X)$, we say that U is *X-dense in V* if it is dense in the finite topology in $\text{Map}(X,X)$, that is, for any $v \in V$ and $x_1, \ldots, x_n \in X$ there exists $u \in U$ such that $u(x_i) = v(x_i)$ for $i = 1, \ldots, n$.

Throughout M will denote a left A-module with $S = {}_A\text{End}(M)$ acting from the right. The finite topology on $\text{Map}(M,M)$ induces the finite topology on $\text{End}_S(M)$. The importance of density in this topology is based on the following facts (cf. [46, 15.7,15.8], [52]).

42.2. Density properties. *Let M be faithful and $B \subset A$ a subring.*

(1) $\sigma[{}_BM] = \sigma[{}_AM]$ if and only if B is M-dense in A.

(2) If M is a generator or a weak cogenerator in $\sigma[M]$, then A is M-dense in $\text{End}_S(M)$.

42.3. The M-adic topology. Let M be an A-module by an algebra morphism $\varphi : A \to \text{End}_R(M)$. Considering $\text{End}_R(M)$ with the finite topology, the topology induced on A by φ is called an *M-adic topology* on A.

Open left ideals. A basis of the filter of open left ideals of A is given by

$$\mathcal{B}_M = \{\text{An}_A(E) \,|\, E \text{ a finite subset of } M\},$$

and the filter of all open left ideals is

$$\mathcal{F}_M = \{I \subset A \,|\, I \text{ is a left ideal and } A/I \in \sigma[M]\}.$$

This filter yields a generator G in $\sigma[M]$ by putting $G = \bigoplus\{A/I \,|\, I \in \mathcal{F}_M\}$. Properties of the filter $\mathcal{F}_M$ correspond to properties of the class $\sigma[M]$ of modules in ${}_A\mathbf{M}$.

Closed left ideals. *For a left ideal $I \subset A$ the following are equivalent:*

(a) I is closed in the M-adic topology;

(b) $I = \mathrm{An}_A(W)$ *for some* $W \in \sigma[M]$;

(c) A/I *is cogenerated by some (minimal) cogenerator of* $\sigma[M]$;

(d) $I = \bigcap_\Lambda I_\lambda$, *where* $A/I_\lambda \in \sigma[M]$ *and finitely cogenerated (cocyclic).*

The last equivalence follows from the fact that the M-injective hulls of simple modules form a cogenerating set in $\sigma[M]$.

Closure of left ideals. *For any left ideal* $J \subset A$ *the closure in the* M*-adic topology is* $\overline{J} = \mathrm{Ke}\,_A\mathrm{Hom}(A/J, K)$, *for some cogenerator* K *in* $\sigma[M]$.

Self-injective modules. *If* M *is self-injective, then every finitely generated right ideal of* S *is closed in the finite topology (e.g., [46, 28.1]).*

For an (left, right) ideal or subring $T \subset A$, any A-module may be considered as a module over the (nonunital) ring T. So some knowledge about modules over rings without units is useful.

Let T be any associative ring (without a unit). A left T-module N is called *s-unital* if $u \in Tu$ for every $u \in N$. T itself is called *left s-unital* if it is s-unital as a left T-module. For an ideal $T \subset A$, every A-module is a T-module and we observe the elementary properties (e.g., [201], [62]):

42.4. s-unital T-modules. *For any subring* $T \subset A$ *the following assertions are equivalent:*

(a) M *is an s-unital* T*-module;*

(b) for any $m_1, \ldots, m_k \in N$, *there exists* $t \in T$ *with* $m_i = tm_i$ *for all* $i \leq k$;

(c) for any set Λ, $N^{(\Lambda)}$ *is an s-unital* T*-module.*

Proof. (a) $\Rightarrow$ (b) We proceed by induction. Assume the assertion holds for $k-1$ elements. Choose $t_k \in T$ such that $t_k n_k = n_k$ and put $a_i = m_i - m_k n_i$, for all $i \leq k$. By assumption there exists $t' \in T$ satisfying $a_i = t'a_i$, for all $i \leq k-1$. Then $t := t' + t_k - t't_k \in T$ is an element satisfying the condition in (b). The remaining assertions are easily verified. □

42.5. Flat factor rings. *For an ideal* $T \subset A$ *the following are equivalent:*

(a) A/T *is a flat right* A*-module;*

(b) for every left ideal I *of* A, $TI = T \cap I$;

(c) every injective left A/T*-module is* A*-injective;*

(d) ${}_{A/T}\mathbf{M}$ *contains an* A*-injective cogenerator;*

(e) for every A*-module* ${}_AL \subset {}_AN$, $TL = TN \cap L$;

(f) T *is left s-unital;*

(g) T *is idempotent and* ${}_AT$ *is a (self-)generator in* $\sigma[{}_AT]$.

Under these conditions T is a flat right A-module, and, for any $N \in_A \mathbf{M}$, the canonical map $T \otimes_A N \to TN$ is an isomorphism.

Proof. The equivalence of (a) and (b) is shown in [46, 36.6].

(a) $\Rightarrow$ (c) Put $D := A/T$. Let N be an injective D-module and $L \subset A$ a left ideal. By (a), the sequence $0 \to D \otimes_A L \to D \otimes_A A$ is exact in ${}_D\mathbf{M}$ and there is a commutative diagram with exact rows and canonical isomorphisms,

$$\begin{array}{ccccc}
{}_D\mathrm{Hom}(D \otimes_A A, N) & \longrightarrow & {}_D\mathrm{Hom}(D \otimes_A L, N) & \longrightarrow & 0 \\
\downarrow \simeq & & \downarrow \simeq & & \\
{}_A\mathrm{Hom}(A, {}_D\mathrm{Hom}(D, N)) & \longrightarrow & {}_A\mathrm{Hom}(L, {}_D\mathrm{Hom}(D, N)) & \longrightarrow & 0 \\
\downarrow \simeq & & \downarrow \simeq & & \\
{}_A\mathrm{Hom}(A, N) & \longrightarrow & {}_A\mathrm{Hom}(L, M) & \longrightarrow & 0 .
\end{array}$$

Since N is an injective D-module, the first row is exact and so are the others, that is, N is injective as an A-module.

(d) $\Rightarrow$ (a) Let N be a cogenerator in ${}_D\mathbf{M}$ that is A-injective. For a left ideal $L \subset A$ there is an exact sequence $0 \to K \to D \otimes_A L \to D \otimes_A A$ in ${}_D\mathbf{M}$, and we want to prove $K = 0$. Consider the exact sequence

$${}_D\mathrm{Hom}(D \otimes_A A, N) \to {}_D\mathrm{Hom}(D \otimes_A L, N) \to {}_D\mathrm{Hom}(K, N) \to 0.$$

Now in the above diagram the bottom row is exact (N is A-injective). This implies that the top row is also exact, that is, ${}_D\mathrm{Hom}(K, N) = 0$. Since N is a cogenerator in ${}_D\mathbf{M}$, we conclude $K = 0$.

(g) $\Rightarrow$ (f) Since T is a self-generator, for any $t \in T$ there is an A-epimorphism $\varphi : T^k \to At$. $T^2 = T$ implies $At = \varphi(T^k) = T\varphi(T^k) = Tt$ and hence $t \in Tt$.

The remaining implications are straightforward to verify. $\square$

42.6. Dense ideals and unital modules. *Let ${}_AM$ be faithful. For an ideal $T \subset A$ the following are equivalent:*

(a) T is M-dense in A;

(b) M is an s-unital T-module;

(c) for every A-submodule $L \subset M$, $L = TL$;

(d) every $N \in \sigma[M]$ is an s-unital T-module;

(e) for every $N \in \sigma[M]$, $N = TN$;

(f) for every $N \in \sigma[M]$, the canonical map $\varphi_N : T \otimes_A N \to N$ is an isomorphism.

If $T \in \sigma[M]$, then (a)–(f) are equivalent to:

(e) $T^2 = T$ and T is a generator in $\sigma[M]$.

Proof. (a) ⇒ (b) Apply the density property for the unit of A.

(b) ⇒ (a) Let $a \in A$ and $m_1, \ldots, m_k \in M$. By 42.4, there exists $t \in T$ such that $am_i = t(am_i) = (ta)m_i$, for $i = 1, \ldots, k$. Since T is an ideal, $ta \in T$, and this shows that T is N-dense in A.

The remaining implications follow from 42.4 and 42.5. □

42.7. Trace ideals. The trace of M in A, $\mathrm{Tr}(M, A)$, is called the *trace ideal of* M, and the trace of $\sigma[M]$ in A, $\mathcal{T}^M(A) = \mathrm{Tr}(\sigma[M], A)$, is called the *trace ideal of* $\sigma[M]$. While $\mathrm{Tr}(M, A)$ is more directly related to properties of the module M, $\mathcal{T}^M(A)$ is better suited to study properties of $\sigma[M]$. Clearly $\mathrm{Tr}(M, A) \subset \mathcal{T}^M(A)$, and the two ideals coincide provided that M is a generator in $\sigma[M]$, or else if A is a left self-injective algebra.

42.8. Trace ideals of M. Denote ${}^*M = {}_A\mathrm{Hom}(M, A)$, and $T = \mathrm{Tr}(M, A) = {}^*M(M)$. Any $f \in {}^*M$ defines an A-linear map

$$\phi_f : M \to S, \quad m \mapsto f(-)m.$$

$\Delta = \sum_{f \in {}^*M} \mathrm{Im}\, \phi_f$ is an ideal in S and $M\Delta \subset TM$.

The following are equivalent:

(a) $M = TM$;

(b) $M = M\Delta$;

(c) for any $L \in {}_A\mathbf{M}$, $\mathrm{Tr}(M, L) = TL$.

If this holds, T and Δ are idempotent ideals and $\Delta = \mathrm{Tr}(M_S, S)$.

Proof. (a) ⇔ (b) are obvious from the definitions.

(a) ⇔ (c) Clearly T is M-generated, and $M = TM$ implies that M-generated A-modules are T-generated.

Assume the conditions hold. By definition, $\Delta \subset \mathrm{Tr}(M_S, S)$. For any S-linear map $g : M \to S$ and $m \in M$, write $m = \sum_i m_i\delta_i$, where $m_i \in M$ and $\delta_i \in \Delta$, to obtain

$$g(m) = g(\sum_i m_i\delta_i) = \sum_i g(m_i)\delta_i \in \Delta,$$

thus showing $\mathrm{Tr}(M_S, S) \subset \Delta$. □

42.9. Canonical map. For any $N \in \mathbf{M}_A$ there is a map

$$\alpha_{N,M} : N \otimes_A M \to \mathrm{Hom}_A({}^*M, N), \; n \otimes m \mapsto [f \mapsto nf(m)],$$

which is injective if and only if

for any $u \in N \otimes_A M$, $(I_N \otimes f)(u) = 0$ for all $f \in {}^*M$, implies $u = 0$.

We are interested in modules M for which $\alpha_{N,M}$ is injective for all $N \in \mathbf{M}_A$.

A module M is said to be *locally projective* if, for any diagram of left A-modules with exact rows,

$$\begin{array}{ccccc} 0 \longrightarrow & F & \xrightarrow{i} & M & \\ & & & \downarrow g & \\ & L & \xrightarrow{f} & N & \longrightarrow 0, \end{array}$$

where F is finitely generated, there exists $h : M \to L$ such that $g \circ i = f \circ h \circ i$.

42.10. Locally projective modules. *With the notation from 42.8, the following are equivalent:*

(a) *M is locally projective;*

(b) *$\alpha_{N,M}$ is injective, for any (cyclic) right A-module N;*

(c) *for each $m \in M$, $m \in {}^*M(m)M$;*

(d) *for any $m_1, \ldots, m_k \in M$ there exist $x_1, \ldots, x_n \in M$, $f_1, \ldots, f_n \in {}^*M$, such that*

$$m_j = \sum_i f_i(m_j)x_i, \quad \textit{for } j = 1, \ldots, k;$$

(e) *$M = TM$, and M is an s-unital right Δ-module;*

(f) *$M = TM$ and S/Δ is flat as a left S-module;*

(g) *$M = TM$ and M_S is a generator in $\sigma[M_S]$.*

Proof. (a) $\Rightarrow$ (d) Put $N = M$ and $L = A^{(\Lambda)}$ in the defining diagram.

(d) $\Rightarrow$ (a) follows by the fact that A is projective as left A-module.

(b) $\Rightarrow$ (c) Assume $\alpha_{N,M}$ to be injective for cyclic right A-modules N. For any $m \in M$ put $J = {}^*M(m)$ and consider the monomorphism

$$\phi : M/JM \simeq A/J \otimes_A M \xrightarrow{\alpha_{N,M}} \mathrm{Hom}_A({}^*M, A/J).$$

For $x \in M$ and $f \in {}^*M$, $\phi(x+JM)(f) = f(x)+J$, and hence $\phi(m+JM) = 0$. By injectivity of ϕ this implies $m \in JM$.

(d) $\Rightarrow$ (b) Let $N \in \mathbf{M}_A$ and let $v = \sum_{j=1}^{r} n_j \otimes m_j \in N \otimes_A M$. Choose $x_1, \ldots, x_n \in M$ and $f_1, \ldots, f_n \in {}^*M$ such that $m_j = \sum_i f_i(m_j)x_i$, for $j = 1, \ldots, r$. Then

$$v = \sum_{i,j} n_j \otimes f_i(m_j)x_i = \sum_i \alpha_{N,M}(v)(f_i) \otimes x_i,$$

and hence $v = 0$ if $\alpha_{N,M}(v) = 0$, that is, $\alpha_{N,M}$ is injective.

(c) $\Leftrightarrow$ (d) $\Leftrightarrow$ (e) For $m \in M$, $m \in {}^*M(m)M$ means that there are $x_1, \ldots, x_n \in M$ and $f_1, \ldots, f_n \in {}^*M$ such that

$$m = \sum_i f_i(m)x_i = m[\sum_i f_i(-)x_i] \in m\Delta,$$

showing that M is an s-unital right Δ-module (see 42.4).

(e) $\Leftrightarrow$ (f) $\Leftrightarrow$ (g) As observed in 42.8, $TM = M$ implies $M\Delta = M$, and hence the equivalences follow by 42.6. □

Modules M with all $\alpha_{N,M}$ injective are termed *universally torsionless (UTL)* modules in [118], in [221] they are called *locally projective*, in [171] *trace modules*, and in [181] *modules plats et strictement de Mittag Leffler.*

42.11. Properties. *Let M be locally projective.*

(1) M is flat and every pure submodule of M is locally projective.

(2) If M is finitely generated or A is left perfect, then M is projective.

(3) For any algebra morphism $A \to B$, $B \otimes_A M$ is a locally projective B-module.

Proof. (1) For monomorphisms $f : L \to N$ in $\mathbf{M}_A$ and $i : U \to M$ in ${}_A\mathbf{M}$, there are commutative diagrams:

$$\begin{array}{ccc} L \otimes_A M & \xrightarrow{\alpha_{L,M}} & \mathrm{Hom}_A({}^*M, L) \\ {\scriptstyle f\otimes I_M}\downarrow & & \downarrow{\scriptstyle \mathrm{Hom}({}^*M,f)} \\ N \otimes_A M & \xrightarrow{\alpha_{N,M}} & \mathrm{Hom}_A({}^*M, N), \end{array} \qquad \begin{array}{ccc} N \otimes_A U & \xrightarrow{\alpha_{N,U}} & \mathrm{Hom}_A({}^*U, N) \\ {\scriptstyle I_N\otimes i}\downarrow & & \downarrow{\scriptstyle \mathrm{Hom}(i^*,N)} \\ N \otimes_A M & \xrightarrow{\alpha_{N,M}} & \mathrm{Hom}_A({}^*M, N). \end{array}$$

Since all $\alpha_{-,M}$ are injective, the left-hand diagram implies that $f \otimes I_M$ is injective and hence M is flat. If $i : U \to M$ is pure, then $I_N \otimes i$ is injective and the right-hand diagram implies that $\alpha_{N,U}$ is injective. Thus U is locally projective.

(2) For M finitely generated the assertion is clear from the definition. Over left perfect rings A, any flat left A-module is projective and so the assertion follows from (1).

(3) The algebra map $A \to B$ yields maps ${}_A\mathrm{Hom}(M, A) \to {}_A\mathrm{Hom}(M, B) \simeq {}_B\mathrm{Hom}(B \otimes_A M, B) = {}^*(B \otimes_A M)$ and, then, for any $N \in {}_B\mathbf{M}$, the commutative diagram

$$\begin{array}{ccc} N \otimes_B (B \otimes_A M) & \xrightarrow{\alpha_{N,B\otimes_A M}} & \mathrm{Hom}_B({}^*(B \otimes_A M), N) \\ \simeq\downarrow & & \downarrow \\ N \otimes_A M & \xrightarrow{\alpha_{N,M}} & \mathrm{Hom}_A({}^*M, N), \end{array}$$

in which $\alpha_{N,M}$ is injective and hence so is $\alpha_{N,B\otimes_A M}$. Thus $B \otimes_A M$ is a locally projective B-module. □

42.12. Right Noetherian rings. *Consider the following conditions:*

(a) M is locally projective;

(b) *the map* $\psi_M : M \to ({}^*M)^* \to A^{{}^*M}$, $k \mapsto [f \mapsto f(k)] \mapsto (f(k))_{f\in {}^*M}$, *is a pure monomorphism;*

(c) *M is a pure submodule of some product* A^Λ, Λ *a set.*

Then (a) $\Rightarrow$ (b) $\Rightarrow$ (c). If A is right Noetherian, then (a) $\Leftrightarrow$ (b) $\Leftrightarrow$ (c).

Proof. The map $\alpha_{N,M}$ can be placed in the commutative diagram

$$\begin{array}{ccc} N\otimes_A M & \xrightarrow{I_N\otimes\psi_M} & N\otimes_A A^{{}^*M} \\ {\scriptstyle \alpha_{N,M}}\downarrow & & \downarrow{\scriptstyle \varphi_N} \\ \mathrm{Hom}_A(M^*,N) & \hookrightarrow & N^{{}^*M}, \end{array} \qquad \begin{array}{ccc} m\otimes k & \longmapsto & m\otimes (f(k))_{f\in{}^*M} \\ \downarrow & & \downarrow \\ [f\mapsto f(k)m] & \longmapsto & (f(k)m)_{f\in{}^*M}, \end{array}$$

where φ_N is the canonical map (see 40.17). So, if ψ_M is a pure monomorphisms then $\alpha_{N,M}$ is injective provided φ_N is injective.

(a) $\Rightarrow$ (b) If $\alpha_{N,M}$ is injective for all $N \in {}_A\mathbf{M}$, then all $I_N\otimes\psi_M$ are injective, which means that ψ_M is a pure monomorphism.

(b) $\Rightarrow$ (c) is trivial.

Now let A be right Noetherian.

(c) $\Rightarrow$ (a) For any index set Λ, there is an injection $\Lambda \to {}^*(A^\Lambda)$ taking any $\lambda\in\Lambda$ to the canonical projection π_λ in ${}^*(A^\Lambda)$. This implies that the resulting map $\psi_{A^\Lambda} : A^\Lambda \mapsto A^{{}^*(A^\Lambda)}$ is a splitting monomorphism and hence A^Λ is locally projective by (a) $\Leftrightarrow$ (b). As noticed in 42.11, any pure submodule of a locally projective module is locally projective. □

42.13. Tensor product with morphisms. *For $Q \in {}_A\mathbf{M}$ consider the map*

$$\nu_M : {}_A\mathrm{Hom}(M,A)\otimes_A Q \to {}_A\mathrm{Hom}(M,Q),\quad h\otimes q \mapsto [m\mapsto h(m)q].$$

If M is locally projective, then Im ν_M *is dense in* ${}_A\mathrm{Hom}(M,Q)$.

Proof. For any $m_1,\ldots,m_k\in M$, there are $x_1,\ldots,x_n\in M$ and $f_1,\ldots,f_n\in {}^*M$ such that $m_j = \sum_i f_i(m_j)x_i$, for $j=1,\ldots,k$. For $h\in {}_A\mathrm{Hom}(M,Q)$, put $u := \sum_i f_i\otimes h(x_i) \in {}^*M\otimes_A Q$. Then

$$\nu_M(u)(m_j) = \textstyle\sum_i f_i(m_j)h(x_i) = h(\sum_i f_i(m_j)x_i) = h(m_j), \text{ for } j=1,\ldots,k,$$

proving that Im ν_M is dense in ${}_A\mathrm{Hom}(M,Q)$. □

42.14. $\sigma[M]$ closed under extensions. *The following are equivalent:*

(a) *$\sigma[M]$ is closed under extensions in* ${}_A\mathbf{M}$;

(b) *for every* $X\in {}_A\mathbf{M}$, $\mathcal{T}^M(X/\mathcal{T}^M(X)) = 0$;

(c) *there exists an A-injective $Q \in {}_A\mathbf{M}$ such that*

$$\sigma[M] = \{N \in {}_A\mathbf{M} \mid {}_A\mathrm{Hom}(N, Q) = 0\}.$$

Notice that Q can be obtained as the product of injective hulls of cyclic modules $X \in {}_A\mathbf{M}$ with $\mathcal{T}^M(X) = 0$.

Proof. For these assertions we refer to [47, 9.2–9.5]. □

Notice that $\sigma[M]$ is closed under extensions in ${}_A\mathbf{M}$ if and only if the M-adic topology is a *Gabriel topology* (see [44, Chapter VI]). In particular, the lattice $\mathcal{F}_M$ is closed under products. Under certain conditions this property characterises the left exactness of $\mathcal{T}^M$. We say that $\mathcal{F}_M$ is of *finite type* if every $J \in \mathcal{F}_M$ contains a finitely generated $\tilde{J} \in \mathcal{F}_M$. This holds if and only if $\sigma[M]$ contains a generating set of modules that are finitely presented in ${}_A\mathbf{M}$. For M locally Noetherian this implies that all $J \in \mathcal{F}_M$ are finitely generated. A filter of left ideals of A is called *bounded* if every $J \in \mathcal{F}_M$ contains a two-sided ideal $\tilde{J} \in \mathcal{F}_M$. The following is shown in [16, Proposition 5.7].

42.15. $\mathcal{F}_M$ closed under products. *Let M be locally Noetherian with $\mathcal{F}_M$ bounded and of finite type. If $\mathcal{F}_M$ is closed under products, then $\sigma[M]$ is closed under extensions.*

42.16. $\mathcal{T}^M$ as an exact functor. *Putting $\tilde{T} = \mathcal{T}^M(A)$, the following assertions are equivalent:*

(a) *the functor $\mathcal{T}^M : {}_A\mathbf{M} \to \sigma[M]$ is exact;*

(b) *$\sigma[M]$ is closed under extensions and the class $\{X \in {}_A\mathbf{M} \,|\, \mathcal{T}^M(X) = 0\}$ is closed under factor modules;*

(c) *for every $N \in \sigma[M]$, $\tilde{T}N = N$;*

(d) *M is an s-unital $\tilde{T}$-module;*

(e) *$\tilde{T}$ is an M-dense subring of A.*

Proof. (a) ⇒ (b) Let $\mathcal{T}^M$ be exact. For any exact sequence in ${}_A\mathbf{M}$ as a bottom row, there is a commutative diagram with exact rows,

$$\begin{array}{ccccccccc}
0 & \longrightarrow & \mathcal{T}^M(K) & \longrightarrow & \mathcal{T}^M(L) & \longrightarrow & \mathcal{T}^M(N) & \longrightarrow & 0 \\
 & & \downarrow & & \downarrow & & \downarrow & & \\
0 & \longrightarrow & K & \longrightarrow & L & \longrightarrow & N & \longrightarrow & 0.
\end{array}$$

In case $\mathcal{T}^M(K) = K$ and $\mathcal{T}^M(N) = N$ this implies $\mathcal{T}^M(L) = L$, showing that $\sigma[M]$ is closed under extensions. Moreover $\mathcal{T}^M(L) = 0$ implies $\mathcal{T}^M(N) = 0$ as required.

(b) ⇒ (c) Since $\sigma[M]$ is closed under extensions, $\mathcal{T}^M(A/T) = 0$. For any $N \in \sigma[M]$, $N/\tilde{T}N$ is generated by $A/\tilde{T}$, and so by (b), $\mathcal{T}^M(N/\tilde{T}N) = 0$ and hence $N = \tilde{T}N$.

(c) ⇒ (a) First observe that the hypothesis implies $\mathcal{T}^M(X) = TX$, for any $X \in {}_A\mathbf{M}$. Consider an exact sequence in ${}_A\mathbf{M}$, $0 \to K \to L \to N \to 0$. Since T_A is flat (see 42.5), tensoring with $T \otimes_A -$ yields an exact sequence $0 \to \tilde{T}K \to \tilde{T}L \to \tilde{T}N \to 0$.

(c) ⇔ (d) ⇔ (e) is shown in 42.6. □

We say $\sigma[M]$ is *closed under small epimorphisms* if, for any epimorphism $f : P \to N$ in ${}_A\mathbf{M}$, where $\mathrm{Ke}\, f \ll P$ and $N \in \sigma[M]$, we obtain $P \in \sigma[M]$.

42.17. Corollary. *Assume that the functor $\mathcal{T}^M : {}_A\mathbf{M} \to \sigma[M]$ is exact.*

(1) $\sigma[M]$ is closed under small epimorphisms.

(2) If P is finitely presented in $\sigma[M]$, then P is finitely presented in ${}_A\mathbf{M}$.

(3) If P is projective in $\sigma[M]$, then P is projective in ${}_A\mathbf{M}$.

Proof. (1) Put $\tilde{T} = \mathcal{T}^M(A)$. Consider an exact sequence $0 \to K \to P \to N \to 0$ in ${}_A\mathbf{M}$, where $K \ll P$ and $N \in \sigma[M]$. From this we obtain the following commutative diagram with exact rows:

$$\begin{array}{ccccccccc}
0 & \longrightarrow & K & \longrightarrow & P & \longrightarrow & N & \longrightarrow & 0 \\
 & & \downarrow & & \downarrow = & & \downarrow & & \\
0 & \longrightarrow & K + \tilde{T}P & \longrightarrow & P & \longrightarrow & P/(K + \tilde{T}P) & \longrightarrow & 0.
\end{array}$$

Clearly $P/(K + \tilde{T}P) \in \sigma[M]$ and by condition 42.16(b), $\tilde{T}(P/(K + \tilde{T}P)) = 0$. This implies $P = K + \tilde{T}P$, that is, $P \in \sigma[M]$.

(2) It is enough to show this for any cyclic module $P \in \sigma[M]$ that is finitely presented in $\sigma[M]$. For this we construct the following commutative diagram with exact rows (applying $\mathcal{T}^M$):

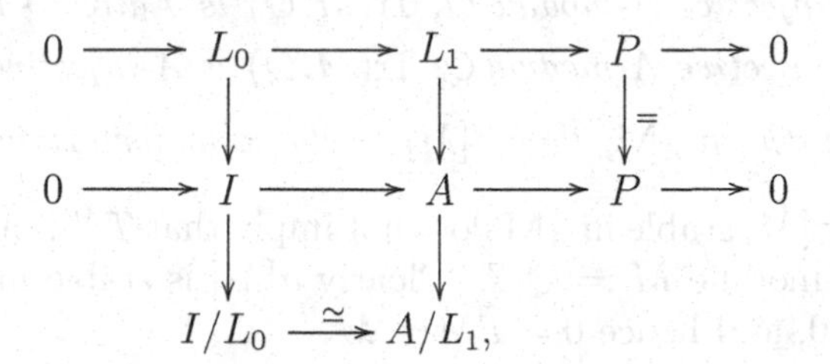

where L_0 and L_1 are suitable finitely generated modules in $\sigma[M]$. So I/L_0 is finitely generated, and hence so is I and P is finitely presented in ${}_A\mathbf{M}$.

(3) This is shown with a similar diagram as in the proof of (2). □

42.18. Corollary. *Suppose that $\sigma[M]$ has a generator that is locally projective in ${}_A\mathbf{M}$. Then $\mathcal{T}^M : {}_A\mathbf{M} \to \sigma[M]$ is an exact functor.*

Proof. Let $P \in \sigma[M]$ be a locally projective generator. Then clearly $\sigma[M] = \sigma[P]$ and $\tilde{T} = \mathcal{T}^M(A) = \mathrm{Tr}(P, A)$. By 42.8 and 42.10, $\tilde{T}^2 = \tilde{T}$ and $\tilde{T}P = P$. So $\tilde{T}$ generates P and 42.16 applies. □

42.19. Projective covers in $\sigma[M]$. *Let $\sigma[M]$ be locally Noetherian and suppose that A is f-semiperfect. Then the following are equivalent:*

(a) *the functor $\mathcal{T}^M : {}_A\mathbf{M} \to \sigma[M]$ is exact;*

(b) *$\sigma[M]$ has a generator that is (locally) projective in ${}_A\mathbf{M}$;*

(c) *there are idempotents $\{e_\lambda\}_\Lambda$ in A such that the Ae_λ are in $\sigma[M]$ and form a generating set in $\sigma[M]$;*

(d) *$\sigma[M]$ is a semiperfect category.*

Proof. (a) ⇒ (c) Let S be any simple module in $\sigma[M]$. S is finitely presented in $\sigma[M]$ and hence in ${}_A\mathbf{M}$ (by 42.17(2)). Since A is f-semiperfect, S has a projective cover P in ${}_A\mathbf{M}$ (see 41.18). By 42.17(1), $P \in \sigma[M]$ and clearly $P \simeq Ae$ for some idempotent $e \in A$. Now a representing set of simple modules in $\sigma[M]$ yields the required family of idempotents.

(c) ⇒ (b) is obvious and (b) ⇒ (a) follows from 42.18.

(c) ⇔ (d) This is clear by 41.16. □

A class of A-modules is said to be *stable* in ${}_A\mathbf{M}$ provided it is closed under essential extensions in ${}_A\mathbf{M}$ (e.g., [207, 5.1]).

42.20. $\sigma[M]$ stable in ${}_A\mathbf{M}$. *For M the following assertions are equivalent:*

(a) *$\sigma[M]$ is stable in ${}_A\mathbf{M}$;*

(b) *$\sigma[M]$ is closed under injective hulls in ${}_A\mathbf{M}$;*

(c) *every M-injective module in $\sigma[M]$ is A-injective;*

(d) *for every injective A-module Q, $\mathrm{Tr}(M, Q)$ is a direct summand in Q;*

(e) *for every injective A-module Q, $\mathrm{Tr}(M, Q)$ is A-injective.*

If $\sigma[M]$ is stable in ${}_A\mathbf{M}$, then $\sigma[M]$ is closed under extensions in ${}_A\mathbf{M}$.

Notice that $\sigma[M]$ stable in ${}_A\mathbf{M}$ does not imply that $\mathcal{T}^M$ is an exact functor. Consider the $\mathbb{Z}$-module $M := \mathbb{Q}/\mathbb{Z}$. Clearly $\sigma[M]$ is stable in $\mathbf{M}_{\mathbb{Z}}$. However, $\tilde{T} := \mathcal{T}^M(\mathbb{Z}) = 0$, and hence $0 = \tilde{T}M \neq M$.

References. Albu and Wisbauer [52]; Berning [62]; Prest and Wisbauer [175]; Wisbauer [46, 207, 209]; Zimmermann [219]; Zimmermann-Huisgen [220, 221].

43 Cogenerating and generating conditions

Throughout this section A is an R-algebra, M is a left A-module and $S = {}_A\mathrm{End}(M)$ denotes its endomorphism ring acting from the right, so that M is an (A, S)-bimodule.

Definitions. M is called a *self-cogenerator* if it cogenerates all its factor modules. $N \in \sigma[M]$ is a *cogenerator in* $\sigma[M]$ if it cogenerates all modules in $\sigma[M]$. M is said to be a *weak cogenerator (in* $\sigma[M]$*)* if, for every finitely generated submodule $K \subset M^n$, $n \in \mathbb{N}$, the factor module M^n/K is cogenerated by M.

It is a most interesting fact that generating (cogenerating) properties of ${}_AM$ imply certain projectivity (injectivity) conditions on M_S, and vice versa (e.g., [46, 48.1, 47.7, 47.8]).

43.1. Weak cogenerators. *For M the following are equivalent:*

(a) ${}_AM$ is a weak cogenerator (in $\sigma[M]$);

(b) M_S is weakly M_S-injective, and every finitely generated submodule of M^n, $n \in \mathbb{N}$, is M-reflexive.

If ${}_AM$ is a finitely generated weak cogenerator, then M_S is FP-injective.

43.2. Injectivity of M_S.

(1) M_S is FP-injective if and only if ${}_AM$ cogenerates the cokernels of morphisms $M^k \to M^n$, $k, n \in \mathbb{N}$.

(2) The following are equivalent:

(a) for any finitely generated left ideal $L \subset S$, the canonical map $\mathrm{Hom}_S(S, M) \to \mathrm{Hom}_S(L, M)$ is surjective;

(b) ${}_AM$ cogenerates the cokernels of morphisms $M \to M^n$, $n \in \mathbb{N}$.

If this holds and ${}_AM$ is linearly compact, then M_S is S-injective.

(3) If ${}_AM$ is a self-cogenerator and M_S is S-injective, then ${}_AM$ is linearly compact.

(4) If ${}_AM$ cogenerates all finitely M-generated modules, then ${}_AM$ is linearly compact if and only if M_S is S-injective.

A weakly M-injective weak cogenerator M is called a *weak quasi-Frobenius (weak QF) module.* From [46, 48.2] we obtain the following characterisation.

43.3. Weak QF modules. *If ${}_AM$ is faithful, the following are equivalent:*

(a) ${}_AM$ is a weak QF module;

(b) (i) ${}_AM$ is weakly ${}_AM$-injective and M_S is weakly M_S-injective, and

(ii) A is dense in $\mathrm{End}_S(M)$;

(c) M_S is a weak QF module and A is dense in $\mathrm{End}_S(M)$*;*

(d) ${}_AM$ and M_S are weak cogenerators in $\sigma[{}_AM]$ and $\sigma[M_S]$, respectively.
For any weak QF module ${}_AM$, $\mathrm{Soc}\,{}_AM = \mathrm{Soc}\,M_S$.

A locally Noetherian weak QF module M is an injective cogenerator in $\sigma[M]$ (by [46, 16.5]). A Noetherian weak QF module is called a *quasi-Frobenius* or *QF module.* A ring is a *QF ring* if it is QF as a left (or right) module. The reader should be warned that "quasi Frobenius" is given varying meanings in the literature.

43.4. Corollary. *Let M be a balanced (A, S)-module, which is locally Noetherian both as a left A-module and as a right S-module. Then the following are equivalent:*

(a) ${}_AM$ is an injective cogenerator in $\sigma[{}_AM]$;

(b) M_S is an injective cogenerator in $\sigma[M_S]$;

(c) ${}_AM$ and M_S are cogenerators in $\sigma[{}_AM]$ and $\sigma[M_S]$, respectively.

43.5. QF modules. *If ${}_AM$ is finitely generated, the following are equivalent (cf [46, 48.14,48.15]):*

(a) M is a self-projective QF module;

(b) M is a Noetherian (Artinian) projective cogenerator in $\sigma[M]$;

(c) M is a Noetherian injective generator in $\sigma[M]$;

(d) M is self-injective and injectives are projective in $\sigma[M]$;

(e) M is a generator and projectives are injective in $\sigma[M]$;

(f) M is a projective generator in $\sigma[M]$ and S is a QF ring.

43.6. QF rings. *For a ring A the following are equivalent:*

(a) ${}_AA$ is a QF module;

(b) A is left Noetherian (Artinian) and a cogenerator in ${}_A\mathbf{M}$;

(c) A is left Noetherian (Artinian) and injective;

(d) all injectives are projective in ${}_A\mathbf{M}$;

(e) all projectives are injective in ${}_A\mathbf{M}$;

(f) $A^{(\mathbb{N})}$ is an injective cogenerator in ${}_A\mathbf{M}$;

(g) A_A is a QF module.

43.7. Cogenerator with commutative endomorphism ring. (Cf. [46, 48.16].) *If $S = {}_A\mathrm{End}(M)$ is commutative, then the following are equivalent:*

(a) M is a cogenerator in $\sigma[M]$;

(b) M is self-injective and a self-cogenerator;

(c) $M \simeq \bigoplus_\Lambda \widehat{E}_\lambda$, *where* $\{E_\lambda\}_\Lambda$ *is a minimal representing set of simple modules in* $\sigma[M]$, *and* $\widehat{E}_\lambda$ *denotes the injective hull of* E_λ *in* $\sigma[M]$;

(d) M *is a direct sum of indecomposable modules* N *that are cogenerators in* $\sigma[N]$.

If (any of) these conditions are satisfied, then:

(1) *Every A-submodule of* M *is fully invariant and hence self-injective and a self-cogenerator.*

(2) *For every* $\lambda \in \Lambda$, *the category* $\sigma[\widehat{E}_\lambda]$ *contains only one simple module (up to isomorphisms).*

(3) *If the* $\widehat{E}_\lambda$ *are finitely generated A-modules, then* M *generates all simple modules in* $\sigma[M]$.

(4) *If* M *is projective in* $\sigma[M]$, *then* M *is a generator in* $\sigma[M]$.

(5) *If* M *is finitely generated, then* M *is finitely cogenerated.*

Let $\sigma_f[M]$ denote the full subcategory of $\sigma[M]$ whose objects are submodules of *finitely* M-generated modules. With this notation $\sigma_f[S_S]$ is the category of submodules of finitely generated right S-modules. This type of category is of particular interest in studying dualities.

43.8. Morita dualities.

(1) *The following are equivalent:*

(a) ${}_A\mathrm{Hom}(-, M)\colon \sigma_f[M] \to \sigma_f[S_S]$ *is a duality;*

(b) ${}_AM$ *is an injective cogenerator in* $\sigma[M]$, *and* M_S *is an injective cogenerator in* $\mathbf{M}_S$;

(c) ${}_AM$ *is linearly compact, finitely cogenerated, and an injective cogenerator in* $\sigma[M]$.

(2) *If* M *is an injective cogenerator in* $\sigma[M]$, *the following are equivalent:*

(a) ${}_AM$ *is Artinian;*

(b) ${}_AM$ *is semi-Artinian and* M_S *is* S*-injective;*

(c) M_S *is a* Σ*-injective cogenerator in* $\mathbf{M}_S$;

(d) S *is right Noetherian.*

Proof. Part (1) is shown in [46, 47.12].

(2) (a) $\Leftrightarrow$ (b) follows by 43.2 and the fact that linearly compact semi-Artinian modules are in fact Artinian (see 41.13(4)).

(a) $\Leftrightarrow$ (d) follows from the fact that descending chains of submodules of ${}_AM$ correspond to ascending chains of right ideals of S.

(a) $\Rightarrow$ (c) Over a right Noetherian ring S, any (FP-)injective module is Σ-injective. Now the assertion is clear by (1).

(c) $\Rightarrow$ (d) Since $\mathbf{M}_S$ has a Σ-injective cogenerator, S is Noetherian (e.g., [46, 27.3]). $\square$

43.9. Hom and tensor functors. (Cf. [46, 45.8].) The functors

$$M \otimes_S - : {}_S\mathbf{M} \to {}_A\mathbf{M}, \quad {}_A\mathrm{Hom}(M, -) : {}_A\mathbf{M} \to {}_S\mathbf{M},$$

form an adjoint pair by the isomorphism, for any $N \in {}_A\mathbf{M}$ and $X \in {}_S\mathbf{M}$,

$${}_A\mathrm{Hom}(M \otimes_S X, N) \to {}_S\mathrm{Hom}(X, {}_A\mathrm{Hom}(M, N)), \quad \delta \mapsto [x \mapsto \delta(- \otimes x)],$$

with inverse map $\varphi \mapsto [m \otimes x \mapsto \varphi(x)(m)]$. Thus the counit and unit of adjunction come out as

$$\begin{aligned} \psi_N &: M \otimes_S {}_A\mathrm{Hom}(M, N) \to N, \quad m \otimes f \mapsto f(m), \\ \eta_X &: X \to {}_A\mathrm{Hom}(M, M \otimes_S X), \quad x \mapsto [m \mapsto m \otimes x]. \end{aligned}$$

Each of the following compositions of maps yields the identity:

$${}_A\mathrm{Hom}(M,N) \xrightarrow{\eta_{\mathrm{Hom}(M,N)}} {}_A\mathrm{Hom}(M, M \otimes_S {}_A\mathrm{Hom}(M, N)) \xrightarrow{\mathrm{Hom}(M,\psi_N)} {}_A\mathrm{Hom}(M,N),$$
$$M \otimes_S X \xrightarrow{id \otimes \eta_X} M \otimes_S {}_A\mathrm{Hom}(M, M \otimes_S X) \xrightarrow{\psi_{M \otimes X}} M \otimes_S X.$$

Notice that the image of the tensor functor $M \otimes_S -$ lies in $\sigma[M]$, and hence by (co-)restriction we obtain the adjoint pair of functors

$$M \otimes_S - : {}_S\mathbf{M} \to \sigma[M], \quad {}_A\mathrm{Hom}(M, -) : \sigma[M] \to {}_S\mathbf{M}.$$

43.10. Static and adstatic modules. An A-module N is called *M-static* if ψ_N in 43.9 is an isomorphism and the class of all M-static A-modules is denoted by $\mathrm{Stat}(M)$. An S-module X is called *M-adstatic* if η_X in 43.9 is an isomorphism and we denote the class of all M-adstatic S-modules by $\mathrm{Adst}(M)$. Clearly, for every M-static module N, ${}_A\mathrm{Hom}(M, N)$ is M-adstatic, and, for each M-adstatic module ${}_SX$, $M \otimes_S X$ is M-static.

43.11. Basic equivalence. *The functor* ${}_A\mathrm{Hom}(M, -)\colon \mathrm{Stat}(M) \to \mathrm{Adst}(M)$ *defines an equivalence with inverse* $M \otimes_S -$.

43.12. M as generator in $\sigma[M]$. *Putting* $B = \mathrm{End}_S(M)$ *and* $\bar{A} = A/\mathrm{An}(M)$*, the following statements are equivalent:*

(a) M is a generator in $\sigma[M]$;

(b) the functor ${}_A\mathrm{Hom}(M, -) : \sigma[M] \to {}_S\mathbf{M}$ is faithful;

(c) M generates every (cyclic) submodule of $M^{(\mathbb{N})}$;

(d) for every (cyclic) submodule $K \subset M^{(\mathbb{N})}$, ψ_K (see 43.9) is surjective;

(e) *M_S is flat and every (finitely) M-generated module N is M-static;*

(f) *M_S is flat and every injective module in $\sigma[M]$ is M-static;*

(g) *M_S is flat, $\bar{A}$ is dense in B, and, for injective modules $V \in \sigma[M]$, the canonical map $M \otimes_S {}_A\mathrm{Hom}(M,V) \to {}_A\mathrm{Hom}(B,V)$ is injective;*

(h) $\mathrm{Stat}(M) = \sigma[M]$.

Proof. The equivalences of (a) to (d) are shown in [46, 15.5 and 15.9].

(a) $\Leftrightarrow$ (h) It suffices to show that $\psi_N : M \otimes_S {}_A\mathrm{Hom}(M,N) \to N$ is injective, for any $N \in \sigma[M]$. Let $u \in \mathrm{Ke}\,\psi_N$, that is,

$$u = \textstyle\sum_{i=1}^k m_i \otimes f_i \in M \otimes_S {}_A\mathrm{Hom}(M,N) \quad \text{with } \sum_i m_i f_i = 0.$$

By assumption, $(m_1,\ldots,m_k) = \sum_j a_j g_j$, for some $a_j \in M$ and morphisms $g_j \in {}_A\mathrm{Hom}(M, \mathrm{Ke}\,(\sum_i f_i))$, where

$$\sum_i f_i : M^k \to N,\ (x_1,\ldots,x_k) \mapsto \sum_i x_i f_i\,.$$

With the canonical projections $\pi_i : M^k \to M$, $\sum_i g_j \pi_i f_i = 0$ for all j, and hence

$$\sum_{i=1}^k m_i \otimes f_i = \sum_i \sum_j a_j g_j \pi_i \otimes f_i = \sum_j a_j \otimes \sum_i g_j \pi_i f_i = 0\,,$$

which shows that ψ_N is injective.

(a) $\Rightarrow$ (e),(f) By [46, 15.9], any generator in $\sigma[M]$ is flat over its endomorphism ring.

(e) $\Rightarrow$ (d) For $K \subset M^k$, $k \in \mathbb{N}$, consider the canonical projection $g : M^k \to M^k/K =: N$. By assumption, we may construct the exact commutative diagram (tensoring over S)

$$\begin{array}{ccccccc}
0 \longrightarrow & M \otimes {}_A\mathrm{Hom}(M,K) & \longrightarrow & M \otimes {}_A\mathrm{Hom}(M,M^k) & \longrightarrow & M \otimes {}_A\mathrm{Hom}(M,N) \\
 & \downarrow \psi_K & & \downarrow \simeq & & \downarrow \psi_N \\
0 \longrightarrow & K & \longrightarrow & M^k & \xrightarrow{\ g\ } & N\,,
\end{array}$$

where ψ_N is an isomorphism and hence ψ_K is surjective.

(f) $\Rightarrow$ (a) For any $K \in \sigma[M]$, there exists an exact sequence $0 \to K \to Q_1 \to Q_2$, where Q_1, Q_2 are injectives in $\sigma[M]$ and hence are M-static. We construct an exact commutative diagram (tensoring over S)

$$\begin{array}{ccccccc}
0 \longrightarrow & M \otimes {}_A\mathrm{Hom}(M,K) & \longrightarrow & M \otimes {}_A\mathrm{Hom}(M,Q_1) & \longrightarrow & M \otimes {}_A\mathrm{Hom}(M,Q_2) \\
 & \downarrow \psi_K & & \downarrow \simeq & & \downarrow \simeq \\
0 \longrightarrow & K & \longrightarrow & Q_1 & \xrightarrow{\ g\ } & Q_2\,,
\end{array}$$

showing that ψ_K is an isomorphism and so K is M-generated.

(f) $\Leftrightarrow$ (g) By the Density Theorem (see [46, 15.7]), $\bar{A}$ is dense in B. This implies ${}_A\mathrm{Hom}(B,V) \simeq \mathrm{Hom}_B(B,V) \simeq V$. $\square$

43.13. Self-progenerator. *For $_AM$ finitely generated, the following assertions are equivalent (cf.[46, 18.5]):*

(a) M is a projective generator in $\sigma[M]$;

(b) M is a generator in $\sigma[M]$ and M_S is faithfully flat;

(c) (i) for every left ideal $J \subset S$, $MJ \neq M$, and
(ii) every finitely M-generated A-module is M-static;

(d) there are functorial isomorphisms

$$I_{S\mathbf{M}} \simeq {}_A\mathrm{Hom}(M, M \otimes_S -) \quad \text{and} \quad M \otimes_S {}_A\mathrm{Hom}(M, -) \simeq I_{\sigma[M]};$$

(e) ${}_A\mathrm{Hom}(M, -) : \sigma[M] \to {}_S\mathbf{M}$ is an equivalence of categories.

References. Wisbauer [46, 208, 209]; Zimmermann-Huisgen [220].

44 Decompositions of $\sigma[M]$

In this section decompositions of closed categories into "direct sums" of closed subcategories are considered. Again let A be an R-algebra and M a left A-module with $S = {}_A\mathrm{End}(M)$.

A submodule $N \subset M$ is said to be *fully invariant* if it is invariant under endomorphisms of M, that is, N is an (A, S)-submodule. The ring of (A, S)-endomorphisms of M is the centre of S (e.g., [47, 4.2]).

44.1. Decompositions and idempotents. Let $M = \bigoplus_\Lambda M_\lambda$ be a decomposition with A-submodules $M_\lambda \subset M$. Then there exists a family of orthogonal idempotents $\{e_\lambda\}_\Lambda$ in S, where $M_\lambda = Me_\lambda$, for each $\lambda \in \Lambda$. If all the M_λ are fully invariant (fully invariant decomposition), then all the e_λ are central idempotents of S.

44.2. Big cogenerators. An M-injective module $Q \in \sigma[M]$ is said to be a *big injective cogenerator* in $\sigma[M]$ if every cyclic module in $\sigma[M]$ is isomorphic to a submodule of $Q^{(\mathbb{N})}$. Clearly every big injective cogenerator in $\sigma[M]$ is a cogenerator as well as a subgenerator in $\sigma[M]$. Such modules always exist:

Let $\{N_\lambda\}_\Lambda$ be a representing set of the cyclic modules in $\sigma[M]$. Then the M-injective hull of $\bigoplus_\Lambda N_\lambda$ is a big injective cogenerator in $\sigma[M]$.

If M is locally Noetherian and $\{E_\lambda\}_\Lambda$ a representing set of the indecomposable M-injectives in $\sigma[M]$, then $\bigoplus_\Lambda E_\lambda$ is a big injective cogenerator in $\sigma[M]$.

If M is locally of finite length, every injective cogenerator in $\sigma[M]$ is big.

In 42.7, for any A-module L, we defined $\mathcal{T}^M(L) := \mathrm{Tr}(\sigma[M], L)$. Recall that every closed subcategory of ${}_A\mathbf{M}$ is of type $\sigma[N]$, for some A-module N.

44.3. Correspondence relations. *Let Q be a big injective cogenerator in the category $\sigma[M]$.*

(1) For every $N \in \sigma[M]$, $\sigma[N] = \sigma[\mathrm{Tr}(N, Q)]$.

(2) The assignment $\sigma[N] \mapsto \mathrm{Tr}(N, Q)$ yields a bijective correspondence between the closed subcategories of $\sigma[M]$ and the fully invariant submodules of Q.

(3) If $\sigma[N]$ is closed under essential extensions (injective hulls) in $\sigma[M]$, then $\mathrm{Tr}(N, Q)$ is an A-direct summand of Q.

(4) If M is locally Noetherian and $\mathrm{Tr}(N, Q)$ is an A-direct summand of Q, then $\sigma[N]$ is closed under essential extensions in $\sigma[M]$.

(5) $N \in \sigma[M]$ is semisimple if and only if $\mathrm{Tr}(N, Q) \subset \mathrm{Soc}({}_AQ)$.

Proof. Since Q is M-injective, $\mathrm{Tr}(\sigma[N], Q) = \mathrm{Tr}(N, Q)$.

(1) $\mathrm{Tr}(N,Q)$ is a fully invariant submodule that, by definition, belongs to $\sigma[N]$. Consider any finitely generated $L \in \sigma[N]$. Then, by assumption, $L \subset Q^k$, for some $k \in \mathbb{N}$, and hence $L \subset \mathrm{Tr}(L,Q)^k \subset \mathrm{Tr}(N,Q)^k$. This implies $N \in \sigma[\mathrm{Tr}(N,Q)]$.

Parts (2) and (5) are immediate consequences of (1).

(3) If $\sigma[N]$ is closed under essential extensions in $\sigma[M]$, then $\mathrm{Tr}(N,Q)$ is an A-direct sumand in Q (and hence is injective in $\sigma[M]$).

(4) Let M be locally Noetherian and $\mathrm{Tr}(N,Q)$ a direct summand of Q. Consider any N-injective module L in $\sigma[N]$. Then L is a direct sum of N-injective uniform modules $U \in \sigma[M]$. Clearly U is (isomorphic to) a direct summand of $\mathrm{Tr}(N,Q)$ and hence of Q; that is, U is M-injective and so L is M-injective, too. □

44.4. Sum and decomposition of closed subcategories. For any $K, L \in \sigma[M]$ we write $\sigma[K] \cap \sigma[L] = 0$, provided $\sigma[K]$ and $\sigma[L]$ have no nonzero module in common. Given a family $\{N_\lambda\}_\Lambda$ of modules in $\sigma[M]$, define

$$\sum\nolimits_\Lambda \sigma[N_\lambda] := \sigma[\bigoplus\nolimits_\Lambda N_\lambda].$$

This is the smallest closed subcategory of $\sigma[M]$ containing all the N_λ. Moreover, we write

$$\sigma[M] = \bigoplus\nolimits_\Lambda \sigma[N_\lambda],$$

provided, for every module $L \in \sigma[M]$, $L = \bigoplus_\Lambda \mathcal{T}^{N_\lambda}(L)$ (internal direct sum). This decomposition of $\sigma[M]$ is known as a *σ-decomposition*. The category $\sigma[M]$ is *σ-indecomposable* provided it has no nontrivial σ-decomposition. In view of the fact that every closed subcategory of ${}_A\mathbf{M}$ is of type $\sigma[N]$, for some A-module N, these constructions describe the decomposition of any closed subcategory into closed subcategories.

44.5. σ-decomposition of modules. *For a decomposition $M = \bigoplus_\Lambda M_\lambda$, the following are equivalent (cf. [211]):*

(a) for any distinct $\lambda, \mu \in \Lambda$, M_λ and M_μ have no nonzero isomorphic subfactors;

(b) for any distinct $\lambda, \mu \in \Lambda$, ${}_A\mathrm{Hom}(K_\lambda, K_\mu) = 0$, where K_λ, K_μ are subfactors of M_λ, M_μ, respectively;

(c) for any distinct $\lambda, \mu \in \Lambda$, $\sigma[M_\lambda] \cap \sigma[M_\mu] = 0$;

(d) for any $\mu \in \Lambda$, $\sigma[M_\mu] \cap \sigma[\bigoplus_{\lambda \neq \mu} M_\lambda] = 0$;

(e) for any $L \in \sigma[M]$, $L = \bigoplus_\Lambda \mathcal{T}^{N_\lambda}(L)$.

If these conditions hold, we call $M = \bigoplus_\Lambda M_\lambda$ a σ-decomposition *and in this case*

$$\sigma[M] = \bigoplus\nolimits_\Lambda \sigma[M_\lambda].$$

44.6. Corollary. *Let $\sigma[M] = \bigoplus_\Lambda \sigma[N_\lambda]$ be a decomposition of $\sigma[M]$. Then:*

(1) *each $\sigma[N_\lambda]$ is closed under essential extensions and small epimorphisms in $\sigma[M]$;*

(2) *any $L \in \sigma[N_\lambda]$ is M-injective if and only if it is N_λ-injective;*

(3) *any $P \in \sigma[N_\lambda]$ is projective in $\sigma[M]$ if and only if it is projective in $\sigma[N_\lambda]$;*

(4) *any $L \in \sigma[N_\lambda]$ is projective in $\sigma[M]$ if and only if it is projective in $\sigma[N_\lambda]$;*

(5) *$M = \bigoplus_\Lambda \mathcal{T}^{N_\lambda}(M)$ is a σ-decomposition of M.*

44.7. Corollary. *Let $\sigma[M] = \bigoplus_\Lambda \sigma[N_\lambda]$ be a σ-decomposition of $\sigma[M]$. Then the trace functor $\mathcal{T}^M$ is exact if and only if the trace functors $\mathcal{T}^{N_\lambda}$ are exact, for all $\lambda \in \Lambda$.*

44.8. Corollary. *If M is a projective generator or an injective cogenerator in $\sigma[M]$, then any fully invariant decomposition of M is a σ-decomposition.*

Proof. Let $M = \bigoplus_\Lambda M_\lambda$ be a fully invariant decomposition. If M is a projective generator in $\sigma[M]$, then every submodule of M_λ is generated by M_λ. Since the M_λ are projective in $\sigma[M]$, any nonzero (iso)morphism between (sub)factors of M_λ and M_μ yields a nonzero morphism between M_λ and M_μ. So the assertion follows from 44.5.

Now suppose that M is an injective cogenerator in $\sigma[M]$. Then every subfactor of M_λ must be cogenerated by M_λ. From this it follows that for $\lambda \neq \mu$, there are no nonzero maps between subfactors of M_λ and M_μ and so 44.5 applies. □

A module M is said to be *σ-indecomposable* if it has no non-trivial σ-decompositions.

44.9. Corollary. *For M the following assertions are equivalent:*

(a) *$\sigma[M]$ is σ-indecomposable;*

(b) *M is σ-indecomposable;*

(c) *any subgenerator in $\sigma[M]$ is σ-indecomposable;*

(d) *an injective cogenerator, which is a subgenerator in $\sigma[M]$, has no non-trivial fully invariant decomposition.*

If there exist projective generators in $\sigma[M]$, then (a)–(d) are equivalent to:

(e) *projective generators in $\sigma[M]$ have no fully invariant decompositions.*

Note that the category ${}_A\mathbf{M}$ is σ-indecomposable if and only if A has no nontrivial central idempotents.

44.10. σ-decomposition when ${}_A\text{End}(M)$ is commutative. (Cf. 43.7) Let M be a cogenerator in $\sigma[M]$ with a commutative endomorphism ring ${}_A\text{End}(M)$. Then $M \simeq \bigoplus_\Lambda \widehat{E}_\lambda$, where $\{E_\lambda\}_\Lambda$ is a minimal representing set of simple modules in $\sigma[M]$. This is a σ-decomposition of M and

$$\sigma[M] = \bigoplus\nolimits_\Lambda \sigma[\widehat{E}_\lambda],$$

where each $\sigma[\widehat{E}_\lambda]$ is σ-indecomposable and has only one simple module.

As an example, consider the $\mathbb{Z}$-module $\mathbb{Q}/\mathbb{Z} = \bigoplus_{p\ prime} \mathbb{Z}_{p^\infty}$ and the decomposition of the category of torsion Abelian groups as a direct sum of the categories of p-groups,

$$\sigma[\mathbb{Q}/\mathbb{Z}] = \bigoplus\nolimits_{p\ prime} \sigma[\mathbb{Z}_{p^\infty}].$$

Notice that, although $\mathbb{Q}/\mathbb{Z}$ is an injective cogenerator in $\mathbf{M}_\mathbb{Z}$ with a nontrivial σ-decomposition, $\mathbf{M}_\mathbb{Z}$ is σ-indecomposable. This is possible since $\mathbb{Q}/\mathbb{Z}$ is not a subgenerator in $\mathbf{M}_\mathbb{Z}$.

To find σ-decompositions of modules, some technical observations are first required.

44.11. Relations on families of modules. Consider any family of A-modules $\{M_\lambda\}_\Lambda$ in $\sigma[M]$. Define a relation $\sim$ on $\{M_\lambda\}_\Lambda$ by putting

$$M_\lambda \sim M_\mu \quad \text{if there exist non-zero morphisms } M_\lambda \to M_\mu \text{ or } M_\mu \to M_\lambda.$$

Clearly $\sim$ is symmetric and reflexive, and we denote by $\approx$ the smallest equivalence relation on $\{M_\lambda\}_\Lambda$ determined by $\sim$, that is,

$$\begin{aligned} M_\lambda \approx M_\mu \quad & \text{if there exist } \lambda_1, \ldots, \lambda_k \in \Lambda, \\ & \text{such that } M_\lambda = M_{\lambda_1} \sim \cdots \sim M_{\lambda_k} = M_\mu\,. \end{aligned}$$

Then $\{M_\lambda\}_\Lambda$ is the disjoint union of the equivalence classes $\{[M_\omega]\}_\Omega$, where $[M_\omega] = \{M_\lambda\}_{\Lambda_\omega}$, $\Lambda = \dot{\bigcup}_\Omega \Lambda_\omega$. In case each $M_\lambda = \widehat{E}_\lambda$, the M-injective hull of some simple module $E_\lambda \in \sigma[M]$, then

$$\widehat{E}_\lambda \sim \widehat{E}_\mu \text{ if and only if } \operatorname{Ext}_M(E_\lambda, E_\mu) \neq 0 \text{ or } \operatorname{Ext}_M(E_\mu, E_\lambda) \neq 0.$$

A decomposition $M = \bigoplus_\Lambda M_\lambda$ is said to *complement direct summands* if, for every direct summand K of M, there exists a subset $\Lambda' \subset \Lambda$ such that $M = K \oplus (\bigoplus_{\Lambda'} M_\lambda)$ (cf. [2, § 12]).

44.12. Lemma. *Let $M = \bigoplus_\Lambda M_\lambda$ be a decomposition that complements direct summands, where all M_λ are indecomposable. Then M has a decomposition $M = \bigoplus_{\alpha \in I} N_\alpha$, where each $N_\alpha \subset M$ is a fully invariant submodule and does not decompose nontrivially into fully invariant submodules.*

Proof. Consider the equivalence relation $\approx$ on $\{M_\lambda\}_\Lambda$ (see 44.11) with the equivalence classes $\{[M_\omega]\}_{\Lambda_\alpha}$ and $\Lambda = \bigcup_\Omega \Lambda_\omega$. Then $N_\omega := \bigoplus_{\Lambda_\omega} M_\lambda$ is a fully invariant submodule of M, for each $\omega \in \Omega$, and

$$M = \bigoplus_{\omega\in\Omega} (\bigoplus_{\Lambda_\omega} M_\lambda) = \bigoplus_{\omega\in\Omega} N_\omega.$$

Assume $N_\omega = K \oplus L$ for fully invariant submodules $K, L \subset N_\omega$. Since the defining decomposition of N_ω complements direct summands, we may assume that Λ_ω has subsets X, Y such that

$$K = \bigoplus_X M_\lambda, \quad L = \bigoplus_Y M_\lambda.$$

By construction, for any $x \in X$, $y \in Y$, we observe that $M_x \approx M_y$, and it is easy to see that this implies the existence of nonzero morphisms $K \to L$ or $L \to K$, contradicting our assumption. So N_ω does not decompose into fully invariant submodules. □

44.13. Proposition. *Let $M = \bigoplus_\Lambda M_\lambda$ with each ${}_A\mathrm{End}(M_\lambda)$ a local ring.*

(1) If M is M-injective, the decomposition complements direct summands.

(2) If M is projective in $\sigma[M]$ and $\mathrm{Rad}(M) \ll M$*, then the decomposition complements direct summands.*

Proof. For the first assertion we refer to [15, 8.13]. The second condition characterises M as semiperfect in $\sigma[M]$ (see 41.14) and the assertion follows from [15, 8.12]. □

44.14. σ-decomposition for locally Noetherian modules. *Let M be a locally Noetherian A-module. Then M has a σ-decomposition $M = \bigoplus_\Lambda M_\lambda$ and*

$$\sigma[M] = \bigoplus_\Lambda \sigma[M_\lambda],$$

where each $\sigma[M_\lambda]$ is σ-indecomposable.

(1) $\sigma[M]$ is σ-indecomposable if and only if, for any indecomposable injectives $K, L \in \sigma[M]$, $K \approx L$ (as defined in 44.11).

(2) If M has locally finite length, then $\sigma[M]$ is σ-indecomposable if and only if, for any simple modules $S_1, S_2 \in \sigma[M]$, $\widehat{S}_1 \approx \widehat{S}_2$ (M-injective hulls).

Proof. Let Q be an injective cogenerator that is also a subgenerator in $\sigma[M]$. Then Q is a direct sum of indecomposable M-injective modules, and this is a decomposition that complements direct summands (by 44.13). By Lemma 44.12, Q has a fully invariant decomposition $Q = \bigoplus_\Lambda Q_\lambda$ such that Q_λ has no nontrivial fully invariant decomposition. Now the assertions follow from Corollaries 44.8, 44.9 and 44.6.

Part (1) is clear from the above discussion, and in (2) – by the assumption – every indecomposable M-injective module is an M-injective hull of a simple module in $\sigma[M]$. □

44.15. σ-decomposition for semiperfect generators. *If M is a projective generator that is semiperfect in $\sigma[M]$, then M (and $\sigma[M]$) has a σ-decomposition $M = \bigoplus_\Lambda M_\lambda$, where each M_λ is σ-indecomposable.*

In particular, every semiperfect ring A has a σ-decomposition $A = Ae_1 \oplus \cdots \oplus Ae_k$, where the e_i are central idempotents of A that are not a nontrivial sum of orthogonal central idempotents.

Proof. By [46, 42.5] and 44.13, M has a decomposition that complements direct summands. By Lemma 44.12, M has a fully invariant decomposition, and the assertions follow from Corollaries 44.8, 44.9 and 44.6. □

References. García, Jara and Merino [116]; Năstăsescu and Torrecillas [165]; Vanaja [202]; Wisbauer [211].

Bibliography

Monographs

[1] Abe, E., *Hopf Algebras*, Cambridge University Press, Cambridge (1977)

[2] Anderson, F., Fuller, K., *Rings and Categories of Modules*, Springer, Berlin (1974)

[3] Borceux, F., *Handbook of Categorical Algebra 2. Categories and Structures*, Cambridge University Press, Cambridge (1994)

[4] Bourbaki, N., *Algebra I*, Addison-Wesley, Reading, MA (1974)

[5] Bourbaki, N., *Algèbre, Chapitres 1 à 3*, Hermann, Paris (1970)

[6] Bourbaki, N., *Algèbre commutative, Chapitres 1 à 9*, Hermann, Paris (1961–83)

[7] Brown, K.A., Goodearl, K.R., *Lectures on Algebraic Quantum Groups*, Birkhäuser, Basel (2002)

[8] Caenepeel, S., *Brauer Groups, Hopf Algebras and Galois Theory*, Kluwer, Dordrecht (1998)

[9] Caenepeel, S., Militaru, G., Zhu, S., *Frobenius and Separable Functors for Generalized Hopf Modules and Nonlinear Equations*, LNM 1787, Springer, Berlin (2002)

[10] Cartan, H., Eilenberg, S., *Homological Algebra*, Princeton University Press, Princeton, NJ (1956)

[11] Chari, V., Pressley, A., *A Guide to Quantum Groups*, Cambridge University Press, Cambridge (1994)

[12] Chase, S.U., Sweedler, M.E., *Hopf Algebras and Galois Theory*, Springer, Berlin-Heidelberg-New York (1969)

[13] Childs, L.N., *Taming Wild Extensions: Hopf Algebras and Local Galois Module Theory*, AMS, Providence, RI (2000)

[14] Dăscălescu, S., Năstăsescu, C., Raianu, S., *Hopf Algebras. An Introduction*, Marcel Dekker, New York-Basel (2001)

[15] Dung, N.V., Huynh, D.V., Smith, P.F., Wisbauer, R., *Extending Modules*, Pitman RN 313, London (1994)

[16] Golan, J.S., *Linear Topologies on a Ring*, Pitman Research Notes 159, London (1987)

[17] Grothendieck, A., *Technique de descente et théorèmes d'existence en géométrie algébrique, I. Généralités, descente par morphismes fidèlement plats*, Séminaire Bourbaki 12, No. 190 (1959/1960)

[18] Grothendieck, A., *Technique de descente et théorèmes d'existence en géométrie algébrique, II. Le théorème d'existence en théorie formelle des modules*, Séminaire Bourbaki 12, No. 195 (1959/1960)

[19] Gumm, H.P., *Elements of the General Theory of Coalgebras*, Preliminary version, Universität Marburg (2000)

[20] Hilton, P.J., Stambach, U., *A Course in Homological Algebra*, Springer, Berlin (1971)

[21] Husemoller, D., *Fibre Bundles*, Springer, Berlin (1992)

[22] Jonah, D.W., *Cohomology of Coalgebras*, Mem. AMS 82 (1968)

[23] Kadison, L., *New Examples of Frobenius Extensions*, AMS, Providence, RI (1999)

[24] Kaplansky, I., *Bialgebras*, Lecture Notes in Math., University of Chicago, Chicago (1975)

[25] Kassel, C., *Quantum Groups*, Springer, Berlin (1995)

[26] Klimyk, A., Schmüdgen, K., *Quantum Groups and Their Representations*, Springer, Berlin (1997)

[27] Kriz, I., May, J.P., *Operads, Algebras, Modules and Motives*, Astérisque 293 (1995)

[28] Knus, M.A., Ojanguren, M., *Théorie de la descente et algèbres d'Azumaya*, LNM 389, Springer, Berlin (1974)

[29] Loday, J.L., Stasheff, J., Voronov, A.A., (Eds.), *Operads: Proceedings of Renaissance Conferences*, Contemp. Math. 202 (1997)

[30] Lusztig, G., *Introduction to Quantum Groups*, Birkhäuser, Basel (1993)

[31] Mac Lane, S., *Categories for the Working Mathematician*, GTM 5, Springer, New York (1998)

[32] Mackenzie, K., *Lie Groupoids and Lie Algebroids in Differential Geometry*, Cambridge University Press, Cambridge (1987)

[33] Majid, S., *Foundations of Quantum Group Theory*, Cambridge University Press, Cambridge (1995)

[34] Majid, S., *A Quantum Group Primer*, Cambridge University Press, Cambridge (2002)

[35] May, J.P., *The Geometry of Iterated Loop Spaces*, LNM 271, Springer, Berlin (1972)

[36] Montgomery, S., *Fixed Rings of Finite Automorphism Groups of Associative Rings*, LNM 818, Springer, Berlin (1980)

[37] Montgomery, S., *Hopf Algebras and Their Actions on Rings*, Reg. Conf. Series in Math., CBMS 82, AMS, Providence RI (1993)

[38] Pareigis, B., *Categories and Functors*, Academic Press, New York-London (1970)

[39] Pierce, R., *Associative Algebras*, Springer, Berlin (1982)

[40] Popescu, N., *Abelian Categories with Applications to Rings and Modules*, Academic Press, London-New York (1973)

[41] Rotman, J.J., *An Introduction to Homological Algebra*, Academic Press, New York (1979)

[42] Schubert, H., *Categories*, Springer, Berlin (1972)

[43] Shnider, S., Sternberg, S., *Quantum Groups. From Coalgebras to Drinfeld Algebras*, International Press, Boston (1993)

[44] Stenström, B., *Rings of Quotients*, Springer, Berlin (1975)

[45] Sweedler, M.E., *Hopf Algebras*, Benjamin, New York (1969)

[46] Wisbauer, R., *Foundations of Module and Ring Theory*, Gordon and Breach, Reading-Paris (1991) (*Grundlagen der Modul- und Ringtheorie*, Verlag Reinhard Fischer, München (1988))

[47] Wisbauer, R., *Modules and Algebras: Bimodule Structure and Group Actions on Algebras*, Pitman Mono. PAM 81, Addison Wesley, Longman, Essex (1996)

Articles

[48] Abuhlail, J.Y., *Dualitätssätze für Hopf-Algebren über Ringen*, PhD Thesis, University of Düsseldorf (2001)

[49] Abuhlail, J.Y., Gómez-Torrecillas, J., Lobillo, F.J., *Duality and rational modules in Hopf algebras over commutative rings*, J. Algebra 240, 165–184 (2001)

[50] Abuhlail, J.Y., Gómez-Torrecillas, J., Wisbauer, R., *Dual coalgebras of algebras over commutative rings*, J. Pure Appl. Algebra 153, 107–120 (2000)

[51] Al-Takhman, K., *Äquivalenzen zwischen Komodulkategorien von Koalgebren über Ringen*, PhD Thesis, University of Düsseldorf (1999)

[52] Albu, T., Wisbauer, R., *M-density, M-adic completion and M-subgeneration*, Rend. Sem. Mat. Univ. Padova 98, 141–159 (1997)

[53] Allen, H.P., Trushin, D., *Coproper coalgebras*, J. Algebra 54, 203–215 (1978)

[54] Allen, H.P., Trushin, D., *A generalized Frobenius structure for coalgebras with applications to character theory*, J. Algebra 62, 430–449 (1980)

[55] Amitsur, S., *Simple algebras and cohomology groups of arbitrary fields*, Trans. Amer. Math. Soc. 90, 73–112 (1959)

[56] Anquela, J.A., Cortés, T., Montaner, F., *Nonassociative coalgebras*, Comm. Algebra 22, 4693–4716 (1994)

[57] Artin, M., *On Azumaya algebras and finite representations of rings*, J. Algebra 11, 532–563 (1969)

[58] Auslander, M., Reiten, I., Smalø, S., *Galois actions on rings and finite Galois coverings*, Math. Scand. 65, 5–32 (1989)

[59] Beattie, M., Dăscălescu, S., Raianu, S., *Galois extensions for co-Frobenius Hopf algebras*, J. Algebra 198, 164–183 (1997)

[60] Beattie, M., Dăscălescu, S., Grünenfelder, L., Năstăsescu, C., *Finiteness conditions, co-Frobenius Hopf algebras, and quantum groups*, J. Algebra 200, 312–333 (1998)

[61] Bergen, J., Montgomery, S., *Smash products and outer derivations*, Israel J. Math. 53, 321–345 (1986)

[62] Berning, J., *Beziehungen zwischen links-linearen Toplogien und Modulkategorien*, Dissertation, University of Düsseldorf (1994)

[63] Blattner, R.J., Montgomery, S., *A duality theorem for Hopf module algebras*, J. Algebra 95, 153–172 (1985)

[64] Böhm, G., *Doi-Hopf modules over weak Hopf algebras*, Comm. Algebra 28, 4687–4698 (2000)

[65] Böhm, G., Nill, F., Szlachányi, K., *Weak Hopf algebras I. Integral theory and C^*-structure*, J. Algebra 221, 385–438 (1999)

[66] Böhm, G., Szlachányi, K., *A coassociative C^*-quantum group with non-integral dimension*, Lett. Math. Phys. 35, 437–456 (1996)

[67] Brzeziński, T., *Translation map in quantum principal bundles*, J. Geom. Phys. 20, 349–370 (1996)

[68] Brzeziński, T., *Quantum homogenous spaces and coalgebra bundles*, Rep. Math. Phys. 40, 179–185 (1997)

[69] Brzeziński, T., *On modules associated to coalgebra-Galois extensions*, J. Algebra 215, 290–317 (1999)

[70] Brzeziński, T., *Frobenius properties and Maschke-type theorems for entwined modules*, Proc. Amer. Math. Soc. 128, 2261–2270 (2000)

[71] Brzeziński, T., *Coalgebra-Galois extensions from the extension theory point of view*, in Hopf Algebras and Quantum Groups, Caenepeel and van Oystaeyen (Eds.), LN PAM 209, Marcel Dekker, New York (2000)

[72] Brzeziński, T., *The cohomology structure of an algebra entwined with a coalgebra*, J. Algebra 235, 176–202 (2001)

[73] Brzeziński, T., *The structure of corings. Induction functors, Maschke-type theorem, and Frobenius and Galois-type properties*, Algebras Rep. Theory 5, 389–410 (2002)

[74] Brzeziński, T., *The structure of corings with a grouplike element*, Preprint ArXiv math.RA/0108117 (2001), to appear in Banach Center Publications

[75] Brzeziński, T., *Towers of corings*, Preprint ArXiv math.RA/0201014 (2002), to appear in Comm. Algebra

[76] Brzeziński, T., Caenepeel, S., Militaru, G., *Doi-Koppinen modules for quantum groupoids*, J. Pure Appl. Alg. 175, 45–62 (2002)

[77] Brzeziński, T., Hajac, P.M., *Coalgebra extensions and algebra coextensions of Galois type*, Comm. Algebra 27, 1347–1367 (1999)

[78] Brzeziński, T., Kadison, L., Wisbauer, R., *On coseparable and biseparable corings*, Preprint ArXiv math.RA/0208122 (2002)

[79] Brzeziński, T., Majid, S., *Quantum group gauge theory on quantum spaces*, Comm. Math. Phys. 157, 591–638 (1993) (Erratum: 167, 235 (1995))

[80] Brzeziński, T., Majid, S., *Coalgebra bundles*, Comm. Math. Phys. 191, 467–492 (1998)

[81] Brzeziński, T., Militaru, G., *Bialgebroids, $\times_A$-bialgebras and duality*, J. Algebra 251, 279–294 (2002)

[82] Caenepeel, S., De Groot, E., *Modules over weak entwining structures*, AMS Contemp. Math. 267, 31–54 (2000)

[83] Caenepeel, S., De Groot, E., Militaru, G., *Frobenius functors of the second kind*, Preprint ArXiv math.RA/0106109 (2001), to appear in Comm. Algebra

[84] Caenepeel, S., Ion, B., Militaru, G., *The structure of Frobenius algebras and separable algebras*, K-Theory 19, 365–402 (2000)

[85] Caenepeel, S., Militaru, G., Zhu, S., *Crossed modules and Doi-Hopf modules*, Israel J. Math. 100, 221–247 (1997)

[86] Caenepeel, S., Militaru, G., Zhu, S., *Doi-Hopf modules, Yetter-Drinfel'd modules and Frobenius type properties*, Trans. Amer. Math. Soc. 349, 4311–4342 (1997)

[87] Caenepeel, S., Raianu, S., *Induction functors for the Doi-Koppinen unified Hopf modules*, in Abelian Groups and Modules, Facchini and Menini (Eds.), Kluwer Academic Publishers, Dordrecht, 73–94 (1995)

[88] Calinescu, C., *Integrals for Hopf algebras*, Dissertation, University of Bucharest (1998)

[89] Cao-Yu, C., Nichols, W.D., *A duality theorem for Hopf module algebras over Dedekind rings*, Comm. Algebra 18(10), 3209–3221 (1990)

[90] Cartier, P., *Cohomologie des coalgèbres*, Sém. Sophus Lie 1955–1956, exp. 5.

[91] Castaño Iglesias, F., Gómez-Torrecillas, J., Năstăsescu, C., *Separable functors in coalgebras. Applications*, Tsukuba J. Math. 21, 329–344, (1997)

[92] Chase, S.U., Harrison, D.K., Rosenberg, A., *Galois theory and Galois cohomology of commutative rings*, Mem. AMS 52, 1–20 (1965)

[93] Chen, J., Hu, Q., *Morita contexts and ring extensions*, Proc. 2nd Japan-China Intern. Symp. on Ring Theory and 28th Symp. on Ring Theory, Okayama (1995)

[94] Cheng, C.C., *Separable semigroup algebras*, J. Pure Appl. Alg. 33, 151–158 (1984)

[95] Chin, W., Montgomery, S., *Basic coalgebras*, in Modular Interfaces. Modular Lie Algebras, Quantum Groups, and Lie Superalgebras, Chari et al. (Eds.) AMS/IP Stud. Adv. Math. 4, 41–47 (1997)

[96] Cipolla, M., *Discesa fedelemente piatta dei moduli*, Rend. Circ. Mat. Palermo (2) 25, 43–46 (1976)

[97] Cohen, M., *A Morita context related to finite automorphism groups of rings*, Pac. J. Math. 98, 37–54 (1982)

[98] Cohen, M., Fishman, D., *Hopf algebra actions*, J. Algebra 100, 363–379 (1986)

[99] Cohen, M., Fischman, D., Montgomery, S., *Hopf-Galois extensions, smash products, and Morita equivalence*, J. Algebra 133, 351–372 (1990)

[100] Cohen, M., Westreich, S., *Central invariants of H-module algebras*, Comm. Algebra 21, 2859–2883 (1993)

[101] Connes, A., *Non-commutative differential geometry*, Inst. Hautes Études Sci. Publ. Math. 62, 257–360 (1985)

[102] Cuadra, J., Gómez-Torrecillas, J., *Idempotents and Morita-Takeuchi theory*, Comm. Algebra 30(5), 2405–2426 (2002)

[103] Cuntz, J., Quillen, D., *Algebra extensions and nonsingularity*, J. Amer. Math. Soc. 8, 251–289 (1995)

[104] Doi, Y., *Homological coalgebra*, J. Math. Soc. Japan 33, 31–50 (1981)

[105] Doi, Y., *On the structure of relative Hopf modules*, Comm. Algebra 11, 243–253 (1983)

[106] Doi, Y., *Unifying Hopf modules*, J. Algebra 153, 373–385 (1992)

[107] Doi, Y., Takeuchi, M., *Cleft comodule algebras for a bialgebra*, Comm. Algebra 14, 801–817 (1986)

[108] Doi, Y., Takeuchi, M., *Hopf-Galois extensions of algebras, the Miyashita-Ulbrich action, and Azumaya algebras*, J. Algebra 121, 488–516 (1989)

[109] Donkin, S., *On Projective Modules for Algebraic Groups*, J. London Math. Soc. 54, 75–88 (1996)

[110] Drinfeld, V.G., *Quantum groups*, in Proceedings of the International Congress of Mathematicians, Vol. 1, 2 (Berkeley, Calif., 1986), AMS, Providence, RI, 798–820 (1987)

[111] El Kaoutit, L., Gómez-Torrecillas, J., *Comatrix corings: Galois corings, Descent Theory, and a Structure Theorem for Cosemisimple corings*, Preprint ArXiv math.RA/0207205 (2002), to appear in Math. Z.

[112] El Kaoutit, L., Gómez-Torrecillas, J., Lobillo, F.J., *Semisimple corings*, preprint (2001), to appear in Algebra Colloquium

[113] Etingof, P., Nikshych, D., *Dynamical quantum groups at roots of 1*, Duke Math. J. 108, 135–168 (2001)

[114] Fang Li, *Weak Hopf algebras and some new solutions of the quantum Yang-Baxter equation*, J. Algebra 208, 72–100 (1998)

[115] García, J.J., del Rio, A., *Actions of groups on fully bounded noetherian rings*, Comm. Algebra 22(5), 1495–1505 (1994)

[116] García, J.M., Jara, P., Merino, L.M., *Decomposition of locally finite modules*, in Rings, Hopf algebras, and Brauer groups, Caenepeel and Verschoren (Eds.), LNPAM 197, Marcel Dekker, New-York (1998)

[117] García, J.M., Jara, P., Merino, L.M., *Decomposition of comodules*, Comm. Algebra 27(4), 1797–1805 (1999)

[118] Garfinkel, G.S., *Universally torsionless and trace modules*, Trans. Amer. Math. Soc. 215, 119–144 (1976)

[119] Gerstenhaber, M., *The cohomology structure of an associative ring*, Ann. Math. 78, 267–288 (1963)

[120] Gerstenhaber, M., Schack, S.D., *Algebras, bialgebras, quantum groups and algebraic deformations*, in Deformation Theory and Quantum Groups with Applications to Mathematical Physics, Gerstenhaber and Stasheff (Eds.), AMS Contemp. Math. 134, 51–92 (1992)

[121] Gómez-Torrecillas, J., *Coalgebras and comodules over a commutative ring*, Rev. Roum. Math. Pures Appl. 43(5-6), 591–603 (1998)

[122] Gómez-Torrecillas, J., *Separable functors in corings*, Int. J. Math. Math. Sci. 30, 203–225 (2002)

[123] Gómez-Torrecillas, J., Năstăsescu, C., *Quasi-co-Frobenius coalgebras*, J. Algebra 174, 909–923 (1995)

[124] Green, J.A., *Locally finite representations*, J. Algebra 41, 137–171 (1976)

[125] Grunenfelder, L., Paré, R., *Families parametrized by coalgebras*, J. Algebra 107, 316–375 (1987)

[126] Guzman, F., *Cointegration and Relative Cohomology for Comodules*, PhD Thesis, Syracuse University, New York (1985)

[127] Guzman, F., *Cointegrations, relative cohomology for comodules, and coseparable corings*, J. Algebra 126, 211–224 (1989)

[128] Hayashi, T., *Quantum group symmetry of partition functions of IRF models and its applications to Jones' index theory*, Comm. Math. Phys. 157, 331–345 (1993)

[129] Heyneman, R.G., Radford, D.E., *Reflexivity and coalgebras of finite type*, J. Algebra 28, 215–246 (1974)

[130] Hirata, K., *Some types of separable extensions of rings*, Nagoya. Math. J. 33, 107–115 (1968)

[131] Hobst, D., Pareigis, B., *Double quantum groups*, J. Algebra 242, 460–494 (2001)

[132] Hochschild, G., *Cohomology groups of an associative algebra*, Ann. Math. 46, 58–67 (1945)

[133] Hochschild, G., *Relative homological algebra*, Trans. Amer. Math. Soc. 82, 246–269 (1956)

[134] Hughes, J., *A study of Categories of Algebras and Coalgebras*, PhD Thesis, Carnegie Mellon University, Pittsburgh (2000)

[135] Jimbo, M., *A q-difference analog of $U(g)$ and the Yang-Baxter equation*, Lett. Math. Phys. 10, 63–69 (1985)

[136] Jones, V.F.R., *Index of subrings of rings*, AMS Contemp. Math. 43, 181–190 (1985)

[137] Kadison, L., *Hopf algebroids and H-separable extensions*, Preprint (2002)

[138] Kadison, L., Szlachányi, K., *Dual bialgebroids with actions on depth two extensions*, Preprint ArXiv math.RA/0108067 (2001)

[139] Kasch, F., *Projektive Frobenius-Erweiterungen*, Sitzungsber. Heidelberger Akad. Wiss. Math.-Natur. Kl. 4, 89–109 (1960/61)

[140] Kleiner, M., *The dual ring to a coring with a grouplike*, Proc. Amer. Math. Soc. 91, 540–542 (1984)

[141] Kluge, L., Paal, E., Stasheff, J., *Invitation to composition*, Comm. Algebra 28, 1405–1422 (2000)

[142] Kontsevich, M., *Deformation quantization of Poisson manifolds I*, Preprint q-alg/9709040 (1997)

[143] Kontsevich, M., *Operads and motives in deformation quantization*, Lett. Math. Phys. 48, 35–72 (1999)

[144] Koppinen, M., *Variations on the smash product with applications to group-graded rings*, J. Pure Appl. Alg. 104, 61–80 (1994)

[145] Koppinen, M., *On twistings of comodule algebras*, Comm. Algebra 25, 2009–2027 (1997)

[146] Kraft, M., *Nicht-koassoziative Koalgebren*, Diploma Thesis, Universität Düsseldorf (2001)

[147] Kreimer, H.F., Takeuchi, M., *Hopf algebras and Galois extensions of an algebra*, Indiana Univ. Math. J. 30, 675–691 (1981)

[148] Larson, R.G., *Coseparable Hopf Algebras*, J. Pure Appl. Algebra 3, 261–267 (1973)

[149] Larson, R.G., Sweedler, M.E., *An associative orthogonal bilinear form for Hopf algebras*, Amer. J. Math. 91, 75–93 (1969)

[150] Lin, B.I., *Morita's theorem for coalgebras*, Comm. Algebra 1, 311–344 (1974)

[151] Lin, B.I., *Products of torsion theories and applications to coalgebras*, Osaka J. Math. 12, 433–439 (1975)

[152] Lin, B.I., *Semiperfect Coalgebras*, J. Algebra 49, 357–373 (1977)

[153] Lomp, Ch., *Primeigenschaften von Algebren in Modulkategorien über Hopf algebren*, PhD Thesis, University of Düsseldorf (2002)

[154] Lu, J.H., *Hopf algebroids and quantum groupoids*, Int. J. Math. 7, 47–70 (1996)

[155] Masuoka, A., *Corings and invertible bimodules*, Tsukuba J. Math. 13(2), 353–362 (1989)

[156] Menini, C., Năstăsescu, C., *When are induction and coinduction functors isomorphic?*, Bull. Belg. Math. Soc. 1, 521–558 (1994)

[157] Menini, C., Torrecillas, B., Wisbauer, R., *Strongly rational comodules and semiperfect Hopf algebras over QF rings*, J. Pure Appl. Algebra 155, 237–255 (2001)

[158] Menini, C., Zuccoli, M., *Equivalence theorems and Hopf-Galois extensions*, J. Algebra 194, 245–274 (1997)

[159] Milnor, J., Moore, J.C., *On the structure of Hopf algebras*, Ann. Math. 81, 211–264 (1965)

[160] Miyamoto, H., *A note on coreflexive coalgebras*, Hiroshima Math. J. 5, 17–22 (1975)

[161] Montgomery, S., *Indecomposable coalgebras, simple comodules, and pointed Hopf algebras*, Proc. Amer. Math. Soc. 123, 2343–2351 (1995)

[162] Morita, K., *Adjoint pairs of functors and Frobenius extensions*, Sci. Rep. Tokyo Kyoiku Daigaku Sect. A, 9, 40–71 (1965)

[163] Müller, E.F., Schneider, H.-J., *Quantum homogeneous spaces with faithfully flat module structures*, Israel J. Math. 111, 157–190 (1999)

[164] Nakajima, A., *Bialgebras and Galois extensions*, Math. J. Okayama Univ. 33, 37–46 (1991)

[165] Năstăsescu, C., Torrecillas, B., *Colocalization on Grothendieck categories with applications to coalgebras*, J. Algebra 185, 108–124 (1996)

[166] Năstăsescu, C., Torrecillas, B., Zhang, Y.H., *Hereditary coalgebras*, Comm. Algebra 24(4), 1521–1528 (1996)

[167] Nichols, W.D., *Cosemisimple Hopf algebras*, in Advances in Hopf Algebras, Montgomery and Bergen (Eds.), Marcel Dekker LNPAM 158, 135–151 (1994)

[168] Nichols, W.D., Sweedler, M., *Hopf algebras and combinatorics*, AMS Contemp. Math. 6, 49–84 (1982)

[169] Nill, F., *Axioms for a weak bialgebra*, Preprint ArXiv math.QA/9805104 (1998)

[170] Nuss, P., *Noncommutative descent and non-Abelian cohomology*, K-Theory 12, 23–74 (1997)

[171] Ohm, J., Bush, D.E., *Content modules and algebras*, Math. Scand. 31, 49–68 (1972)

[172] Ocneanu, A., *Quantized groups, string algebras and Galois theory for algebras*, in Operators algebras and applications 2, Evans et al. (Eds.), Cambridge University Press, Cambridge (1988)

[173] Pareigis, B., *When Hopf algebras are Frobenius algebras*, J. Algebra 18, 588–596 (1971)

[174] Pareigis, B., *On the cohomology of modules over Hopf algebras*, J. Algebra 22, 161–182 (1972)

[175] Prest, M., Wisbauer, R., *Finite presentation and purity in categories $\sigma[M]$*, preprint (2002)

[176] Radford, D.E., *Coreflexive coalgebras*, J. Algebra 26, 512–535 (1973)

[177] Radford, D.E., *Finiteness conditions for a Hopf algebra with a nonzero integral*, J. Algebra 46, 189–195 (1977)

[178] Radford, D.E., *On the structure of pointed coalgebras*, J. Algebra 77, 1–14 (1982)

[179] Radford, D.E., Towber, J., *Yetter-Drinfeld categories associated to an arbitrary algebra*, J. Pure Appl. Algebra 87, 259–279 (1993)

[180] Rafael, M.D., *Separable functors revisited*, Comm. Algebra 18, 1445–1459 (1990)

[181] Raynaud, M., Gruson, L., *Critère de platitude et de projectivité*, Inv. Math. 13, 1–89 (1971)

[182] Rojter, A.V., *Matrix problems and representations of BOCS's*, LNM 831, Springer, Berlin, 288–324 (1980)

[183] Schauenburg, P., *Bialgebras over noncommutative rings and structure theorems for Hopf bimodules*, Appl. Categorical Structures 6, 193–222 (1998)

[184] Schauenburg, P., *Doi-Koppinen Hopf modules versus entwined modules*, New York J. Math. 6, 325–329 (2000)

[185] Schauenburg, P., *Weak Hopf algebras and quantum groupoids*, Preprint ArXiv math.QA/0204180 (2002)

[186] Schneider, H.-J., *Principal homogeneous spaces for arbitrary Hopf algebras*, Israel J. Math. 72, 167–195 (1990)

[187] Schneider, H.-J., *Representation theory of Hopf Galois extensions*, Israel J. Math. 72, 196–231 (1990)

[188] Schneider, H.-J., *Normal basis and transitivity of crossed products for Hopf algebras*, J. Algebra 152, 289–312 (1992)

[189] Shudo, T., *A note on coalgebras and rational modules*, Hiroshima Math. J. 6, 297–304 (1976)

[190] Shudo, T., Miyamoto, H., *On the decomposition of coalgebras*, Hiroshima Math. J. 8, 499–504 (1978)

[191] Sullivan, J.B., *The uniqueness of integrals for Hopf algebras and some existence theorems of integrals for commutative Hopf algebras*, J. Algebra 19, 426–440 (1971)

[192] Sweedler, M.E., *Groups of simple algebras*, Inst. Hautes Études Sci. Publ. Math. 44, 79–189 (1975)

[193] Sweedler, M.E., *The predual theorem to the Jacobson-Bourbaki theorem*, Trans. Amer. Math. Soc. 213, 391–406 (1975)

[194] Taft, E.J., *Reflexivity of algebras and coalgebras*, Am. J. Math. 94, 1111–1130 (1972)

[195] Takeuchi, M., *A correspondence between Hopf ideals and sub-Hopf algebras*, Manuscripta Math. 7, 251–270 (1972)

[196] Takeuchi, M., *Groups of algebras over $A \otimes \bar{A}$*, J. Math. Soc. Japan 29, 459–492 (1977)

[197] Takeuchi, M., *Morita theorems for categories of comodules*, J. Fac. Sci. Univ. Tokyo Sect. IA Math. 24, 629–644 (1977)

[198] Takeuchi, M., *Relative Hopf modules - equivalences and freeness criteria*, J. Algebra 60, 452–471 (1979)

[199] Tamarkin, D.E., *Another proof of M. Kontsevich formality theorem for* $\mathbf{R}^n$, Preprint ArXiv math.QA/9803025 (1998)

[200] Tambara, D., *The coendomorphism bialgebra of an algebra*, J. Fac. Sci. Univ. Tokyo Sect. IA, Math. 37, 425–456 (1990)

[201] Tominaga, H., *On s-unital rings*, Math. J. Okayama Univ. 18, 117–134 (1976)

[202] Vanaja, N., *All finitely generated M-subgenerated modules are extending*, Comm. Algebra 24(2), 543–572 (1996)

[203] Wang, Z., *Actions of Hopf algebras on fully bounded noetherian rings*, Comm. Algebra 24, 3117–3120 (1996)

[204] Wilke, B., *Zur Modulstruktur von Ringen über Schiefgruppenringen*, Algebra-Berichte 73, München (1994)

[205] Wisbauer, R., *Localization of modules and the central closure of rings*, Comm. Algebra 9, 1455–1493 (1981)

[206] Wisbauer, R., *Local-global results for modules over algebras and Azumaya rings*, J. Algebra 135, 440–455 (1990)

[207] Wisbauer, R., *On module classes closed under extensions*, in Rings and Radicals, Gardner et al. (Eds.), Pitman RN 346, 73–97 (1996)

[208] Wisbauer, R., *Tilting in module categories*, in Abelian Groups, Module theory, and Topology, Dikranjan and Salce (Eds.), Marcel Dekker LNPAM 201, 421–444 (1998)

[209] Wisbauer, R., *Static modules and equivalences*, in Interactions Between Ring Theory and Representations of Algebras, van Oystaeyen and Saorin (Eds.), Marcel Dekker, 423–449 (2000)

[210] Wisbauer, R., *Semiperfect coalgebras over rings*, in Algebras and Combinatorics, ICA'97, Hong Kong, Shum et al. (Eds.), Springer Singapore, 487–512 (1999)

[211] Wisbauer, R., *Decompositions of modules and comodules*, in Algebra and Its Applications, Huynh et al. (Eds.), AMS Contemp. Math. 259, 547–561 (2000)

[212] Wisbauer, R., *Weak corings*, J. Algebra 245, 123–160 (2001)

[213] Wischnewsky, M.B., *On linear representations of affine groups I*, Pac. J. Math. 61, 551–572 (1975)

[214] Woronowicz, S.L., *Compact matrix pseudogroups*, Comm. Math. Phys. 111, 613–665 (1987)

[215] Xu, P., *Quantum groupoids*, Comm. Math. Phys. 216, 539–581 (2001)

[216] Yamanouchi, T., *Duality for generalized Kac algebras and a characterisation of finite groupoid algebras*, J. Algebra 163, 9–50 (1994)

[217] Yetter, D.N., *Quantum groups and representations of monoidal categories*, Math. Proc. Camb. Phil. Soc. 108, 261–290 (1990)

[218] Zhang, Y., *Hopf Frobenius extensions of algebras*, Comm. Algebra 20 (7), 1907–1915 (1992)

[219] Zimmermann, W., *Modules with chain conditions for finite matrix subgroups*, J. Algebra 190, 68–87 (1997)

[220] Zimmermann-Huisgen, B., *Endomorphism rings of self-generators*, Pac. J. Math. 61 587–602 (1975)

[221] Zimmermann-Huisgen, B., *Pure submodules of direct products of free modules*, Math. Ann. 224, 233–245 (1976)

Index

Printed in the United States
By Bookmasters